The Industrial Electronics Handbook
SECOND EDITION

FUNDAMENTALS OF
INDUSTRIAL ELECTRONICS

The Industrial Electronics Handbook
SECOND EDITION

FUNDAMENTALS OF INDUSTRIAL ELECTRONICS

POWER ELECTRONICS AND MOTOR DRIVES

CONTROL AND MECHATRONICS

INDUSTRIAL COMMUNICATION SYSTEMS

INTELLIGENT SYSTEMS

The Electrical Engineering Handbook Series

Series Editor
Richard C. Dorf
University of California, Davis

Titles Included in the Series

The Avionics Handbook, Second Edition, Cary R. Spitzer
The Biomedical Engineering Handbook, Third Edition, Joseph D. Bronzino
The Circuits and Filters Handbook, Third Edition, Wai-Kai Chen
The Communications Handbook, Second Edition, Jerry Gibson
The Computer Engineering Handbook, Vojin G. Oklobdzija
The Control Handbook, Second Edition, William S. Levine
CRC Handbook of Engineering Tables, Richard C. Dorf
Digital Avionics Handbook, Second Edition, Cary R. Spitzer
The Digital Signal Processing Handbook, Vijay K. Madisetti and Douglas Williams
The Electric Power Engineering Handbook, Second Edition, Leonard L. Grigsby
The Electrical Engineering Handbook, Third Edition, Richard C. Dorf
The Electronics Handbook, Second Edition, Jerry C. Whitaker
The Engineering Handbook, Third Edition, Richard C. Dorf
The Handbook of Ad Hoc Wireless Networks, Mohammad Ilyas
The Handbook of Formulas and Tables for Signal Processing, Alexander D. Poularikas
Handbook of Nanoscience, Engineering, and Technology, Second Edition,
    William A. Goddard, III, Donald W. Brenner, Sergey E. Lyshevski, and Gerald J. Iafrate
The Handbook of Optical Communication Networks, Mohammad Ilyas and
    Hussein T. Mouftah
The Industrial Electronics Handbook, Second Edition, Bogdan M. Wilamowski
    and J. David Irwin
The Measurement, Instrumentation, and Sensors Handbook, John G. Webster
The Mechanical Systems Design Handbook, Osita D.I. Nwokah and Yidirim Hurmuzlu
The Mechatronics Handbook, Second Edition, Robert H. Bishop
The Mobile Communications Handbook, Second Edition, Jerry D. Gibson
The Ocean Engineering Handbook, Ferial El-Hawary
The RF and Microwave Handbook, Second Edition, Mike Golio
The Technology Management Handbook, Richard C. Dorf
Transforms and Applications Handbook, Third Edition, Alexander D. Poularikas
The VLSI Handbook, Second Edition, Wai-Kai Chen

The Industrial Electronics Handbook
SECOND EDITION

FUNDAMENTALS OF INDUSTRIAL ELECTRONICS

Edited by
Bogdan M. Wilamowski
J. David Irwin

CRC Press
Taylor & Francis Group
Boca Raton London New York

CRC Press is an imprint of the
Taylor & Francis Group, an **informa** business

CRC Press
Taylor & Francis Group
6000 Broken Sound Parkway NW, Suite 300
Boca Raton, FL 33487-2742

First issued in paperback 2017

© 2011 by Taylor and Francis Group, LLC
CRC Press is an imprint of Taylor & Francis Group, an Informa business

No claim to original U.S. Government works

ISBN-13: 978-1-4398-0279-3 (hbk)
ISBN-13: 978-1-138-07439-2 (pbk)

Library of Congress Cataloging-in-Publication Data

Fundamentals of industrial electronics / editors, Bogdan M. Wilamowski and J. David Irwin.
    p. cm.
  "A CRC title."
  Includes bibliographical references and index.
  ISBN 978-1-4398-0279-3 (alk. paper)
  1. Industrial electronics. I. Wilamowski, Bogdan M. II. Irwin, J. David. III. Title.

TK7881.F86 2010
621.381--dc22
    2010019980

Visit the Taylor & Francis Web site at
http://www.taylorandfrancis.com

and the CRC Press Web site at
http://www.crcpress.com

Contents

PART I   Circuits and Signals

PART II   Devices

PART III Digital Circuits

PART IV Digital and Analog Signal Processing

PART V Electromagnetics

Preface

The field of industrial electronics covers a plethora of problems that must be solved in industrial practice. Electronic systems control many processes that begin with the control of relatively simple devices like electric motors, through more complicated devices such as robots, to the control of entire fabrication processes. An industrial electronics engineer deals with many physical phenomena as well as the sensors that are used to measure them. Thus, the knowledge required by this type of engineer is not only traditional electronics but also specialized electronics, for example, that required for high-power applications. The importance of electronic circuits extends well beyond their use as a final product in that they are also important building blocks in large systems, and thus the industrial electronics engineer must also possess knowledge of the areas of control and mechatronics. Since most fabrication processes are relatively complex, there is an inherent requirement for the use of communication systems that not only link the various elements of the industrial process but are also tailor-made for the specific industrial environment. Finally, the efficient control and supervision of factories require the application of intelligent systems in a hierarchical structure to address the needs of all components employed in the production process. This is accomplished through the use of intelligent systems such as neural networks, fuzzy systems, and evolutionary methods. The Industrial Electronics Handbook addresses all these issues and does so in five books outlined as follows:

1. *Fundamentals of Industrial Electronics*
2. *Power Electronics and Motor Drives*
3. *Control and Mechatronics*
4. *Industrial Communication Systems*
5. *Intelligent Systems*

The editors have gone to great lengths to ensure that this handbook is as current and up to date as possible. Thus, this book closely follows the current research and trends in applications that can be found in *IEEE Transactions on Industrial Electronics*. This journal is not only one of the largest engineering publications of its type in the world, but also one of the most respected. In all technical categories in which this journal is evaluated, its worldwide ranking is either number 1 or number 2 depending on category. As a result, we believe that this handbook, which is written by the world's leading researchers in the field, presents the global trends in the ubiquitous area commonly known as industrial electronics.

Fundamentals of Industrial Electronics deals with the fundamental areas that form the basis for the field of industrial electronics. Because of the breadth of this field, the knowledge required spans a wide spectrum of technology, which includes analog and digital circuits, electronics, electromagnetic machines, and signal processing. The knowledge gained here is then applied in *Power Electronics and Motor Drives*, *Control and Mechatronics*, *Industrial Communication Systems*, and *Intelligent Systems*, and in total form the Industrial Electronics Handbook.

For MATLAB® and Simulink® product information, please contact

The MathWorks, Inc.
3 Apple Hill Drive
Natick, MA, 01760-2098 USA
Tel: 508-647-7000
Fax: 508-647-7001
E-mail: info@mathworks.com
Web: www.mathworks.com

Acknowledgments

The editors wish to express their heartfelt thanks to their wives Barbara Wilamowski and Edie Irwin for their help and support during the execution of this project.

Editorial Board

Editors

Bogdan M. Wilamowski received his MS in computer engineering in 1966, his PhD in neural computing in 1970, and Dr. habil. in integrated circuit design in 1977. He received the title of full professor from the president of Poland in 1987. He was the director of the Institute of Electronics (1979–1981) and the chair of the solid state electronics department (1987–1989) at the Technical University of Gdansk, Poland. He was a professor at the University of Wyoming, Laramie, from 1989 to 2000. From 2000 to 2003, he served as an associate director at the Microelectronics Research and Telecommunication Institute, University of Idaho, Moscow, and as a professor in the electrical and computer engineering department and in the computer science department at the same university. Currently, he is the director of ANMSTC—Alabama Nano/Micro Science and Technology Center, Auburn, and an alumna professor in the electrical and computer engineering department at Auburn University, Alabama. Dr. Wilamowski was with the Communication Institute at Tohoku University, Japan (1968–1970), and spent one year at the Semiconductor Research Institute, Sendai, Japan, as a JSPS fellow (1975–1976). He was also a visiting scholar at Auburn University (1981–1982 and 1995–1996) and a visiting professor at the University of Arizona, Tucson (1982–1984). He is the author of 4 textbooks, more than 300 refereed publications, and has 27 patents. He was the principal professor for about 130 graduate students. His main areas of interest include semiconductor devices and sensors, mixed signal and analog signal processing, and computational intelligence.

Dr. Wilamowski was the vice president of the IEEE Computational Intelligence Society (2000–2004) and the president of the IEEE Industrial Electronics Society (2004–2005). He served as an associate editor of *IEEE Transactions on Neural Networks*, *IEEE Transactions on Education*, *IEEE Transactions on Industrial Electronics*, the *Journal of Intelligent and Fuzzy Systems*, the *Journal of Computing*, and the *International Journal of Circuit Systems and IES Newsletter*. He is currently serving as the editor in chief of *IEEE Transactions on Industrial Electronics*.

Professor Wilamowski is an IEEE fellow and an honorary member of the Hungarian Academy of Science. In 2008, he was awarded the Commander Cross of the Order of Merit of the Republic of Poland for outstanding service in the proliferation of international scientific collaborations and for achievements in the areas of microelectronics and computer science by the president of Poland.

J. David Irwin received his BEE from Auburn University, Alabama, in 1961, and his MS and PhD from the University of Tennessee, Knoxville, in 1962 and 1967, respectively.

In 1967, he joined Bell Telephone Laboratories, Inc., Holmdel, New Jersey, as a member of the technical staff and was made a supervisor in 1968. He then joined Auburn University in 1969 as an assistant professor of electrical engineering. He was made an associate professor in 1972, associate professor and head of department in 1973, and professor and head in 1976. He served as head of the Department of Electrical and Computer Engineering from 1973 to 2009. In 1993, he was named Earle C. Williams Eminent Scholar and Head. From 1982 to 1984, he was also head of the Department of Computer Science and Engineering. He is currently the Earle C. Williams Eminent Scholar in Electrical and Computer Engineering at Auburn.

Dr. Irwin has served the Institute of Electrical and Electronic Engineers, Inc. (IEEE) Computer Society as a member of the Education Committee and as education editor of *Computer*. He has served as chairman of the Southeastern Association of Electrical Engineering Department Heads and the National Association of Electrical Engineering Department Heads and is past president of both the IEEE Industrial Electronics Society and the IEEE Education Society. He is a life member of the IEEE Industrial Electronics Society AdCom and has served as a member of the Oceanic Engineering Society AdCom. He served for two years as editor of *IEEE Transactions on Industrial Electronics*. He has served on the Executive Committee of the Southeastern Center for Electrical Engineering Education, Inc., and was president of the organization in 1983–1984. He has served as an IEEE Adhoc Visitor for ABET Accreditation teams. He has also served as a member of the IEEE Educational Activities Board, and was the accreditation coordinator for IEEE in 1989. He has served as a member of numerous IEEE committees, including the Lamme Medal Award Committee, the Fellow Committee, the Nominations and Appointments Committee, and the Admission and Advancement Committee. He has served as a member of the board of directors of IEEE Press. He has also served as a member of the Secretary of the Army's Advisory Panel for ROTC Affairs, as a nominations chairman for the National Electrical Engineering Department Heads Association, and as a member of the IEEE Education Society's McGraw-Hill/Jacob Millman Award Committee. He has also served as chair of the IEEE Undergraduate and Graduate Teaching Award Committee. He is a member of the board of governors and past president of Eta Kappa Nu, the ECE Honor Society. He has been and continues to be involved in the management of several international conferences sponsored by the IEEE Industrial Electronics Society, and served as general cochair for IECON'05.

Dr. Irwin is the author and coauthor of numerous publications, papers, patent applications, and presentations, including *Basic Engineering Circuit Analysis*, 9th edition, published by John Wiley & Sons, which is one among his 16 textbooks. His textbooks, which span a wide spectrum of engineering subjects, have been published by Macmillan Publishing Company, Prentice Hall Book Company, John Wiley & Sons Book Company, and IEEE Press. He is also the editor in chief of a large handbook published by CRC Press, and is the series editor for Industrial Electronics Handbook for CRC Press.

Dr. Irwin is a fellow of the American Association for the Advancement of Science, the American Society for Engineering Education, and the Institute of Electrical and Electronic Engineers. He received an IEEE Centennial Medal in 1984, and was awarded the Bliss Medal by the Society of American Military Engineers in 1985. He received the IEEE Industrial Electronics Society's Anthony J. Hornfeck Outstanding Service Award in 1986, and was named IEEE Region III (U.S. Southeastern Region) Outstanding Engineering Educator in 1989. In 1991, he received a Meritorious Service Citation from the IEEE Educational Activities Board, the 1991 Eugene Mittelmann Achievement Award from the IEEE Industrial Electronics Society, and the 1991 Achievement Award from the IEEE Education Society. In 1992, he was named a Distinguished Auburn Engineer. In 1993, he received the IEEE Education Society's McGraw-Hill/Jacob Millman Award, and in 1998 he was the recipient of the

IEEE Undergraduate Teaching Award. In 2000, he received an IEEE Third Millennium Medal and the IEEE Richard M. Emberson Award. In 2001, he received the American Society for Engineering Education's (ASEE) ECE Distinguished Educator Award. Dr. Irwin was made an honorary professor, Institute for Semiconductors, Chinese Academy of Science, Beijing, China, in 2004. In 2005, he received the IEEE Education Society's Meritorious Service Award, and in 2006, he received the IEEE Educational Activities Board Vice President's Recognition Award. He received the Diplome of Honor from the University of Patras, Greece, in 2007, and in 2008 he was awarded the IEEE IES Technical Committee on Factory Automation's Lifetime Achievement Award. In 2010, he was awarded the electrical and computer engineering department head's Robert M. Janowiak Outstanding Leadership and Service Award. In addition, he is a member of the following honor societies: Sigma Xi, Phi Kappa Phi, Tau Beta Pi, Eta Kappa Nu, Pi Mu Epsilon, and Omicron Delta Kappa.

Contributors

Michael E. Baginski
Department of Electrical and Computer
  Engineering
Auburn University
Auburn, Alabama

R. Jacob Baker
Department of Electrical and Computer
  Engineering
Boise State University
Boise, Idaho

Carlotta A. Berry
Department of Electrical and Computer
  Engineering
Rose-Hulman Institute of Technology
Terre Haute, Indiana

Carles Cané
National Microelectronics Center
Barcelona, Spain

Luis Castañer
Department of Electronic Engineering
Polytechnic University of Catalonia
Catalonia, Spain

John M. Dell
Microelectronics Research Group
University of Western Australia
Perth, Western Australia, Australia

Lorenzo Faraone
Microelectronics Research Group
University of Western Australia
Perth, Western Australia, Australia

Montserrat Fernández-Bolaños
Ecole Polytechnique Fédérale de Lausanne
Lausanne, Switzerland

Stephen M. Haddock
Department of Electrical and Computer
  Engineering
Auburn University
Auburn, Alabama

James A. Heinen
Department of Electrical and Computer
  Engineering
Marquette University
Milwaukee, Wisconsin

Tina Hudson
Department of Electrical and Computer
  Engineering
Rose-Hulman Institute of Technology
Terre Haute, Indiana

Francisco Ibáñez
European Commission
Brussels, Belgium

Adrian Ionescu
Ecole Polytechnique Fédérale de Lausanne
Lausanne, Switzerland

J. David Irwin
Department of Electrical and Computer
    Engineering
Auburn University
Auburn, Alabama

Marcin Jagiela
Faculty of Applied Informatics
University of Information Technology
    and Management in Rzeszów
Rzeszów, Poland

Tyler N. Killian
Department of Electrical and Computer
    Engineering
Auburn University
Auburn, Alabama

Ernest M. Kim
Department of Engineering
University of San Diego
San Diego, California

Alicja Konczakowska
Faculty of Electronics, Telecommunications
    and Informatics
Gdansk University of Technology
Gdansk, Poland

Robert Lempkowski
Motorola Applied Research and Technology Center
Schaumburg, Illinois

Sin Ming Loo
Department of Electrical and Computer
    Engineering
Boise State University
Boise, Idaho

Antonio Luque
Department of Electronic Engineering
University of Seville
Sevilla, Spain

Dalton S. Nelson
Department of Electrical and Computer
    Engineering
The University of Alabama at Birmingham
Birmingham, Alabama

Victor P. Nelson
Department of Electrical and Computer
    Engineering
Auburn University
Auburn, Alabama

Russell J. Niederjohn (deceased)
Department of Electrical and Computer
    Engineering
Marquette University
Milwaukee, Wisconsin

Guofu Niu
Department of Electrical and Computer
    Engineering
Auburn University
Auburn, Alabama

Nam Pham
Department of Electrical and Computer
    Engineering
Auburn University
Auburn, Alabama

Arlen Planting
Department of Electrical and Computer
    Engineering
Boise State University
Boise, Idaho

José M. Quero
Department of Electronic Engineering
University of Seville
Sevilla, Spain

Sadasiva M. Rao
Department of Electrical and Computer
    Engineering
Auburn University
Auburn, Alabama

Angel Rodríguez
Department of Electronic Engineering
Polytechnic University of Catalonia
Catalonia, Spain

Juan J. Rodriguez-Andina
Department of Electronic Technology
University of Vigo
Vigo, Spain

Vishal Saxena
Department of Electrical and Computer
   Engineering
Boise State University
Boise, Idaho

Thomas F. Schubert, Jr.
Department of Engineering
University of San Diego
San Diego, California

Jianjian Song
Department of Electrical and Computer
   Engineering
Rose-Hulman Institute of Technology
Terre Haute, Indiana

John W. Steadman
College of Engineering
University of South Alabama
Mobile, Alabama

Eduardo de la Torre
Center of Industrial Electronics
Polytechnic University of Madrid
Madrid, Spain

David R. Voltmer
Department of Electrical and Computer
   Engineering
Rose-Hulman Institute of Technology
Terre Haute, Indiana

Deborah J. Walter
Department of Electrical and Computer
   Engineering
Rose-Hulman Institute of Technology
Terre Haute, Indiana

Buren Earl Wells
Department of Electrical and Computer
   Engineering
The University of Alabama in Huntsville
Huntsville, Alabama

Edward Wheeler
Department of Electrical and Computer
   Engineering
Rose-Hulman Institute of Technology
Terre Haute, Indiana

Bogdan M. Wilamowski
Department of Electrical and Computer
   Engineering
Auburn University
Auburn, Alabama

Tiantian Xie
Department of Electrical and Computer
   Engineering
Auburn University
Auburn, Alabama

I

Circuits and Signals

1

DC and Transient Circuit Analysis

Carlotta A. Berry
Rose-Hulman Institute of Technology

Deborah J. Walter
Rose-Hulman Institute of Technology

1.1 Introduction

Direct current (DC) circuit analysis is the study of circuits with a constant voltage or current source. The most popular example of a DC circuit is a battery and a light bulb. A DC circuit contains an active circuit element (i.e., battery) capable of generating electric energy. These electric sources convert nonelectric energy to electric energy (i.e., a voltage or current). Independent electric sources produce a constant voltage or current in the circuit regardless of the current through or voltage across the source. The symbols for an ideal DC voltage and current source are shown in Figure 1.1. It should be noted that an ideal voltage and current source can deliver or absorb power to an electric circuit. An example of an ideal voltage source absorbing power is a rechargeable battery.

Dependent sources establish a voltage or current in a circuit that is based upon the value of a voltage or current elsewhere in the circuit. One use of dependent sources is to model operational amplifiers and transistors. Table 1.1 presents a summary of the four types of dependent sources.

A passive circuit element models devices that cannot generate electric energy such as a light bulb. The most common passive circuit elements are inductors, capacitors, and resistors. The voltage–current relationships for these devices will be described in the subsequent section.

1.1.1 Ohm's Law

Ohm's law states that the voltage (V) difference across a resistor is linearly related to the current (I) through the resistor (see Equation 1.1):

$$V = IR \tag{1.1}$$

where R is the resistance of the resistor in Ohms (Ω). The conductance (G) of a resistor is the inverse of the resistance ($1/R$) and is in units of Siemens (S). Resistors always absorb power, so the standard way to represent a resistive element is to draw the resistor in the passive sign convention (see Figure 1.2). If the resistor is not drawn in the passive sign convention, then $V = -IR$.

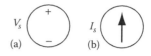

FIGURE 1.1 Ideal DC sources. (a) Voltage source. (b) Current source.

1.1.2 Inductors and Capacitors

As previously stated, the other two passive circuit elements are inductors and capacitors. Both the inductor and the capacitor have the ability to store energy. Inductors store energy in the form of current and capacitors store energy in the form of voltage. The energy stored in these elements is released back into the circuit when a DC source is removed. Therefore, these two elements exhibit behavior that is a function of time. The analysis of these types of circuits is transient analysis that will be addressed later in this chapter. Table 1.2 describes the current–voltage relationship for inductors and capacitors where the inductance (L) is in henrys (H), capacitance (C) is in farads (F), and time (t) is in seconds (s).

1.1.3 Kirchhoff's Current Law

The law of conservation of energy states that energy can neither be created nor destroyed, only transferred. Another way to state this law is for any electric circuit, the total power delivered by the elements must be equal to the total power absorbed by the elements. Kirchhoff's current law (KCL) is based upon the law of conservation of energy. A node in a circuit is any point at which two or more circuit elements are connected. KCL states that the sum of currents entering a node is zero (i.e., current in = current out). KCL can be applied to any node in a closed circuit. The circuit in Figure 1.3 has three branch currents: I_1, I_2, and I_3. Since all of these currents are leaving Node A, KCL at this node yields Equation 1.6:

$$I_1 + I_2 + I_3 = 0 \tag{1.2}$$

TABLE 1.1 Summary of Dependent Sources

Element	Description	Symbols
Current-controlled current source (CCCS)	Establishes a current in the circuit based upon the value of controlling variable, I_x, and the gain α	αI_x
Voltage-controlled voltage source (VCVS)	Establishes a voltage in the circuit based upon the value of controlling variable, V_x, and the gain β	βV_x
Voltage-controlled current source (VCCS)	Establishes a current in the circuit based upon the value of controlling variable, V_x, and the gain μ	μV_x
Current-controlled voltage source (CCVS)	Establishes a voltage in the circuit based upon the value of controlling variable, I_x, and the gain ρ	ρI_x

1.1.4 Kirchhoff's Voltage Law

Kirchhoff's voltage law (KVL) is also based upon the law of conservation
of energy. A loop is any closed path in a circuit. KVL states that the sum
of the voltages around a loop is zero (i.e., sum of the voltage drops =
sum of the voltage rises). KVL is applied to the loop shown in Figure 1.4.
Note that the direction of the loop goes from the negative terminal to the

FIGURE 1.2 Resistor.

TABLE 1.2 Inductor and Capacitor Current–Voltage Relationships

Element	Circuit Symbol	Relationship	
Inductor		$v = L\dfrac{di}{dt}$	(1.3)
		$i = \dfrac{1}{L}\displaystyle\int_0^t v\,dt + i(0)$	(1.4)
Capacitor		$i = C\dfrac{dv}{dt}$	(1.5)
		$v = \dfrac{1}{C}\displaystyle\int_0^t i\,dt + v(0)$	(1.6)

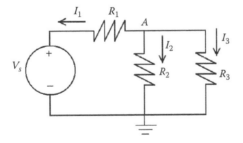

FIGURE 1.3 KCL at Node A.

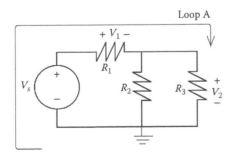

FIGURE 1.4 KVL applied around Loop A.

positive terminal on the voltage source, which indicates it is a voltage rise. For the KVL expression in Equation 1.7, voltage rises are negative and voltage drops are positive:

$$-V_s + V_1 + V_2 = 0 \tag{1.7}$$

Example 1.1: DC Circuit Analysis with Independent Sources

For the circuit shown in Figure 1.5, apply Ohm's law, KVL, and KCL to solve for the labeled voltages and currents.

The first step in the analysis is to apply KCL at Node A and KVL at the left and right loop. These equations are provided in Equations 1.8 through 1.10:

$$\text{KCL at Node A:} \quad -I_s + I_2 + I_3 = 0 \tag{1.8}$$

$$\text{KVL at left loop:} \quad -120 + V_1 + V_2 = 0 \tag{1.9}$$

$$\text{KVL at right loop:} \quad -V_2 + V_3 + V_4 = 0 \tag{1.10}$$

Next, use Ohm's law to rewrite Equations 1.9 and 1.10 in terms of the branch currents and resistor values. These equations are shown in Equations 1.11 and 1.12:

$$\text{KVL at left loop:} \quad 50I_s + 100I_2 = 120 \tag{1.11}$$

$$\text{KVL at right loop:} \quad -100I_2 + 20I_3 + 80I_3 = 0 \tag{1.12}$$

Solving the simultaneous set of equations, (1.8), (1.11), and (1.12) yields

$$I_s = 1.2\,\text{A}, \quad I_2 = 0.6\,\text{A}, \quad I_3 = 0.6\,\text{A} \tag{1.13}$$

The results in (1.13) and Ohm's law can be used to find the unknown voltages:

$$V_1 = 50I_s = 60\,\text{V} \tag{1.14}$$

$$V_2 = 100I_2 = 60\,\text{V} \tag{1.15}$$

$$V_3 = 20I_3 = 12\,\text{V} \tag{1.16}$$

$$V_4 = 80I_3 = 48\,\text{V} \tag{1.17}$$

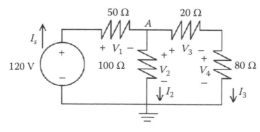

FIGURE 1.5   DC circuit with independent sources.

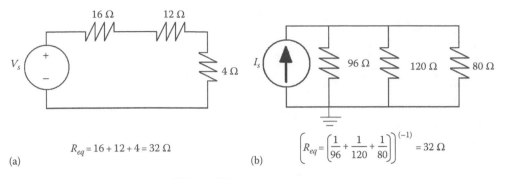

$$R_{eq} = 16 + 12 + 4 = 32\ \Omega$$

(a)

$$\left(R_{eq} = \left[\frac{1}{96} + \frac{1}{120} + \frac{1}{80}\right]^{(-1)}\right) = 32\ \Omega$$

(b)

FIGURE 1.6 Resistors in (a) series and (b) parallel.

1.1.5 Series and Parallel Relationships

At times, it is useful to simplify resistive networks by combining resistors in series and parallel into an equivalent resistance. Exactly two resistors that are connected at a single node share the same current and are said to be connected in series. It is important to note that the equivalent resistance of series resistors is larger than each of the individual resistances. Resistors that are connected together at a pair of nodes ("single node pair") have the same voltage and are said to be connected in parallel. The equivalent conductance of resistors in parallel is the sum of the conductances of the individual resistors. Therefore, the reciprocal of the equivalent resistance is the sum of the individual conductances. Note that the equivalent resistance of parallel resistors is smaller than each of the individual resistances. Figure 1.6a provides an example of a circuit with series resistors and the equivalent resistance seen by the voltage source. Figure 1.6b provides an example of a circuit with parallel resistors and the equivalent resistance seen by the current source.

Example 1.2: Analysis of Example 1.1 by Combining Resistors

It is possible to analyze the circuit in Example 1.1, to find the source current, I_s. The first step is to recognize that the 80 and 20 Ω resistors are in series and combine to yield 100 Ω. This simplified circuit is shown in Figure 1.7.

The next step is to note that the two 100 Ω resistors are in parallel. Combine these two resistors to yield the equivalent resistance of 50 Ω (see Figure 1.8).

The last simplification is to note that the 50 Ω resistors in Figure 1.8 are in series and yield the equivalent resistance of 100 Ω (see Figure 1.9).

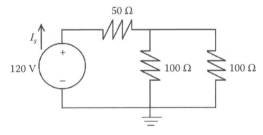

FIGURE 1.7 Circuit in Example 1.1 simplified by putting 80 W in series with 20 W.

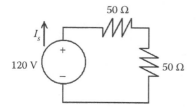

FIGURE 1.8 Circuit in Figure 1.7 simplified by putting 100 W resistors in parallel.

Finally, the last step is to use Ohm's law to solve I_s, which yields

$$I_s = \frac{120}{100} = 1.2\,\text{A} \tag{1.18}$$

Note that this result is consistent with the answer to Example 1.1.

1.1.6 Voltage and Current Divider Rule

Given a set of series resistors with a voltage sourced across them, the voltage across each individual resistor divides in direct proportion to the value of the resistor. This relationship is referred to as the voltage divider rule and it can be derived from KVL. Given a set of parallel resistors with a current sourced through them, the current through each individual resistor divides inversely proportional to the value of the resistor. This relationship is defined as the current divider rule and it can be derived from KCL. These two rules are shown for the circuits in Figure 1.6 and are shown in Figure 1.10.

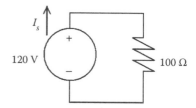

FIGURE 1.9 Circuit in Figure 1.8 simplified by putting 50 W resistors in series.

Example 1.3: Analysis of Example 1.1 Using Voltage and Current Divider

For the circuit in Figure 1.5, given that $I_s = 1.2$ A, use the current divider to find I_2 and the voltage divider to find V_4. The first step in the analysis is to recognize that the 100 Ω resistor is in parallel with the 80 and 20 Ω series combination. The current divider relationship to find I_2 is shown in Equation 1.19:

$$I_2 = \frac{100\|(80+20)}{100} I_s = 0.6\,\text{A} \tag{1.19}$$

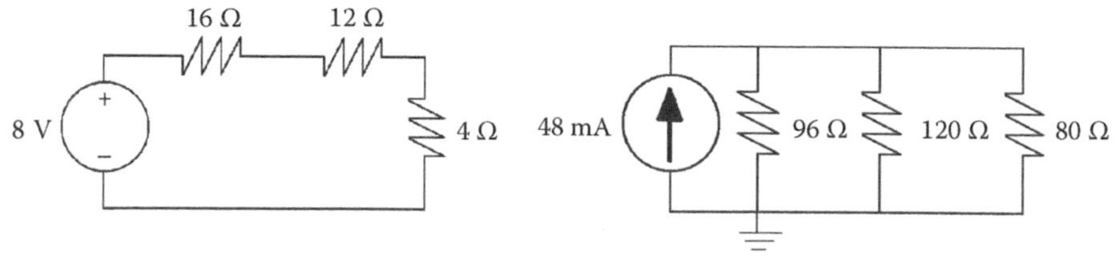

$$V_{16\Omega} = \frac{16}{(16+12+4)8} = 4\,\text{V}$$

$$V_{12\Omega} = \frac{16}{(16+12+4)8} = 3\,\text{V}$$

$$V_{4\Omega} = \frac{4}{(16+12+4)8} = 1\,\text{V}$$

(a)

$$I_{96\Omega} = \frac{96\|120\|80}{96}\,48 = 16\,\text{mA}$$

$$I_{120\Omega} = \frac{96\|120\|80}{120}\,48 = 12.8\,\text{mA}$$

$$I_{80\Omega} = \frac{96\|120\|80}{80}\,48 = 19.2\,\text{mA}$$

(b)

FIGURE 1.10 Voltage and current divider rule for circuits in Figure 1.6. (a) Series circuit (voltage divider). (b) Parallel circuit (current divider).

TABLE 1.3　Delta–Wye (Δ–Y) Transformations

Δ Configuration	Y Configuration

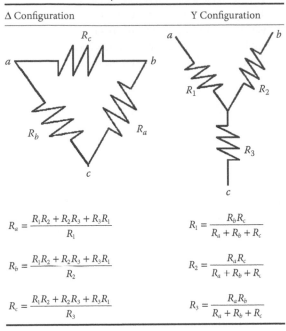

$$R_a = \frac{R_1 R_2 + R_2 R_3 + R_3 R_1}{R_1}$$

$$R_b = \frac{R_1 R_2 + R_2 R_3 + R_3 R_1}{R_2}$$

$$R_c = \frac{R_1 R_2 + R_2 R_3 + R_3 R_1}{R_3}$$

$$R_1 = \frac{R_b R_c}{R_a + R_b + R_c}$$

$$R_2 = \frac{R_a R_c}{R_a + R_b + R_c}$$

$$R_3 = \frac{R_a R_b}{R_a + R_b + R_c}$$

Ohm's law can be used to find the voltage, V_2, across the 100 Ω resistor, $V_2 = 100 I_2 = 60$ V. The voltage divider can be used to find the voltage, V_4, as shown in Equation 1.20:

$$V_4 = \frac{80}{80 + 20} V_2 = 48 \text{ V} \tag{1.20}$$

Note that these results are consistent with the solution to Example 1.1.

1.1.7 Delta–Wye (Δ–Y) Transformations

There are some resistance configurations that are neither in series or parallel. These special configurations are referred to as delta ("Δ") or wye ("Y") interconnections. These two configurations are equivalent based upon the relationships shown in Table 1.3. Equivalence means that both configurations have the same voltage and current characteristics at terminals a, b, however internal to the network, the values may not be the same.

1.2 Systematic Circuit Analysis Techniques

There are two general approaches to solving circuits using systematic techniques. The systematic techniques are the node-voltage method based on KCL and the mesh-current method based on KVL. These techniques are used to derive the minimum number of linearly independent equations necessary to find the solution.

1.2.1 Node-Voltage Method

The node-voltage method is a general technique that can be applied to any circuit. An independent KCL equation can be written at every essential node (nodes with three or more elements connected) except for one. The standard practice is to choose the ground node as the reference node and omit the ground node from the set of equations. Next, each essential node is labeled with a voltage variable (V_1, V_2, etc.). The node voltage represents the positive voltage difference at the labeled node with respect to the reference node. A KCL equation is written summing the currents leaving the node in terms of the unknown node voltages. Lastly, this set of linearly independent equations is solved for the unknown node voltages. Finally, the node voltages can be used to find any current in the circuit.

Example 1.4: Node-Voltage Method with Independent Sources

Given the circuit in Figure 1.11, use the node-voltage method to find the power delivered by each source.

    Recall that the first step in the analysis was to label the essential nodes. The four essential nodes in Figure 1.11 have already been labeled as V_1, V_2, V_3, and ground (0 V). Since V_1 is the voltage at that node with respect to the reference node ("ground node"), it is tied to the 200 V source so $V_1 = 200$ V. The node voltages V_2 and V_3 are unknown, thus KCL must be performed to find these values. In order to simplify analysis, the KCL equations are derived such that the current is drawn leaving the node if it is not given. The KCL equations at V_2 and V_3 are given in Equations 1.21 and 1.22:

$$\text{KCL at } V_2: \quad I_{500\,\Omega} + I_{250\,\Omega} + I_{400\,\Omega} = \frac{V_2 - V_1}{500} + \frac{V_2}{250} + \frac{V_2 - V_3}{400} = 0 \tag{1.21}$$

$$\text{KCL at } V_3: \quad I_{100\,\Omega} + I_{400\,\Omega} - 1 = \frac{V_3 - V_1}{100} + \frac{V_3 - V_2}{400} - 1 = 0 \tag{1.22}$$

By substituting $V_1 = 200$ into Equations 1.21 and 1.22, and solving the simultaneous system of equations yields

$$V_1 = 200 \text{ V}, \quad V_2 = 125 \text{ V}, \quad V_3 = 265 \text{ V} \tag{1.23}$$

Using the results of Equation 1.23, it is possible to find the power associated with the 1 A current source. Since the voltage across the current source is V_3, and it is not in the passive sign convention, the power is

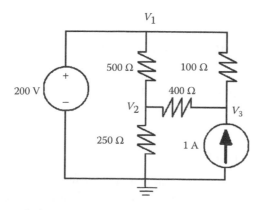

FIGURE 1.11   Node-voltage method circuit.

$P = -V_3\,(1) = -265\,W$ *or* 265 W *delivered.* In order to find the current through the 200 V source, it is necessary to use KCL at V_1. The KCL equation at V_1 is given in Equation 1.24:

$$\text{KCL at } V_1: \quad I_s + I_{500\,\Omega} + I_{100\,\Omega} = I_s + \frac{V_1 - V_2}{500} + \frac{V_1 - V_3}{100} = 0 \tag{1.24}$$

$$I_s = 500\,\text{mA} \tag{1.25}$$

Since the 200 V source obeys the passive sign convention, the power is $P = 100I_s = 100\,W$ absorbed. In order to check that the analysis is correct, the law of conservation can be used to verify that the sum of all of the power delivered equals the sum of all of the power absorbed.

Example 1.5: Analysis of Example 1.4 with Δ-Y Transformations

For the circuit in Figure 1.11, use Δ-Y transformations to find the power associated with the 200 V source. The first step in the analysis is to identify that the 500, 100, and 400 Ω resistors form a Δ configuration as R_a, R_b, and R_c, respectively. This circuit can be simplified by converting the Δ configuration to a Y configuration. Equations 1.26 through 1.28 are used to find the resistor values in the Y configuration. The simplified circuit is shown in Figure 1.12.

$$R_1 = \frac{R_b R_c}{R_a + R_b + R_c} = \frac{(400)(100)}{500 + 100 + 400} = 40\,\Omega \tag{1.26}$$

$$R_2 = \frac{R_c R_a}{R_a + R_b + R_c} = \frac{(400)(500)}{500 + 100 + 400} = 200\,\Omega \tag{1.27}$$

$$R_3 = \frac{R_a R_b}{R_a + R_b + R_c} = \frac{(500)(100)}{500 + 100 + 400} = 50\,\Omega \tag{1.28}$$

In order to find the power associated with the 200 V source, perform KCL at essential Node A. The equation and solution are shown in Equations 1.29 and 1.30:

$$\text{KCL at } V_A: \quad I_s + I_{200+250\,\Omega} - I = \frac{V_A - V_1}{50} + \frac{V_A}{450} - 1 = 0 \tag{1.29}$$

$$V_A = 225\,\text{V} \tag{1.30}$$

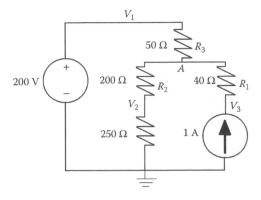

FIGURE 1.12 Circuit in Figure 1.11 simplified.

Using the result in Equation 1.30 to find the current through the 200 V source yields

$$I_s = \frac{V_A - 200}{50} = 500 \text{ mA}$$

(1.31)

Thus, the power absorbed by the 200 V source is 100 W, consistent with the prior solution.

Example 1.6: Node-Voltage Method with Dependent Sources

The circuit in Figure 1.13 models an operational amplifier. An operational amplifier is an active circuit element used to perform mathematical operations such as addition, subtraction, multiplication, division, differentiation, and integration. This electronic unit is an integrated circuit that can be modeled as a VCVS. The gain of the op amp is the ratio of the output voltage to the input voltage, (V_o/V_s). Use KCL to determine the gain of the circuit in Figure 1.13.

The KCL equations at Nodes A and B are shown in Equations 1.32 and 1.33:

$$\text{KCL at } V_A: \quad I_{10\,k\Omega} + I_{2\,M\Omega} + I_{20\,k\Omega} = \frac{V_A - V_s}{10\,k} + \frac{V_A - V_B}{20\,k} + \frac{V_A}{2\,M} = 0$$

(1.32)

$$\text{KCL at } V_B: \quad I_{50\,\Omega} + I_{20\,k\Omega} = \frac{V_B - V_A}{20\,k} + \frac{V_B - 2 \times 10^5 V_d}{50} = 0$$

(1.33)

Note that the dependent source introduces a constraint equation based upon the relationship between the node voltage and the controlling voltage, V_d. This relationship is $V_A = -V_d$. This produces two equations and two unknowns that can be solved for the gain shown in Equation 1.34:

$$\frac{V_o}{V_s} \approx \frac{-20\,k}{10\,k} = -2$$

(1.34)

A special case of the node-voltage method is when there is a voltage source between two nonreference essential nodes (see Figure 1.14).

In this case, an additional unknown variable must be introduced to describe the current in the branch with the voltage source. To minimize the number of unknowns, an alternate method to introducing another variable is to label the voltage source and any element in parallel with it as a supernode. The supernode in Figure 1.14 is denoted by the superimposed oval. The node-voltage method with supernodes involves deriving a KCL and KVL equation at the supernode as well as KCL equations at any other essential nodes where the voltage is unknown and solving the simultaneous system of equations.

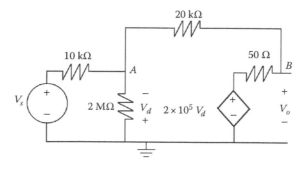

FIGURE 1.13 DC circuit with dependent sources (operational amplifier model).

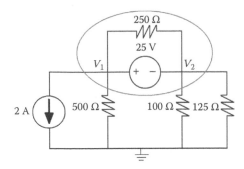

FIGURE 1.14 Node-voltage method with a supernode.

Example 1.7: Node-Voltage Method with Supernodes

Use the node-voltage method on the circuit in Figure 1.14 to find the current through the voltage source. The first step in the analysis is to label the node voltages and supernode. These have already been labeled in the circuit in Figure 1.14. Next, KCL at the supernode yields Equation 1.35, and KVL at the supernode yields Equation 1.36:

$$\text{KCL at supernode:}\quad 2 + I_{500\,\Omega} + I_{100\,\Omega} + I_{125\,\Omega} = 2 + \frac{V_1}{500} + \frac{V_2}{100} + \frac{V_2}{125} = 0 \tag{1.35}$$

$$\text{KVL at supernode:}\quad -V_1 + 25 + V_2 = 0 \tag{1.36}$$

Solving these two equations and two unknowns yields

$$V_1 = -77.5\ \text{V} \tag{1.37}$$

$$V_2 = -102.5\ \text{V} \tag{1.38}$$

To find the current through the voltage source, it is necessary to perform KCL at V_1 or V_2. Since the 2 A current source is connected to V_1, this selection will have one less term with a voltage variable. Assuming the current through the voltage source, I_s, flows from right to left and applying KCL at V_1 yields the following equation:

$$\text{KCL at supernode:}\quad I_s = 2 + I_{500\,\Omega} + I_{250\,\Omega} = 2 - 155\ \text{m} + 100\ \text{m} = 1.945\ \text{A} \tag{1.39}$$

1.2.2 Mesh-Current Method

The goal of the mesh-current method is to determine all of the unknown mesh currents in a circuit. A mesh is a loop in a circuit that does not contain any other loops. The mesh-current method is only applicable to planar circuits, circuits that can be drawn on a plane with no crossing branches. The first step in the analysis is to label all of the mesh currents in a circuit. The mesh currents are fictitious currents that circulate in a mesh. Note that a mesh current may or may not be a branch current, but all of the branch currents can be found from the mesh currents. The next step is to write KVL equations summing voltage drops around the mesh in terms of the unknown mesh currents. For *n* meshes, there will be *n* linearly independent mesh-current equations to solve.

Example 1.8: Mesh-Current Method on Example 1.6

For the circuit in Figure 1.15, use the mesh-current method to determine the output voltage V_o if the input voltage $V_s = 3\,V$.

The first step in the analysis is to label the two mesh currents, and this has been done in Figure 1.15. The second step is to write the KVL equations around meshes 1 and 2 in terms of the mesh currents, I_1 and I_2 (see Equations 1.40 and 1.41):

$$\text{KVL at mesh 1:} \quad 3 + 10\,kI_1 + 2\,M\,(I_1 - I_2) = 0 \tag{1.40}$$

$$\text{KVL at mesh 2:} \quad 2\,M\,(I_2 - I_1) + 20\,kI_2 + 50I_2 + 2\times10^5\,V_d = 0 \tag{1.41}$$

Similar to the prior analysis, the dependent source introduces the following constraint equation:

$$\text{constraint:} \quad V_d = 2\,M\,(I_2 - I_1) \tag{1.42}$$

Solving these three simultaneous equations for the mesh currents yields

$$I_1 = 299.997\ \mu A, \quad I_2 = 300.002\ \mu A, \quad V_d = -30.075\ \mu V \tag{1.43}$$

Using the mesh current value to find V_o yields

$$V_o = 50I_2 + 2 \times 10^5\,V_d = -6\,V \tag{1.44}$$

The reader should verify that this gain is consistent with Example 1.6.

A special case of the mesh-current method occurs when a current source is shared between two meshes. In this case, it is necessary to introduce another variable to describe the voltage across the current source in order to write the KVL equation. An alternate approach is to define the two meshes that include the current source and anything in series with it as a *supermesh*. The 6 A current source in series with the 1 Ω resistor in Figure 1.16 creates a supermesh denoted by the superimposed rectangle.

In order to analyze a supermesh, it is necessary to perform KVL and KCL at the supermesh. Lastly, write a KVL equation for any other unknown mesh currents in the circuit and solve the simultaneous system of equations. This method will be demonstrated on the circuit in Figure 1.16.

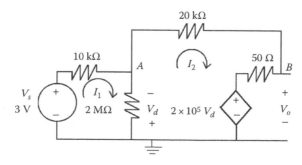

FIGURE 1.15   Mesh-current method example.

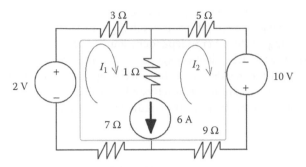

FIGURE 1.16 Mesh-current method with a supermesh.

Example 1.9: Mesh-Current Method with a Supermesh

Use the mesh-current method to find the power associated with the 6 A current source. The first step in the analysis is to label the supermesh and mesh currents. These have already been labeled in the circuit in Figure 1.16. The next step is to derive the KVL and KCL equations at the supermesh and these are shown in Equations 1.45 and 1.46:

$$\text{KVL at supermesh:} \quad -2 + 3I_1 + 5I_2 - 10 + 9I_2 + 7I_1 = 0 \tag{1.45}$$

$$\text{KCL at supermesh:} \quad I_1 - I_2 = 6 \tag{1.46}$$

Solving this simultaneous set of equations yields

$$I_1 = 4\,\text{A}, \quad I_2 = -2\,\text{A} \tag{1.47}$$

In order to determine the power associated with the 6 A current source, it is necessary to perform KVL at the left or right mesh to find the voltage across the current source. Assuming the voltage across the current source, V_s, is positive on top and applying KVL at the left mesh yields

$$\text{KVL at mesh 1:} \quad V_s = 1(I_2 - I_1) - 3I_1 + 2 - 7I_1 = -44\,\text{V} \tag{1.48}$$

The 6A current source is drawn in the passive sign convention and since V_s is negative, the power associated with this source is $P = +(-44)(6) = -264\,\text{W}$ or 264 W delivered.

1.2.3 Superposition

Superposition applies to linear circuits that have multiple independent sources. The principle of superposition states that the electrical quantities, voltage or current, due to all the sources acting at the same time is equal to the sum of the same quantity due to each source acting alone. The method to solve for an unknown variable in a circuit involves solving for the variable of interest for one source acting alone by deactivating all the other independent sources, then sum the results for each source acting alone. To deactivate an independent voltage source, replace the voltage source with a short circuit (0 V). To deactivate an independent current source, replace the current source with an open circuit (0 A). Dependent sources are never deactivated ("turned off"). The benefit in applying the principle of superposition is that many times, the circuit with the deactivated source is simpler to solve for the unknown value.

Example 1.10: Circuit Analysis Using Superposition

For the circuit in Figure 1.16, apply the principle of linear superposition to solve for the unknown branch current I_1 (see Figure 1.17).

The first step in the analysis is to disable the 6 A and 10 V sources and use KVL to calculate I_1. The solution to this analysis is shown in Equation 1.49. The variable of interest is given a prime to denote that it is due to one source acting alone (see Figures 1.18 through 1.20).

$$\text{KVL at mesh 1:} \quad -2 + 3I_1 + 5I_1 + 9I_1 + 7I_1 = 0$$

$$I_1' = 83.33 \text{ mA} \tag{1.49}$$

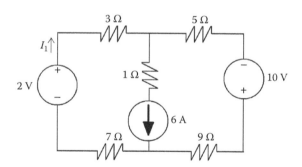

FIGURE 1.17 Superposition circuit for Example 1.10.

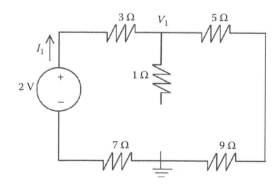

FIGURE 1.18 Superposition circuit with 2 V source activated.

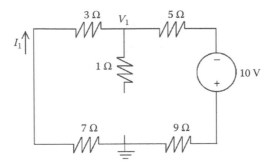

FIGURE 1.19 Superposition circuit with 10 V source activated.

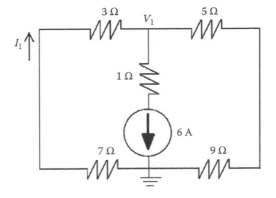

FIGURE 1.20 Superposition circuit with 6 A source activated.

In the next step, disable the 2 V and 6 A sources and use KVL to calculate I_1. The solution to this analysis is shown in Equation 1.50:

$$\text{KVL at mesh 1:} \quad 3I_1 + 5I_1 - 10 + 9I_1 + 7I_1 = 0$$

$$I_1'' = 416.67 \, \text{mA} \tag{1.50}$$

In the next step, disable the 2 and 10 V sources and use KCL at V_1 to calculate I_1. The solution to this analysis is shown in Equation 1.51:

$$\text{KCL at } V_1: \quad \frac{V_1}{3+7} + 6 + \frac{V_1}{5+9} = 0$$

$$V_1 = -35 \, \text{V}$$

$$I_1''' = \frac{0 - V_1}{7+3} = 3.5 \, \text{A} \tag{1.51}$$

Applying the principle of superposition yields

$$I_1 = I_1' + I_1'' + I_1''' = 4A \tag{1.52}$$

Note that the value for I_1 is consistent with the solution to Example 1.9.

1.3 Circuit Modeling Techniques

Just like it is possible to model the behavior of multiple resistors connected in parallel and series with a single equivalent resistance, it is also possible to model resistive circuits containing sources and resistors as either a Thevenin or Norton equivalent model. These models are useful simplifying techniques when only the circuit behavior at a single port is of interest.

1.3.1 Source Transformations

Source transformations are another simplifying technique for circuit analysis. Source transformations are based upon the concept of equivalence of the voltage and current terminal characteristics at a single port. A voltage source in series with a resistor can be replaced by a current source in parallel with a resistor if they have the relationships given in Figure 1.21.

1.3.2 Thevenin and Norton Equivalent Circuits

A simple resistive circuit can be simplified to an independent voltage source in series with a resistor and this is referred to as the Thevenin equivalent circuit. The voltage source is referred to as the Thevenin voltage, V_{TH}, and the resistor is the Thevenin resistance, R_{TH}. In addition, a simple resistive circuit can be simplified to an independent current source in parallel with a resistor and this is referred to as the Norton equivalent circuit. The current source is the Norton current, I_N, and the resistance is the same as the Thevenin resistance. These are important simplification techniques when the values of interest are the port characteristics such as the voltage, current, or power delivered to a load placed across the terminals. The method to find the Thevenin voltage is to determine the open circuit voltage across terminals a and b. The method to find the Norton current is to find the short circuit current between terminals a and b. There are several techniques to find the Thevenin equivalent resistance. When there are only independent sources, one of the more popular methods is to deactivate all independent sources and find the equivalent resistance of the network across terminals a and b. Alternately, the Thevenin resistance can be calculated by using the following formula:

$$R_{TH} = \frac{V_{oc}}{I_{sc}} = \frac{V_{TH}}{I_N}$$

(1.53)

where

 V_{oc} represents the open circuit voltage across terminals a and b
 I_{sc} represents the short circuit current between terminals a and b

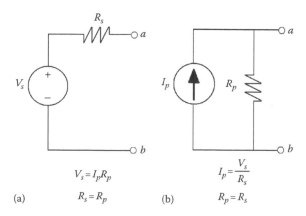

$$V_s = I_p R_p$$
$$R_s = R_p$$

$$I_p = \frac{V_s}{R_s}$$
$$R_p = R_s$$

(a)   (b)

FIGURE 1.21 Source transformation relationships. (a) Series circuit, $V_s = I_p R_p$, $R_s = R_p$ and (b) parallel circuit, $I_p = V_s/R_s$, $R_p = R_s$.

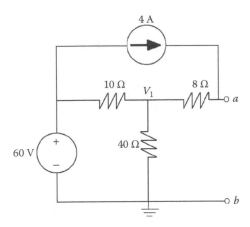

FIGURE 1.22 Circuit for Thevenin equivalent example.

Note that when there are dependent sources in the circuit, there is a third technique based upon deactivating all independent sources and using a test voltage or current at terminals *a* and *b* to find the equivalent resistance. The reader is encouraged to review this technique for future study.

Example 1.11: Thevenin Equivalent Resistance

For the circuit in Figure 1.22, determine the Thevenin equivalent resistance to the left of terminals *a* and *b*.

In order to find the Thevenin equivalent resistance, deactivate the two independent sources and find the equivalent resistance to the left of terminals *a* and *b*. The circuit in Figure 1.22 is shown in Figure 1.23 with the sources deactivated.

In the circuit in Figure 1.23, the 10 and 40 Ω resistors are in parallel. This parallel combination is in series with the 8 Ω resistor. Equation 1.54 shows the derivation of the Thevenin equivalent resistance, R_{TH}:

$$R_{TH} = 10 \parallel 40 + 8 = 16 \, \Omega \tag{1.54}$$

Example 1.12: Thevenin and Norton Equivalent Circuits

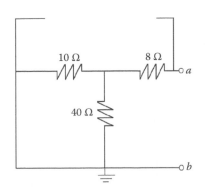

FIGURE 1.23 Circuit in Figure 1.22 with independent sources deactivated.

For the circuit in Figure 1.22, find the Thevenin and Norton equivalent circuit to the left of terminals *a* and *b*. The first step in the analysis is to find the open circuit voltage between terminals *a* and *b* ($V_{TH} = V_{oc} = V_a$). Either the node-voltage or mesh-current method would be an acceptable technique to find this value, however the node-voltage method was used by writing the KCL equation at V_1 and V_a and these are shown in Equation 1.55:

$$\text{KCL at } V_1: \quad \frac{V_1 - 60}{10} + \frac{V_1}{40} + \frac{V_1 - V_a}{8} = 0 \tag{1.55}$$

$$\text{KCL at } V_a: \quad \frac{V_a - V_1}{8} = 4$$

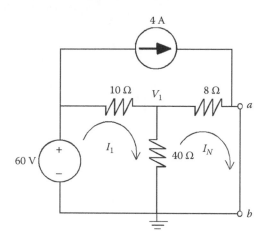

FIGURE 1.24   Circuit in Figure 1.22 with terminals *a* and *b* short-circuited.

$$V_1 = 80\,V, \qquad V_{TH} = V_{oc} = V_a = 112\,V$$

The next step in the analysis is to find the short circuit current between terminals *a* and *b* ($I_N = I_{sc} = I_{ab}$). The mesh-current method will be used to determine short circuit current, I_{sc}, as shown in Figure 1.24. The result of the analysis is shown in Equation 1.56:

$$\text{KVL at } I_1: \quad -60 + 10(I_1 - 4) + 40\,(I_1 - I_N) = 0$$

$$\text{KVL at } I_N: \quad 40\,(I_N - I_1) + 8(I_N - 4) = 0 \tag{1.56}$$

$$I_1 = 7.6\,A, \quad I_N = I_{sc} = I_{ab} = 7\,A$$

The Thevenin equivalent resistance can also be found from

$$R_{TH} = \frac{V_{oc}}{I_{sc}} = \frac{V_{TH}}{I_N} = \frac{112}{7} = 16\,\Omega \tag{1.57}$$

Note that this Thevenin resistance is consistent with Example 1.11. The final step in the result is to draw the Thevenin and Norton equivalent circuits to the left of terminals *a* and *b*. These are shown in Figure 1.25.

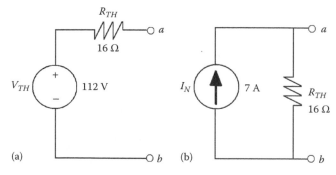

FIGURE 1.25   Thevenin and Norton equivalent of circuit in Figure 1.22. (a) Thevenin equivalent. (b) Norton equivalent.

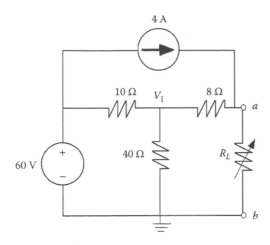

FIGURE 1.26 Circuit for maximum power transfer example.

1.3.3 Maximum Power Transfer

One benefit and purpose for determining the Thevenin (Norton) equivalent of a circuit is to determine the power delivered to a load placed across terminals *a* and *b*. With the knowledge of the Thevenin and Norton equivalent, it is possible to design a circuit or select a load for maximum power transfer to the load. It can be shown that if the load resistance is equal to the Thevenin equivalent resistance, then maximum power is transferred to the load. Therefore, the condition for maximum power transfer is to set the load resistance equal to the Thevenin equivalent resistance. When the load resistance is equal to the Thevenin equivalent resistance, the maximum power delivered to the load is

$$P_L = \frac{V_{TH}^2}{4R_{TH}} \tag{1.58}$$

Example 1.13: Maximum Power Transfer

For the circuit shown in Figure 1.22, determine the value of a load resistor placed across terminals *a* and *b* for maximum power transfer and calculate the value of the power for the load selected (see Figure 1.26).

Since the Thevenin equivalent resistance of this circuit is 16 Ω, select $R_L = R_{TH} = 16$ Ω for maximum power transfer. Finally, the value of the power delivered to the 16 Ω is calculated as follows:

$$P_L = \frac{V_{TH}^2}{4R_{TH}} = \frac{112^2}{4(16)} = 196 \text{ W} \tag{1.59}$$

1.4 Transient Analysis

Transient analysis describes a circuit's behavior as a function of time. Since capacitors and inductors store energy in the electric or magnetic field, transient analysis focuses on the current and voltage values in circuits where energy is either stored by or released by an inductor or a capacitor.

1.4.1 First-Order Circuits

First-order circuits contain resistors and either capacitors or inductors, but not both. These configurations are either RL circuits or RC circuits based upon whether they have resistors and capacitors or resistors and inductors, respectively. RL and RC circuits are known as first-order circuits because the equations that describe these circuits are first-order ordinary differential equations. If a voltage or current source is suddenly applied to a first-order circuit (i.e., a switch), then energy will begin to store in the capacitor as an electric field or in the inductor as a magnetic field. When a source is instantaneously applied, the time-dependent current or voltage in the circuit is called the step response. If the source of energy is suddenly removed, then the time-dependent current or voltage in the circuit is called the natural response. It is important to note that the voltage across a capacitor cannot change instantaneously and the current through an inductor cannot change instantaneously. The natural and step response of first-order circuits can be found by using circuit analysis techniques such as KVL and KCL to derive the first-order differential equation that describes the circuit. Using the initial conditions and differential equations, these equations can be solved for voltage and current. In order to find the initial conditions for a first-order circuit, it is necessary to draw the circuit under DC conditions before the switching occurs. Note that under DC or steady state conditions, inductors can be modeled as short circuits and capacitors can be modeled as open circuits. The general form of the solution for a first-order circuit is the sum of the transient response and the steady-state response. The transient response is the portion of the response that decays over time. The steady-state response is the portion of the response that remains after a long time. Furthermore, the general form of the solution can be described as the sum of the natural response and the forced response. The forced response is the portion due to the independent sources and the natural response is due to the energy stored in the circuit. The general solution for natural and step responses for first-order circuits is given in Equation 1.60:

$$x(t) = x_{steady\text{-}state} + x_{transient} = x(t \to \infty) + [x(t = 0^+) - x(t \to \infty)]e^{-t/\tau}$$

$$x(t) = x_{step} + x_{natural} = \left[x(t \to \infty) - x(t \to \infty)e^{-t/\tau} \right] + x(t = 0^+)e^{-t/\tau}$$

(1.60)

where

$\tau = R_{TH}C$ or $\tau = L/R_{TH}$
R_{TH} is equivalent resistance across the inductor or the capacitor
$x(t)$ is either voltage or current

The time constant, τ, is in units of seconds and describes how fast the transient signal settles. At a time equal to 5τ, the transient solution will settle to within 1% of the steady-state or final value. The general solution can be applied to any RL or RC circuit provided that the initial and final conditions and equivalent resistance seen by the inductor or capacitor can be found.

Example 1.14: Natural Response of an RL Circuit

For the circuit in Figure 1.27, assume that the switch is in position *a* for a long time and moves to position *b* at *t* = 0. Find the current through the inductor, *i*(*t*), and the voltage across the inductor, *v*(*t*), for *t* > 0.

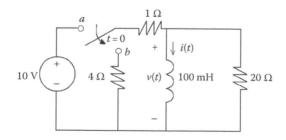

FIGURE 1.27   RL circuit for Example 1.14.

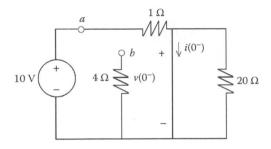

FIGURE 1.28 RL circuit at $t = 0^-$.

The first step in the analysis is to find the initial conditions for the circuit in Figure 1.27. In order to find the initial conditions, redraw the circuit at $t = 0^-$ (before the switching occurs) under DC conditions. The circuit to find the initial conditions is shown in Figure 1.28. Since an inductor under DC or steady-state conditions is modeled as a short circuit, the initial voltage, $v(0^-)$, is 0V. It is modeled as a short circuit because the current is constant with time; therefore, the time rate of change of the current (di_L/dt) is zero and the voltage ($v_L = L(di_L/dt)$) over the inductor is 0V. Using Ohm's law on the circuit in Figure 1.28, it is possible to find the initial current, $i(0^-)$ as shown in Equation 1.61. Note that because the inductor is a short circuit, the 20 Ω resistor is shorted out and has no affect on the circuit.

$$i(0^-) = \frac{10}{1} = 10 \text{ A} \tag{1.61}$$

Since the voltage across an inductor can change instantaneously, the circuit must also be analyzed immediately after switching occurs at $t = 0^+$. For this analysis, model the inductor as a 10 A current source because current cannot change instantaneously so $i(0^+) = i(0^-) = 10$ A. Next, use KCL to find the voltage across the inductor. The circuit is shown in Figure 1.29 and the analysis in Equation 1.62:

$$v(0^+) = -[(4+1)\|20]10 = -40 \text{ V} \tag{1.62}$$

The circuit in Figure 1.29 can also be used to find the time constant, $\tau = L/R_{TH}$, where R_{TH} is the equivalent resistance seen by the inductor. Since the 4 and 1 Ω resistors are in series and they are in parallel with the 20 Ω resistor, R_{TH} is given by Equation 1.63:

$$R_{TH} = (4+1)\|20 = 4 \ \Omega \tag{1.63}$$

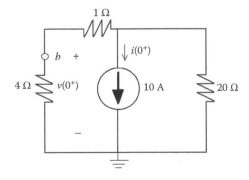

FIGURE 1.29 RL circuit at $t = 0^+$.

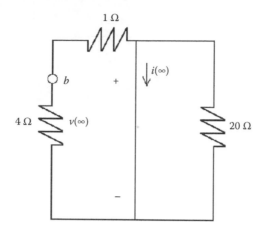

FIGURE 1.30   RL circuit at $t = \infty$.

The time constant is given by Equation 1.64.

$$\tau = \frac{L}{R_{TH}} = \frac{0.1}{4} = 25\,\text{ms} \tag{1.64}$$

In order to find the final value for the current and voltage, analyze the circuit under steady-state conditions a long time after switching occurs. This circuit is shown in Figure 1.30; note that once again the 20 Ω resistor is shorted out. Since, it is assumed that the circuit has been in this state for a long time, the current through the inductor and voltage across the inductor can be represented as $i(\infty)$ and $v(\infty)$, respectively. Since the inductor is still modeled as a short circuit, thus $v(\infty) = 0\,\text{V}$ and since there are no sources, $i(\infty)$ is 0 A.

These values make sense because for the natural response of a first-order circuit, the inductor has stored energy in the form of current and over time, it discharges until it is eventually 0 A. Using the general solution equation in (1.60) yields

$$i(t) = i(\infty) + [i(0^+) - i(\infty)]e^{-t/\tau} = 10e^{-40t}\ \text{A}, \quad t > 0$$

$$v(t) = v(\infty) + [v(0^+) - v(\infty)]e^{-t/\tau} = -40e^{-40t}\ \text{V}, \quad t > 0$$

$$\tag{1.65}$$

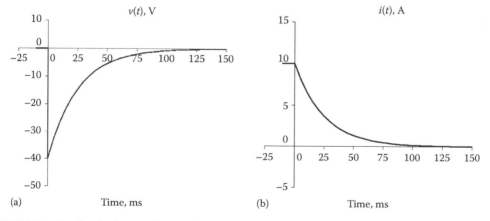

(a)   Time, ms   (b)   Time, ms

FIGURE 1.31   Graphs of voltage and current for the inductor in Example 1.14. (a) Voltage across the inductor and (b) current through the inductor.

The graphs of $v(t)$ ad $i(t)$ are shown in Figure 1.31; note that these are exponentially decaying functions to represent the natural response and the fact that the inductor is discharging. Also notice that there is a discontinuous step at $t = 0$ for the voltage across the inductor to denote the change from storing or stored energy to releasing energy.

Example 1.15: Step response of an RC Circuit

For the circuit in Figure 1.32, the switch has been in position a for a long time, and at $t = 0$, it moves to position b Find the voltage across the capacitor, $v(t)$, and the current through the capacitor, $i(t)$, for $t > 0$.

The first step in the analysis is to find the initial conditions by analyzing the circuit under steady-state conditions before switching occurs to find $i(0^-)$ and $v(0^-)$. Since a capacitor under DC or steady-state conditions is modeled as an open circuit, the initial current, $i(0^-)$, is 0 A. It is modeled as an open circuit because the voltage is constant with time, therefore, the time rate of change of the voltage (dv_C/dt) is zero and the current ($i_c = Cdv_C/dt$) over the inductor is 0 A. The circuit is shown in Figure 1.33 and the analysis using Ohm's law yields Equation 1.66:

$$v(0^-) = (1\,m)(1\,k) = 1\,V \tag{1.66}$$

Since the current through a capacitor can change instantaneously, the circuit must be analyzed right after switching occurs to find $i(0^+)$. However, since voltage across a capacitor cannot change instantaneously, after switching occurs, the capacitor can be modeled as a 1 V source (i.e., $v(0^-) = v(0^+) = 1\,V$). This circuit is shown in Figure 1.34. KCL can be used to analyze this circuit to find the current through the capacitor as shown in Equation 1.67:

$$i(0^+) = \frac{-1}{12\,k} + \frac{30-1}{6\,k} = 4.75\,mA \tag{1.67}$$

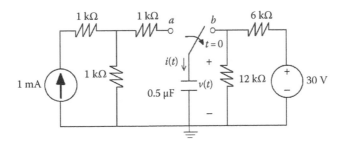

FIGURE 1.32   RC circuit for Example 1.15.

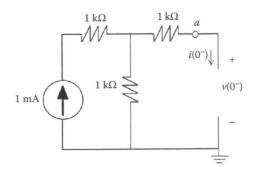

FIGURE 1.33   RC circuit for initial conditions ($t = 0^-$).

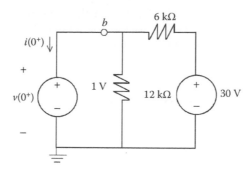

FIGURE 1.34 RC circuit for initial conditions ($t = 0^+$).

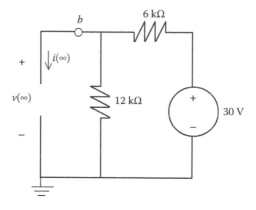

FIGURE 1.35 RC circuit for final conditions ($t = \infty$).

The circuit in Figure 1.34 can also be used to find the time constant, $\tau = R_{TH}C$. The equivalent resistance across the capacitor can be found by disabling the 30 V source. After deactivating the 30 V source, the 6 kΩ resistor is in parallel with the 12 kΩ resistor, thus $R_{TH} = 4\,k\Omega$. The time constant is $\tau = R_{TH}C = (4\,k)$ $(0.5\,\mu) = 2\,ms$. Finally, to find the steady-state voltage and current for the capacitor, analyze the circuit under steady-state conditions a long time after switching occurs. Since the capacitor is modeled as an open circuit under steady-state conditions, $i(\infty) = 0\,A$. The circuit is shown in Figure 1.35. The final value of the capacitor voltage, $v(\infty)$, can be found by using the voltage divider as shown in Equation 1.68:

$$v(\infty) = \frac{12}{12+6}(30) = 20\,V \tag{1.68}$$

Using the general solution for a first-order circuit in Equation 1.60 yields the following equations for $v(t)$ and $i(t)$:

$$i(t) = i(\infty) + [i(0^+) - i(\infty)]e^{-t/\tau} = 4.75e^{-500t}\,mA, \quad t > 0$$

$$v(t) = v(\infty) + [v(0^+) - v(\infty)]e^{-t/\tau} = 20 - 19e^{-500t}\,V, \quad t > 0 \tag{1.69}$$

The graphs of the current and voltage for the capacitor are shown in Figure 1.36. Note that since this is a step response and the capacitor voltage is charging, the graph is an exponentially increasing function. Also, since current through a capacitor can change instantaneously, there is a discontinuous step in the graph at $t = 0$ when the capacitor begins charging from 1 to 20 V.

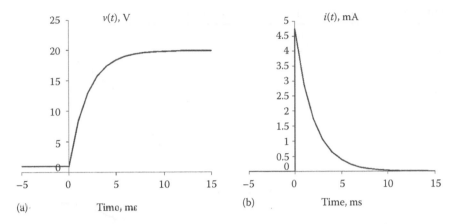

FIGURE 1.36 Graphs of voltage and current for the capacitor in Example 1.15. (a) Voltage across the capacitor and (b) current through the capacitor.

1.4.2 Second-Order Circuits

Second-order circuits contain resistors, capacitors, and inductors. These circuits are known as second-order circuits because the equations that describe these circuits are second-order ordinary differential equations. Second-order circuits also exhibit a natural and step response based upon whether the capacitor and inductor are storing or releasing energy. To solve these types of circuits, use circuit analysis to generate the second-order differential equation governing the behavior and then use differential equations and initial conditions to solve for the solution. One approach to solving the differential equation describing the transient response is to guess a solution and plug into the differential equation. Then the initial and final values can be used to determine the time response. For example, the general differential equation for these types of circuits is given by

$$\frac{d^2x}{dt^2} + 2\alpha \frac{dx}{dt} + \omega_o^2 x = K \tag{1.70}$$

where
   x represents the voltage or current
   α is the Neper frequency in rad/s
   ω_o is the resonant frequency in rad/s
   K is related to the steady-state value of the variable of interest

The damping factor and resonant frequency can be identified from the second-order equation derived for the circuit. To solve these equations, assume that the solution is of the form $x(t) = Ae^{st}$. Substitute this value into Equation 1.71 to yield the following:

$$Ae^{st}(s^2 + 2\alpha s + \omega_o^2) = K \tag{1.71}$$

Thus, the characteristic equation of any second-order circuit is $s^2 + 2\alpha s + \omega_o^2$. The general form of the solution to the second-order differential equation in (1.72) is given by

$$x(t) = A_1 e^{s_1 t} + A_2 e^{s_2 t} + x(t \to \infty) \tag{1.72}$$

The roots of the characteristic equation, s_1 and s_2, can be used to determine the type of response. There are three types of responses for second-order circuits: overdamped, critically damped, and underdamped.

TABLE 1.4 Summary of Second-Order Circuit Responses

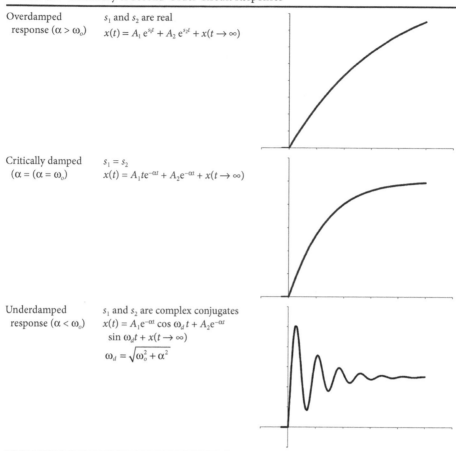

Overdamped response ($\alpha > \omega_o$)	s_1 and s_2 are real $x(t) = A_1 e^{s_1 t} + A_2 e^{s_2 t} + x(t \to \infty)$
Critically damped ($\alpha = (\alpha = \omega_o)$)	$s_1 = s_2$ $x(t) = A_1 t e^{-\alpha t} + A_2 e^{-\alpha t} + x(t \to \infty)$
Underdamped response ($\alpha < \omega_o$)	s_1 and s_2 are complex conjugates $x(t) = A_1 e^{-\alpha t} \cos \omega_d t + A_2 e^{-\alpha t}$ $\sin \omega_d t + x(t \to \infty)$ $\omega_d = \sqrt{\omega_o^2 + \alpha^2}$

The overdamped response has a slow response and long settling time. The critically damped response has a fast response and short settling time. The underdamped response has the fastest response and a long settling time. Table 1.4 presents the relationship between the three responses, the roots of the characteristics equation, the Neper frequency, resonant frequency, form of the solution, and the graph.

Example 1.16: Natural Response of an RLC Circuit

For the circuit in Figure 1.37, assume that the switch opens instantaneously at $t = 0$, what is the voltage, $v(t)$ across the capacitor and current, $i(t)$ through the capacitor.

The first step in the analysis is to determine the initial conditions or the energy stored in the inductor and capacitor. In order to find these values, analyze the circuit under steady-state conditions right before switching occurs ($t = 0^-$). This circuit is shown in Figure 1.38. As previously stated, in this circuit, the inductor is modeled as a short circuit and the capacitor is modeled as an open circuit. Since the capacitor is an open circuit, $i(0^-)$ is 0 V and since it is in parallel with a short circuit, $v(0^-)$ is 0 V. Finally, since the inductor is a short circuit and current follows the path of least resistance, $i_L(0^-) = 2$ A.

Next, the circuit must be analyzed right after switching occurs to find $i(0^+)$ (see Figure 1.39). In this circuit, the inductor is modeled as a 2 A current source and the capacitor is modeled as a 0 V voltage source or a wire. This circuit is shown in Figure 1.39. Since current is continuous for inductors and voltage is continuous for capacitors, these values do not change. However, the current through the capacitor changes to $i(0^+) = -2$ A.

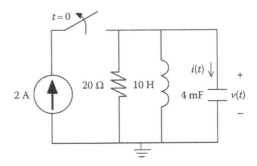

FIGURE 1.37 Natural response of a parallel RLC circuit.

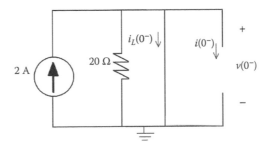

FIGURE 1.38 Parallel RLC circuit at $t = 0^-$.

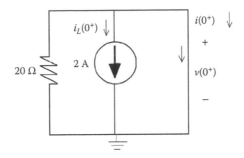

FIGURE 1.39 Parallel RLC circuit at $t = 0^+$.

The next step in the analysis is to analyze the circuit at some point after switching occurs to derive the second-order differential equation. This circuit is shown in Figure 1.40, and the derivation of the equation using KCL is shown in (1.73):

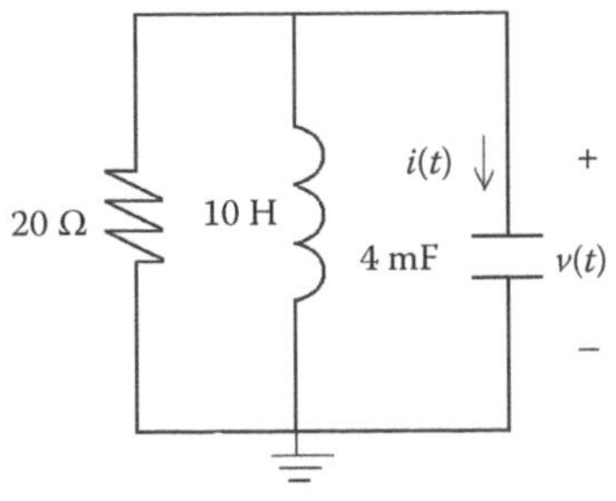

$$i_R + i_L + i_C = 0$$

$$\frac{v}{20} + \frac{1}{10}\int_0^t v\,dt + i(0^+) + 4\,\text{m}\frac{dv}{dt} = 0$$

$$\frac{1}{20}\frac{dv}{dt} + \frac{1}{10}v + 4\,\text{m}\frac{d^2v}{dt^2} = 0$$

$$\frac{d^2v}{dt^2} + 12.5\frac{dv}{dt} + 25v = \frac{d^2v}{dt^2} + 2\alpha\frac{dv}{dt} + \omega_0^2 v = 0 \qquad (1.73)$$

FIGURE 1.40 Parallel RLC circuit for Example 1.16.

From examination of Equation 1.73, it is evident that $\alpha = 6.25$ and $\omega_o = 5$ rad/s. Since $\alpha > \omega_o$, the voltage and current response are overdamped. The roots of the characteristic equation $(s^2 + 12.5s + 25)$ are -2.5 and -10, and the general form of the response, $v(t)$, is given in Equation 1.74:

$$v(t) = A_1 e^{s_1 t} + A_2 e^{s_2 t} + v(t \to \infty) = A_1 e^{-2.5t} + A_2 e^{-10t} \qquad (1.74)$$

Note that since this is a natural response and the capacitor and inductor are discharging, $v(\infty)$ and $i(\infty)$ are zero. In order to find the values of A_1 and A_2, use the circuit's initial conditions. The evaluation of Equation 1.74 and its first derivative at $t = 0^+$ yields

$$v(0^+) = A_1 + A_2 \qquad (1.75)$$

$$\frac{dv(0^+)}{dt} = -2.5A_1 - 10A_2$$

Using the initial current through the inductor and the initial voltage across the capacitor with the results of (1.75), the values of A_1 and A_2 can be found as shown in Equation 1.76:

$$v(0^+) = A_1 + A_2 = 0 \qquad (1.76)$$

$$i_R(0^+) + i_L(0^+) + i_C(0^+) = \frac{v(0^+)}{20} + i_L(0^+) + 4\,\mathrm{m}\frac{dv(0^+)}{dt} = 0$$

$$\qquad (1.77)$$

$$\frac{dv(0^+)}{dt} = -\frac{v(0^+)}{(20)(4\,\mathrm{m})} - \frac{i_L(0^+)}{4\,\mathrm{m}} = -2.5A_1 - 10A_2 = -500$$

Solving the simultaneous set of Equations 1.76 and 1.77 yields $A_1 = -66.7$ and $A_2 = 66.7$. Finally, the solutions to the example are

$$v(t) = -66.7e^{-2.5t} + 66.7e^{-10t}\ \mathrm{V}, \quad t > 0$$

$$\qquad (1.78)$$

$$i(t) = 4\,\mathrm{m}\frac{dv}{dt} = 667e^{-2.5t} - 2.667e^{-10t}\ \mathrm{mA}, \quad t > 0$$

The reader is encouraged to verify that the solution for the transient response of the capacitors' voltage and current does indeed obey the initial conditions.

Example 1.17: Natural Response of an RLC Circuit

For the circuit in Figure 1.41, assume the switch has reached steady-state before the switch moves from position *a* to position *b*. If at time $t = 0$ the switch moves to position b, calculate $i(t)$ and $v(t)$ for $t > 0$.

Similar to Example 1.17, the first step in the solution process is to determine the stored energy in the inductor and capacitor. The circuit in Figure 1.42 illustrates the circuit under steady-state conditions before the switch moves from position "*a*" ($t = 0^-$); the inductor is modeled as a short circuit ($v(0^-) = 0\,\mathrm{V}$) and the capacitor is modeled as an open circuit. By observation, the voltage across the capacitor is also $0\,\mathrm{V}$ and the current through the inductor can be found from Ohm's law, $i(0^-) = 50/10 = 5\,\mathrm{A}$.

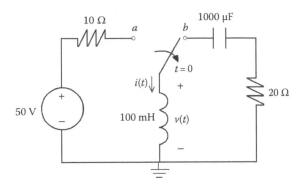

FIGURE 1.41   Natural response of a series RLC circuit.

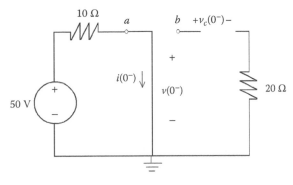

FIGURE 1.42   Series RLC circuit at $t = 0^-$.

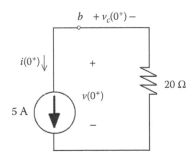

FIGURE 1.43   Series RLC circuit at $t = 0^+$.

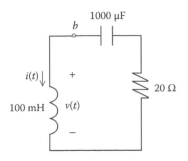

FIGURE 1.44   Series RLC circuit at $t = 0^+$.

The next step in the analysis is to use the values found at $t = 0^-$ to model the initial conditions in the circuit right after switching occurs (see Figure 1.43). Since current cannot change instantaneously through the inductor, it is modeled as a 5 A current, and because voltage cannot change instantaneously across a capacitor, it is modeled as a 0 V source or wire.

Since the 20 Ω resistor is in parallel with the 5 A current source in Figure 1.43, they have the same voltage ($v(0^+) = -(20)(5) = -100\,\text{V}$). At any instant of time after the switch moves, observe that this circuit is a series RLC circuit and KVL can be used to derive the second-order differential equation that describes it (see Figure 1.44 and Equation 1.79):

$$v_L + v_R + v_C = 0$$

$$100\,\text{m}\frac{di}{dt} + 20i + \frac{1}{1000\mu}\int_0^t i\,dt + v(0^+) = 0$$

$$100\,\text{m}\frac{d^2i}{dt^2} + 20\frac{di}{dt} + \frac{1}{1000\mu}i = 0 \qquad (1.79)$$

$$\frac{d^2i}{dt^2} + 200\frac{di}{dt} + 10{,}000i = \frac{d^2i}{dt^2} + 2\alpha\frac{di}{dt} + \omega_0^2 i = 0$$

From the examination of Equation 1.79, the Neper frequency, $\alpha = 100\,\text{rad/s}$ and the resonant frequency, $\omega_o = 100\,\text{rad/s}$. By reviewing Table 1.4, since $\alpha = \omega_o$, this is a critically damped circuit. The characteristic equation is $s^2 + 200s + 10{,}000$ and there is one repeated root, 100. The general form of the solution for the current through the inductor is given in Equation 1.80. Note that since there is no active source on the circuit after the switch moves to position "b," the inductor is discharging ($i(t \to \infty) = 0\,\text{A}$):

$$i(t) = A_1 t e^{-\alpha t} + A_2 e^{-\alpha t} + i(t \to \infty) = A_1 t e^{-100t} + A_2 e^{-100t} \tag{1.80}$$

Once again, it is necessary to use the initial conditions to determine the values of A_1 and A_2. In order to do this, the equation in 1.80 and its first derivative must be evaluated at $t = 0^+$. These equations are given in (1.81) and (1.82):

$$i(0^+) = A_2 \tag{1.81}$$

$$\frac{di(0^+)}{dt} = A_1 - 100A_2 \tag{1.82}$$

Next, using the initial conditions found from Figures 1.42 and 1.43, KVL, and Equations 1.81 and 1.82, it is possible to solve for A_1 and A_2 (see Equations 1.83 and 1.84):

$$i(0^+) = A_2 = 5 \tag{1.83}$$

$$v_L(0^+) + v_R(0^+) + v_C(0^+) = 0$$

$$100\,\text{m}\frac{di(0^+)}{dt} + 20i(0^+) + v_C(0^+) = 0$$

$$\frac{di(0^+)}{dt} = -\frac{20i(0^+)}{100\,\text{m}} - \frac{v_C(0^+)}{100\,\text{m}} = A_1 - 100A_2 = -1000 \tag{1.84}$$

By solving Equations 1.83 and 1.84 yields, $A_1 = -500$ and $A_2 = 5$. The specific solution for the inductor's current and voltage are shown in Equations 1.85 and 1.86:

$$i(t) = -500t e^{-100t} + 5e^{-100t}\;\text{A}, \quad t > 0 \tag{1.85}$$

$$v(t) = 100\,\text{m}\frac{di}{dt} = -100e^{-100t} + 5000t e^{-100t}\;\text{V}, \quad t > 0 \tag{1.86}$$

It is left to the reader to verify that Equations 1.85 and 1.86 are consistent with the initial conditions found from the circuit in Figure 1.43.

Example 1.18: Step Response of an RLC Circuit

The circuit in Figure 1.45 has been in position *a* for a long time, and at $t = 0$ it moves to position b. Determine the voltage, $v(t)$, across and current, $i(t)$, through the 25 Ω resistor.

In order to find the initial conditions for the inductor, capacitor, and resistor, the circuit in Figure 1.46 must be analyzed. Since the inductor is a short circuit and the resistor does not have any current because of the break in the circuit $v(0^-)$, $i(0^-)$, and $i_L(0^-)$ are zero. However, the capacitor is an open circuit and it is the same as the voltage across the 40 Ω resistor. Thus, the voltage can be found by using the voltage divider, $v_C(0^-) = (40/50)(25) = 20\,V$.

The next step in the analysis is to model the initial voltage across the capacitor and the initial current through the inductor as independent sources right after it moves to position *b* (see Figure 1.47). The

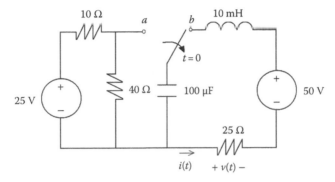

FIGURE 1.45 Step response of an RLC circuit.

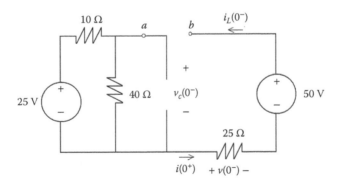

FIGURE 1.46 Series RLC circuit at $t = 0$.

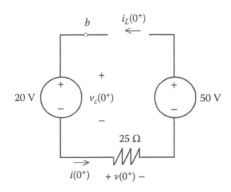

FIGURE 1.47 Series RLC circuit at $t = 0^+$.

capacitor is modeled as a 20 V voltage source, the inductor had an initial current of 0 A, so this is modeled as an open circuit. The resistor model remains the same because voltage across and current through a resistor can change instantaneously. However, since there is a break in the circuit, the current, $i(0^+)$, and the voltage, $v(0^+)$, are still zero.

Since this circuit is a step response and the inductor and/or capacitor are storing energy, it is also necessary to analyze the circuit after the switch has been in position "b" for a long time. It is assumed that this circuit has reached steady-state conditions and the circuit model is shown in Figure 1.48. The analysis of this circuit indicates that after a long time, the capacitor has charged to 50 V ($v_C(\infty) = 50$). The current through the inductor and resistor is 0 A because of the open circuit. Finally, the voltage across the resistor is $v(\infty) = 0$ V because there is no current flow.

The next step in the analysis is to derive the second-order differential equation that describes this series RLC circuit after the switch moves to position "b." The circuit is shown in Figure 1.49 and it is analyzed in Equation 1.87:

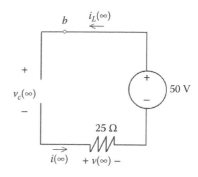

FIGURE 1.48 Series RLC circuit at $t = \infty$.

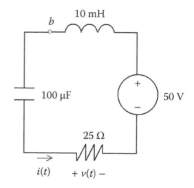

FIGURE 1.49 Series RLC circuit.

$$v_L + v_R + v_C = 50$$

$$10\,m\frac{di}{dt} + 25i + v_c = 50$$

$$\text{substitute} \quad i = 100\,\mu\frac{dv_c}{dt}$$

$$(10\,m)(100\,\mu)\frac{d^2v_C}{dt} + (25)(100\,\mu))\frac{dv_C}{dt} + v_C = 50 \tag{1.87}$$

$$\frac{d^2v_C}{dt} + 2500\frac{dv_C}{dt} + 1Mv_C = 50M$$

The Neper frequency can be found from Equation 1.87 to be $\alpha = 1.25$ krad/s and the resonant frequency is $\omega_o = 1$ krad/s. Since the Neper frequency is greater than the resonant frequency, this is also an overdamped response. The roots of the characteristics equation are −500 and −2000. The equation for the voltage across the capacitor is given by Equation 1.88:

$$v_C(t) = A_1 e^{-500t} + A_2 e^{-2000} + 50 \tag{1.88}$$

The variables, A_1 and A_2, will be found for the capacitor voltage and this equation will be used to find the resistor current and voltage. The initial conditions found earlier will be used to find A_1 and A_2, as shown in (1.89) and (1.90):

$$v_C(0^+) = 50 + A_1 + A_2 = 20 \tag{1.89}$$

$$i_L(0^+) = 100\,\mu\frac{dv_C(0^+)}{dt}$$

$$\frac{dv_c(0^+)}{dt} = \frac{i_L(0^+)}{100\,\mu} = 0$$

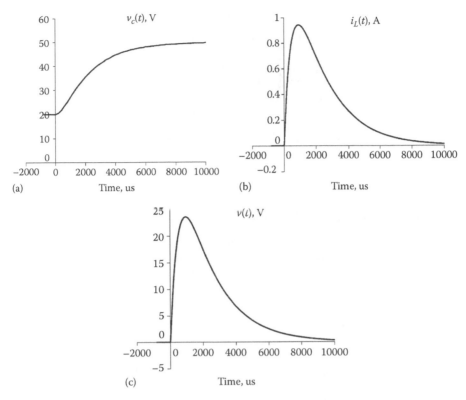

FIGURE 1.50 Graph of voltages and currents in Example 1.18. (a) Voltage across the capacitor, (b) current through the inductor, and (c) voltage across the 25Ω resistor.

$$\frac{dv_C(0^+)}{dt} = -500A_1 - 2000A_2 = 0 \tag{1.90}$$

Solving the simultaneous system of equations, (1.89) and (1.90), yields $A_1 = -40$ and $A_2 = 10$. Finally, the solution for all values is given in (1.91) through (1.93):

$$v_C(t) = -40e^{-500t} + 10e^{-2000t} + 50\,\text{V}, \quad t > 0 \tag{1.91}$$

$$i(t) = i_L(t) = i_C(t) = 100\,\mu\frac{dv_C}{dt} = 2e^{-500t} - 2e^{-2000t}\,\text{A}, \quad t > 0 \tag{1.92}$$

$$v(t) = 25i(t) = 50e^{-500t} - 50e^{-2000t}\,\text{V}, \quad t > 0 \tag{1.93}$$

The initial conditions for the solution can be verified by observing the graphs of the functions shown in Figure 1.50.

Example 1.19: Step Response of an RLC Circuit

The switch in the circuit in Figure 1.51 has been opened for a long time and the circuit has reached steady-state conditions. If the switch closes at $t = 0$, determine $v(t)$ and $i(t)$ for $t > 0$.

The first step in the analysis is to determine the initial conditions. In order to simplify the analysis, a source transformation was performed on the 18 V source and 6 Ω resistor. The circuit is shown in Figure 1.52.

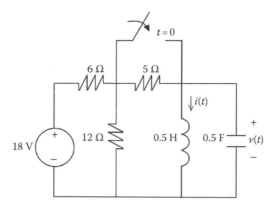

FIGURE 1.51 Step response of a parallel RLC circuit.

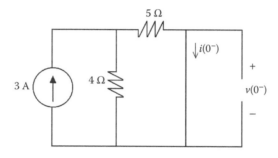

FIGURE 1.52 Parallel RLC circuit at $t = 0^-$.

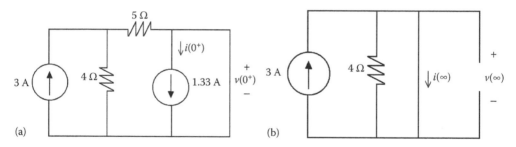

(a)   (b)

FIGURE 1.53 Parallel RLC circuit initial and final values. (a) Parallel RLC circuit at $t = 0^+$ and (b) parallel RLC circuit as $t \to \infty$.

Since the capacitor is an open circuit across the shorted out inductor, $v(0^-) = 0\,V$. The current divider can be used to find the current through the inductor $i(0^-) = [(4\|5)5] = 1.33$ A. Since current through inductors is continuous and voltage across capacitors is continuous, these values remain the same after the switch closes ($t = 0^+$). The next step in the analysis is to analyze the circuit in Figure 1.53 to determine the final values for the voltage and current. When the switch closes, it shorts out the 4 Ω resistor and it no longer has an effect on the circuit. Since current always follows the path of least resistance, $i(\infty) = 3$ A and the voltage across the capacitor $v(\infty) = 0\,V$.

In order to derive the second-order equation that describes the response, the parallel RLC circuit in Figure 1.54 will be analyzed using KCL. This derivation is given in Equation 1.94.

$$i_L + i_R + i_C = 3$$

$$i + \frac{v}{4} + (0.5)\frac{dv}{dt} = 3$$

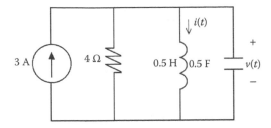

FIGURE 1.54 Parallel RLC circuit.

$$\text{substitute} \quad v = 0.5 \frac{di}{dt}$$

$$i + \frac{0.5}{4} \frac{di}{dt} + (0.5)(0.5) \frac{d^2 i}{dt} = 3 \tag{1.94}$$

$$\frac{d^2 i}{dt} + 0.5 \frac{di}{dt} + 4i = 3$$

From examination of Equation 1.94, this circuit's behavior exhibits an underdamped response because $\alpha = 0.25$ rad/s is less than $\omega_o = 2$ rad/s. The general form for this solution is given in Equation 1.95:

$$\omega_d = \sqrt{\omega_0^2 - \alpha^2} = 1.98 \, \text{rad/s}$$

$$i(t) = 3 + A_1 e^{-0.25t} \cos 1.98t + A_2 e^{-0.25t} \sin 1.98t, \quad t > 0 \tag{1.95}$$

It is necessary to use the initial conditions to solve for the constants, A_1 and A_2 in Equation 1.95. The analysis to find the values for these constants is shown in (1.96) and (1.97):

$$i(0^+) = 3 + A_1 = 1.33 \tag{1.96}$$

$$v(0^+) = 0.5 \frac{di(0^+)}{dt} \tag{1.97}$$

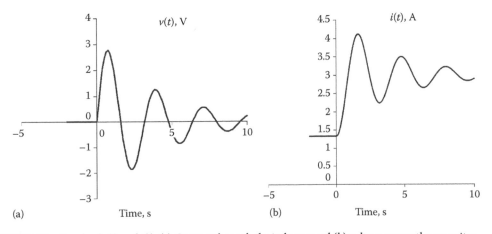

(a) Time, s   (b) Time, s

FIGURE 1.55 Graph of $v(t)$ and $i(t)$. (a) Current through the inductor and (b) voltage across the capacitor.

$$\frac{di(0^+)}{dt} = \frac{v(0^+)}{0.5} = 0$$

$$\frac{di(0^+)}{dt} = -0.25A_1 + 1.98A_2 = 0$$

Solving this system of equations ((1.96), (1.97)) yields $A_1 = -1.67$ and $A_2 = -0.211$. The final solution for $v(t)$ and $i(t)$ are given as (1.98) and (1.99):

$$i(t) = 3 - 1.67e^{-0.25t} \cos 1.98t - 0.211e^{-0.25t} \sin 1.98t \text{ A}, \quad t > 0 \qquad (1.98)$$

$$v(t) = 0.5\frac{di}{dt} = -280\mu e^{-0.25t} \cos 1.98t + 3.36e^{-0.25t} \sin 1.98t \text{ V}, \quad t > 0 \qquad (1.99)$$

Finally, the graphs of these two equations are shown in Figure 1.55; the reader is encouraged to verify that they do indeed satisfy the initial conditions.

This section has presented the transient analysis of first- and second-order circuits using ordinary differential equations. It should be noted that there is an alternate method for solving these types of circuits. As the circuits become more complex, it may be advisable to use complex frequency ($s = \sigma + j\omega$) or Laplace analysis. This analysis technique converts the circuits to the complex frequency domain and simplifies the mathematics by using algebra with complex numbers to solve.

1.5 Conclusions

This chapter has presented the fundamental concepts related to DC circuit analysis, including ideal sources, active and passive circuit elements, the law of conservation of energy, and analysis techniques. Transient analysis for first-order circuits was also presented, including the step and natural responses to model the storing and releasing of energy for inductors and capacitors. Finally, transient analysis was presented for second-order circuits and the three types of responses, overdamped, underdamped, and critically damped, were reviewed. For further study, the reader is encouraged to review AC circuit analysis, frequency-selective circuits, and operational amplifiers.

Bibliography

J.W. Nilsson and S.E. Riedel, *Electric Circuits*, 8th edition, Upper Saddle River, NJ: Prentice Hall, 2007, 880 pp.
M.N.O. Sadiku and C.K. Alexander, *Fundamentals of Electric Circuits*, 3rd edition, New York: McGraw-Hill, 2007, 901 pp.

2

AC Circuit Analysis

Carlotta A. Berry
*Rose-Hulman Institute
of Technology*

Deborah J. Walter
*Rose-Hulman Institute
of Technology*

2.1 Introduction

Alternating current (AC) circuits are important to the field of electronics. AC signals can be found in power distribution systems or in common household electrical systems. The household wall outlet delivers AC power to loads such as lamps, televisions, refrigerators, stoves, washers, and dryers. A step-down transformer transmits AC power from the power plant to the house.

AC signals are based upon a sinusoidal source. A sinusoidal source produces a voltage or current that has the form of a cosine. The equation to represent the sinusoidal voltage source is

$$v(t) = V_m \cos(\omega t + \phi) V \tag{2.1}$$

From Equation 2.1, it is evident that the source varies with time, t, in seconds; has maximum amplitude, V_m, in volts; angular frequency, ω, in rad/s; and a phase, ϕ, in radians. Note that the angular frequency, ω, can also be related to the cyclic frequency, f, in hertz and period, T, in seconds, as shown in Equation 2.2:

$$\omega = 2\pi f = \frac{2\pi}{T} \tag{2.2}$$

Figure 2.1 illustrates a phase-shifted sinusoidal voltage source V_1 with respect to the positive cosine voltage source, V_2. The amplitude, period, frequency, and phase are shown in the figure. Note that if V_1 is shifted to the left with respect to V_2, then the phase $\phi > 0$ and V_1 is leading V_2. If V_1 is shifted to the right with respect to V_2, then $\phi < 0$ and V_1 is lagging with respect to V_2.

Amplitude, $V_m = 1$ V

Period, $T = 6.28$ s $= 2\pi$ s

Angular frequency, $\omega = 2\pi/T = 1$ rad/s

Cyclic frequency, $f = 1/T = 0.159$ Hz

Phase, $\phi = (1.57 - 0.79)\,\omega = 0.79 = \pi/4$ radians

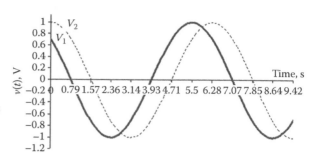

FIGURE 2.1 Sinusoidal voltage source, $v(t)$.

A phasor is a complex number used to represent a sinusoid. The phasor representation includes the amplitude, V_m, and phase, ϕ, of the sinusoid. Phasor representation of a sinusoid is based upon Euler's identity:

$$e^{j\phi} = \cos\phi + j\sin\phi, \tag{2.3}$$

where j is the imaginary number $\sqrt{-1}$. The real part of the identity is $\cos\phi$ and $e^{j\phi}$ can be used to represent the sinusoid in Equation 2.1 as

$$v(t) = V_m\cos(\omega t + \phi) \rightarrow V = V_m e^{j\phi} \tag{2.4}$$

Note that the phasor representation holds the magnitude and phase information but not the frequency. There are three forms of phasor representation: exponential form, polar form, and rectangular form. The exponential form is given in Equation 2.4, the polar form is shown in (2.5), and the rectangular form in (2.6):

$$V = V_m \angle \phi \tag{2.5}$$

$$V = a + jb \tag{2.6}$$

The formulas in (2.7) and (2.8) give the conversion between the exponential, angle, and rectangular forms:

$$V_m = \sqrt{a^2 + b^2} \quad \phi = \tan^{-1}\left(\frac{b}{a}\right) \tag{2.7}$$

$$a = V_m\cos\phi \quad b = V_m\sin\phi \tag{2.8}$$

Another way to think of a phasor is as a vector representation of a complex number where the angle with respect to the real axis is ϕ and the magnitude of the vector is V_m (see Figure 2.2).

The phasor representation of the sinusoid V_1 in Figure 2.1 in all three forms is shown in (2.9):

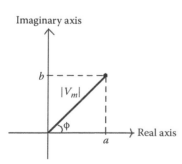

FIGURE 2.2 Phasor representation of a complex number.

$$V = 1\angle 45° = 1e^{j\pi/4} = 0.707 + j0.707 \text{ V} \tag{2.9}$$

2.2 Circuit Elements

2.2.1 Passive Circuit Elements

The three passive circuit elements are resistors, inductors, and capacitors. A passive circuit element is incapable of delivering power to a circuit. However, inductors store energy in the form of current and capacitors store energy in the form of voltage so they can release energy previously stored back to the circuit.

How can we represent the voltage and current relationship for these three passive elements as phasors? Examine the circuit in Figure 2.3 to find the voltage and current for the resistor, inductor, and capacitor.

The first step in the analysis is to use the voltage and current relationships to find the voltage across each element. These equations in the time domain are given in (2.10) through (2.12):

$$v_R(t) = iR = RI_m \cos(\omega t + \phi) \tag{2.10}$$

$$v_L(t) = L\frac{di}{dt} = -LI_m \omega \sin(\omega t + \phi) = LI_m \omega \cos(\omega t + \phi + 90°) \tag{2.11}$$

$$v_C(t) = \frac{1}{C}\int id\tau = \frac{I_m}{\omega C} \sin(\omega t + \phi) = \frac{I_m}{\omega C} \cos(\omega t + \phi - 90°) \tag{2.12}$$

The phasor relationships for the resistor, inductor, and capacitor are derived from the prior equations as follows:

$$V_R = RI_m \; e^{j\phi} = RI \tag{2.13}$$

$$V_L = LI_m \omega e^{j\phi} e^{j\pi/2} = j\omega L I_m e^{j\phi} = j\omega L I \tag{2.14}$$

$$V_C = \frac{I_m}{\omega C} e^{j\phi} e^{-j\pi/2} = \frac{-j}{\omega C} I_m e^{j\phi} = \frac{-j}{\omega C} I \tag{2.15}$$

Table 2.1 summarizes the voltage and current relationships for resistors, inductors, and capacitors.

Based upon the prior derivation, in the phasor domain, the relationship between voltage and current is $V = IZ$, where Z is the impedance of the resistor, inductor, or capacitor in ohms (Ω).

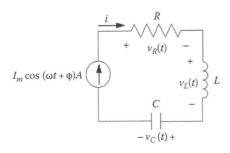

FIGURE 2.3 Voltage and current phasor relationships for R, L, and C.

TABLE 2.1 Phasors and Passive Circuit Elements

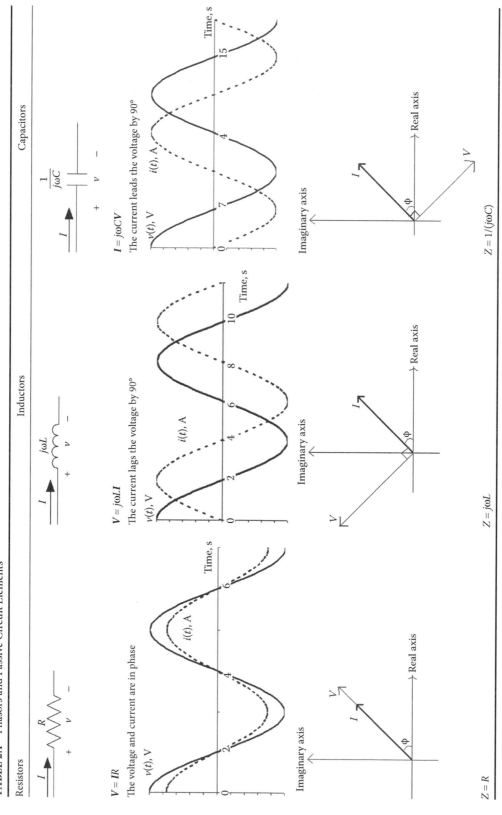

2.2.2 Mutual Inductance

Another important circuit element is based upon the concept of mutual inductance. When the AC current flows through a coil (inductor), it creates a changing magnetic field. Self-inductance, L, is the parameter that relates induced voltage across a coil to the current flowing through it ($v = L \, di/dt$). Since the time-varying current through an inductor creates a magnetic field, if a secondary coil is placed next to the primary coil then a portion of the magnetic flux will be coupled to the secondary coil thereby inducing a voltage across the secondary coil. Mutual inductance, M, is the parameter that relates the voltage induced in the second coil due to the current flowing through the first coil. Just like self-inductance, the units for mutual inductance are also henrys (H). Figure 2.4 illustrates a mutual inductance circuit with two coils.

The relationship between the induced voltage in the second coil and the current flowing through the first coil is given in Equation 2.16:

$$v_2(t) = \pm M \frac{di_1}{dt} \tag{2.16}$$

In the phasor domain, the voltage across the secondary coil is given by

$$V_2 = (j\omega M)I \tag{2.17}$$

Therefore, the impedance for the mutual inductance is $Z = V/I = j\omega M$. Note that the relationship in Equation 2.16 can be positive or negative and the polarity is based upon the dot convention for mutual inductors. The dot convention states that when the current enters the coil through the dot on the first coil, then the voltage induced on the second coil is *positive* at the dotted terminal. If the current leaves the coil through the dot on the first coil, then the voltage induced on the second coil is *negative* at the dotted terminal. Table 2.2 presents a summary of the two possible configurations of the voltages and currents.

The mutual inductance parameter, M, can also be described in terms of the coefficient of coupling, k. The range of values for k is between 0 and 1, where 0 represents no coupling and 1 represents a very tight coupling (i.e. transformer). The relationship between mutual inductance and the coefficient of coupling is shown in Equation 2.18:

$$M = k\sqrt{L_1 L_2} \tag{2.18}$$

In terms of phasor impedance Equation 2.18 becomes

$$j\omega M = k\sqrt{(j\omega L_1)(j\omega L_2)} \tag{2.19}$$

The total energy stored in the coupled inductors is given by

$$w(t) = 0.5L_1 \left(i_1(t)\right)^2 + 0.5L_2 \left(i_2(t)\right)^2 \pm M i_1(t) i_2(t) \tag{2.20}$$

In the energy expression, use $+M i_1(t) i_2(t)$ if both currents enter or leave inductors through the dotted terminal, otherwise use the minus sign.

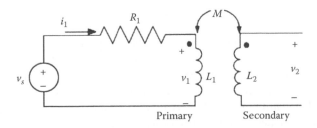

FIGURE 2.4   Mutual inductance circuit.

TABLE 2.2   Dot Convention Configurations

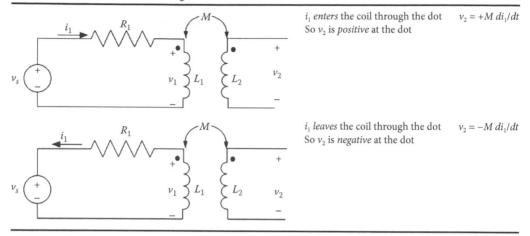

i_1 *enters* the coil through the dot $v_2 = +M \, di_1/dt$
So v_2 is *positive* at the dot

i_1 *leaves* the coil through the dot $v_2 = -M \, di_1/dt$
So v_2 is *negative* at the dot

2.2.3 Ideal Transformer

One of the primary applications of ideal transformers is for energy transmission. The step-up transformer is used to increase the voltage for the power lines from the power plant. As previously mentioned, the step-down transformer is used to reduce the voltage for delivery to household outlets at 240 or 120 V_{AC}. An ideal transformer consists of two magnetically coupled coils with the following characteristics:

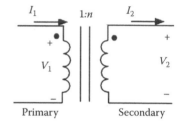

FIGURE 2.5   Ideal transformer circuit.

1. The coils are perfectly coupled; therefore $k = 1$.
2. The self-inductance of each coil is infinite ($L_1 = L_2 = \infty$).
3. Coil losses due to resistance are negligible.

Figure 2.5 is the circuit symbol for an ideal transformer.

The left coil and its circuit are called the primary side of the transformer. The right coil and its circuit are called the secondary side of the transformer. The turns ratio, n, of a transformer is a ratio of the turns on the secondary coil to the turns on the primary coil:

$$n = \frac{N_2}{N_1} \tag{2.21}$$

There are two characteristics of the terminal behavior of an ideal transformer:

$$1. \; V_2 = \pm n V_1 \tag{2.22}$$

$$2. \; I_1 = \pm n I_2 \tag{2.23}$$

Note that the relationship in Equations 2.22 and 2.23 can be positive or negative and the polarity is based upon the dot convention for ideal transformers. For Equation 2.22, if the coil voltages V_1 and V_2 are both positive or both negative at the dotted terminals, use a *plus* sign. If one voltage is positive and one voltage is negative at the dotted terminals, use a *minus* sign. For Equation 2.23, if both of the currents I_1 and I_2 enter or leave through the dotted terminal, use a *minus* sign. If one current enters and one current leaves through the dotted terminal then use a *plus* sign.

Table 2.3 presents two possible configurations for the dot convention.

TABLE 2.3 Ideal Transformer Dot Convention

Circuit	Voltage Relationship	Current Relationship
	$V_2 = +nV_1$	$I_1 = +nI_2$
	$V_2 = -nV_1$	$I_1 = +nI_2$

(Circuit diagrams: Primary/Secondary transformer windings with dot conventions, $1:n$ turns ratios, labeled I_1, V_1, I_2, V_2.)

2.2.4 Autotransformer

An autotransformer is a single winding ideal transformer with a center tap. Alternately, the primary and secondary windings are connected in series to form an autotransformer. If a center tap is used, it is adjustable such that the turns ratio is variable to step up or step down to the desired voltage. In this configuration, there is an electrical coupling between the primary and secondary windings. The primary benefit of this configuration is that it is possible to deliver a larger apparent power to the load or to step a voltage up or down by a small amount. Apparent power will be described in Section 2.4 and an example shown with the autotransformer. However, one of the disadvantages of the autotransformer is the fact that there is no electrical isolation between the primary and secondary windings. Figure 2.6 presents three different examples of autotransformer configurations and the voltage and current

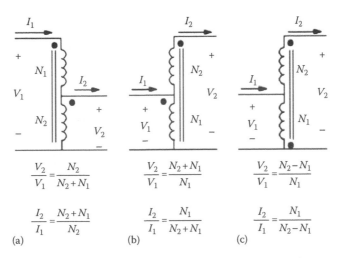

$$\frac{V_2}{V_1} = \frac{N_2}{N_2 + N_1} \qquad \frac{V_2}{V_1} = \frac{N_2 + N_1}{N_1} \qquad \frac{V_2}{V_1} = \frac{N_2 - N_1}{N_1}$$

$$\frac{I_2}{I_1} = \frac{N_2 + N_1}{N_2} \qquad \frac{I_2}{I_1} = \frac{N_1}{N_2 + N_1} \qquad \frac{I_2}{I_1} = \frac{N_1}{N_2 - N_1}$$

(a)   (b)   (c)

FIGURE 2.6 Autotransformer configurations. (a) Step down, (b) step up, and (c) subtractive.

relationships. The first two represent a traditional step-down and step-up transformation. The last configuration demonstrates the subtractive connection of an autotransformer.

2.3 Analysis Techniques

2.3.1 Phasor Analysis

The prior sections introduced sinusoidal sources, phasors, and the phasor representation of some common circuit elements. Phasor analysis also referred to as sinusoidal steady-state analysis involves representing voltage and sinusoidal waveforms as phasors in circuit analysis. Note that in sinusoidal steady-state analysis, all of the voltages and currents will have the same frequency; therefore, it is only necessary to find amplitudes and phase shifts. The steps of phasor analysis are

1. Convert the circuit to the phasor domain (sources to phasors, elements to impedances).
2. Use circuit analysis techniques to find relevant voltages and currents.
3. Convert the phasor values back to the time domain.

The phasor analysis technique will be demonstrated on several types of circuits.

Example 2.1: Mesh-Current Method

For the circuit shown in Figure 2.7, use phasor analysis to find the current through the voltage source. This will be achieved by using the mesh-current method with phasor analysis to find the three unknown mesh currents.

The first step in phasor analysis is to convert the circuit to the phasor domain or the sinusoidal steady state. Note that the angular frequency, ω, of the source is $2\pi 60 = 377$ rad/s. Using this frequency, convert all of the sources and passive circuit elements to phasors and impedances. The voltage source becomes $170\angle 0°$ V by using the polar form. The next step would be to convert all of the passive circuit elements to impedances. Recall that for resistors the impedance is the same in the time domain and phasor domain. The inductor and capacitor impedances become

$$Z_L = j\omega L = j(377)(3) = j1131\,\Omega \tag{2.24}$$

$$Z_C = \frac{-j}{\omega C} = \frac{-j}{(377)(2.2\mu)} = -j1206\,\Omega \tag{2.25}$$

Using the results of (2.24), (2.25), the circuit in Figure 2.6 redrawn in the phasor domain is shown in Figure 2.8.

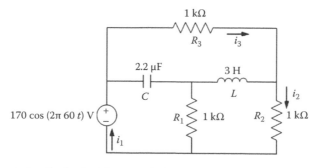

FIGURE 2.7   Mesh-current method using phasor analysis.

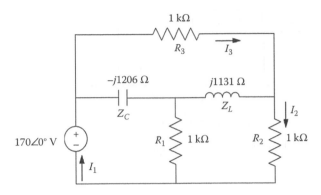

FIGURE 2.8 Mesh-current circuit in phasor domain.

The next step is to use the Kirchhoff's voltage law (KVL) to solve for I_1, I_2, and I_3. The KVL equations for meshes 1, 2, and 3 are given in Equations 2.26 through 2.28, respectively:

$$(1000 - j1206)I_1 - 1000I_2 + j1206I_3 = 170 \tag{2.26}$$

$$-1000I_1 + (2000 + j1131)I_2 - j113I_3 = 0 \tag{2.27}$$

$$j1206I_1 - j1131I_2 + (1000 - j75)I_3 = 0 \tag{2.28}$$

Solving Equations 2.26 through 2.28 yields the following solution for I_1, I_2 and I_3:

$$I_1 = 150.16 \angle 36.25° \text{ mA} \tag{2.29}$$

$$I_2 = 101.36 \angle 18.17° \text{ mA} \tag{2.30}$$

$$I_3 = 80.19 \angle -23.21° \text{ mA} \tag{2.31}$$

The final step involves converting the phasors I_1 and I_2 back to the time domain:

$$i_1(t) = 150.16 \cos(2\pi 60t + 36.25°) \text{ mA} \tag{2.32}$$

$$i_2(t) = 101.36 \cos(2\pi 60t + 18.17°) \text{ mA} \tag{2.33}$$

$$i_3(t) = 80.19 \cos(2\pi 60t - 23.21°) \text{ mA} \tag{2.34}$$

Finally, the current through the voltage source is $i_1(t)$.

Example 2.2: Example 2.1 Revisited Using *T*-π Transformations

An alternate method to find the unknown voltage and currents in an AC circuit is to use circuit simplification along with the Ohm's law, voltage divider, current divider, and Kirchhoff's current (KCL) and voltage (KVL) laws. Circuit simplification involves combining impedances by using series/parallel combinations and *T*-π transformations to reduce the number of circuit elements. Figure 2.9 shows the relationship between the *T* and π configuration. It should be noted that since these configurations are equivalent, the voltage and current characteristics at the three terminals are the same. The terminals are the only locations where the voltage and current characteristics are the same.

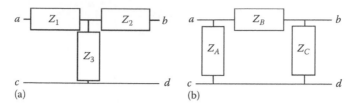

FIGURE 2.9   (a) T and (b) π transformation configurations.

To convert from the π configuration to the T configuration, use the following relationships:

$$Z_1 = \frac{Z_A Z_B}{Z_A + Z_B + Z_C}$$

$$Z_2 = \frac{Z_B Z_C}{Z_A + Z_B + Z_C} \tag{2.35}$$

$$Z_3 = \frac{Z_C Z_A}{Z_A + Z_B + Z_C}$$

The conversion from the T configuration to the π configuration are given by the following relationships:

$$Z_A = \frac{Z_1 Z_2 + Z_2 Z_3 + Z_3 Z_1}{Z_2}$$

$$Z_B = \frac{Z_1 Z_2 + Z_2 Z_3 + Z_3 Z_1}{Z_3} \tag{2.36}$$

$$Z_C = \frac{Z_1 Z_2 + Z_2 Z_3 + Z_3 Z_1}{Z_1}$$

In order to use a T to π transformation to simplify the circuit in Example 2.1, let $Z_1 = -j1206\ \Omega$, $Z_2 = j1131\ \Omega$, and $Z_3 = R_1 = 1000\ \Omega$. Using the formulas in Equation 2.35 yields $Z_A = -66 - j1206\ \Omega$, $Z_B = 1364 - j75\ \Omega$, and $Z_C = 62 + j1131\ \Omega$. The simplified circuit is shown in Figure 2.10.

Figure 2.10 can be simplified by combining Z_B in parallel with R_3 and Z_C in parallel with R_2. Equations show the values of these impedances after the parallel combinations:

$$Z_B' = \frac{Z_B R_3}{Z_B + R_3} = 577 - j13\ \Omega \tag{2.37}$$

$$Z_C' = \frac{Z_C R_2}{Z_C + R_2} = 559 + j470\ \Omega \tag{2.38}$$

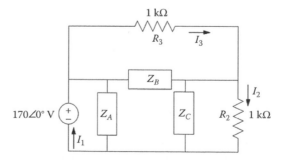

FIGURE 2.10   Figure 2.8 simplified using a T to π transformation.

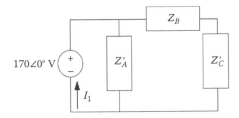

FIGURE 2.11   Figure 2.8 simplified using parallel combinations.

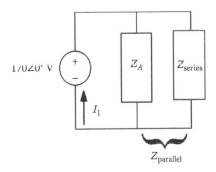

FIGURE 2.12   Figure 2.8 simplified to a single node pair.

Figure 2.11 shows the simplified circuit after the parallel combinations. By combining Z_B' and Z_C' in series, the circuit is reduced to a voltage source in parallel with Z_A and the series combination ($Z_B' + Z_C'$). Finally, the last two impedances are put in parallel with the voltage source and this simplified circuit is used to find the source current (see Figure 2.12).

The solution for the final series and parallel combinations are given by

$$Z_{\text{series}} = Z_B' + Z_C' = 1136 + j456 \ \Omega \tag{2.39}$$

$$Z_{\text{parallel}} = \frac{Z_A R_{\text{series}}}{Z_A + Z_{\text{series}}} = 913 - j669 \ \Omega \tag{2.40}$$

Finally, the value for the source current is found by using Ohm's law as shown in Equation 2.41 and it is consistent with Equation 2.29:

$$I_1 = \frac{170}{Z_{\text{parallel}}} = 150.16 \angle 36.25° \ \text{mA} \tag{2.41}$$

Example 2.3: Example 2.1 Revisited Using the Node-Voltage Method

For the circuit in Figure 2.7, use the node-voltage method to find the current through R_2. The first step is to label all of the node voltages and the modified circuit is shown in Figure 2.13.

Performing KCL at nodes V_2 and V_3 yields

$$V_1 = 170 \tag{2.42}$$

$$\frac{V_2 - V_1}{-j1206} + \frac{V_2}{1000} + \frac{V_2 - V_3}{j1131} = 0 \tag{2.43}$$

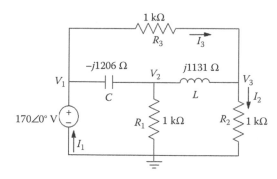

FIGURE 2.13   Node-voltage circuit in the phasor domain.

$$\frac{V_3}{1000} + \frac{V_3 - V_2}{j1131} + \frac{V_3 - V_1}{1000} = 0 \tag{2.44}$$

Solving Equations 2.42 through 2.44 for the node voltages yields

$$\boldsymbol{V_2} = 62.33 \angle 66.55° \, V \tag{2.45}$$

$$\boldsymbol{V_3} = 101.36 \angle 18.17° \, V \tag{2.46}$$

Converting the phasors in Equations 2.45 and 2.46 to the time domain yields the following:

$$v_2(t) = 62.33 \cos (2\pi 60 t \pm 66.55°) \, V \tag{2.47}$$

$$v_3(t) = 101.36 \cos (2\pi 60 t \pm 18.17°) \, V \tag{2.48}$$

To confirm that the node-voltage method and the mesh-current method produce the same results use the relationship in Equation 2.49:

$$\boldsymbol{I_2} = \frac{V_3}{1000} = 101.36 \angle 118.17° \, mA \tag{2.49}$$

Since $\boldsymbol{I_2}$ is the same as Equation 2.30, the two methods are consistent.

Example 2.4: Mutual Inductance

For the mutual inductance circuit shown in Figure 2.14, phasor analysis and the mesh-current method (KVL) will be used to find the primary and secondary currents.

The first step in the analysis is to convert the circuit to the phasor domain. To find the mutual inductance, M, from the coupling coefficient use the following formula:

$$M = k\sqrt{L_1 L_2} = 0.8\sqrt{(4700\mu)(5600\mu)} = 4104 \, \mu H \tag{2.50}$$

In terms of phasor impedance, Equation 2.50 becomes

$$j\omega M = k\sqrt{(j\omega L_1)(j\omega L_2)} = j1231\,\Omega \tag{2.51}$$

The circuit in Figure 2.10 converted to the phasor domain is shown in Figure 2.15.

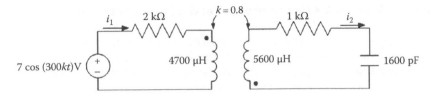

FIGURE 2.14 Mutual inductance using phasor analysis.

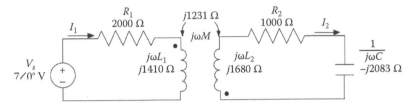

FIGURE 2.15 Mutual inductance circuit in the phasor domain.

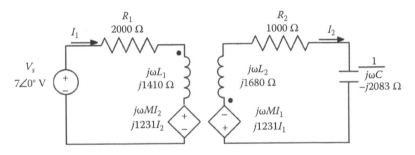

FIGURE 2.16 Mutual inductance circuit in the phasor domain.

The next step in the analysis is to model the voltage induced by the mutual inductance as a current-controlled voltage source. Current I_1 enters the primary coil at the dot and induces a voltage equal to $j1231\,I_1$ across the secondary coil, which is positive at the dot. Current I_2 enters the secondary coil through the dot and induces a voltage equal to $j1231\,I_2$ across the secondary coil that is positive at the dot. Figure 2.16 shows the mutual inductance circuit with the induced voltages modeled as dependent voltage sources.

The KVL equations for the two meshes are

$$\text{Mesh 1:}\quad (R_1 + j\omega L_1) + j\omega M I_2 = (2000 + j1410)I_1 + j1231 I_2 = 7 \tag{2.52}$$

$$\text{Mesh 2:}\quad j\omega M I_1 + (R_2 + j\omega L_2 - j/(\omega C))I_2 = j1231 I_1 + (1000 - j403)I_2 = 0 \tag{2.53}$$

Solving Equations 2.52 and 2.53 yields

$$I_1 = 1.83\angle -30.36° \text{ mA} \tag{2.54}$$

$$I_2 = 2.09\angle -98.41° \text{ mA} \tag{2.55}$$

The final step involves converting the phasors I_1 and I_2 back to the time domain:

$$i_1(t) = 1.83 \cos(300kt - 30.36°) \text{ mA} \tag{2.56}$$

$$i_2(t) = 2.09 \cos(300kt - 98.41°) \text{ mA} \tag{2.57}$$

Example 2.5: Example 2.4 Revisited with *T* and π Equivalent Circuits

An alternate approach to the dependent source model for mutual inductance is *T* or π equivalent configuration. It should be noted that these configurations can only be used if the primary and secondary sides of the network have a common node, typically the reference or ground. In addition, the following analysis assumes that there is no energy initially stored in the circuit. For the circuit in Figure 2.11, it is possible to write the mesh equations as
The KVL equations for the two meshes are

$$\text{Mesh 1:} \quad (R_1 + j\omega L_1 + j\omega M)I_1 - j\omega M (I_1 - I_2) = V_s \tag{2.58}$$

$$\text{Mesh 2:} \quad -j\omega M (I_2 - I_1) + \left(R_2 + j\omega L_2 + j\omega M - \frac{j}{\omega C} \right) I_2 = 0 \tag{2.59}$$

Note that these equations are equivalent to Equations 2.52 and 2.53 and thus the circuit in Figure 2.17 can also be used to solve for the mesh currents.

Since this is a *T* equivalent circuit, you can use the relationships in Equation 2.36 to create the π equivalent circuit. In order to create the π equivalent circuit, let $Z_1 = R_1 + j\omega (L_1 + M)\ \Omega$, $Z_2 = R_2 + j\omega (L_2 + M)\ \Omega$, and $Z_3 = -j\omega M\ \Omega$ (Figure 2.18). Substitution into Equation 2.36 yields the following:

$$Z_A = \frac{Z_1 Z_2 + Z_2 Z_3 + Z_3 Z_1}{Z_2} = 1587 + j151\ \Omega$$

$$Z_B = \frac{Z_1 Z_2 + Z_2 Z_3 + Z_3 Z_1}{Z_3} = -3875 + j931\ \Omega \tag{2.60}$$

$$Z_C = \frac{Z_1 Z_2 + Z_2 Z_3 + Z_3 Z_1}{Z_1} = 1357 + j593\ \Omega$$

It is important to note that this configuration is for modeling purposes only and a negative resistance is physically impossible.

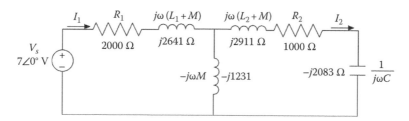

FIGURE 2.17 *T* equivalent of mutual inductance circuit in Figure 2.15.

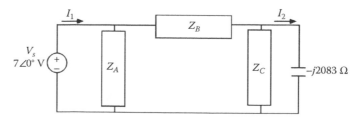

FIGURE 2.18 π equivalent of mutual inductance circuit in Figure 2.15.

Example 2.6: Ideal Transformer

For the ideal transformer circuit in Figure 2.19, find the voltage, V_L, delivered to the load. The load voltage can be found by using phasor analysis, reflection, and the mesh-current method.

Typically, the simplest way to analyze an ideal transformer circuit is to use reflection to simplify the circuit to a single mesh. In this case, the steps would include reflecting from the secondary to the middle or feeder, and reflecting from the middle to the primary. The characteristic equations for the secondary transformer are

$$V_4 = -nV_3 \qquad (2.61)$$

$$I_L = -\frac{I_2}{n} \qquad (2.62)$$

Therefore, the impedance as seen from the feeder is the ratio of Equations 2.61 and 2.62:

$$Z_3 = \frac{V_3}{I_2} = \frac{V_4/I_L}{n^2} = \frac{32 - j40}{(0.1)^2} = 3200 - j4000 \ \Omega \qquad (2.63)$$

The secondary reflected to the feeder is shown in Figure 2.20.

The next step will be to reflect the feeder to the primary using a similar approach to Equation 2.63:

$$Z_1 = \frac{V_1}{I_1} = \frac{V_2/I_2}{n^2} = \frac{5200 - j4000}{(2)^2} = 1300 - j1000 \ \Omega \qquad (2.64)$$

The feeder and secondary reflected to the primary side is shown in Figure 2.21.

Finally, use KVL around the single loop to find I_1 and V_1, and then use the transformer voltage and current relationships to find the voltages and currents in the feeder and secondary sides. This analysis is shown below:

$$I_1 = \frac{7}{400 + 500 + 800 - j1000} = 3.55\angle30.47° \ \text{mA} \qquad (2.65)$$

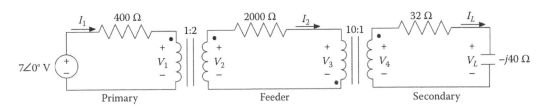

FIGURE 2.19 Ideal transformer using phasor analysis.

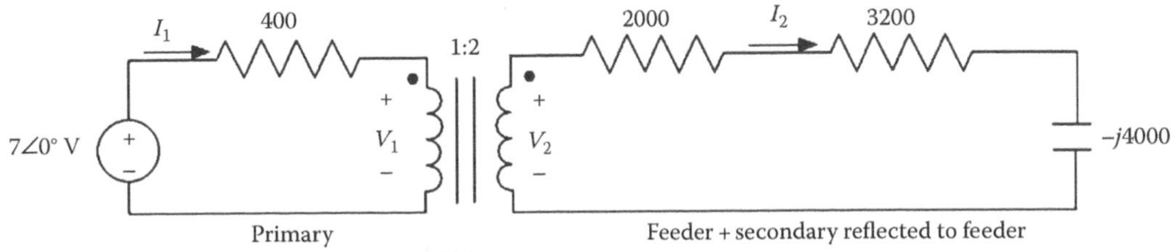

FIGURE 2.20 Ideal transformer reflection to the middle.

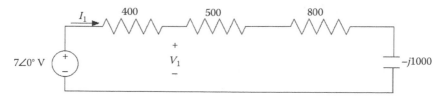

FIGURE 2.21 Ideal transformer reflection to the primary.

$$V_1 = (500 + 800 - j1000)I_1 = 5.82\angle -7.1° \, V \tag{2.66}$$

$$V_2 = nV_1 = 2V_1 = 11.64 \angle -7.1° \, V \tag{2.67}$$

$$I_2 = \frac{I_1}{n} = \frac{I_1}{2} = 1.77\angle 30.47° \, mA \tag{2.68}$$

$$V_3 = V_2 - (2000)I_2 = 9.09 \angle -20.87° \, V \tag{2.69}$$

$$V_4 = -nV_3 = -0.1\,V_3 = 909 \angle 159.13° \, mV \tag{2.70}$$

$$I_L = -\frac{I_2}{n} = -10I_2 = 17.75\angle -149.53° \, mA \tag{2.71}$$

$$V_L = \frac{-j40}{32 - j40}V_4 = 710\angle 120.47° \, mV \tag{2.72}$$

Example 2.7: Autotransformer

For the autotransformer in Figure 2.22, find the voltage V_L delivered to the load. This transformer is in a step-up subtractive configuration and it can also be analyzed by using KCL and the ideal transformer relationships. Equations 2.73 and 2.74 present the results of the analysis:

$$V_L = \frac{N_2 - N_1}{N_1}V_1 = \frac{100 - 27}{27}(120) = 324 \, V \tag{2.73}$$

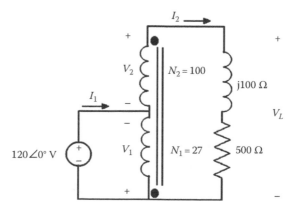

FIGURE 2.22 Subtractive step-up autotransformer.

$$I_2 = \frac{V_L}{500 + j100} = 635\angle -11° \,\text{mA} \qquad (2.74)$$

In conclusion, this section has presented the basic theory of phasor analysis and demonstrated the technique on several types of circuits. The next section will also consider the transient response as well as the steady-state response. In order to analyze a circuit to get the transient and steady-state response, it is necessary to use Laplace (complex frequency, $s = \sigma + j\omega$) analysis.

2.3.2 Frequency Response (Laplace) Analysis

Phasor analysis is used to find the sinusoidal steady-state response of a circuit. The steady-state sinusoidal response is the forced response based upon the sinusoidal source. The transient and sinusoidal steady-state response of the circuit can be found by using frequency response (Laplace) analysis. The transient response is based upon the sudden application or removal of a source and the initial conditions of the passive circuit elements. The transfer function of a circuit is the ratio of the output to the input of a circuit assuming zero initial conditions. Frequency response analysis is used to determine the behavior of the circuit as a function of frequency variation. This analysis is based upon using the Laplace transform of the differential equation that describes the circuit. The first step in the Laplace analysis is to convert the circuit from the time domain to the complex frequency or s-domain, where the complex frequency is represented by $s = \sigma + j\omega$. The benefit of the Laplace analysis is that it transforms differential equations to algebraic equations. Similar to phasor analysis, the next step is to use circuit analysis techniques to find relevant voltages and currents in the circuit. The final step is to convert the s-domain values back to the time domain.

2.3.2.1 Impedance

The first step in the Laplace analysis of circuits is to convert the circuit to the s-domain where $V(s) = \mathcal{L}\{v(t)\}$ and $I(s) = \mathcal{L}\{i(t)\}$. Table 2.4 provides a summary of the impedance conversions from the time domain to the complex frequency domain.

Note that the frequency domain circuits for the inductor and capacitor including initial conditions represent the Thevenin and Norton equivalent circuit with respect to the terminals. Because they are equivalent circuits, it is possible to convert from one to the other by using source transformations. Similar to phasor analysis, the voltage current relationship, $V = IZ$, still holds.

2.3.2.2 Initial and Final Conditions

It is evident from Table 2.4 that to model a circuit in the s-domain it is necessary to find the initial conditions. In order to find the initial voltage across a capacitor and the initial current through an inductor it is necessary to understand the properties of inductors and capacitors. The properties of an inductor are

1. The current through and inductor cannot change instantaneously.
2. The voltage across an inductor can change instantaneously.
3. At low frequencies (or DC conditions), an inductor models a short circuit.
4. At high frequencies, an inductor models an open circuit.

The properties of a capacitor are

1. The voltage across a capacitor cannot change instantaneously.
2. The current through a capacitor can change instantaneously.
3. At low frequencies (or DC conditions), a capacitor models an open circuit.
4. At high frequencies, a capacitor models a short circuit.

TABLE 2.4   Impedance Relationships for Laplace Analysis

Element	Time Domain	Voltage Current Relationships (Time Domain)	Laplace Transform (s-Domain)	Voltage Current Relationships (s-Domain)
Resistor	$v = iR$ $i = v/R$		$\mathbf{V} = \mathbf{I}R$ $\mathbf{I} = \mathbf{V}/R$ $Z = R$	
Inductor	$v = \mathcal{L}\dfrac{di}{dt}$ $i = \dfrac{1}{\mathcal{L}}\displaystyle\int_0^\tau v\,d\tau + I_o$		$\mathbf{V} = s\mathcal{L}\mathbf{I} - \mathcal{L}I_o$ $\mathbf{I} = \dfrac{v}{s\mathcal{L}} + \dfrac{I_o}{s}$ $Z = s\mathcal{L}$	
Capacitor	$i = C\dfrac{dv}{dt}$ $v = \dfrac{1}{C}\displaystyle\int_0^\tau i\,d\tau + V_o$		$\mathbf{I} = sC\mathbf{V} - CV_o$ $\mathbf{V} = \dfrac{I}{sC} + \dfrac{V_o}{s}$ $Z = \dfrac{1}{sC}$	

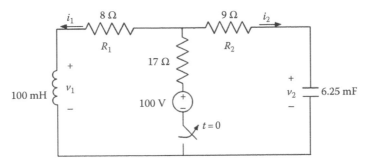

FIGURE 2.23 Laplace analysis circuit.

Example 2.8: Initial and Final Conditions

The process of finding initial conditions and using Laplace analysis to find voltages and currents will be demonstrated on the circuit in Figure 2.23.

The switch under the 100 V source in Figure 2.23 opens at $t = 0$. Therefore, to find the initial conditions for the elements, analyze the circuit just before and right after the switch opens. At the instance of time right before the switch moves ($t = 0^-$), assume that the circuit is in a steady-state or DC condition. In addition, assume as t approaches ∞ the circuit is in a steady-state or DC condition. Therefore, at $t = 0^-$, the inductor acts like a short circuit (0 V) and the capacitor acts like an open circuit (0 A). At the instance of time right after the switch moves ($t = 0^+$), the capacitor voltage and inductor current can be modeled as an independent source that is equal to the initial conditions. The three values for v_1, v_2, i_1, and i_2 are shown in Figure 2.24.

Example 2.9: Laplace Circuit Analysis

The circuit in Figure 2.16 can be redrawn in the s-domain by using the initial conditions and the equivalent impedances. This process involves modeling the sources, passive circuit elements, and their initial values using the $t = 0^+$ values. Figure 2.25 presents the results of the circuit conversion.

It is possible to use KVL to solve for V_1, V_2, I_1, and I_2 and this is shown Equation 2.75:

$$0.1sI_1 + 17I_1 + \frac{160}{s}I_1 = 0.4 + \frac{32}{s} \tag{2.75}$$

Solving (2.75) for I_1 yields

$$I_1 = \frac{4s + 320}{s^2 + 170s + 1600} = \frac{4s + 320}{(s+10)(s+160)} \tag{2.76}$$

The inverse Laplace transform of (2.76) yields

$$i_1(t) = (1.87e^{-10t} + 2.13e^{-160t}) u(t)A = -i_2(t) \tag{2.77}$$

It is also possible to find V_1 and V_2 by using I_1, I_2, and Ohm's law, and these are shown below:

$$V_1 = I_1 Z - LI_o = (0.1s)I_1 - 0.4 = \frac{0.4s^2 + 32s}{(s+10)(s+160)} - 0.4 \tag{2.78}$$

$$V_2 = I_2 Z + \frac{V_o}{s} = \left(\frac{160}{s}\right)I_2 + \frac{32}{s} = -\frac{640s + 51,200}{s(s+10)(s+160)} + \frac{32}{s} \tag{2.79}$$

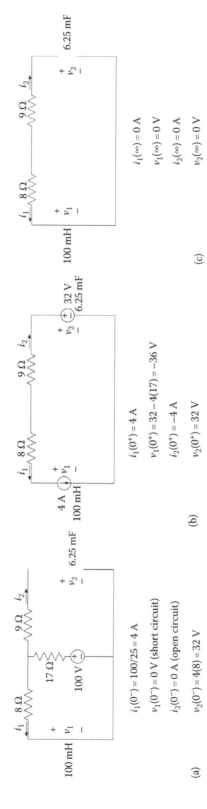

FIGURE 2.24 Initial conditions analysis circuits. (a) $t = 0^-$, (b) $t = 0^+$, and (c) $t \to \infty$.

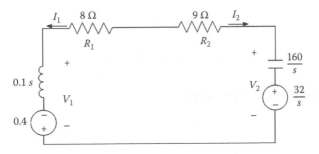

FIGURE 2.25   Figure 2.23 in the *s*-domain.

Finding the inverse Laplace transform of (2.78) and (2.79) yields

$$v_1(t) = (-34.13e^{-160t} - 1.87e^{-10t})u(t) \, V \tag{2.80}$$

$$v_2(t) = (29.87e^{-10t} + 2.13e^{-160t})u(t) \, V \tag{2.81}$$

It can be confirmed that the solutions do obey the initial and final values shown in Figure 2.17. If the circuit was given in the *s*-domain instead of the time domain, it is also possible to find the initial and final value for voltages and currents by using the initial and final value theorems. These theorems are given in Equations 2.82 and 2.83. Note that the solutions to these equations are also consistent with the conditions presented in Figure 2.24.

$$\text{Initial value theorem}: \quad \lim_{t \to 0} f(t) = \lim_{s \to \infty} sF(s) \tag{2.82}$$

$$\text{Final value theorem}: \quad \lim_{t \to \infty} f(t) = \lim_{s \to \infty} sF(s) \tag{2.83}$$

Example 2.10: Laplace Circuit Analysis

Now let us examine what happens if the 9 Ω resistor in Figure 2.16 is replaced with a wire. Notice that this resistor changes only one initial condition, $v_1(0^+)$ is 0 V. Based upon this change to the circuit, the values of I_1, I_2, and V_1, and V_2 also change:

$$I_1 = \frac{4s + 320}{s^2 + 80s + 1600} = \frac{4s + 320}{(s + 40)^2} \tag{2.84}$$

$$V_1 = I_1 Z - LI_o = (0.1s)I_1 - 0.4 == \frac{0.4s^2 + 32s}{(s + 40)^2} - 0.4 \tag{2.85}$$

$$V_2 = I_2 Z + \frac{V_o}{s} = \left(\frac{160}{s}\right)I_2 + \frac{32}{s} = -\frac{640s + 51{,}200}{s(s + 40)^2} + \frac{32}{s} \tag{2.86}$$

The inverse Laplace transform of Equations 2.87 through 2.89 are completely different forms when compared to the prior analysis. These values are presented in Equations 2.87 through 2.89:

$$i_1(t) = 4e^{-40t}(1 + 40t)u(t) \, A = -i_2(t) \tag{2.87}$$

$$v_1(t) = (-640te^{-40t})u(t) \text{ V} \qquad (2.88)$$

$$v_2(t) = 32e^{-40t}(1+20t)u(t) \text{ V} \qquad (2.89)$$

Example 2.11: Laplace Circuit Analysis

Now let's examine what happens if the 9 Ω resistor is still a wire and the 8 Ω resistor in Figure 2.16 is replaced with a 4.8 Ω resistor. The circuit now looks like that shown in Figure 2.26.

Notice that the removal of this resistor changes more than just the value of $v_1(0^+)$. The new initial and final conditions and the new values are summarized in Table 2.5.

Based upon these changes to R_1 and R_2, the values of $I_1, I_2,$ and $V_1,$ and V_2 become

$$I_1 = \frac{4.59s + 220}{s^2 + 48s + 1600} = \frac{4.59s + 220}{(s + 24 - j32)(s + 24 + j32)} = -I_2 \qquad (2.90)$$

$$V_1 = I_1Z - LI_o = (0.1s)I_1 - 0.459 = \frac{0.459s^2 + 22s}{(s + 24 - j32)(s + 24 + j32)} - 0.459 \qquad (2.91)$$

$$V_2 = I_2Z + \frac{V_o}{s} = \left(\frac{160}{s}\right)I_2 + \frac{32}{s} = -\frac{73.4s + 35,200}{s(s + 24 - j32)(s + 24 + j32)} + \frac{32}{s} \qquad (2.92)$$

The inverse Laplace transform of Equations 2.90 through 2.92 are completely different forms of responses than the prior analysis. These values are presented in Equations 2.93 through 2.95:

$$i_1(t) = 5.73e^{-24t}(\cos(32t - 36.79°)u(t) \text{ A} = -i_2(t) \qquad (2.93)$$

$$v_1(t) = 22.93e^{-24t}(\cos(32t + 90°)u(t) \text{ V} \qquad (2.94)$$

$$v_2(t) = 26.19e^{-24t}(\cos(32t - 32.85°)u(t) \text{ V} \qquad (2.95)$$

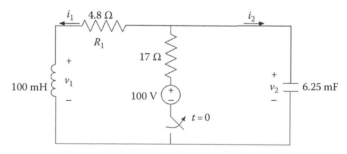

FIGURE 2.26 Modified Figure 2.16 ($R_1 = 4.8$ Ω, $R_2 = 0$ Ω).

TABLE 2.5 New Initial Conditions for Figure 2.26

$t = 0^-$	$t = 0^+$	$t = \infty$
$i_1(0^-) = 100/21.8 = 4.59$ A	$i_1(0^+) = 4.59$ A	$i_1(\infty) = 0$ A
$v_1(0^-) = 0$ V (short circuit)	$v_1(0^+) = 22 - 4.49(4.8) = 0$ V	$v_1(\infty) = 0$ V
$i_2(0^-) = 0$ A (open circuit)	$i_2(0^+) = -4.59$ A	$i_2(\infty) = 0$ A
$v_2(0^-) = 4.8(4.59) = 22$ V	$v_2(0^+) = 22$ V	$v_2(\infty) = 0$ V

Example 2.12: Transfer Functions

Assuming zero initial conditions, the *transfer function, H(s)*, of a system or circuit is the ratio of the output, *Y(s)*, to the input, *X(s)*, in the *s*-domain. The transfer function is very useful for characterizing the circuit behavior, stability, and responses. The transfer function is defined as

$$H(s) = \frac{\text{output}}{\text{input}} = \frac{Y(s)}{X(s)} \tag{2.96}$$

Figure 2.27 presents a system description of the transfer function, input, and output.

The transfer function is a rational function of *s*, and the denominator of the transfer function is the *characteristic equation* of the system. The roots of this characteristic equation can be used to determine the system stability. The roots of the characteristic equation are called the *poles*. The roots of the numerator of the transfer function are called *zeros*. The poles are used to identify the frequencies where the system will grow without being bound or becoming unstable. The graph of the poles of a system on the s-plane can be used to quickly identify whether a system is stable. On the s-plane, the zeros are presented as O's and the poles are represented as X's. If all of the poles of the system are on the open left-hand plane, then the system is stable. Another way of stating this criteria is that "if the real part of the pole is negative," then the system will be *stable*. If the poles are purely imaginary, then the system is *marginally stable*, otherwise it is *unstable*. Figure 2.28 illustrates the graphical relationship between system poles and stability.

2.3.2.2.1 Types of Responses

It is evident from the examples of the Laplace analysis that there are three forms of the solutions for *second-order* circuits. Recall that a *second-order circuit* is one that can be described by a second order differential equation and these circuits have all three passive circuit elements: resistors, inductors, and capacitors. The three types of responses are *overdamped*, *underdamped*, and *critically damped*. The *overdamped response* has a slow response and a long settling time. The *critically damped response* has a fast response and a short settling time. The *underdamped response* has the fastest response and long settling time. It is possible to determine the type of response a circuit will have by examining the roots of the *characteristic equation*. Equation 2.77 is an example of an *overdamped response*. Equation 2.87 is an example of a *critically damped response*. Equation 2.93 is an example of an *underdamped* response. Table 2.6 presents a summary of the three types of responses and roots of the characteristic equations as they relate to the Laplace analysis examples.

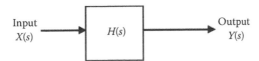

FIGURE 2.27   Transfer function representation of a system.

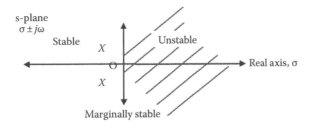

FIGURE 2.28   Poles on the s-plane and stability.

TABLE 2.6 Types of Second Order Responses

Overdamped (Example 2.9)	Critically Damped (Example 2.10)	Underdamped (Example 2.11)
Slow rise time, long settling time	Fast rise time, short settling time	Fast rise time, long settling time
Characteristic equation has two distinct real roots, i.e., $(s + 160)(s + 10)$	Characteristic equation has repeated real roots, i.e., $(s + 40)^2$	Characteristic equation has two complex conjugate roots, i.e., $(s + 24 + j32)(s + 24 - j32)$

2.3.2.2.2 Impulse Response

From Equation 2.96, it is evident that the output of a system can be found from the transfer function by using

$$Y(s) = H(s)X(s) \tag{2.97}$$

The *impulse response* of a system is the output, $y(t)$, when the input is an impulse function, $x(t) = \delta(t)$. Since the Laplace transform of $\delta(t)$ is 1, $X(s) = 1$ so

$$Y(s) = H(s) \tag{2.98}$$

$$y(t) = L^{-1}\{H(s)\} = h(t) \tag{2.99}$$

2.3.2.2.3 Step Response

The *step response* of a system is the output, $y(t)$, when the input is a step function, $x(t) = u(t)$. Since the Laplace transform of $u(t)$ is $1/s$, $X(s) = 1/s$ so

$$Y(s) = \left(\frac{1}{s}\right)H(s) \tag{2.100}$$

$$y(t) = L^{-1}\left\{\left(\frac{1}{s}\right)H(s)\right\} \tag{2.101}$$

2.3.2.2.4 Sinusoidal Steady-State

Recall from the phasor analysis section that the steady-state sinusoidal response is the output when the input is $x(t) = A \cos(\omega t + \phi)$. The Laplace transform of the input, $x(t)$, is

$$X(s) = \frac{As\cos\phi}{s^2 + \omega^2} - \frac{A\omega\sin\phi}{s^2 + \omega^2} = \frac{A(s\cos\phi - \omega\sin\phi)}{s^2 + \omega^2} \tag{2.102}$$

The output is

$$Y(s) = H(s)X(s) = H(s)\frac{A(s\cos\phi - \omega\sin\phi)}{s^2 + \omega^2} \tag{2.103}$$

and using partial fraction expansion,

$$Y(s) = \text{terms from poles of } H(s) + \frac{K_1}{s - j\omega} + \frac{K_1^*}{s + j\omega} \tag{2.104}$$

The complete response $y(t)$ can be separated into the steady-state and transient response, or the natural response and forced response:

$$y(t) = \text{steady-state} + \text{transient response} \tag{2.105}$$

$$y(t) = \text{natural} + \text{forced response} \tag{2.106}$$

The *natural response* is due to the stored energy in the circuit being released and the *forced response* is due to the input or independent source suddenly applied. The *transient response* is the changing part of the response that decays exponentially as time increases. The *steady-state response* is the part that remains after all of the transient decays off. Therefore, only the input, $X(s)$, terms in Equation 2.104 contribute to the steady-state response. The terms due to the transfer function, $H(s)$, are transient and approach 0 as t approaches ∞. The poles of the transfer function, $H(s)$, are in the open left-half of the s-plane (they have to be for stability!). The poles of the sinusoidal input, $X(s)$, are on the imaginary axis of the s-plane because they oscillate forever. Using partial fraction expansion to solve for K_1 in Equation 2.104 yields

$$K_1 = \left. \frac{H(s)A(s\cos\phi - \omega\sin\phi)}{s + j\omega} \right|_{s=j\omega} = \frac{1}{2}H(j\omega)Ae^{j\phi} = \frac{A}{2}\left|H(j\omega)\right|e^{j[\theta(\omega)+\phi]} \tag{2.107}$$

From these results, the response due to the steady-state is

$$Y_{ss}(j\omega) = (H(j\omega))(A\angle\phi) = A\left|H(j\omega)\right|e^{j(\phi+\theta(\omega))} \tag{2.108}$$

The steady-state solution for $y(t)$ or the steady-state sinusoidal response is

$$y_{ss}(t) = A\left|H(j\omega)\right|\cos[\omega t + \phi + \theta(\omega)] \tag{2.109}$$

The amplitude of the solution is equal to the amplitude of the source, A, times the magnitude of the transfer function, $|H(j\omega)|$. The phase angle of the response is equal to the phase angle of the source, ϕ, plus the phase angle of the transfer function, $\theta(\omega)$, at the frequency of the source, ω.

2.3.2.2.5 Complete Response

It is also possible to find the complete response (steady-state and transient) due to an excitation by finding the Laplace transform of the input and solving for the product $H(s)X(s)$. The complete response, $y(t)$, is the inverse Laplace transform of $Y(s)$.

2.3.3 Impulse Response Example

For the circuit shown in Figure 2.29a, find the transfer function $H(s) = V_o(s)/V_S(s)$ and the impulse response. Recall that the transfer function assumes zero initial conditions. The first step in the solution is to convert the circuit to the s-domain and this is shown in Figure 2.29b.

Next, use the node-voltage method to find the voltage across the resistor:

$$\frac{sV_o}{1G} + \frac{V_o}{1000} + \frac{1000V_o}{s} = \frac{1000V_s}{s} \tag{2.110}$$

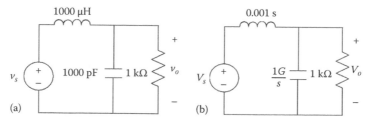

FIGURE 2.29 Transfer function example circuit. (a) Time domain and (b) s domain.

Solving for V_o yields

$$V_o = \frac{1T}{s^2 + 1Ms + 1T} V_s \tag{2.111}$$

Finally solve for the transfer function, $H(s) = V(s)/V_S(s)$,

$$H(s) = \frac{V_o}{V_S} = \frac{1T}{s^2 + 1Ms + 1T} \tag{2.112}$$

Recall that the *impulse response* is the output, $y(t)$, when the input $x(t) = \delta(t)$, this is found from (2.113):

$$V_o(s) = H(s)V_s(s) = H(s)(1) = \frac{1T}{s^2 + 1Ms + 1T} \tag{2.113}$$

The inverse Laplace of (2.113) yields the *impulse response*

$$h(t) = v_o(t) = (1.15M\, e^{-500kt} \cos(866kt - 90°))\, u(t) \text{ V} \tag{2.114}$$

2.3.4 Step Response Example

It is also possible to use the transfer function in the equation in (2.112) to find the *step response*. If the input is $v_s(t) = u(t)$, then

$$V_o(s) = H(s)V_s(s) = H(s)\left(\frac{1}{s}\right) = \frac{1T}{s(s^2 + 1Ms + 1T)} \tag{2.115}$$

The inverse Laplace of (2.115) yields the *step response*

$$v_o(t) = 1 + 1.15e^{-500kt} \cos(866kt - 90°)u(t) \text{ V} \tag{2.116}$$

2.3.5 Sinusoidal Steady-State Example

To use the transfer function, $H(s)$, to find the sinusoidal steady-state response, assume that the input $v_s(t) = 5 \cos(700kt)$ V. The first step in the solution is to find the value of the transfer function $H(j\omega)$ at $\omega = 700$ krad/s:

$$H(j\omega) = H(j700k) = \frac{1T}{(j700k)^2 + 1M(j700k) + 1T} = 1.15\angle - 53.92° \tag{2.117}$$

Using Equation 2.108,

$$V_{oss}(j\omega) = (H(j\omega))(A\angle\phi) = (1.15\angle - 53.92°)(5\angle 0°) = 5.75\angle - 53.92° \text{ V} \tag{2.118}$$

Finally, the answer can be found from Equation 2.109:

$$v_{oss}(t) = 5.75 \cos(700kt - 53.92°)u(t) \text{ V} \tag{2.119}$$

2.3.6 Complete Response Example

It is also possible to find the complete response (steady-state and transient), $v_o(t)$, due to the input $v_s(t) = 5\cos(700kt)$ V. The first step is to find the Laplace transform of $v_s(t)$:

$$V_s(s) = \frac{5s}{s^2 + (700k)^2} \tag{2.120}$$

Then to find the output, $V_o(s)$, use the following equation:

$$V_o(s) = H(s)V_s(s) = \left(\frac{1T}{s^2 + 1Ms + 1T}\right)\left(\frac{5s}{s^2 + (700k)^2}\right) \tag{2.121}$$

In order to find $v_o(t)$, it is necessary to find the constants, C_1 and K_1, of the partial fraction expansion shown in (2.122):

$$V_o(s) = \frac{C_1}{s + 500k - j866k} + \frac{C_1^*}{s + 500k + j866k} + \frac{K_1}{s - j700k} + \frac{K_1^*}{s + j700k} \tag{2.122}$$

The constants for this expression are

$$C_1 = \left(\frac{1T}{s^2 + 500k + j866k}\right)\left(\frac{5s}{s^2 + (700k)^2}\right)\Bigg|_{s=-500k+j866k} = 3.333\angle 120.66° \tag{2.123}$$

$$K_1 = \left(\frac{1T}{s^2 + 1Ms + 1T}\right)\left(\frac{5s}{s + j700k}\right)\Bigg|_{s=j700k} = 2.887\angle -53.92°$$

The answer for the complete response is found by using the formula in (2.124):

$$v_o(t) = 2|C_1|e^{-\alpha t}\cos(\omega t + \angle\theta_{C1}) + 2|K_1|\cos(\omega t + \angle\theta_{K1}) \tag{2.124}$$

Finally the answer for the complete response is

$$v_o(t) = 6.67e^{-500kt}\cos(866kt + 120.66°) + 5.77\cos(100kt - 53.92°)\,u(t)\,\text{V} \tag{2.125}$$

Notice that this equation is indeed the sum of the transient and steady-state response.

2.4 Complex Power

2.4.1 Instantaneous, Average, and Reactive Power

The instantaneous power for an AC circuit is given by the equation $p = vi$, where the units are watts (W). Since the source is a sinusoidal voltage or current source, this yields

$$p(t) = \frac{1}{2}V_m I_m(\cos(\theta_v - \theta_i) + \cos(\theta_v - \theta_i)\cos(2\omega t) - \sin(\theta_v - \theta_i)\sin(2\omega t)) \tag{2.126}$$

Recall that V_m and I_m represent the maximum values or magnitude of the voltage and current sinusoids, respectively. This relationship is found by using trigonometric properties to rearrange the product of the two sinusoids. Note that this expression includes a constant term plus the sum of two sinusoids at frequencies double the voltage and current frequency. Figure 2.30 illustrates the plot of the voltage, current,

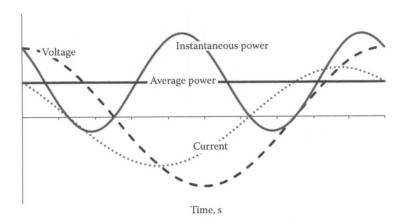

FIGURE 2.30 Instantaneous and average power.

and instantaneous power waveforms. During the positive portion of the power waveform, energy is being stored in an inductor and/or capacitor, and during the negative portion, energy is being released back into the circuit. Also, notice that the waveform is offset by a real positive number that represents the average (real) power, P. The average power is the useful output of a device typically transformed from electric to nonelectric energy. The reactive power, Q, represents the energy stored in an inductor or capacitor. The reactive power is the energy that is transferred back and forth in the circuit but is not converted into nonelectric energy.

The instantaneous power expression in (2.126) can be written in terms of the real and reactive power as

$$p(t) = P + P\cos(2\omega t) - Q\sin(2\omega t) \text{ [W]} \tag{2.127}$$

where

$$P = \frac{1}{2}V_m I_m \cos(\theta_v - \theta_i) \text{ [W]} \tag{2.128}$$

$$Q = \frac{1}{2}V_m I_m \sin(\theta_v - \theta_i) \text{ [VAR]} \tag{2.129}$$

The angle, $\theta_v - \theta_i$, in the instantaneous power expression is the power factor angle. The cosine of the power factor angle is the power factor (pf $= \cos(\theta_v - \theta_i)$). The power factor represents the effectiveness of a load to absorb the average power from a source. In other words, the higher the power factor, the more power will be delivered to the load. A power factor is lagging if the current lags the voltage (i.e., current peaks after the voltage). The current is leading if the current peaks before the voltage. Table 2.7 presents a summary of the relationships between instantaneous, average, reactive power, and power factor for purely resistive, inductive, and capacitive circuits.

2.4.2 Effective or RMS Value

The RMS (root mean square) value of a periodic function is given by

$$X_{rms} = \sqrt{\frac{1}{T}\int_{t_o}^{t_o+T} x(t)^2 \, dt} \tag{2.130}$$

TABLE 2.7 Ideal Transformer Dot Convention

	Purely Resistive	Purely Inductive	Purely Capacitive
Circuit			
Waveforms			
Instantaneous power	$p(t) = P + P\cos(2\omega t)$	$p(t) = -Q\sin(2\omega t)$	$p(t) = -Q\sin(2\omega t)$
Average power, P	$P > 0$	$P = 0$	$P = 0$
Reactive power, Q	$Q = 0$	$Q > 0$	$Q < 0$
Power factor	pf = 1 $\theta_v - \theta_i = 0°$	pf lagging $\theta_v - \theta_i > 0°$	pf leading $\theta_v - \theta_i < 0°$

However, if the periodic function is a sinusoid, then this expression reduces to

$$X_{rms} = \sqrt{\frac{1}{T}\int_{t_o}^{t_o+T} X_m^2 \cos^2(\omega t + \phi)dt} = \frac{X_m}{\sqrt{2}},$$

(2.131)

where T is the period.

The RMS value is also referred to as the effective value of a sinusoidal current or voltage source. The term effective value is based upon the fact that given an equivalent resistive load and an equivalent time period, the RMS value of a source delivers the same energy to the resistor as a DC source of the same value. Thus, the average and reactive power expression in Equations 2.128 and 2.129 can also be written in terms of the RMS value:

$$P = V_{rms}I_{rms}\cos(\theta_v - \theta_i) \text{ [W]}$$

(2.132)

$$Q = V_{rms}I_{rms}\sin(\theta_v - \theta_i) \text{ [VAR]}$$

(2.133)

Also, if a sinusoidal voltage or current source is supplied to a resistor, then the average power delivered to the resistor is

$$P = \frac{V_{rms}^2}{R} = I_{rms}^2 R \text{ [W]}$$

(2.134)

2.4.3 Complex and Apparent Power

Complex Power, S, is the complex sum of the average and reactive power given by

$$S = P + jQ = \frac{1}{2}V_m I_m^* = V_{rms}I_{rms}^* \text{ [VA]}$$

(2.135)

As is evident from this expression, the complex power can also be found from the voltage and current phasors in the circuit. The magnitude of the complex power is the apparent power, $|S|$, and can be found from the following expression:

$$|S| = \sqrt{P^2 + Q^2} \text{ [VA]}$$

(2.136)

The apparent power of a device represents the volt-ampere capacity to supply the average power. This value may be more useful in designing a device than the average power that represents the useful output of an energy-converting device. Thus, the power factor indicates the capability of a device to convert the apparent power supplied to the useful output or average power. To summarize the units of AC power are watts (W) for instantaneous and average power, Volts–Amperes Reactive (VARS) for reactive power, and volts–amperes (VA) for complex power and apparent power.

There is also a geometric relationship between average, reactive, apparent, complex power, power factor, and impedance. This relationship can be represented by the impedance and power triangles that are useful when performing complex power calculations. The power and impedance triangles are shown in Figure 2.31.

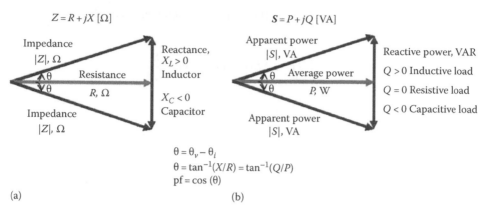

FIGURE 2.31   (a) Impedance and (b) power triangles.

2.4.4 Maximum Average Power Transfer

When designing a power delivery system, the design goal is to transfer the maximum amount of voltage from the source to the load. However, in some communication and power electronic circuits, it is desired to design a system to deliver the maximum power to the load. This can be achieved with impedance matching or by setting the load impedance equal to the conjugate of the Thevenin equivalent impedance ($Z_L = Z_{th}^*$) of the source circuit. The Thevenin equivalent impedance, Z_{th}, of the source is the net equivalent impedance seen looking into the terminals of the source when all independent sources are disabled. The value of the maximum power transferred can be found by using the load resistance and the Thevenin equivalent voltage, V_{th}. The Thevenin equivalent voltage is the open circuit voltage found when looking back into the terminals of the source. The value for the maximum average power is given by

$$P_{max} = \frac{|V_{th}|^2}{4R_L} \text{ (rms)} = \frac{|V_{th}|^2}{8R_L} \text{ (peak) [W]} \tag{2.137}$$

Sometimes there are restrictions on the load and it is not possible to set this value to the conjugate of the Thevenin equivalent impedance. One solution is to design the load impedance magnitude to equal the magnitude of the Thevenin impedance, $|Z_L| = |Z_{th}|$. Alternately, if there are restrictions on the values of the resistance or reactance, it is possible to design X_L as close as possible to X_{th} and then adjust R_L as close as possible to $\sqrt{R_{Th}^2 + (X_L + X_{Th})^2}$. Lastly, when the load impedance is set and cannot be adjusted at all, an ideal transformer can be used to design for *impedance matching*. Impedance matching is the process of designing a system to match the source and load impedance in order to maximize the average power delivered to the load.

2.4.5 Power Factor Correction

Most industrial loads are inductive with a lagging power factor. Power factor correction is the method used to increase the power factor of a load for more effective delivery of the average power from the source. Increasing the load power factor can be achieved by placing a capacitor in parallel with the load. The addition of this capacitor will decrease the phase angle between the voltage and the current supplied to the load but the average power will not be changed. The load with the capacitor draws less current which reduces the average power loss in transmission. The ideal solution is to reduce the reactive power because this is the energy that is transferred back and forth, and that is not lost. A shunt capacitor across

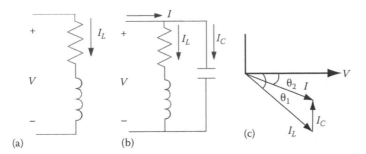

FIGURE 2.32 Power factor correction. (a) Original load, (b) load with capacitor, and (c) phasor diagram.

the load will decrease this oscillating energy. Figure 2.32 presents a circuit and phasor diagram for an inductive load before and after the power factor correction.

The process of the power factor correction can also be illustrated using the power triangle where θ_1 represents the original power factor angle and θ_2 represents the corrected power factor angle. It is evident from Figure 2.33 that the addition of the capacitor only changes the reactive power and leaves the average power unchanged.

It is possible to use the power triangle in Figure 2.33 to determine the value of the capacitor that is necessary to adjust the load to the desired power factor angle of θ_2. The equations found from the figure are

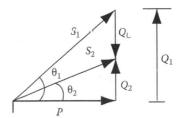

FIGURE 2.33 Power factor correction and the power triangle.

$$Q_1 = P \tan \theta_1 \tag{2.138}$$

$$Q_2 = P \tan \theta_2 \tag{2.139}$$

$$Q_C = Q_1 - Q_2 = P(\tan \theta_1 - \tan \theta_2) \tag{2.140}$$

Finally, the value of the shunt capacitance can be found from (2.133) and is given by

$$C = \frac{Q_C}{\omega V_{rms}^2} \tag{2.141}$$

Example 2.13: RMS and Average Power

The waveform in Figure 2.34 represents the voltage across a 10 Ω resistor. What is the average power absorbed by the resistor?

The first step in the solution is to find the RMS value of the voltage waveform by using Equation 2.130:

$$V_{rms} = \sqrt{\frac{1}{2}\int_0^1 120^2 t^2 \, dt + \frac{1}{2}\int_1^2 120^2 \, dt} = 120\sqrt{\frac{2}{3}} \tag{2.142}$$

Finally, the average power delivered to the resistor is found from Equation 2.134:

$$P = \frac{V_{rms}^2}{R} = 960 \text{ W} \tag{2.143}$$

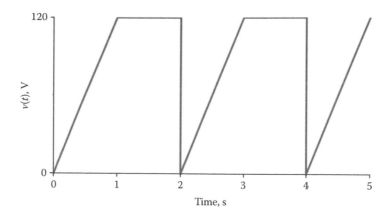

FIGURE 2.34   Voltage waveform.

Example 2.14: Law of Conservation of Energy (Power Triangle)

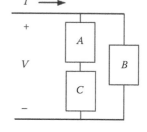

For the system shown in Figure 2.35, the load consists of three devices. Device A includes ten 60 W light bulbs, device B is a 3 kW kitchen range with a 0.8 pf leading, and device C is a 700 VA television with a 0.4 pf lagging. Determine the complex power supplied by the source.

It is possible to create a power triangle for each of the devices and then use trigonometry to find the complex power delivered to each of them (see Figure 2.36). Based upon the law of conservation of energy, the complex power supplied by the source will be the sum of the complex power delivered to the load.

FIGURE 2.35   Real and reactive power.

Since device A has only average power, there is no reactive power. In order to find Q_B, P_C, and Q_C, use the following equations:

$$\theta_B = \cos^{-1}(0.8) = -36.87° (\theta_B < 0 \text{ for leading pf}) \tag{2.144}$$

$$\theta_C = \cos^{-1}(0.4) = 66.42° (\theta_C > 0 \text{ for lagging pf}) \tag{2.145}$$

$$Q_B = P_B \tan(\theta_B) = -2.25 \text{ kVAR} \tag{2.146}$$

$$P_C = S_C \cos(\theta_C) = 280 \text{ W} \tag{2.147}$$

$$Q_C = S_C \sin(\theta_C) = 641.56 \text{ VAR} \tag{2.148}$$

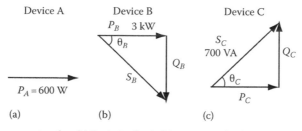

FIGURE 2.36   Device power triangles. (a) Resistive load, (b) capacitive load, and (c) inductive load.

Therefore, summing the complex power absorbed by each of the devices yields

$$S = S_A + S_B + S_C = P_A + P_B + P_C + j(Q_A + Q_B + Q_C) = 3.88 - j1.61 \text{ kVA} = 4.2\angle -22.52° \text{kVA} \qquad (2.149)$$

Example 2.15: Average Power Line Losses

In order to compare power line losses for a power delivery system with and without an ideal transformer, the circuit in Figure 2.37 will be analyzed. If the power delivered to the load is 153 W at 120 V_{rms}, what is the average power lost in the line? This can be found by solving for the current, I_L, using the power rating and voltage V_L:

$$I_L = \frac{120}{7.5 + j3.8} = 14.28\angle -26.87° \text{ A} \qquad (2.150)$$

The average power lost in the line is given by

$$P = |I_L|^2 (0.2) = 40.74 \text{ W} \qquad (2.151)$$

However, how does this compare to the power delivery system with the ideal transformer? The circuit in Figure 2.38 models a power delivery system with a source, transmission line, and load. One benefit of using a transformer is that it reduces the average power loss in the line by stepping up the voltage to the line and reducing the current. This method also saves costs because smaller wires can be used to carry the current. For the following figure, determine the average power lost in the line, the power delivered to the load, and the power factor. Finally, compare the average power lost in the line to a similar system without the transformer.

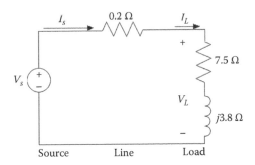

FIGURE 2.37 Power delivery system without a transformer.

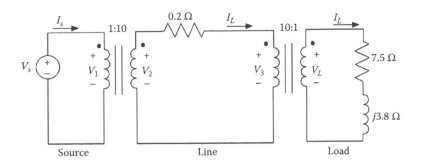

FIGURE 2.38 Power delivery system with a transformer.

The first in the step would be to use reflection to determine the line current in the feeder:

$$I_1 = (0.1)I_L = 0.1(14.28\angle -26.87°) = 1.43\angle -26.87° \text{ A} \tag{2.152}$$

Notice that the line current is greatly reduced with the voltage stepped up from the source. The average power lost in the line is now given by

$$P = |I_1|^2 (0.2) = 407.41 \text{ mW} \tag{2.153}$$

The loss of power in the line is significantly reduced by a magnitude of 100 as compared with the circuit in Figure 2.37. However, the power delivered to the load is still 153 W at 120 V_{rms} at 0.892 pf lagging.

Example 2.16: Apparent Power (Autotransformers)

In order to compare the performance of an autotransformer and an ideal transformer with the same turns ratio, the circuits in Figures 2.39 and 2.40 will be analyzed to determine the voltage and apparent power delivered to the load. The process to find the voltage and apparent power involves using the ideal transformer terminal equations and KCL. For the circuit in Figure 2.39, reflection from the primary to the

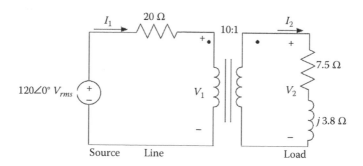

FIGURE 2.39   Ideal transformer apparent power.

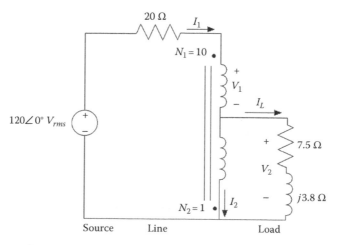

FIGURE 2.40   Autotransformer apparent power.

secondary is used to find the load current and the voltage, and apparent power. These values are shown in (2.154) through (2.156):

$$I_2 = \frac{12}{0.2 + 7.5 + j3.8} = 1.4\angle -26.27° \text{ A} \tag{2.154}$$

$$V_2 = (7.5 + j3.8)I_2 = 11.75\angle 0.6° \text{ V} \tag{2.155}$$

$$S_L = V_2 I_2^* = 16.42\angle 26.87° \text{ VA} = 14.65 + j7.422 \text{ VA} \tag{2.156}$$

Therefore, the load has 11.75 V at 14.65 W delivered to it. The derivation of the load voltage and the apparent power for the autotransformer in Figure 2.40 involves creating the ideal transformer terminal relationships, performing KVL around the left and right mesh, and KCL at the autotransformer tap. This configuration is a step-down transformer with a subtractive connection. It is not possible to perform reflection on this circuit because the sides are not decoupled. This analysis yields the following five equations:

$$20I_1 + V_1 + V_2 = 120 \tag{2.157}$$

$$(7.5 + j3.8)I_L - V_2 = 0 \tag{2.158}$$

$$I_1 = I_2 + I_L \tag{2.159}$$

$$V_2 = -0.1V_1 \tag{2.160}$$

$$I_1 = 0.1I_2 \tag{2.161}$$

Solving this system of five equations yields

$$I_L = 1.55 \angle 153.87° \text{ V} \tag{2.162}$$

$$V_2 = 13\angle -179.26° \text{ V} \tag{2.163}$$

$$S_L = V_2 I_2^* = 20.08\angle 26.87° \text{ VA} = 17.91 + j9.07 \text{ VA} \tag{2.164}$$

Observe that the same transformer in the autotransformer configuration delivers a slightly higher voltage, power, and apparent power with the same power factor.

Example 2.17: Power Factor Correction

In order to design a system for maximum efficiency of power delivery, it may be necessary to design the load to have a certain power factor. For example, the higher the power factor, the smaller the reactive power and the larger the average power delivered to the load. Since the reactive power, Q, represents energy that is exchanged back and forth between the source and the load, it would be ideal to reduce this. The power factor of a load can be corrected by putting a capacitor in parallel with it. This will be demonstrated by designing the circuit in Figure 2.41 to have a power factor of 0.95.

The first step would be to determine the power factor and average power delivered to the load of the original circuit. This can be done by using the following equations:

$$\theta_{old} = \tan^{-1}(3.8 / 7.5) = 26.87° \tag{2.165}$$

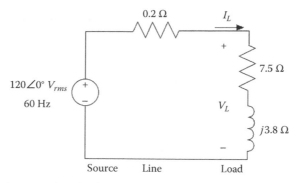

FIGURE 2.41   Power factor correction circuit.

$$pf = \cos(\theta_{old}) = 0.892 \tag{2.166}$$

$$I_L = 120/(0.2+7.5+j3.8) = 13.98\angle-26.27°\,\text{A} \tag{2.167}$$

$$V_L = I_L(7.5+j3.8) = 117.5\angle0.6°\,\text{V} \tag{2.168}$$

$$S_L = V_L I_L^* = 1642\angle26.87° = 1465+j742.17\,\text{VA} \tag{2.169}$$

Thus, the average power delivered to the load is 1.47 kW at a 0.89 pf lagging. In order to increase the power factor without changing the average power, a shunt capacitor must be placed in parallel with the load. The formula to determine the value of the capacitor is

$$C = \frac{Q_C}{\omega V_{rms}^2} = \frac{P(\tan\theta_{old} - \tan\theta_{new})}{\omega V_{rms}^2} \tag{2.170}$$

The desired power factor can be used to determine the new power factor angle and this value can be substituted into (2.169) to find the value of the capacitor:

$$\theta_{new} = \cos^{-1}(0.95) = 18.19° \tag{2.171}$$

$$C = \frac{P(\tan\theta_{old} - \tan\theta_{new})}{\omega V_{rms}^2} = \frac{1465(\tan(26.87°) - \tan(18.19°))}{(2\pi60)(117.5)^2} = 50\,\mu\text{F} \tag{2.172}$$

The redesigned circuit using power factor correction is shown in Figure 2.42.

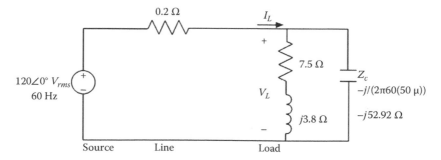

FIGURE 2.42   Power factor correction circuit with capacitor.

The final step in the analysis involves verifying that the new circuit has the new power factor and the same average power delivered to the load:

$$Z_L = (7.5 + j3.8) \| (-j52.92) = 8.51 + j2.8\,\Omega \tag{2.173}$$

$$I_L = \frac{120}{0.2 + Z_L} = 13.12\angle{-17.8°}\ A \tag{2.174}$$

$$V_L = I_L Z_L = 117.5\angle 0.39°\,V \tag{2.175}$$

$$S_L = V_L I_L^* = 1542\angle 18.19° = 1465 + j481.31\,\text{VA} \tag{2.176}$$

Thus, the average power delivered to the load is 1.47 kW at 0.95 pf lagging, which confirms that the design meets the specifications.

Example 2.18: Maximum Average Power Transfer (Impedance Matching)

In order to deliver the maximum average power to a load in a circuit, it is necessary to perform impedance matching. Impedance matching is the process of designing the system such that the load impedance is matched to the Thevenin equivalent impedance of the source. When there are limitations on the source or load impedance, this can be accomplished by using an ideal transformer and designing for the appropriate turns ratio. This process will be demonstrated with the following two examples. Design the circuit in Figure 2.43 to deliver the maximum average power to the load and determine the value of the power delivered.

In order to deliver maximum power, determine the turns ratio to match the load impedance as close as possible to the source impedance:

$$n = \sqrt{\frac{R_L}{R_s}} = 20 \tag{2.177}$$

$$P = (1.38)^2(75) = 143.48\ W \tag{2.178}$$

Design the circuit in Figure 2.44 to deliver the maximum average power to the load when there are no limitations on the load. Finally, determine the average power delivered to the load selected.

The first step in the analysis is to determine the Thevenin equivalent voltage and impedance as seen from the load. The Thevenin equivalent voltage is the open circuit voltage across terminals a and b.

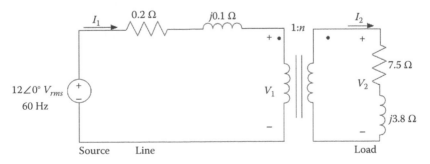

FIGURE 2.43   Impedance matching.

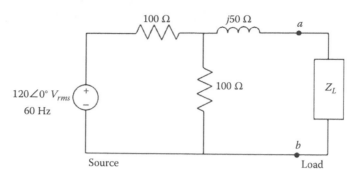

FIGURE 2.44   Maximum average power transfer.

The Thevenin equivalent impedance is the impedance across terminals a and b when the independent voltage source is disabled. These values are found using formulas (2.179) and (2.180):

$$Z_{th} = 100 \,\|\, 100 + j50 = 50 + j50 \,\Omega \tag{2.179}$$

$$V_{th} = \frac{1}{2}(120) = 60\angle 0° V_{rms} \tag{2.180}$$

Therefore, for impedance matching, set the load $Z_L = Z_{th}^*$ and determine the maximum average power delivered to the load:

$$Z_{th} = 50 - j50 \,\Omega \tag{2.181}$$

$$P = \frac{|V^{th}|^2}{4R_{th}} = 18 \, W \tag{2.182}$$

2.5 Conclusions

This chapter has presented a brief summary of the fundamental concepts of AC circuits. These concepts included mutual inductance, transformers, phasor analysis, Laplace analysis, and complex power. Since many of these concepts were summarized, the reader is encouraged to extend their study to other topics related to AC circuits, including filters, operational amplifiers, three-phase power, and Fourier analysis.

Bibliography

J.W. Nilsson and S.E. Riedel, *Electric Circuits*, 8th edition, Upper Saddle River, NJ: Prentice Hall, 2007, 880 pp.
M.N.O. Sadiku and C.K. Alexander, *Fundamentals of Electric Circuits*, 3rd edition, New York: McGraw-Hill, 2007, 901 pp.

3

Computational Methods in Node and Loop Analyses

Stephen M. Haddock
Auburn University

J. David Irwin
Auburn University

In this section, node and loop analysis are introduced as methods for computing voltages and currents in electrical circuits. Readers are assumed to have no previous knowledge of these methods. Therefore, the basic principles are first explained and applied to simple dc circuits. Following a few examples, ac circuits are introduced as well. As will become apparent, circuit analysis can quickly become too arduous when many nodes and loops are present. For this reason, a brief introduction to circuit simulation with MATLAB®, PSPICE, and Multisim concludes the section.

3.1 Node Analysis

In a node analysis, the node voltages are the variables in a circuit, and KCL is the vehicle used to determine them. Once a node in the network is selected as a reference node, then all other node voltages are defined with respect to that particular node. This reference node is typically referred to as ground using the symbol ($\perp$), indicating that it is at ground-zero potential. Consider the network shown in Figure 3.1. The network has three nodes, and the node at the bottom of the circuit has been selected as the reference node. Therefore, the two remaining nodes, labeled V_1 and V_2, are measured with respect to this reference node.

Suppose that the node voltages, V_1 and V_2, have somehow been determined, i.e., $V_1 = 4\,V$ and $V_2 = -4\,V$. Once these node voltages are known, Ohm's law can be used to find all branch currents. For example,

$$I_1 = \frac{V_1 - 0}{2} = 2\ \text{A}$$

$$I_2 = \frac{V_1 - V_2}{2} = \frac{4 - (4)}{2} = 4\ \text{A}$$

$$I_3 = \frac{V_2 - 0}{1} = -\frac{4}{1} = -4\ \text{A}$$

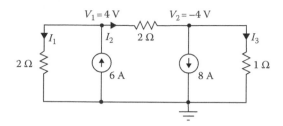

FIGURE 3.1 A three-node network.

Note that KCL is satisfied at every node, i.e.,

$$I_1 - 6 + I_2 = 0$$

$$-I_2 + 8 + I_3 = 0$$

$$-I_1 + 6 - 8 - I_3 = 0$$

Therefore, as a general rule, if the node voltages are known, all branch currents in the network can be immediately determined.

In order to determine the node voltages in a network, we apply KCL to every node in the network except the reference node. Therefore, given an N-node circuit, we employ $N - 1$ linearly independent simultaneous equations to determine the $N - 1$ unknown node voltages. Graph theory can be used to prove that exactly $N - 1$ linearly independent KCL equations are required to find the $N - 1$ unknown node voltages in a network.

Let us now demonstrate the use of KCL in determining the node voltages in a network. For the network shown in Figure 3.2, the bottom node is selected as the reference and the three remaining nodes, labeled V_1, V_2, and V_3, are measured with respect to that node. All unknown branch currents are also labeled. The KCL equations for the three non-reference nodes are

$$I_1 + 4 + I_2 = 0$$

$$-4 + I_3 + I_4 = 0$$

$$-I_1 - I_4 - 2 = 0$$

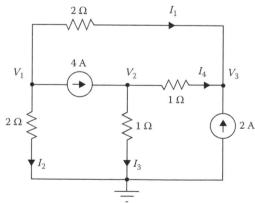

FIGURE 3.2 A four-node network.

Using Ohm's law, these equations can be expressed as

$$\frac{V_1 - V_3}{2} + 4 + \frac{V_1}{2} = 0$$

$$-4 + \frac{V_2}{1} + \frac{V_2 - V_3}{1} = 0$$

$$-\frac{V_1 - V_3}{2} - \frac{V_2 - V_3}{1} - 2 = 0$$

Solving these equations, using any convenient method, yields $V_1 = -8/3\,\mathrm{V}$, $V_2 = 10/3\,\mathrm{V}$, and $V_3 = 8/3\,\mathrm{V}$. Applying Ohm's law, we find that the branch currents are $I_1 = -16/6\,\mathrm{A}$, $I_2 = -8/6\,\mathrm{A}$, $I_3 = 20/6\,\mathrm{A}$, and $I_4 = 4/6\,\mathrm{A}$. A quick check indicates that KCL is satisfied at every node.

The circuits examined thus far have contained only current sources and resistors. In order to expand our capabilities, we next examine a circuit containing voltage sources. The circuit shown in Figure 3.3 has three non-reference nodes labeled V_1, V_2, and V_3. However, we do not have three unknown node voltages. Since known voltage sources exist between the reference node and nodes, V_1 and V_3, these two node voltages are known, i.e., $V_1 = 12\,\mathrm{V}$ and $V_3 = -4\,\mathrm{V}$. Therefore, we have only one unknown node voltage, V_2. The equations for this network are then

$$V_1 = 12$$

$$V_2 = -4$$

and

$$-I_1 + I_2 + I_3 = 0$$

The KCL for node V_2 written using Ohm's law is

$$-\frac{12 - V_2}{1} + \frac{V_2}{2} + \frac{V_2 - (-4)}{2} = 0$$

Solving this equation yields $V_2 = 5\,\mathrm{V}$, $I_1 = 7\,\mathrm{A}$, $I_2 = 5/2\,\mathrm{A}$, and $I_3 = 9/2\,\mathrm{A}$. Therefore, KCL is satisfied at every node.

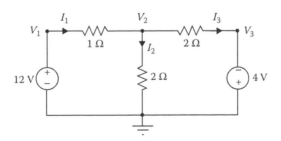

FIGURE 3.3   A four-node network containing voltage sources.

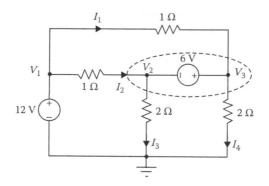

FIGURE 3.4   A four-node network used to illustrate a supernode.

Thus, the presence of a voltage source in the network actually simplifies a node analysis. In an attempt to generalize this idea, consider the network in Figure 3.4. Note that, in this case, $V_1 = 12\,V$ and the difference between node voltages, V_3 and V_2, is constrained to be $6\,V$. Hence, two of the three equations needed to solve for the node voltages in the network are

$$V_1 = 12$$

$$V_3 - V_2 = 6$$

To obtain the third required equation, we form what is called a supernode, indicated by the dotted enclosure in the network. Just as KCL must be satisfied at any node in the network, it must be satisfied at the supernode as well. Therefore, summing all the currents leaving the supernode yields the equation:

$$\frac{V_2 - V_1}{1} + \frac{V_2}{2} + \frac{V_3 - V_1}{1} + \frac{V_3}{2} = 0$$

The three equations yield the node voltages, $V_1 = 12\,V$, $V_2 = 5\,V$, and $V_3 = 11\,V$, and therefore $I_1 = 1\,A$, $I_2 = 7\,A$, $I_3 = 5/2\,A$, and $I_4 = 11/2\,A$.

3.2 Mesh Analysis

In a mesh analysis, the mesh currents in the network are the variables and KVL is the mechanism used to determine them. Once all the mesh currents have been determined, Ohm's law will yield the voltages anywhere in the circuit. If the network contains N independent meshes, then graph theory can be used to prove that N independent linear simultaneous equations will be required to determine the N mesh currents.

The network shown in Figure 3.5 has two independent meshes. They are labeled I_1 and I_2 as shown. If the mesh currents are known to be $I_1 = 7\,A$ and $I_2 = 5/2\,A$, then all voltages in the network can be calculated.

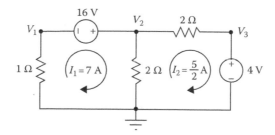

FIGURE 3.5   A network containing two independent meshes.

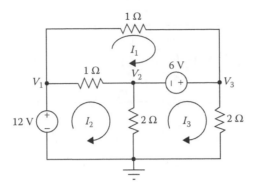

FIGURE 3.6 A three-mesh network.

For example, the voltage V_1, i.e., the voltage across the 1 Ω resistor, is $V_1 = -I_1R = -(7)(1) = -7$ V. Likewise, $V_2 = (I_1 - I_2)R = (7 - 5/2)(2) = 9$ V. Furthermore, we can check our analysis by showing that KVL is satisfied around every mesh. Starting at the lower left-hand corner and applying KVL to the left-hand mesh, we obtain

$$-(7)(1) + 16 - \left(7 - \frac{5}{2}\right)(2) = 0$$

where we have assumed that increases in energy level are positive and decreases in energy level are negative.

Consider now the network in Figure 3.6. Once again, if we assume that an increase in energy level is positive and a decrease in energy level is negative, the three KVL equations for the three meshes defined are

$$-I_1(1) - 6 - (I_1 - I_2)(1) = 0$$

$$+12 - (I_2 - I_1)(1) - (I_2 - I_3)(2) = 0$$

$$-(I_3 - I_2)(2) + 6 - I_3(2) = 0$$

These equations can be written as

$$2I_1 - I_2 = -6$$

$$-I_1 + 3I_2 - 2I_3 = 12$$

$$-2I_2 + 4I_3 = 6$$

Solving these equations using any convenient method yields $I_1 = 1$ A, $I_2 = 8$ A, and $I_3 = 5.5$ A. Any voltage in the network can now be easily calculated, e.g., $V_2 = (I_2 - I_3)(2) = 5$ V and $V_3 = I_3(2) = 11$ V.

Just as in the node analysis discussion, we now expand our capabilities by considering circuits that contain current sources. In this case, we will show that for mesh analysis, the presence of current sources makes the solution easier.

The network in Figure 3.7 has four meshes that are labeled I_1, I_2, I_3, and I_4. However, since two of these currents, i.e., I_3 and I_4, pass directly through a current source, two of the four linearly independent equations required to solve the network are

$$I_3 = 4$$

$$I_4 = -2$$

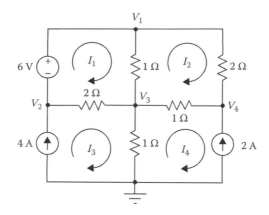

FIGURE 3.7   A four-mesh network containing current sources.

The two remaining KVL equations for the meshes defined by I_1 and I_2 are

$$+6-(I_1-I_2)(1)-(I_1-I_3)(2)=0$$

$$-(I_2-I_1)(1)-I_2(2)-(I_2-I_4)(1)=0$$

Solving these equations for I_1 and I_2 yields $I_1 = 54/11$ A and $I_2 = 8/11$ A. A quick check will show that KCL is satisfied at every node. Furthermore, we can calculate any node voltage in the network. For example, $V_3 = (I_3 - I_4)(1) = 6$ V and $V_1 = V_3 + (I_1 - I_2)(1) = 112/11$ V.

3.3 An AC Analysis Example

Both node analysis and mesh analysis have been presented and discussed. Although the methods have been presented within the framework of dc circuits with only independent sources, the techniques are applicable to ac analysis and circuits containing dependent sources.

To illustrate the applicability of the two techniques to ac circuit analysis, consider the network in Figure 3.8. All voltages and currents are phasors and the impedance of each passive element is known.

In the node analysis case, the voltage, V_4, is known and the voltage between V_2 and V_3 is constrained. Therefore, two of the four required equations are

$$V_4 = 12\angle 0°$$

$$V_2 + 6\angle 0° = V_3$$

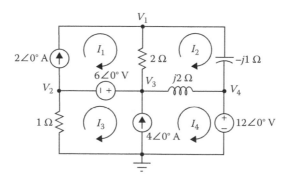

FIGURE 3.8   A network containing five nodes and four meshes.

KCL for the node labeled V_1 and the supernode containing the nodes labeled V_2 and V_3 is

$$\frac{V_1 - V_3}{2} + \frac{V_1 - V_4}{-j1} = 2\angle 0°$$

$$\frac{V_2}{1} + 2\angle 0° + \frac{V_3 - V_1}{2} + \frac{V_3 - V_4}{-j2} = 2\angle 0°$$

Solving these equations yields the remaining unknown node voltages:

$$V_1 = 11.9 - j0.88 = 11.93\angle -4.22° \text{ V}$$

$$V_2 = 3.66 - j1.07 = 3.91\angle -16.34° \text{ V}$$

$$V_3 = 9.66 - j1.07 = 9.72\angle -6.34° \text{ V}$$

In the mesh analysis case, the currents, I_1 and I_3, are constrained to be

$$I_1 = 2\angle 0°$$

$$I_4 - I_3 = -4\angle 0°$$

The two remaining KVL equations are obtained from the mesh defined by mesh current, I_2, and the loop that encompasses the meshes defined by mesh currents, I_3 and I_4:

$$-2(I_2 - I_1) - (-j1)I_2 - j2(I_2 - I_4) = 0$$

$$-I_3 + 6\angle 0° - j2(I_4 - I_2) - 12\angle 0° = 0$$

Solving these equations yields the remaining unknown mesh currents

$$I_2 = 0.88\angle -6.34° \text{ A}$$

$$I_2 = 3.91\angle 163.66° \text{ A}$$

$$I_4 = 1.13\angle 72.35° \text{ A}$$

As a quick check we can use these currents to compute the node voltages. For example, if we calculate

$$V_2 = -1(I_3)$$

and

$$V_1 = -j(I_2) + 12\angle 0°$$

we obtain the answer computed earlier.

Since both node and mesh analyses will yield all currents and voltages in a network, which technique should be used? The answer to this question depends upon the network to be analyzed. If the network contains more voltage sources than current sources, node analysis might be the easier technique. If, however, the network contains more current sources than voltages sources, mesh analysis may be the easiest approach.

3.4 Computer Simulation of Networks

While any network can be analyzed using mesh or nodal techniques, the required calculations are cumbersome for more than three loops or nodes. In these cases, computer simulation is an attractive alternative. As an example, we will solve for the current, I_0, in the circuit in Figure 3.9 using first MATLAB, then PSPICE, and finally Multisim.

MATLAB requires a matrix representation of the network. Nodal analysis yields the equations

$$V_1\left(\frac{1}{Z_1}+\frac{1}{Z_2}\right) + V_2\left(-\frac{1}{Z_1}\right) + V_3\left(-\frac{1}{Z_2}-\frac{1}{Z_4}\right) + V_4\left(\frac{1}{Z_4}+\frac{1}{Z_5}+\frac{1}{Z_6}+\alpha\frac{1}{Z_6}\right) + V_5\left(-\frac{1}{Z_5}\right) = 0$$

$$V_2 = V_{s2}$$

$$V_1(-\beta) + V_2(\beta) + V_3 = 0$$

$$-V_1 + V_4 = V_{s1}$$

$$V_4\left(-\frac{1}{Z_5}-\alpha\frac{1}{Z_6}\right) + V_5\left(\frac{1}{Z_5}+\frac{1}{Z_7}\right) = 0$$

Note the supernode between V_1 and V_2 and also that V_2 is obtained from the voltage source, V_{S2}. The values of these components have been left to variables because of the flexibility it will afford in MATLAB analysis. If the circuit is defined in this way in code, then changing a single variable will result in a universal change through the equations. The matrix equation that solves for the node voltages is

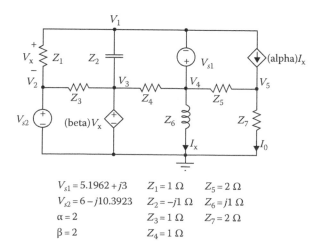

$$V_{s1} = 5.1962 + j3 \qquad Z_1 = 1\ \Omega \qquad Z_5 = 2\ \Omega$$
$$V_{s2} = 6 - j10.3923 \qquad Z_2 = -j1\ \Omega \qquad Z_6 = j1\ \Omega$$
$$\alpha = 2 \qquad\qquad\qquad Z_3 = 1\ \Omega \qquad Z_7 = 2\ \Omega$$
$$\beta = 2 \qquad\qquad\qquad Z_4 = 1\ \Omega$$

FIGURE 3.9 A network containing six nodes and six loops. Computer simulation is helpful in this analysis.

$$
\begin{bmatrix} V_1 \\ V_2 \\ V_3 \\ V_4 \\ V_5 \end{bmatrix} = \begin{bmatrix} \dfrac{1}{Z_1}+\dfrac{1}{Z_2} & -\dfrac{1}{Z_1} & -\dfrac{1}{Z_2}-\dfrac{1}{Z_4} & \dfrac{1}{Z_4}+\dfrac{1}{Z_5}+\dfrac{1}{Z_6}+\alpha\dfrac{1}{Z_6} & -\dfrac{1}{Z_5} \\ 0 & 1 & 0 & 0 & 0 \\ -\beta & \beta & 1 & 0 & 0 \\ -1 & 0 & 0 & 1 & 0 \\ 0 & 0 & 0 & -\dfrac{1}{Z_5}-\alpha\dfrac{1}{Z_6} & \dfrac{1}{Z_5}+\dfrac{1}{Z_7} \end{bmatrix}^{-1} \begin{bmatrix} 0 \\ V_{s2} \\ 0 \\ V_{s1} \\ 0 \end{bmatrix}
$$

Though MATLAB can be used at a command prompt, there are advantages to writing m-files that describe systems that need to be solved. One in particular is that changes can be made quickly. An m-file capable of solving this circuit for the current, I_0, is given at the conclusion of this section. The results given by MATLAB are

Io_mag = 10.4756e+000
Io_phase = −152.0373e+000

In PSPICE, we draw the circuit using one of the accompanying schematic entry tools, either *Schematics* or *Capture*. The resulting *Schematics* file is shown in Figure 3.10. Note that capacitors and inductors must be specified in Farads and Henries, respectively. Therefore, any excitation frequency can be chosen with the L and C values calculated from the known impedances. The most convenient frequency is $\omega = 1$ rad/s or 0.1591 Hz. Figure 3.11 shows the ac sweep settings to produce a single frequency analysis at 0.1591 Hz. Also, the IPRINT part is required to load the simulation results into the OUTPUT file.

Finally, to preserve the clarity of the schematic, the BUBBLE part is used for connecting the controlling parameters into the dependent sources. This is accomplished by naming connecting BUBBLE parts with the same name. Figure 3.12 shows the relevant portion of the output file produced by PSPICE. PSPICE produces an answer that equals that given by MATLAB.

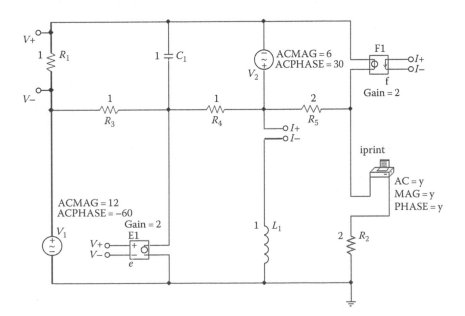

FIGURE 3.10 Circuit drawn in PSPICE.

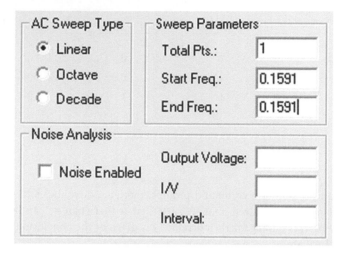

FIGURE 3.11 AC Sweep setup window.

```
FREQ      IM(V_PRINT1) IP(V_PRINT1)

1.591E-01  1.048E+01 −1.520E+02
```

FIGURE 3.12 Excerpt from the PSPICE output file.

$10.481|-152.0°$A, which matches the results given by MATLAB.

Multisim affords another method of computing the desired current in the circuit shown in Figure 3.9. The interface for Multisim is similar to the one presented in PSPICE. The capacitors and inductors must again be input in units of Farads and Henries, respectively. Figure 3.13 shows the circuit drawn in Multisim.

The frequency $\omega = 1, f = 0.1591$ Hz, is also used. Just as with PSPICE, Multisim must be told where to make measurements, so a measurement probe is placed on the branch where the current is to be

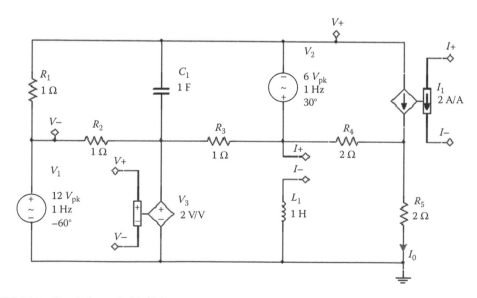

FIGURE 3.13 Circuit drawn in Multisim.

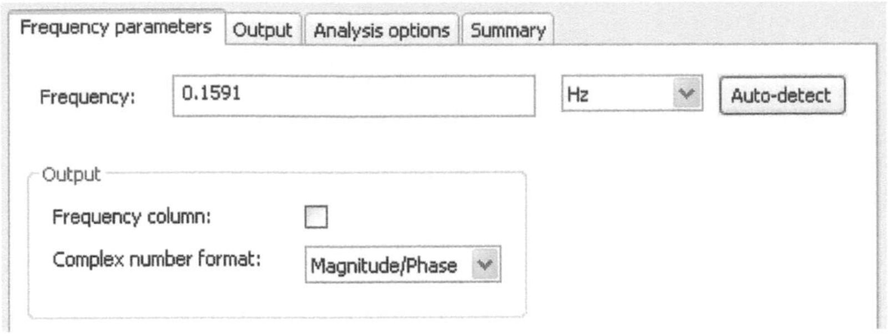

FIGURE 3.14 Single Frequency AC Analysis configuration window.

Single Frequency AC Analysis @ 0.1591 Hz

	AC Frequency Analysis	Magnitude	Phase (deg)
1	I(Io)	10.47515	-152.04829

FIGURE 3.15 Output from Multisim.

measured. There is no special part for connecting the dependent sources to their controlling values in Multisim; just name the connecting nodes with the same name to connect them. To find the current, a single frequency ac analysis will be employed. The menu settings for this analysis are shown in Figure 3.14. Results are shown in Figure 3.15.

$I_0 = 10.475\underline{|-152.05°}$A. All simulation results match to four significant digits.

3.5 MATLAB® m-File

The following MATLAB code was used to compute the output current for the circuit in Figure 3.9.

```
% These lines clear all the previously used
% variables, close all figures, and clear the
% prompt window
clear all
close all
clc

% Use engineering numeric formatting
format short eng

% Define circuit values
    % Resistors
R1 = 1;    G1 = 1/R1;
R2 = -1j;  G2 = 1/R2;
R3 = 1;    G3 = 1/R3;
R4 = 1;    G4 = 1/R4;
R5 = 2;    G5 = 1/R5;
R6 = 1j;   G6 = 1/R6;
R7 = 2;    G7 = 1/R7;
```

```
    % Voltage sources
Vs1 = 5.1962 +3j;
Vs2 = 6 - 10.3923j;

    % Gain terms
alpha = 2;
beta = 2;

% Setup the G matrix
G =    [G1+G2  -G1    -G2-G4   G4 + G5 + G6 + alpha*G6   -G5;
         0      1      0       0                          0;
        -beta   beta   1       0                          0;
        -1      0      0       1                          0;
         0      0      0       -G5-alpha*G6               G5 + G7];

% Setup the I matrix
I = [0;
     Vs2;
     0;
     Vs1;
     0];

% Compute the V matrix
V = G\I;

% Use v5 from the V matrix to solve for Io
Io = V(5)*G7;

% display Io
Io_mag = abs(Io)
Io_phase = angle(Io)*180/pi
```

Defining Terms

ac: An abbreviation for alternating current.

dc: An abbreviation for direct current.

Kirchhoff's current law (KCL): This law states that the algebraic sum of the currents either entering or leaving a node must be zero. Alternatively, the law states that the sum of the currents entering a node must be equal to the sum of the currents leaving that node.

Kirchhoff's voltage law (KVL): This law states that the algebraic sum of the voltages around any loop is zero. A loop is any closed path through the circuit in which no node is encountered more than once.

MATLAB, PSPICE, Multisim: Computer-aided analysis techniques.

Mesh analysis: A circuit analysis technique in which KVL is used to determine the mesh currents in a network. A mesh is a loop that does not contain any loops within it.

Node analysis: A circuit analysis technique in which KCL is used to determine the node voltages in a network.

Ohm's law: A fundamental law that states that the voltage across a resistance is directly proportional to the current flowing through it.

Reference node: One node in a network that is selected to be a common point, and all other node voltages are measured with respect to that point.

Supernode: A cluster of nodes, interconnected with voltage sources, such that the voltage between any two nodes in the group is known.

Bibliography

J.D. Irwin and R.M. Nelms, *Basic Engineering Circuit Analysis*, 9th edn., New York: John Wiley & Sons, 2008.

4

Transistor Operation and Modeling

Tina Hudson
Rose-Hulman Institute of Technology

4.1 Introduction

Transistors are the fundamental building blocks of microelectronic circuits. They are used in both analog and digital circuits. In digital circuits, they can be thought of as voltage-gated switches. In analog circuits, they can be thought of as transconductance (voltage-to-current) amplifiers. There are many different types of transistors, determined by the material it has been made from, the underlying physics of operation, and the primary carrier creating current in the device. However, they can all be used in the same types of circuits. The most commonly used transistors are made out of silicon: the metal-oxide-semiconductor field-effect transistor (MOSFET) and the bipolar-junction transistor (BJT). *Field-effect transistor* (FET) operation is based upon an input generating an *electromagnetic field* to turn the device on and off. A *junction transistor* is controlled by turning a *p–n junction diode* within the device on and off. Both types of transistors have two subtypes determined by the dominant current carrier, holes (p-type), or electrons (n-type), resulting in four transistor types: MOSFETs (nFET and pFET) and BJTs (the npn and the pnp).

The rest of this chapter will discuss more details about these four transistor types, including the device operation, characteristic equations and curves, and modeling. While the underlying physics of these devices are different, the basic operation and circuits for which they are used are very similar. Discussion concerning the trade-offs between these devices are embedded throughout the chapter.

4.2 Transistor Operation and Characterization

4.2.1 MOSFET Operation and Characterization

The MOSFET contains four terminals: the gate (G), drain (D), source (S), and bulk/body/substrate (B). There are several common schematic symbols used for this device, shown in Figure 4.1a for a nFET and Figure 4.1b for a pFET with current directions labeled. Figure 4.2 shows the cross-sectional view of the nFET and pFET. The drain and source regions form a p–n junction diode with the substrate. A dielectric, typically a thin layer of oxide, is placed between the gate, typically made from polysilicon, and the substrate forming a capacitor at the gate. As a result, the FET will exhibit zero gate current, as shown in Figure 4.1. The bulk is placed at a potential to guarantee that the drain and source p–n junction diodes are reverse biased so that excess current does not flow into the bulk.

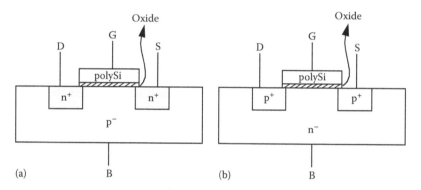

FIGURE 4.1   Schematic representations of (a) the nFET and (b) the pFET.

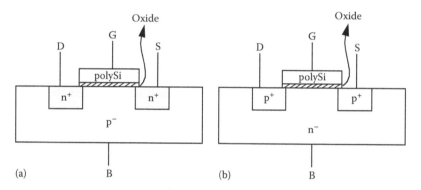

FIGURE 4.2   Cross-sectional views of (a) the nFET and (b) the pFET.

The operation of the nFET is shown in Figure 4.3. The nFET is turned on by placing a voltage on the gate, G, relative to the bulk. The gate voltage attracts electrons to the gate. When enough charge has been attracted to the gate, a channel is formed, which creates a low-resistive connection between the drain and the source (Figure 4.3a). A higher gate voltage places more charge on the gate, resulting in more electron attraction, more charge in the channel, and thus more current through the channel (Figure 4.3b). Current flows through the channel when the drain voltage is greater than the source voltage, attracting electrons to the drain, resulting in positive current flow to the source (Figure 4.3c). As drain voltage increases, more electrons are attracted to the drain resulting in more current to the source. However, an increase in drain voltage also decreases the gate-drain voltage, which causes less channel charge on the drain side of the channel. When the gate-drain voltage is large enough, the channel pinches off at the drain limiting the channel current (Figure 4.3d).

The operation of the pFET is similar to the nFET as shown in Figure 4.4. However, because the dominate carriers in a pFET are holes, the gate voltage must be lower than the substrate voltage to attract holes to the gate (Figure 4.4a and b). Similarly, the drain voltage must be lower than the source voltage to attract holes to the drain through the channel (Figure 4.4c and d). Pinch-off occurs for the same reason as the nFET: a lower drain-gate voltage relative to the source-gate voltage, which lowers the channel charge at the drain. Like the nFET, when the pFET is in pinch-off, the channel current is limited.

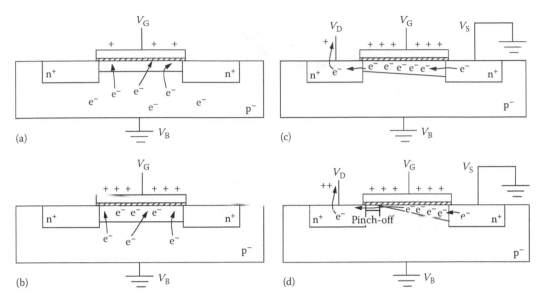

FIGURE 4.3 Device operation of the nFET.

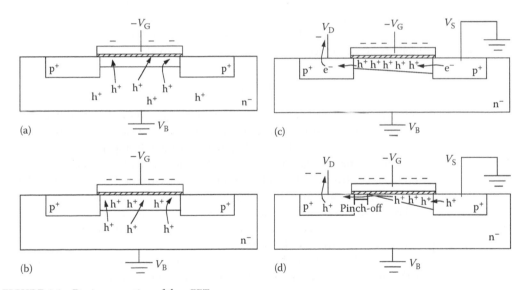

FIGURE 4.4 Device operation of the pFET.

The device operation leads to the characteristic equations of the device. To turn the nFET on, the gate voltage must be large enough to generate the channel. The voltage necessary to generate the channel is called the threshold voltage (V_t). If the gate-bulk voltage, V_{GB}, is larger than the threshold voltage, the device is considered to be on and can allow current to flow through the channel. A larger V_{GB} allows more current. However, even if $0 < V_{GB} < V_t$, the current through the channel is very small, but not zero, which can be a source of leakage current in some circuits.

Although V_{GB} generates the channel, the characteristic curves typically use the gate-source voltage, V_{GS}, in the equations to allow for more flexibility in the circuit designs. In discrete devices, the source is often connected to the bulk internal to the device, making $V_{GB} = V_{GS}$. In integrated circuits, the substrate is not always connected to the source. In these cases, this discrepancy is accounted for by increasing the

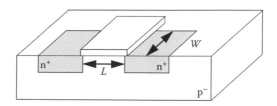

FIGURE 4.5 Three-dimensional sketch of FET showing W and L.

threshold voltage as a function of the source-bulk voltage, V_{SB}. As a result, the threshold voltage may be larger for devices whose bulk is not at the same potential as the substrate.

As a consequence of the device operation, the channel current, or current from the drain to the source (I_{DS}), depends on $V_{GS} - V_t$. A larger V_{GS} will generate a larger current by drawing more charge into the channel. If the channel is not pinched-off, the device is in the *linear region* and the current will also increase as a function of the drain-source voltage, V_{DS}, as follows:

$$I_{DS} = K'\left(\frac{W}{L}\right)\left[(V_{GS} - V_t)V_{DS} - \frac{V_{DS}^2}{2}\right] \tag{4.1}$$

In this equation, W is the width of the transistor (see Figure 4.5) and L is the length of the transistor. A larger transistor *width* acts like a wider river allowing more water (or current) to flow. A larger transistor *length* causes more resistance in the channel, resisting current flow.

K' depends on the mobility of the carrier and the capacitance of the oxide:

$$K' = \mu_n C_{OX} \tag{4.2}$$

K' determines how many electrons will be in the channel for a given V_{GS} by indicating the effectiveness of the gate in generating the channel. A larger oxide capacitance will generate a larger electromagnetic field, pulling more charge into the channel for the same V_{GS}. The carrier mobility, μ, indicates how many carriers will flow through the channel for a given electromagnetic field. Higher mobility results in more carrier movement for the same electromagnetic field, resulting in more current. However, the mobility is a fixed value for a given carrier type and material. Electrons always have a higher mobility than holes, resulting in larger K' values for nFETs than pFETs that are otherwise the same. Additionally, mobility varies with material type: e.g., GaAs transistors have higher electron mobility than silicon, which is why many of the high-speed electronic devices use GaAs devices.

When V_{DS} is large enough to pinch-off the channel, the FET is in the *saturation region*, where the current saturates (a further increase in V_{DS} has little effect on the current) as follows:

$$I_{DS} = \frac{K'}{2}\left(\frac{W}{L}\right)(V_{GS} - V_t)^2 \tag{4.3}$$

The voltage that places the channel on the edge of pinch-off is called the saturation voltage ($V_{DS,sat}$), and defined as follows:

$$V_{DS,\,sat} = V_{GS} - V_t \tag{4.4}$$

In the saturation region, there is a small increase in channel current due to an increase in V_{DS} due to channel-length modulation. As V_{DS} increases, the pinch-off region grows, effectively making the channel length shorter and producing a small increase in drain current. To account for this second-order effect, the saturation equation can be modified as follows; however, its most significant impact is felt in producing a finite output resistance for the transistor.

$$I_{DS} = \frac{K'}{2}\left(\frac{W}{L}\right)(V_{GS} - V_t)^2(1 + \lambda V_{DS})$$ (4.5)

In this equation, λ is the channel-length modulation parameter given in units of V^{-1}.

The characteristic equations are depicted graphically in Figure 4.6. In the saturation region, the channel current increases as a square-law once $V_{GS} > V_t$ (Figure 4.6a). This square-law relationship determines how well the transistor converts a change in voltage into a change in current when the device is used in an amplifier circuit. When the transistor is being used as an amplifier, V_{GS} typically changes by a small amount (a *small signal*). As a result, the change in drain current, I_D, is approximately linear and equal to the derivative of the square-law relationship at a particular point (shown in Figure 4.6a). This *small-signal parameter*, called the transconductance (g_m) of the FET, is a direct measure of the amplification capability of the transistor. The transconductance is calculated as follows, where I_{DQ} is the tangent point to the nonlinear curve. The derivation assumes that λV_{DS} is small and can be ignored.

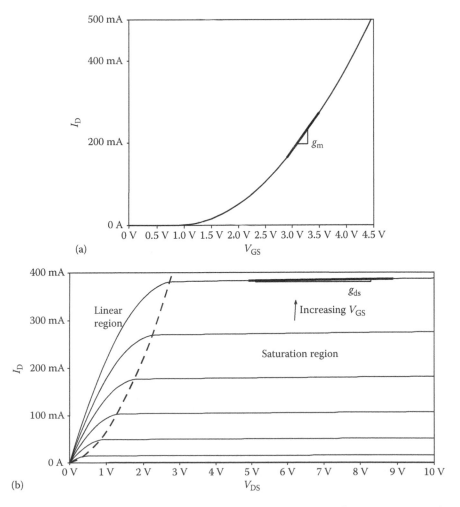

FIGURE 4.6 (a) I_D versus V_{GS} characteristics curves for a FET. (b) I_D versus V_{DS} characteristic curves for a FET.

$$g_{\text{mFET}} = \left.\frac{\partial I_{\text{DS}}}{\partial V_{\text{GS}}}\right|_{\text{IDQ}} \cong \left.\frac{\partial\left[\frac{K'}{2}\left(\frac{W}{L}\right)(V_{\text{GS}} - V_t)^2\right]}{\partial V_{\text{GS}}}\right|_{\text{IDQ}} = \left.2\frac{K'}{2}\left(\frac{W}{L}\right)(V_{\text{GS}} - V_t)\right|_{\text{IDQ}} \quad (4.6)$$

By noting that $\sqrt{I_{\text{DQ}}} = \sqrt{(K'/2)(W/L)}(V_{\text{GSQ}} - V_t)$, the equation can be rewritten in terms of the drain current as follows:

$$g_{\text{mFET}} = \sqrt{2K'\frac{W}{L}I_{\text{DQ}}} \quad (4.7)$$

Figure 4.6b depicts the drain current as V_{GS} and V_{DS} change, showing the linear and saturation regions. Note that the transition from the linear to the saturation region changes with V_{GS}. When the FET is being used as an amplifier, the goal is to keep it in the saturation region. When the FET is being used as a switch, the goal is to make the transistor switch from the saturation region to the linear region and vice versa.

Figure 4.6b also demonstrates the channel-length modulation in the saturation region, where I_{D} increases slightly with V_{DS}. The slope of the line in this region at a given point represents the output conductance of the transistor (g_{ds}), which indicates how the drain current changes when V_{DS} changes. This parameter assumes that the input, V_{GS}, is changing by a small amount (*small signal*), which keeps the transistor in the saturation region and thereby the slope of the line linear. The small-signal value of g_{ds} can be calculated in a similar manner as g_{m}, except using the relationship between I_{DS} and V_{DS}. Clearly, to include the V_{DS} relationship, the I_{D} equation must include the channel-length modulation parameter:

$$g_{\text{ds}} = \left.\frac{\partial I_{\text{DS}}}{\partial V_{\text{DS}}}\right|_{\text{IDQ}} = \left.\frac{\partial\left[\frac{K'}{2}\left(\frac{W}{L}\right)(V_{\text{GS}} - V_t)^2(1 + \lambda V_{\text{DS}})\right]}{\partial V_{\text{DS}}}\right|_{\text{IDQ}} = \lambda I_{\text{DQ}} \quad (4.8)$$

Many designers prefer to use the inverse of this relationship, r_{ds}, which models the output resistance of the transistor when the input is a small signal. Ideally, a designer would like the $r_{\text{ds}} = \infty$. The predominant impact of a finite output resistance on an amplifier circuit includes a small decrease in voltage gain and a limited accuracy and reproducibility of current values in current mirrors.

The characteristic equations of the pFET are similar to the characteristic equations of the nFET, since the only difference in the basic operation is the polarity of the voltages required to account for the reverse in carrier types. The equations can be written identically with the polarity reversal as follows:

$$I_{\text{SD}} = K'\left(\frac{W}{L}\right)\left[\left(V_{\text{SG}} - |V_t|\right)V_{\text{SD}} - \frac{V_{\text{SD}}^2}{2}\right] \quad (4.9)$$

$$I_{\text{SD}} = \frac{K'}{2}\left(\frac{W}{L}\right)\left(V_{\text{SG}} - |V_t|\right)^2 \quad (4.10)$$

$$V_{\text{SD,sat}} = V_{\text{SG}} - |V_t| \quad (4.11)$$

$$I_{SD} = \frac{K'}{2}\left(\frac{W}{L}\right)(V_{SG} - |V_t|)^2(1 + \lambda V_{SD}) \tag{4.12}$$

The absolute value of the threshold voltage is used since the threshold voltage for pFETs is generally negative, reflecting that the gate voltage must be brought lower than the bulk to generate a channel. It is important to keep in mind that a large V_{SG} producing a large drain current is generated by a low gate voltage relative to ground, and a small V_{SG} producing small drain current is generated with a high gate voltage relative to ground.

Since the characteristic equations are identical to the n-devices with only polarity reversals, the derivative of these equations will result in the same small-signal equations as follows:

$$g_{mFET} = \left.\frac{\partial I_{SD}}{\partial V_{SG}}\right|_{IDQ} = \sqrt{2K'\frac{W}{L}I_{DQ}} \tag{4.13}$$

$$g_{sd} = \left.\frac{\partial I_{SD}}{\partial V_{SD}}\right|_{IDQ} = \lambda I_{DQ} \tag{4.14}$$

4.2.2 BJT Operation and Characterization

The BJT contains three terminals, the base (B), the collector (C), and the emitter (E), and uses the schematic symbols shown in Figure 4.7a for an npn and Figure 4.7b for a pnp with current directions labeled. Figure 4.8 shows the cross-sectional view of the npn and pnp. Figure 4.8a shows the standard back-to-back p–n junction cross section, while Figure 4.8b presents the typical integrated circuit implementation. In the integrated circuit implementation, the collector of one transistor is isolated from the collectors of other transistors by placing the transistors into a substrate of opposite doping type. The substrate is held at a potential to keep the collector–substrate junction reverse-biased.

To turn on the BJT, the base–emitter junction diode must be forward-biased, using a positive V_{BE} voltage for npn's and a positive V_{EB} voltage for pnp's. If the base–collector junction diode is also forward-biased, the transistor is in the *saturation region*, and base current is shared between the two forward-biased diodes (see Figure 4.9a). If the base–emitter diode and the base–collector diode are on equally strong and the width of the base is small enough that electrons will not recombine, equal numbers of electrons will diffuse through the base region from the collector to the emitter and from the emitter to the collector, resulting in net current equal to zero. Holes will still diffuse from the base to both the collector and the emitter in equal numbers, thereby requiring base current in this condition; however, since net hole movement is equal in each case, net current from emitter to collector is still zero. This condition occurs in the npn when $V_{CE} = 0$, which causes $V_{BE} = V_{BC}$, and in the pnp when $V_{EC} = 0$, which causes $V_{EB} = V_{CB}$.

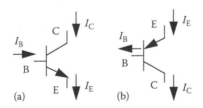

FIGURE 4.7   Schematic representations of (a) the npn and (b) the pnp.

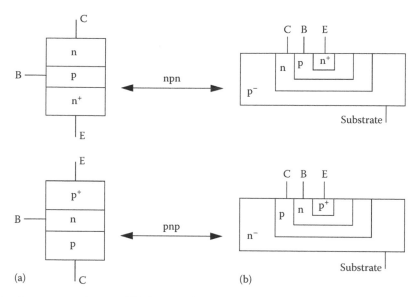

FIGURE 4.8 Cross-sectional views of (a) theoretical BJT (the npn) and (b) integrated circuit BJT (the pnp).

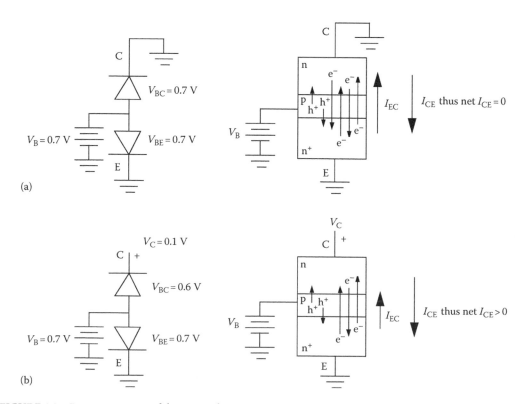

FIGURE 4.9 Device operation of the npn in the saturation region.

As V_{CE} increases (Figure 4.9b), the base–collector diode is on less strongly, resulting in more electrons diffusing from the emitter to the collector than from the collector to the emitter. Since positive current flow is designated in the opposite direction of electron flow, this results in net positive current from the collector to the emitter. The number of electrons diffusing from the collector to the emitter drops exponentially as the collector voltage increases (basic diode relationship), so the net current from the collector to the emitter increases exponentially.

At the saturation point of the transistor, which usually occurs when V_{CE} for an npn and V_{EC} for a pnp is between 0.2 and 0.3 V, the base–collector diode becomes reverse biased and the BJT moves into the *active region*. In this region of operation, the device exhibits current amplification: small base current results in large collector current. In the active region, the reverse-biased junction (V_{BC}) generates a depletion region that is significant in size relative to the base width (see Figure 4.10). The forward-biased base–emitter junction causes holes to diffuse from the base into the emitter and electrons to diffuse from the emitter into the base. However, if the base width is small enough, electrons will be swept into the collector by the electric field in the collector–base depletion region. As a result, few electrons recombine in the base. These carrier movements result in net positive current from the collector to the emitter. The hole diffusion current from the base to the emitter dominates the base current. The electron carrier movement from the emitter to the collector dominates the collector current. The emitter sees holes coming in from the base and electrons leaving for the collector, so it is a sum of the base and collector current that results in KCL applying to the devices as in Equation 4.15 with the current directions shown in Figure 4.8a and b:

$$I_E = I_C + I_B \qquad (4.15)$$

In manufacturing, the emitter is more highly doped than the base (shown in the figures with the n$^+$ as compared to the p). As a result of the higher doping, more electrons diffuse into the collector than holes diffuse from the base. Therefore, $I_C \gg I_B$, resulting in current amplification. The number of carriers swept into the collector for every base carrier is called the forward common-emitter current gain, (β), and relates the base current to the collector current as follows:

$$I_C = \beta * I_B \qquad (4.16)$$

The percentage of carriers received by the collector relative to the current in the emitter is called the common-base current gain (α), and relates the emitter current to the collector current as follows:

$$I_C = \alpha * I_E \qquad (4.17)$$

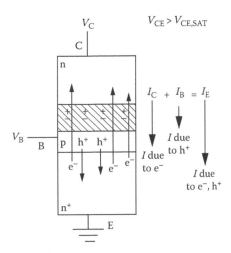

FIGURE 4.10 Device operation of the npn in the active region.

Fundamentals of Industrial Electronics

Many designers use these current amplification equations to design their circuits. However, to understand the operation of many circuits, it is important to keep in mind that the base–emitter voltage controls the base–emitter junction diode that controls the device. As a result, the base current changes exponentially with changes in the base–emitter voltage. Since the collector current is a scaled version of the base current, the collector current also changes exponentially with the base–emitter voltage, resulting in the following current–voltage relationship:

$$I_{CE} = \beta I_{B0}(e^{V_{BE}/V_T} - 1) \quad \text{for npn} \tag{4.18}$$

$$I_{EC} = \beta I_{B0}(e^{V_{EB}/V_T} - 1) \quad \text{for pnp} \tag{4.19}$$

In this equation, I_{B0} is the reverse-bias current of the base–emitter diode and V_T is the thermal voltage, equal to approximately 26 mV at room temperature. The thermal voltage is more exactly defined by the following equation:

$$V_T = \frac{kT}{q} \tag{4.20}$$

In this equation, k/q is Boltzmann's constant and equal to 8.61738×10^{-5} eV/K, and T is the temperature in Kelvin. Similar to the FET equations, since the only difference in the basic operation of the npn and the pnp is the carrier type, the equations can be written identically using the reverse polarity on the voltages and current directions.

Due to the exponential relationship between I_C and V_{BE}, I_C must change by a factor of 100 for V_{BE} to change by 0.1 V. In normal amplifiers with typical rails of ±15 V or less, I_C does not typically change by that large a factor. Therefore, the base–emitter voltage appears to be constant; however, it does change by a small amount.

Similar to FETs, in the active region, there is a small increase in collector current due to an increase in V_{CE} for npn's and V_{EC} for pnp's due to base-width modulation, or Early effect. As V_{CE} increases, the reverse-bias voltage on the base–collector diode increases, resulting in a larger depletion region. This larger depletion region further narrows the base width and allows fewer carries to recombine in the base, producing a small increase in collector current. To account for this second-order effect, the active region equations can be modified as follows; however, similar to the FET, it's most significant impact is felt in producing a non-infinite output resistance for the transistor:

$$I_{CE} = \beta I_{B0}\left(e^{V_{BE}/V_T} - 1\right)\left(1 + \frac{V_{CE}}{V_A}\right) \quad \text{for npn} \tag{4.21}$$

$$I_{EC} = \beta I_{B0}\left(e^{V_{EB}/V_T} - 1\right)\left(1 + \frac{V_{EC}}{V_A}\right) \quad \text{for pnp} \tag{4.22}$$

The characteristic equations of the BJT are depicted graphically in Figure 4.11. In the active region, the channel current increases exponentially as a function of V_{BE} (Figure 4.11a). This exponential relationship determines how well the transistor converts a change in voltage into a change in current when the device is used in an amplifier circuit. Just like the FET, when the transistor is being used as an amplifier, V_{BE} typically changes by a small amount (a *small signal*). As a result, the change in collector current, I_C, is approximately linear and equal to the derivative of the exponential relationship at a particular point (shown in Figure 4.11a). This small-signal parameter, called the transconductance of the BJT,

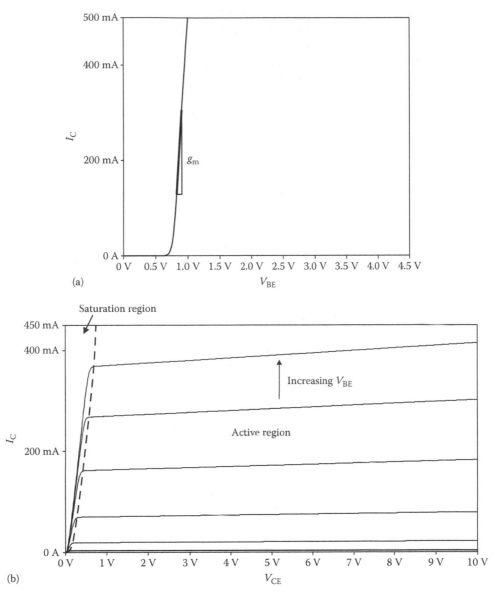

FIGURE 4.11 (a) I_C versus V_{BE} characteristic curves for an npn. (b) I_C versus V_{CE} characteristic curves for an npn.

is calculated for the BJT as follows, where I_{CQ} is the tangent point. The derivation assumes that V_{CE}/V_A is small and can be ignored, and that $e^{V_{BE}/V_T} \gg 1$, resulting in $I_{CE} \cong \beta I_{B0} e^{V_{BE}/V_T}$.

$$g_{mBJT} = \left. \frac{\partial I_{CE}}{\partial V_{BE}} \right|_{I_{CQ}} \cong \left. \frac{\partial \left[\beta I_{B0}(e^{V_{BE}/V_T}) \right]}{\partial V_{BE}} \right|_{I_{CQ}} = \frac{\beta I_{B0}(e^{V_{BE}/V_T})}{V_T} = \frac{I_{CQ}}{V_T} \tag{4.23}$$

This equation is true for both npn and pnp devices, since the polarity reversal is embedded in the definition of I_{CQ}.

Figure 4.11b depicts the collector current as V_{BE} and V_{CE} change, showing the saturation and active regions. Note that the transition from the saturation to the active region is approximately constant for varying V_{BE} values. This effect occurs because the change in current is exponential with V_{CE}. Identical to the FET, when the BJT is being used as an amplifier, the goal is to keep it in the active region. When the BJT is being used as a switch, the goal is to make the transistor switch from the active region to the saturation region and vice versa.

Figure 4.11b also demonstrates the base-width modulation in the active region, where I_C increases slightly with V_{CE}. Similar to the FET, the slope of the line in this region at a given point represents the output conductance of the transistor (g_{ce}), which indicates how the collector current changes when V_{CE} changes. This parameter assumes that the input, V_{BE}, is changing by a small amount (*small signal*), which keeps the transistor in the active region and thereby the slope of the line linear. The small-signal value of g_{ce} can be calculated in a similar manner as g_m, including the Early effect in the I_C equation:

$$g_{ce} = \left. \frac{\partial I_{CE}}{\partial V_{CE}} \right|_{I_{CQ}} \cong \left. \frac{\partial \left[\beta I_{B0} \left(e^{V_{BE}/V_T} \right) (1 + (V_{CE}/V_A)) \right]}{\partial V_{CE}} \right|_{I_{CQ}} = \frac{I_{CQ}}{V_A} \tag{4.24}$$

Like the FETs, many designers prefer to use the inverse of this relationship, r_{ce}, which models the output resistance of the transistor when the input is a small signal. While ideally $r_{ce} = \infty$, a finite output resistance has the same impact on BJT circuits as FET circuits, including a small decrease in voltage gain and limited accuracy and reproducibility of current values in current mirrors.

4.2.3 FETs versus BJTs

Although the FET and BJT operation are based on different physical principles, the net result is similar: a small change in input voltage (FET = V_{GS}, BJT = V_{BE}) produces a large change in output current (FET = I_{DS}, BJT = I_{CE}). This is the fundamental principle that governs the use of these devices in amplifiers and switches. Note that both devices have a region of current saturation (FET = saturation region, BJT = active region), which is used in voltage amplification. In this region, the change in current is dominated by the change in the input voltage ($g_m \gg g_{out} = g_{ds}, g_{ce}$), which helps to keep the voltage amplification linear. Additionally,

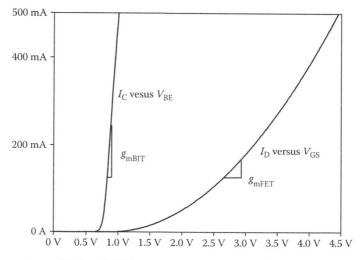

FIGURE 4.12 Comparison of FET and BJT input voltage–output current characteristic curves.

both devices have a region in which the current changes with the output voltage (FET = linear region, BJT = saturation region), which is used to turn current on and off in use as a switch.

While FETs and BJTs work similarly in the abstract sense, the difference in characteristic equations produce a difference in how well each of these devices performs these operations. The difference between the devices is depicted in Figure 4.12. This figure shows the current–voltage relationship for an FET (I_D versus V_{GS}) overlaid upon the current–voltage relationship of a BJT (I_C versus V_{BE}). Note how much more sharply the BJT curve changes than the FET curve. This graph demonstrates that a small change in input voltage will generate a much larger change in output current for the BJT than the FET (e.g., $g_{mBJT} > g_{mFET}$). This difference in operation between the two devices in the saturation region can be seen by looking at Figures 4.7 and 4.12. In these figures, the change in output current (FET = I_D, BJT = I_C) is shown as a function of output voltage (FET = V_{DS}, BJT = V_{CE}). Notice how the saturation voltage for a BJT is approximately constant while the saturation voltage for the FET changes with V_{GS}. This will impact how the saturation voltage is approximated in Chapter 5.

5

Application of Operational Amplifiers

Carlotta A. Berry
*Rose-Hulman Institute
of Technology*

Deborah J. Walter
*Rose-Hulman Institute
of Technology*

5.1 Introduction

An **operational amplifier (op amp)** is an electronic device or integrated circuit (IC) made of transistors and resistors used to amplify or attenuate inputs. The op amp can be used for mathematical operations such as addition, subtraction, multiplication, division, differentiation, integration, and filtering. Op amp applications include audio preamplifiers, voltage comparators, rectifiers, regulators, analog-to-digital converters, oscillators, sensor transducers, and waveform generators. The five terminals of interest associated with an op amp are as follows:

- Inverting input (V_n)
- Non-inverting input (V_p)
- Output (V_o)
- Positive power supply (V^+)
- Negative power supply (V^-)

Figure 5.1 shows the op amp circuit symbol with the five terminals of interest labeled.

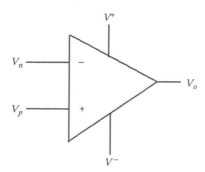

FIGURE 5.1   Op amp IC symbol.

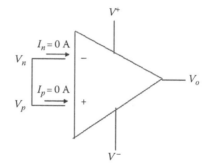

FIGURE 5.2   Op amp assumptions.

5.1.1 Ideal Op Amp Assumptions

The model for an ideal op amp has infinite gain ($V_o/(V_n - V_p)$), infinite input impedance, and zero output impedance. Based upon this model, there are two assumptions made for analysis of an ideal op amp:

The *infinite input impedance* assumption states that no current enters the op amp from the positive or negative terminal ($I_p = I_n = 0$ A).

The *virtual short circuit* assumption states that the voltage at the positive terminal is the same as the voltage at the negative terminal ($V_n = V_p$).

Figure 5.2 provides an illustration of the two ideal assumptions.

Note that these two assumptions are only valid when the op amp is used in a closed-loop configuration where the output is fed back to the inverting input through a passive circuit element. This configuration is referred to as *negative feedback*. With the negative feedback, the closed-loop gain is decreased based upon a portion of the output subtracting from the input so the infinite gain does not cause the output to grow without bound to the positive or negative power supply.

5.1.2 Linear Range of Operation

The output (V_o) of an ideal op amp is a product of the gain (A) and the input voltage (V_{in}). However, the output is only linear when the output is in the range between the positive (V^+) and negative (V^-) power supplies. When the output is outside of this range, the output maximizes or minimizes at the power supply voltage, and the op amp is saturated. Figure 5.3 illustrates the linear and nonlinear region for an op amp. The curve in Figure 5.3 is also called the saturation curve.

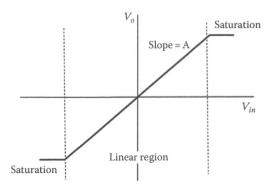

FIGURE 5.3 Op amp saturation curve.

5.2 Node Voltage Analysis of Op Amp Circuits

The most common analysis technique for an ideal op amp circuit is Kirchhoff's current law (KCL). By using KCL and the two ideal op amp assumptions, it is possible to find the output voltage and current for any op amp circuit. The steps to analyze the circuit are as follows:

1. If the voltage is unknown, use the *infinite input impedance* assumption to write the KCL equation at the positive (V_p) and/or negative (V_n) terminals.
2. Use the *virtual short circuit* assumption to simplify the KCL equations and solve for the output voltage (V_o) or any other required values.

This analysis technique will be demonstrated on several op amp circuits using DC and AC circuit analysis techniques.

5.2.1 Inverting Amplifier

The op amp circuit shown in Figure 5.4 represents an *inverting amplifier*. Note that R_f is the feedback resistor that feeds a portion of the output V_o back to the inverting input V_n. It is an amplifier because the value of the output is based upon the ratio of the feedback resistor R_f and the input resistor R_i. It is an inverting amplifier because the polarity of the output voltage is the reverse of the polarity of the input voltage.

The first step in the analysis is to note that V_p is tied to ground, therefore $V_p = 0\,\text{V}$.

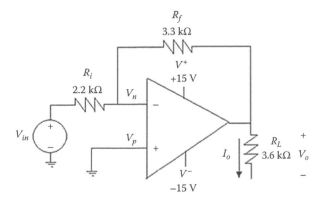

FIGURE 5.4 Inverting amplifier.

Using the *infinite input impedance* assumption to create the KCL equation at V_n yields

$$\frac{V_{in} - V_n}{R_i} = \frac{V_n - V_o}{R_f} \tag{5.1}$$

Using the fact that $V_p = 0\,\text{V}$ and the *virtual short circuit assumption* so $V_n = 0\,\text{V}$, it is possible to solve Equation 5.1 for V_o.

$$V_o = \left(-\frac{R_f}{R_i}\right)V_{in} = \left(-\frac{3.3}{2.2}\right)V_{in} = 1.5V_{in} \tag{5.2}$$

$$A = \frac{V_o}{V_{in}} = \left(-\frac{R_f}{R_i}\right) = -1.5 \tag{5.3}$$

From Equation 5.3, it is evident that the gain is the ratio of the feedback resistor to the input resistor, and the output is inverted. Note that this circuit can also be an attenuator by making the value of R_f less than R_i.

5.2.2 Non-Inverting Amplifier

The op amp circuit shown in Figure 5.5 represents a *non-inverting amplifier.*

The first step in the analysis is to note that $V_p = V_{in}$. Then, using the *infinite input impedance* assumption to write the KCL equation at V_n yields

$$\frac{V_n - 0}{R_i} = \frac{V_o - V_n}{R_f} \tag{5.4}$$

Using the fact that $V_p = V_{in}$ and the *virtual short circuit assumption*, it is possible to solve Equation 5.4 for V_o and the gain, A.

$$V_o = \left(1 + \frac{R_f}{R_i}\right)V_{in} = \left(1 + \frac{5.6}{2.0}\right)V_{in} = 3.8V_{in} \tag{5.5}$$

$$A = \frac{V_o}{V_{in}} = \left(1 + \frac{R_f}{R_i}\right) = 3.8 \tag{5.6}$$

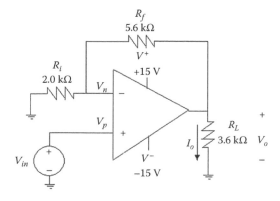

FIGURE 5.5 Non-inverting amplifier.

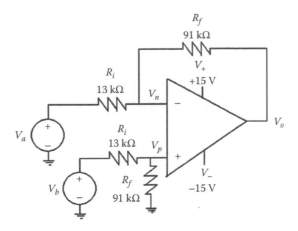

FIGURE 5.6 Difference amplifier.

Note that this gain, A, is positive and therefore the op amp configuration in Figure 5.5 produces a *non-inverting* amplifier. Also note that from Equation 5.6 it is not possible for this configuration of a non-inverting amplifier to attenuate the input voltage.

5.2.3 Difference Amplifier

The circuit shown in Figure 5.6 is a difference amplifier and it will be analyzed using KCL.

$$V_n: \quad \frac{V_n - V_a}{R_i} + \frac{V_n - V_o}{R_f} = 0$$

$$V_p: \quad \frac{V_p - V_b}{R_i} + \frac{V_p - 0}{R_f} = 0$$

(5.7)

Using the virtual short circuit condition and solving Equation 5.7 yields

$$V_o = \left(\frac{R_f}{R_i}\right)(V_b - V_a) = 7(V_b - V_a)$$

(5.8)

The differential amplifier has a differential input mode defined as the difference between the two input voltages. The second feature of a difference amplifier is the common mode input, the average of the two input voltages. Ideally, the common mode gain should be zero and only the differential input should be amplified. Typically, the common mode signal represents the noise found in most electric signals, which should be suppressed at the output. To design for this feature, the resistors used in the op amp must be well matched. The **common mode rejection ratio (CMRR)** is used to measure how well a difference amplifier performs. The CMRRR is the ratio of the differential mode gain to the common mode gain.

5.2.4 Weighted Difference Amplifier

The circuit shown in Figure 5.6 can be changed to a weighted difference amplifier by modifying the values of the resistors. This modified circuit is shown in Figure 5.7 and it will be analyzed using superposition.

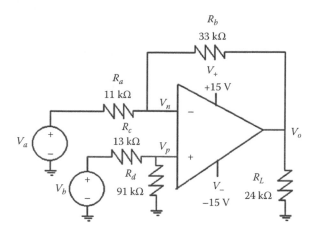

FIGURE 5.7 Weighted difference amplifier.

Superposition involves turning on each of the voltage inputs one at a time and determining the output due only to that source. The principal of linearity then states that the total output voltage will be the sum of the voltage due to each source acting alone. Note that to turn off a voltage source, replace it with a short circuit that represents 0 Ω and 0 V. If only the voltage V_a is turned on, then V_b becomes 0 V. Note that this configuration is the same as Figure 5.4 and is an inverting amplifier. So from Equation 5.2,

$$V_o' = \left(-\frac{R_b}{R_a}\right)V_a = \left(-\frac{33}{11}\right)V_a = -3.0V_a \tag{5.9}$$

Now, if only the voltage V_b is turned on, then V_a becomes 0 V. This configuration is similar to Figure 5.5 and is a non-inverting amplifier. V_p can be found by using the voltage divider and substituting into Equation 5.5. This yields the following relationship:

$$V_p = \left(\frac{R_d}{R_c + R_d}\right)V_b = \left(\frac{91}{13 + 91}\right)V_b = \frac{7}{8}V_b \tag{5.10}$$

$$V_o'' = \left(1 + \frac{R_b}{R_a}\right)\left(\frac{R_d}{R_c + R_d}\right)V_b = \left(1 + \frac{33}{11}\right)\left(\frac{7}{8}\right)V_b = 3.5V_b \tag{5.11}$$

Therefore, using the principle of superposition, the total output V_o would be the sum of the voltages given in Equations 5.9 and 5.11:

$$V_o = V_o' + V_o'' = 3.5V_b - 3.0V_a \tag{5.12}$$

5.2.5 Output Current

For the difference amplifier shown in Figure 5.7, if $V_a = V_b = 5$ V, the output voltage, V_o, can be found from Equation 5.12 as

$$V_o = 3.5V_a - 3.0V_b = 3.5(5) - 3.0(5) = 2.5 \text{ V} \tag{5.13}$$

The output current, I_o, can be found by using the following equation:

$$I_o = \frac{V_o}{R_L} = \frac{2.5}{24k} = 104 \text{ mA} \tag{5.14}$$

One benefit of an op amp is the fact that the addition of the load does not change the output voltage, V_o. However, it is important when designing an op amp circuit to insure that the output current does not exceed the output current specifications for the specific device. This can be accomplished with proper selection of the load, feedback, and input resistors, typically in the kΩ range.

5.2.6 Saturation

For the circuit shown in Figure 5.5, if we assume that the non-inverting amplifier saturates exactly at the positive and negative power supply, the saturation curve is shown in Figure 5.8. This curve has a positive slope of 3.8 with a maximum voltage of 15 V and a minimum voltage of –15 V. The saturation curve for an inverting amplifier would be similar except with a negative slope.

5.2.7 Differentiator

The circuit shown in Figure 5.9 is a differentiator and can also be analyzed using KCL.

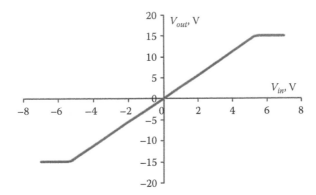

FIGURE 5.8   Saturation curve for a non-inverting amplifier.

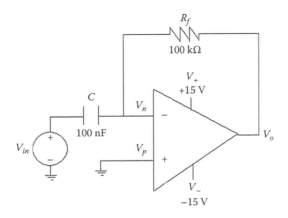

FIGURE 5.9   Differentiator.

The first step in the analysis is to note that $V_p = 0\,V$ and therefore $V_n = 0\,V$ based upon the *virtual short circuit assumption*. Using the *infinite input impedance* assumption to create the KCL equation at V_n yields

$$C\frac{d(V_{in} - V_n)}{dt} = \frac{V_n - V_o}{R_f} \qquad (5.15)$$

Substituting the numeric value of $V_n = 0\,V$ into Equation 5.15 yields

$$C\frac{dV_{in}}{dt} = -\frac{V_o}{R_f} \qquad (5.16)$$

Solving Equation 5.16 for V_o yields

$$V_o = -R_f C\frac{dV_{in}}{dt} = -0.01\frac{dV_{in}}{dt} \qquad (5.17)$$

From Equation 5.17, note that the output is the inverse of the derivative of the input with a gain of $A = -R_f C$.

5.2.8 Bandpass Filter

The circuit shown in Figure 5.10 is a *broadband* bandpass filter.

A broadband bandpass filter is defined as one where $\omega_{c2} \gg \omega_{c1}$. *Broadband* means that the upper cutoff frequency is at least a magnitude or two greater than the lower cutoff frequency. It is possible to identify this circuit as a bandpass filter by using qualitative analysis. Intuitively, at low frequencies (D.C. conditions), a capacitor models an open circuit, and at high frequencies it models a short circuit. Therefore at low frequencies, the output of the circuit in Figure 5.9 is zero volts because the input is zero volts. At high frequencies, the output of the circuit in Figure 5.9 is zero because of the

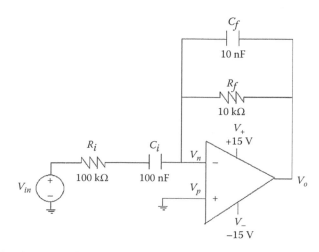

FIGURE 5.10   Broadband bandpass filter.

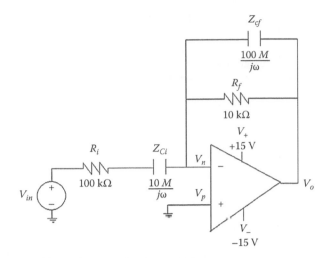

FIGURE 5.11 Broadband bandpass filter (frequency domain).

virtual short circuit assumption. At the resonance frequency, $(\omega_o = \sqrt{\omega_{c1}\omega_{c2}})$, the transfer function is purely real and the gain of the circuit is $\cong R_f/R_i$.

This circuit can also be analyzed quantitatively by using KCL and steady-state analysis. This yields an equation similar to Equation 5.2, because this circuit is still an inverting amplifier but with a feedback impedance and input impedance instead of resistors.

Recall that the impedance of a capacitor in the frequency domain is $Z = 1/(j\omega C)$. The circuit in Figure 5.10 redrawn in the phasor domain is shown in Figure 5.11.

Use the inverting amplifier relationship in Equation 5.2 to derive the output for Figure 5.10. The relationship in terms of the input and feedback impedance is shown in Equation 5.18, where Z_i is the series combination of R_i and Z_{Ci}, and Z_f is the parallel combination of R_f and Z_{Cf}.

$$V_o = \left(-\frac{Z_f}{Z_i}\right) V_{in}, \tag{5.18}$$

The expressions for Z_i and Z_f are given in Equations 5.19 and 5.20, respectively:

$$Z_i = R_i + \frac{1}{j\omega C_i} = \frac{j\omega C_i R_i + 1}{j\omega C_i} = \frac{100kj\omega + 10M}{j\omega} \tag{5.19}$$

$$Z_f = R_f \parallel \frac{1}{j\omega C_f} = \frac{R_f}{1 + j\omega R_f C_f} = \frac{10k}{1 + j\omega 100\mu} \tag{5.20}$$

By substituting Equations 5.19 and 5.20 into 5.18, the output equation becomes

$$V_o = \left(-\frac{Z_f}{Z_i}\right) V_{in} = \left(-\frac{R_f/(j\omega C_f R_f + 1)}{(j\omega C_i R_i + 1)/j\omega C_i}\right) V_{in} = \left(-\frac{j\omega C_i R_f}{(j\omega C_i R_i + 1)(j\omega C_f R_f + 1)}\right) V_{in} \tag{5.21}$$

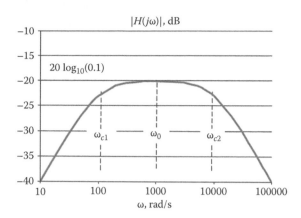

FIGURE 5.12 Bandpass filter Bode diagram.

Equation 5.21 can be written as a transfer function in terms of the gain (K), bandwidth (β), and resonant frequency (ω_o) as

$$H(j\omega) = \frac{V_o}{V_i} = \left(-\frac{(R_f/R_i)(1/C_f R_f)j\omega}{(j\omega + (1/R_i C_i))(j\omega + (1/R_f C_f))}\right) = \left(-\frac{K\omega_{c2}j\omega}{(j\omega + \omega_{c1})(j\omega + \omega_{c2})}\right)$$

$$= \left(-\frac{K\beta j\omega}{-\omega^2 + \beta j\omega + \omega_o^2}\right) = \left(-\frac{(0.1)(10k)j\omega}{(j\omega + 100)(j\omega + 10k)}\right) \tag{5.22}$$

From examination of Equation 5.22, the gain K is $R_f/R_i = 0.1$, the lower cutoff frequency ω_{c1} is $1/(R_i C_i) = 100$ rad/s, the upper cutoff frequency ω_{c2} is $1/(R_f C_f) = 10$ krad/s, the resonant frequency ω_o is $\sqrt{\omega_{c1}\omega_{c2}} = 1000$ rad/s, and the bandwidth β is $\omega_{c2} + \omega_{c1} = 10.1$ krad/s. Note that the numerator is not exactly the bandwidth, thus the requirement that the upper cutoff frequency be much larger than the lower cutoff frequency. The Bode diagram for the magnitude of the gain of the bandpass filter is shown in Figure 5.12. This is the ideal Bode diagram because it does not consider gain-bandwidth limitations; these and other practical considerations of op amp design will be addressed in Section 5.4.

5.2.9 Phase-Shift Oscillator

The circuit shown in Figure 5.13 is a phase-shift oscillator.

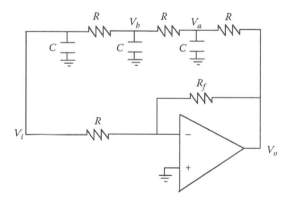

FIGURE 5.13 Phase-shift oscillator.

This design of this system is based upon the control theory that a feedback system such as an op amp will oscillate when it becomes marginally stable. It will become marginally stable when the poles of the transfer function are purely imaginary and the gain is 1 or 0 dB. Another way to express this requirement is that the phase shift at a certain frequency must be −180° or the frequency where the transfer function is purely real. The RC network will be used to get a phase shift of −180° or −60° per stage. The negative feedback gain (R_f/R_δ) of the inverter will be used to compensate for the attenuation in the RC network to meet the gain specification of 1.

The first step in the design of the phase-shift oscillator is to find the transfer function of the RC network, $V_i(s)/V_o(s)$. The transfer function of the RC network is the feedback gain of the closed-loop control system. This gain can be found by applying KCL at the nodes V_a, V_b, and V_i. The results of this analysis are shown in Equation 5.23.

$$V_a: \quad \frac{V_a - V_b}{R} + \frac{V_a - V_o}{R} + sCV_a = 0$$

$$V_b: \quad \frac{V_b - V_i}{R} + \frac{V_b - V_a}{R} + sCV_b = 0 \tag{5.23}$$

$$V_i: \quad \frac{V_i - V_b}{R} + \frac{V_i - 0}{R} + sCV_i = 0$$

Solving the three equations in (Equation 5.23) yields

$$H(s) = \frac{V_i(s)}{V_o(s)} = \frac{1}{s^3 R^3 C^3 + 6s^2 R^2 C^2 + 10sRC + 4} \tag{5.24}$$

In order to design for the −180° phase shift, express the transfer function in Equation 5.24 as a function of $j\omega$:

$$H(j\omega) = \frac{V_i(j\omega)}{V_o(j\omega)} = \frac{1}{4 - 6\omega^2 R^2 C^2 + j(10\omega RC - \omega^3 R^3 C^3)} \tag{5.25}$$

The phase shift for the transfer function in Equation 5.25 is

$$\phi = 0^0 - \tan^{-1}\left(\frac{10\omega RC - \omega^3 R^3 C^3}{4 - 6\omega^2 R^2 C^2}\right) \tag{5.26}$$

In order for the phase shift to be −180°, the imaginary part of the denominator must be equal to zero so that the denominator is purely real. Solving this expression for ω yields

$$\omega = \frac{\sqrt{10}}{RC} \tag{5.27}$$

In order to find the gain of the inverting amplifier to satisfy the 0 dB gain specification, find the magnitude of the gain for the transfer function $H(j\omega_R)$, where $\omega_R = \sqrt{10}/RC$.

$$|H(j\omega_R)| = \frac{1}{4 - 6\omega_R^2 R^2 C^2} = \frac{1}{56} \tag{5.28}$$

To compensate for the attenuation in the RC network, the gain of the inverting amplifier should be the inverse of the gain in (5.28). Therefore, set

$$\frac{R_f}{R} = 56 \tag{5.29}$$

5.3 Common Op Amp Circuits

Using KCL and the two ideal op amp assumptions it is possible to analyze and design many types of op amps. Table 5.1 presents a summary of some common op amp circuits and their output relationships.

5.4 Circuit Design with Op Amps

The prior section has summarized the analysis of several op amp circuits using the two ideal assumptions. However, in op amp design, these ideal assumptions may not always hold true and there are several practical considerations including selecting reasonable resistor values, finding input resistance, gain-bandwidth compensation, balancing inputs, input offset voltages, input bias and offset currents, slew rate, and frequency response. This section will discuss the practical consideration and design of op amp applications and the subsequent section will discuss the features of a more realistic op amp model.

5.4.1 Practical Considerations

One of the most fundamental considerations when designing an op amp circuit is the selection of the input, feedback, and load resistors. It is to select resistor sizes that do not require too much current from the op amp but supply sufficient current to drive a load. Small resistors may dissipate a great deal of power; however, larger resistors exhibit more noise. Therefore for op amp design, reasonable values of resistors would be between $1\,k\Omega$ and $1\,M\Omega$. In addition, because capacitor values may be more limiting, it is important to select resistors first in a design and then choose standard capacitor values in the μF to pF range.

5.4.2 Practical Applications

One practical application for op amps is in audio systems. An op amp can be used to drive a speaker. The microphone converts acoustic pressure to a voltage typically in the mV range. The op amp is used to amplify the voltage to drive a headphone or speaker. In order for the amplifier not to load the output of the microphone, it must have high input impedance. This high input impedance can be accomplished by using a non-inverting amplifier that has higher input impedance than an inverting amplifier. Op amps can also be used for signal conditioning to amplify and/or convert the signal from sensors such as temperature, force, pressure, acceleration, and light intensity. For a photocell amplifier, the output of a photocell is connected to a current-to-voltage converter to produce a value proportional to incident light. This solution works because it provides a constant terminal voltage to the photocell typically 0 V and low load impedance. To measure strain, a strain gauge can be used in a Wheatstone bridge to linearize the output resistance with respect to the strain. The output of the bridge is a voltage in the mV range that can be input into an instrumentation amplifier with high input impedance and low output impedance. The instrumentation amplifier is used to amplify the output, the voltage proportional to strain. Lastly, op amps can be used to filter out unwanted noise in communication signals, for example that caused by the power line in telemetry. In this case, a notch or band stop filter can be used to remove a 60 Hz signal.

5.4.3 Gain Bandwidth Limitation

An op amp typically has a closed-loop gain that rolls off at 20 dB/decade at a certain frequency. The op amp has an internal compensation network built in to improve performance. The fixed compensation modifies the open-loop frequency response and is typically a passive RC low-pass filter. The transfer function for this RC compensation network is $G(s) = G_o/(1 + s/\omega_o)$, where G_o is the open-loop gain and ω_o is the cutoff frequency. The *gain-bandwidth product* (GBP) is a constant that is found by

TABLE 5.1 Common Op Amp Circuits

Amplifier Type	Circuit Configuration	Relationships
Voltage follower		$V_o = V_{in}$
Inverting		$V_o = \left(-\dfrac{R_f}{R_i} \right) V_{in}$
Non-inverting		$V_o = \left(1 + \dfrac{R_f}{R_i} \right) V_{in}$
Summing		$V_o = \left(-\dfrac{R_f}{R_a} \right) V_a + \left(-\dfrac{R_f}{R_b} \right) V_b$ $+ \left(-\dfrac{R_f}{R_c} \right) V_c$
Difference		$V_o = \left(\dfrac{R_1 + R_2}{R_3 + R_4} \right) \left(\dfrac{R_4}{R_1} \right) V_2 - \left(\dfrac{R_2}{R_1} \right) V_1$

(continued)

TABLE 5.1 (continued)　Common Op Amp Circuits

Amplifier Type	Circuit Configuration	Relationships
Instrumentation Amplifier (Type I)		$V_o = \left(\dfrac{R_2}{R_1} \right)(V_2 - V_1)$
Instrumentation Amplifier (Type II)		$V_o = \left(1 + \dfrac{2R}{R_G} \right)(V_2 - V_1)$
Differentiator		$V_o = -RC \dfrac{dV_{in}}{dt}$
Integrator		$V_o = -\dfrac{1}{RC} \displaystyle\int V_{in} d\tau$
Lowpass filter		$H(s) = \dfrac{V_o(s)}{V_{in}(s)} = \dfrac{(R_f/R_i)((1/R_f C_f))}{s + (1/R_f C_f)}$

TABLE 5.1 (continued) Common Op Amp Circuits

Amplifier Type	Circuit Configuration	Relationships
Butterworth Lowpass filter	C_1, R, R, C_2, V_{in}, V_o	$H(s) = \dfrac{V_o(s)}{V_{in}(s)}$ $= \dfrac{1/C_1C_2R^2}{s^2 + (2/C_1R)s + (1/C_1C_2R^2)}$
Highpass filter	R_f, R_i, C_i, V_{in}, V_o	$H(s) = \dfrac{V_o(s)}{V_{in}(s)} = \dfrac{(R_f/R_i)s}{s + (1/R_iC_i)}$
Butterworth Highpass filter	R_1, C, C, R_2, V_{in}, V_o	$H(s) = \dfrac{V_o(s)}{V_{in}(s)}$ $= \dfrac{s^2}{s^2 + (2/R_2C)s + (1/R_1R_2C^2)}$
Broadband Bandpass filter $\omega_{c2} \gg \omega_{c1}$ $Q < 0.5$	C_f, R_f, R_i, C_i, V_{in}, V_o	$H(s) = \dfrac{V_o(s)}{V_{in}(s)}$ $= \dfrac{(R_f/R_i)(1/C_fR_f)s}{(s + (1/R_iC_i))(s + (1/R_fC_f))}$
Narrowband Bandpass filter	C, R_3, R_1, C, R_2, V_{in}, V_o	$H(s) = \dfrac{K\beta s}{s^2 + \beta s + \omega_o^2}$ $H(s) = \dfrac{-s/(R_1C)}{s^2 + 2/(R_3C)s + 1/(R_{eq}R_3C^2)}$ $\beta = 2/(R_3C)$ $\omega_o^2 = 1/(R_{eq}R_3C^2)$ $R_{eq} = R_1 \| R_2$ $K\beta = -1/(R_1C)$

(continued)

TABLE 5.1 (continued)   Common Op Amp Circuits

Amplifier Type	Circuit Configuration	Relationships
Narrowband Bandreject filter		$H(s) = \dfrac{\left(s^2 + \omega_o^2\right)}{s^2 + \beta s + \omega_o^2}$ $H(s) = \dfrac{s^2 + 1/(R^2C^2)}{s^2 + (4(1-\sigma)/RC)\,s + 1\,(R^2C^2)}$ $\beta = 4(1 - \sigma)/(RC)$ $\omega_o^2 = 1/(R^2C^2)$ $\sigma = 1 - \beta/(4\omega_o)$
I–V Converter		$V_o = -R_f I_{in}$
V–I Converter		$i_L = \dfrac{V_a - V_b}{R}$
Square-wave oscillator		Period, $T = (2R_1 C) \ln\left(\dfrac{1+\lambda}{1-\lambda}\right)$ s $\lambda = \dfrac{R_2}{R_2 + R_3}$

TABLE 5.1 (continued) Common Op Amp Circuits

Amplifier Type	Circuit Configuration	Relationships
Phase-shift oscillator	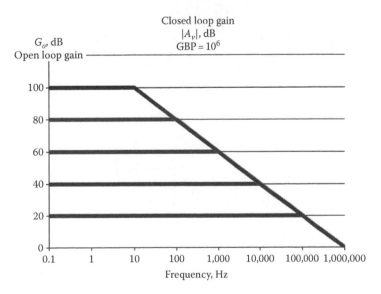	$H(s) = \dfrac{v_i(s)}{v_o(s)}$ $= \dfrac{1}{s^3 R^3 C^3 + 6s^2 R^2 C^2 + 10 s R C + 4}$ $\omega = \left(\dfrac{\sqrt{10}}{RC}\right), \quad \dfrac{R_f}{R} = -56$
Gyrator		$Z = \dfrac{V}{I} = s \dfrac{C R_1 R_3 R_4}{R_2}$ $L = \dfrac{C R_1 R_3 R_4}{R_2} H$

multiplying the open-loop gain G_o, with the bandwidth ω_o. Figure 5.14 presents the gain-bandwidth curves for an op amp with a GBP of 10^6. The vertical axis is the open-loop gain G_o, and each curve is a different closed-loop gain A. For design purposes, the op amp closed-loop gain should be 1/20th of the open-loop gain at a given frequency to avoid performance issues based upon gain-bandwidth limitations.

FIGURE 5.14 Example closed-loop gain-bandwidth curves.

5.4.4 Design Examples

Example 5.1: Displacement Signal Conditioning

A potentiometer is used to measure the location and position of boxes on a conveyor system. The displacement of the box moves the wiper of the potentiometer. The potentiometer converts the linear change in displacement to a resistance. The model for the system is shown in Figure 5.15.

If a 100 kΩ potentiometer is used to measure the displacement of a box from 0 to 50 feet, design a signal-conditioning circuit with a sensitivity of 0.25 V/ft by using an inverting amplifier. Note that the potentiometer could be used in a voltage divider or bridge circuit to supply the input to the op amp; however, in both of these methods the linear relationship between displacement and resistance becomes nonlinear. Therefore, the most feasible approach would be to use the potentiometer in the feedback loop of the op amp. In this configuration, the resistance of the potentiometer is directly proportional to the gain of the op amp. The resolution of the potentiometer is found by using Equation 5.30.

$$\Delta = \frac{100\,k\Omega}{50\,ft} = 2000\,\Omega/ft \qquad (5.30)$$

The inverting amplifier configuration was selected because the relationship between the feedback resistor and the input is nonlinear for a non-inverting amplifier. Assume the input to the inverting amplifier is $V_i = -1\,V$. When the box moves 1 foot, the potentiometer output is $R_f = 2\,k\Omega$ based upon the required sensitivity of 0.25 V/ft. R_i can be found from the following equation:

$$R_i = -\frac{R_f}{V_o}V_i = -\frac{2000}{0.25}(-1) = 8\,k\Omega \qquad (5.31)$$

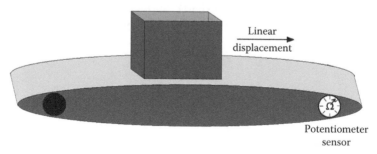

FIGURE 5.15 Box conveyor system.

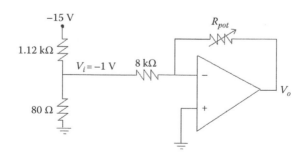

FIGURE 5.16 Inverting amplifier example.

Figure 5.16 presents the final design of the displacement signal-conditioning circuit with V_i as the input from a voltage divider. Note that the resistor values for the voltage divider were selected in order to limit loading effects of the op amp input resistance.

To confirm that this design meets the specification equation it is necessary to find the potentiometer output, op amp output, and displacement sensitivity assuming a box moves 25 ft. The output of the potentiometer is

$$R_{pot} = (25 \text{ ft})(2 \text{k}\Omega/\text{ft}) = 50 \text{ k}\Omega \tag{5.32}$$

The output of the op amp is

$$V_o = -\frac{R_{pot}}{R_i}V_{in} = -\frac{50k}{8k}(-1) = 6.25 \text{ V} \tag{5.33}$$

$$\text{Sensitivity} = \frac{6.25 \text{ V}}{25 \text{ ft}} = 0.25 \text{ V/ft} \tag{5.34}$$

Therefore, this design does meet the specification.

Example 5.2: Current–Voltage (*I–V*) Converter Example

It is possible to design a photodiode signal-conditioning circuit with a sensitivity of 30 V/fc by using an *I–V* converter. Photodiodes are used for camera flash controls, headlight dimmers, bar code scanners, and laser printers. Assume that the photodiode outputs a small current that is proportional to the incident light intensity at a range of 30 µA to 30 pA for 1000–0.001 fc. Select resistor values for the circuit in Figure 5.17 to satisfy the design constraints.

The first step in the design is to determine the gain of each of the stages. If the sensitivity is 30 V/fc, then the sensitivity is (30 V/fc) (1000 fc/30 µA) = 1 mV/pA. This means that the overall gain for the conditioning circuit is 1×10^9. It is possible to implement this gain with one stage, but that would require a 1000 MΩ resistor, which is very noisy. Instead, a viable solution would be to convert the current to a voltage in nV and then use the inverting amplifier to amplify the signal to mV. Therefore,

$$\frac{V_{o1}}{I_d} = 1 \text{M}\Omega = R_f \tag{5.35}$$

$$\frac{V_o}{V_{o1}} = -100k = -\frac{R_2}{R_1} \tag{5.36}$$

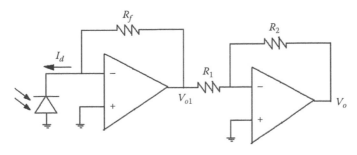

FIGURE 5.17   Current–voltage converter example.

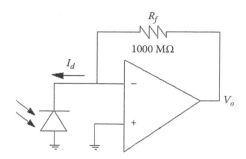

FIGURE 5.18   Single-stage current–voltage converter.

one set of reasonable resistor values that satisfy these design equations are $R_f = 1\,M\Omega$, $R_2 = 100\,M\Omega$, and $R_1 = 1\,k\Omega$. The one-stage solution for the design is shown in Figure 5.18.

The final step would be to confirm that the circuit designed satisfies the 30 V/fc sensitivity requirement. If the light intensity is 0.001 fc, then the photodiode current is 30 pA. The output of the *I–V* converter would be (30 pA) (10 kV/A) = 300 nV/A. The output of the inverting amplifier would be (300 nV)(−100k) = −30 mV, so the design constraint is satisfied.

Example 5.3: Difference Amplifier

It is possible to design a strain gauge signal-conditioning circuit with sensitivity 3 mV/(μm/m) by using a Wheatstone bridge and difference amplifier. Input buffers are added to the input of the difference amplifier to create an *instrumentation amplifier* that resolves the problem of low input impedance to the op amp. The 3 kΩ strain gauge is placed in one branch of the Wheatstone bridge and at full scale deflection, 3030 Ω indicates a 5000 μm/m strain. Therefore the strain or deflection is proportional to a change in resistance, ΔR. Note that when there is no strain, the bridge is at null where all four legs are 3 kΩ. Figure 5.19 presents the strain gauge signal-conditioning circuit.

Since a deflection of 30 Ω indicates a 5000 μm/m strain, the output voltage of the signal-conditioning circuit should be

$$(3\ mV/\mu m/m)(5000\ \mu m/m) = 15\ V \tag{5.37}$$

It is possible to find the output of the Wheatstone bridge when the strain gauge outputs 3030 Ω by using the voltage divider. Equations 5.38 and 5.39 present the result of this analysis.

$$V_a = \frac{R}{2R}15 = \frac{3k}{6k}15 = 7.5\,V \tag{5.38}$$

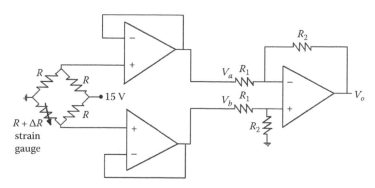

FIGURE 5.19   Difference (instrumentation) amplifier example.

$$V_b = \frac{R + \Delta R}{2R + \Delta R} 15 = \frac{3.03k}{6.03k} 15 = 7.537 \, V \qquad (5.39)$$

Using the results of Equations 5.38 and 5.39, the differential voltage V_d is obtained:

$$V_d = V_a - V_b = 0.037 \, V \qquad (5.40)$$

The differential voltage can be used to find the op amp gain, which is

$$A = \frac{15}{0.037} = 402 = \frac{R_2}{R_1} \qquad (5.41)$$

If we let $R_2 = 400\,k\Omega$ and $R_1 = 1\,k\Omega$, this satisfies the design constraint. The final step would be to confirm that the circuit design satisfies the 3 mV/μm/m sensitivity requirement. If the strain is 1 μm/m, then the output of the gauge is 3000 + (30/5000)(1) = 3000.006 Ω. The output of the difference amplifier is V_d = 15(0.5000005−0.5) = 7.5 μV. The output of the difference amplifier is (400)(7.5 μV) = 3 mV so the design constraint is satisfied.

Example 5.4: Broadband Bandpass Filter Design

If a communication signal is 1 V_{rms} at 1 kHz with 100 mV_{rms} 60 Hz power line noise and 100 mV_{rms} 100 kHz machine vibration noise, design a filter to amplify the communication signal to 20 dB with at least a 40 dB signal to noise ratio (SNR) for the power line and machine vibration noise. In order to meet this design specification, it is necessary to use the broadband bandpass filter shown in Figure 5.20.

The resonant frequency should be ω_o = 2π(1000) = 6283 rad/s with a gain of 20 dB or $K = 10^{20/20} = 10$ in order to meet the design specifications. These gain design specifications yield the following constraint equation:

$$|K| = \frac{R_f}{R_i} = 10 \qquad (5.42)$$

The other two constraint equations are set by the lower and upper cutoff frequencies. In order to attenuate the noise signals, set the lower cutoff ω_{c1} = 2π(500) = 3.14 krad/s and the upper cutoff to ω_{c2} = 2π(10k) = 62.83 krad/s. These two requirements yield the following two constraint equations:

$$\omega_{c1} = \frac{1}{R_i C_i} = 3.14 \, krad/s \qquad (5.43)$$

$$\omega_{c2} = \frac{1}{R_f C_f} = 62.83 \, krad/s \qquad (5.44)$$

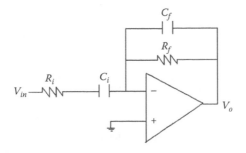

FIGURE 5.20 Broadband bandpass filter example.

Since there are four unknown component values and three constraint equations, choose one of the values. Since capacitors are typically more restrictive in the available values, set $C_f = 1$ nF. Using Equation 2.35 yields $R_f = 15.92$ kΩ. Substituting R_f into Equation 5.42 yields $R_i = 1.59$ kΩ. Lastly, substitute the value for R_i into Equation 5.43 to calculate C_i. The calculation for C_i is

$$C_i = \frac{1}{(3.14k)(1.59k)} = 0.2\,\mu F \tag{5.45}$$

The final step would be to confirm that the design meets the specification. The transfer function for the designed filter is

$$H(s) == -\frac{(R_f/R_i)(1/C_f R_f)s}{(s+1/R_i C_i)(s+(1/R_f C_f))} = -\frac{(628.3k)s}{(s+3.14k)(s+62.83k)} \tag{5.46}$$

In order to verify the gain and SNR, substitute the frequencies of interest into Equation 5.46, which yields (5.47) through (5.49):

$$H(j2\pi 60) = 1.19\angle -97° \tag{5.47}$$

$$H(j2\pi 1000) = 8.9\angle -160° \tag{5.48}$$

$$H(j2\pi 100k) = 0.995\angle 96° \tag{5.49}$$

To verify the gain at resonance, use the result in (5.48):

$$20\log_{10}(8.9/1) = 19\,dB \tag{5.50}$$

The SNR for the power line noise is found from the result of (5.47):

$$SNR = 20\log_{10}(8.9/0.119) = 37.5\,dB \tag{5.51}$$

The SNR for the power line noise is found from the result of (5.49):

$$SNR = 20\log_{10}(8.9/0.0995) = 39\,dB \tag{5.52}$$

Although not exact, the filter gain error is only 1 dB lower than specification and the low frequency noise SNR is 2 dB lower than specification. Figure 5.21 shows the signal before and after passing through the Bode diagram of the filter.

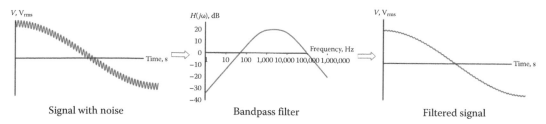

FIGURE 5.21 Communication signal before and after filtering.

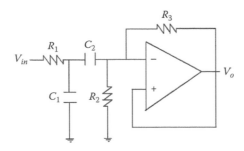

FIGURE 5.22 Sallen–Key bandpass filter.

An alternate solution for this problem would be to use a high-Q (narrowband) filter. The Sallen–Key filter shown in Figure 5.22 is a second-order high-Q filter. This filter has a quality factor ($Q > 1/2$), and unlike the prior broadband filter it does not have discrete real poles. It can be designed to meet the bandwidth and resonant frequency specifications and cascaded with an amplifier to meet the gain specification.

The transfer function for the filter is

$$H(s) = \frac{(1/(R_1C_1))s}{s^2 + \beta s + \omega_o^2}, \quad \text{where} \tag{5.53}$$

$$\omega_o^2 = \frac{R_1 + R_3}{R_1R_2R_3C_1C_2} \quad \text{and} \quad \beta = \frac{1}{R_2C_2} + \frac{1}{R_1C_1} + \frac{1}{R_2C_1} \tag{5.54}$$

In order to simplify the analysis, let $R_2 = 2R_1$, $C_1 = C_2 = 0.1\ \mu F$, and $Q = 1$. Using this analysis yields $R_1 = 3.2\ k\Omega$, $R_2 = 6.4\ k\Omega$, and $R_3 = 455\ \Omega$. The final step would be to confirm that the design meets the specification. The transfer function for the designed high-Q filter cascaded with an amplifier with a gain of 20 is

$$H(s) = \frac{(1/(R_1C_1))s}{s^2 + \beta s + \omega_o^2} \times \frac{R_f}{R_i} = \frac{(3125)s}{s^2 + 6250s + 40M} \times 20 \tag{5.55}$$

In order to verify the gain and SNR, substitute the frequencies of interest into Equation 5.55, which yields (5.56) through (5.58):

$$H(j2\pi60) = 0.6\angle87° \tag{5.56}$$

$$H(j2\pi1000) = 10\angle0.8° \tag{5.57}$$

$$H(j2\pi100k) = 0.1\angle-89° \tag{5.58}$$

To verify the gain at resonance, use the result in (5.48):

$$20\log_{10}(10/1) = 20\ dB \tag{5.59}$$

The SNR for the power line noise is found from the result of (5.60):

$$SNR = 20\log_{10}(10/0.06) = 44\ dB \tag{5.60}$$

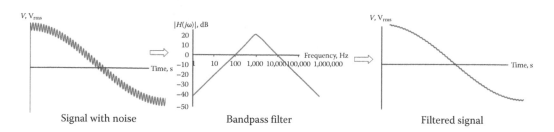

Signal with noise   Bandpass filter   Filtered signal

FIGURE 5.23   Communication signal before and after filtering (Sallen–Key).

The SNR for the machine vibration noise is found from the result of (5.61):

$$SNR = 20\log_{10}(10/0.01) = 60\,dB \qquad (5.61)$$

The communications signal before and after the Sallen–Key filter is shown in Figure 5.23. It is evident from the results that this is a steeper and more accurate design.

Example 5.5: Proportional-Integral-Derivative Controller

Op amps can be used for signal conditioning and also for controller design for a process control system (Figure 5.24). A proportional-integral-derivative (PID) controller is one type of controller that can be used to maintain a system at a given set point while reducing oscillations, offset error, and accumulation error (Figure 5.25). Figures 5.24 and 5.25 provide the block diagram of this system and the circuit schematic, respectively.

For example, if a thermostat is set to a certain temperature, a PID controller can be used to send an output to the heater based upon error between the set point and the room's temperature. Assume that a thermostat is used to represent the temperature set point as a voltage from 0 to 5 V and that the output of the controller is a 0 to 10 V signal to the heater. Design a PID controller with a proportional gain of 3%/%, an integral gain of 10%/(% − min), and a derivative gain of 1%/(%/min) using 1000 μF capacitors.

The first step in the design is to determine the gain of the proportional, integral, and derivative controllers. Equations 5.62 through 5.64 finds the gain for the three controllers by using the ratio of the percentage change of the output to the percentage change of the input multiplied by the gain.

$$K_p = \frac{3(0.01)(10)}{(0.01)(5)} = 6 \qquad (5.62)$$

$$K_I = \frac{10(0.01)(10)}{(0.01)(5)} = 20\,\text{min}^{-1} = 0.333\,\text{s}^{-1} \qquad (5.63)$$

$$K_D = \frac{1(0.01)(10)}{(0.01)(5)} = 2\,\text{min} = 120\,\text{s} \qquad (5.64)$$

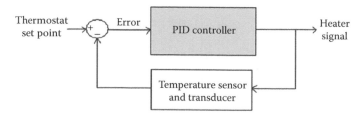

FIGURE 5.24   Process control system.

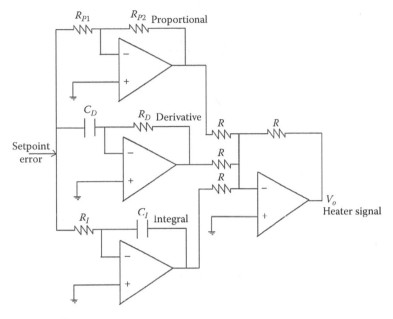

FIGURE 5.25 PID controller.

Since $C_D = C_I = 1000\,\mu\text{F}$, R_D and R_I are found from the following formulas:

$$R_I = \frac{1}{K_I C_I} = 3\,\text{k}\Omega \tag{5.65}$$

$$R_D = \frac{K_D}{C_D} = 120\,\text{k}\Omega \tag{5.66}$$

If we select $R = R_{P1} = 1\,\text{k}\Omega$, then R_{P2} can be found from

$$R_{P2} = K_p R_{p1} = 6\,\text{k}\Omega \tag{5.67}$$

It is also possible to implement the PID controller with one op amp, and this is shown in Figure 5.26. This design would be more cost effective although it has the constraint that the controller gains are not decoupled, which may force the selection of nonstandard resistors or capacitors values.

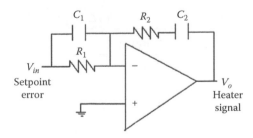

FIGURE 5.26 PID controller (one op amp).

The gains for the one op amp PID controller are

$$K_P = \frac{R_2}{R_1} + \frac{C_2}{C_1} \tag{5.68}$$

$$K_I = \frac{1}{R_1 C_1} \tag{5.69}$$

$$K_D = C_1 R_2 \tag{5.70}$$

By selecting $C_2 = 1000\,\mu F$, $R_1 = 3\,k\Omega$, $R_2 = 17.6\,k\Omega$, and $C_1 = 6.8\,mF$ it is possible to meet the design constraints of the previous example.

Example 5.6: Gyrator

The final example will use an op amp circuit to model an inductor or gyrator. Inductors can be very expensive to use in a design; therefore, if an engineer wishes to develop a circuit that produces a 5 H equivalent inductance using only 0.01 μF capacitors, resistors, and op amps, the solution to this problem is to use the gyrator circuit. The gyrator was shown in Table 5.1 and the equivalent inductance is repeated here in Equation 5.71.

$$L = \frac{C R_1 R_3 R_4}{R_2} H \tag{5.71}$$

Solving equation for the resistances yields

$$500M = \frac{R_1 R_3 R_4}{R_2} \tag{5.72}$$

By selecting $R_1 = R_2 = 1\,k\Omega$, the equation in (5.72) reduces to

$$500M = R_3 R_4 \tag{5.73}$$

Finally, let $R_3 = 20\,k\Omega$ and $R_4 = 25\,k\Omega$ to satisfy the 5 H constraint. The final design for the gyrator is shown in Figure 5.27.

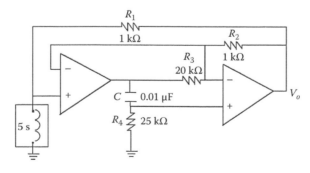

FIGURE 5.27   Gyrator design.

5.5 Realistic Op Amp Model

It should be noted that all of the designs in this chapter have assumed ideal op amps with the virtual short circuit and infinite input impedance assumptions. However, a more realistic model has a finite input impedance R_i, a finite open-loop gain G, and a nonzero output resistance R_o. The realistic op amp model is shown in Figure 5.28. Typical values for the input resistance, output resistance, and open-loop gain are $2\,\mathrm{M\Omega}$, $75\,\Omega$, and 10^5, respectively.

5.5.1 Inverting Amplifier

How does the more realistic model of the op amp affect the analysis of the inverting amplifier in Figure 5.4? The inverting amplifier with the more realistic model is shown in Figure 5.29. The KCL equations at the output and inverting terminals of this op amp are given in Equation 5.74.

$$\text{KCL @ } V_n: \quad \frac{V_n - V_i}{R_1} + \frac{V_n - V_o}{R_2} + \frac{V_n}{R_i} = 0$$

$$\text{KCL @ } V_o: \quad \frac{V_o - V_n}{R_2} + \frac{V_o - G(-V_n)}{R_o} = 0$$

(5.74)

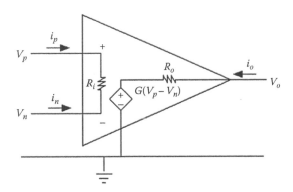

FIGURE 5.28 Realistic op amp model.

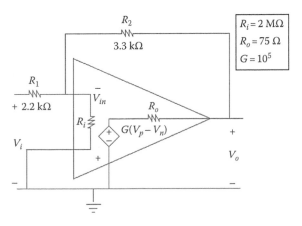

FIGURE 5.29 Inverting amplifier realistic model.

Note that there was not an equation at the output node for the ideal op amp analysis. Solving the set of equations in (5.74) yields

$$A = \frac{V_o}{V_i} = \left(-\frac{G(R_2/R_1) + (R_o/R_i)}{(G+1+(R_2/R_1)) + (R_2/R_i) + (R_o/R_1) + (R_o/R_i)} \right) = -1.5 \tag{5.75}$$

Since the gain for this model is the same as for the ideal model, it can be noted that as long as the input resistance and open-loop gain are relatively large and the output resistance is relatively small, the simplifying assumptions are valid. However, the source resistance and load resistance have to be considered based upon realistic values of the input and output resistance.

5.5.2 Non-Inverting Amplifier

How does the more realistic model of the op amp affect the analysis of the non-inverting amplifier in Figure 5.5? The non-inverting amplifier with the more realistic model is shown in Figure 5.30.

$$\text{KCL @ } V_n: \quad \frac{V_n}{R_1} + \frac{V_n - V_o}{R_2} + \frac{V_n - V_i}{R_i} = 0$$

$$\text{KCL @ } V_o: \quad \frac{V_o - V_n}{R_2} + G\frac{V_o - G(V_i - V_n)}{R_o} = 0 \tag{5.76}$$

Solving the KCL equations at the inverting terminal and output terminal yields the gain as

$$A = \frac{V_o}{V_i} = \left(\frac{G(1 + (R_2/R_1)) + (R_o/R_i)}{(G+1+(R_2/R_1)) + (R_2/R_i) + (R_o/R_1) + (R_o/R_i)} \right) = 3.8 \tag{5.77}$$

If the load resistor from Figure 5.5 is considered in the derivation, then the gain changes to

$$A = \frac{V_o}{V_i} = \left(-\frac{G(1 + (R_2/R_1)) + (R_o/R_i)}{(G+1+(R_2/R_1)) + ((R_2/R_i) + (R_o/R_1) + (R_o/R_i)) + ((R_2R_o/R_iR_L) + (R_2R_o/R_1R_L) + (R_o/R_L))} y \right) = 3.8 \tag{5.78}$$

Based upon a similar result, the same analysis conclusion found for the inverting amplifier holds here.

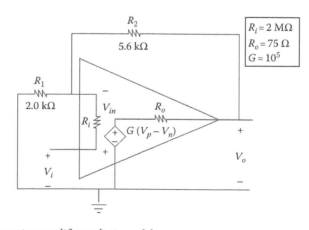

FIGURE 5.30   Non-inverting amplifier realistic model.

5.5.3 Input Offset Voltage

Unlike an ideal op amp, when the input voltage to the practical op amp is zero, the output is not zero. The **input offset voltage** is the differential input voltage that is needed to make the output zero when the input is zero. A typical value for the input offset voltage is 2 mV. The output when the input is zero is the output dc offset voltage. The way to compensate for this value in a realistic design is to add a small voltage source at the inverting or non-inverting amplifier of the opposite magnitude and polarity. It may be necessary to use a potentiometer with the op amp input set to zero to find the exact value to cancel out the input offset voltage.

5.5.4 Input Bias and Offset Currents

Unlike an ideal op amp, the current into the input terminals is not zero. This current value is typically 2 μA. The **bias current** is the average of the current into the positive and negative input terminals. The **input offset current** is the difference between the input bias currents. The bias current has the most effect when the source impedance is large because it produces a large voltage. When the source impedance is small, the bias current can be neglected. If the op amp is designed such that the DC resistance from the positive terminal to ground is the same as the DC resistance from the negative terminal to ground, then the effects of the input bias current are negligible. The way to create a design to resolve input bias currents is to select resistors to cancel out the output error. The smaller the resistor values, the smaller the error; however, there must be a tradeoff between resistor size and current and power requirements for the op amp. For design considerations, in order to cancel out input bias current, you could select the input resistance on the non-inverting terminal to be equivalent to the parallel combination of the feedback resistance(s) and input resistance(s) on the inverting input. In order to resolve the input offset current, you can follow a similar procedure to the input offset voltage and use a voltage source and potentiometer to cancel out the effects on the output.

5.5.5 Frequency Response

As previously stated, the GBP is a constant that determines the limitations of the op amps performance for a certain gain at a certain frequency. This constant must be considered when designing a filter to insure that the closed-loop gain is within the proper range for the given frequency.

5.5.6 Slew Rate

Another characteristic of a practical op amp is that it exhibits slew rate limiting. This means that the output due to a step input is not a perfect step function. This is due to the fact that the response is frequency dependent as described in the prior section. Slewing happens when the input voltage to the op amp is so large that it causes it to saturate and the output curve cannot rise fast enough.

5.6 Conclusion

This chapter has presented a general introduction to the design of op amps including ideal assumptions, practical considerations, analysis, applications, and a more realistic model. Further study is encouraged in the component circuitry that make up the op amp IC. The basic building block of this circuitry is the differential amplifier. This study will also introduce a more in depth discussion of topics such as common mode and differential mode gains, level shifters, power supply rejection ratio, balanced inputs and outputs, and coupling between multiple inputs.

Additionally, there are many other op amp applications such as power audio op amps, bridge oscillators, Chebyshev filters, and nonlinear circuits. There are also feedback limiter configurations that constrain a signal to be above or below a specific breakpoint. Op amps can also be used as comparators to determine which of the two voltages is larger. The next phase in comparators are Schmitt triggers, which provide feedback for steeper and more rapid transitions where the saturation states are held until a certain input is reached. It is also possible to design digital-to-analog and analog-to-digital converters using op amps. Table 5.1 presents a summary of some of these op amp configurations.

Bibliography

C.D. Johnson, *Process Control and Instrumentation Technology*, 8th edition, Upper Saddle River, NJ: Prentice Hall, 2006, 704 pp.

J.W. Nilsson and S.E. Riedel, *Electric Circuits*, 8th edition, Upper Saddle River , NJ: Prentice Hall, 2007, 880 pp.

M.N.O. Sadiku and C.K. Alexander, *Fundamentals of Electric Circuits*, 3rd edition, New York: McGraw-Hill, 2007, 901 pp.

6

Frequency Response and Bode Diagrams

Thomas F.
Schubert, Jr.
University of San Diego

Ernest M. Kim
University of San Diego

6.1 Introduction

The response of a system to an input can be characterized in a variety of formats. Most commonly, the response is given in either the time domain, by the impulse response, or in the frequency domain, by the frequency response. Of interest here is the frequency response, a measure of the relationship of a system's output to a sinusoidal input as a function of the input frequency. While most commonly applied to electronic/electrical systems, frequency-response techniques are often used in mechanical, as well as biological, systems.

Frequency-response characteristics can commonly be found in the consumer market with respect to audio equipment, in particular, audio amplifiers, loudspeakers, and microphones. Radio-frequency response can refer to a variety of items such as amplifiers, receivers and transmitters, antennas, coaxial cables, video switchers, etc. Subsonic frequency-response characterizations can include earthquakes and brain waves.

Frequency-response techniques are most valuable when dealing with linear, time-invariant systems, where a sinusoidal input produces a sinusoidal output of the same frequency, where the response to a sum of inputs is the sum of the individual responses to each input, and where the response of the system does not depend on when the input was applied. Practically, a small amount of variation is allowed in these system constraints and is typically treated as distortion.

6.2 Theoretical Relationships

Linear, time-invariant systems can be characterized by the impulse response or the frequency response. These two response characterizations are directly related.

In the time domain, the response, $x_o(t)$, of such a system to an arbitrary input, $x_i(t)$, can be characterized, by a convolution of the input, $x_i(t)$ with the system-impulse response, $h(t)$:

$$x_o(t) = \int_0^t h(t - \tau) x_i(\tau) \, d\tau.$$

While this representation is useful in many situations, a frequency-domain representation of the system characteristics is also typical and highly useful. The frequency-domain spectrum of a system, $X_o(\omega)$, can be derived from the impulse response expression by performing the Fourier transform on the output, $x_o(t)$,

$$X_o(\omega) = \{x_o(t)\} = \int_{-\infty}^{\infty} x_o(t)\, e^{-j\omega t}\, dt$$

This operation reduces to the frequency-domain relationship,

$$X_o(\omega) = H(\omega)\, X_i(\omega).$$

Simply stated, the spectrum of the output, $X_o(\omega)$, is the product of the input spectrum, $X_i(\omega)$, and the frequency-domain transfer function of the system, $H(\omega)$. The impulse response and the frequency-domain transfer function of a system are related by the Fourier transform,

$$H(\omega) = \mathfrak{F}\{h(t)\}.$$

$H(\omega)$ is usually represented by its fundamental parts in polar form:

- $|H(\omega)|$—the magnitude response
- $\angle H(\omega)$—the phase response

6.3 Measurement of the Frequency Response

The frequency response of a system can be determined by measurements either in the time domain or in the frequency domain. Time-domain measurements will result in the system-impulse response, which is then transformed to the frequency response through the Fourier transform. Time-domain measurements are obtained by either:

- Applying an impulse to the system and measuring its response, $h(t)$, or
- Applying a signal with a wide frequency spectrum (e.g., digitally generated maximum length sequence noise, a step function, or analog-filtered white noise equivalent, like pink noise), and calculating the impulse response by a deconvolution of this input signal and the output signal of the system.

Frequency-domain measurements are more direct and consist of

- Sweeping a constant-amplitude pure sinusoid through the frequency range of interest and measuring the output level and phase shift relative to the input.

While various devices exist for obtaining $H(\omega)$ automatically, manual determination is common. Such a determination requires two basic pieces of equipment:

- A source with sinusoidal output of variable frequency
- A device capable of measuring and/or displaying both amplitude and phase shift

An oscilloscope often serves well as the measurement device in electronic applications. A typical oscilloscope display of the input and output signals for a system under test is shown in Figure 6.1. At this particular frequency (f), the magnitude of the system response is given by

$$H(f) = \left| \frac{V_2}{V_1} \right|$$

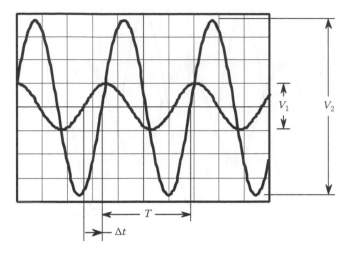

FIGURE 6.1 Gain and phase shift measurement.

The phase shift (in degrees) is determined as

$$\theta = \frac{\Delta t}{T}(360°) = f\Delta t(360°),$$

where
    T is the period of the sinusoidal voltage
    Δt is the delay time between the input (V_1) wave, as reference, and the output (V_2) wave as observed
      on the oscilloscope

Phase is positive if the output (V_2) wave is to the left of (leads) the input V_1 wave and is negative if V_2 lags behind the V_1 wave. Assuming the two signals shown in Figure 6.1 are on the same scale, the magnitude of system response at this frequency is approximately 3.6 (11.13 dB) and the phase shift is approximately −70° (−1.22 rad).

Measurements are typically performed in three stages: a broad frequency scan (often only one measurement in each frequency decade) to determine the general frequency range of interest, a systematic frequency scan (often at as few as three frequencies in each decade) to determine the general shape of the response, and a focused frequency scan near points of interest. For the systematic frequency scan, multipliers of 1, 2, and 5 divide each frequency decade into approximately equal thirds and are easily located on a logarithmic grid.

In a noisy environment, automated measurement devices may give spurious estimations of amplitude and phase. For example, for the signals shown in Figure 6.2, a peak-to-peak detector will output a value approximately 18% larger than the proper value for the noisy signal shown. All attempts must be made to visually filter out the noise in parameter estimation. For the signals shown, the gain is approximately 0.50 (−3.0 dB) and the phase shift about +150° (+2.6 rad).

6.4 Displaying the Frequency Response—The Bode Diagram

In the late 1930s, Hendrik Wade Bode (1905–1982) pioneered presenting the two fundamental response quantities, $|H(\omega)|$ and $\angle H(\omega)$, as a pair of plots [1]. The typical format for these plots is the magnitude (typically on a decibel scale, but occasionally on a linear scale) or the phase (in either radians or degrees) as the ordinate and frequency (on a logarithmic scale) as the abscissa. Each of these two plots is identified as a ***Bode plot*** in commemoration of Bode's early work. The two plots together

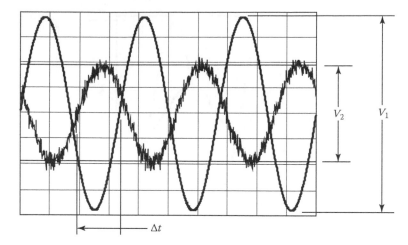

FIGURE 6.2   Gain and phase shift measurement: noisy output signal.

form a **Bode diagram**. Originally used in the design of feedback amplifiers, magnitude, and phase plots, Bode diagrams have provided engineers with an intuitive tool useful in the design and analysis of a large variety of systems. In particular, Bode's introduction of the concepts of gain margin and phase margin, each easily interpreted on a Bode plot, provided engineers with an especially valuable viewpoint in the investigation of the stability of systems [2,3].

   While many computer simulation programs exist for the efficient creation of such plots directly from the frequency-response equations of the system, much insight into the functioning of a system (or any electronic circuit) and the dependence of the responses to individual system parameter variation can be gained by the manual creation of simple straight-line asymptotic magnitude and approximate phase plots. Similarly, a good approximation to system response can be generated by fitting straight-line plots to experimental or simulation data.

6.4.1 Mathematical Derivations

The response of a linear, time-invariant system to a sinusoidal input of varying frequency can, in most cases, be described as the ratio of two polynomials in frequency ω,[*]

$$H(\omega) = K_0 \frac{Z_m(\omega)}{P_n(\omega)}. \tag{6.1}$$

Here, $Z_m(\omega)$ is an mth-order polynomial whose roots identify the m zeroes of the response. The roots of the nth-order polynomial, $P_n(\omega)$, identify the n poles. For all real systems, the response must not become infinite as frequency approaches infinity; thus, there must be at least as many poles as zeroes, $n \geq m$. Equation 6.1 can be rewritten in the format

$$H(\omega) = K_0 \frac{1 + (j\omega)a_1 + (j\omega)^2 a_2 + \cdots + (j\omega)^m a_m}{1 + (j\omega)b_1 + (j\omega)^2 b_2 + \cdots + (j\omega)^n b_n}$$

---

[*] Many sources prefer to express the system response in Laplace domain, $H(s)$. In the case of sinusoidal frequency response, $H(s)$ reduces to $H(j\omega)$ by replacing s with $j\omega$. Similarly, frequency can be expressed as either angular frequency, ω (in rad/s), or temporal frequency, f (in Hz), without any loss of generality.

The numerator and denominator polynomials, $Z_m(\omega)$ and $P_n(\omega)$ respectively, can each be written as a factored product of multiples of four types of simple function, identified as the *factors*, $\{F(\omega)\}$:

1. K—a constant
2. $j\omega$—a root at the origin
3. $1 + \dfrac{j\omega}{\omega_o}$ — a simple root at $\omega = \omega_o$
4. $1 + 2\zeta\dfrac{j\omega}{\omega_o} + \left(\dfrac{j\omega}{\omega_o}\right)^2$ —a complex conjugate root pair

Each of these four factors has a straight-line approximate Bode representation. The use of decibels (a logarithmic function) as the vertical scale for the magnitude plot and a linear vertical scale for phase converts the product of the factors into the *sum* of the factor magnitudes (in dB) and the *sum* of the factor phases: division becomes subtraction. Therefore, the total system response plot is obtained by algebraically summing the plots of the individual simple factors describing the system.

6.4.2 Bode Plots of the Factors

The Bode *magnitude* straight-line representation of each of the four simple factors is unique. The Bode *phase* straight-line representation is unique for two of the factors (a constant and a root at the origin) and universally standardized for a third (a simple root). However, a variety of approximate representations for the phase of a complex conjugate root pair exists in the literature [4–7]. The representation for a constant and for a root at the origin is exact: the other two factor representations are asymptotic in the case of magnitude, and approximate in the case of phase. The straight-line representation of each Bode factor follows; a summary can be found in Table 6.1.

6.4.2.1 Constant

A constant is the most simple of the factors:

$$F_C(\omega) = K.$$

The magnitude plot is constant at $20 \log |K|$ if the factor is in the numerator, or $-20 \log |K|$ if in the denominator. The phase plot is also constant at $\theta = 0°$ or at $\theta = \pm 180°$ depending on the mathematical sign of K. Each plot is an exact representation of the factor.

6.4.2.2 Root at the Origin

Roots at the origin:

$$F_0(\omega) = j\omega,$$

also have a simple, exact, Bode representation. The magnitude plot is a straight line with a slope of $\pm 20\,\text{dB/decade}^*$:

$$|F_0(\omega)|_{dB} = 20\log|j\omega| = 20\log(\omega)$$

If the root is in the numerator (indicating a zero), the line has positive slope. Denominator roots (poles) have negative slope. In each case, the line passes through the point $\{0\,\text{dB}, 1\,\text{rad/s}\}$.

---

* A decade is a change in frequency by a factor of 10. A slope of 20 dB/decade is also identified in some sources as 6 dB/ octave (actually $\approx 6.0206\,\text{dB/octave}$), where "octave" is derived from the eight-tone musical scale and is a frequency change by a factor of 2.

TABLE 6.1   Bode Factor Magnitude and Phase Plots

Factor	Bode Magnitude Plot	Bode Phase Plot
$F_C = K_0$	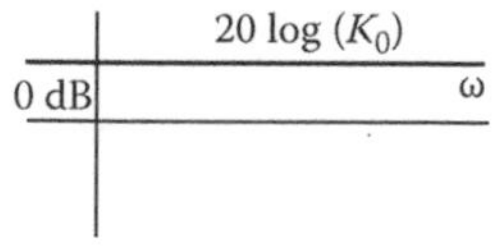 $20 \log(K_0)$; 0 dB	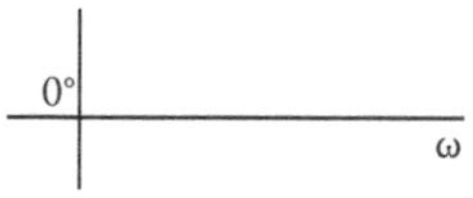 $0°$
$F_0(\omega) = j\omega$	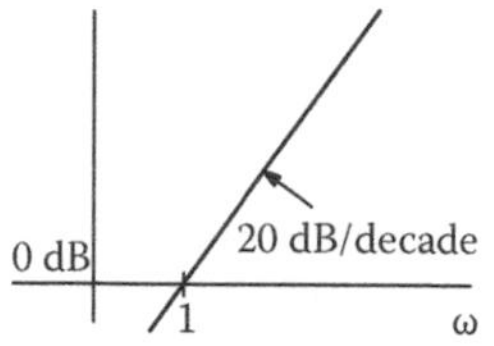 0 dB; 20 dB/decade; 1	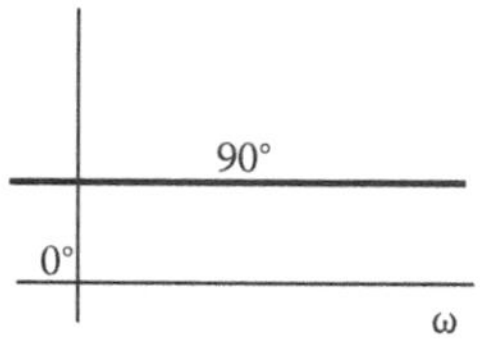 $90°$; $0°$
$F_1(\omega, \omega_0) = 1 + \dfrac{j\omega}{\omega_0}$	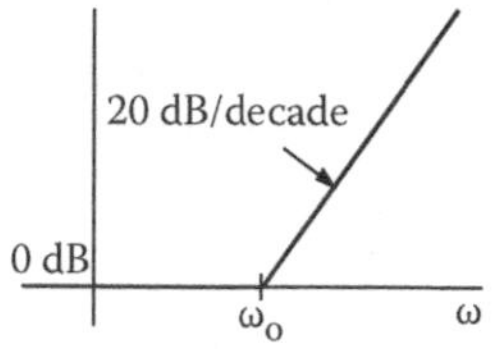 20 dB/decade; 0 dB; ω_0	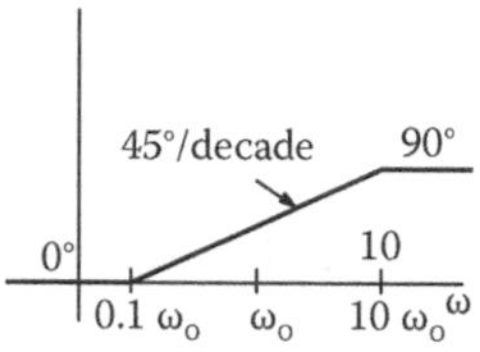 $45°$/decade; $90°$; $0°$; $0.1\,\omega_0$; ω_0; $10\,\omega_0$; 10
$F_2(\omega, \omega_0, \zeta) =$ $1 + 2\zeta\dfrac{j\omega}{\omega_0} + \left(\dfrac{j\omega}{\omega_0}\right)^2$	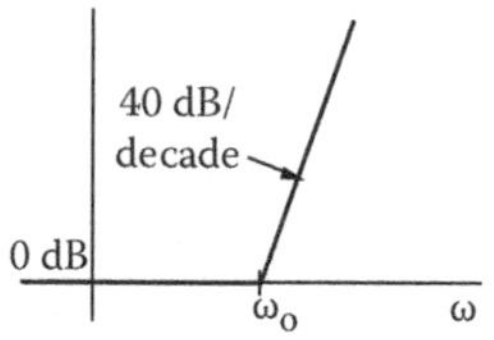 40 dB/decade; 0 dB; ω_0	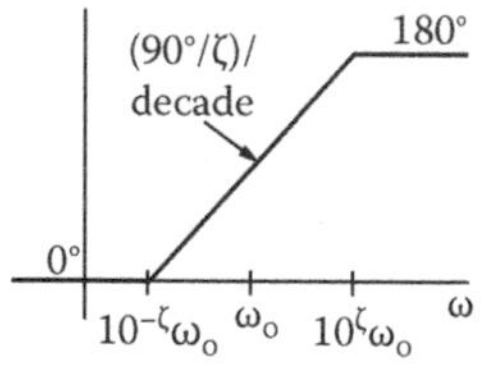 $(90°/\zeta)/$decade; $180°$; $0°$; $10^{-\zeta}\omega_0$; ω_0; $10^{\zeta}\omega_0$

Note: Factors are in the numerator, representing zeroes.

The phase of a root at the origin is constant at $\theta = 90°$. Consequently, the phase plot for a zero will be at $+90°$ and at $-90°$ for a pole.

6.4.2.3 Simple Root at $\omega = \omega_0$

Simple roots not at the origin have more complex plots. The factor for a simple root is

$$F_1(\omega, \omega_0) = 1 + \frac{j\omega}{\omega_0},$$

where ω_0 is the frequency of the root. The magnitude of the factor (in dB) is given by

$$\left| F_1(\omega, \omega_0) \right|_{dB} = 20 \log \sqrt{1 + \left(\frac{\omega}{\omega_0}\right)^2} = 10 \log \left(1 + \left(\frac{\omega}{\omega_0}\right)^2 \right).$$

The factor phase is given by

$$\angle F_1(\omega, \omega_0) = \tan^{-1}\left(\frac{\omega}{\omega_0}\right)$$

These two components of the factor are shown in Figure 6.3. Also shown in the figure are the standard Bode straight-line approximations. A factor in the denominator (representing a pole) will have plots that are the vertical mirror image of a factor in the numerator (representing a zero).

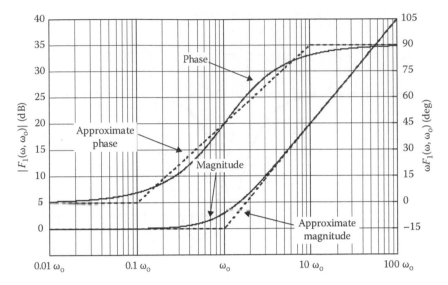

FIGURE 6.3 Simple root factor plots.

The Bode approximate plot of the magnitude is an asymptotic approximation and comprises two intersecting straight lines that form a piecewise continuous plot. If $\omega \ll \omega_o$, $F_1(\omega, \omega_o) \approx 1$ and the magnitude plot is asymptotically constant at 0 dB. If $\omega \gg \omega_o$,

$$\left.\left| F_1(\omega, \omega_o) \right|\right|_{dB} \approx 10\log\left(\left(\frac{\omega}{\omega_o}\right)^2\right) = 20\log(\omega) - 20\log(\omega_o)$$

This asymptote is a straight line with a slope of 20 dB/decade that intersects the 0 dB line at the root frequency, $\omega = \omega_o$. The Bode approximation transitions between the two asymptotic lines at the root frequency, where the approximation has its greatest error ($20\log(2) \approx 3.01$ dB).

As can be seen in Figure 6.3, the simple root phase never exceeds 90° and essentially all phase change takes place within ± one decade of the root frequency, $\omega_o(\angle F_1(0.1\omega_o, \omega_o) = 5.71°$ and $\angle F_1(10\omega_o, \omega_o) = 84.29°)$. The universally adopted Bode phase approximation is a continuous, three-segment, straight-line plot that uses ± one decade as the transition points between segments. Beyond one decade from the root, ω_o, the phase is approximated by a constant:

$$\omega < 0.1\,\omega_o \implies \angle F_1(\omega, \omega_o) = 0°$$

$$\omega > 10\,\omega_o \implies \angle F_1(\omega, \omega_o) = 90°$$

Within one decade of ω_o, the phase is approximated by a straight line of slope 45°/decade:

$$0.1\,\omega_o \leq \omega \leq 10\,\omega_o \implies \angle F_1(\omega, \omega_o) = 45°[\log(10\omega/\omega_o)]$$

This straight-line Bode approximation has a maximum error of ~5.71°.

The error introduced by the Bode straight-line approximations has distinct symmetry about ω_o (Figure 6.4). Any corrections to the plots, if necessary, are accomplished by interpolation and comparison to standard plots. Some data points that are helpful in making accurate corrections are

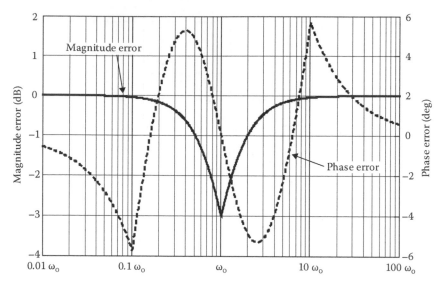

FIGURE 6.4   Bode straight-line errors: simple root.

Magnitude plot

- The magnitude error at ω_o is 3.01 dB
- The magnitude error at ± one octave ($0.5\omega_o$ and $2\omega_o$) is 0.97 dB

Phase plot

- Slope at ω_o is $\approx 66°$/decade
- The phase error is zero at $0.159\omega_o$, ω_o, and $6.31\omega_o$
- The phase error at ± one decade is ±5.71°

6.4.2.4 Complex Conjugate Root Pair

Complex conjugate root pairs,

$$F_2(\omega, \omega_o, \zeta) = 1 + 2\zeta \frac{j\omega}{\omega_o} + \left(\frac{j\omega}{\omega_o}\right)^2,$$

have a more complex relationship. Again, ω_o is identified as the resonant frequency; ζ is identified as the damping factor.* A plot of the magnitude and phase of this factor with the damping factor as a parameter ($0.2 \le \zeta \le 0.9$ in increments of 0.1) is shown in Figure 6.5. The Bode plots for a complex conjugate pair of poles will take the same shapes but will be vertical mirror images.

The variation of the plot with damping coefficient interferes with the accuracy of two-segment, straight-line magnitude approximations near the resonant frequency. Beyond half of a decade from the resonant frequency, asymptotic approximations of the magnitude plot have a high degree of validity. If $\omega \ll \omega_o$, $F_2(\omega, \omega_o, \zeta) \approx 1$; the asymptotic magnitude plot is constant at 0 dB. If $\omega \gg \omega_o$,

$$\left|F_2(\omega, \omega_o, \zeta)\right|_{dB} \approx 20\log\left|\left(\frac{j\omega}{\omega_0}\right)^2\right| = 40\log(\omega) - 40\log(\omega_o).$$

In this region, the asymptotic magnitude plot is a line with a slope of 40 dB/decade and intersects to other asymptote at the resonant frequency, $\omega = \omega_o$ (as was the case for a simple root). Near the resonant

---

* The damping factor lies in the range, $0 \le \zeta < 1$, for complex conjugate root pairs.

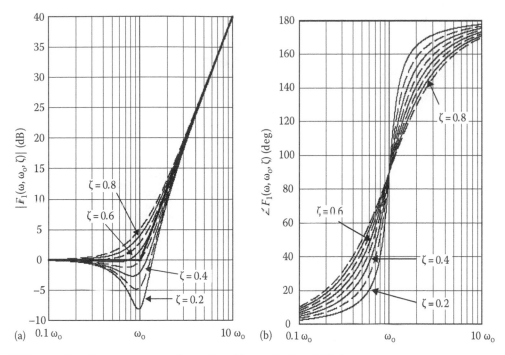

FIGURE 6.5 Complex conjugate root factor plots: (a) magnitude and (b) phase.

frequency, there can be a large difference between the approximate Bode magnitude plot and the true magnitude plot. Depending on the application, corrections to the curve may be necessary. These corrections can usually be accomplished with the addition of only a few data points and interpolation. Two helpful data points are the magnitude at the resonant frequency and the magnitude of the peak in the curve (if one exists). The magnitude at the resonant frequency is given by

$$\left| F_2(\omega_o, \omega_o, \zeta) \right|_{dB} = 20 \log(2\zeta).$$

For damping coefficients less than $1/\sqrt{2}$, a valley occurs in the magnitude of the factor. This valley will occur at the maximum difference between the true magnitude plot and the Bode straight-line approximation (Figure 6.6). Interestingly, this valley in the factor magnitude plot is usually encountered with the factor in the denominator and is consequently identified in the literature as a "peak." If only the factor is considered, the valley (peak) occurs at a frequency somewhat lower than the resonant frequency:

$$\omega_{peak} = \omega_o \sqrt{1 - 2\zeta^2}.$$

The magnitude of this valley (peak) is

$$\left| F_2(\omega_{peak}, \omega_o, \zeta) \right|_{dB} = 20 \log\left(2\zeta\sqrt{1 - \zeta^2}\right).$$

In cases where a complex conjugate pair of poles (or zeroes) cancels lower-frequency zeroes (or poles) so that the asymptotic plot is a constant for frequencies higher than the resonant frequency (see Example 6.3), the peak (valley), necessarily of the same magnitude, occurs at a frequency somewhat higher than the resonant frequency:

$$\omega_{peak} = \frac{\omega_o}{\sqrt{1 - 2\zeta^2}}.$$

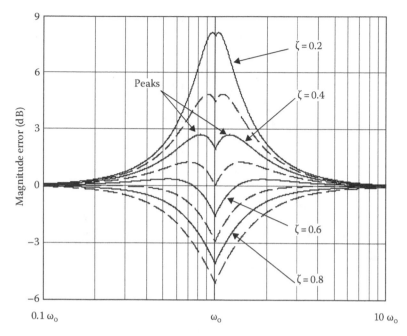

FIGURE 6.6   Bode straight-line magnitude error: complex conjugate root pair.

The Bode phase plot for a complex root pair is also simplified into a three-segment, straight-line plot. However, the damping coefficient complicates the location of the transition between segments. At frequencies much lower than the resonant frequency, the phase is near constant at 0°; at frequencies much higher than ω_o, the phase is near constant at 180°. Near ω_o the Bode approximate curve is a straight line joining the other segments.

The location of the transition points between the segments of the approximate Bode phase plot is not uniformly described in the literature; a ζ-independent approximation [4,5] and at least two approximations that depend on the value of ζ are to be found. The ζ-independent approximation chooses transition points at ± one decade, while the most prevalent of the ζ-dependent approximations vary the transition points (as a fraction of a decade) linearly with damping coefficient, ζ. Each approximation has its strong features. For example, the tangent-of-phase approximation [6] is useful in that it identifies the slope of the phase plot, m_{phase}, at the resonant frequency, ω_o, as

$$m_{phase} = \left(\frac{180}{\pi}\right)\left(\frac{\ln(10)}{\zeta}\right) \approx \frac{131.928}{\zeta} \approx \frac{132}{\zeta} \text{ °/decade.}$$

However, it has been shown that the decade-fraction approximation [7] more closely approximates the phase plot variation with damping coefficient under a variety of criteria [8].

The decade-fraction approximation identifies the transition frequencies as lying $\pm\zeta$ decades from the resonant frequency. The constant phase regions are a function of ζ and are described as

$$\omega < \omega_o 10^{-\zeta} \implies \angle F_2(\omega, \omega_o, \zeta) = 0°$$

$$\omega > \omega_o 10^{\zeta} \implies \angle F_2(\omega, \omega_o, \zeta) = 180°$$

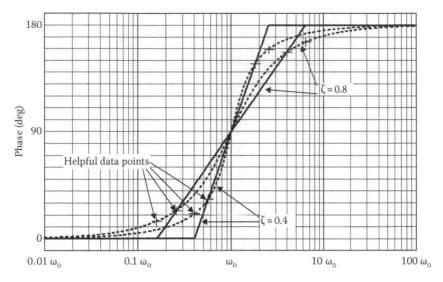

FIGURE 6.7 The decade-fraction ζ-dependent Bode phase approximation.

The phase between the two constant phase regions is approximated by a straight line of slope $(90/\zeta)°/$ decade passing through the point ω_o, 90°:

$$\omega_o 10^{-\zeta} \le \omega \le \omega_o 10^{\zeta} \Rightarrow \angle F_2(\omega, \omega_o, \zeta) = \frac{90°}{\zeta}\left(\log\left(\frac{\omega}{\omega_o}\right)\right) + 90°$$

As was the case in first-order factors, the error introduced by the Bode straight-line approximations has distinct symmetry about ω_o. Any corrections to the plots, if necessary, are accomplished by interpolation and comparison to standard plots. Some data points that are helpful in making accurate corrections to the phase plot are

- Phase plot slope at ω_o is $\approx 132/\zeta$ °/decade
- The phase error goes to zero at
  - $\omega = \omega_o$
  - $\omega = 10^{\pm(0.3\zeta^2 + 0.49\zeta + 0.01)}$

The magnitude of the phase error at the transition points $(\omega = \omega_o 10^{\pm\zeta})$ is

- $\approx -15.8\zeta + 27.2$ for $\zeta \ge 0.4$
- $= 10\zeta^3 - 22.8\zeta^2 - 0.7\zeta + 23.5$ (all values of ζ)

The decade-fraction ζ-dependent Bode phase approximation is shown in Figure 6.7 for two values of ζ along with the actual phase plots. Helpful data points, as identified above are marked with a "+" sign.

6.4.3 Time Delay

Systems with an inherent time delay, t_d, will experience a fifth factor, a phase shift that is linear with frequency

$$F_D(\omega) = e^{-j\omega t_d}.$$

The delay factor has a Bode magnitude plot that is constant at 0 dB; the system Bode magnitude plot is unchanged by the factor. Unfortunately, the phase is linear with frequency $\{\phi = -t_d\omega\}$ and does not have

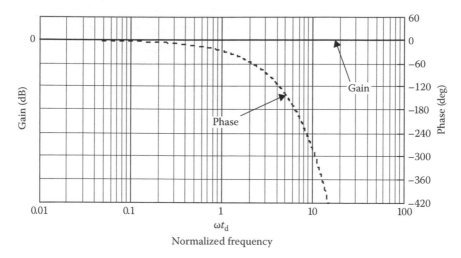

FIGURE 6.8   Time delay factor Bode diagram.

a "nice" plot on a logarithmic frequency scale; each decade in frequency experiences ten times the phase variation of the decade directly lower in frequency. Typically, any system delay is treated separately from the Bode phase plots; in systems where the time delay is small compared to the period of the highest frequency present, delay effects are usually considered insignificant and, consequently, ignored. A Bode diagram of the delay factor is shown in Figure 6.8.

6.4.4 Temporal Frequency versus Angular Frequency

Theoretical derivations of Bode factors tend to utilize angular frequency (ω: rad/s). However, in the real world, it is likely that systems will be described using temporal frequency (f: Hz).

The functional form of the response, $H(\omega)$ or $H(f)$, is identical and conversions can be made using $\omega = 2\pi f$. The only other difference appears in a difference in the constant factor F_C for a particular system. If there are simple roots at the origin ($F_0 = j\omega$), the constant terms in $H(\omega)$ and $H(f)$ for the system {$K\omega$ and K_f, respectively} will differ by a factor

$$K_f = (2\pi)^{m-n} K_\omega$$

where $m-n$ is the number of simple roots at the origin in the numerator minus those in the denominator.

Example 6.1: Drawing Bode Plots

Draw the Bode diagram for the following transfer function:

$$H(\omega) = \frac{[1.7 \times 10^{-3}][j\omega]^2}{[1+(j\omega/120)][1+(j\omega/700)][1+(j\omega/500\times10^3)+((j\omega)^2/160\times10^9)]}$$

Solution

The Bode diagram consists of the Bode magnitude plot and the Bode phase plot. The given transfer function contains all of the primary factor types:

- A constant
- Two zeroes at the origin
- Two simple poles
- A complex conjugate pair of poles

The Bode magnitude plot for the constant is a horizontal line at

$$20\log[1.7\times10^{-3}] = -55.39\,\text{dB}$$

The Bode magnitude plot for the two zeroes at the origin has a slope of $+40\,\text{dB/decade}$ ($2 \times 20\,\text{dB/decade}$) and passes through ($0\,\text{dB}$, $1\,\text{rad/s}$).

The simple poles each introduce a slope increment of $-20\,\text{dB/decade}$ beginning at the pole frequencies

$$\omega_{p1} = 120\,\text{rad/s} \quad \text{and} \quad \omega_{p2} = 700\,\text{rad/s}$$

The complex pair of poles will introduce a slope increment of $-40\,\text{dB/decade}$ at the resonant frequency

$$\omega_o^2 = 160\times10^9 \quad \Rightarrow \quad \omega_o = 400\times10^3\,\text{rad/s}$$

The pole pair damping coefficient is calculated to be

$$\frac{2\zeta}{\omega_o} = \frac{1}{500\times10^3} \quad \Rightarrow \quad \zeta = 0.4$$

The magnitude at the resonant frequency can then be (optionally) corrected by determining

$$\left|F_2(\omega_o,\,\omega_o,\,0.4)\right| = 2\times0.4 = 0.8 \quad \Rightarrow \quad -1.94\,\text{dB}$$

Since the damping coefficient is less than $1/\sqrt{2}$ and the factor cancels out lower-frequency zeroes, a peak in the magnitude response exists above the resonant frequency at

$$\omega_{peak} = \omega_o\sqrt{1-2\zeta^2} = 400\times10^3(0.825) = 329.9\times10^3\,\text{rad/s.}$$

Consequently, the other optional correction point has value

$$\left|F_2(\omega_{peak},\,\omega_0,\,0.4)\right| = 2(0.4)\sqrt{1-0.4^2} = 0.733 \quad \Rightarrow \quad -2.79\,\text{dB}$$

Since the complex conjugate pair are poles, the factor lies in the denominator, the slope is $-40\,\text{dB/decade}$, and the signs of the magnitude corrections are reversed (making them both, in this case, positive).

The magnitude plot is constructed as follows:

- A low-frequency starting point is found, where the sum of all factors is known and below any pole (or zero) frequencies (other than those at the origin). Here $\omega = 10$ is a good choice (evaluate the constant and the zeroes at the origin):

$$|H(10)| \approx -55.39 + 20\,\text{dB} + 20\,\text{dB} = -15.39\,\text{dB.}$$

- The slope of the plot at the above point is 20 dB/decade × (number of poles at origin), i.e., 40 dB/decade.
- The pole and zero frequencies are located and marked: simple poles (down) and zeroes (up) by an arrow. Similarly, complex conjugate pairs are marked by a double arrow.
- The plot slope is incremented at the arrow frequencies by 20 dB/decade in the direction of the arrow. Multiple arrows at the same frequency indicate a multiplicative change in slope.
- Any higher-order corrections are then made (if desired).

The resultant uncorrected Bode straight-line magnitude plot, four optional correction points (marked by "+"), and the exact magnitude plot for this system are shown in Figure 6.9.

The Bode phase plot for the constant and each of the zeroes at the origin are simple horizontal lines at 0° and 90°, respectively. Each simple pole will increment the phase by −45°/decade at one decade below the pole frequency and decrement the phase by the same quantity one decade above the pole frequency. For the complex conjugate pair phase plot, the frequency range where the phase changes must be calculated by determining the quantity

$$10^\zeta = 10^{0.4} = 2.51$$

The transition frequencies for the phase change are then determined:

$$\frac{\omega_o}{2.51} < \omega < 2.51\omega_o \quad \Rightarrow \quad 120 \times 10^3 < \omega < 750 \times 10^3$$

Since the factor is in the denominator, the phase plot between the transition frequencies has a slope increment of

$$\frac{-90}{\zeta} = -225°/\text{decade}$$

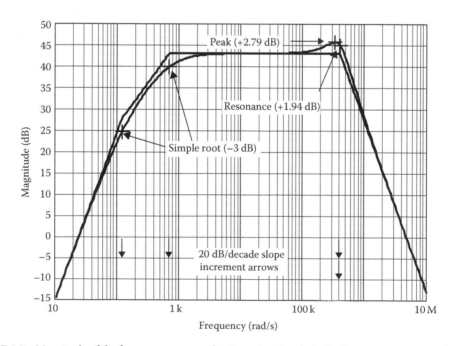

FIGURE 6.9   Magnitude of the frequency response for Example 6.1 with the Bode approximation overlay.

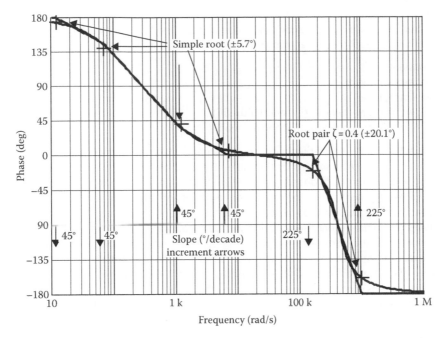

FIGURE 6.10 Phase of the frequency response for Example 6.1 with the Bode approximation overlay.

The phase plot is constructed as follows:

- A low-frequency starting point is found where the sum of all factors is known and at least one decade below any pole (or zero) frequencies (other than those at the origin); here $\omega = 0.1$ is a good choice: $\angle H(0.1) \approx 0° + 90° + 90° - 0° - 0° - 0° = 180°$.
- The phase-plot transition frequencies are located and marked: simple poles and zeroes by opposing arrow pairs at one tenth and ten times the pole or zero frequency (12 and 1200; 70 and 7000); complex conjugate root pairs by opposing arrow pairs at the calculated slope transition frequencies (120×10^3 and 750×10^3).
- The phase-plot slope is changed at the arrow frequencies by the appropriate amount in the direction of the arrow.
- Any higher-order corrections are then made (if desired).

The system uncorrected Bode straight-line approximate phase plot and the exact phase plot are shown in Figure 6.10. Six optional correction points (marked by "+") are also shown. Note that while these optional correction points improve the curve in regions where roots are far apart, in regions where the root transition regions overlap (i.e., at 70 and 1200 rad/s), their usage may not improve the overall curve.

Example 6.2: Drawing Bode Plots—Alternate Method

Draw the Bode diagram for the following transfer function:

$$H(\omega) = \frac{-625 \times 10^{-6} [j\omega]^2}{[1 + (j\omega/160)][1 + (j\omega/1000)][1 + (j\omega/200 \times 10^3)][1 + (j\omega/900 \times 10^3)]}$$

Solution

The Bode diagram consists of the Bode magnitude plot and the Bode phase plot. The given transfer function contains three of the primary factor types:

- A constant
- Two zeroes at the origin
- Four simple poles
    - Two at low frequencies (160 and 1000 rad/s)
    - Two at high frequencies (200 and 900 krad/s)

This particular form of the transfer function is quite common in electronic applications. Low-frequency poles (in this case, two) are canceled by the same number of zeroes at the origin resulting in a middle range of frequencies (the midband region) where the transfer function is essentially constant, and regions of decreasing gain as frequency varies from the midband region for both higher and lower frequencies. Similarly, the phase is relatively constant in the midband region. In such a situation, the Bode plot can be begun in the midband region and can progress outward. One should note that this method is not dependent on the poles being simple; the presence of complex conjugate pole pairs does not change the method described as long as there is a midband region of essentially constant gain or phase.

The value of the transfer function in the midband region can be determined by assuming a midband frequency, ω_{mid}, that is conceptually much larger than the largest of the low-frequency poles, but much smaller than the smallest of the high-frequency poles. Under that assumption, the pole factors take on a simpler form and result in an approximate transfer function value in the midband region:

$$H(\omega_{mid}) \approx \frac{-625 \times 10^{-6}[j\omega_{mid}]^2}{\left[(j\omega_{mid}/160)\right]\left[(j\omega_{mid}/1000)\right][1][1]} = -100$$

The Bode magnitude plot for this example in the midband region becomes a horizontal line at

$$20 \log|-100| = 40\,dB.$$

The magnitude plot is constructed as follows:

- The plot begins in the midband region with a horizontal line at the midband gain value; here, 40 dB.
- The pole and zero frequencies are located and marked: simple poles (down) and zeroes (up) by an arrow. Similarly, complex conjugate pairs are marked by a double arrow.
- The plot slope is incremented at the arrow frequencies by 20 dB/decade in the direction of the arrow. Multiple arrows at the same frequency indicate a multiplicative change in slope.
- Any higher-order corrections are then made (if desired).

The resultant uncorrected Bode straight-line magnitude plot and the exact magnitude plot for this system are shown in Figure 6.11.

The Bode phase plot can be similarly constructed from the midband outward. The phase plot is constructed as follows:

- The plot begins in the midband with a horizontal line at the midband phase value; here, $\angle H(\omega_{mid}) \approx \angle(-100) = \pm 180°$.
  Since poles introduce negative angles, $-180°$ is more commonly chosen.
- The phase-plot transition frequencies are located and marked: simple poles and zeroes by opposing arrow pairs at one tenth and ten times the pole or zero frequency; complex conjugate root pairs by opposing arrow pairs at the calculated slope transition frequencies with the appropriate slope increment noted.
- The phase plot slope is changed at the arrow frequencies by the appropriate amount in the direction of the arrow.
- Any higher-order corrections are then made (if desired).

The system uncorrected Bode straight-line approximate phase plot and the exact phase plot are shown in Figure 6.12.

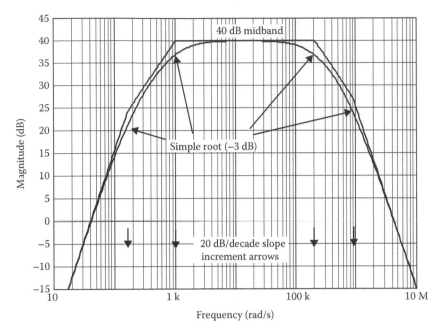

FIGURE 6.11 Magnitude of the frequency response for Example 6.2 with the Bode approximation overlay.

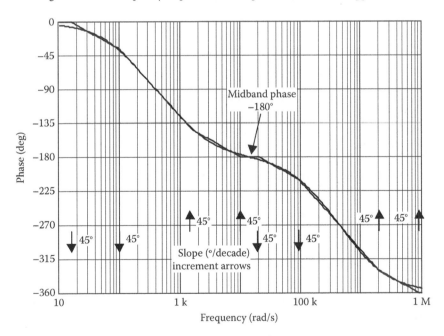

FIGURE 6.12 Phase of the frequency response for Example 6.2 with the Bode approximation overlay.

Example 6.3: Determining the System Response Using Bode Straight-Line Plots

In order to determine the system response, $H(\omega)$, the response of the system was determined through test as shown in Table 6.2. Initial tests indicated that the frequency range of interest was from 1 Hz to 10 kHz. Further tests at three samples per decade yielded the data shown. The region from 5 to 10 Hz was determined to be a region of interest and two additional data points were taken at 6 and 8 Hz.

TABLE 6.2 Experimental Data for Example 6.3

f (Hz)	$|H(f)|$	Phase (°)	$|H(f)|_{db}$
1	0.8	175	−1.9
2	3.4	170	10.6
5	28	145	28.9
10	85.5	55	38.6
20	58.5	15	35.3
50	52	5	34.3
100	51	−5	34.2
200	50	−10	34.0
500	45	−30	33.1
1k	34	−45	30.6
2k	21	−65	26.4
5k	9	−80	19.1
10k	4.5	−85	13.1
6	45	135	33.1
8	85	90	38.6

Solution

The data need to be plotted on an appropriate set of axis. The experimental data need to be converted to decibels and that column is added to Table 6.2. Working in temporal frequency will lead directly to $H(f)$ which will be converted to $H(\omega)$.

Observations $\Rightarrow$ Conclusions:

- The magnitude slope is +40 dB/decade at low frequencies
  $\Rightarrow$ There are two zeroes at the origin
- The response levels out between 50 and 200 Hz at ~50 (34 dB)
  $\Rightarrow$ There are two low-frequency poles in the system
- There is only one low-frequency change in slope that is a multiple of 20 dB/decade
  $\Rightarrow$ The system is described by a complex pole pair
- The low-frequency asymptote intersects the 34 dB line at $f \approx 8$ Hz
  $\Rightarrow f_o = 8$ Hz ($\omega_o = 2\pi f_o = 16\pi$)
  $\Rightarrow$ The system constant can be determined as (50 $\Leftrightarrow$ 34 dB)

$$50 = \lim_{f \to \infty} \left(\frac{K_f (jf)^2}{\left(1 + \frac{jf}{8}\right)^2} \right) \quad \Rightarrow \quad K_f = \frac{50}{8^2} \approx 0.781$$

$\Rightarrow$ There is a peak in the magnitude of somewhat above +4.6 dB; assume it to be 4.7 dB (a factor of 1.72):
  $\Rightarrow 2\zeta\sqrt{1-\zeta^2} = 1/1.72 \Rightarrow \zeta \approx 0.30$
- At the resonant frequency, $f = 8$ Hz, the magnitude is +4.6 dB over the nominal value of 34 dB (a factor of 1.7)
  $\Rightarrow 2\zeta = 1/1.7 \Rightarrow \zeta \approx 0.295$
- The tangent-of-phase characterization can provide another approximation for the damping coefficient by estimating the phase slope at the factor resonant frequency. The slope of the phase at the resonant frequency (Hz) is given by

$$m_{phase} \approx \frac{135 - (50)}{\log(10) - \log(6)} = 383.2$$

$$\Rightarrow \zeta \approx \frac{132}{383.2} \approx 0.34$$

- At high frequencies the magnitude slope is $-20\,\text{dB/decade}$ and the phase is $\approx -90°$
  $\Rightarrow$ There is a single high-frequency pole in the system
- The high-frequency asymptote intersects the low-frequency asymptote at $f \approx 900\,\text{Hz}$
  $\Rightarrow f_\text{o} = 900\,\text{Hz}$ ($\omega_\text{o} = 2\pi f_\text{o} = 1800\pi$)

While there is a slight variation in the three estimates for the damping coefficient, the system response can be reasonably represented as an appropriate product of the factors. If one chooses the damping coefficient to be the average of the estimates ($\zeta = 0.31$),

$$H(f) = \frac{0.781(jf)^2}{\left[1+2(0.31)\dfrac{jf}{8}+((jf)^2/(8)^2)\right]\left[1+(jf/900)\right]}$$

When converting to angular frequency (ω), the resonant frequencies are modified to be in radians per second and the system constant is altered ($m-n = 2$):

$$K_\omega = \frac{K_f}{(2\pi)^2} \approx 0.0198$$

These changes yield the final system response:

$$H(\omega) = \frac{0.0198(j\omega)^2}{\left[1+2(0.31)(j\omega/16\pi)+((j\omega)^2/(16\pi)^2)\right]\left[1+(j\omega/1800\pi)\right]}$$

Figure 6.13 is a plot of the data points, the uncorrected Bode straight-line approximate plots, and the experimentally derived $H(f)$, all on a temporal frequency (Hz) scale.

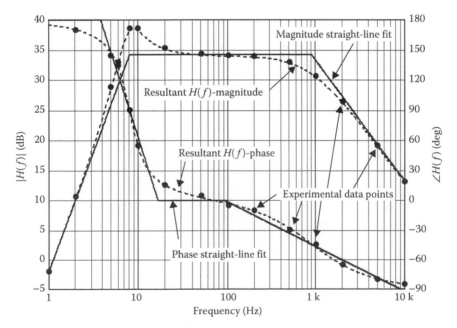

FIGURE 6.13 System Bode diagram—Example 6.3.

References

1. H.W. Bode, *Network Analysis and Feedback Amplifier Design*, Van Nostrand, Princeton, NJ, 1945.
2. T.J. Cavicchi, *Phase Margin Revisited: Phase-Root Locus, Bode Plots, and Phase Shifters, IEEE Transactions on Education*, 46(1), 168–176, February 2003.
3. G.F. Franklin, J.D. Powell, and A. Emami-Naeini, *Feedback Control of Dynamic Systems*, 3rd edn., Addison-Wesley, Reading, MA, 1994.
4. N.S. Nise, *Control System Engineering*, 4th edn., John Wiley & Sons, New York, 2003.
5. R.C. Dorf and R.H. Bishop, *Modern Control Systems*, 9th edn., Prentice Hall, Upper Saddle River, NJ, 2001.
6. J.W. Nilsson and S.A. Riedel, *Electric Circuits*, 8th edn., Prentice Hall, Upper Saddle River, NJ, 2008.
7. T.F. Schubert Jr. and E.M. Kim, *Active and Non-Linear Electronics,* John Wiley & Sons, New York, 1996.
8. T.F. Schubert Jr., A quantitative comparison of three Bode straight-line phase approximations for second-order, underdamped systems, *IEEE Transactions on Education*, 40(2), 135–138, May 1997.

7

Laplace Transforms

Dalton S. Nelson
*The University of Alabama
at Birmingham*

7.1 Introduction

In practice, many engineering and technology problems may be represented as integro-differential equations that need to be solved. Depending on the order of the equation, the solution may be simple or very complex, especially where the standard techniques of integral and differential calculus are used. An alternative is to use the *Laplace transform*. The Laplace transform allows us to carry out differentiation and integration using purely algebraic manipulations of addition and subtraction, respectively. This transformation is analogous to taking the logarithm of a function [B82]. So, where we had multiplication and division, after "taking logs" of the function, the multiplication becomes addition, and division becomes subtraction. Then, after the various algebraic manipulations in the log form, the results can be transformed back to get the final result. This logarithmic transformation allows for ease and convenience of algebraic manipulation, and the same will be true for the Laplace transform for integro-differential equations. The Laplace transform itself is named after Pierre Simon Laplace (1749–1827), a French mathematician and astronomer, who first used such transforms [KH00].

The Laplace transform is defined for positive values of time, t (most often used independent variable in electrical engineering), and denoted as the function, $f(t)$, with known initial conditions. The Laplace transform [S68] of $f(t)$ is formally defined by

$$L\{f(t)\} = F(s) = \int_0^\infty e^{-st} f(t)\,dt \tag{7.1}$$

provided that the integral exists.

Note that this integral is defined in the range, $0 \le t < \infty$, and that s may be real or complex. For most realizable systems in engineering and technology, s will be real [G88]. $L\{f(t)\}$ is the *Laplace transform* of $f(t)$, and $F(s)$, is the *Laplace transformation operator*. The Laplace transform is a mathematical tool that can be used to solve linear initial-value problems that often occur in the form of ordinary and partial differential equations [G88]. By convention, the function of time is written in lower-case letters (e.g., $f(t)$),

TABLE 7.1   Laplace Transforms Pairs (Except for Impulse Functions, $u(t)$ Is Understood)

$f(t)$	$F(s) = L\{f(t)\}$
1. $\delta(t)$, unit impulse, at $t = 0$	1
2. $\delta(t - T)$, unit impulse, at $t = T$	e^{-sT}
3. 1 or $u(t)$, unit step, *at $t = 0$*	$\dfrac{1}{s}$
4. $u_T(t)$, unit step, at $t = T$	$\dfrac{e^{-sT}}{s}$
5. t, unit ramp, at $t = 0$	$\dfrac{1}{s^2}$
6. t^n, nth order ramp, at $t = 0$	$\dfrac{n!}{s^{n+1}}$
7. e^{-at}, exponential decay	$\dfrac{1}{s+a}$
8. $1 - e^{-at}$, exponential growth	$\dfrac{a}{s(s+a)}$
9. te^{-at}	$\dfrac{1}{(s+a)^2}$
10. $t^n e^{-at}$	$\dfrac{n!}{(s+a)^{n+1}}$
11. $\dfrac{1}{(n-1)!}t^{n-1}$	$\dfrac{1}{s^n}$
12. Rectangular pulse, magnitude M, duration a	$\dfrac{M}{s}(1-e^{-as})$
13. Triangular pulse, magnitude M, duration $2a$	$\dfrac{M}{as^2}(1-e^{-as})^2$
14. Sawtooth pulse, magnitude M, duration a	$\dfrac{M}{as^2}[1-(as+1)e^{-as}]$
15. Sinusoidal pulse, magnitude M, duration $\dfrac{\pi}{a}$	$\dfrac{Ma}{s^2+a^2}[1+e^{-\pi s/a}]$
16. $\sin \omega t$, a sine wave	$\dfrac{\omega}{s^2+\omega^2}$
17. $\cos \omega t$, a cosine wave	$\dfrac{s}{s^2+\omega^2}$
18. $e^{-at}\sin \omega t$, a damped sine wave	$\dfrac{\omega}{(s+a)^2+\omega^2}$
19. $e^{-at}\cos \omega t$, a damped cosine wave	$\dfrac{s+a}{(s+a)^2+\omega^2}$
20. $1 - \cos \omega t$	$\dfrac{\omega^2}{s(s^2+\omega^2)}$
21. $t\sin \omega t$	$\dfrac{2\omega s}{(s^2+\omega^2)^2}$
22. $\sin(\omega t + \theta)$	$\dfrac{\omega\cos\theta+s\sin\theta}{s^2+\omega^2}$
23. $\cos(\omega t + \theta)$	$\dfrac{s\cos\theta+\omega\sin\theta}{s^2+\omega^2}$
24. $e^{-at}\sin(\omega t + \theta)$	$\dfrac{(s+a)\sin\theta+\omega\cos\theta}{(s+a)^2+\omega^2}$
25. $e^{-at}\cos(\omega t + \theta)$	$\dfrac{(s+a)\cos\theta-\omega\sin\theta}{(s+a)^2+\omega^2}$

while the transformed values, that are now functions of *s*, are written in upper-case letters (e.g., *F*(*s*)). The real variable, *s*, for the Laplace transform is now in the *s-domain* or *s-plane* (also called the *complex frequency* domain). While the original function of time, *t*, is said to be in the *time-domain*. Assuming that we will be using the Cartesian coordinate system, there are two components in the *s*-domain: the *y*-component and the *x*-component. These are, respectively, the imaginary, or *j*ω component; and the real part, or sigma, σ component. The transformation *maps* the time-domain function to the *s*-domain, and this process is reversible, so it is also possible to map an *s*-domain function to the corresponding time-domain description [GV91].

In Table 7.1, except for the *impulse function* item numbers, it should be understood that each overall function of *t* is multiplied by *u*(*t*).

7.2 Properties of the Laplace Transform

The fundamental mathematical properties of the Laplace transform are outlined in the following.

7.2.1 Linearity

The Laplace transform, *L*{*f*(*t*)} = *F*(*s*), is a linear operation. Therefore, if two time functions, *f*(*t*) and *g*(*t*), have Laplace transforms, then the transform of the algebraic sum of the time functions is the algebraic sum of the two separate Laplace transforms. That is,

$$L\{af(t)+bf(t)\} = aL\{f(t)\}.$$

where *a* and *b* are constants.

7.2.2 s-Domain Shifting

This property is also referred to as the *First shifting property*, and it is used to determine the Laplace transform of functions that include exponential factors. If *L*{*f*(*t*)} = *F*(*s*) then

$$L\{e^{at}f(t)\} = F(s-a)$$

Example 7.1

Find the Laplace transform, *L*{$e^{-at}t^n$}. From Table 7.1, Item #6, the Laplace transform pair is given as $t^n \leftrightarrow n!/s^{n+1}$, therefore the Laplace transform for the function given will be

$$L\{e^{-at}t^n\} = \frac{n!}{(s-a)^{n+1}}$$

7.2.3 Time-Domain Shifting

This property is also referred as the *second shift theorem*, and is used when a signal is shifted in time by a factor of *T*. The result is that the Laplace transform is multiplied by e^{-sT}. If *F*(*s*) is the Laplace transform of *f*(*t*), then

$$L\{f(t-T)u(t-T)\} = e^{-st}F(s)$$

This time-domain shifting property can be applied to all Laplace transforms, and can be particularly useful in signal-processing applications.

7.2.4 Periodic Functions

Given a function $f(t)$, which is periodic with period T, the Laplace transform of the function is

$$L\{f(t)\} = \frac{1}{1 - e^{-sT}} F_1(s)$$

where $F_1(s)$ is the Laplace transform of the function for the first period.

7.2.5 Initial and Final Value Theorems

The *Initial Value Theorem* states that *if a function of time f(t) has a Laplace transform F(s) then, in the limit, as time tends to zero, the value of that function is given by the expression*

$$\lim_{t \to 0} f(t) = \lim_{t \to \infty} sF(s)$$

The *Final Value Theorem* states that *if a function of time f(t) has a Laplace transform F(s) then in the limit as time tends toward infinity, the value of the function is given by the expression*

$$\lim_{t \to \infty} f(t) = \lim_{t \to 0} sF(s)$$

Example 7.2

For the function below, find (a) the initial value and (b) the final value of the function

$$X(s) = \frac{3}{s(s+3)}$$

Solution

   a. Initial value:

$$X(0) = \lim_{s \to \infty} sX(s) = \lim_{s \to \infty} s\left[\frac{3}{s(s+3)}\right] = \lim_{s \to \infty}\left[\frac{3}{s+3}\right] = 0$$

   b. Final value:

$$X(\infty) = \lim_{s \to 0} sX(s) = \lim_{s \to 0} s\left[\frac{3}{s(s+3)}\right] = \lim_{s \to 0}\left[\frac{3}{s+3}\right] = \frac{3}{3} = 1$$

7.2.6 Integrals (Integration)

The Laplace transform of the integral function, $f(t)$, that has a transform, $F(s)$, is given by

$$L\left\{\int_0^t f(t)\,dt\right\} = \frac{1}{s}F(s)$$

This expression suggests that the original integral function of t, is simply the Laplace transformed version of $f(t)$, multiplied by the reciprocal of s, $1/s$. The multiplicative "$1/s$" term is the equivalent of the first integral in the time domain. This property can be generalized to use higher integral functions. For example, the Laplace transform of a second integral is given by

$$L\left\{\int\int_0^t f(t)\,dt\right\} = \frac{1}{s^2}F(s)$$

And the Laplace transform of a triple integral is given by

$$L\left\{\int\int\int_0^t f(t)\,dt\right\} = \frac{1}{s^3}F(s)$$

Generally, it can be observed that the number of integrals used in the time domain description, n, is used as the power of s in the Laplace transform of the coefficient term $1/s^n$. Each $1/s$ term corresponds to a time-domain integral.

7.2.7 Derivatives (Differentiation)

The Laplace transform of the differential function time, $f(t)$, that has a transform, $F(s)$, is given by

$$L\left\{\frac{d}{dt}f(t)\right\} = sF(s) - f(0)$$

where $f(0)$ is the value of the function when $t = 0$ (often referred to as the *initial conditions*).

This expression suggests that the original differential function of t is the Laplace transform of the time function, multiplied by s. The multiplicative "s" term is the first differential equivalent in the time-domain description. This property can be generalized for higher-order derivatives. For example, for a second derivative, we obtain

$$L\left\{\frac{d^2}{dt^2}f(t)\right\} = s^2F(s) - sf(0) - \frac{d}{dt}f(0)$$

where $df(0)/dt$ is the value of the first derivative at $t = 0$.

The nth derivative is given by

$$L\left\{\frac{d^n}{dt^n}f(t)\right\} = s^nF(s) - s^{n-1}f(0) - s^{n-2}f(0) - \cdots - sf^{n-2}(0) - f^{n-1}(0)$$

where the $f^i(0)$ terms are the initial conditions of the function, with $n - 1 \leq i \leq 0$.

If the initial conditions are zero, it becomes even simpler to state in general terms that

$$L\left\{\frac{d^n}{dt^n}f(t)\right\} = s^nF(s)$$

Of course, we must be very careful that these conditions are met, before using this form of the expression.

7.3 The Inverse Transform

The inverse Laplace transform allows us to convert the s-domain function, $F(s)$, back to the corresponding time-domain function, $f(t)$. This operation can be expressed in algebraic terms as

$$L^{-1}\{F(s)\} = f(t)$$

Most often, the inverse can be found by simply using a table of Laplace transform pairs, as shown in Table 7.1. However, there is no standardized table of the pairs, and so you must either get a "good table," or work out the value from first principles. Because of the linearity property of Laplace transforms, it is possible to work out the results for several parts of a sum of values, invert them, and sum the partial results. Hence, we have that

$$L^{-1}\{aF(s) + bG(s)\} = aL^{-1}F(s) + bL^{-1}G(s)$$

where a and b are constants.

7.3.1 Partial Fractions

The *partial fractions* technique is a means of separating out the various components of a Laplace transform expression, and transforming each term back to the time domain. This technique exploits the fact that the Laplace transform is a linear operation, and as such, the sum of the individual terms will be the overall result. Whether these results are obtained from a table of Laplace transform pairs, as in Table 7.1, or worked out from first principles, this is nearly always required, and familiarity with the partial fraction technique is therefore essential.

If a given function, $F(s)$, is a ratio of two polynomials, and is easily identifiable from a given table of Laplace transform pairs, say Table 7.1, then algebraic manipulations can be made to "make" the terms "look" like those in the table, with the appropriate fractional forms. This is the process of partial fractions. A major constraint on the technique is that the numerator term should not exceed the denominator term, i.e., if n is the numerator term, and m the denominator term, then $n < m$. The degree of the polynomial is, by definition, the highest power of s in the overall expression.

Five general forms of the partial fraction can be readily identified, as outlined below:

1. Those that contain non-repeated linear factors, such as a denominator of the form $(s + a)$, $(s + b)$, $(s + c)$,...

2. Those that contain repeated linear factors of form $(s - a)^r$ in the denominator that corresponds to the partial fraction

$$\frac{A_1}{(s-a)} + \frac{A_2}{(s-a)^2} + \frac{A_3}{(s-a)^3} + \cdots + \frac{A_r}{(s-a)^r}$$

3. Those that include a denominator that contains quadratic factors, but which also, when factorized, contains imaginary terms. For example, if $a_1 = \sigma + j\omega$ is complex, then $\omega \neq 0$, and there will be imaginary terms that have complex conjugates. So, there will be $a_1 = \sigma + j\omega$ and the complex conjugate $\bar{a}_1 = \sigma + j\omega$. If we take this complex conjugate to be the equivalent of the coefficient term A_2 and the pole location, a_2, as $\bar{a}_1$, then the partial fraction expansion can proceed as follows:

$$\frac{A_1}{(s-a_1)} + \frac{\bar{A}_1}{(s-\bar{a}_1)} + \frac{A_3}{(s-a_3)} + \cdots + \frac{A_n}{(s-a_n)}$$

4. Those contain a non-repeated quadratic factor $(s^2 + as + b)$ in the denominator, which corresponds to the form

$$\frac{As + B}{s^2 + as + b}$$

5. Those that include a repeated quadratic factor $(s^2 + as + b)^r$ in the denominator, that correspond to partial fractions of the form

$$\frac{A_1 s + B_1}{(s^2 + as + b)} + \frac{A_2 s + B_2}{(s^2 + as + b)^2} + \frac{A_3 s + B_3}{(s^2 + as + b)^3} + \cdots + \frac{A_r s + B_r}{(s^2 + as + b)^r}$$

7.4 Miscellaneous Examples

Several examples are shown below that will illustrate the use of Table 7.1, and some of the properties of the Laplace transform.

Example 7.3

Find the Laplace transform of

$$e^{-10t} \cos(3t-1)u(t)$$

Solution

Recall that $\cos(\omega t + \theta) = \cos(\omega t) \cdot \cos(\theta) - \sin(\omega t) \cdot \sin(\theta)$, so that the expression becomes

$$e^{-10t} \cos(3t - t)u(t) = \left[e^{-10t} \left(\cos(3t) \cdot \cos(1) + \sin(3t) \cdot \sin(1) \right) \right] u(t)$$

$$= \left[e^{-10t} \left(\cos(3t) \cdot \cos(1) \right) + e^{-10t} \left(\sin(3t) \cdot \sin(1) \right) \right] u(t)$$

Using Items #18 and #19 from Table 7.1, we obtain

$$= \left[\frac{(s+10)}{(s+10)^2 + (3)^2} \right] \cdot \cos(1) + \left[\frac{3}{(s+10)^2 + (3)^2} \right] \cdot \sin(1)$$

$$= \left[\frac{\cos(1)(s+10)}{(s+10)^2 + 9} \right] + \left[\frac{3\sin(1)}{(s+10)^2 + 9} \right]$$

$$\therefore \quad L\left\{ e^{-10t} \cos(3t - 1)u(t) \right\} = \left[\frac{\cos(1)(s+10) + 3\sin(1)}{(s+10)^2 + 9} \right]$$

Example 7.4

Find the inverse transform of the following Laplace expression, using partial fraction expansion.

$$F(s) = \frac{2(s^2 + 2s + 6)}{(s+1)(s+2)(s+3)}$$

Solution

Expanding the numerator, the result is

$$F(s) = \frac{2s^2 + 4s + 12}{(s+1)(s+2)(s+3)}$$

This function has non-repeating linear terms, and so the expansion appears as follows:

$$F(s) = \frac{A}{s+1} + \frac{B}{s+2} + \frac{C}{s+3}$$

Substituting to find the respective numerator terms

$$A: \quad \frac{2s^2 + 4s + 12}{(s+2)(s+3)}\bigg|_{s=-1} = 5$$

$$B: \quad \frac{2s^2 + 4s + 12}{(s+1)(s+3)}\bigg|_{s=-2} = -12$$

$$C: \quad \frac{2s^2 + 4s + 12}{(s+1)(s+2)}\bigg|_{s=-3} = 9$$

$$F(s) = \frac{5}{s+1} - \frac{12}{s+2} + \frac{9}{s+3}$$

Using Item #7 from Table 7.1, along with each coefficient, the final result is obtained as

$$\therefore \quad L^{-1}\{F(s)\} = f(t) = 5e^{-t} - 12e^{-2t} + 9e^{-3t}, \quad \text{for } t \geq 0$$

Example 7.5

Find the *transfer function* (i.e., the ratio of $Y(s)/U(s)$, and by definition, the initial conditions are always 0) of the differential equation below.

$$\ddot{y} + 5\dot{y} + 6y = 2\dot{u} + 3u$$

Solution

Writing in differential notation, we obtain

$$\frac{d^2 y(t)}{dt^2} + 5\frac{dy(t)}{dt} + 6y(t) = 2\frac{du(t)}{dt} + 3u(t)$$

Using the Laplace transform for each term (see property in Section 7.2.7), the result is

$$s^2 Y(s) + 5sY(s) + 6Y(s) = 2sU(s) + 3U(s)$$

Factorizing

$$Y(s)[s^2 + 5s + 6] = U(s)[2s + 3]$$

$$\frac{Y(s)}{U(s)} = \frac{2s + 3}{s^2 + 5s + 6}$$

$$\therefore \quad \frac{Y(s)}{U(s)} = \frac{2(s + (3/2))}{(s + 3)(s + 2)}$$

Example 7.6

Solve the following differential equation using Laplace transforms:

$$\frac{dy(t)}{dt} - 3y(t) = u(t), \quad \text{given that, } y(0) = 1$$

Solution
Using the Laplace transform for each term (see property in Section 7.2.7), and Item #3 from Table 7.1, the result is

$$[sY(s) - y(0)] - 3Y(s) = \frac{1}{s}$$

$$sY(s) - 3Y(s) = \frac{1}{s} + y(0)$$

Substituting the initial value, and factoring, we obtain

$$(s - 3)Y(s) = \frac{1}{s} + 1$$

$$Y(s) = \frac{1}{s(s - 3)} + \frac{1}{(s - 3)}$$

By partial fraction expansion

$$Y(s) = \frac{1/3}{(s - 3)} - \frac{1/3}{s} + \frac{1}{(s - 3)}$$

Taking the inverse transform from Table 7.1, Item #7, the solution is found as

$$y(t) = L^{-1}[Y(s)] = \left[\frac{1}{3}e^{-3t} - \frac{1}{3} + e^{-3t}\right]u(t)$$

$$\therefore \quad y(t) = \left[\frac{1}{3}(4e^{-3t} - 1)\right]u(t), \quad \text{for } t \geq 0$$

Example 7.7

Solve the following differential equation using Laplace transforms:

$$\frac{dy(t)}{dt} + 10y(t) = 8e^{-10t}u(t), \quad \text{given that, } y(0) = 0$$

Solution

Using Item #7 from Table 7.1, and Property in Section 7.2.7, the following expression is obtained:

$$[sY(s) - y(0)] + 10Y(s) = \frac{8}{s+10}$$

Factorizing

$$(s+10)Y(s) = \frac{8}{s+10}$$

And solving for Y(s)

$$Y(s) = \frac{8}{((s+10)(s+10))} = \frac{8}{(s+10)^2}$$

Finding the inverse transform, using Item #10 from Table 7.1, we obtain the following result:

$$\therefore \quad y(t) = L^{-1}[Y(s)] = 8te^{-10t}u(t), \quad \text{for } t \geq 0$$

Example 7.8

Solve the following differential equation using Laplace transforms:

$$\frac{dy(t)}{dt} + 10 = 4\sin(2t)u(t), \quad \text{given that, } y(0) = 1$$

Solution

Using Items #16 from Table 7.1, and property in Section 7.2.7 the following expression is obtained:

$$[sY(s) - y(0)] + 10Y(s) = \frac{8}{(s^2+4)}$$

Since

$$(s+10)Y(s) = \frac{8}{(s^2+4)} + y(0)$$

Substituting for y(0), the result is

$$(s+10)Y(s) = \frac{8}{(s^2+4)} + 1$$

Solving for $Y(s)$

$$Y(s) = \frac{8}{(s+10)(s^2+4)} + \frac{1}{(s+10)}$$

By partial fraction expansion, the results are

$$Y(s) = \frac{14/13}{(s+10)} - \frac{1}{13} \cdot \frac{s}{(s^2+4)} + \frac{10}{26} \frac{2}{(s^2+4)}$$

Finding the inverse transform, using Items #16 and #17 from Table 7.1, the solution is then complete.

$$\therefore \ L^{-1}[Y(s)] = y(t) = \left[\frac{14}{13} e^{-10t} - \frac{1}{13}\cos(2t) + \frac{10}{26}\sin(2t) \right] u(t), \quad \text{for } t \geq 0$$

Example 7.9

Solve the following differential equation using Laplace transforms:

$$\frac{d^2 y(t)}{dt^2} + 6\frac{dy(t)}{dt} + 8y = u(t), \quad \text{given that, } y(0)=0; \ \dot{y}(t)=1$$

Solution

Using Items #16 from Table 7.1, and property in Section 7.2.7, the following expression is obtained:

$$[s^2 Y(s) - sy(0) - \dot{y}(0)] + 6[sY(s) - y(0)] + 8Y(s) = \frac{1}{s}$$

Substituting values of the initial conditions

$$[s^2 Y(s) - 0s - 1] + 6[sY(s) - 0] + 8Y(s) = \frac{1}{s}$$

$$s^2 Y(s) - 1 + 6sY(s) + 8Y(s) = \frac{1}{s}$$

$$[s^2 + 6s + 8]Y(s) = \frac{1}{s} + 1$$

Factorizing and solving for $Y(s)$

$$Y(s) = \frac{1}{(s+4)(s+2)s} + \frac{1}{(s+4)(s+2)}$$

By partial fractions

$$Y(s) = \frac{-(3/8)}{(s+4)} + \frac{1/4}{(s+2)} + \frac{1/8}{s}$$

Finding the inverse transform using Item #7 from Table 7.1, the following solution is obtained:

$$\therefore \; L^{-1}[Y(s)] = y(t) = \left[-\frac{3}{8}e^{-4t} + \frac{1}{4}e^{-2t} + \frac{1}{8} \right] u(t), \quad \text{for } t \geq 0$$

References

[B82] J. A. Bogart, *Laplace Transforms and Control Systems Theory for Technology*, John Wiley & Sons, New York, 1982.

[G88] M. D. Greenberg, *Advanced Engineering Mathematics*, Prentice-Hall, Upper Saddle River, NJ, 1988.

[GV91] J. Golten and A. Verwer, *Control System Design and Simulation*, McGraw-Hill International (UK) Ltd., London, U.K., 1991.

[KH00] E. W. Kamen and B. S. Heck, *Fundamentals of Signals and Systems Using the Web and MATLAB* (2nd edition), Prentice-Hall, Upper Saddle River, NJ, 2000.

[S68] M. R. Spiegel, *Schaum's Outline Series—Mathematical Handbook of Formulas and Tables*, McGraw-Hill, New York, 1968.

II

Devices

8

Semiconductor Diode

Bogdan M.
Wilamowski
Auburn University

The *pn* semiconductor junctions exhibit nonlinear current–voltage characteristics, and they are used to rectify and shape electrical signals ([1],[3],[4],[5]). Exponential current–voltage characteristics are sometimes used to build logarithmic amplifiers. The thickness of the depletion layer depends on applied reverse voltage, and the voltage-dependent capacitance can be used to tune frequency characteristics of electronic circuits [2]. There are over 20 different types of diodes using different properties of *pn* junctions or metal–semiconductor junction properties [7]. Some of these special diodes are described in Section 8.6.

8.1 Nonlinear Static *I–V* Characteristics

Typical *I–V* diode characteristics are shown in Figure 8.1. In the case of common silicon diode, the forward direction current increases exponentially at first, and then it is limited by an ohmic resistance of the structure. A very small current in the reverse direction at first increases slightly with applied voltage and then starts to multiply near the breakdown voltage. The current at the breakdown is limited by the ohmic resistances of the structure. In germanium (small energy gap) diodes, the recombination–generation component of the current is much smaller than the diffusion components, and in a wide range of reverse voltages the current is almost constant. In the case of silicon diodes (larger energy gap), the diffusion component of the reverse current is negligibly small, and the reverse current is caused by the recombination–generation phenomena, and the current is proportional to the size of the depletion layer (which increases slightly with the voltage). The diode equation (8.6) is not valid for silicon diodes in the reverse direction. Typical reverse characteristics of germanium and silicon diodes are shown in Figure 8.2.

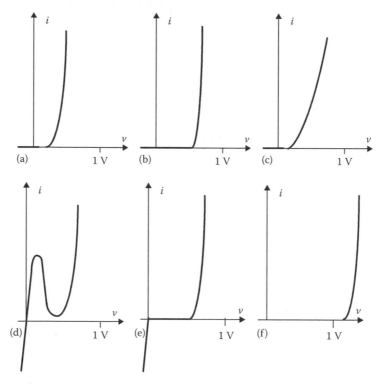

FIGURE 8.1 Forward current–voltage characteristics of various types of diodes: (a) germanium diode, (b) silicon diode, (c) Schottky diode, (d) tunnel diode, (e) backward diode, and (f) LED.

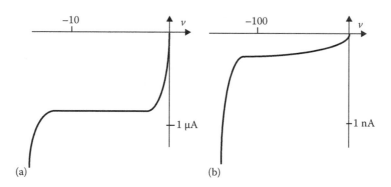

FIGURE 8.2 Reverse current–voltage characteristics: (a) germanium diode, (b) silicon diode.

8.1.1 *pn* Junction Equation

The n-type semiconductor material has a positive impurity charge attached to the crystal lattice structure. This fixed positive charge is compensated by free moving electrons with negative charges. Similarly, the p-type semiconductor material has a lattice with a negative charge, which is compensated by free moving holes, as is shown in Figure 8.3. The number of majority carriers (electrons in p-type and holes in n-type materials) are approximately equal to the donor or acceptor impurity concentrations, i.e., $n_n = N_D$ and $p_p = N_A$. The number of minority carriers (electrons in p-type and holes in n-type) can be found using the equations

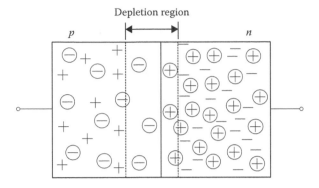

FIGURE 8.3 Illustration of the *pn* junction.

$$n_p = \frac{n_i^2}{p_p} \approx \frac{n_i^2}{N_A} \quad p_n = \frac{n_i^2}{n_n} \approx \frac{n_i^2}{N_D} \tag{8.1}$$

The intrinsic carrier concentration, n_i, is given by

$$n_i^2 = \xi T^3 \exp\left(-\frac{V_g}{V_T}\right); \quad V_T = \frac{kT}{q} \tag{8.2}$$

where
    $V_T = kT/q$ is the thermal potential ($V_T = 25.9\,\text{mV}$ at $300\,\text{K}$)
    T is the absolute temperature in K
    $q = 1.6 \times 10^{-16}\,\text{C}$ is the electron charge
    $k = 8.62 \times 10^{-5}\,\text{eV/K}$ is the Boltzmann's constant
    V_g is the potential gap ($V_g = 1.1\,\text{V}$ for silicon)
    ξ is a material constant

For silicon the intrinsic concentration n_i is given by

$$n_i = 3.88 \times 10^{16} T^{3/2} \exp\left(-\frac{7000}{T}\right) \tag{8.3}$$

For silicon at $300\,\text{K}$, $n_i = 1.5 \times 10^{10}\,\text{cm}^{-2}$.

When a *pn* junction is formed, the fixed electrostatic lattice charges form an electrical field at the junction. Electrons are pushed by electrostatic forces deeper into the *n*-type region and holes into the *p*-type region, as illustrated in Figure 8.4. Between *n*-type and *p*-type regions there is a depletion layer with a built-in potential that is a function of impurity doping level and intrinsic concentration n_i:

$$V_{pn} = V_T \ln\left(\frac{N_A N_D}{n_i^2}\right) = V_T \ln\left(\frac{n_n p_p}{n_i^2}\right) = V_T \ln\left(\frac{n_n}{n_p}\right) = V_T \ln\left(\frac{p_p}{p_n}\right) \tag{8.4}$$

The junction current as a function of biasing voltage is described by the diode equation

$$i = I_s\left[\exp\left(\frac{v}{V_T}\right) - 1\right] \tag{8.5}$$

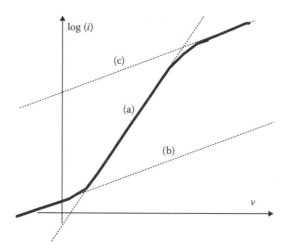

FIGURE 8.4 Current–voltage characteristics of the *pn* junction in forward direction: (a) diffusion current, (b) recombination current, and (c) high-level injection current.

where

$$I_s = Aqn_i^2 V_T \left(\frac{\mu_p}{\int_0^{L_p} n_n dx} + \frac{\mu_n}{\int_0^{L_n} p_p dx} \right)$$ (8.6)

where

$n_n \approx N_D$

$p_p \approx N_A$

μ_n and μ_p are the mobility of electrons and holes

L_n and L_p are the diffusion length for electron and holes

A is the device area

In the case of diodes made of silicon or other semiconductor material with a high energy gap, the reverse-biasing current cannot be calculated from the diode equation (8.5). This is due to the carrier generation–recombination phenomenon. Lattice imperfection and most impurities are acting as generation–recombination centers. Therefore, the more imperfections there are in the structure, the larger the deviation from ideal characteristics.

8.1.2 Forward *I–V* Diode Characteristics

The diode equation (8.5) was derived with an assumption that injected carriers are recombining on the other side of the junction. The recombination within the depletion layer was neglected. In forward-biased diode, electrons and holes are injected through the depletion region, and they may recombine there. The recombination component of the forward-biased diode is given by

$$i_{rec} = qwA \frac{n_i}{2\tau_0} \exp\left(\frac{v}{2V_T} \right) = I_{ro} \exp\left(\frac{v}{2V_T} \right)$$ (8.7)

where

w is the depletion layer thickness

τ_0 is the carrier lifetime in depletion region

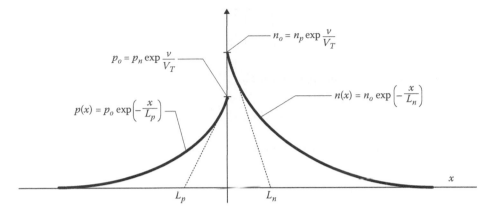

FIGURE 8.5 Minority carrier distribution in the vicinity of the *pn* junction biased in forward direction.

The total diode current $i_T = i_D + i_{rec}$, where i_D and i_{rec} are defined by Equations 8.5 and 8.7. The recombination component dominates at low current levels, as Figure 8.5 illustrates.

Also, in very high current levels the diode equation (8.5) is not valid. Two phenomena cause this deviation. First, there is always an ohmic resistance that plays an important role for large current values. Second, deviation is due to high concentration of injected minority carriers. For very high current levels, the injected minority carrier concentrations may approach, or even become larger than, the impurity concentration. An assumption of the quasi-charge neutrality leads to an increase of the majority carrier concentration. Therefore, the effective diode current is lower, as can be seen from Equation 8.6. The high current level in the diode follows the relation

$$i_h = I_{ho} \exp\left(\frac{v}{2V_T}\right) \tag{8.8}$$

Figure 8.4 shows the diode *I–V* characteristics, which include generation–recombination, diffusion, and high current phenomena. For modeling purposes, the forward diode current can be approximated by

$$i_D = I_o \exp\left(\frac{v}{\eta V_T}\right) \tag{8.9}$$

where η has a value between 1.0 and 2.0. Note that the η coefficient is a function of current, as shown in Figure 8.5. It has a larger value for small and large current regions and it is close to unity in the medium current region.

8.1.3 Reverse *I–V* Characteristics

The reverse leakage current in silicon diodes is mainly caused by the electron-hole generation in the depletion layer. This current is proportional to the number of generation–recombination centers. These centers are formed either by a crystal imperfection or deep impurities, which create energy states near the center of the energy gap. Once the reverse voltage is applied, the size of the depletion region and the number of generation–recombination centers increase. Thus, the leakage current is proportional to the thickness of the depletion layer $w(v)$. For a step-abrupt junction

$$w = \sqrt{\frac{2\varepsilon\varepsilon_o \left(V_{pn} - v\right)}{qN_{eff}}} \qquad (8.10)$$

For other impurity profiles, w can be approximated by

$$w = K\left(V_{pn} - v\right)^{1/m} \qquad (8.11)$$

The reverse diode current for small and medium voltages can therefore be approximated by

$$i_{rev} = Aw(v)\frac{qn_i}{2\tau_o} \qquad (8.12)$$

where n_i is given by Equation 8.2 and w by Equation 8.10 or 8.11. The reverse current increases rapidly near the breakdown voltage. This is due to the avalanche multiplication phenomenon. The multiplication factor is often approximated by

$$M = \frac{1}{1 - \left(v/BV\right)^m} \qquad (8.13)$$

where
  BV stands for the breakdown voltage
  m is an exponent chosen experimentally

Note that for the reverse biasing, both v and BV have negative values and the multiplication factor M reaches an infinite value for $v = BV$.

8.2 Diode Capacitances

Two types of capacitances are associated with a diode junction. The first capacitance, known as diffusion capacitance, is proportional to the diode current. This capacitance exists only for the forward-biased condition and has the dominant effect there. The second capacitance, known as the depletion capacitance, is a weak function of the applied voltage.

8.2.1 Diffusion Capacitance

In a forward-biased diode, minority carriers are injected into opposite sides of the junction. Those minority carriers diffuse from the junction and recombine with the majority carriers. Figure 8.6 shows the distribution of minority carriers in the vicinity of the junction of uniformly doped n-type and p-type regions. The electron charge stored in the p-region corresponds to the area under the curve, and it is equal to $Q_n = qn_oL_n$. Similarly, the charge of stored holes $Q_p = qp_oL_p$. The storage charge can be also expressed as $Q_n = I_n\tau_n$ and $Q_p = I_p\tau_p$, where I_n and I_p are electron and hole currents at the junction, τ_n and τ_p are the lifetimes for minority carriers. Assuming $\tau = \tau_n = \tau_p$ and knowing that $I = I_p + I_n$ the total storage charge at the junction is $Q = I\tau$. The diffusion capacitance can then be computed as

$$C_{dif} = \frac{dQ}{dv} = \frac{d}{dv}\left[\tau I_o \exp\left(\frac{v}{\eta V_T}\right)\right] = \frac{\tau I_B}{\eta V_T} \qquad (8.14)$$

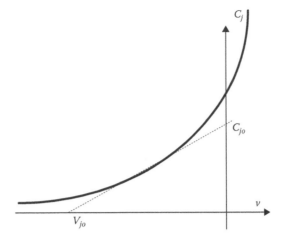

FIGURE 8.6 Capacitance–voltage characteristics for reverse-biased junction.

As one can see, the diffusion capacitance C_{dif} is proportional to the storage time τ and to the diode biasing current I_B. Note that the diffusion capacitance does not depend on the junction area, but it only depends on the diode current. The diffusion capacitances may have very large values. For example, for 100 mA current and $\tau = 1\,\mu s$, the junction diffusion capacitance is about $4\,\mu F$. Fortunately, this diffusion capacitance is connected in parallel to the small-signal junction resistance $r = \eta V_T/I_B$, and the time constant rC_{dif} is equal to the storage time τ.

8.2.2 Depletion Capacitance

The reversed-biased diode looks like a capacitor with two "plates" formed of p-type and n-type regions and the dielectric layer (depletion region) between them. The capacitance of a reversed-biased junction can then be written as

$$C_{dep} = A\frac{\varepsilon}{w}$$

(8.15)

where
  A is a junction area
  ε is the dielectric permittivity of semiconductor material
  w is the thickness of the depletion layer

The depletion layer thickness w is a weak function of the applied reverse-biasing voltage. In the simplest case, with step-abrupt junction, the depletion capacitance is

$$C_j = \sqrt{\frac{qN_{eff}\varepsilon\varepsilon_o}{2(V_{pn}-v)}}; \quad \frac{1}{N_{eff}} = \frac{1}{N_D} + \frac{1}{N_A}$$

(8.16)

The steepest capacitance–voltage characteristics are in pn^+n diodes with the impurity profiles shown in Figure 8.1f. In general, for various impurity profiles at the junction, the depletion capacitance C_j can be approximated by

$$C_j = C_{jo}(V_{pn}-v)^{1/m}$$

(8.17)

or using linear approximation, as shown in Figure 8.6

$$C_j = C_{jo}\left(1 - \frac{v}{V_{jo}}\right) \qquad (8.18)$$

8.3 Diode as a Switch

The switching time of the *pn* junction is limited mainly by the storage charge of injected minority carriers into the vicinity of the junction (electrons injected in *p*-type region and holes injected in *n*-type region). When a diode is switched from forward to reverse direction, these carriers may move freely through the junction. Some of the minority carriers recombine with time. Others are moved away to the other side of the junction. The diode cannot recover its blocking capability as long as a large number of the minority carriers exist and can flow through the junction. An example of the current–time characteristics of a diode switching from forward to reverse direction is shown in Figure 8.7. Few characteristics that are shown in the figure are for the same forward current and different reverse currents. Just after switching, these reverse currents are limited only by external circuitry. In this example, shown in Figure 8.7, most of the minority carriers are moved to the other side of the junction by the reverse current and the recombination mechanism is negligible. Note that the larger the reverse current flows after switching, the shorter time is required to recover the blocking capability. This type of behavior is typical for commonly used high-voltage diodes.

In order to shorten the switching time, diodes sometimes are doped with gold or other deep-level impurities to create more generation centers and to increase the carrier recombination. This way, the minority carrier lifetimes of such switching diodes are significantly reduced. The switching time is significantly shorter, but it is almost independent of the reverse diode current after switching, as Figure 8.8 shows.

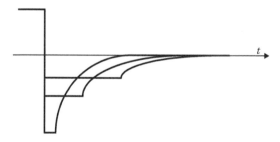

FIGURE 8.7   Currents in diode with large minority carrier lifetimes after switching from forward to reverse direction.

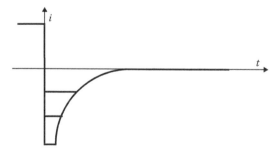

FIGURE 8.8   Currents in diode with small minority carrier lifetimes after switching from forward to reverse direction.

This method of artificially increasing recombination rates has some severe disadvantages. Such switching diodes are characterized by very large reverse leakage current and small breakdown voltages.

The best switching diodes utilize metal–semiconductor contacts. They are known as the Schottky diodes. In such diodes there is no minority carrier injection phenomenon, therefore, these diodes recover the blocking capability instantaneously. The Schottky diodes are also characterized by a relatively small (0.2–0.3 V) voltage drop in the forward direction. However, their reverse leakage current is larger, and the breakdown voltage rarely exceeds 20–30 V. Lowering the impurity concentration in the semiconductor material leads to slightly larger breakdown voltages, but at the same time, the series diode resistances increase significantly.

8.4 Temperature Properties

Both forward and reverse diode characteristics are temperature dependent. These temperature properties are very important for correct circuit design. The temperature properties of the diode can be used to compensate for the thermal effects of electronic circuits. Diodes can be used also as accurate temperature sensors. The major temperature effect in a diode is caused by the strong temperature dependence of the intrinsic concentration n_i (Equations 8.2 and 8.3) and by the exponential temperature relationship of the diode equation (8.7). By combining Equations 8.2 and 8.7 and assuming the temperature dependence of carrier mobilities, the voltage drop on the forward-biased diode can be written as

$$v = \eta \left[V_T \ln \left(\frac{i}{\xi T^\alpha} \right) + V_g \right]$$ (8.19)

or diode current

$$i = I_o \left(\frac{T}{T_o} \right)^\alpha \exp \left(\frac{T_o}{T} \frac{(v/\eta) - V_g}{V_{To}} \right)$$ (8.20)

where
   V_g is the potential gap in semiconductor material
   $V_g = 1.1$ V for silicon and $V_g = 1.4$ V for GaAs
   α is a material coefficient ranging between 2.5 and 4.0

The temperature dependence of the diode voltage drop dv/dT can be obtained by calculating the derivative of Equation 8.19

$$\frac{dv}{dT} = \frac{v - \eta(V_g + \alpha V_T)}{T}$$ (8.21)

For example, in the case of the silicon diode with a 0.6 V drop and assuming $\eta = 1.1$, $\alpha = 3.0$, and $T = 300$ K, the $dV/dT = 1.87$ mV/°C.

The reverse diode current is a very strong function of the temperature. For diodes made of the semiconductor materials with a small potential gap, such as germanium, the diffusion component dominates. In this case, the reverse current is proportional to

$$i_{rev} \propto T^\alpha \exp \left(-\frac{q V_g}{kT} \right)$$ (8.22)

For diodes made of silicon and semiconductors with a higher energy gap, the recombination is the dominant mechanism. In this case, reverse leakage current is proportional to

$$i_{rev} \propto T^{\alpha/2} \exp\left(-\frac{qV_g}{2kT}\right)$$

(8.23)

Using Equation 8.23, one may calculate that for silicon diodes at room temperatures, the reverse leakage current doubles for about every 10°C.

The breakdown voltage is also temperature dependent. The tunneling effect dominates in diodes with small breakdown voltages. This effect is often known in literature as the Zener breakdown. In such diodes, the breakdown voltage decreases with the temperature. The avalanche breakdown dominates in diodes with large breakdown voltages. When the avalanche mechanism prevails then the breakdown voltage increases 0.06%–0.1% per °C. For medium range breakdown voltages, one phenomenon compensates the other, and the temperature-independent breakdown voltage can be observed. This zero temperature coefficient exists for diodes with breakdown voltages equal to about $5V_g$. In the case of the silicon diode, this breakdown voltage, with a zero temperature coefficient, is equal to about 5.6 V.

8.5 Piecewise Linear Model

The nonlinear diode characteristics are often approximated by the piecewise linear model. There are a few possible approaches to linearize the diode characteristics, as shown in Figure 8.9. The parameters of the most accurate linearized diode model are shown in Figure 8.10a, and the linearized diode equivalent circuit is shown in Figure 8.10b

The modified diode equation (8.9) also can be written as

$$v = \eta V_T \ln\left(\frac{i}{I_o}\right)$$

(8.24)

For the biasing point V_B and I_B, the small-signal diode resistance dv/di can be computed from Equation 8.24 as

$$r = \frac{dv}{di} = \frac{\eta V_T}{I_B}; \quad V_{tho} = V_B - V_T$$

(8.25)

and it is only the function of the thermal potential V_T and the biasing current I_B. Note that the small-signal diode resistance is almost independent on the diode construction or semiconductor material used. If one requires that this linearized diode have the I_B current for the V_B voltage, then the piecewise diode characteristics should be as in Figure 8.10. The equivalent Thevenin and Norton circuits are

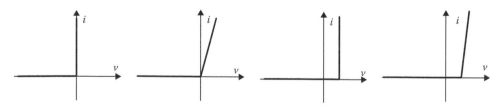

FIGURE 8.9 Various ways of linearizing diode characteristics.

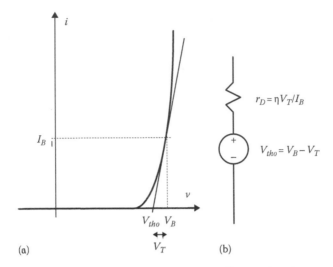

FIGURE 8.10 Linearization of the diode: (a) diode characteristics, (b) equivalent diagram.

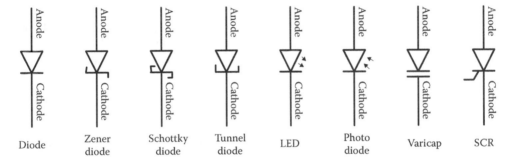

| Diode | Zener diode | Schottky diode | Tunnel diode | LED | Photo diode | Varicap | SCR |

FIGURE 8.11 Commonly used symbols for various diodes.

shown in Figure 8.11. In a case of large signal operation, the diode can be approximated by shifting the characteristics to the left by ΔV. In this case, the threshold voltage becomes $V_{tho} = V_B - 2V_T$ instead of $V_{tho} = V_B - V_T$.

8.6 Different Types of Diodes

Using different phenomena in semiconductors, it is possible to develop many different types of diodes with specific characteristics. Different diodes have different symbols, as shown in Figure 8.11. Various types of diodes are briefly described in this section.

8.6.1 Switching Diodes

Switching diodes are usually small-power *pn* junction diodes that are designed for fast switching. In order to reduce the storage time (and diffusion capacitances), the lifetime of electron and holes were purposely reduced by introducing deep-level impurities such as gold or platinum.

8.6.2 Zener Diodes

Zener diodes use the reverse-breakdown voltage to stabilize voltages in electronic circuits. The breakdown voltage of *pn* junction decreases with an increase of the impurity level. When junction is heavily doped, the breakdown voltage is controlled by the tunneling mechanism, and it decreases with temperature. With lightly doped junction and high breakdown voltages, the avalanche breakdown is the dominant mechanism and the breakdown voltage increases with temperature. In other words, Zener diodes for small voltages have negative temperature coefficient, and Zener diodes for large voltages have positive temperature coefficient. It is worth noticing that for voltages equal about 5 energy gaps (about 5.6 V for silicon) Zener diodes have close to zero temperature coefficients. Such temperature-compensated Zener diodes are known as reference diodes.

8.6.3 Tunnel Diodes (Esaki Diodes)

When both sides of the junction are very heavily doped, then for small forward-biasing voltages (0.1–0.3 V) a large tunneling current may occur. For larger forward voltages (0.4–0.5 V), this tunneling current vanishes. This way, the current–voltage characteristic has a negative resistance region somewhere between 0.2 and 0.4 V (Figure 8.1d). The germanium and other than silicon semiconductors are used to fabricate tunnel diodes.

8.6.4 Backward Diodes

The backward diode has slightly lower impurity concentrations than the tunnel diode, and the tunneling current in forward direction does not occur (Figure 8.1e). The backward diode is characterized by a very sharp knee near 0 V, and it is used for detection (rectifications) of signals with very small magnitudes.

8.6.5 PIN Diodes

Diodes with high breakdown voltage have the *pin* structure with an impurity profile shown in Figure 8.12d. A similar *pin* structure is also used in microwave circuits as a switch or as an attenuating resistor. For reverse biasing, such microwave *pin* diode represents an open circuit with a small parasitic junction capacitance. In the forward direction, this diode operates as a resistor whose conductance is proportional to the biasing current. At very high frequencies, electrons and holes will oscillate rather than flow. Therefore, the microwave *pin* diode exhibits linear characteristics even for large modulating voltages.

8.6.6 Schottky Diodes

The switching time of a *pn* junction from forward to reverse direction is limited by the storage time of minority carriers injected into the vicinity of the junction. Much faster operation is possible in the Schottky diode, where minority carrier injection does not exist. Another advantage of the Schottky diode is that the forward voltage drop is smaller (0.2–0.3 V) than in the silicon *pn* junction. This diode uses the metal–semiconductor contact for its operation. Schottky diodes are characterized by relatively small reverse-breakdown voltage, rarely exceeding 50 V.

8.6.7 Super Barrier Diodes

The major drawback of Schottky diodes is their low breakdown voltage. By combining *pn* junctions with Schottky contact, it is possible to make super barrier diodes where voltage drop in forward direction is determined by Schottky contact, but for reverse direction, specially profiled *pn* junction [6] lower electrical field Schottky contact, resulting in much larger breakdown voltages.

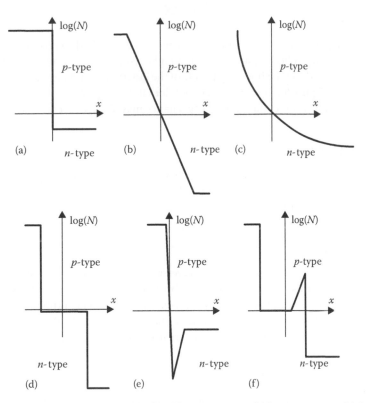

FIGURE 8.12 Impurity profiles for various diodes: (a) step junction, (b) linear junction, (c) diffusion junction, (d) *pin* junction, (e) hyperabrupt junction, and (f) *pipn* junction.

8.6.8 Step-Recovery Diodes

When *pn* junction is biased in forward detection, then minority carriers are injected to both sides of the junction. Holes are injected into the *n*-type region and electrons are injected into the *p*-type region. When diode switches into reverse directions, injected minority carriers flow back through the junction creating large temporary reverse current. This current decays with time as the number of minority carriers decreases with time. The step-recovery diodes have such impurity profile that the built-in potentials push minority carriers away from the junction, so large reverse current flows only for a very short time after switching and then this current drops very rapidly generating very sharp current pulse. When RF signal is applied to this diode, many higher harmonic frequencies are generated. These step-recovery diodes are used for frequency multiplication.

8.6.9 Avalanche Diodes

The destructive thermal breakdown in high-power diodes occur when the leakage current, which is an exponential function of the temperature, starts to increase. Locally, larger current creates large local heat dissipation, which leads to further temperature increase and, as a consequence, the destructive thermal breakdown. When the breakdown is controlled by avalanche mechanism, the breakdown voltage increases locally with temperature. In the hotter part of the diode, the breakdown voltage increases, the current stops to flow in this region, and the region cools down. Therefore, high-power avalanche diodes can sustain large reverse currents without destruction. The avalanche mechanism is also used to generate truly random noise and, of course, these are low-power avalanche diodes.

8.6.10 Varicaps

The thickness of the depletion layer in *pn* junction depends on the applied voltage, and this phenomenon can be used as voltage-controlled capacitor to tune high-frequency electronic circuits. Capacitance–voltage relationship depends on the impurity profile of the junction. In a typical junction with an impurity profile as shown in Figure 8.12a or b, capacitance does not change much; but with the hyperabrupt junction, as shown in Figure 8.12e, the value of capacitance may change several times.

8.6.11 Solar Batteries

The semiconductor *pn* junction illuminated by light will generate a voltage on its terminals. Such a diode is known as a solar battery or photovoltaic cell. The maximum voltage of a solar cell is limited by forward *pn* junction characteristics.

8.6.12 Photodiodes

If *pn* junction is eliminated with light, then the reverse junction current is proportional to the light intensity at the junction. This phenomenon is used in photodiodes. If the reversely biased collector junction is illuminated with light, then this photocurrent is amplified beta times by the transistor and as a result phototransistor is beta times more sensitive than the photodiode. Phototransistors are slower than photodiodes, and as a result only photodiodes are used in optical communications.

8.6.13 LEDs

If a diode is biased in the forward direction, it may emit light. In order to obtain high emission efficiency, the light emitting diode (LED) should be made out of a semiconductor material with a direct energy band structure. This way electrons and holes can recombine directly between valence and conduction bands. The silicon diodes do not emit light because the silicon has indirect band structure and the probability of a direct band-to-band recombination is very small. Typically, LEDs are fabricated using various compositions of $Ga_yAl_{1-y}As_xP_{1-x}$. The wavelength of generated light is inversely proportional to the potential gap of junction material.

8.6.14 Laser Diodes

When in LEDs the light intensity is enhanced by the addition of micromirrors, then laser action may occur. Laser diodes have a better efficiency and they generate coherent light.

8.6.15 Gun Diodes

These are other microwave diodes that generate microwave signals. The gun diode uses material like gallium arsenide, where for certain electrical field electron velocity decreases with electrical field. This phenomenon leads to grouping moving electrons in packs and to the generation of microwave frequencies on its terminals.

8.6.16 IMPATT Diodes

Another interesting "diode" structure having the impurity profile is shown in Figure 8.12f. When reverse biasing exceeds the breakdown voltage, this element generates a microwave signal with a frequency related to the electron transient time through structure. Such a diode is known as an IMPATT (IMPact Avalanche Transit Time) diode.

8.6.17 Peltier Diodes

Moving electrons and holes are also carrying thermal energy. In order to efficiently transfer heat, the semiconductor material must have large mobility and as small as possible thermal conductivity. Peltier diodes are composed of *p*-type and *n*-type bars connected in such a way that currents through *p*-type and *n*-type bars flow in opposite directions. Since holes move in the same direction as current, and electrons move in the opposite direction of current, the majority of carriers (both electron and holes) move in the same direction in different bars carrying heat. Peltier diodes are used for thermoelectric cooling.

There are many other two-terminal bulk semiconductor devices that can be considered diodes. Thermistors are made out of semiconductors with small energy gap (0.2–0.3 eV), and their conductance increases exponentially with temperature. Photoresistors are made from semiconductors with large minority carrier lifetime and their resistances change with illumination. Photoresistors have very slow response. Piezoresistors are sensitive to the induced stress. They can be over 100 times more sensitive than thin film tensometers. Magnetoresistors change their resistance with magnetic field. The most popular magnetoresistor structure is the Corbino ring.

References

1. S. M. Sze, *Modern Semiconductor Device Physics*, John Wiley & Sons, New York, 1998.
2. G. W. Neudeck, *The PN Junction Diode*, Vol. II, *Modular Series on Solid-State Devices*, Addison-Wesley, Reading, MA, 1983.
3. R. S. Muller and T. I. Kamins, *Device Electronics for Integrated Circuits*, 2nd edn., John Wiley & Sons, New York, 1986.
4. B. G. Streetman, *Solid State Electronic Devices*, 4th edn., Prentice Hall, Englewood Cliffs, NJ, 1995.
5. D. A. Neamen, *Semiconductor Physics and Devices*, Irwin, Boston, MA, 1992.
6. B. M. Wilamowski, *Solid-State Electronics*, 26(5), 491–493, 1983.
7. B. L. Anderson and R. L. Anderson, *Fundamentals of Semiconductor Devices*, McGraw-Hill, Burr Ridge, IL, 2005.

9

Bipolar Junction Transistor

Bogdan M.
Wilamowski
Auburn University

Guofu Niu
Auburn University

The bipolar junction transistor (BJT) is historically the first solid-state analog amplifier and digital switch, and formed the basis of integrated circuits (ICs) in the 1970s [6]. Starting in the early 1980s, the MOSFET had gradually taken over; particularly for mainstream digital ICs. However, in the 1990s, the invention of silicon–germanium base heterojunction bipolar transistor (SiGe HBT) brought the bipolar transistor back into high-volume commercial production, mainly for the now widespread wireless and wire line communications applications. Today, SiGe HBTs are used to design radio-frequency (RF) ICs and systems for cell phones, wireless local area network (WLAN), automobile collision avoidance radar, wireless distribution of cable television, millimeter wave radios, and many more applications, due to its outstanding high-frequency performance and ability to integrate with CMOS for realizing digital, analog, and RF functions on the same chip.

Below, we will first introduce the basic concepts of BJT using a historically important equivalent circuit model, the Ebers–Moll model [1]. Then the Gummel–Poon model [2] will be introduced, as it is widely used for computer-aided design, and is the basis of modern BJT models like the VBIC, Mextram, and HICUM models. Current gain, high current phenomena, fabrication technologies, and SiGe HBTs will then be discussed.

9.1 Ebers–Moll Model

An NPN BJT consists of two closely spaced PN junctions connected back to back sharing the same p-type region, as shown in Figure 9.1a. The drawing is not to scale. The emitter and base layers are thin, typically less than $1\,\mu m$, and the collector is much thicker to support a high output voltage swing. For forward mode operation, the emitter–base (EB) junction is forward biased, and the collector-base (CB) junction is reverse biased. Minority carriers are injected from the emitter to base,

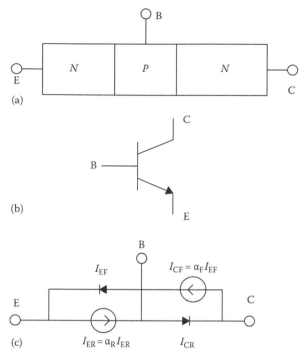

FIGURE 9.1 (a) Cross-sectional view of an NPN BJT, (b) circuit symbol, and (c) the Ebers–Moll equivalent circuit model.

travel across the base, and are then collected by the reverse biased CB junction. Therefore, the collector current is transported from the EB junction, and thus is proportional to the EB junction current. In the forward–active mode, the current–voltage characteristic of the EB junction is described by the well-known diode equation

$$I_{EF} = I_{E0} \left[\exp\left(\frac{V_{BE}}{V_T} \right) - 1 \right] \tag{9.1}$$

where
  I_{E0} is the EB junction saturation current
  $V_T = kT/q$ is the thermal potential (about 25 mV at room temperature)

The collector current is typically smaller than the emitter current is, $I_{CF} = \alpha_F I_{EF}$, where α_F is the forward current gain.

Under reverse mode operation, the CB junction is forward biased and the EB junction is reverse biased. Like in the forward mode, the forward-biased CB junction current gives the collector current

$$I_{CF} = I_{C0} \left[\exp\left(\frac{V_{BC}}{V_T} \right) - 1 \right] \tag{9.2}$$

where I_{C0} is the CB junction saturation current. Similarly $I_{ER} = \alpha_R I_R$, where α_R is the reverse current gain. Under general biasing conditions, it can be proven that to first order, a superposition of the above-described forward and reverse mode equivalent circuits can be used to describe the transistor operation, as shown in Figure 9.1b. The forward transistor operation is described by Equation 9.1, and the reverse transistor operation is described by Equation 9.2. From the Kirchoff's current law, one can write $I_C = I_{CF} - I_{CR}$, $I_E = I_{EF} - I_{ER}$, and $I_B = I_E - I_C$. Using Equations 9.1 and 9.2 the emitter and collector currents can be described as

$$I_E = a_{11}\left(\exp\frac{V_{BE}}{V_T} - 1\right) - a_{12}\left(\exp\frac{V_{BC}}{V_T} - 1\right)$$

$$I_C = a_{21}\left(\exp\frac{V_{BE}}{V_T} - 1\right) - a_{22}\left(\exp\frac{V_{BC}}{V_T} - 1\right)$$

(9.3)

which are known as the Ebers–Moll equations [1]. The Ebers–Moll coefficients a_{ij} are given as

$$a_{11} = I_{E0} \quad a_{12} = \alpha_R I_{C0} \quad a_{21} = \alpha_F I_{E0} \quad a_{22} = I_{C0}$$

(9.4)

The Ebers–Moll coefficients are a very strong function of the temperature

$$a_{ij} = K_x T^m \exp\frac{V_{go}}{V_T}$$

(9.5)

where

K_x is proportional to the junction area and is independent of the temperature

$V_{go} = 1.21\,\text{V}$ is the bandgap voltage in silicon (extrapolated to $0\,\text{K}$)

m is a material constant with a value between 2.5 and 4

When both EB and CB junctions are forward biased, the transistor is called to be working in the saturation region. Current injection through the collector junction may activate the parasitic transistors in ICs by using a p-type substrate, where the base acts as the emitter, the collector as the base, and the substrate as the collector. In typical ICs, bipolar transistors must not operate in saturation. Therefore, for the integrated bipolar transistor the Ebers–Moll equations can be simplified to the form

$$I_E = a_{11}\left(\exp\frac{V_{BE}}{V_T} - 1\right)$$

$$I_C = a_{21}\left(\exp\frac{V_{BE}}{V_T} - 1\right)$$

(9.6)

where $a_{21}/a_{11} = \alpha_F$. This equation corresponds to the circuit diagram shown in Figure 9.1c.

9.2 Gummel–Poon Model

In real bipolar transistors, the current–voltage characteristics are more complex than those described by the Ebers–Moll equations. Typical current–voltage characteristics of the bipolar transistor, plotted in semi-logarithmic scale, are shown in Figure 9.2. At small-base emitter voltages, due to the generation–recombination phenomena, the base current is proportional to

$$I_{BL} \propto \exp\frac{V_{BE}}{2V_T}$$

(9.7)

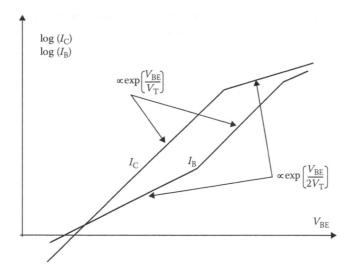

FIGURE 9.2 Collector and base currents as a function of base–emitter voltage.

Also, due to the base conductivity modulation at high-level injections, the collector current for larger voltages can be expressed by the similar relation

$$I_{CH} \propto \exp\frac{V_{BE}}{2V_T} \tag{9.8}$$

Note that the collector current for a wide range is given by

$$I_C = I_s \exp\frac{V_{BE}}{V_T} \tag{9.9}$$

The saturation current is a function of device structure parameters

$$I_s = \frac{qAn_i^2 V_T \mu_B}{\displaystyle\int_0^{w_B} N_B(x)dx} \tag{9.10}$$

where
  $q = 1.6 \times 10^{-19}$ C is the electron charge
  A is the EB junction area
  n_i is the intrinsic concentration ($n_i = 1.5 \times 10^{10}$ at 300 K)
  μ_B is the mobility of the majority carriers in the transistor base
  w_B is the effective base thickness
  $N_B(x)$ is the distribution of impurities in the base

Note that the saturation current is inversely proportional to the total impurity dose in the base. In the transistor with the uniform base, the saturation current is given by

$$I_s = \frac{qAn_i^2 V_T \mu_B}{w_B N_B} \tag{9.11}$$

When a transistor operates in the reverse-active mode (emitter and collector are switched), then the current of such a biased transistor is given by

$$I_E = I_s \exp \frac{V_{BC}}{V_T} \qquad (9.12)$$

Note that the I_s parameter is the same for forward and reverse modes of operation. The Gummel–Poon transistor model [2] was derived from the Ebers–Moll model using the assumption that $a_{12} = a_{21} = I_s$. For the Gummel–Poon model, Equations 9.3 are simplified to the form

$$I_E = I_s \left(\frac{1}{\alpha_F} \exp \frac{V_{BE}}{V_T} - \exp \frac{V_{BC}}{V_T} \right)$$

$$I_C = I_s \left(\exp \frac{V_{BE}}{V_T} - \frac{1}{\alpha_R} \exp \frac{V_{BC}}{V_T} \right) \qquad (9.13)$$

These equations require only three coefficients, while the Ebers–Moll requires four. The saturation current I_s is constant for a wide range of currents. The current gain coefficients α_F and α_R have values smaller, but close to unity. Often instead of using the current gain as $\alpha = I_C/I_E$, the current gain β as a ratio of the collector current to the base current $\beta = I_C/I_B$ is used. The mutual relationships between α and β coefficients are given by

$$\alpha_F = \frac{\beta_F}{\beta_F + 1} \quad \beta_F = \frac{\alpha_F}{1 - \alpha_F} \quad \alpha_R = \frac{\beta_R}{\beta_R + 1} \quad \beta_R = \frac{\alpha_R}{1 - \alpha_R} \qquad (9.14)$$

The Gummel–Poon model was implemented in SPICE [3] and other computer programs for circuit analysis. To make the equations more general, the material parameters η_F and η_R were introduced:

$$I_C = I_s \left[\exp \frac{V_{BE}}{\eta_F V_T} - \left(1 + \frac{1}{\beta_R} \right) \exp \frac{V_{BC}}{\eta_R V_T} \right] \qquad (9.15)$$

The values of η_F and η_R vary from one to two.

9.3 Current Gains of Bipolar Transistors

The transistor current gain β, is limited by two phenomena: base transport efficiency and emitter injection efficiency. The effective current gain β can be expressed as

$$\frac{1}{\beta} = \frac{1}{\beta_I} + \frac{1}{\beta_T} + \frac{1}{\beta_R} \qquad (9.16)$$

where
  β_I is the transistor current gain caused by emitter injection efficiency
  β_T is the transistor current gain caused by base transport efficiency
  β_R is the recombination component of the current gain

As one can see from Equation 9.16, smaller values of β_I, β_T, and β_R dominate. The base transport efficiency can be defined as a ratio of injected carriers into the base, to the carriers that recombine within the base.

This ratio is also equal to the ratio of the minority carrier's lifetime to the transit time of carriers through the base. The carrier transit time can be approximated by an empirical relationship

$$\tau_{\text{transit}} = \frac{w_{\text{B}}^2}{V_{\text{T}}\mu_{\text{B}}(2+0.9\eta)}; \quad \eta = \ln\left(\frac{N_{\text{BE}}}{N_{\text{BC}}}\right) \tag{9.17}$$

where
μ_{B} is the mobility of the minority carriers in base
w_{B} is the base thickness
N_{BE} is the impurity doping level at the emitter side of the base
N_{BC} is the impurity doping level at the collector side of the base

Therefore, the current gain due to the transport efficiency is

$$\beta_{\text{T}} = \frac{\tau_{\text{life}}}{\tau_{\text{transit}}} = (2+0.9\eta)\left(\frac{L_{\text{B}}}{w_{\text{B}}}\right)^2 \tag{9.18}$$

where $L_{\text{B}} = \sqrt{V_{\text{T}}\mu_{\text{B}}\tau_{\text{life}}}$ is the diffusion length of minority carriers in the base.
　　The current gain β_{I}, due to the emitter injection efficiency, is given by

$$\beta_{\text{I}} = \frac{\mu_{\text{B}}\displaystyle\int_0^{w_{\text{E}}} N_{\text{Eeff}}(x)dx}{\mu_{\text{E}}\displaystyle\int_0^{w_{\text{B}}} N_{\text{B}}(x)dx} \tag{9.19}$$

where
μ_{B} and μ_{E} are minority carrier mobilities in the base and in the emitter
$N_{\text{B}}(x)$ is the impurity distribution in the base
N_{Eeff} is the effective impurity distribution in the emitter

The recombination component of the current gain β_{R} is caused by the different current–voltage relationship of base and collector currents as can be seen in Figure 9.2. The slower base current increase is due to the recombination phenomenon within the depletion layer of the base-emitter junction. Since the current gain is a ratio of the collector current to that of the base current the relation for β_{R} can be found as

$$\beta_{\text{R}} = K_{\text{R0}}I_{\text{C}}^{1-(1/\eta_{\text{R}})} \tag{9.20}$$

As can be seen from Figure 9.2, the current gain β is a function of the current. This gain-current relationship is illustrated in Figure 9.3. The range of a constant current gain is wide for bipolar transistors with a technology characterized by a lower number of generation–recombination centers.
　　With an increase of the CB voltage, the depletion layer penetrates deeper into the base. Therefore, the effective thickness of the base decreases. This leads to an increase of transistor current gain with applied collector voltages. Figure 9.4 illustrates this phenomenon known as Early's effect. The extensions of transistor characteristics (dotted lines in Figure 9.4) are crossing the voltage axis at the point—V_{A}, where V_{A} is known as the Early voltage. The current gain β, as a function of the collector voltage is usually expressed by using the relation

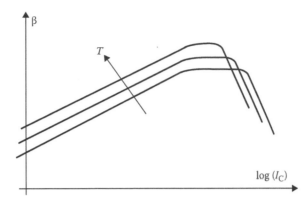

FIGURE 9.3 The current gain β as the function of the collector current.

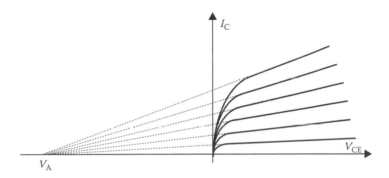

FIGURE 9.4 Current–voltage characteristics of a bipolar transistor.

$$\beta = \beta_0 \left(1 + \frac{V_{CE}}{V_A} \right)$$ (9.21)

A similar equation can be defined for the reverse mode of operation.

9.4 High Current Phenomena

The concentration of minority carriers increases with the rise of transistor currents. When the concentration of moving carriers exceeds a certain limit, the transistor property degenerates. Two phenomena are responsible for this limitation. The first is related to the high concentration of moving carriers (electrons in the npn transistor) in the base–collector depletion region. This is known as the Kirk effect. The second phenomenon is caused by a high level of carriers injected into the base. When the concentration of injected minority carriers in the base exceeds the impurity concentration there, then the base conductivity modulation limits the transistor's performance.

To understand the Kirk effect, consider the NPN transistor in the forward-active mode with the base–collector junction reversely biased. The depletion layer consists of the negative lattice charge of the base region and the positive lattice charge of the collector region. Boundaries of the depletion layer are such that the total positive and negative charges are equal. When a collector current that carries the negatively charged electrons flows through the junction, the effective negative charge on the base side of junction increases. Also, the positive lattice charge of the collector side of the junction is compensated by the negative charge of moving electrons. This way, the CB space charge region moves toward the collector, resulting in a thicker effective base. With a large current level, the thickness of the base may

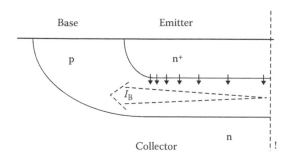

FIGURE 9.5 Current crowding effect.

be doubled or tripled. This phenomenon, known as the Kirk effect, becomes very significant when the charge of moving electrons exceeds the charge of the lightly doped collector N_C. The threshold current for the Kirk effect is given by

$$I_{max} = qAv_{sat}N_C$$

(9.22)

where v_{sat} is the saturation velocity for electrons ($v_{sat} = 10^7$ cm/s for silicon).

The conductivity modulation in the base, or high-level injection, starts when the concentration of injected electrons into the base exceeds the lowest impurity concentration in the base N_{Bmin}. This occurs for the collector current I_{max} given by

$$I_{max} < qAN_{Bmax}v = \frac{qAV_T\mu_BN_{Bmax}(2+0.9\eta)}{w_B}$$

(9.23)

The above equation is derived using (9.17), for the estimation of the base transient time.

The high current phenomena are significantly enlarged by the current crowding effect. The typical cross section of the bipolar transistor is shown in Figure 9.5. The horizontal flow of the base current results in the voltage drop across the base region under the emitter. This small voltage difference on the base–emitter junction causes a significant difference in the current densities at the junction. This is due to the very nonlinear junction–current–voltage characteristics. As a result, the base–emitter junction has very nonuniform current distribution across the junction. Most of the current flows through the part of the junction closest to base contact. For transistors with larger emitter areas, the current crowding effect is more significant. This nonuniform transistor current distribution makes the high current phenomena, such as the base conductivity modulation and the Kirk effect, start for smaller currents than given by Equations 9.22 and 9.23. The current crowding effect is also responsible for the change of the effective base resistance with a current. As the base current increases, the larger part of the emitter current flows closer to the base contact, and the effective base resistance decreases.

9.5 Small Signal Model

Small signal transistor models are essential for the design of an AC circuit. The small signal equivalent circuit of the bipolar transistor is shown in Figure 9.6a. The lumped circuit shown in Figure 9.6a is only an approximation. In real transistors, resistances and capacitances have a distributed character. For most design tasks, this lumped model is adequate, or even the simple equivalent transistor model shown in Figure 9.6b can be considered. The small signal resistances, r_π and r_o, are inversely proportional to the transistor currents, and the transconductance g_m is directly proportional to the transistor currents.

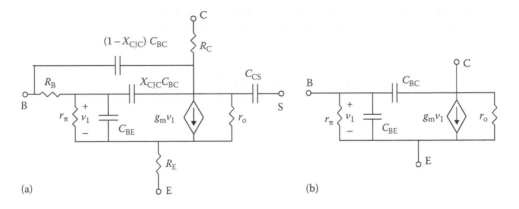

FIGURE 9.6 Bipolar transistor equivalent diagrams: (a) SPICE model and (b) simplified model.

$$r_\pi = \frac{\eta_F V_T}{I_B} = \frac{\eta_F V_T \beta_F}{I_C}; \quad r_0 = \frac{V_A}{I_C}; \quad g_m = \frac{I_C}{\eta_F V_T} \tag{9.24}$$

where

η_F is the forward emission coefficient, ranging form 1.0 to 2.0
V_T is the thermal potential ($V_T = 25\,\text{mV}$ at room temperature)

Similar equations to (9.24) can be written for the reverse transistor operation as well.

The series base, emitter, and collector resistances R_B, R_E, and R_C are usually neglected for simple analysis (Figure 9.6b). However, for high-frequency analysis it is essential to use at least the base series resistance R_B. The series emitter resistance R_E usually has a constant and bias independent value. The collector resistance R_C may significantly vary with the biasing current. The value of the series collector resistance may lower by one or two orders of magnitude if the collector junction becomes forward biased. A large series collector resistance may force the transistor into the saturation mode. Usually however, when the collector–emitter voltage is large enough, the effect of the collector resistance is not significant. The SPICE model assumes a constant value for the collector resistance R_C.

The series base resistance R_B may significantly limit the transistor performance at high frequencies. Due to the current crowding effect and the base conductivity modulation, the series base resistance is a function of the collector current I_C [4]

$$R_B = R_{Bmin} + \frac{R_{B0} - R_{Bmin}}{0.5 + \sqrt{0.25 + (I_C/I_{KF})}} \tag{9.25}$$

where

I_{KF} is β_F high-current roll-off current
R_{B0} is the base resistance at very small currents
R_{Bmin} is the minimum base resistance at high currents

Another possible approximation of the base series resistance R_B, as the function of the base current I_B, is [4]

$$R_B = 3(R_{B0} - R_{Bmin})\frac{\tan z - z}{z \tan^2 z} + R_{Bmin} \quad z = \frac{\sqrt{1 + (1.44\, I_B/\pi^2 I_{RB})} - 1}{(24/\pi^2)\sqrt{(I_B/I_{RB})}} \tag{9.26}$$

where I_{RB} is the base current for which the base resistance falls halfway to its minimum value.

The base–emitter capacitance C_{BE} is composed of two terms: the diffusion capacitance, which is proportional to the collector current, and the depletion capacitance, which is a function of the base–emitter voltage V_{BE}. The C_{BE} capacitance is given by

$$C_{BE} = \tau_F \frac{I_C}{\eta_F V_T} + C_{JE0} \left(1 - \frac{V_{BE}}{V_{JE0}}\right)^{-m_{JE}} \tag{9.27}$$

where

V_{JE0} is the base–emitter junction potential
τ_F is the base transit time for the forward direction
C_{JE0} is base–emitter zero-bias junction capacitance
m_{JE} is the base–emitter grading coefficient

The base–collector capacitance C_{BC} is given by a similar expression as Equation 9.27. In the case when the transistor operates in the forward-active mode, it can be simplified to

$$C_{BC} = C_{JC0} \left(1 - \frac{V_{BC}}{V_{JC0}}\right)^{-m_{JC}} \tag{9.28}$$

where

V_{JC0} is the base–collector junction potential
C_{JC0} is the base–collector zero-bias junction capacitance
m_{JC} is the base–collector grading coefficient

In the case when the bipolar transistor is in the integrated form, the collector–substrate capacitance C_{CS} has to be considered

$$C_{CS} = C_{JS0} \left(1 - \frac{V_{CS}}{V_{JS0}}\right)^{-m_{JS}} \tag{9.29}$$

where

V_{JS0} is the collector–substrate junction potential
C_{JS0} the collector–substrate zero-bias junction capacitance
m_{JS} is the collector–substrate grading coefficient

When the transistor enters saturation, or it operates in the reverse-active mode, Equations 9.27 and 9.28 should be modified to

$$C_{BE} = \tau_F \frac{I_S \exp(V_{BE}/\eta_F V_T)}{\eta_F V_T} + C_{JE0} \left(1 - \frac{V_{BE}}{V_{JE0}}\right)^{-m_{JE}} \tag{9.30}$$

$$C_{BC} = \tau_R \frac{I_S \exp(V_{BC}/\eta_R V_T)}{\eta_R V_T} + C_{JC0} \left(1 - \frac{V_{BC}}{V_{JC0}}\right)^{-m_{JC}} \tag{9.31}$$

9.6 Technologies

The bipolar technology was used to fabricate the first ICs more than 40 years ago. A similar standard bipolar process is still used. In recent years, for high performance circuits and for BiCMOS technology, the standard bipolar process was modified by using the thick selective silicon oxidation instead of the p-type isolation diffusion. Also, the diffusion process was substituted by the ion implantation process, low temperature epitaxy, and CVD.

9.6.1 Integrated NPN Bipolar Transistor

The structure of the typical integrated bipolar transistor is shown in Figure 9.7. The typical impurity profile of the bipolar transistor is shown in Figure 9.8. The emitter doping level is much higher than the base doping, so large current gains are possible (see Equation 9.19). The base is narrow and it has an impurity gradient, so the carrier transit time through the base is short (see Equation 9.17). Collector concentration near the base collector junction is low, therefore, the transistor has a large breakdown voltage, large Early voltage V_{AF}, and CB depletion capacitance is low. High impurity concentration in the buried layer leads to a small collector series resistance. The emitter strips have to be as narrow as technology allows, reducing the base series resistance and the current crowding effect. If large emitter area is required, many narrow emitter strips interlaced with base contacts have to be used in a single transistor. Special attention has to be taken during the circuit design, so the base–collector junction is not forward biased. If the base–collector junction is forward biased, then the parasitic PNP transistors activate. This leads to an undesired circuit operation. Thus, the integrated bipolar transistors must not operate in reverse or in saturation modes.

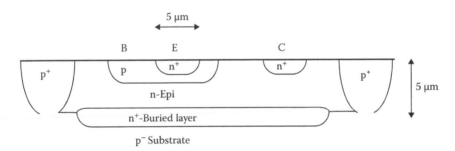

FIGURE 9.7 NPN bipolar structure.

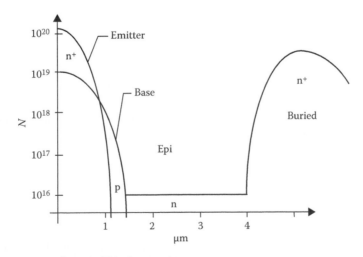

FIGURE 9.8 Cross section of a typical bipolar transistor.

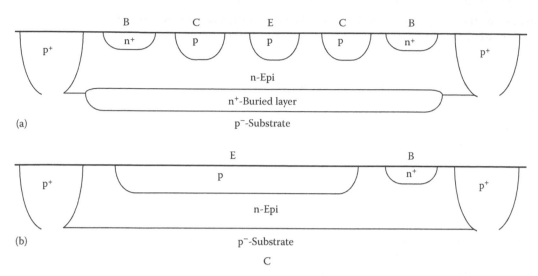

FIGURE 9.9   Integrated PNP transistors: (a) lateral PNP transistor and (b) substrate PNP transistor.

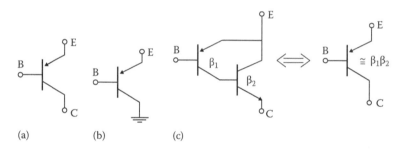

FIGURE 9.10   Integrated PNP transistors: (a) lateral transistor, (b) substrate transistor, and (c) composed transistor.

9.6.2 Lateral and Vertical PNP Transistors

The standard bipolar technology is oriented for the fabrication of the NPN transistors with the structure shown in Figure 9.7. Using the same process, other circuit elements, such as resistors and PNP transistors, can be fabricated as well.

The lateral transistor, shown in Figure 9.9a, uses the base p-type layer for both the emitter and collector fabrications. The vertical transistor, shown in Figure 9.9b, uses the p-type base layer for the emitter, and the p-type substrate as the collector. This transistor is sometimes known as the substrate transistor. In both transistors, the base is made of the n-type epitaxial layer. Such transistors with a uniform and thick base are slow. Also, the current gain β of such transistors is small. Note that the vertical transistor has the collector shorted to the substrate as Figure 9.10b illustrates. When a PNP transistor with a large current gain is required, then the concept of the composite transistor can be implemented. Such a composite transistor, known also as the super-beta transistor, consists of a PNP lateral transistor, and the standard NPN transistor is connected, as shown in Figure 9.10c. The composed transistor acts as the PNP transistor and it has a current gain β approximately equal to $\beta_{pnp}\beta_{npn}$.

9.7 Model Parameters

It is essential to use proper transistor models in the computer-aided design tools. The accuracy of simulation results depends on the model's accuracy, and on the values of the model parameters used. In this section, the thermal sensitivity and second order effects in the transistor model are discussed. The SPICE bipolar transistor model parameters are also discussed.

9.7.1 Thermal Sensitivity

All parameters of the transistor model are temperature dependent. Some parameters are very strong functions of temperature. To simplify the model description, the temperature dependence of some parameters is often neglected. In this chapter, the temperature dependence of the transistor model is described based on the model of the SPICE program [3–5]. Deviations from the actual temperature dependence will also be discussed. The temperature dependence of the junction capacitance is given by

$$C_J(T) = C_J \left\{ 1 + m_J \left[4.0 \times 10^{-4} (T - T_{\text{NOM}}) + \left(1 - \frac{V_J(T)}{V_J} \right) \right] \right\} \tag{9.32}$$

where T_{NOM} is the nominal temperature, which is specified in the SPICE program in the .OPTIONS statement. The junction potential $V_J(T)$ is a function of temperature

$$V_J(T) = V_J \frac{T}{T_{\text{NOM}}} - 3V_T \ln\left(\frac{T}{T_{\text{NOM}}} \right) - E_G(T) + E_G \frac{T}{T_{\text{NOM}}} \tag{9.33}$$

The value of 3, in the multiplication coefficient of above equation is from the temperature dependence of the effective state densities in the valence and conduction bands. The temperature dependence of the energy gap is computed in the SPICE program from

$$E_G(T) = E_G - \frac{7.02 \times 10^{-4} T^2}{T + 1108} \tag{9.34}$$

The transistor saturation current as a function of temperature is calculated as

$$I_s(T) = I_s \left(\frac{T}{T_{\text{NOM}}} \right)^{X_{TI}} \exp\left[\frac{E_G (T - T_{\text{NOM}})}{V_T T_{\text{NOM}}} \right] \tag{9.35}$$

where E_G is the energy gap at the nominal temperature. The junction leakage currents I_{SE} and I_{SC} are calculated using

$$I_{SE}(T) = I_{SE} \left(\frac{T}{T_{\text{NOM}}} \right)^{X_{TI} - X_{TB}} \exp\left[\frac{E_G (T - T_{\text{NOM}})}{\eta_E V_T T_{\text{NOM}}} \right] \tag{9.36}$$

and

$$I_{SC}(T) = I_{SC} \left(\frac{T}{T_{\text{NOM}}} \right)^{X_{TI} - X_{TB}} \exp\left[\frac{E_G (T - T_{\text{NOM}})}{\eta_C V_T T_{\text{NOM}}} \right] \tag{9.37}$$

The temperature dependence of the transistor current gains β_F and β_R are modeled in the SPICE as

$$\beta_F(T) = \beta_F \left(\frac{T}{T_{\text{NOM}}} \right)^{X_{TB}} \quad \beta_R(T) = \beta_R \left(\frac{T}{T_{\text{NOM}}} \right)^{X_{TB}} \tag{9.38}$$

The SPICE model does not give accurate results for the temperature relationship of the current gain β at high currents. For high current levels, the current gain decreases sharply with the temperature, as can be seen from Figure 9.3. Also, the knee current parameters IKF, IKR, IKB are temperature dependent, and this is not implemented in the SPICE program.

9.7.2 Second Order Effects

The current gain β is sometimes modeled indirectly by using different equations for the collector and base currents [4,5]

$$I_C = \frac{I_s(T)}{Q_b}\left(\exp\frac{V_{BE}}{\eta_F V_T} - \exp\frac{V_{BC}}{\eta_R V_T}\right) - \frac{I_s(T)}{\beta_R(T)}\left(\exp\frac{V_{BC}}{\eta_R V_T} - 1\right) - I_{SC}(T)\left(\exp\frac{V_{BC}}{\eta_C V_T} - 1\right) \tag{9.39}$$

where

$$Q_b = \frac{1+\sqrt{1+4Q_X}}{2\left(1-(V_{BC}/V_{AF})-(V_{BE}/V_{AR})\right)} \tag{9.40}$$

and

$$Q_X = \frac{I_s(T)}{I_{KF}}\left(\exp\frac{V_{BE}}{\eta_F V_T} - 1\right) + \frac{I_s(T)}{I_{KR}}\left(\exp\frac{V_{BC}}{\eta_R V_T} - 1\right) \tag{9.41}$$

$$I_B = \frac{I_s}{\beta_F}\left(\exp\frac{V_{BE}}{\eta_F V_T} - 1\right) + I_{SE}\left(\exp\frac{V_{BE}}{\eta_E V_T} - 1\right) + \frac{I_s}{\beta_R}\left(\exp\frac{V_{BC}}{\eta_R V_T} - 1\right) + I_{SC}\left(\exp\frac{V_{BC}}{\eta_C V_T} - 1\right) \tag{9.42}$$

where
   I_{SE} is the base–emitter junction leakage current
   I_{SC} is the base–collector junction leakage current
   η_E is the base–emitter junction leakage emission coefficient
   η_C is the base–collector junction leakage emission coefficient

The forward transit time τ_F is a function of biasing conditions. In the SPICE program, the τ_F parameter is computed by using

$$\tau_F = \tau_{F0}\left[1 + X_{TF}\left(\frac{I_{CC}}{I_{CC}+I_{TF}}\right)\exp\frac{V_{BC}}{1.44V_{TF}}\right] \quad I_{CC} = I_S\left(\exp\frac{V_{BE}}{\eta_F V_T} - 1\right) \tag{9.43}$$

At high frequencies, the phase of the collector current shifts. This phase shift is computed in the SPICE program in the following way:

$$I_C(\omega) = I_C \exp\left(j\omega P_{TF}\tau_F\right) \tag{9.44}$$

where P_{TF} is a coefficient for excess phase calculation.

Noise is usually modeled as the thermal noise for parasitic series resistances, and as shot and flicker noise for collector and base currents

$$\overline{i_R^2} = \frac{4kT\Delta f}{R} \tag{9.45}$$

where K_F and A_F are the flicker noise coefficients. More detailed information about noise modeling

$$\overline{i_B^2} = \left(2qI_B + \frac{K_F I_B^{A_F}}{F} \right) \Delta f \tag{9.46}$$

$$\overline{i_C^2} = 2qI_C \Delta f \tag{9.47}$$

is given in the bipolar noise chapter of this handbook.

9.7.3 SPICE Model of the Bipolar Transistor

The SPICE model of the bipolar transistor uses similar or identical equations as described in this chapter [3–5]. Table 9.1 shows the parameters of the bipolar transistor model and its relation to the parameters used in this chapter.

The SPICE (Simulation Program with Integrated Circuit Emphasis [3]) was developed mainly for the analysis of ICs. During the analysis, it is assumed that the temperatures of all circuit elements are the same. This is not true for power ICs where the junction temperatures may differ by 30 K or more. This is obviously not true for circuits composed of the discrete elements where the junction temperatures may differ by 100 K and more. These temperature effects, which can significantly affect the analysis results, are not implemented in the SPICE program.

Although the SPICE bipolar transistor model uses more than 40 parameters, many features of the bipolar transistor are not included in the model. For example, the reverse junction characteristics are described by Equation 9.32. This model does not give accurate results. In the real silicon junction, the leakage current is proportional to the thickness of the depletion layer, which is proportional to $V^{1/m}$. Also, the SPICE model of the bipolar transistor assumes that there is no junction breakdown voltage. A more accurate model of the reverse junction characteristics is described in the diode section of this handbook. The reverse transit time τ_R is very important to model the switching property of the lumped bipolar transistor, and it is a strong function of the biasing condition and the temperature. Both phenomena are not implemented in the SPICE model.

9.8 SiGe HBTs

The performance of the Si bipolar transistor can be greatly enhanced with proper engineering of the base bandgap profile using a narrower bandgap material, SiGe, an alloy of Si and Ge. Structure wise, a SiGe HBT is essentially a Si BJT with a SiGe base. Its operation and circuit level performance advantages can be illustrated with the energy band diagram in Figure 9.11 [7]. Here the Ge content is linearly graded from the emitter toward the collector to create a large accelerating electric field that speeds up the minority carrier transport across the base, thus making the transistor speed much faster and the cutoff frequency much higher. Everything else being the same, the potential barrier for electron injection into the base is reduced, thus exponentially enhancing the collector current. The base current is the same for SiGe HBT and Si BJT, as the emitter is typically made the same. Beta is thus higher in SiGe HBT. Figure 9.12 confirms these expectations experimentally with the data from a typical first generation SiGe HBT technology. The measured doping and Ge profiles are shown in Figure 9.13. The metallurgical base width is only 90 nm, and the neutral base width is around 50 nm. Figure 9.14 shows the experimental cutoff frequency f_T improvement from using a graded SiGe base, which also directly translates into maximum oscillation frequency f_{max} improvement.

TABLE 9.1   Parameters of SPICE Bipolar Transistor Model

Name Used	Equations	SPICE Name	Parameter Description	Unit	Typical Value	SPICE Default
I_s	10, 11	IS	Saturation current	A	10^{-15}	10^{-16}
I_{SE}	39	ISE	B–E leakage saturation current	A	10^{-12}	0
I_{SC}	39	ICS	B–C leakage saturation current	A	10^{-12}	0
β_F	14, 16, 21	BF	Forward current gain	—	100	100
β_R	14, 16, 21	BF	Reverse current gain	—	0.1	1
η_F	15, 24, 30, 31, 39–41	NF	Forward current emission coefficient	—	1.2	1.0
η_R	15, 24, 30 31, 39–42	NR	Reverse current emission coefficient	—	1.3	1.0
η_E	39	NE	B–E leakage emission coefficient	—	1.4	1.5
η_C	39	NC	B–C leakage emission coefficient	—	1.4	1.5
V_{AF}	21, 40	VAF	Forward Early voltage	V	200	∞
V_{AR}	21, 40	VAR	Reverse Early voltage	V	50	∞
I_{KF}	22, 23, 40	IKF	β_F high current roll-off corner	A	0.05	∞
I_{KR}	22, 23, 40	IKR	β_R high current roll-off corner	A	0.01	∞
I_{RB}	26	IRB	Current where base resistance falls by half	A	0.1	∞
R_B	25, 26	RB	Zero base resistance	Ω	100	0
R_{Bmin}	25, 26	RBM	Minimum base resistance	Ω	10	RB
R_E	Figure 9.6	RE	Emitter series resistance	Ω	1	0
R_C	Figure 9.6	RC	Collector series resistance	Ω	50	0
C_{JE0}	27	CJE	B–E zero-bias depletion capacitance	F	10^{-12}	0
C_{JC0}	28	CJC	B–C zero-bias depletion capacitance	F	10^{-12}	0
C_{JS0}	29	CJS	Zero-bias collector–substrate capacitance	F	10^{-12}	0
V_{JE0}	27	VJE	B–E built-in potential	V	0.8	0.75
V_{JC0}	28	VJC	B–C built-in potential	V	0.7	0.75
V_{JS0}	29	VJS	Substrate junction built-in potential	V	0.7	0.75
m_{JE}	27	MJE	B–E junction exponential factor	—	0.33	0.33
m_{JC}	28	MJC	B–C junction exponential factor	—	0.5	0.33
m_{JS}	29	MJS	Substrate junction exponential factor	—	0.5	0
X_{CJC}	Figure 9.6	XCJC	Fraction of B–C capacitance connected to internal base node (see Figure 9.6)	—	0.5	0
τ_F	17, 28, 30, 42	TF	Ideal forward transit time	s	10^{-10}	0
τ_R	31	TR	Reverse transit time	s	10^{-8}	0
X_{TF}	43	XTF	Coefficient for bias dependence of τ_F	—		0
V_{TF}	43	VTF	Voltage for τ_F dependence on V_{BC}	V		∞
I_{TF}	43	ITF	Current where $\tau_F = f(I_C, V_{BC})$ starts	A		0
P_{TF}	44	PTF	Excess phase at *freq* $= 1/(2\pi\tau_F)$ Hz	deg		0
X_{TB}	38	XTB	Forward and reverse beta temperature exponent			0
E_G	34	EG	Energy gap	eV	1.1	1.11
X_{TI}	35–37	XTI	Temperature exponent for effect on I_s	—	3.5	3
K_F	46	KF	Flicker-noise coefficient	—		0
A_F	46	AF	Flicker-noise exponent	—		1
F_C		FC	Coefficient for the forward-biased depletion capacitance formula	—	0.5	0.5
T_{NOM}	32–38	TNOM	Nominal temperature specified in OPTION statement	K	300	300

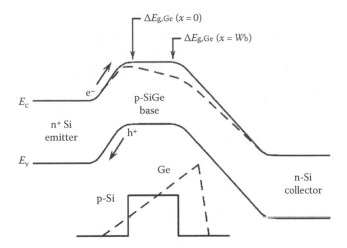

FIGURE 9.11 Energy band diagram of a graded base SiGe HBT and a comparably constructed Si BJT.

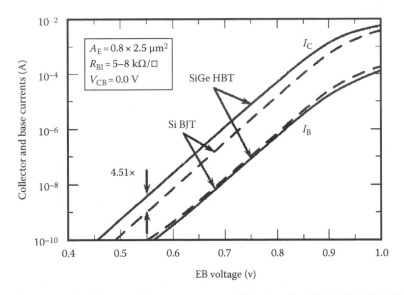

FIGURE 9.12 Experimental collector and base currents versus EB voltage for SiGe HBT and Si BJT.

9.8.1 Operation Principle and Performance Advantages over Si BJT

In modern transistors, particularly with the use of polysilicon emitter, beta may be sufficient. If so, the higher beta potential of SiGe HBT can then be traded for reduced base resistance, using higher base doping. The unique ability of simultaneously achieving high beta, low base resistance, and high cutoff frequency makes SiGe HBT attractive for many RF circuits. Broadband noise is naturally reduced, as low base resistance reduces transistor input noise voltage, and high beta as well as high f_T reduces transistor input noise current [7]. Experimentally, $1/f$ noise at the same base current was found to be approximately the same for SiGe HBT and Si BJT [8]. Consequently, $1/f$ noise is often naturally reduced in SiGe HBT circuits for the same biasing collector current, as the base current is often smaller due to a higher beta, as shown in Figure 9.15, by using the corner frequency as a figure-of-merit.

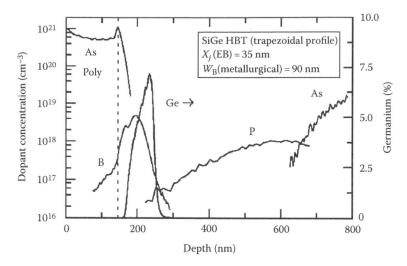

FIGURE 9.13   Measured doping and Ge profiles of a modern SiGe HBT.

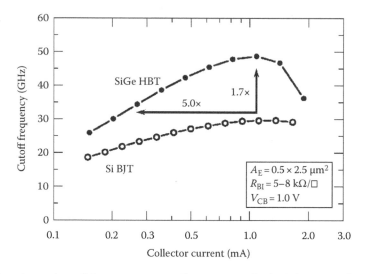

FIGURE 9.14   Experimental cutoff frequency versus collector current for the SiGe HBT and Si BJT.

These, together with circuit level optimization can lead to excellent low phase noise oscillators and frequency synthesizers suitable for both wireless and wire line communication circuits. Another less obvious advantage from grading Ge is the collector side of the neutral base that has less impact on the collector current than the emitter side of the neutral base. Consequently, as the collector voltage varies and the collector side of the neutral base is shifted toward the emitter due to increased CB junction-depletion layer thickness, the collector current is increased to a much lesser extent than in a comparably constructed Si BJT, leading to a much higher output impedance or Early voltage. The $\beta \times V_A$ product is thus much higher in SiGe HBT than in Si BJT.

9.8.2 Industry Practice and Fabrication Technology

The standard industry practice today is to integrate SiGe HBT with CMOS to form a SiGe BiCMOS technology. The ability to integrate with CMOS is also a significant advantage of SiGe HBT over III–V HBT. Modern SiGe BiCMOS combines the analog and RF performance advantages of the SiGe HBT,

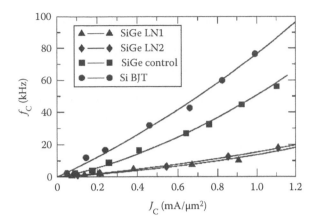

FIGURE 9.15 Experimentally measured corner frequency as a function of collector current density for three SiGe HBTs with different base SiGe designs, and a comparatively constructed Si BJT.

and the lower power logic, high integration level, and memory density of Si CMOS, into a single cost effective system-on-chip (SoC) solution. Typically, SiGe HBTs with multiple breakdown voltages are offered through the selective collector implantation, to provide more flexibility in the circuit design.

The fabrication process of SiGe HBT and its integration with CMOS has been constantly evolving in the past two decades, and varies from company to company. Below are some common fabrication elements and modules shared by many, if not all, commercial first generation (also most widespread in manufacturing at present) SiGe technologies:

1. A starting N^+ subcollector around 5 Ω/square on a p-type substrate at 5×10^{15}/cm^3, typically patterned to allow CMOS integration.
2. A high temperature, lightly doped n-type collector, around 0.4–0.6 µm thick at 5×10^{15}/cm^3.
3. Polysilicon filled deep trenches for isolation from adjacent devices, typically 1 µm wide and 7–10 µm deep.
4. Oxide filled shallow trenches or LOCOS for local device isolation, typically 0.3–0.6 µm deep.

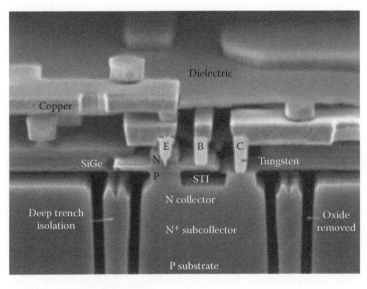

FIGURE 9.16 Structure of a modern SiGe HBT.

5. An implanted collector reaches through to the subcollector, typically at 10–20 $\Omega\mu m^2$.
6. A composite SiGe epilayer consisting of a 10–20 nm Si buffer, a 70–100 nm boron-doped SiGe active layer, with or without C doping to help suppress boron out diffusion, and a 10–30 nm Si cap. The integrated boron dose is typically 1–3 × 10^{13}/cm^2.
7. A variety of EB self-alignment scheme, depending on the device structure and SiGe growth approach. All of them utilize some sort of spacer that is 100–300 nm wide.
8. Multiple self-aligned collector implantations to allow multiple breakdown voltages on the same chip.
9. Polysilicon extrinsic base, usually formed during SiGe growth over shallow trench oxide, and additional self-aligned extrinsic implantation to lower base resistance.
10. A silicided extrinsic base.
11. A 100–200 nm thick heavily doped (>5 × 10^{20}/cm^3) polysilicon emitter, either implanted or *in situ* doped.
12. A variety of multiple level back-end-of-line metallization schemes using Al or Cu, typically borrowed from the parent CMOS process.

These technological elements can also be seen in the electronic image of a second generation SiGe HBT shown in Figure 9.16.

References

1. J. J. Ebers and J. M. Moll, Large signal behavior of bipolar transistors. *Proceedings IRE* 42(12), 1761–1772, December 1954.
2. H. K. Gummel and H. C. Poon, An integral charge-control model of bipolar transistors. *Bell System Technical Journal* 49, 827–852, May 1970.
3. L. W. Nagel and D. O. Pederson, SPICE (Simulation Program with Integrated Circuit Emphasis). University of California, Berkeley, ERL Memo No. ERL M382, April 1973.
4. P. Antognetti and G. Massobrio, *Semiconductor Device Modeling with SPICE*, McGraw-Hill, New York, 1988.
5. A. Vadimiresku, *The SPICE Book*, John Wiley & Sons, Hoboken, NJ, 1994.
6. R. S. Muller and T. I. Kamins, *Device Electronics for Integrated Circuits*, 2nd edn., John Wiley & Sons, New York, 1986.
7. J. D. Cressler and G. Niu, *Silicon–Germanium Heterojunction Bipolar Transistor*, Artech House, Boston, MA, 2003.
8. G. Niu, Noise in SiGe HBT RF technology: Physics, modeling and circuit implications. *Proceedings of the IEEE*, 93(9), 1583–1597, September 2005.

10

Field Effect Transistors

Bogdan M.
Wilamowski
Auburn University

J. David Irwin
Auburn University

10.1 Introduction

There are several different types of field effect transistors (FETs), each of which has a different operational principle. For example, there are metal oxide semiconductor (MOS) transistors, junction field effect transistors (JFETs), static induction transistors (SITs), the punch-through transistors (PTTs), and others. All of these devices employ the flow of majority carriers. The most popular one among this group is the MOS transistor, which is primarily used in integrated circuits [T99,N06,S05]. In contrast, the JFET is not suitable for integration and so it is primarily fabricated as an individual device [E97,R99]. All FETs have very large input resistance on the order of $10^{12}\ \Omega$. The MOS transistor typically operates with very small currents [N02], and thus for power electronics applications, thousands of MOS transistors are connected in parallel. A JFET usually operates with larger currents. Both JFET and MOS transistors have relatively small transconductances, and this means that they cannot control current flow as effectively as bipolar junction transistors (BJTs). Since the parasitic capacitors are of the same order of magnitude, BJTs can charge and discharge these capacitors much faster and so BJTs are more suitable for high-frequency operations. Because current flow in MOS transistors is very close to the silicon surface where surface states can fluctuate with time, MOS devices have a relatively higher noise level, especially at low frequencies.

10.2 MOS Transistor

The MOS transistor can be considered a capacitor in which the applied voltage to the gate G would attract carriers (elections in NMOS and holes in PMOS) from the semiconductor substrate. This layer of accumulated carriers near the surface conducts current between source and drain [T99]. If the voltage on the gate is increased, then more carries (electrons or holes) will be accumulated near the surface, causing a larger current to flow, as indicated in Figure 10.1. In order to better understand the process of carrier accumulation under the gate, the MOS structure must be analyzed in detail.

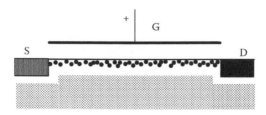

FIGURE 10.1   The principle of operation for an NMOS transistor, where electrons are accumulated by the positive gate voltage.

10.2.1 MOS Structure and Threshold Voltage

Figure 10.2 shows the cross section of the MOS band structure with a p-type silicon substrate. Note that the Fermi level has a different location in every material. In metals, there is no forbidden energy gap, and the Fermi level E_{Fm} is on the edge of the conduction band. Albert Einstein received his Nobel Prize for the photoelectric effect, in which he was able to measure the energy required to free an electron from metal to the vacuum. This energy is now known as the work function ϕ_m. Work functions for various materials are shown in Table 10.1.

It is important to note that the Fermi levels in semiconductors may depend on the doping level N and on the type of impurities present. In the n-type material, the Fermi level E_{Fs} is above the center of the energy gap E_i and in the p-type material, the Fermi level E_{Fs} is below E_i, as shown in Figure 10.2. The work function ϕ_s for intrinsic, i.e., undoped, silicon is 3.8 eV, as listed in Table 10.1, and the energy needed to free an electron from the p-type silicon is

$$\phi_s = 3.8 + \phi_F \quad \text{where } \phi_F = V_T \ln\left(\frac{N_A}{n_i}\right) \tag{10.1}$$

and this energy is dependent on the acceptor doping level N_A and intrinsic carrier concentration n_i. At room temperature in silicon $n_i = 10^{10}$ cm^{-3}. Similarly, in the n-type silicon:

$$\phi_s = 3.8 - \phi_F \quad \text{where } \phi_F = V_T \ln\left(\frac{N_D}{n_i}\right) \tag{10.2}$$

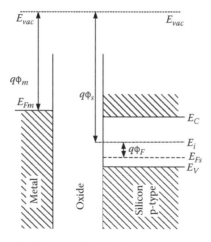

FIGURE 10.2   The location of Fermi levels in metal and in the silicon.

TABLE 10.1   Work Functions ϕ_m for Various Materials

	Si	p⁺Si	n⁺Si	Al	Mo	Au	Cu
ϕ_m (eV)	3.8	4.5	3.05	3.2	3.95	4.1	3.8

and N_D is the donor doping level. When positive voltage V_G is applied to the gate, the metal's band structure will move down by qV_G, as shown in Figure 10.3, and the depletion layer in the silicon will be formed.

Figure 10.3 shows the case with the gate (metal) biasing exactly at the threshold as the accumulation layer of carriers at the silicon surface is just being formed. This particular state, called strong inversion, is one in which the silicon surface is now an n-type level with the same electron concentration as the hole concentration in the bulk p-type silicon. It also means the voltage drop on the depletion layer is

$$V_d = 2\phi_F \tag{10.3}$$

Knowing the voltage drop V_d, the thickness w of the depletion layer can be found from

$$w = \sqrt{2\frac{\varepsilon_o\varepsilon_{Si}V_d}{qN}} = \sqrt{4\frac{\varepsilon_o\varepsilon_{Si}\phi_F}{qN}} \tag{10.4}$$

where

$$\varepsilon_{Si} = 11.8 \quad \varepsilon_o = 8.85\cdot10^{-14}\,\text{F/cm} \tag{10.5}$$

and N is the impurity concentration in the silicon substrate. The charge of ionized impurities in the depletion layer is

$$Q_d = qNw = \sqrt{4\varepsilon_{Si}\varepsilon_o\phi_F qN} = \sqrt{6.68\cdot10^{-31}\phi_F N} \tag{10.6}$$

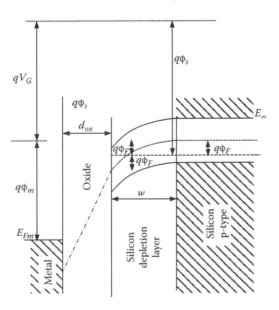

FIGURE 10.3   MOS band structure with positive voltage on the gate V_G in p-type silicon.

TABLE 10.2   Surface Charges in Silicon for Various Crystal Orientations

Crystal Orientation	$\langle 111 \rangle$	$\langle 110 \rangle$	$\langle 100 \rangle$
Number of surface states Q_{ss}/q (cm^{-2})	$1.5 \cdot 10^{11}$	10^{11}	$2 \cdot 10^{10}$
Surface charge Q_{ss} (C)	$2.4 \cdot 10^{-8}$	$1.6 \cdot 10^{-8}$	$3.2 \cdot 10^{-9}$

On top of the depletion layer charge, there exists a silicon surface charge Q_{ss} that depends on the silicon crystal orientation as indicated in Table 10.2.

Since the MOS structure can be considered as a capacitor, knowledge of the charge in silicon $Q_d + Q_{ss}$ yields the corresponding voltage

$$V = \frac{Q_d + Q_{ss}}{C_{ox}} \tag{10.7}$$

The MOS structure unit capacitance is

$$C_{ox} = \frac{\varepsilon_{ox}\varepsilon_o}{d_{ox}} = \frac{3.45 \cdot 10^{-9}}{d_{ox}\,(\mu m)} \; -> (F/cm^2) \tag{10.8}$$

where d_{ox} is the thickness of the oxide and $\varepsilon_{ox} = 3.9$. In addition to the electrical charges, the voltage drop on the depletion layer $V_d = 2\phi_F$ and the difference in the work functions should also be considered in determining the threshold voltage

$$\phi_{ms} = \phi_m - \phi_s = \phi_m - \left(3.8 + s\phi_F\right) = \phi_m - 3.8 - s\phi_F \tag{10.9}$$

where the symbol s indicates the sign, which is

$$s = \begin{pmatrix} 1 & \text{for n channel (p-type impurities)} \\ -1 & \text{for p channel (n-type impurities)} \end{pmatrix} \tag{10.10}$$

In conclusion, the threshold voltage for a MOS structure is given by

$$V_{th} = \frac{sQ_d - Q_{ss} + sqF_{imp}}{C_{ox}} + s2\phi_F + \phi_{ms} \tag{10.11}$$

Note that the effective charge can be controlled by ion implantation using F_{imp} dose (cm^{-2}), which can be made from p-type or n-type impurities, and as a result, the threshold voltage can be properly adjusted.

Example 10.1

Calculate the threshold voltage V_{th} for a MOS transistor with a p$^+$ polysilicon gate and a p-type substrate with $N_A = 10^{16}$ cm^{-3}. Assume a <100> crystal orientation and a $d_{ox} = 0.1$ μm. Find the implantation dose required for adjusting the threshold voltage to $V_{th} = +2$ V. Specify if boron or phosphor should be used for implantation. The calculations required for this example are

$$\phi_F = V_T \ln \frac{N}{n_i} = 0.0258 \ln \frac{10^{16}}{1.5_{10}10} = 0.347 \quad 2\phi_F = 0.694$$

$$s = \begin{pmatrix} 1 & \text{for n channel (p-type substrate)} \\ -1 & \text{for p channel (n-type substrate)} \end{pmatrix} = +1$$

$$\phi_{ms} = \phi_m - \phi_s \quad \phi_s = 3.8 + s\phi_F$$

$$\phi_{ms} = 4.5 - (3.8 + s\phi_F) = 4.5 - (3.8 + 0.347) = 0.353$$

$$Q_d = qNw = \sqrt{4\varepsilon_{Si}\varepsilon_o\phi_F qN} = \sqrt{6.68\cdot 10^{-31}\phi_F N} = \sqrt{6.68\cdot 10^{-31}\cdot 0.347\cdot 10^{16}} = 4.8145\cdot 10^{-8}$$

$$Q_{ss} \xrightarrow{<100>} 3.3\cdot 10^{-9}$$

$$C_{ox} = \frac{\varepsilon_{ox}\varepsilon_o}{d_{ox}} = \frac{3.45\cdot 10^{-9}}{d_{ox}(\mu m)} = \frac{3.45\cdot 10^{-9}}{0.1} = 3.45\cdot 10^{-8}\ (F/cm^2)$$

$$V_{th} = s\frac{Q_d}{C_{ox}} + s2\phi_F + \phi_{ms} - \frac{Q_{ss}}{C_{ox}} = \frac{4.8145 - 0.32}{3.45} + 0.694 + 0.353 = 2.35$$

Since the requirement specifies a $V_{th} = 2$ V, this value must be lowered by 0.35 V:

$$\Delta V_{th} = \frac{qF_{imp}}{C_{ox}} \quad F_{imp} = \frac{\Delta V_{th}C_{ox}}{q} = \frac{0.35\cdot 3.45\cdot 10^{-8}}{1.6\cdot 10^{-19}} = 7.54\cdot 10^{10}\ atm/cm^2$$

The threshold voltage obtained from Equation 10.11 is valid for the case in which the substrate has the same potential as the source for the MOS transistor. If the substrate is biased with an additional voltage V_{SB}, then due to the substrate biasing, the threshold voltage will change by

$$\Delta V_{th} = \frac{\sqrt{2\varepsilon_{Si}\varepsilon_o qN}}{C_{ox}}\left(\sqrt{2\phi_F + |V_{SB}|} - \sqrt{2\phi_F}\right) = \gamma\left(\sqrt{2\phi_F + |V_{SB}|} - \sqrt{2\phi_F}\right) \tag{10.12}$$

10.2.2 MOS Transistor Current Characteristics

Depending on the value of the drain-source voltage V_{DS}, the MOS transistor characteristics are described by different formulas. For small values of V_{DS}, known as the "linear" or "triode" region, the current is a strong function of the drain voltage, as shown in Figure 10.4. For large values of V_{DS} known as the "current saturation" or "pentode" region, the current is almost independent of the drain voltage. The current–voltage characteristics are often approximated by the formula:

$$I_D = \begin{cases} K\left[(\Delta)V_{DS} - 0.5V_{DS}^2\right](1 + \lambda V_{DS}) & \text{for } V_{DS} \leq \Delta \\ 0.5K(\Delta)^2(1 + \lambda V_{DS}) & \text{for } V_{DS} \geq \Delta \end{cases} \tag{10.13}$$

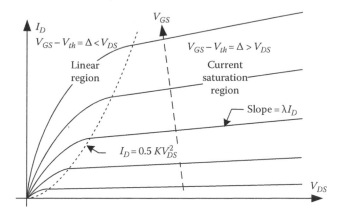

FIGURE 10.4   Output characteristics for MOS transistors.

where Δ shows how much the gate voltage V_{GS} exceeds the threshold voltage V_{th}, i.e.,

$$\Delta = V_{GS} - V_{th} \tag{10.14}$$

$$K = K' \frac{W}{L} = \mu C_{ox} \frac{W}{L} \tag{10.15}$$

The λ parameter describes the slope of the output characteristics in the current saturation region. The typical values of the λ parameter are 0.02 to 0.04 [V⁻¹]. For small signal analysis, a MOS transistor in the current saturation region can be described by two parameters r_m and r_o:

$$r_m = \frac{1}{g_m} = \frac{\Delta}{2I_D} = \frac{1}{K\Delta} = \frac{1}{\sqrt{2KI_D}} \tag{10.16}$$

$$r_o = \frac{1}{\lambda I_D} \tag{10.17}$$

Figure 10.5 shows a small-signal equivalent model of the MOS transistor. For the voltage-controlled circuit, the input capacitances need not to be included, i.e., input capacitance is a part of the previous stage. Assuming that the loading capacitance C is the capacitance of the identical transistor of the next stage $C = C_{ox}WL$, the maximum frequency of operation is

$$f_{max} = \frac{1}{2\pi(r_m \| r_o)C} \approx \frac{\sqrt{2\mu C_{ox}\frac{W}{L}I_D}}{2\pi C_{ox}WL} = \frac{\sqrt{\frac{2\mu I_D}{C_{ox}W}}}{2\pi L} \tag{10.18}$$

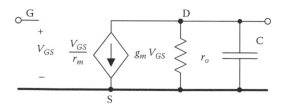

FIGURE 10.5   Small-signal equivalent model for a MOS transistor.

Example 10.2

Consider the NMOS transistor described in Example 10.1 with $V_{th} = +2$ V. Neglecting the channel-length modulation (), and assuming the following parameters: electron mobility $\lambda = 0.03$, $\mu = 600$ cm²/Vs, $L = 2\,\mu$m and $W = 20\,\mu$m, calculate

a. Drain current for $V_{GS} = 4$ V and $V_{DS} = 1$ V
b. Drain current for $V_{GS} = 4$ V and $V_{DS} = 10$ V
c. Maximum frequency for $V_{GS} = 4$ V and $V_{DS} = 10$ V

$$a. \quad K = \mu C_{ox}\frac{W}{L} = 600 \cdot 3.45 \cdot 10^{-8}\frac{20}{2} = 0.2 \cdot 10^{-3} - 200\,\mu A / V^2$$

$$I_D = K\left[(V_{GS} - V_{th})V_{DS} - \frac{V_{DS}^2}{2}\right](1 + \lambda V_{DS}) = 200 \cdot 10^{-6}\left[2 \cdot 1 - \frac{1}{2}\right](1 + 0.03) = 309\,\mu A$$

$$b. \quad I_D = \frac{K}{2}(V_{GS} - V_{th})^2(1 + \lambda V_{DS}) = \frac{200 \cdot 10^{-6}}{2}(4 - 2)^2(1 + 0.3) = 520\,\mu A$$

$$c. \quad r_m = \frac{1}{\sqrt{2KI_D}} = \frac{1}{\sqrt{2 \cdot 200\mu \cdot 520\mu}} = 2.19\text{k}\Omega$$

$$r_o = \frac{1}{\lambda I_D} = \frac{1}{0.03 \cdot 520\mu} = 64.1\text{k}\Omega$$

$$C = C_{ox}WL = 3.45 \cdot 10^{-8} \cdot 2 \cdot 10^{-4} \cdot 20 \cdot 10^{-4} = 13.8\,\text{fF}$$

$$f_{max} = \frac{1}{2\pi(r_m \| r_o)C} = \frac{1}{2\pi(2.19\text{k}\Omega \| 64.1\text{k}\Omega)13.8\,\text{fF}} = 5.45\,\text{GHz}$$

Note that only the gate oxide capacitance was included. Since all other junction parasitic capacitances were ignored, the calculated maximum frequency f_{max} is significantly larger than the actual one.

10.2.3 Second-Order Effects on a MOS Transistor

There are a number of second-order effects that significantly affect the operation of a MOS transistor, such as channel-length modulation, carrier velocity limitation, surface mobility degradation, subthreshold conduction, etc. [T99]. These effects will now be described with some detail.

10.2.3.1 Channel-Length Modulation

The effect of channel-length modulation is shown in Figure 10.6. The thickness of the depletion layer d depends on the drain-substrate voltage and is described by an equation that is similar to Equation 10.4:

$$d = \sqrt{2\frac{\varepsilon_o \varepsilon_{Si} V_D}{qN}} \tag{10.19}$$

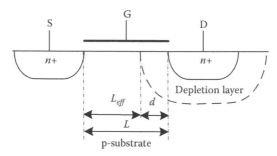

FIGURE 10.6 Channel-length modulation.

As a consequence, the effective channel length L_{eff} is shorter than the distance L between the implanted source and drain regions, and, therefore, instead of Equation 10.15, the transconductance coefficient K should be expressed as

$$K = K'\frac{W}{L_{eff}} = K'\frac{W}{L - \sqrt{2\frac{\varepsilon_o\varepsilon_{Si}V_D}{qN}}} = K'\frac{W}{L}\frac{1}{\left(1 - \frac{\eta}{L}\sqrt{V_D}\right)} \approx K'\frac{W}{L}\left(1 + \frac{\eta}{L}\sqrt{V_D}\right) \tag{10.20}$$

As indicated in Figure 10.6, the effect of channel-length modulation becomes more significant with the reduction in channel length. Thus, the textbook formula, i.e., Equation 10.13, using the λ parameter to describe the effect of channel-length modulation, and also employed in the basic Spice MOS transistor models (Level 2 and Level 3), is not correct. Equation 10.13 implies that for a given drain voltage, the ratio between d and L_{eff} is always the same; however, clearly this cannot be true for different channel lengths. With longer transistor channels, the effect of channel-length modulation is less significant and the output resistance actually increases with the channel-length L.

As Equation 10.19 indicates, the channel-length modulation can be reduced only by increasing the impurity concentration N in the silicon. Without a significant increase in impurities, there would be a punch-through effect in a short channel transistor. Punch-through occurs when the depletion layer thickness d becomes equal to the channel length L, as illustrated in Figure 10.6. Unfortunately, a larger impurity concentration in the substrate leads to large parasitic capacitances, and therefore a reduction in the transistor size does not result in a similar reduction of the parasitic capacitances. For example, it is interesting to note that a significant reduction in the size of transistors in the last decade from 0.3 μm to 0.05 μm did not result in any noticeable increase in the computer clock frequencies. Of course, there are also other factors that are limiting clock frequencies. One of the most important limitations is an ability to dissipate heat, since power dissipation in computer chips is proportional to clock frequency.

10.2.3.2 Effect of Carrier Velocity Saturation

Most of the textbook equations are derived with an assumption that the mobility of carriers is constant and independent of the electrical field. Actually, in silicon, the maximum carrier velocity for both electrons and holes is $v_{sat} = 10^7$ cm/s $= 10^{11}$ μm/s. As indicated in Figure 10.7, the critical field in which the velocity saturates is about 1 V/μm. Note that the channel length in modern MOS transistors is significantly smaller than 1 μm. Therefore, even reducing the supply voltage below 2 V produces an average electrical field in MOS transistors, which is significantly larger than 1 V/μm. Fortunately, large electrical fields exist near the drain and significantly smaller electrical fields are near the source, and these smaller electric fields shape the transistor current characteristics, so Equations 10.13 are

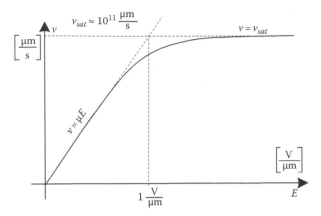

FIGURE 10.7   Carriers' velocity as a function of the electrical field in silicon.

still in use even if the drain current in the current saturation region no longer is described by a quadratic equation:

$$I_D \sim (V_{GS} - V_{th})^n \tag{10.21}$$

where n has value between 1 and 2.

10.2.3.3 Carrier Mobility Degradation near the Surface

The key feature of MOS transistor operation is the fact that most of the current flows near the silicon surface and, as a result of crystal imperfections, carrier mobilities near the surface are reduced. This effect becomes even more significant with increased gate voltage when a large electrical field is created perpendicular to the direction of current flow. As a consequence, with larger gate voltages, a larger number of carriers are accumulated near the surface. However, these carriers are moving slower due to surface mobility degradation and the fact that the drain current is not increasing as fast as would be predicted by Equation 10.13 with a quadratic relationship. In addition to mobility degradation in the transverse electric field, i.e., the gate voltage, there is a strong degradation due to the longitudinal electric field, i.e., the drain voltage. As a result, experimental characteristics for short-channel MOS transistors exhibit almost linear dependence with gate voltage.

10.2.3.4 Subthreshold Conduction

As illustrated in Section 2.1, as the gate voltage increases, carriers gradually accumulate near the surface. The assumption, that suddenly there is a strong surface inversion where the concentration of minority carriers near the surface is exactly the same as the concentration of carriers in the substrate, is very artificial. Below and near the threshold, the drain current in a MOS transistor is described by an exponential relationship:

$$I_D = I_{ON} \exp\left(\frac{V_{GS} - V_{th} - \eta V_T}{\eta V_T} \right) \quad \text{for } V_{GS} < V_{th} + \eta V_T \tag{10.22}$$

where

$$I_{ON} = \frac{K}{2}(\eta V_T)^2 \tag{10.23}$$

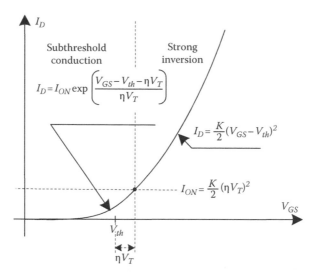

FIGURE 10.8 Drain current versus the gate-source voltage in subthreshold conduction and the strong inversion regions.

$V_T = kT/q$ is the thermal potential and η depends on the device geometry and lies between 1.5 and 2.5 (Figure 10.8). The subthreshold conduction is the reason why MOS transistors are actually never completely turned OFF and there is always some current leak through MOS transistors. When the popular CMOS technology was first developed, one of the underlying assumptions was that along the power path, there would never be even one MOS transistor in the OFF state, so power would not be taken from power supply. While a transistor can be in the OFF state, there is a leakage current caused by the subthreshold conduction in CMOS VLSI circuits, which can be very significant in situations where the number of MOS transistors exceeds one billion.

10.3 Junction Field Effect Transistor

The principle of operation for a JFET is quite different than that of a MOS transistor [S05,N06]. The current flows through a thin semiconductor layer that is surrounded by a gate made of semiconductor material of the opposite type, as shown in Figure 10.9. The gate–channel (source) junction is biased in the reverse direction, so there is no gate current. The thickness of the depletion regions controls current flow between source S and drain D. The thickness of the depletion regions is the function of the gate voltage:

$$d = \sqrt{2 \frac{\varepsilon_o \varepsilon_{Si} V_{channel}}{qN}} \tag{10.24}$$

As indicated in Figure 10.9a, the thickness of the depletion layer d is constant only if there is no voltage applied between source and drain. With applied drain voltage, this issue becomes much more complicated because the channel width is a nonlinear function of distance, and there is a nonlinear voltage drop along the channel length. As a consequence, there is a different gate-channel voltage and a different depletion layer thickness d along the channel, as shown in Figure 10.9b. With any further increase in the gate voltage, the channel is pinched off and only a depletion layer exists between point x in Figure 10.9c and the drain D. In that region, the operation of a JFET is very interesting. In the n-type JFET case shown in Figure 10.9, electrons in the channel close to the source are being pushed away by the negative gate voltage. This is why the depletion layer is formed. But the large positive drain voltage creates an electric field between point x and drain, and this electric field swipes all electrons that

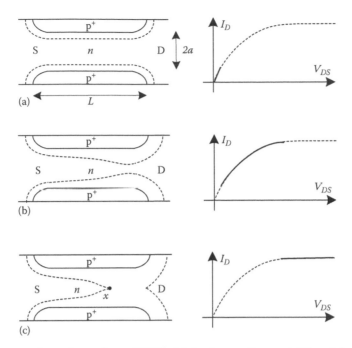

FIGURE 10.9 Cross sections of an n-channel JFET with fixed gate voltage and different drain-source voltages. Characteristics on the right side of the figure show the corresponding modes of operations.

have reached point x through the pinch-off region near the drain. Interestingly, when in this mode of operation, the drain voltage does not have a direct effect on the drain current, and the drain current is determined by the triangular shape of the channel region to the left of point x. Of course, in a manner analogous to that of the MOS transistor, the JFET also experiences second-order effects from the channel-length modulation. With an increase in drain voltage, the point x is moved to the left, which reduces the effective resistance of the "triangular-like channel region" and results in a slight increase in drain current.

The geometry of the JFET is usually more complicated than the one shown in Figure 10.9. Also, in most JFET devices, there is a nonlinear impurity distribution in the channel. Therefore, a derivation of the current–voltage characteristics would be either too simplistic or too complicated. As a consequence, various approximation formulas are being used. The most popular equations for determining JFET drain currents are

$$\text{if } V_{GS} < V_P \quad I_D = 0 \tag{10.25}$$

$$\text{if } V_{GS} \geq V_P \quad I_D = \begin{cases} 2I_{DSS}\left[\left(\dfrac{V_{GS}}{V_P}-1\right)\dfrac{V_{DS}}{V_P}-0.5\left(\dfrac{V_{DS}}{V_P}\right)^2\right](1+\lambda V_{DS}) & \text{for } V_{DS} \leq V_{GS}-V_P \\[2em] I_{DSS}\left(\dfrac{V_{GS}}{V_P}-1\right)^2(1+\lambda V_{DS}) & \text{for } V_{DS} \geq V_{GS}-V_P \end{cases} \tag{10.26}$$

where V_P is the pinch-off voltage and I_{DSS} is the drain current for the case in which the gate is connected with source and a relatively large drain voltage is applied. Both V_P and I_{DSS} can be related to the JFET geometry as shown in Figure 10.9.

The pinch-off voltage V_P can be calculated as

$$V_P = \frac{qa^2 N}{2\varepsilon\varepsilon_0} \tag{10.27}$$

and the drain current I_{DSS} can be calculated from Ohm's law by dividing V_P by the resistance of the channel between source and drain. The nonlinear distribution of the resistance is usually assumed to be many times larger than the resistance without the applied drain voltage, as shown in Figure 10.9a:

$$R_{\text{channel}} = 3\frac{\rho L}{2aW} \tag{10.28}$$

where a is half of the thickness of the channel, L is the channel length, and W is the channel width:

$$I_{DDS} = \frac{V_P}{3}\frac{2aW}{\rho L} = V_P\frac{2aW}{3L}\sigma = V_P\frac{2aW}{3L}q\mu N \tag{10.29}$$

Example 10.3

An *n*-channel junction JFET has a uniformly doped channel 2 µm thick, 20 µm wide, and 20 µm long with $N_D = 0.5 \cdot 10^{16}$ cm^{-3}. Determine the pinch-off voltage as well as the drain current for $V_{GS} = -1$ V and $V_{DS} = 10$ V. Assume $\mu_n = 500$ cm^2/Vs. Neglect the channel-length modulation effects.

Since the gate concentration is not given, the increase in potential is neglected:

$$V_P = \frac{qa^2 N_D}{2\varepsilon\varepsilon_0} = \frac{1.6\cdot10^{-19}(10^{-4})^2 0.5\cdot10^{16}}{2\cdot11.8\cdot8.85\cdot10^{-14}} = 3.83\text{V}$$

$$I_{DDS} = V_P\frac{2aW}{3L}q\mu N = \frac{2\cdot10^{-4}\cdot20\cdot10^{-4}}{20\cdot10^{-4}}(1.6\cdot10^{-19}\cdot500\cdot0.5\cdot10^{16})\frac{3.83}{3} = 1.0\text{ mA}$$

For $V_{GS} = -1$ V in an n-channel JFET, the pinch-off voltage must be negative, e.g., $V_P = -3.83$ V, and neglecting the channel-length modulation effect $\lambda = 0$:

$$I_D = I_{DDS}\left(\frac{V_{GS}}{V_P} - 1\right)^2 = 1.0\text{ mA}\left(\frac{-1-(-3.83)}{-3.83}\right)^2 = 0.55\text{ mA}$$

Note the similarities in Equations 10.13 for the MOS transistor and Equation 10.26 for the JFET. Actually, if the pinch-off voltage V_P and the drain initial current I_{DSS} are replaced by

$$V_{th} = V_P \quad \text{and} \quad K = \frac{2I_{DSS}}{V_P^2} \tag{10.30}$$

then, to calculate the drain current in a JFET, we need not employ the JFET equations (10.26) and use instead the well-known equations for MOS transistors working in the depletion mode:

$$I_D = \begin{cases} K[(V_{GS}-V_{th})V_{DS}-0.5V_{DS}^2](1+\lambda V_{DS}) & \text{for } V_{DS} \le V_{GS}-V_{th} \\ 0.5K(V_{GS}-V_{th})^2(1+\lambda V_{DS}) & \text{for } V_{DS} \ge V_{GS}-V_{th} \end{cases} \tag{10.31}$$

Because current flow in JFETs is far from the surface, JFETs have a significantly smaller $1/f$ noise level and they are a preferred choice for low-noise amplifiers. Another advantage of the JFET is that it is relatively safe to exceed gate-source or gate-drain break voltages. In the case of MOS transistors, large gate voltages may result in a permanent break-through of the oxide, leading to the destruction of the transistor. It is however very difficult to integrate several JFETs into one chip. Another disadvantage of the JFET is that the gate voltage should always have a polarity opposite that of the drain voltage, and this makes it almost impossible to fabricate digital circuits using JFETs.

10.4 Static Induction Transistor

Static induction devices were invented by J. Nishizawa [NTS75]. The device has characteristics similar to that of the vacuum triode. Its fabrication is relatively difficult, and Japan is actually the only country where a family of static induction devices was successfully fabricated [W99].

The cross section of the static induction transistor (SIT) is shown in Figure 10.10. In this *n*-channel structure, the gate is biased with a negative potential and the drain has a significantly large positive potential. There are two reverse-biased junctions; one between gate and source, and second between gate and drain. Because n⁻ regions have a very low concentration of impurities (10^{14} cm^{-3} or below) with very small applied voltages or simply the junctions' built-in potential, these n⁻ regions are depleted of carriers. As a consequence, gate-drain voltages form a relatively complex potential surface. Samples of such surfaces for a gate voltage equal −10 V and a drain voltage equal +50 V are shown in Figure 10.11. Note that gate and drain voltages create opposite electric fields near the source. With an increase in gate voltage, the height of the potential barrier increases, as shown in Figure 10.11, while the larger drain

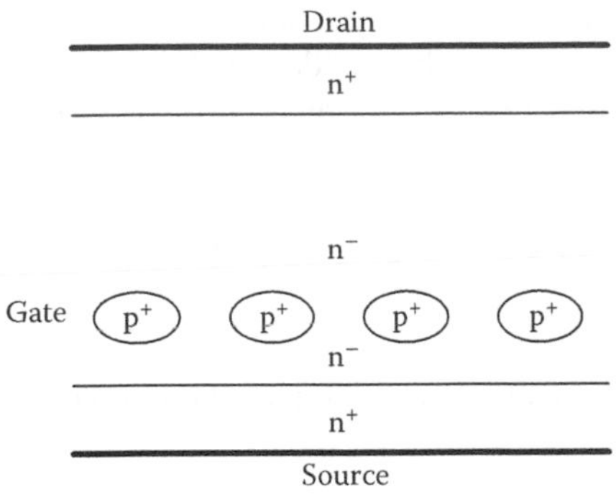

FIGURE 10.10   Cross section of the SIT.

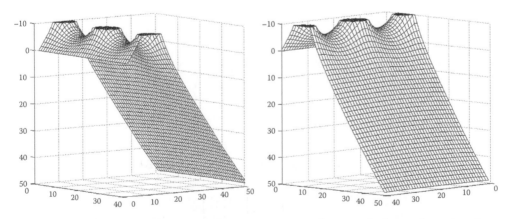

FIGURE 10.11   Potential distribution in the SIT with a gate voltage of −10 V and a drain voltage of +50 V.

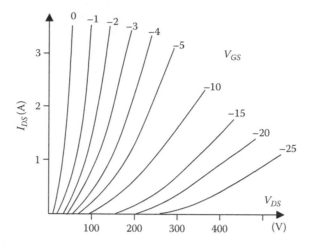

FIGURE 10.12   Typical current–voltage characteristics for the SIT.

voltage leads to a lowering of the potential barrier. Typical current–voltage characteristics for the SIT are shown in Figure 10.12.

10.4.1 Theory of SIT Operation for Small Currents

Consider first a derivation of the formula for the one-dimensional electron current flow through a potential with a parabolic shape. In the *n*-channel device, the electron current is described by a differential equation that includes both drift and diffusion of carriers [N02,WSM92]:

$$J_n = -qn(x)\mu_n \frac{d\phi(x)}{dx} + qD_n \frac{dn(x)}{dx} \tag{10.32}$$

where $D_n = \mu_n V_T$ and $V_T = \dfrac{kT}{q}$. By multiplying both sides of the equation by [PW80]

$$\exp\left(-\frac{\phi(x)}{V_T}\right) \tag{10.33}$$

and rearranging

$$J_n \exp\left(-\frac{\phi(x)}{V_T}\right) = qD_n \frac{d}{dx}\left[n(x)\exp\left(-\frac{\phi(x)}{V_T}\right)\right] \tag{10.34}$$

After integration from source x_S to drain x_D:

$$J_n = qD_n \frac{n(x_D)\exp\left(-\dfrac{\phi(x_D)}{V_T}\right) - n(x_S)\exp\left(-\dfrac{\phi(x_S)}{V_T}\right)}{\displaystyle\int_{x_S}^{x_D} \exp\left(-\frac{\phi(x)}{V_T}\right)dx} \tag{10.35}$$

By inserting

$$\phi(x_S) = 0 \quad n(x_S) = N_S$$
$$\phi(x_D) = V_D \quad n(x_D) = N_D \tag{10.36}$$

Equation 10.35 reduces to

$$J_n = \frac{qD_nN_S}{\displaystyle\int_{x_S}^{x_D} \exp\left(-\frac{\phi(x)}{V_T}\right)dx} \tag{10.37}$$

Equation 10.37 is very general and it describes the current flow over a potential barrier with a shape given by $\varphi(x)$. This equation can be used not only in SIT devices but also in bipolar transistors or it can be used to calculate the subthreshold conduction in a MOS transistor.

Note that because of this exponential relationship, only the shape of the potential distribution near the top of the potential barrier is important, and in the SIT, this shape can be approximated (see Figure 10.11) by quadratic equations along the x direction and across the y direction of the channel:

$$\phi(y,x) = \Phi\left(\left(\frac{x}{L}\right)^2 - \left(\frac{y}{W}\right)^2\right) \tag{10.38}$$

where L is the effective channel length, W is the effective channel width, and Φ is the height of the potential barrier in the center of the channel. Using (10.38) and integrating (10.37) first along the channel and then across it leads to a simple formula for the drain current of a SIT as a function of the height of potential barrier:

$$I_D = qD_pN_SZ\frac{W}{L}\exp\left(\frac{\Phi}{V_T}\right) \tag{10.39}$$

where Φ is the potential barrier height in reference to the source potential, and N_S is the electron concentration at the source. The W/L ratio describes the shape of the potential saddle in the vicinity of the barrier, and Z is the length of the source strip.

Since the barrier height Φ can be a linear function of gate and drain voltages:

$$I_D = qD_pN_SZ\frac{W}{L}\exp\left(\frac{aV_{GS} + bV_{DS} + \Phi_0}{V_T}\right) \tag{10.40}$$

The actual characteristics of the SIT device on a logarithmic scale are shown in Figure 10.13. Indeed, for a small current range, the characteristics have an exponential character, but this is not true for large currents where the space charge for moving electrons affects the potential distribution.

10.4.2 Theory of SIT for Large Currents

For large current levels, the SIT current is controlled by the space charge of the moving carriers [PW81,MW83]. In the one-dimensional case, the potential distribution is described by the Poisson equation:

$$\frac{d^2\phi}{dx_2} = -\frac{\rho(x)}{\varepsilon_{Si}\varepsilon_o} = \frac{I_{DS}}{Av(x)} \tag{10.41}$$

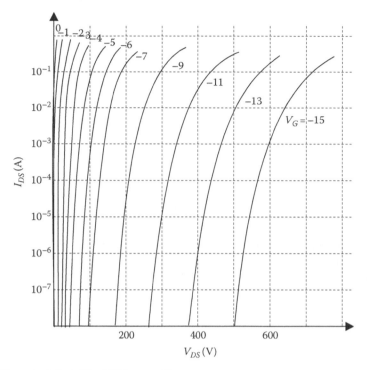

FIGURE 10.13   Characteristics of the SIT for a small current range.

where A is the effective device cross section and $v(x)$ is the carrier velocity. For a small electrical field, $v(x)=\mu E(x)$ and the solution of (10.41) is

$$I_D = \frac{9}{8}V_{DS}^2\mu\varepsilon_{Si}\varepsilon_o\,\frac{A}{L^3} \tag{10.42}$$

For a large electrical field, $v(x) = v_{sat}$ and

$$I_D = 2V_{DS}v_{sat}\varepsilon_{Si}\varepsilon_o\,\frac{A}{L^2} \tag{10.43}$$

where L is the channel length and $v_{sat} \approx 10^{11}$ μm/s is the carrier saturation velocity. In practical devices, the current–voltage relationship is described by an exponential relationship (10.9) for small currents, a quadratic relationship (10.11), and finally for large voltages, by an almost linear relationship (10.12). SIT characteristics drawn in linear and logarithmic scales are shown in Figures 10.12 and 10.13, respectively.

10.4.3 Bipolar Mode of Operation of the SIT

The bipolar mode of operation for the SIT was first reported in 1977 [NW77a]. Several complex theories for the bipolar mode of operation were developed [NTT86,NOC82], but actually the simple formula (10.37) works well not only for the typical mode of SIT operation, but also for the bipolar mode as well. Furthermore, the same formula works very well for classical bipolar transistors. Typical characteristics of the SIT, operating in normal and in bipolar modes, are shown in Figure 10.14.

In a SIT, a virtual base is formed, not by impurity doping but rather by a potential barrier that is induced by the gate voltage. As a consequence, in the bipolar mode of operation, the SIT may have a

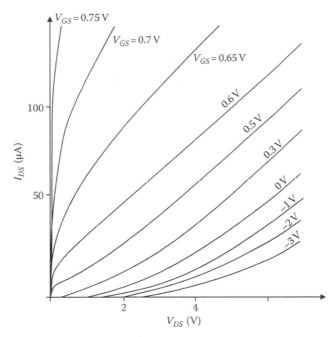

FIGURE 10.14 SIT transistor characteristic operating in both normal and the bipolar modes, $I_D = f(V_{DS})$ with V_{GS} as parameter.

very large current gain β. Also, the SIT operates with a very low level of impurity concentration and its parasitic capacitances are very low. When a bipolar transistor in integrated injection logic (I^2L) was replaced by a SIT, the time-delay product of such a device was reduced almost 100 times [NW77]. Such a drastic improvement in the power-delay product is possible because the SITL structure has a significantly smaller junction parasitic capacitance, and, furthermore, the voltage swing is reduced.

Another interesting application of the SIT is a replacement for Schottky diodes in the protection of a bipolar transistor against saturation, leading to a much faster switching time. The use of a SIT [WMS84] is more advantageous than that obtained with a Schottky diode since it does not require additional area on a chip and it does not introduce additional capacitance between the base and the collector.

10.5 Lateral Punch-Through Transistor

The fabrication of SITs is a very challenging endeavor. The channel area requires very low impurity concentration ($N < 10^{14}$ cm^{-3}), and, at least, a part of the channel near the source has to be made using an epitaxial layer, as shown in Figure 10.10, which should be about 100 times more pure than that which is considered an epitaxial layer with low impurity concentration ($N = 10^{16}$ cm^{-3}). The second difficult issue is the creation of a buried gate region. With high temperature epitaxial growth and subsequent diffusion processes, it is extremely difficult to concentrate gate impurity in one place without spreading it into the channel area and actually closing the channel. Only a couple of Japanese companies (Yamaha and Sony) were able to develop a fabrication process for SIT devices.

The lateral punch-through transistor (LPTT), which has characteristics that are similar to the SIT, can be fabricated with a very simple process [WJ82]. The cross section of the LPTT device is shown in Figure 10.15, and its characteristics are shown in Figure 10.16. The LPTT device, in contrast to SIT device, must use the same type of the impurity in the channel as is used in the gate. For the n-type channel, p$^-$ must be used instead of n$^-$. With an increase in positive drain voltage, the thickness of the drain depletion region increases. Once the depletion layer reaches the source, the punch-through current will start to flow between source and drain. The current can be stopped by applying a negative voltage on the

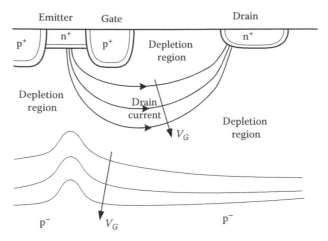

FIGURE 10.15 Cross section of an LPTT.

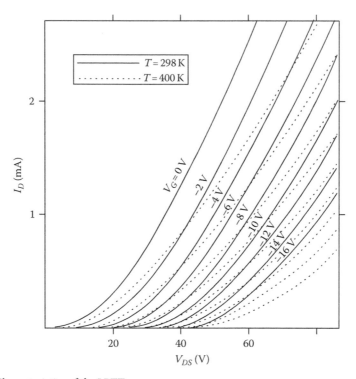

FIGURE 10.16 Characteristics of the LPTT.

gate. Eventually, the electrical field near the source will be affected by both the gate and source voltages. Because of its proximity to the source, the gate voltage can be much more effective in controlling the device current. Typical current–voltage characteristics are shown in Figure 10.16.

The concept of the LPTT can be extended to a punch-through MOS transistor (PTMOS), where the current is controlled by the MOS gate [W83a,WJF84]. The principle of operation of such a device is shown in Figure 10.17. In this device, instead of the implanted p-type gate, the gate is formed by an accumulation p-type layer under the MOS gate, as shown in Figure 10.17a, and the punch-through current can be controlled by a potential applied to the adjacent p-type region, which is normally

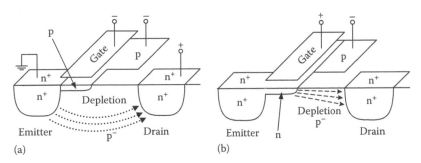

FIGURE 10.17 Punch-through MOS transistor: (a) transistor in the punch-through mode for a negative gate potential and (b) transistor in the on-state for a positive gate potential.

biased with a large negative potential. With positive voltage on the MOS gate, the n-type accumulation layer is formed under the gate and the current may flow easily from source to drain, as indicated in Figure 10.17b. Note that in typical CMOS technology, in order to prevent a punch-through phenomenon between source and drain, a large impurity doping in the substrate must be used, which in turn leads to larger parasitic capacitances. In a MOS transistor with shorter channels, a larger impurity concentration must be used to prevent the punch-through phenomenon, and parasitic capacitances are also larger. The PTMOS takes advantage of the punch-through phenomenon and the substrate impurity concentration can be very low, which leads to very small parasitic capacitances and a significant reduction in power consumption for digital circuits operating with very large clock frequencies.

The PTMOS transistor has several advantages over the traditional MOS transistor:

1. The gate capacitance is very small.
2. The carrier move with a velocity close to saturation velocity.
3. The substrate doping is much lower and the existing depletion layer leads to a much smaller drain capacitance.

The device operates in a fashion that is similar to that of the MOS transistor in subthreshold conditions, but this process occurs at much higher current levels. Such a "bipolar mode" of operation may have many advantages in VLSI applications.

10.6 Power MOS Transistors

MOS transistors have a relatively small transconductance in comparison to bipolar transistors. Therefore, in power electronic applications, the integrated device structures usually should consist of thousands of transistors connected in parallel. There are two types of power MOS transistors: VMOS (shown in Figure 10.18a) and DMOS (shown in Figure 10.18b). In the VMOS structure, MOS gates and channels are formed on etched surfaces. This way many transistors can be efficiently connected together. VMOS uses the silicon surface very efficiently.

The DMOS transistor does not use the chip area as efficiently as the VMOS transistor, but it can be fabricated with much larger breakdown voltages. In DMOS, a fragile MOS structure is protected from a large electric field by a concept borrowed from the SIT [NTS75] and the high-voltage Schottky diode [W83]. The n⁻ area under the gate, as illustrated in Figure 10.18b, is depleted from carriers, and neighboring p-type regions work as electrostatic screens as is done in SIT devices. As a result, this transistor may withstand much larger drain voltages and also the effect of channel-length modulation is significantly reduced. The latter effect leads to larger output resistances of the transistor. Therefore, the drain current is less sensitive to drain voltage variations. In fact, the DMOS structure can be considered as a composition of the MOS transistor and the SIT, as shown in Figure 10.19.

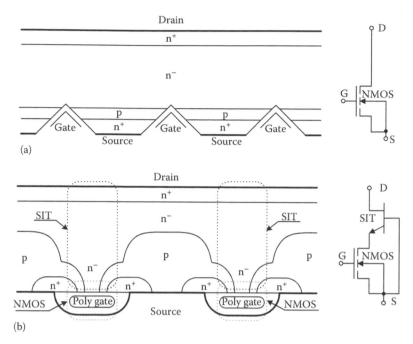

FIGURE 10.18   Cross section of power MOS transistors: (a) VMOS and (b) DMOS.

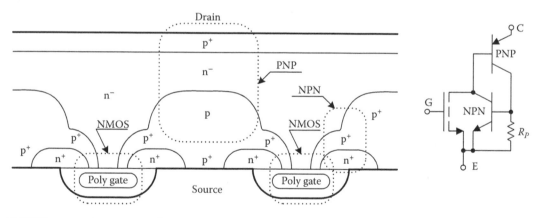

FIGURE 10.19   Cross section of an IGBT and its internal equivalent circuit.

The major disadvantage of power MOS transistors is the relatively larger drain series resistance and much smaller transconductance in comparison to bipolar transistors. Both of these parameters can be improved dramatically if the n^+ layer near the drain is replaced by p^+ layer as is shown in Figure 10.19. This way an integrated structure is being built where its equivalent diagram consists of a MOS transistor integrated with a bipolar transistor, as shown in Figure 10.19. Such a structure has a transconductance that is β times larger, where β is the current gain of the PNP bipolar transistor, and a much smaller series resistance due to the conductivity modulation effect caused by holes injected into the lightly doped drain region. Such device is known as insulated gate bipolar transistor (IGBT). An IGBT can work with large currents and voltages. Its main disadvantage is a large switching time that is limited primarily by the poor switching performance of the bipolar transistor. Another difficulty is related to a possible latch-up action of four layer $n^+pn^-p^+$ structure. This undesired effect could be suppressed by using a heavily doped p^+ region in the base of the NPN structure, which leads to a significant reduction in the current gain of this parasitic transistor, shown in Figure 10.19. The gain of the PNP transistor must be kept large so the transconductance of the entire device can be large. IGBT transistors may have breakdown voltages over 1000 V, with turn-off times in the range from 0.1 to 0.5 µs. In addition, they may operate with currents above 100 A with a forward voltage drop of about 3 V.

References

[E97] R. Enderlein, *Fundamentals of Semiconductor Physics and Devices*, World Scientifics, New York, 1997.

[MW83] R. H. Mattson and B. M. Wilamowski, Punch-through devices operating in space-charge-limited modes, *IEEE International Workshop on the Physics of Semiconductor Devices*, Delhi, India, December 5–10, 1983.

[N02] Kwok K. Ng, *The Complete Guide to Semiconductor Devices*, Wiley-IEEE Press, New York, 2002.

[N06] D. Neamen, *An Introduction to Semiconductor Devices*, McGraw-Hill, New York, 2006.

[NOC82] J. Nishizawa, T. Ohmi, and H. L. Chen, Analysis of static characteristics of a bipolar-mode SIT (BSIT), *IEEE Transactions on Electron Devices*, **29**, 8, 1233–1244, 1982.

[NTS75] J. Nishizawa, T. Terasaki, and J. Shibata, Field-effect transistor versus analog transistor (static induction transistor), *IEEE Transactions on Electron Devices*, **22**, 4, 185–197, 1975.

[NTT86] Y. Nakamura, H. Tadano, M. Takigawa, I. Igarashi, and J. Nishizawa, Experimental study on current gain of BSIT, *IEEE Transactions on Electron Devices*, **33**, 6, 810–815, 1986.

[NW77] J. Nishizawa and B. M. Wilamowski, Integrated logic—State induction transistor logic, *International Solid State Circuit Conference*, Philadelphia, PA, pp. 222–223, 1977.

[NW77a] J. Nishizawa and B. M. Wilamowski, Static induction logic—A simple structure with very low switching energy and very high packing density, *International Conference on Solid State Devices*, Tokyo, Japan, pp. 53–54, 1976; and *Journal of Japanese Society of Applied Physics*, **16-1**, 158–162, 1977.

[PW80] P. Plotka and B. M. Wilamowski, Interpretation of exponential type drain characteristics of the SIT, *Solid-State Electronics*, **23**, 693–694, 1980.

[PW81] P. Plotka and B. M. Wilamowski, Temperature properties of the static induction transistor, *Solid-State Electronics*, **24**, 105–107, 1981.

[R99] D. J. Roulston, *Introduction to the Physics of Semiconductor Devices*, Oxford University Press, New York, 1999.

[S05] B. Streetman, *Solid State Electronic Devices*, Prentice Hall, New York, 2005.

[T99] Y. Tsividis, *Operation and Modeling the MOS Transistor*, McGraw-Hill, New York, 1999.

[W83] B. M. Wilamowski, Schottky diodes with high breakdown voltage, *Solid-State Electronics*, **26**, 5, 491–493, 1983.

[W83a] B. M. Wilamowski, The punch-through transistor with MOS controlled gate, *Physica Status Solidi (a)*, **79**, 631–637, 1983.

[W99] B. M. Wilamowski, High speed, high voltage, and energy efficient static induction devices, *12 Symposium of Static Induction Devices—SSID'99*, Tokyo, Japan, pp. 23–28, April 23, 1999.

[WJ82] B. M. Wilamowski and R. C. Jaeger, The lateral punch-through transistor, *IEEE Electron Device Letters*, **3**, 10, 277–280, 1982.

[WJF84] B. M. Wilamowski, R. C. Jaeger, and J. N. Fordemwalt, Buried MOS transistor with punch-through, *Solid State Electronics*, **27**, 8/9, 811–815, 1984.

[WMS84] B. M. Wilamowski, R. H. Mattson, and Z. J. Staszak, The SIT saturation protected bipolar transistor, *IEEE Electron Device Letters*, **5**, 263–265, 1984.

[WSM92] B. M. Wilamowski, Z. J. Staszak, and R. H. Mattson, An electrical network approach to the analyses of semiconductor devices, *IEEE Transactions on Education*, **35**, 2, 144–152, 1992.

11

Noise in Semiconductor Devices

Alicja
Konczakowska
*Gdansk University
of Technology*

Bogdan M.
Wilamowski
Auburn University

11.1 Introduction

Noise (a spontaneous fluctuation in current or in voltage) is generated in all semiconductor devices. The intensity of these fluctuations depends on device type, its manufacturing process, and operating conditions. The resulted noise, as a superposition of different noise sources, is defined as an inherent noise. The equivalent noise models (containing all noise sources) are created for a particular device: for example, bipolar transistor (BJT), junction field effect transistor (JFET), or metal oxide semiconductor field effect transistor (MOSFET).

The inherent noise of semiconductor devices is considered as an undesired effect and sometimes is referred to as a useful signal. It is specially important for input (front-end) stages of electronic systems. However, the inherent noise can also be used for the quality assessment of semiconductor devices. Quite often it has been used as an important factor during the development of the production process of new semiconductor devices. Inherent noise is also used for the classification of semiconductor devices into groups with different quality and reliability.

The most important sources of noise are thermal noise, shot noise, generation-recombination noise, $1/f$ noise (flicker noise), $1/f^2$ noise, burst noise or random telegraph signal (RTS) noise, and avalanche noise. Detailed description of noise sources is presented in references [1–6].

11.2 Sources of Noise in Semiconductor Devices

11.2.1 Thermal Noise

Thermal noise is created by random motion of charge carriers due to thermal excitation. This noise is sometimes known as the Johnson noise. In 1905, Einstein presented his theory of fluctuating movement of charges in thermal equilibrium. This theory was experimentally verified by Johnson in 1928. The thermal motion of carriers creates a fluctuating voltage on the terminals of each resistive element.

The average value of this voltage is zero, but the power on its terminals is not zero. The internal noise voltage source or current source is described by the Nyquist equation

$$\bar{v}_{th}^2 = 4kTR\Delta f \quad \bar{i}_{th}^2 = \frac{4kT\Delta f}{R} \tag{11.1}$$

where
  k is the Boltzmann constant
  T is the absolute temperature
  $4kT$ is equal to $1.61 \cdot 10^{-20}\,\text{V} \cdot \text{C}$ at room temperature

The thermal noise is proportional to the frequency bandwidth Δf. It can be represented by the voltage source in series with resistor R or by the current source in parallel to the resistor R. The maximum noise power can be delivered to the load when $R_L = R$. In this case, maximum noise power in the load is $kT\Delta f$. The noise power density, $dP_n/df = kT$, is independent of frequency. Thus, the thermal noise is the white noise. The RMS noise voltage and the RMS noise current are proportional to the square root of the frequency bandwidth Δf. The thermal noise is associated with every physical resistor in the circuit.

The spectral density function of the equivalent voltage and current thermal noise are given by

$$S_{thvR} = 4kTR \tag{11.2}$$

or

$$S_{thiG} = 4kTG \tag{11.3}$$

These noise spectral densities are constant up to 1 THz and they are proportional to temperature and to resistance of elements, and as such can be used to indirectly measure the following:

- The device temperature
- The base-spreading resistance of BJT
- The quality of contacts and connections

11.2.2 Shot Noise

Shot noise is associated with a discrete structure of electricity and the individual carrier injection through the *pn* junction. In each forward-biased junction, there is a potential barrier that can be overcome by the carriers with higher thermal energy. This is a random process and the noise current is given by

$$\bar{i}_{sh}^2 = 2qI\Delta f \tag{11.4}$$

The noise spectral density function of the shot noise is temperature independent (white noise) and it is proportional to the junction current

$$S_{shi} = 2qI \tag{11.5}$$

where
  q is the electron charge
  I is the forward junction current

Shot noise is usually considered as a current source connected in parallel to the small signal junction resistance. The measurement of shot noise in modern nanoscale devices is relatively difficult since measured values of current are in the range of 10–100 fA.

Shot noise has to be proportional to the current and any deviation from this relation can be used to evaluate parasitic leaking resistances. It can be used for diagnosis of photodiodes, Zener diodes, avalanche diodes, and Schottky diodes.

11.2.3 Generation-Recombination Noise

Generation-recombination noise is caused by the fluctuation of number of carriers due to existence of the generation-recombination centers. Variation of number of carriers leads to changes of device conductance. This type of noise is a function of both temperature and biasing conditions. The spectral density function of the generation-recombination noise is described by

$$\frac{S_{g-r}(f)}{N^2} = \frac{(\overline{\Delta N})^2}{N^2} \cdot \frac{4\tau}{1 + (2\pi f \cdot \tau)^2} \tag{11.6}$$

where
$(\overline{\Delta N})^2$ is the variance of the number of carriers N
τ is the carrier lifetime

Spectral density is constant up to the frequency $f_{g-r} = 1/(2\pi\tau)$, and after that it decreases proportionally to $1/f^2$.

In the case, when there are several types of generation-recombination centers with different carrier life time, the resultant noise spectrum will be a superposition of several distributions described by (11.6). Therefore, the spectral distribution of noise can be used to investigate various generation-recombination centers. This is an alternative method to deep-level transient spectroscopy (DLTS) to study generation-recombination processes in semiconductor devices.

11.2.4 1/*f* Noise

1/*f* noise is the dominant noise in the low frequency range and its spectral density function is proportional to 1/*f*. This noise is present in all semiconductor devices under biasing. This noise is usually associated with material failures or with imperfection of a fabrication process. Most of research results conclude that this noise exists even for very low frequencies up to 10^{-6} Hz (frequency period of several weeks). This noise is sometimes used to model fluctuation of device parameters with time. There are two major models of 1/*f* noise:

- Surface model developed by McWhorter in 1957 [7]
- Bulk model developed by Hooge in 1969 [8]

The simplest way to obtain 1/*f* characteristics is to superpose many different spectra of generation-recombination noise, where free carriers are randomly trapped and released by centers with different life times. This was the basic concept behind the McWhorter model where it was assumed that

- In the silicon oxide near the silicon surface there are uniformly distributed trap centers.
- The probability of the carrier penetration to trap centers is decreasing exponentially with the distance from the surface.
- Time constants of trap centers increases with the distance from the surface.
- Trapping mechanisms by separate centers are independent.

The resulted noise spectral density function is given by

$$S_{1/f} \propto (\overline{\Delta N})^2 \int_{\tau_1}^{\tau_2} \frac{1}{\tau} \frac{4\tau}{1 + \omega\tau^2} \cdot d\tau = (\overline{\Delta N})^2 \cdot \frac{1}{f} \quad \text{for} \quad \frac{1}{\tau_2} \ll \omega \ll \frac{1}{\tau_1} \tag{11.7}$$

The spectral density function is constant up to frequency $f_2 = 1/(2\pi\tau_2)$, then it is proportional to $1/f$ between f_2 and $f_1 = 1/(2\pi\tau_1)$, and from frequency f_1 it is proportional to $1/f^2$. The McWhorter model is primarily used for MOSFET devices.

For BJT, the Hooge bulk model is more adequate. In this noise model, Hooge uses in the carrier transport two scattering mechanisms of carries: scattering on the silicon lattice and scattering on impurities. He assumed that only scattering on the crystal lattice is the source of the $1/f$ noise, while scattering on the impurities has no effect on noise level. All imperfections of the crystal lattice leads to large $1/f$ noise.

The noise spectral density function for the Hooge model is

$$S_{1/f} = \frac{\alpha_H \cdot I^\alpha}{f^\gamma \cdot N} \tag{11.8}$$

where

$\alpha_H = 2 \cdot 10^{-3}$ is the Hooge constant [8]
α and γ are material constants
N is the number of carriers

Later [9], Hooge proposed to use α_H as variable parameter, which in the case of silicon devices may vary from $5 \cdot 10^{-6}$ to $2 \cdot 10^{-3}$.

The $1/f$ noise is increasing with the reduction of device dimensions and as such is becoming a real problem for devices fabricated in nanoscale. The level of $1/f$ noise is often used as the measure of the quality of devices and its reliability. Devices fabricated with well-developed technologies usually have a much smaller level of $1/f$ noise. The $1/f$ noise (flicker noise) sometimes is considered to be responsible for the long-term device parameter fluctuation.

11.2.5 Noise $1/f^2$

Noise $1/f^2$ is a derivative of $1/f$ noise and it is observed mainly in metal interconnections of integrated circuits. It has become more evident for very narrow connections where there is a possibility of electromigration due to high current densities. In aluminum, the electromigration begins at current densities of $200\,\mu A/\mu m^2$ and noise characteristics changes from $1/f^2$ to $1/f^\gamma$, where $\gamma > 2$. Also the noise level increases proportionally to the third power of the biasing current:

$$S_{1/f^2}(f) = \frac{C \cdot J^\beta}{f^\gamma \cdot T} \cdot \exp\left(\frac{-E_a}{k \cdot T}\right) \tag{11.9}$$

where

$\beta \geq 3, \gamma \geq 2$
C is the experimentally found constant
E_a is the activation energy of the electromigration
$k = 8.62 \cdot 10^{-5}\,eV/K$ is the Boltzmann constant

The degeneration of the metallic layer is described by

$$\nu_d \propto J^n \exp\left(\frac{-E_a}{k \cdot T}\right) \tag{11.10}$$

Since Equations 11.9 and 11.10 have a similar character, the $1/f^2$ noise can be used as the measure of the quality of metal interconnections. This is a relatively fast and accurate method to estimate reliability of metal interconnections.

11.2.6 Burst Noise/RTS Noise

Burst noise is another type of noise at low frequencies. Recently, this noise was described as RTS noise. With given biasing condition of a device, the magnitude of pulses is constant, but the switching time is random. The burst noise looks, on an oscilloscope, like a square wave with the constant magnitude, but with random pulse widths (see Figure 11.1). In some cases, the burst noise may have not two but several different levels.

Noise spectral density function of the RTS noise has a similar form like generation-recombination noise:

$$S_{RTS}(f) = C\frac{4 \cdot (\Delta I)^2}{1 + (2\pi f/f_{RTS})^2} \tag{11.11}$$

where

$$C = \frac{1}{(\overline{\tau_l} + \overline{\tau_h}) \cdot f_{RTS}^2}$$

$$f_{RTS} = \frac{1}{\tau} = \frac{1}{\overline{\tau_l}} + \frac{1}{\overline{\tau_h}} = \frac{\overline{\tau_l} + \overline{\tau_h}}{\overline{\tau_h} \cdot \overline{\tau_l}}$$ RTS noise corner frequency, below this frequency spectrum of the RTS noise is flat. τ is observation time, $\overline{\tau_l}$ is average time of pulses at low level, and $\overline{\tau_h}$ is average time of pulses at high level

$$\overline{\tau_l} = \frac{1}{P}\sum_{i=1}^{P}\tau_{l,p}, \ \overline{\tau_h} = \frac{1}{S}\sum_{j=1}^{S}\tau_{h,s}$$

The intensity of the RTS noise depends on the location of the trap center with reference to the Fermi level. Only centers in the vicinity of Fermi levels are generating the RTS noise. These trapping centers, which are a source for RTS noise, are usually the result of silicon contamination with heavy metals or lattice structure imperfections.

In the SPICE program the burst noise is often approximated by

$$\overline{i}_{RTS}^2 = K_B\frac{I^{A_B}}{1 + (f/f_{RTS})^2}\Delta f \tag{11.12}$$

where K_B, A_B, and f_{RTS} are experimentally chosen parameters, which usually vary from one device to another. Furthermore, a few different sources of the burst noise can exist in a single transistor. In such a case, each noise source should be modeled by separate Equation 11.11 with different parameters (usually different noise RTS corner frequency f_{RTS}).

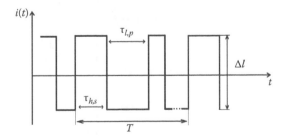

FIGURE 11.1   The RTS noise.

Kleinpenning [10] showed that RTS noise exists with devices with small number of carriers, where a single electron can be captured by a single trapping center. RTS noise is present in submicrometer MOS transistors and in BJTs with defective crystal lattice. It is present in modern SiGe transistors.

This noise has significant effect at low frequencies. It is a function of temperature, induced mechanical stress, and also radiation. In audio amplifiers, the burst noise sounds as random shots, which are similar to the sound associated with making popcorn. Obviously, BJTs with large burst noise must not be used in audio amplifiers and in other analog circuitry. The burst noise was often observed in epiplanar BJTs with large β coefficients. It is now assumed that devices fabricated with well-developed and established technologies do not generate the RTS noise. This is unfortunately not true for modern nanotransistors and devices fabricated with other than silicon materials.

11.2.7 Avalanche Noise

Avalanche noise in semiconductor devices is associated with reverse-biased junctions. For large reverse junction voltages, the leakage current can be multiplied by the avalanche phenomenon. Carriers in the junctions gain energies in a high electrical field and then they collide with the crystal lattice. If the energy gained between collisions is large enough, then during collision another pair of carriers (electron and hole) can be generated. This way the revised biased current can be multiplied. This is a random process and obviously the noise source is associated with the avalanche carrier generation. The intensity of the avalanche noise is usually much larger than any other noise component. Fortunately, the avalanche noise exists only in the *pn* junction biased with a voltage close to the breakdown voltage. The avalanche phenomenon is often used to build the noise sources.

Noise spectral density function of the avalanche noise is frequency independent:

$$S_{av}(f) = \frac{2qI}{(2\pi f \cdot \tau)^2} \qquad (11.13)$$

where *I* is an average value of the reverse biasing current.

An avalanche phenomenon is in most cases reversible. Therefore, semiconductor devices, where the avalanche breakdown took place, are regaining their low noise properties once devices are no longer working at avalanche region.

11.3 Noise of BJTs, JFETs, and MOSFETs

11.3.1 Noise of BJTs

In Figure 11.2, the equivalent diagram of BJT with noise sources is presented. These are as follows: thermal noise of base-spreading resistance r_b, shot noise and $1/f$ type noise of base bias current I_B, and

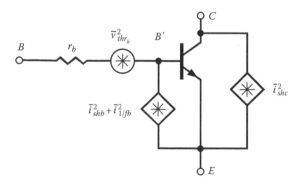

FIGURE 11.2 Equivalent diagram of BJT with noise sources.

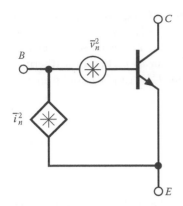

B

$\bar{v}_n^2$

$\bar{i}_n^2$

C

E

FIGURE 11.3 The $v_n^2 - i_n^2$ amplifier model of the bipolar transistor.

shot noise of collector current I_C. Spreading resistance is shown as external resistor r_b (noiseless resistor) between internal base B' and external base B.

The intensities (mean-square values) of noise sources are given by the following relations:

$$\bar{v}_{r_b}^2 = 4kTr_b \cdot \Delta f \text{—thermal noise of base spreading resistance } r_b \quad (11.14)$$

$$\bar{i}_{shb}^2 = 2qI_B \cdot \Delta f \text{—shot noise of base bias current } I_B \quad (11.15)$$

$$\bar{i}_{1/fb}^2 = \frac{k_f \cdot I_B^\alpha \cdot \Delta f}{f^\gamma} \text{—flicker noise of base bias current } I_B \quad (11.16)$$

$$\bar{i}_{shc}^2 = 2qI_C \cdot \Delta f \text{—shot noise of collector bias current } I_C \quad (11.17)$$

Coefficients α i γ in properly fabricated BJTs are close to 1. In silicon BJTs, the noise $1/f$ is caused by fluctuation of the recombination current in the depletion region of the base-emitter junction near the silicon surface. The *npn* transistors have usually higher levels of $1/f$ than *pnp* transistors.

Figure 11.3 shows the $v_n^2 - i_n^2$ amplifier model of the BJT with the equivalent input noise voltage and current $\bar{v}_n^2$ and $\bar{i}_n^2$, respectively.

The equivalent input noise voltage and current (mean-square values) can be expressed by [4]

$$\bar{v}_n^2 = 4kTr_b \cdot \Delta f + \left(2q \cdot I_B \cdot \Delta f + \frac{K_f \cdot I_B^\alpha}{f^\gamma} \Delta f\right) \cdot r_b^2 + 2qI_C \cdot \Delta f \left(\frac{r_b}{\beta} + \frac{V_T}{I_C}\right)^2 \quad (11.18)$$

$$\bar{i}_n^2 = 2qI_B \cdot \Delta f + \frac{K_f \cdot I_B^\alpha}{f^\gamma} \cdot \Delta f + \frac{2qI_C}{\beta^2} \cdot \Delta f \quad (11.19)$$

where β is the common-emitter current gain, $V_T = k \cdot T_0/q$, for $T_0 = 290\,\text{K}$ and $V_T = 25\,\text{mV}$.

In practice, the intensities of these noise sources versus frequency f have to be taken into account, and of course the $1/f$ noise sources in low frequency range are the main ones. For this reason, the flicker noise corner frequency f_{cor} is one of the important parameters. The flicker noise corner frequency f_{cor} is understood as being the frequency for which the $1/f$ noise and the white noise (thermal, shot) are equal to each other (see Figure 11.4).

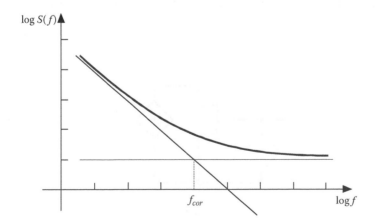

FIGURE 11.4 Noise power spectral density function $S(f)$ versus frequency f.

The shape of noise power spectral density function $S(f)$ as a function of frequency f, representing intensity of input noise voltage or current sources $\overline{v}_n^2/\Delta f$, $\overline{i}_n^2/\Delta f$, respectively, is the same. Function $S(f)$ should have $-10\,\text{dB/decade}$ slope in the low frequency range below flicker noise corner frequency f_{cor}. For the $f < f_{cor}$ the $1/f$ noise is the dominant component and for $f > f_{cor}$ the white (thermal and shot) noise is prevailing. The flicker noise corner frequency f_{cor} for $\overline{i}_n^2/\Delta f$ can be found from the relation [4]

$$f_{cor} = \frac{K_f}{2q(1+1/\beta)} \tag{11.20}$$

where parameters K_f and β are measured experimentally.

For BJTs, the f_{cor} is in the range from tenth of Hz to several kHz. The value of the flicker noise corner frequency can be evaluated separately for both equivalent input noise sources, for equivalent input noise voltage, and equivalent input noise current. The values of f_{cor} are not the same.

For evaluating a noise behavior of BJTs in a high and a very high frequency range the noise factor F can be applied. The noise factor F is given by the relation

$$F = \frac{\overline{v}_{ni}^2}{4kTR_S\Delta f} \tag{11.21}$$

where
$\overline{v}_{ni}^2$ is the mean-square equivalent noise input voltage for the CE or the CB configuration of BJT
R_S is the noise source resistance [4]

One way to reduce the thermal noise level of the base spreading resistance, r_b, is the connection of several (N) BJTs in parallel and to assure that the total current of all transistors is the same as for one transistor. By this way, the level of shot noise stays on the same level and the thermal noise is reduced to r_b/N.

11.3.2 Noise of JFETs

In Figure 11.5, the equivalent diagram of JFET with attached noise sources is presented. These are as follows: thermal noise of drain current I_D, $1/f$ noise of drain current I_D, and shot noise of gate current I_G. At the normal operating conditions, the gate-source junction is reverse biased and the shot noise of gate current, I_G, can be neglected.

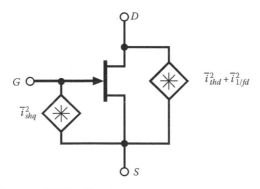

FIGURE 11.5 Equivalent diagram of JFET with noise sources.

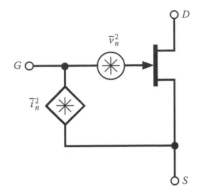

FIGURE 11.6 The $v_n^2 - i_n^2$ amplifier model of the JFET.

The intensities (mean-square values) of noise sources are given by the following relations:

$$\overline{i_{shg}^2} = 2qI_G \cdot \Delta f \text{—shot noise of gate current } I_G \tag{11.22}$$

$$\overline{i_{thd}^2} = 4kT\left(\frac{2g_m}{3}\right) \cdot \Delta f \text{—thermal noise of drain current } I_D \tag{11.23}$$

$$\overline{i_{1/fd}^2} = \frac{K_f \cdot I_D^\alpha}{f^\gamma} \cdot \Delta f \text{—flicker noise of drain current } I_D \tag{11.24}$$

Coefficients α i γ in properly fabricated JFETs are close to 1.

Figure 11.6 shows the $v_n^2 - i_n^2$ amplifier model of the JFET with the equivalent input noise voltage and current $\overline{v_n^2}$ and $\overline{i_n^2}$, respectively.

The equivalent input noise voltage and current (mean-square values) can be expressed by [4]

$$\overline{v_n^2} = \frac{\overline{i_{thd}^2} + \overline{i_{1/fd}^2}}{g_m^2} = 4kT\left(\frac{2}{3g_m}\right) \cdot \Delta f + \frac{K_f \cdot I_D^\alpha}{g_m^2 \cdot f^\gamma} \cdot \Delta f = \frac{4kT \cdot \Delta f}{3\sqrt{\beta \cdot I_D}} + \frac{K_f \cdot \Delta f}{4\beta \cdot f} \tag{11.25}$$

$$\overline{i_n^2} = \overline{i_{shg}^2} = 2qI_G \cdot \Delta f \tag{11.26}$$

where β is the transconductance coefficient.

As for BJTs, in the low frequency range, the flicker noise corner frequency f_{cor} is one of the important parameter. The frequency f_{cor} can be evaluated only for the power spectral density function $S(f)$ representing the intensity of input noise voltage $\overline{v_n^2}/\Delta f$, because the equivalent input noise current does not include the $1/f$ noise. For JFETs, the flicker noise frequency f_{cor} is understood as being the frequency for which the $1/f$ noise and thermal noise of $\overline{v_n^2}/\Delta f$ are equal to each other (see Figure 11.4).

Noise power spectral density function $S(f)$ as function of frequency f (representing intensity of input noise voltage source) should have −10 dB/decade slope in the low frequency range below flicker noise corner frequency f_{cor}. For the $f < f_{cor}$, the $1/f$ noise of drain current is the dominant component; and for $f > f_{cor}$, the thermal noise of drain current is prevailing.

The flicker noise corner frequency f_{cor} for $\overline{v_n^2}/\Delta f$ can be calculated from the relation

$$f_{cor} = \frac{3 \cdot K_f}{16 \cdot kT}\sqrt{\frac{I_D}{\beta}} \tag{11.27}$$

The typical flicker noise corner frequency f_{cor} for JFETs is in the range of several kHz.

For high and very high frequency, the noise factor F for JFETs can be calculated using relation (11.21), where $\overline{v_{ni}^2}$ is the mean-square equivalent noise input voltage for the CS or the CG configuration of JFET, and R_S is the noise source resistance [4].

11.3.3 Noise of MOSFETs

In Figure 11.7, the equivalent diagram of MOSFET with attached noise sources is presented. These are as follows: thermal noise of drain current I_D and $1/f$ noise of drain current I_D.

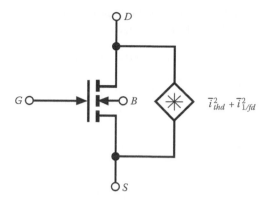

FIGURE 11.7   Equivalent diagram of MOSFET with noise sources.

The intensities (mean-square values) of noise sources are given by the following relations:

$$\overline{i}_{thd}^2 = 4kT\left(\frac{2g_m}{3}\right)\cdot \Delta f \text{—thermal noise of drain current } I_D \tag{11.28}$$

$$\overline{i}_{1/fd}^2 = \frac{K_f \cdot I_D^\alpha}{L^2 C_{ox} f^\gamma}\cdot \Delta f \text{—flicker noise of drain current } I_D \tag{11.29}$$

where
  L is the channel length
  C_{ox} is the gate oxide capacitance per unit area
  Coefficients α i γ in properly fabricated MOSFETs are close to 1

Figure 11.8 shows the $v_n^2 - i_n^2$ amplifier model of the MOSFET with the equivalent input noise voltage and current $\overline{v}_n^2$ and $\overline{i}_n^2$, respectively.

The equivalent input noise voltage and current (mean-square values) can be expressed by [4]

$$\overline{v}_n^2 = \frac{\overline{i}_{thd}^2 + \overline{i}_{1/fd}^2}{g_m^2} = \frac{4kT\cdot \Delta f}{3\sqrt{K\cdot I_D}} + \frac{K_f \cdot \Delta f}{4KL^2 C_{ox}\cdot f^\gamma} \tag{11.30}$$

$$\overline{i}_n^2 = 0 \tag{11.31}$$

where K is the transconductance coefficient.

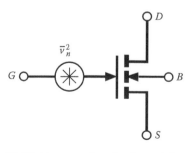

FIGURE 11.8   The $v_n^2 - i_n^2$ amplifier model of the MOSFET.

From the noise power spectral density function $S(f)$ versus frequency f (representing intensity of input noise voltage source), the flicker noise corner frequency f_{cor} can be found (see Figure 11.4). This noise corner frequency f_{cor} for $\overline{v}_n^2/\Delta f$ can be evaluated from the relation

$$f_{cor} = \frac{3\cdot K_f}{16\cdot kTL^2 C_{ox}}\cdot \sqrt{\frac{I_D}{K}} \tag{11.32}$$

For the $f < f_{cor}$, the 1/f noise of drain current is the dominant component; and for $f > f_{cor}$, the thermal noise of drain current is prevailing.

The typical values of f_{cor} in MOSFETs could be even larger than 10 MHz. The noise level at very high frequencies is very low.

11.3.4 Low Noise Circuits for Low Frequency Range

There are special semiconductor devices named "noiseless" that have very low levels of noise, especially in the low frequency range. These are transistors (bipolar and unipolar), transistor pairs, specially matched transistors, and amplifiers. For these devices, the equivalent input noise voltage source or the equivalent input noise current source (see Figure 11.3 for BJTs, Figure 11.6 for JFETs, Figure 11.8 for MOSFETs) at low frequency is given in technical data by manufacturers. Typically, this information is given for 1 kHz (sometimes for 10 Hz) at the given value of the device current. For these devices, the $1/f$ noise intensity and flicker noise corner frequency are important.

For low noise system, the input (front-end) stages are very important. For small source resistances, the BJTs are the preferred devices for these stages, and typically they have about 10 times lower level of equivalent input noise voltage than JFETs. In the selection of the BJT, the large value of the current gain β and the small value of the base spreading resistance r_b is important. For example, BJTs *npn* 2SD786 i *pnp* 2SB737 of Japanese company ROHM have $\beta = 400$ and $r_b = 4\,\Omega$. Transistors MAT-2 from Analog Devices (monolithic transistor pair) have $\beta = 500$ and r_b below 1 Ω. Noise parameters of transistors MAT 02E are as follows: intensity of equivalent input noise voltage source at collector current of 1 mA at frequency 10 Hz is equal to 1.6 / 2 nV/$\sqrt{\text{Hz}}$, at frequency 100 Hz is 0.9 / 1 nV/$\sqrt{\text{Hz}}$, and at frequency 1–100 kHz is 0.85 / 1 nV/$\sqrt{\text{Hz}}$.

For high source resistances, the JFETs are the preferred choice. It is important that JFET transistors have large transconductance g_m and small gate capacitance. For transistors 2N5515 made by INTENSIL, the intensity of equivalent input noise current source at 10 Hz and 1.6 mA is smaller than 1 fA/$\sqrt{\text{Hz}}$, and is the same up to 10 kHz. Whereas, the intensity of equivalent input noise voltage source at drain current of 600 µA is 10 nV / $\sqrt{\text{Hz}}$. This transistor has $f_{cor} = 10$ kHz at drain current of 600 µA; it means that at 10 kHz the intensity of $1/f$ noise is equal to intensity of white noise.

For low frequencies, MOSFETs should not be used because of the high level of $1/f$ noise.

There are also special operational amplifiers for low noise applications. One such amplifier is manufactured by Precision Monolithics Inc. and it has the intensity of equivalent input noise voltage source in the range 3.5/5.5 nV/$\sqrt{\text{Hz}}$ at 10 Hz and 3/3.8 nV/$\sqrt{\text{Hz}}$ at 1 kHz. Whereas, the intensity of equivalent current noise source is 1.7/4 pA/$\sqrt{\text{Hz}}$ at 10 Hz, and 0.4/0.6 pV/$\sqrt{\text{Hz}}$ at 1 kHz. This low noise OPAMP has very small flicker noise corner frequency, which is equal to 2.7 Hz for equivalent input noise voltage source and 140 Hz for equivalent input noise current source. This amplifier is specially suited for small source resistances ($R_S < 1$ kΩ). For input resistances larger than 1 kΩ, better noise property have amplifiers OP-07 i OP-08. For very large input resistances, the better choice is OPA-128, which has intensity of equivalent input noise voltage source equal to 27 nV/$\sqrt{\text{Hz}}$ at 1 kHz, and intensity of equivalent input noise current source is 0.12 fA/$\sqrt{\text{Hz}}$ in the frequency range from 0.1 Hz to 20 kHz. The low noise amplifiers AD 797 from Analog Devices has the intensity of equivalent input noise voltage source equal to 1.7 nV/$\sqrt{\text{Hz}}$ at 10 Hz and 0.9 nV/$\sqrt{\text{Hz}}$ at 1 kHz. Similar properties have low noise amplifiers LT 1028/LT 1128 from LINEAR TECHNOLOGY. They have the intensity of equivalent input noise voltage source equal to 1 nV/$\sqrt{\text{Hz}}$ at 10 Hz and 1.1 nV/$\sqrt{\text{Hz}}$ at 1 kHz. The flicker noise corner frequency f_{cor} is very low and it is equal to 3.5 Hz.

A special low noise amplifier for sources with large resistances is TLC 2201. It has, at 100 Hz, the intensity of equivalent input noise voltage source of 10 nV/$\sqrt{\text{Hz}}$, and the intensity of equivalent input noise current source of 0.6 fA/$\sqrt{\text{Hz}}$.

In practical applications, for very low noise circuits, usually in the first stage of the system, low noise transistor is applied and then at the next stages low noise amplifiers are used. Special care should also be taken for proper design of power supplies.

References

1. Ambrozy A. *Electronic Noise*. Akademiai Kiadó, Budapest, Hungary, 1982.
2. Konczakowska A. *Szumy z zakresu małych częstotliwości*. Akademicka Oficyna Wydawnicza EXIT, Warszawa, Poland, 2006.
3. Lukyanchikova N. *Noise Research in Semiconductor Devices*. B. K. Jones (Ed.), Gordon and Breach Science Publisher, Amsterdam, the Netherlands, 1996.
4. Marshall Leach W. Jr. *Fundamentals of Low-Noise Electronics*. Georgia Institute of Technology, School of Electrical and Computer Engineering, Atlanta, GA, 1999–2008.
5. Motchenbacher C. D., Fitchen F. C. *Low-Noise Electronic System Design*. A Wiley-Interscience Publication, John Wiley & Sons, Inc., New York, 1993.
6. Van der Ziel A. *Noise in Solid State Devices and Circuits*. John Wiley & Sons, New York, 1986.
7. McWhorter A. L. 1/f noise and germanium surface prosperities. In *Semiconductor Surface Physics*. R. H. Kingdton (Ed.), University of Pennsylvania Press, Philadelphia, PA, 1957, pp. 207–228.
8. Hooge F. N. 1/f noise is no surface effect. *Physics Letters*, 29A (3), 1969, 139–140.
9. Hooge F. N. The relation between 1/f noise and number of electrons. *Physica B*, 162, 1990, 334–352.
10. Kleinpenning T. G. M. On 1/f noise and random telegraph noise in very small electronic devices. *Physica B*, 164, 1990, 331–334.

12

Physical Phenomena Used in Sensors

Tiantian Xie
Auburn University

Bogdan M.
Wilamowski
Auburn University

12.1 Introduction **12**-1
12.2 Piezoresistive Effect...**12**-1
12.3 Thermoelectric Effect **12**-5
12.4 Piezoelectric Effect **12**-5
12.5 Pyroelectric Effect...**12**-6
12.6 Photoelectric Effect in Semiconductors **12**-8
12.7 Photoelectric Effect in p-n Junctions..............................**12**-9
12.8 Temperature Effect in p-n Junctions **12**-9
12.9 Hall Effect **12**-11
12.10 Conclusion **12**-12
References..**12**-12

12.1 Introduction

A sensor is used to transform a nonelectrical stimulation to an electrical response that is suitable to be processed by electrical circuits [W91]. Sensors are related with everyday life, such as automobiles, airplanes, radios, and countless other applications [BR90,RPSW01]. Several transformation steps are required before the electric output signal can be generated. These steps involve changes of types of energy where the final step must produce an electrical signal of a desirable format. There are several physical effects that cause generation of electric signals in response to nonelectrical influences. In this chapter, the physical effects behind various sensor applications that can be used for conversion of stimuli into electric signals are introduced, including piezoresistive effect, thermoelectric effect, piezoelectric effect, pyroelectric effect, temperature effect in p-n junction, and Hall effect.

12.2 Piezoresistive Effect

Piezoresistive effect describes the changes of electrical resistance when the material is mechanically deformed. It occurs in crystals that have no polar axes and is well represented in semiconductors. Physically, piezoresistance comes from the anisotropic distribution of energy levels in the k-space of the angular wave vector. This phenomenon is successfully employed in sensors that are sensitive to stress.

The relationship between relative changes in resistance $\Delta R/R$ (where R is specific resistivity) and the mechanical stress applied is given by

$$\frac{\Delta R}{R} = \frac{\pi\sigma}{E} \qquad (12.1)$$

where π is the so-called piezoresistive coefficient, which is dependent on the crystal orientation and the conditions of measurement, for example volume constancy.

The relationship between the stress σ and deformation of the material can be presented as

$$\sigma = E\frac{\Delta L}{L} \tag{12.2}$$

where E is Young's modulus of the material.

It is known that the resistance of a conductor can be calculated by

$$R = \frac{L}{W}R_s \tag{12.3}$$

where

R_s is the unit surface resistance of the material
L and W are the length and the width of the area, respectively

Considering the volume of the material as a constant, Equation 12.3 can be rewritten as

$$R = \frac{L^2}{A}R_s \tag{12.4}$$

where A is the area of the material.

Differentiating Equation 12.4, the following is obtained:

$$\frac{\Delta R}{\Delta L} = 2R_s \frac{L}{A} \tag{12.5}$$

By combining Equations 12.2, 12.4, and 12.5, the normalized resistance change of the wire can be rewritten as a linear function of the stress σ:

$$\frac{\Delta R}{R} = 2\frac{\Delta L}{L} = 2\frac{\sigma}{E} \tag{12.6}$$

In reality, the piezoresistive coefficients π contain 21 components. In the cubic system, only three of the components, π_{11}, π_{12}, and π_{44} are independent of each other. The same is true of monocrystalline silicon. The values of these coefficients depend on the type of conductor and the dosing level.

Let us derive the equations for $\langle 100 \rangle$ and $\langle 111 \rangle$ surface in a silicon wafer. For $\langle 100 \rangle$ surface of silicon, the changes of resistance can be measured by

$$\frac{\Delta R}{R} = 0.5\big[(\pi_{11} + \pi_{12} + \pi_{44})\sigma_{11} + (\pi_{11} + \pi_{12} - \pi_{44})\sigma_{22}\big]\cos^2 \phi$$

$$+ 0.5\big[(\pi_{11} + \pi_{12} - \pi_{44})\sigma_{11} + (\pi_{11} + \pi_{12} + \pi_{44})\sigma_{22}\big]\sin^2 \phi$$

$$+ \pi_{12}\sigma_{33} + (\pi_{11} - \pi_{12})\sigma_{12}\sin(2\phi) + \alpha T \tag{12.7}$$

where

π_{11}, π_{12}, and π_{44} are geometry-dependent constants (see Table 12.1)
α is the temperature coefficient
σ_{11}, σ_{22}, σ_{33}, and σ_{12} are geometry-dependent stresses

TABLE 12.1   Parameters for Silicon Wafer

	n-Type [1/TPa]	p-Type [1/TPa]
π_{11}	−1022	66
π_{12}	534	−11
π_{44}	−136	1381
$\pi_{11} + \pi_{12}$	−488	55
$\pi_{11} - \pi_{12}$	−1556	77
B_1	−312	718
B_2	297	−228
B_3	61	−446
C_1	−7	245
C_2	−305	473
C_3	670	615

It can be simplified to

$$\frac{\Delta R}{R} = 0.5(\pi_{11} + \pi_{12})(\sigma_{11} + \sigma_{22}) + \pi_{12}\sigma_{33} + \alpha T$$

$$+ 0.5\pi_{44}(\sigma_{11} - \sigma_{22})\cos(2\phi) + (\pi_{11} - \pi_{12})\sigma_{12}\sin(2\phi) \tag{12.8}$$

For n-type resistors, using the data in Table 12.1,

$$\frac{\Delta R}{R} = -244(\sigma_{11} + \sigma_{22}) + 534\sigma_{33} + \alpha T - 68(\sigma_{11} - \sigma_{22})\cos(2\phi)$$

$$- 1556\sigma_{12}\sin(2\phi) \tag{12.9}$$

For p-type resistors, using the data in Table 12.1,

$$\frac{\Delta R}{R} = 27(\sigma_{11} + \sigma_{22}) - 11\sigma_{33} + \alpha T + 690(\sigma_{11} - \sigma_{22})\cos(2\phi)$$

$$+ 77\sigma_{12}\sin(2\phi) \tag{12.10}$$

A sum of two perpendicular resistors is not a function of angular location and

$$\frac{\Delta R_1}{R_1} + \frac{\Delta R_2}{R_2} = 0.5(\pi_{11} + \pi_{12})(\sigma_{11} + \sigma_{22}) + \pi_{12}\sigma_{33} + \alpha T \tag{12.11}$$

A difference is a function of angular location:

$$\frac{\Delta R_1}{R_1} - \frac{\Delta R_2}{R_2} = \pi_{44}(\sigma_{11} - \sigma_{22})\cos(2\phi) + 2(\pi_{11} + \pi_{12})\sigma_{12}\sin(2\phi) \tag{12.12}$$

To measure $\sigma_{11} - \sigma_{22}$, $\phi = 0°$ and $90°$ should be used (preferably p-type); while to measure σ_{12}, $\phi = -45°$ and $45°$ should be used (preferably n-type).

For $\langle 111 \rangle$ surface of silicon,

$$\frac{\Delta R}{R} = \left[B_1 \sigma_{11} + B_2 \sigma_{22} + B_3 \sigma_{33} + 2\sqrt{2}(B_2 - B_3)\sigma_{23} \right] \cos^2 \phi$$

$$+ \left[B_2 \sigma_{11} + B_1 \sigma_{22} + B_3 \sigma_{33} - 2\sqrt{2}(B_2 - B_3)\sigma_{23} \right] \sin^2 \phi$$

$$+ \left[2\sqrt{2}(B_2 - B_3)\sigma_{13} + (B_1 - B_2)\sigma_{12} \right] \sin(2\phi) + \alpha T \qquad (12.13)$$

where

$$B_1 = \frac{\pi_{11} + \pi_{12} + \pi_{44}}{2}, \quad B_2 = \frac{\pi_{11} + 5\pi_{12} - \pi_{44}}{6}, \quad B_3 = \frac{\pi_{11} + 2\pi_{12} - \pi_{44}}{3} \qquad (12.14)$$

It can be simplified to

$$\frac{\Delta R}{R} = C_1(\sigma_{11} + \sigma_{22}) + B_3 \sigma_{33} + \alpha T + [C_2(\sigma_{11} - \sigma_{22}) + C_3 \sigma_{23}]\cos(2\phi)$$

$$+ [2C_2 \sigma_{12} + C_3 \sigma_{13}]\sin(2\phi) \qquad (12.15)$$

where

$$C_1 = \frac{2\pi_{11} + 4\pi_{12} + \pi_{44}}{6}, \quad C_2 = \frac{\pi_{11} - \pi_{12} + 2\pi_{44}}{6}, \quad C_3 = 2\sqrt{2}\frac{-\pi_{11} + \pi_{12} + \pi_{44}}{6} \qquad (12.16)$$

For n-type resistors, using the data in Table 12.1,

$$\frac{\Delta R}{R} = -7(\sigma_{11} + \sigma_{22}) + 61\sigma_{33} + \alpha T + [-305(\sigma_{11} - \sigma_{22}) + 670\sigma_{23}]\cos(2\phi)$$

$$+ [-610\sigma_{12} + 670\sigma_{13}]\sin(2\phi) \qquad (12.17)$$

For p-type resistors, using the data in Table 12.1,

$$\frac{\Delta R}{R} = 245(\sigma_{11} + \sigma_{22}) - 446\sigma_{33} + \alpha T + [473(\sigma_{11} - \sigma_{22}) + 750\sigma_{23}]\cos(2\phi)$$

$$+ [94\sigma_{12} + 750\sigma_{13}]\sin(2\phi) \qquad (12.18)$$

The sum of two perpendicular resistors is

$$\frac{\Delta R_1}{R_1} + \frac{\Delta R_2}{R_2} = C_1(\sigma_{11} + \sigma_{22}) + B_3 \sigma_{33} + \alpha T \qquad (12.19)$$

For n-type resistor on $\langle 111 \rangle$ surface

$$\frac{\Delta R_1}{R_1} + \frac{\Delta R_2}{R_2} = -7(\sigma_{11} + \sigma_{22}) + 61\sigma_{33} + 2000\Delta T \qquad (12.20)$$

where σ is in MPa and ΔT is in °C.

For p-type resistor on $\langle 111 \rangle$ surface

$$\frac{\Delta R_1}{R_1} + \frac{\Delta R_2}{R_2} = 245(\sigma_{11} + \sigma_{22}) - 446\sigma_{33} + 2000\Delta T \qquad (12.21)$$

As shown above, the piezoresistive effect in silicon can be several orders of magnitudes larger than in metals, making it a good member for piezoresistive sensors. However, the silicon is very sensitive to temperature. Additional methods should be adopted to counteract the temperature effect to make the sensor more accurate.

12.3 Thermoelectric Effect

Thermoelectric effect is also called Seebeck effect in honor of East Prussian scientist Thomas Johann Seebeck (1770–1831) [AE92]. He observed that an electrical current was present in a series circuit of two different metals that were contacted and at different temperatures.

When different conductors A and B are connected together, free electrons behave like an ideal gas. For different materials, the energies and densities of free electrons are different. Kinetic energy of electrons is a function of the temperature. At the same temperature, when two different materials contact, free electrons diffuse through the junction (contacting point). The electric potential of the material accepting electrons becomes more negative at the interface, while the material emitting electrons becomes more positive. Different electronic concentrations across the junction set up an electric field that balances the diffusion process until equilibrium is established. When the two materials are at different temperatures, the voltage at the junction can be presented as (Figure 12.1)

$$\frac{dV_{AB}}{dT} = S_A - S_B \qquad (12.22)$$

where S_A and S_B are the absolute Seebeck coefficients of the conductors A and B, respectively.

Conversely, when a voltage is applied to the conductors A and B in Figure 12.1, it creates a temperature difference between side A and side B.

The differential Seebeck coefficient $S_{AB} = S_A - S_B$ is called the sensitivity of a thermocouple junction. The Seebeck coefficient is independent on the characteristic of the junction, but only related with the materials. Therefore, to achieve the best sensitivity, the differential Seebeck coefficient of those junction materials should be as large as possible.

12.4 Piezoelectric Effect

Piezoelectric effect is the property of certain crystals that can generate a voltage subjected to a pressure and conversely generate a pressure due to an applied voltage [DN92]. The effect exists in crystals that do not have a symmetrical center.

Each molecule in piezoelectric crystal is polarized: one side is negatively charged while the other side is positively charged, which is also named as a dipole. This is due to the different atoms that make up the molecule, and the way in which the molecules are shaped. The polar axis runs through the center of both charges in the molecule and the molecule is electrically neutral under non-stress conditions. When external force is applied, the lattice is deformed and the electric field is built up; conversely, when the piezoelectric

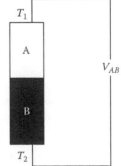

FIGURE 12.1 Voltage of two conductors with different temperatures connected in series.

crystal is under a strong electric field, most of the dipoles in the crystal are forced to line up in nearly the same direction, which results in mechanical stress.

Piezoelectricity is the combined effect of the electrical behavior of the crystal:

$$D = \varepsilon E \tag{12.23}$$

where

    D is the electric charge density displacement (electric displacement)
    ε is the permittivity
    E is the electric field strength

Using Hooke's law,

$$L = s\sigma \tag{12.24}$$

where

    L is strain
    s is compliance
    σ is the stress

By combining Equations 12.7 and 12.8, the piezoelectric effect can be described as

$$\{\mathbf{L}\} = [\mathbf{s}^E]\{\boldsymbol{\sigma}\} + [\mathbf{d}^T]\{\mathbf{E}\}$$

$$\{\mathbf{D}\} = [\mathbf{d}]\{\boldsymbol{\sigma}\} + [\boldsymbol{\varepsilon}^\sigma]\{\mathbf{E}\} \tag{12.25}$$

where

    {} and [] standard for vector and matrix separately
    d is the piezoelectric coefficient
    [$\mathbf{d}$] is the matrix for the direct piezoelectric effect
    transposed [$\mathbf{d}^T$] is the matrix for the converse piezoelectric effect
    the superscript E indicates under a zero or constant electric field
    the superscript σ indicates under a zero or constant stress field

Piezoelectric crystals perform direct conversion between mechanical and electrical energy. Efficiency of the conversion is defined by the coupling coefficients k:

$$k = \sqrt{d \times h} \tag{12.26}$$

where h is the gradient of electric field E multiplied by Young's modulus.

The k-coefficient is an important characteristic for applications where energy efficiency is of prime importance, like in acoustics and ultrasonics.

12.5 Pyroelectric Effect

Pyroelectric effect is the phenomenon of generating a temporary electrical potential when materials are heated or cooled [ZL78]. Different from thermoelectric devices that produce steady voltages, pyroelectric devices generate dynamical charges corresponding to the change of temperatures. So a pyroelectric device is usually used as a heat flow detector rather than a heat detector.

Pyroelectric effect is very tightly connected to the piezoelectric effect. There are several mechanisms that will result in pyroelectricity. Temperature changes cause shortening or elongation of individual dipoles. It may also affect the randomness of the dipole orientations due to thermal agitation. This is the primary pyroelectricity. The second pyroelectricity is induced by the strain in material caused by thermal expansion, which may be described as a result of the piezoelectric effect. Of the classic 32 crystal structures, 10 of these exhibit pyroelectric properties.

The pyroelectric charge coefficient, P_Q, is defined as

$$P_Q = \frac{\partial P_S}{\partial T} \tag{12.27}$$

and the pyroelectric voltage coefficient, P_V, as

$$P_V = \frac{\partial E}{\partial T} \tag{12.28}$$

where
  P_S is the spontaneous polarization
  E is the electric field strength
  T is the absolute temperature

By combining Equations 12.27 and 12.28,

$$\frac{P_Q}{P_V} = \frac{\partial P_S}{\partial E} = \varepsilon = \varepsilon_r \varepsilon_0 \tag{12.29}$$

where
  ε_0 is the electric permittivity of vacuum
  ε_r is the electric permittivity of the materials

If a pyroelectric material is exposed to a heat source, its temperature rises by ΔT and the corresponding charge and voltage changes can be calculated by

$$\Delta Q = P_Q A \Delta T$$
$$\Delta V = P_V L \Delta T \tag{12.30}$$

where A and L is the area and thickness of the material separately.

Since the capacitance is defined as

$$C = \frac{\Delta Q}{\Delta V} = \varepsilon_r \varepsilon_0 \frac{A}{L} \tag{12.31}$$

Integrating Equations 12.13 through 12.15, the relationship between ΔV and ΔT can be described as

$$\Delta V = \Delta T \frac{P_Q L}{\varepsilon_0 \varepsilon_r} \tag{12.32}$$

One may notice that the output voltage is proportional to the sensor's temperature change, pyroelectric charge coefficient, and its thickness.

The equivalent electrical circuit of the pyroelectric sensor is shown in Figure 12.2. It consists of three components: the current source generating a heat induced current *i*; the parasitic capacitance *C*, and the leakage resistance *R*.

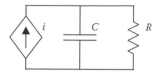

FIGURE 12.2 Equivalent circuit of pyroelectric sensor.

12.6 Photoelectric Effect in Semiconductors

A photoelectric effect is any effect in which light energy is converted to electricity. When light strikes certain light-sensitive materials, it may cause them to give electrons, or change their ability to conduct electricity, or may cause them to develop an electrical potential or voltage across two surfaces.

When a surface is exposed to electromagnetic radiation above a certain threshold frequency, the radiation is absorbed and electrons are emitted. This is called photoelectric effect, which was discovered by A. Einstein. The required photon energy must equal or exceed the energy of a single photon, which is given by

$$E = h\nu \tag{12.33}$$

where
ν is the frequency of light
h is Planck's constant equal to 6.63×10^{-34} J·s

Considering the relationship between frequency and the wavelength, the energy in Equation 12.33 can be written as the function of wavelength

$$E = \frac{hc}{\lambda} = \frac{1.24 \times 10^{-7}}{\lambda} \text{ [eV} \cdot \text{s]} \tag{12.34}$$

where
c is the light speed in materials
λ is the wavelength of light

For Si, the value of E at room temperature is 1.12 eV. However, in a real semiconductor crystal other excitation mechanisms are possible. These include absorption through transitions between the allowed bands and absorption through high levels of distortion in the forbidden band. However, the greatest excitation effect is still provoked by band-to-band absorption.

The electromagnetic radiation can be classified from low frequency to high frequency as infrared, visible, and ultraviolet, respectively. The electromagnetic frequency spectrum is shown in Figure 12.3.

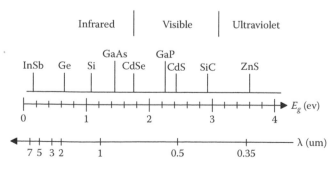

FIGURE 12.3 Electromagnetic frequency spectrum.

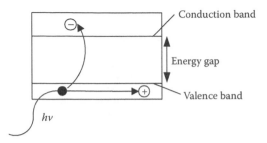

FIGURE 12.4 Photoconductive effect in semiconductor.

When light is absorbed by a semiconductor, a current can be induced and thus cause the change of resistance of the material. As shown in Figure 12.4, the semiconductor in thermal equilibrium contains free electrons and holes. The optical field to be detected is incident on and absorbed in the crystal, thereby exciting electrons into the conduction band or, in p-type semiconductors, holes into the valence band. The electronic deficiency thus created is acted upon by the electric field, and its drift along the field direction gives rise to the signal current.

12.7 Photoelectric Effect in p-n Junctions

When light strikes a semiconductor p-n junction, its energy is absorbed by electrons. Electrons and holes generated by light in the p-n junction are swept by the junction electrical field. As a result, the current flows though the junction. Figure 12.5 shows the process of generating current in photo diode.

Figure 12.5a is the p-n junction with a depletion layer without light. When the p-n junction is exposed to light, the impinging photons create electron–hole pairs everywhere in n-type area, p-type area, and depletion layer. In the depletion layer, the electric field accelerates electrons toward the n-layer and the holes toward the p-layer (Figure 12.5b).

For the electron–hole pairs generated in the n-layer, the electrons, along with electrons that have arrived from the p-layer, are left in the n-layer conduction band. The holes at this time are being diffused through the n-layer up to the depletion layer while being accelerated, and collected in the p-layer valence band. In this manner, electron–hole pairs that are generated in proportion to the amount of incident light are collected in the n- and p-layers. This results in a positive charge in the p-layer and a negative charge in the n-layer. If an external circuit is connected between the p- and n-layers, electrons will flow away from the n-layer, and holes will flow away from the p-layer toward the opposite respective electrodes (Figure 12.5c). The current is thus generated.

12.8 Temperature Effect in p-n Junctions

The temperature effect of p-n junction can be utilized as a temperature sensor. The principal sensor is straightforward. The *I–V* characteristic of the diode is as follows:

$$I_D = I_S \left(\exp\left(\frac{qV_D}{kT} \right) - 1 \right) \tag{12.35}$$

where
    V_D is the applied voltage
    q is electronic charge equal to 1.60×10^{-19} C
    k is Boltzmann's constant equal to 1.38×10^{-23} J/K
    T is absolute temperature
    I_S is the reverse saturation current of diode

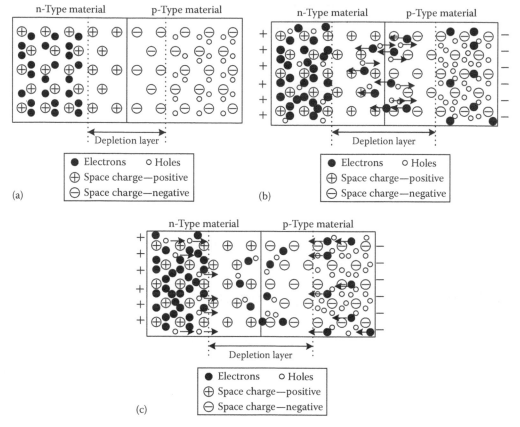

FIGURE 12.5 Photo diode: (a) p-n junction; (b) carrier generation in depletion layer associated with carrier sweep in electric field of depletion layer (fast process); (c) carrier generation in bulk material associated with minority carrier diffusion toward junction (slow process).

$$I_S = Aqn_i^2 \left(\frac{1}{\int_0^{x_n} (N/D_p)dx} + \frac{1}{\int_0^{x_p} (N/D_n)dx} \right) \tag{12.36}$$

where n_i is calculated by

$$n_i^2 = N_v N_c T^3 \exp\left(-\frac{E_g}{kT} \right) \tag{12.37}$$

By combining Equations 12.12 through 12.14, the voltage of the diode can be calculated by

$$V_D = \frac{kT}{q} \ln\left(\frac{I_D}{I_S} + 1 \right) \cong \frac{kT}{q} \ln\left(\frac{I_D}{I_S} \right) \tag{12.38}$$

Since I_S is proportional to n_i^2 (Equation 12.13), taking the derivative with respect to temperature yields

$$\frac{\partial V_D}{\partial T} = \frac{k}{q} \ln\left(\frac{I_D}{I_S} \right) - \frac{kT}{q} \frac{1}{I_S} \frac{\partial I_S}{\partial T} = \frac{V_D - (3kT/q) - (E_g/q)}{T} \tag{12.39}$$

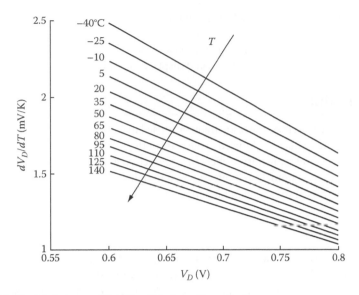

FIGURE 12.6 Relationship between dV_D/dT and V_D for a silicon diode.

where E_g is the semiconductor bandgap energy. The relationship between dV_D/dT and V_D for a silicon diode is shown in Figure 12.6. dV_D/dT decreases with temperature increasing.

With a constant current applied, the voltage across a diode or p-n junction will decrease by approximately 1–2 mV/°C. The diode voltage vs. temperature can be characterized by placing the amplifier in a temperature chamber with a constant current applied to the diode junction.

12.9 Hall Effect

The Hall effect was discovered in 1879 in Johns Hopkins University by E.H. Hall. The effect is based on the interaction between moving electric carriers and an external magnetic field. When an electron moves through a magnetic field, it acts as a sideways force

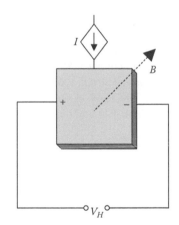

FIGURE 12.7 Principle of Hall effect.

$$F = qvB \tag{12.40}$$

where
    q is an electronic charge
    v is the speed of an electron
    B is the magnetic field

When the electric current source I is perpendicular to the magnetic field B, as shown in Figure 12.7, the so called Hall voltage, V_H, is produced in the direction perpendicular to both B and I. V_H is given by

$$V_H = \frac{IB}{qnd} \tag{12.41}$$

where
    d is the thickness of the hall plate
    n is the carrier density

One very important feature of the Hall effect is that it differentiates between positive charges moving in one direction and negative charges moving in the opposite.

Hall effect devices produce a very low signal level and thus require amplification. The Hall sensors can be used to detect magnetic fields, position, and displacement of objects.

12.10 Conclusion

The physical phenomena described above are frequently used in various sensors. For example, piezoresistive effect and piezoelectric effect can be used to measure both pressure and acceleration. Thermoelectric effect, pyroelectric effect, and temperature effect in p-n junction are usually applied in temperature sensors, bolometers, and so on. Photoelectric effect is used in light detectors such as photodiodes and thermal detectors. Hall effect is always used in measuring magnetic fields and sensing position and motion.

References

[AE92] Boyer, A. and Cisse, E., Properties of thin films thermoelectric materials: Application to sensors using the Seebeck effect, *Materials Science & Engineering*, 13, 103–111, 1992.

[BR90] Barshan, B. and Kuc, R., Differentiating sonar reflections from corners and planes by employing an intelligent sensor, *IEEE Transactions on Pattern Analysis and Machine Intelligence*, 12, 560–569, 1990.

[DN92] Damjanovic, D. and Newnham, R. E., Electrostrictive and piezoelectric materials for actuator applications, *Journal of Intelligent Material Systems and Structures*, 3, 190–208, 1992.

[RPSW01] Leonhard, M. R., Alfred, P., Gerd, S., and Robert, W., SAW-based radio sensor systems, *IEEE Sensors Journal*, 1, 69–78, 2001.

[W91] White, R. W., A sensor classification scheme, In *Microsensors*, IEEE Press, New York, pp. 3–5, 1991.

[ZL78] Zook, J. D. and Liu, S. T., Pyroelectric effects in thin film, *Journal of Applied Physics*, 49, 4604, 1978.

13

MEMS Devices

José M. Quero
University of Seville

Antonio Luque
University of Seville

Luis Castañer
*Polytechnic University
of Catalonia*

Angel Rodríguez
*Polytechnic University
of Catalonia*

Adrian Ionescu
*Ecole Polytechnique
Fédérale de Lausanne*

Montserrat
Fernández-Bolaños
*Ecole Polytechnique
Fédérale de Lausanne*

Lorenzo Faraone
*University of Western
Australia*

John M. Dell
*University of Western
Australia*

13.1 Introduction

Since the appearance of microelectromechanical systems (MEMS) technologies in the early 1950s, a traditional classification in two main classes, sensors and actuators, regarding the direction of its interaction with their surroundings, has been accepted. However, the large increase of new concepts and applications make this rather minimalist, and the increase in their functionality deserves a wider description.

It can be stated that MEMS functional development has two main limitations: actual technological constraints, which are improving quite fast, and engineering creativity. But we should not forget that, in the end, the market decides if a smart design should become a successful product.

In the case of sensor devices, miniaturization is quite a positive feature because the interface with the real world is greatly reduced. For example, only picoliters of samples and analytes are required in a lab-on-chip (LoC), making the mass and energy transportation more precise and faster. But the

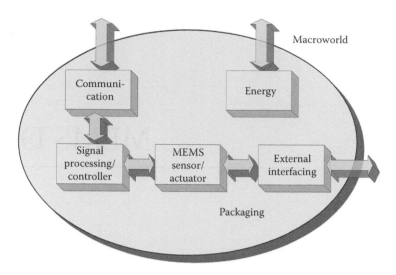

FIGURE 13.1   Block diagram of a full MEMS device.

inherent small amount of energy that a miniature device can handle makes the interfacing with the external world a key issue in actuator design. A sort of amplification scheme is needed to make micro effects to be significative in the macroworld. This way of interaction explains the success of devices like displays or barcode readers based on micromirrors that deflect a laser beam creating large projections in the macroworld, or the aggregated action of fast multiple actuations, like those in ink-jet printers whose small drops become visible to the human eye.

Major MEMS sensing technologies include piezoresistive, capacitive, thermoelectric, or piezoelectric. The use of structures in resonant mode is commonly employed because it simplifies the circuitry and increases the sensitivity of the transducer [1–2].

A generic definition of MEMS actuators defines these devices as integrated energy converters, exchanging energy from one physical domain to another. The basic actuation principles are electrostatic, magnetic, thermal, piezoelectric, and optical, but nowadays we also include other devices like microbatteries and energy scavengers into this class.

In the following sections, a brief description of the main MEMS devices is included, with special emphasis on their basic functioning principles. However, we should keep in mind the basic MEMS definition that conceives these devices as smart systems that comprise all necessary building blocks to be self-contained. For this reason, actual research focuses on the increase of the integration level to include in a common package the MEMS device, controller, communication interface, and energy, and thus creates a full MEMS device (Figure 13.1).

13.2 Sensing and Measuring Principles

13.2.1 Capacitive Sensing

Two conductors placed at a given distance electrically isolated from each other form a capacitor. If the capacitor is formed by two flat, parallel conducting plates, i.e., rectangular of length L and width W separated by a distance d, L and W being much bigger than d, the capacitance is given by

$$C = \varepsilon \frac{W \cdot L}{d} \tag{13.1}$$

where ε is the dielectric constant of the insulating material in between the plates.

Capacitors are frequently used in MEMS sensors as they can provide high sensitivity and linearity, and they provide useful, easy to handle electrical signals. Any of the parameters of the capacitor can be used for sensing if adequately linked to the physical magnitude to detect. A variety of methods allow to precisely measure the capacitances, providing the information of the evolution of that physical magnitude.

An example of this is the capacitive monitorization of a mechanical vibration, where one of the plates of a capacitor is attached to the vibrating point, and the second one is fixed closely, i.e., in the direction of the vibration. The vibration changes the distance *d* between the plates and, therefore, the measurement of *C* gives precise information of the instantaneous position of the moving plate.

When a voltage *V(t)* is applied to a capacitor whose capacitance varies with time, the current of the device can be expressed as

$$i = C\frac{\partial V}{\partial t} + V\frac{\partial C}{\partial t} \tag{13.2}$$

The first term is the current of a constant capacitor, while the second term provides information about the capacitance variation, and therefore of the parameter under study.

Usual values of capacitance of MEMS devices range from femto to pico Farads. The measurement of these low-capacitance values may involve the apparition of noise. The approximation of the capacitance assuming flat parallel plates is very poor and more complicated expressions and simulations have to be used.

13.2.2 Resistive Sensing

Metallic and semiconductor resistors are frequently used in MEMS technology. They can be fabricated by deposition and lithography of the resistor material onto an insulating layer. Semiconductor resistors can also be obtained by selective doping of a semiconductor substrate [3].

The piezoresistive effect is the dependence of a resistor on the strain of the material in which it is fabricated. Both metals and semiconductors can be used to make piezoresistors. Piezoresistors are frequently used in MEMS devices to measure stress, deformation, or bending, i.e., in vibrating structures such as membranes, bridges, and cantilevers.

In metals, piezoresistance is due to the geometrical changes caused by stress. In semiconductors, piezoresistance can be much higher than in metals because the stress induces changes in the band diagram and this has a strong effect on the amount of free carriers available for conduction.

In semiconductors, piezoresistance depends on the type and dose of doping, and in crystalline semiconductors, the piezoresistance effect also depends on the crystal orientation of the strain and of the resistor [4].

In a resistor deformed by a stress, the relative variation of resistance due to a deformation *dL* can be expressed as

$$\frac{dR}{R} = (1 + 2\nu)\frac{dL}{L} + \frac{d\rho}{\rho} \tag{13.3}$$

where
  ν is the Poison ratio
  $d\rho/\rho$ is the relative variation of the resistivity of the material due to the deformation

The gauge factor is the measure of the strain sensitivity of a material. For a thin film device of length *L*, the gauge factor *G* is defined as the ratio

$$G = \frac{\Delta R / R}{\Delta L / L} = 1 + 2\nu + \pi E \tag{13.4}$$

where

E is the strain

π is the piezoresistance coefficient that relates the stress τ and the variation of resistivity of the material ($\delta\rho/\rho = \pi\tau$)

To account for the three-dimensional components of the stress, the crystalline orientation and the crystal anisotropy π is in general a tensor of 21 parameters, but thanks to the symmetries in the case of crystalline silicon, it can be reduced to only three independent elements: $\pi 11$, $\pi 12$, and $\pi 44$ [5].

The stresses and the piezoresistance coefficients can be defined as longitudinal π_l (along the direction of the current flow), and transversal π_t (perpendicular to the flow), and π_s due to shear stress.

$$\Delta = \frac{\Delta R}{R}, \quad \Delta = \pi_l \sigma_l + \pi_t \sigma t + \pi_s \sigma_s \tag{13.5}$$

Commonly, the resistor is a thin film, whose thickness is much smaller than its length L and its width W, and the shear component can be neglected.

Usually, piezoresistors are integrated as components of Wheatstone bridges initially designed to be balanced and therefore providing zero output voltage. The variation of one or more of the resistors of the bridge implies one output voltage related to the variation.

13.2.3 Piezoelectric Sensing

The piezoelectric transduction is an efficient way to convert mechanical vibrations into electrical signals and vice versa. The piezoelectric effect consists on the generation of electric charge by a material when subjected to a mechanical deformation. This is an anisotropic effect. The redistribution of electrical charges in the piezoelectric material causes variations of the electrical field inside the material. Electrical voltages that are functions of the deformation appear in the surfaces of the crystal. In most occasions, this is a linear effect. The reverse piezoelectric effect consists on the appearance of a mechanical deformation as reaction to the application of externally applied voltages.

Even though structures like diodes or bipolar transistors show piezoelectric properties, silicon is not a piezoelectric material; therefore, layers of piezoelectric materials are deposited to obtain this functionality in integrated devices. Piezoelectric materials used in sensors are zinc oxide (ZnO), aluminium nitride (AlN), PZT, etc. They are deposited by RF sputtering, evaporation, etc.

Mechanical waves are easily produced in the piezoelectric material by using sinusoidal voltage excitation. This gives place to mechanical resonances at high enough frequencies. Mechanical resonances originate resonance peaks in the measured admittance between the electrodes of the capacitor.

Acoustic sensing frequently uses piezoelectric transducers. They can be used in viscous liquids. This is used to make microbalances and other types of sensors. In bulk acoustic wave (BAW), the mechanical wave propagates through the material. Resonances are found at frequencies at which the material thickness is an integer number of half wavelengths.

Piezoelectric resonators often have high-quality factors Q, what is a measure of the sharpness of the resonance, defined as the quotient of the resonance frequency over the pass band width.

BAW resonators frequently work at moderate frequencies allowing the use of low-cost electronics. An important application of these devices is gravimetric sensing.

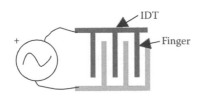

FIGURE 13.2 Layout of an IDT.

Gravimetric sensors measure variations of mass deposited on the surfaces of the resonator. Sawerbrey demonstrated that an increase of mass (Δm) causes a shift (Δf) of the resonant frequency (f_{res}):

$$\frac{\Delta f}{f_{res}} = S_m \Delta m \tag{13.6}$$

Changes of mass are then translated into frequency variations that can be easy and precisely measured. S_m is a negative number that depends on the material, physical dimensions, and operating frequency.

In surface acoustic wave (SAW) devices, mechanical waves propagate in regions close to the surface of the material. Usually, SAW devices work at RF frequencies. Woking as gravimetric sensors, they provide higher frequencial variations than BAW sensors. The fact that the energy of the propagating wave is concentrated near the surface makes these devices more sensitive to variations in the surface than BAW resonators. In SAW fabrication, wave emitters and receivers are placed in the surface of the piezoelectric material. Fundamentally, their working mechanism is the same as in the case of BAW devices. SAW devices can be fabricated on piezoelectric and non-piezoelectric substrates, for instance, silicon or GaAs.

Different types of surface waves can be excited: Raileigh, Lamb, Love, STW, etc. The type of acoustic wave generated in a piezoelectric material depends mainly on the substrate material properties, the crystal cut, and the structure of the electrodes.

Wave excitation and detection is usually done with interdigitated electrodes, and then called interdigitated transducers (IDTs). They consist of a high number of parallel line electrodes (fingers) disposed perpendicularly to the desired direction of the propagation of the wave. Alternating fingers are connected together so that half of them are connected to one contact, and the other half to the other, as in Figure 13.2.

Applying a sinusoidal voltage to the IDT generates a mechanical wave in the interdigital space between two neighbor electrodes. As this wave propagate interferes with the waves generated by other electrodes. Adequate electrode spacing makes generated waves interfere constructively. Therefore, the transduction efficiency is proportional to the number of fingers of the IDT. IDTs are fabricated by depositing a metallic film onto the piezoelectric, and lithographically patterned to the desired shape.

Piezoelectric coefficients are temperature dependent. Temperature variation in sensors has to be compensated. Often, SAW sensors use pairs of devices, one of them actually sensing and the other acting as reference. Even though semiconductors like silicon are not piezoelectric, SAW devices can be fabricated on silicon substrates.

Piezoelectric devices are often applied to sensors like gas or biosensors and to RF devices like filters. Designers have to pay attention to wave reflections at the borders of the device as they can introduce undesired resonances in the transfer function of the device. Reflecting structures can be added to the device to direct the generated waves to the desired regions.

13.2.4 Thermal Transducers

Different applications use thermal sensors. For instance, chemical or biological processes generate heat and, therefore, can be monitored by measuring their temperature; these thermal sensors are called Termistors. Bolometers are thermal transducers that measure radiation by absorbing the wavelengths of interest in small volumes of material well isolated thermally from the surroundings. MEMS technology makes possible the integration of these devices in very small volumes and their thermal isolation; therefore, low powers may produce precisely measurable temperature variations.

13.2.4.1 Metallic Thermoresistors

The resistivity of metals varies with temperature. This is used to fabricate thermal sensors. Over a broad range of temperatures, the dependence of the resistivity of a metal film with T is almost linear. This dependence is mainly due to the dependence of the mobility of free carriers on T. Due to lattice vibration scattering, the free carriers suffer from high scattering and their mobility decreases as temperature increases, therefore increasing the resistivity. This is the dominant scattering process for pure metals at temperatures higher than about 100 K. At low temperatures, scattering is dominated by impurities, which are almost temperature independent.

Platinum is the most used metal in metallic thermoresistor fabrication as it provides highly precise and reliable devices. Alternative metals are tungsten, copper, or nickel, among others. The variation of platinum resistance is very linear with temperature dependence of up to 500°C.

In the range of temperatures from about 75 K to over 1100 K, the resistance of a platinum sensor is usually expressed as

$$R(T) = R_0(1 + AT + BT^2 + CT^3(T - 100°C)) \tag{13.7}$$

where
$A = 3.908 \times 10^{-3} \text{ K}^{-1}$
$B = -5.775 \times 10^{-7} \text{ K}^{-2}$
$C = -4.183 \times 10^{-12} \text{ K}^{-4}$
T in °C

Since in metals the resistively is low, to obtain useful devices, resistors are usually thin and long. Commonly, they are designed with meander shapes to concentrate on a given part of the chip. Common thicknesses are several tenths of micron.

The fabrication process involves the deposition (by sputtering or evaporation) of a thin film of the desired metal onto an oxidized silicon wafer, or other type of substrate, provided there is adequate electrical isolation from the bulk. Lithography and etching are used to shape the resistor. Their precise values can be tuned by laser trimming.

13.2.4.2 Semiconductor Thermoresistors

Semiconductor resistors may present positive or negative dependence with temperature depending on the semiconductor, doping concentration, and temperature of work. The conductivity of a semiconductor is a function of the concentration of free carriers (electrons and holes) and of their mobilities. In general, carrier mobilities decrease with temperature due to the increase in scattering; this alone would provide positive thermal dependence to the resistance. On the other hand, carrier concentrations initially clamped to nearly the doping concentration at high enough temperatures rise exponentially with T, as the intrinsic carrier concentration dominates over doping, then the resistivity decreases as temperature increases. Semiconductor resistors can be obtained with polycrystalline or crystalline materials. In crystalline materials, resistivities are somewhat dependent on crystallographic orientation.

13.2.4.3 Semiconductor P-N Junction and BJT Thermal Sensors

Semiconductor P-N junctions or bipolar junction transistors (BJTs) have current–voltage relationships strongly dependent on temperature. This has been used for a long time in integrated electronics to thermally compensate integrated circuits (ICs) or to measure temperature. The integration of these devices is generally easy, not only in control circuitry but also in membranes or cantilevers of a MEMS device.

Some circuital configurations are of special interest for temperature measurement. If the current of a diode is imposed by a current source, the voltage of the diode is related to the temperature of the junction. This voltage represents a very linear measurement of the temperature of the junction. Typically, for a silicon diode operated in the range from microamperes to several miliamperes, the voltage variation per degree of temperature has a value from −1 to −3 mV/°C. This sensitivity is somewhat dependent on the bias current. Semiconductor diodes have been used for a long time in ICs to regulate the temperature.

The thermal behavior of bipolar transistors and diodes is similar. Temperature sensing using bipolar transistors is based on the emitter current-to-voltage characteristic of the base-emitter junction. When base and collector terminals are short circuited, the transistor basically behaves as a diode. The use of several matched transistors in a circuit can give place to precise sensors such as the classical PTAT (proportional to absolute temperature).

13.2.4.4 Seebeck Effect

The thermoelectric Seebeck effect consists of the fact that a conductor subjected to a thermal gradient develops an electric voltage between its hot and cold sides. This voltage depends on the material and on the difference of temperatures of both sides. The voltage drop developed in a given material per unit of temperature is the Seebeck coefficient $\alpha(T)$ of that material. In semiconductors, Seebeck coefficients can be positive or negative depending on the dominant type of carriers (electrons or holes). Typical Seebeck coefficients are of several μV/°C.

A thermocouple is formed by two different conductors (metals or semiconductors) electrically connected at the hot side at temperature $T1$, while the nonconnected ends of both conductors are at another temperature, $T0$. The thermal gradient in each conductor produces a different voltage variation between hot and cold points; therefore, a voltage can be measured between both conductors at the cold ends, and these are the two electrodes of the thermocouple.

$$V = \int_{T0}^{T1} (\alpha_1(T) - \alpha_2(T)) dT \tag{13.8}$$

In the integration of thermocouples, special attention has to be paid to the thermal isolation between hot and cold pints. Thermocouples can be used as thermoelectric converters by connecting them to a load. To increase the output voltage, several thermocouples can be connected in a series to form thermopiles.

13.2.5 Optical Sensors

The optical properties of materials often involve measuring reflection, transmission, or optical absorbance in the films under characterization. Light detection can be done using different devices. The most common are photodiodes, phototransistors, and photoresistors [6].

Photodiodes consist of a semiconductor P–N junction. For their use, they are reversely biased; therefore, just a very small reverse saturation current flows. When light incides with adequate wavelength in the surface of the semiconductor, the transmitted part propagates in the semiconductor. Light in the semiconductor creates electron–hole pairs. Electrons in the conduction band start flowing to the positive biased terminal, while photogenerated holes flow to the negative biased terminal. This causes a noticeable increase of current of the diode proportional to the incident optical intensity.

Phototransistors are bipolar transistors. Commonly, bias is applied between collector and emitter, leaving the base terminal unconnected. In the absence of light, there is no base current. When light incides on the device, photogenerated carriers near the collector-base junction, one type of carriers is swept to the collector and the other to the base regions of the device, thus creating the base current. The base current causes injection from the emitter to the collector. Therefore, phototransistors provide a current gain and have a much higher sensitivity than photodiodes.

Photoresistors are semiconductor devices whose resistance changes when light incides onto the device. The dark resistance of the material is high because few free carriers are available for conduction. The absorption of light provokes the transition of electrons and holes to the bands, thus increasing the conductivity of the material.

13.2.6 Magnetic Sensors

13.2.6.1 Hall Effect

Hall effect is the consequence of the interaction of a moving charge carrier and a magnetic field. The force acting on a charge q traversing a position in which a magnetic field $\vec{B}$ exists is

$$\vec{F} = q \cdot \vec{v} \times \vec{B} \tag{13.9}$$

where
$\vec{v}$ being the speed of the charge
$\vec{F}$ is a lateral force deflecting the charge toward one side

For instance, in the case of a vertically placed metallic film, were charge carriers are electrons, if the magnetic field is perpendicular to the film directed to it, and the electron current flows in the downward direction, the magnetic force on the electrons pushes them to the right side. If contacts are placed in the right and left sides of the strip, a difference of potential V can be measured between them:

$$V = KIB \sin \alpha \tag{13.10}$$

where
I is the current flowing
α is the angle between the magnetic field vector and the plate
K is a coefficient sensitivity that depends on the material, geometry, and temperature

Typically, a Hall sensor has four terminals. A current flow is set between the two excitation terminals and a differential voltage proportional to the Hall effect is read between the two output ones.

A way to integrate a Hall sensor is by simply defining a square N-type doped region on a P substrate. Contacts are placed on the sides of this region. Two opposite contacts serve to establish the current flow; the other two are the output.

13.2.6.2 Magnetotransistors

These devices are bipolar transistors in which the collector current is modulated by an external magnetic field. Often, they consist of a bipolar transistor with a double collector. In the absence of magnetic field, the current of both collectors is the same. The magnetic field deflects carriers favoring their collection in one of the two contacts unbalancing the collector currents. This difference in current provides a measure of the electric field.

13.2.6.3 Magnetoresistance

It is the property of a material carrying an electric current to change its electrical resistance in the presence of an external magnetic field. Ordinary magnetoresistance may produce resistance variations up to 5% of their original values due to the presence of the magnetic field. It is proportional to the square of the intensity of the magnetic field at low intensities. It is observed with H both parallel to and transversal to the current flow. In semiconductors, magnetoresistance is large and is dependent on the relative angle between the field direction and the current flow. Stacks of alternating layers of magnetic and nonmagnetic metals may present very large, negative values of magnetoresistance. This is called giant magnetoresistance (GMR).

13.3 MEMS Actuation Principles

13.3.1 Introduction

It is recognized that main MEMS actuator devices are based on one of the following actuation principles: electrostatics, thermal, piezoelectric, and magnetic [8].

13.3.2 Electrostatic Actuation

Electrostatic actuation is based upon the application of an electric field between electrodes, one of which at least, can move in some direction, and normally working against a mechanical restoring force, $F_m = -k(g_0 - g)$ with a stiffness coefficient k. Figure 13.3 schematically shows a diagram of two electrodes of area A parallel plate geometry, with an initial gap g_0 between them. A DC voltage V is applied between the plates and the resulting force, normal to the plates' surface, makes the upper plate move, changing the gap to g. The electrostatic force is given by [7]

$$F_e = \frac{V^2 C(g)}{2g} = \frac{\varepsilon A V^2}{2g^2} \tag{13.11}$$

As can be seen, the force is always attractive, as it is independent of the sign of the applied voltage, and it increases nonlinearly as the gap narrows. Applying static and dynamic equilibrium conditions, it can be shown that there is a minimum gap that can be stably reached: $g_{min} = (2/3)g_0$. An important parameter resulting from this analysis is the well known "pull-in voltage," V_{PI}, beyond which the movable electrode is unstable and collapses toward the fixed electrode [9]. It is given by

$$V_{PI} = \sqrt{\frac{8kg_0^3}{27\varepsilon A}} \tag{13.12}$$

where ε is the permittivity of the gas filling the gap. For practical designs, the values for V_{PI} are generally larger than the standard supply voltage of ICs, hence requiring specific step-up converters. Pull-in maximum stable deflection limits the value of the ratio of ON/OFF capacitances in tunable analogue MEMS varactors [10].

The same geometry shown in Figure 13.3 can also be driven by directly injecting a charge to the MEMS capacitor and the movement becomes, ideally, stable beyond the pull-in instability [11]. A number of ways of achieving this operation mode have been reported and the effect of parasitics discussed [12]. It turns out that the transient dynamics developing after injecting a current pulse is equivalent to placing a capacitance in series with the MEMS switch [13].

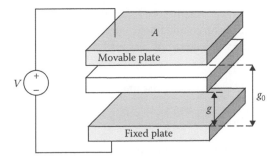

FIGURE 13.3 Diagram of a two-electrode parallel plate geometry.

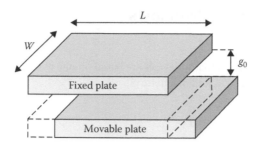

FIGURE 13.4   Diagram of a two-electrode plate geometry with lateral displacement.

Lateral movement (Figure 13.4) can also be achieved electrostatically and, as the gap remains constant, the pull-in instability is avoided. The electrostatic force is independent of the relative placement of the electrodes, the sign always tends to achieve maximum overlapping area and it is given by

$$F_e = \frac{\varepsilon W V^2}{2g_0} \tag{13.13}$$

This tangential force is usually much smaller than the normal force [14]. This operation mode is used to build lateral comb actuators and electrostatic motors. When the motor is a rotating motor, the torque is related to the angle by [15]

$$T(\theta) = \frac{V^2}{2} \frac{dC(\theta)}{d\theta} \tag{13.14}$$

13.3.3 Thermal Actuation

Heat can be used in different ways to provide thermally actuated MEMS devices, such as thermal expansion of beams or cantilevers, shape memory alloy (SMA), thermo neumatic expansion of a gas, or bimetal effect of two-layered materials of different thermal expansion coefficients. Among them, the simplest principle is the displacement that can be achieved from the thermal expansion of materials. Metals are the ones with larger expansion coefficient and hence capable to provide larger forces and larger displacements compared with silicon-based microdevices.

Using a simple one-dimensional model, the thermal expansion $\Delta L/L$ can be made proportional to the temperature increment ΔT. This strain relates to the thermal expansion stress through the Young modulus (E) of the material: $\sigma = E\Delta L/L$ and, finally, the stress is converted in the force by multiplying by the cross section area: $F = A\sigma$.

Being a thermal process, heating is usually very fast as the mass involved in microstructures is very small, but cooling depends basically on conduction to supports and ambient air and to a smaller extent on convection. Figure 13.5 shows a diagram of a typical thermal actuator structure, where a large

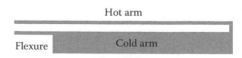

FIGURE 13.5   Diagram of a typical thermal actuator structure.

difference can be seen in the transversal dimension of the hot and cold arms in order to get most of the Joule power dissipated in the hot arm. Despite these design precautions, in practice, a large part of the power is also dissipated in the combination of the flexure and cold arm, hence leading to typically low efficiency. The device shown in Figure 13.5 creates a nonlinear displacement of the tip that can be converted to linear by changing the mechanical design of supporting legs, as, e.g., buckle-beam device [16].

The movement can be enhanced either by using layered structures of different thermal expansion coefficients (bimetal actuators) or by using SMA, which consists on NiTi wire, which undergoes a phase transition from martensitic to austenitic, entailing several percent of longitudinal change and producing forces in the newton range [17].

13.3.4 Piezoelectric Actuation

The piezoelectric effect is present in several natural materials such as quartz, but also in piezo ceramic materials with improved characteristics based on polyvinylidene fluoride (PVDF) and lead zirconate titanate (PZT). Zinc oxide has also been used to provide movement for AFM stages.

The displacement is a function of the applied electric field and the strain produced can be written as

$$\frac{\Delta L}{L} = d_{ij}E \tag{13.15}$$

where
d_{ij} is the piezoelectric strain coefficients
d_{33} stands for the strain parallel to the polarization plane
d_{31} for the orthogonal.

d_{33} is positive and d_{31} is negative, thereby indicating contraction. Strain values up to 0.2% are achievable. To a first-order approximation, the actuator displacement is proportional to the stored charge and hence to the electrical capacitance, larger capacitances can be manufactured by stacking several layers of material connected in parallel. These actuators are capable of producing very large forces and small displacements. Typical applications are nanopositioning stages from some tens to some hundredths of micron travel range. They can also be used in tunneling and atomic force microscopes for nanometer-range resolution.

13.3.5 Magnetic Actuation

Magnetic actuation in the microworld faces challenging issues because of the scaling rules for the Lorentz force, the difficulties of providing coils using mainstream micromachining techniques and the material compatibility. The two main families of magnetic actuators are based on the electromagnetic force and on magnetostrictive films.

The actuators having a movable cantilever or beam being attracted by a magnetic field created by an electrical current in a coil exert a force on it given by [18]

$$F = \frac{1}{2}\left(\frac{NI}{\Re}\right)^2 \frac{d\Re}{dl} \tag{13.16}$$

where
$\Re$ is the reluctance of the gap and core
N is the number of spires of the coil
I is the electrical current

The consequences of Equation 13.16 are that large values of the product Amp-turns and low reluctances are required to provide significant force. These requirements have to be faced with careful considerations of the device dimensions, material permeability values to reach forces in the range of several hundred µN, and also the application of specific fabrication techniques for 2D or multilayer coils using micromachining or LIGA processes.

Magnetostriction stands for the deformation of an object when subject to an external magnetic field. This can be used to build remote actuators by depositing magnetostrictive material on top of cantilever or membrane structures. The action of the magnetic field makes the cantilever bend due to the existing mismatch between the two materials [19].

13.4 MEMS Devices

13.4.1 Inertial Sensors

Inertial sensors measure the state of movement of an object. Accelerometers measure linear acceleration, while gyroscopes measure angular rate movement. Historically, accelerometers were among the first MEMS devices to be successfully commercialized, mainly in the automotive industry. Nowadays, accelerometers and gyroscopes are being used in positioning and guidance systems in a variety of sectors.

13.4.1.1 Accelerometers

Accelerometers have a mass proof, which moves relatively to a fixed frame due to the acceleration of this frame, which is rigidly linked to the object whose acceleration is to be measured. By sensing the displacement between the mass proof and the fixed frame, the desired acceleration measurement can be obtained. Depending on the method used to sense the displacement, accelerometers are classified into capacitive, piezoresistive, thermal, or resonating. There exist accelerometers able to measure three-dimensional movements, but most are limited to one or two axis. The most important accelerometer characteristics are sensitivity, bandwidth, operation range, resolution, and linearity.

Capacitive accelerometers, like the one depicted in Figure 13.6, have very good sensitivity: low-drift, low-noise, and low-temperature sensitivity. They are also inexpensive to fabricate. They can, however, be sensitive to EMI if properly shielded or not. A typical capacitive accelerometer can have a range of 50 g, 10 bits of resolution, 10 kHz of bandwidth and 0.2% of nonlinearity.

Piezoresistive accelerometers sense the displacement of the proof mass by measuring the change in electrical resistance in the cantilevers that support it. They are simple to fabricate and have a low cost, but their sensitivity is not as good as that of capacitive devices. Range of 5 g, sensitivity of 5 mV/g and 5 kHz of resonant frequency are typical values for a piezoresistive accelerometer.

Different types of thermal accelerometers have been developed. One of the most widely used is the based on a bubble movement. The position of the bubble is sensed by measuring the amount of heat transferred from a heater to the liquid where the bubble is located.

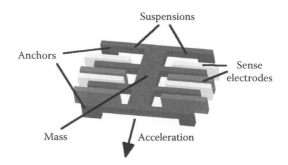

FIGURE 13.6   Structure of a horizontal capacitive accelerometer.

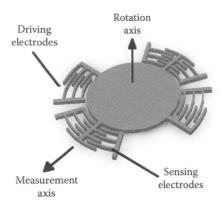

FIGURE 13.7 Structure of a vibrating electrostatic gyroscope.

Resonating accelerometers transfer the mass-proof movement to a series of beams, changing their oscillating frequency. They are most expensive than other types, but feature high resolution and sensitivity.

13.4.1.2 Gyroscopes

The principle of measuring angular rate variation in a gyroscope is to have a rotating or vibrating piece, which is affected by Coriolis pseudo-forces and thus has a relative movement with respect to the fixed frame. Like accelerometers, gyroscopes are classified according to the physical effect used to measure this displacement (piezoresistive, capacitive, etc.). In addition, the effect used to generate the movement of the proof mass can also differentiate gyroscopes, and it can be piezoelectricity, electromagnetism, electrostatics, etc.

Defining performance characteristics for gyroscopes are angle random walk, full scale range, bandwidth, zero rate output, temperature sensitivity, and parasitic effects.

Vibrating (Figure 13.7), rotating, and levitating gyroscopes have been constructed using bulk- and surface-micromachining processes. Quartz and silicon are the most commonly used materials.

A typical rate-grade gyroscope could have an angle random walk of $0.6°/\sqrt{h}$, a range of 500°/s, and a bandwidth of about 100 Hz.

13.4.2 Pressure Sensors

Nowadays, pressure sensors are MEMS devices with the greatest commercial success. Their application fields include automobile, medicine, and industry. We can find pressure sensors everywhere in the car: for fuel distribution, exhaust gases, tires, seats, etc. Very compact sensors are applied to patients to measure internal ocular or cranial pressure.

MEMS pressure sensors have been intensively studied because the manufacturing of membranes with an accurate thickness was developed in the beginning of these technologies. Bulk micromachining, as it has been presented in the previous chapter, is a process that produces membranes with different shapes and widths very precisely. Additionally, there is a wafer that is bonded to the previous one to form a cavity (see Figure 13.8a). The deflection of a membrane due to the difference of pressures applied in their sides is the working principle that most pressure sensors use. This kind of device is called differential or relative pressure sensor.

When one of the pressures is fixed in a closed cavity as a reference, we have an absolute pressure sensor. Normally, this reference pressure is vacuum but if the reference pressure is the atmospheric one, this sensor is called a gauge. Maintaining this reference pressure inside this cavity is difficult due to the permeability of the materials and bonding methods that are employed in its fabrication.

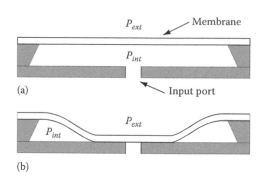

FIGURE 13.8 (a) Pressure sensor based on the deflection of a membrane manufactured with MEMS technology. (b) Pressure sensor working in touch mode.

Diaphragm sensors may work in two modes [20]: a normal mode when the membrane freely deflects depending on the difference of pressures, and touch mode that happens when the diaphragm stays in contact with the substrate, as depicted in Figure 13.8b. In the first mode, the device is more sensitive, but it presents higher linearity in the second mode.

But the detection of the deformation of a membrane is not straightforward, and an indirect phenomenon is used [21]. In the next paragraphs, the most relevant MEMS structures will briefly be described.

The main parameters to evaluate the performance of pressure sensors are sensitivity, linearity, dynamic range, and reliability. The sensitivity is the most relevant feature and it is defined as the change of the output signal (v) with respect to a pressure change, depending on the signal value itself:

$$s = \frac{1}{v}\frac{\partial v}{\partial P} \tag{13.17}$$

The dynamic range in these devices depends on the weight of the membrane and the gap. Due to the fabrication of very small gaps, these sensors are designed for a narrow pressure band.

One of the most important drawbacks related with the MEMS processes that affect the performance of these devices is the residual stress that may appear when a membrane is released, because this permanent stress would initially deflect it. This effect is even worse in multilayer membranes.

13.4.2.1 Capacitive Pressure Sensors

A capacitive pressure sensor uses the membrane as one plate of a capacitor, while the substrate is the other plate, as shown in Figure 13.9.

Obviously, a dielectric layer should be between the membrane and the substrate. This layer can be deposited or generated before the membrane layer, and it can be also used as a stopping layer for the bulk micromachining process. For instance, we can create an insulator layer of SiO_2 on top of a silicon

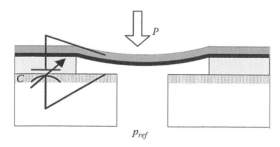

FIGURE 13.9 Working principle of a capacitive pressure sensor.

wafer and then deposit polysilicon as a conductive layer. Many alternative fabrication processes have been proposed since the 1970s when they were first developed, but all of them make use of this basic scheme, which is very simple and can be easily implemented and scaled and also it has no DC power consumption. However, it suffers from several limitations: it has a nonlinear response because of the $1/g$ dependence of the capacitance value respect to the gap g; the existence of parasitic capacitance parallel to the sensing capacitance makes the relative changes of the total capacitance difficult to be measured. The auxiliary electronics that is employed for the capacitance-to-voltage conversion consist of a conventional integrator or a switched capacitor charge integrator.

13.4.2.2 Piezoresistive Pressure Sensors

Nowadays, most of the commercially available pressure MEMS sensors are based on this transducing method. It basically consists of the fabrication of piezoresistors on the membrane by selectively doping special areas, as presented in Figure 13.10. These piezoresistors are placed in a symmetrical disposition with respect to the edges of the membrane.

These piezoresistors can be configured in a Wheatstone bridge. The piezoresistors' resistance has a low value, thus permitting long connections to the readout circuit with minimum perturbation due to the impedance of the wires. But on the other hand, it also represents rather high continuous power consumption.

13.4.2.3 Compensated Pressure Sensors

As it has already been mentioned, the main drawback of pressure sensors based on membrane deflection is that they are nonlinear transducers with a rather low dynamic range. One way to circumvent this problem is the use of a variable reference pressure that balances the pressure on the other side of the membrane. An electronic servo-controller could feedback the signal loop to achieve this objective. However, this compensation scheme requires the application of an external force onto the membrane to be defected in the opposite way than the external pressure to be measured. Normally, an electrostatic actuator is used for this compensation because it generates attractive forces between the membrane and the substrate. This approach complicates the MEMS structure and adds external auxiliary electronics. Besides, the maximum applied voltage should be inferior to the pull-in voltage to prevent the membrane from collapsing.

13.4.2.4 Resonant Pressure Sensors

As it has been explained in the previous chapter, frequency measurements of a resonant structure are quite simple and accurate. For this reason, accurate pressure sensors based on this principle have a beam whose supports are on the membrane vibrating in resonant mode. When the pressure is applied, the membrane deflects creating a change in the resonant frequency due to the stress induced in the beam,

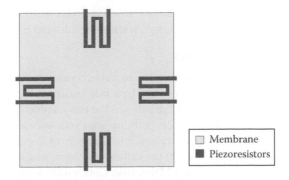

FIGURE 13.10 Piezoresistive pressure sensor layout.

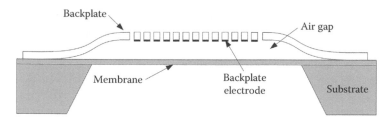

FIGURE 13.11   Structure of a condenser microphone.

thus providing an indirect measurement of the applied differential pressure. The main problem with this approach is that the manufacturing process is much more complicated than in previous sensors.

13.4.2.5 Other Pressure Sensors

There are other alternatives to pressure sensor transducing that should be mentioned. For instance, optical pressure sensors detect the deflection of the membrane by measuring the reflection angle of a laser beam. Other optical schemes include a membrane in the end of an optic fiber, acting the device as a Fabry–Perot interferometer. PCB-MEMS techniques also provide methods for the manufacturing of pressure sensors [22].

13.4.2.6 Microphones

A special case of pressure sensor is a microphone like the one depicted in Figure 13.11. The membrane is released via anisotropic etch of the substrate, while a backplate is formed by surface micromachining. The suspended backplate has holes to facilitate the vibration of the membrane. The capacitive transducer is formed by the membrane and the electrode backplate. This structure allows for the realization of commercial low-cost miniature microphones.

13.4.3 Radio Frequency MEMS: Capacitive Switches and Phase Shifters

13.4.3.1 Introduction

MEMS are very attractive for wireless applications due to their excellent radiofrequency (RF) properties, high linearity, low power consumption, and low cost. A large variety of devices exist under the umbrella of RF MEMS, being categorized according to their main functionality: (1) capacitive and contact switches, (2) tunable or programmable passives (capacitors and inductors), (3) resonators (mimicking full circuit functions) for oscillator and mixer applications, and (4) micromachined transmission lines and antennas. In principle, any true RF MEMS device has a mechanically movable part using one of the four fundamental actuation principles: (1) electrostatic, (2) thermal, (3) magnetic, and (4) piezoelectric. The most used actuation principles, essentially because of their reduced power consumption, are the electrostatic and piezoelectric ones. In this section, we present the state-of-the-art in electrostatically actuated RF MEMS capacitive switches and their use in phase shifters for applications beyond 10 GHz; we illustrate the chapter with some recent fabrication results and measured performances.

13.4.3.2 RF MEMS Capacitive Switches

The RF MEMS capacitive switch is made based on a movable conductive membrane suspended over an air-gap and another conductive electrode covered by a thin insulator. The movable membrane can be actuated electrostatically by applying a voltage between the two conductive electrodes. Figure 13.12 describes the principle of such a capacitive switch: in the up-state (non-actuated switch) the capacitance between the two electrodes is very small (Coff, typically in the order of few fFs) and there is a good signal isolation, in the down-state (actuate switch) the value of the capacitance is substantially increased (Con, typically at least one order of magnitude higher than Coff, values of interest being in the order of hundreds of fFs to pFs) and a high-frequency signal can be coupled between the two electrodes.

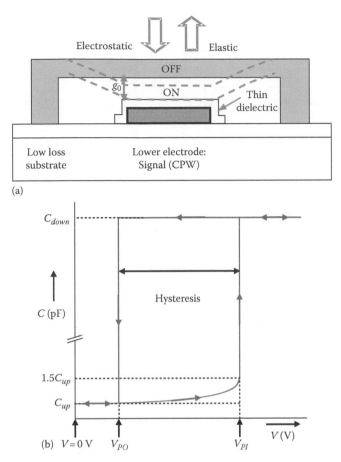

FIGURE 13.12 (a) Capacitive RF MEMS switch with electrostatic actuation and the OFF and ON stable states. (b) Capacitance characteristics of electrostatically actuated RF MEMS capacitive switch showing ON state (C_{down}) and OFF state (C_{up}) capacitances as well as V_{PI} and V_{PO}.

The capacitance characteristic of an RF MEMS capacitive is qualitatively represented in Figure 13.13. One can distinguish as key parameters the actuation voltages, called pull-in (V_{PI}) and pull-out (V_{PO}) voltages. Their expressions can be analytically calculated as a function of the switch geometrical design parameter

$$V_{PI} = \sqrt{\frac{8}{27} \frac{k g_{eff}^3}{\varepsilon_0 A}}, \quad V_{PO} = \sqrt{\frac{2 k g_0 g_\varepsilon^2}{\varepsilon_r^2 \varepsilon_0 A}} \tag{13.18}$$

where
  k is the spring elastic constant
  g_0 is the zero-voltage gap spacing
  g_ε is the dielectric thickness
  g_{eff} is the effective insulator thickness ($=g_0 + g_\varepsilon$)
  A is the membrane area

RF MEMS switches are used in wireless communication systems with two different device functionalities that determine two different capacitive MEMS approaches that are described as it follows.

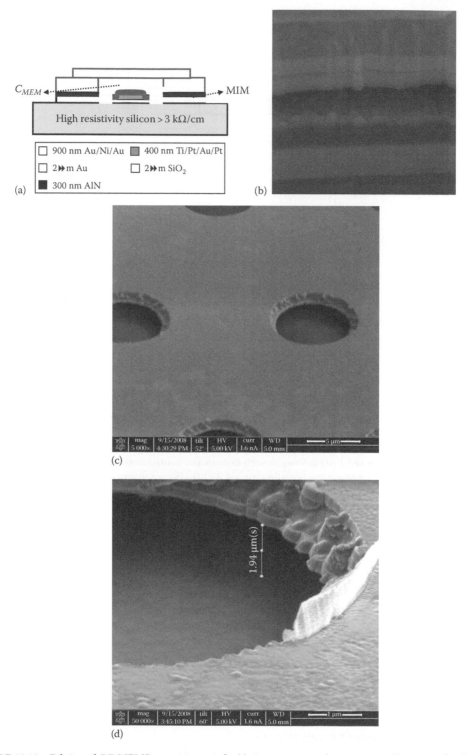

FIGURE 13.13 Fabricated RF MEMS capacitive switch: (a) Cross section schematic view illustrating the different materials and thicknesses. CMEM and CMIM denoted the MEMS and the metal–insulator–metal capacitors. (b) Focused Ion Beam (FIB) cross section of the fabricated device. (c) A view of the Au/Ni/Au membrane showing the releasing holes. (d) A detail of the membrane showing the 2 μm air gap.

The first approach is the RF MEMS capacitive switch or analog MEMS capacitor with a high capacitance ratio (Con/Coff > 50) is used as a high isolation switch for signal routing purpose in reconfigurable front-end or for antenna orientation. To achieve a high isolation, some successful capacitive switches based on shunt and series switches have been reported by Goldsmith et al. [23] and Muldavin and Rebeiz [24]. Other works concern switches with low spring constant meander anchoring [25,26] for low-voltage operation, inductively tuned switches [27], and single or cascaded multiple devices.

The second approach is the MEMS switch capacitor or digital MEMS capacitor that can have a low-to-moderate capacitance ratio (<10) and it is generally used as a part of a tunable system as filters, phase shifters or reconfigurable networks with two well-defined capacitive states. This type of MEMS switch capacitor has been introduced by Hayden and Rebeiz [28] being based on a MEMS capacitive switch in series with a fix small capacitance; the implementations of this type of switch have been reported in metal–insulator–metal (MIM) capacitor, metal–air–metal (MAM) capacitor or interdigitated capacitor.

Today, reliability is still one of the main problems of RF MEMS switches, conditioning their final success for commercialization. Reliability is directly related with the dielectric charging of silicon-dioxide and silicon-nitride layers, and it is the main responsible mechanism for failures in capacitive switches. Degradation mechanisms have been studied by Herfst et al. [29] and different solutions have been proposed to reduce or avoid the charging effect: the dielectric-free capacitive switch architectures [30], the use of leaky dielectric as PZT or AlN [31] or switch miniaturizing to mitigate the effect of the dielectric, as well as decreasing the switching time [32].

In Figure 13.13, we show a recent RF MEMS capacitive switch fabricated in a seven-mask process on high resistivity silicon substrates (>3 kΩ/cm). The silicon substrate is isolated with a thick 2 μm thermal oxide insulator to avoid the effect of free charge carriers in the substrate interface. The RF MEMS switches are loaded on a 78 Ω-high impedance coplanar waveguide (CPW) line consisting of 2.5 μm electroplated gold conductor, except under the MEMS bridge where a thin under-path layer of a sandwich of metals (Ti/Pt/Au/Pt) is employed instead. The thin metal signal line is covered in the actuation region by 300 nm of non-piezoelectric aluminum nitride (AlN), whose dielectric constant is 9.8. The total CPW width is 300 μm with G/W/G of 125/50/125 μm.

The suspended membrane is 900 nm thick consisting of AuNiAu (200/500/200 nm) multilayer and 15 μm thick Ni for the MEMS meander or spring-type anchoring, supports, and stiff bars for avoiding membrane deformation due to multi-metal membrane fabrication–induced stress. The final air gap is around 2 μm.

Figure 13.13 depicts SEM images of three designed and fabricated capacitive shunt switches with the process previously described.

As a typical example, Table 13.1 reports the measured electrostatic and RF performances for the three devices fabricated and depicted in Figure 13.14; their performances are similar. However, the actuation voltage of the external bias switch (Figure 13.14c) is much higher (2.5 times higher) than the other two

TABLE 13.1 Mechanical and RF Performances for Capacitive Switch Devices Shown in Figure 13.14

	Analogue Capacitive Shunt RF MEMS Switch				
	String Type	Meander Type			External Actuation Spring Type
Actuation voltage, V_{PI} (V)	15	13	20	40	33
Spring constant, k (N/m)	13	9			63
Insertion loss @ 40 GHz (dB)	−1.5	−1.1			−0.65
Return loss @ 40 GHz (dB)	−10	−12.5	−11.5	−10	−13
Isolation @ 40 GHz (dB)	−35	−30	−37.5	−35	−30
@ 20 GHz (dB)	−13.5	−12.5	−15	−17	−12

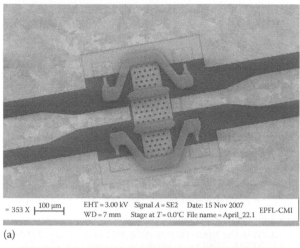

(a)

(b)

(c)

FIGURE 13.14 SEM images of three fabricated capacitive RF MEMS switches: (a) string-type, (b) meander-type anchoring with a DC voltage in line with the RF signal, and (c) stiffer spring-type anchoring with external DC voltage actuation.

(Figure 13.14a and b) due to a higher stiffness of the membrane. The achieved insertion loss of the external DC bias string-type switch is relatively lower. Device characterization from 6 to 40 GHz is usually carried out on using Vector Analyzer and a calibration is performed using an external SOLT commercial calibration kit.

13.4.3.3 Distributed MEMS Phase Shifters

RF MEMS capacitors have been investigated for application in distributed MEMS transmission lines (DMTLs) phase shifters, both analog [33] and digital approaches [34]. The idea is to periodically load a CPW with voltage-controlled varactors (two-state switch capacitors for digital DMTLs [35], in order to tune the distributed capacitance, the phase velocity and the propagation delay in the line. The main advantage of this kind of phase shifter is its constant time delay behavior over a wide band of frequencies.

The first optimized loaded-line phase shifter was implemented using diodes [36]. However, diodes present degraded performances at high frequencies (RF), especially in terms of losses, tuning linearity and intermodulation distortion. RF MEMS capacitors overcome these problems, although, they might have limited tuning range and slower switching; however, these are not limiting factors for most phase-shifter applications. Analog MEMS DMTL phase shifter are commonly limited by its low capacitance ratio (<1.5). Digital solutions based on MEMS switch capacitors achieved higher capacitance ratio (~3) with the advantages of size reduction, lower insertion loss, less dependence on the fabrication technology, but with a higher actuation voltage. For multistate phase shifter, the standard solution is to add bits to individually actuate groups of switch capacitor, which implies different actuation control commands as well as increasing substantially the complexity and usually the size of the device [37,38].

Reliability and temperature sensitivity in phase shifter have rarely been studied [39] as important issues for airborne and space applications, which require operation within a large range of temperature.

RF-distributed MEMS phase shifter concept, DMTL [40] for beam-steering has the main advantage of achieving a true-time delay (TTD) over a wide band. DMTL phase shifters will play important roles in future airborne applications, adding many advantages over traditional mechanical steering phase shifters. These include decreased size, cost, and weight; increased flexibility; lower insertion loss and dispersion; and a good matching over a large bandwidth. Electronically, beam-steering the antenna involves dynamically switching the beam configurations of an antenna array by controlling the relative phases at the input of each element in order to change the orientation of the whole antenna beam [41].

Figure 13.15 illustrates a fabricate DMTL phase shifter [42] by the authors of this chapter; the performances of the measured 5 mm digital DMTL phase shifter are very good in terms of repeatability, fabrication uniformity, reliability, and robustness after different tests in vacuum and in air and at various temperatures (from −130°C to 50°C). The pull-in of all the switches is around 24 V (slightly higher than a single digital capacitor to ensure the actuation of all 13 MEMS capacitors). Figure 13.16 depicts the typical $C-V$ curve of a single digital capacitor and a full array of capacitors loaded on a CPW, namely, the DMTL phase shifter. One can observe a quasi-identical VPI for all the switches (~22V) and a symmetrical curve around $V = 0$, which indicates that either no mobile residual charges exits on the dielectric or that they are compensated. The pull-out is slightly higher than 10 V. The $C-V$ curve appears to be symmetrical because the polarity of the applied voltage has no influence on the actuation since the electrostatic force is proportional to the voltage squared.

Figure 13.16a shows the measured S-parameter data for the phase shifter and Figure 13.16b illustrates the full characterization analysis of the reported DMTL. First, a constant TTD of 25 ps is achieved for a wide frequency band, this time delay corresponding to a phase delay of 125° at 14 GHz. Second, the characteristic impedance of the loaded line in the Up and Down states results in a quasi-adapted matching line of 55 and 45 Ω, respectively. Both values are very close to the desired impedance reference of 50 Ω resulting in a return loss (matching) better than −10 dB for all the states.

The reported results confirm the maturity of RF MEMS technology for phase-shifter applications and its promise for high-frequency applications.

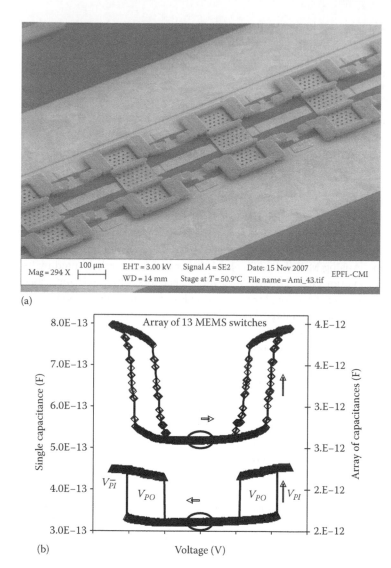

(a)

(b)

FIGURE 13.15 (a) SEM image of the fabricated distributed MEMS digital phase shifter using 13-loaded RF MEMS digital capacitive switches (only four switches are shown in this figure). (b) Measured *C–V* curves of a single digital capacitor and the array of parallel loaded 13 MEMS capacitors forming the phase shifter.

13.4.4 Microfluidic Components

Microfluidic components are used in chemical, biomedical, and energy applications. A microfluidic system is composed of several components that extract, transport, pump, or process fluids at small scales [43,44].

13.4.4.1 Valves

Passive valves (also known as check-valves or one-way valves) are usually built by placing a membrane that can deflect in the direction of flow, but not in the opposite one. When the flow is perpendicular to wafer surface, they are usually fabricated using a combination of bulk micromachining and wafer bonding, like in Figure 13.17. If the channel for fluid is coplanar with the wafer or chip, fabrication is more difficult and can involve several layers of structural photoresist, such as SU-8.

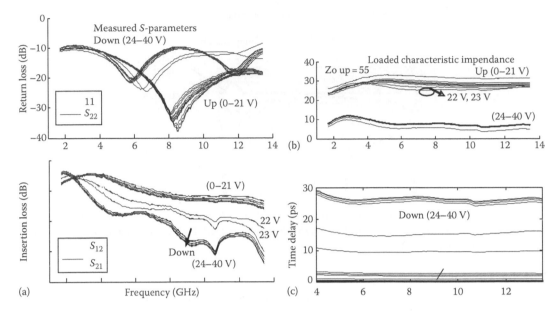

FIGURE 13.16 (a) Measured *S*-parameters of the digital DMTLs of 13 nickel bridges spaced at 294 μm (total length 5 mm) with an actuation voltage of 24 V. (b) Illustration of the real impedance of the loaded digital phase shifter in all the states. (c) The time delay extracted from measurements at different actuation states.

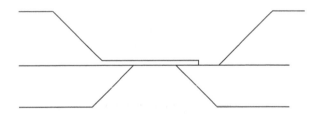

FIGURE 13.17 Passive microvalve by bulk micromachining.

Active microvalves include an external actuation to open or close them. The physical effect used for actuation can be any one of those described in Section 13.2.3. In particular, electrostatic or piezoelectricity actuations are commonly used for microfluidic valves [45].

13.4.4.2 Nozzles and Diffusers

Nozzles and diffusers are easily built using surface micromachining or micromolding, and are used to speed up or slow down the flow at a particular point. Nozzles, e.g., can be used to accelerate a stream until it breaks down into droplets. This technique is used in many inkjet printer cartridges. A combination of a nozzle and a diffuser is used in Venturi devices, which are used in microfluidics to generate a low pressure from a higher one, and employ it to extract or pump a fluid.

13.4.4.3 Pumps

Displacement micropumps work by having a movable component that displaces the fluid. The type of actuation used to move the component can again be any one of the methods presented in Section 13.2.3. The most commonly employed ones are electrostatic (see Figure 13.18), piezoelectric, pneumatic, or thermopneumatic.

Other methods of providing impulsion to fluids have been tried. These include electroosmosis, electrowetting, magnetohydrodynamics, and acoustic streaming [47].

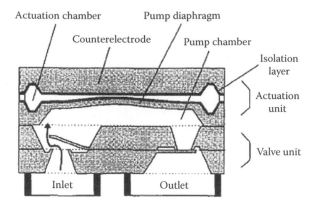

FIGURE 13.18 Bidirectional silicon micropump. (Modified from Zengerle, R. et al., A bidirectional silicon micropump, in *Proceedings of the Microelectromechanical Systems (MEMS'95)*, Amsterdam, the Netherlands, January 29–February 2, 1995.)

13.4.4.4 Flow Sensors

Sensors that measure flow rate can rely on different effects. Force flow sensors measure the force exerted by the fluid over a bendable cantilever; this force is measured using the change in electrical resistance suffered by a piezoresistive element included in the cantilever. Pressure flow sensors use the change in fluid pressure between two points in the flow path; again, pressure is measured using a piezoresistor. Thermal flow sensors provide a known amount of heat to the fluid using a resistor and measure the change in temperature at a second point downstream; measurement of temperature change is usually done with temperature-dependent resistors, or with thermistors.

All these types of flow sensors have been successfully micromachined, usually in silicon, and are commercially sold (see e.g., [48]).

13.4.4.5 Mixers

The difficulty of mixing at the microscale resides in the low Reynolds numbers that are usually present in the flow, due to the small length and size. The transition from laminar to turbulent regime (where mixing occurs most efficiently) is not easily accomplished. Complex geometries have to be employed in order to generate the turbulence. Examples can be found in [49] and [50].

13.4.5 Optical Devices

The development of microscale, reconfigurable MEMS optical components have opened a wide range of applications. The list of optical devices that have been realized is vast, and it is not possible to cover all of the structures that have been fabricated. Rather, this section will examine the core components of any MEMS-based optical system. These are mirrors, lenses, optical gratings, and optical filters. Other components such as variable optical attenuators, optical switches and many optical sensors are variants of these structures. Important devices that are not covered are micro-optic components such as waveguides, resonators, couplers, and micro-optical benches. In the following, the references given are examples rather than exhaustive.

13.4.5.1 Mirrors

Two major types of mirrors have been realized in MEMS structures; planar mirrors that allow tip, tilt, or piston motion (or a combination of all three); and deformable mirrors, primarily for aberration correction. Mirror dimensions that have been realized extend from a few tens of microns to millimeters. The most complex MEMS devices to date are based on large arrays of micromirrors used for digital projection [51]. Actuation of planar mirror structures is most commonly electrostatic [51], piezoelectric [52], or electromagnetic [53] (current through a coil in a permanent magnetic field) because of the high

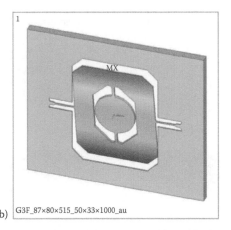

(a)

(b) G3F_87×80×515_50×33×1000_au

FIGURE 13.19 MEMS two-axis tilt mirror designed for use in a portable projector system. This system uses electromagnetic actuation with the structure placed in a magnetic field and current passed through the coil. The actuation of both axes is achieved by driving currents through the coils at frequencies corresponding to the resonant frequency of central mirror and the outside gimbal: (a) An optical micrograph of the device. (b) Modeling when the device is driven at the resonant frequency of the centre mirror. (Modified from Davis, W.O. et al., MEMS-based pico projector display, in *2008 IEEE/LEOS International Conference on Optical MEMs and Nanophotonics*, Freiburg, Germany, pp. 31–32, August, 11–14, 2008.)

speeds needed for most applications of these devices. Deformable mirrors are usually formed using flexible membranes, which are either actuated at a discrete number of locations across the mirror [54] or in a spatially continuous manner (either around the edge or over the entire surface). Local displacement at an array of points can be used to form very complicated shapes for wavefront correction, but requires closely spaced actuators to give surfaces that are smooth on a scale comparable to the wavelength of light. Continuous actuation allows smooth mirror shapes such as variable focal length parabolic mirrors [55], but cannot form mirrors with complicated shapes. A wide range of actuation techniques have been used for deformable mirrors including electrostatic, piezoelectric, pneumatic, electromagnetic, and thermal. Figure 13.19 shows an example of a planar mirror structure that has both tip and tilt, for a miniature projector suitable for consumer electronics applications [53].

Figure 13.20 shows an example of a deformable mirror structure [56]. It consists of a continuous membrane mirror attached to an array of electrostatic actuators allowing the device to correct wavefront errors, improving focusing, and overcoming aberrations in optical systems. Such devices are used in biomedical, astronomical, and scientific instrumentation to improve resolution and decrease dispersion.

13.4.5.2 Lenses

MEMS lenses can be classified as being fixed focus, movable with fixed focus, variable focus, and movable with variable focus. Fixed focus microlenses [57] have been commercially available for many years and will not be discussed here. Virtually all variable focus and movable with variable focus microlenses rely on liquids, using either thermo/pneumatic actuation of microfluidic devices incorporating membranes [58], liquid droplets actuated using electrowetting [59], or a droplet of liquid crystal [60]. All of these technologies are being commercialized, with a major driver being the multilayer, high-density optical data storage applications. The principle of microfluidic lenses [58] is shown in Figure 13.21. Liquid is introduced into a micro-chamber covered by a thin, deformable membrane. By controlling the pressure in the chamber, the extent of deformation of the membrane, and hence the focal length of the lens formed is controlled. By combining several membranes with different liquids, excellent aberration control can be achieved.

Figure 13.22 shows the principle of electrowetting and an example of an electrowetting-based microlens [60]. When a drop of liquid is placed on a solid, the shape of the droplet is generally spherical. The spherical shape of the liquid can then be used to form a lens, and since the surface of the lens is flexible, there exist the possibility of changing the focal length of the lens, by changing the contact angle between the solid

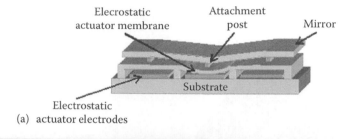

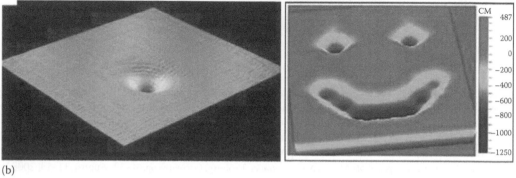

FIGURE 13.20 (a) A cross section of a deformable mirror structure showing the elements of a 1 × 3 actuator array within an electrostatically actuated MEMS deformable mirror (DM). (b) In this image of an optical surface measurement of a 144-element DM, a single pixel is actuated (left). The influence of the single element deflection extends only to its immediate neighbors, leaving the rest of the mirror surface unchanged. This local influence characteristic of the DM allows for high-order and otherwise arbitrary shapes on the DM (right). (Modified from Bierden, P. et al., MEMS deformable mirrors for high-performance AO applications, in *Adaptive Optics for Industry and Medicine: Proceedings of the Sixth International Workshop*, Dainty, C. (ed), Imperial College Press, London, U.K., 2008.)

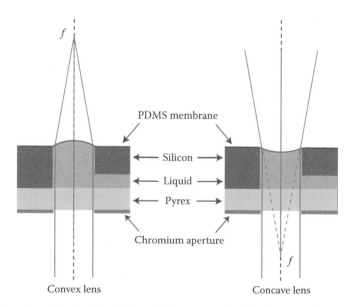

FIGURE 13.21 Operation of a microfluidic lens, which consists of a chamber covered with a flexible membrane into which liquid is introduced. If the pressure is increased above ambient, a convex lens results, as shown on the left. In contrast, if the pressure in the chamber is less than ambient, the membrane is deformed so as to form a concave lens. (Courtesy of Prof. Hans Zappe, IMTEK, University of Freiburg, Freiburg, Germany.)

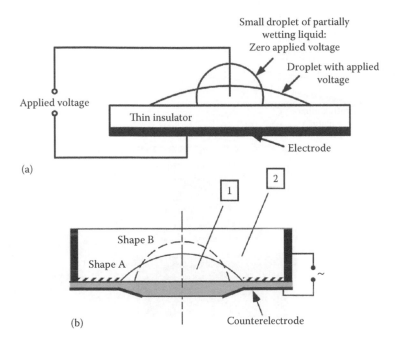

FIGURE 13.22 Operation of tunable microlenses based on electrowetting: (a) The principle of operation in which applying a voltage changes the surface tension of the liquid with respect to the planar surface on which it rests. (b) Shows how this effect can be used to form a variable focal length lens. The cell is filled with water 2. A drop 1 of an insulating and nonpolar liquid is deposited on the bottom wall, which is made of an insulating and transparent material, in gray. The central disc on the bottom wall surface is hydrophobic, in order to trap the drop. The outer zone, hatched in the figure, is hydrophilic. The optical axis is shown as a dashed line. (Modified from Berge, B. and Peseux, J., *Eur. Phys. J. E*, 3, 159, 2000.)

and liquid. The contact angle between the solid and liquid is determined by the values of surface tension at the liquid–solid, liquid–vapor, and solid–vapor interfaces. Electrowetting is a physical effect in which the surface tension between a liquid and a solid can be controlled by a voltage applied across the interface.

Practical devices use a thin insulating layer between the electrode and the liquid, and often use two immiscible liquids, as shown in Figure 13.22. The use of two fluids allows one fluid to be used as the lens, while the other is used to control the surface tension around the lens, obviating the need for a contact to the lens liquid. If the region below the insulator is patterned with an array of electrodes, it also possible to move the lens across the surface by applying potentials at the edge of the lens liquid, making it energetically favorable for the lens droplet to move over the active electrode.

13.4.5.3 Gratings

Micro-optical gratings used to diffract or steer optical signals have been commercially available for some time, and are key components of many systems, most notably optical data storage. When combined with MEMS actuation, normally fixed gratings can be made tunable allowing controllable beam steering, tunable spectral separation, and wavefront control. The angle at which a particular incident wavelength is diffracted by a grating is controlled by the incident angle and the pitch of the grating. Gratings in which the angle of incidence is controlled have been realized by mounting the grating on a movable structure similar to tip/tilt mirrors [62]. Gratings with adjustable pitch have been fabricated by either defining the grating on a flexible substrate and stretching the material to increase the separation of the lines [63], by concertinaing the lines of the grating [64], or by using fine-pitch arrays in which lines are removed from the field of view of the incoming optical signal [65]. An example of the tip-tilt approach is shown in Figure 13.23, while an example of the later stretched grating approach is shown in Figure 13.24.

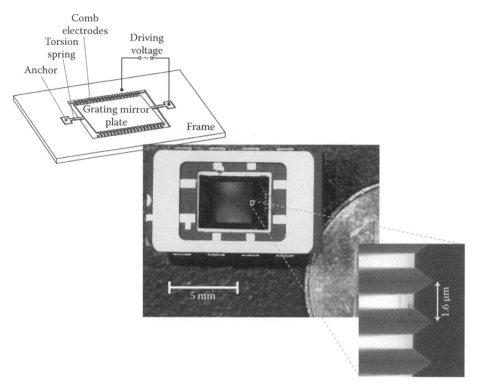

FIGURE 13.23   Movable grating showing the electrostatic comb-drive actuation, allowing the grating to be tilted. This structure was manufactured for use in microspectrometer. (Courtesy of Kraft, M. et al., *Eur. Phys. J. E*, 3, 159, 2000.)

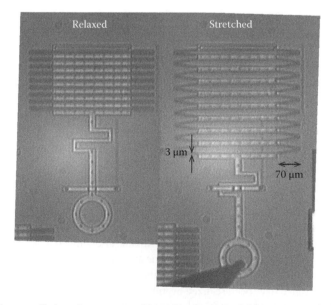

FIGURE 13.24   Pitch-controlled grating structure that realizes an expandable grating in which the pitch of the grating is changed, thanks to the movement of the end point. In the figure a probe has been used to extend the accordion-like structure. In the actual device, a thermal actuator drives a pawl and ratchet mechanism to control the extension of the structure. These structures were developed for use in large astronomical telescopes. (Courtesy of Konidaris, N. et al., *Appl. Opt.*, 42(4), 621, February 2003.)

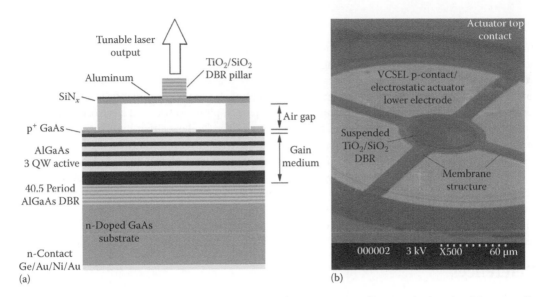

FIGURE 13.25 Tunable vertical cavity surface-emitting laser structure. Changing the length of the optically resonant cavity, the emission wavelength can be controlled: (a) A cross-sectional schematic of the structure. (b) An SEM image of the structure. (Courtesy of Bond, T. et al., Two-state optical filter based on micromechanical diffractive elements, in *2007 IEEE/LEOS International Conference on Optical MEMs and Nanophotonics*, Hualien, Taiwan, pp. 167–168, August 12–July 16, 2007.)

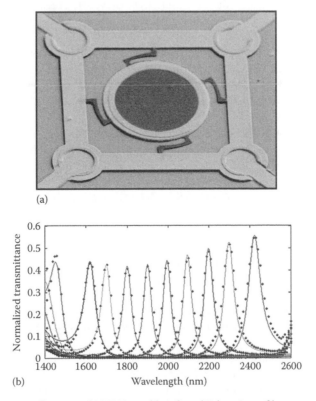

FIGURE 13.26 Electrostatically actuated MEMS tunable infrared Fabry–Perot filter structure: (a) An SEM image of the tunable filter (the circular filter area is 100 μm in diameter) and (b) the spectral characteristics of the device as the voltage is tuned. The pass transmission is limited because the substrate is not antireflection coated.

13.4.5.4 Filters

Tunable filter structures include variable attenuators, polarization filters, and spectral filters. All of these devices have been realized in MEMS using a variety of approaches. Of particular importance are tunable MEMS-based Fabry–Perot spectral filters because of their widespread use in tunable lasers and detectors. In general, these devices use distributed Bragg reflectors for the mirrors that form the Fabry–Perot resonant cavity, and tuning has been achieved using thermal, piezoelectric, and electrostatic actuation. Compromises are made in the design of the Fabry–Perot filter to allow either high spectral resolution but limited tuning range (e.g., as required for dense wavelength-division multiplexing in optical fiber communications) or wide tuning range of up to one octave in wavelength, but limited spectral resolution (e.g., for application in near infrared spectroscopy). The compromise is needed because of mirror flatness requirements and the need to operate in first order for wide spectral coverage. Figure 13.25 shows an example of a tunable vertical cavity surface-emitting laser diode that incorporates a tunable mirror structure [66]. Figure 13.26 shows a tunable filter for spectroscopic applications in the near infrared region [67].

References

1. Sze, S. M., *Semiconductor Sensors*, Wiley, New York, 1994.
2. Korvink, J. and Paul, O., *MEMS A Practical Guide to Design, Analysis and Applications*, William Andrew Inc., Norwich, NY, 2006.
3. Smith, C. S., Piezoresistance effect in germanium and silicon, *Phys. Rev.*, 94(1), 42–49, 1954.
4. Y. Kanda, Piezoresistance effect of silicon, *Sens. Actuators A*, 28(2), 83–91, 1991.
5. Middelhoek, S. and Audet, S. A., *Silicon Sensors*, Delft University Press, Delft, the Netherlands, 1994.
6. Fraden, J., *Handbook of Modern Sensors, Physics, Designs, and Applications*, Springer-Verlag, New York, 1996.
7. Senturia, S. D., *Microsystem Design*, Kluwer Academic Publishers, Norwell, MA, 2001.
8. Bell, D. J., Lu, T. J., Fleck, N. A., and Spearing, S. M., MEMS actuators and sensors: Observations on their performance and selection for purpose, *J. Micromech. Microeng.*, 15, S153–S164, 2005.
9. Nathanson, H. C., Newell, W. E., Wickstrom, R. A., and Davis, J. R., The resonant gate transistor, *IEEE Trans. Electon. Devices*, 14(3), 117–133, 1967.
10. Rebeiz, G. M., *RF MEMS: Theory, Design, and Technology*, Wiley, New York, 2003.
11. Castañer, L., Pons, J., Nadal-Guardia, R., and Rodríguez, A., Analysis of the extended operation range of electrostatic actuators by current-pulse drive, *Sens. Actuators: Phys. A*, 90(3), 181–190, May 2001.
12. Seeger, J. I. and Crary, S. B., Stabilization of electrostatically actuated mechanical devices, in: *Proceedings of the International Conference on Solid State Sensors and Actuators (Transducers '97)*, Vol. 2, Chicago, IL, pp. 1133–1136, June 16–19, 1997.
13. Pons, J., Rodriguez, A., and Castañer, L., Voltage and pull-in time in current drive of electrostatic actuators, *J. Microelectromech. Syst.*, 11(3), 196–205, June 2002.
14. Bao, M.-H., *Handbook of Sensors and Actuators*, Vol. 8, Elsevier, New York, 2000.
15. Fan, L.-S., Tai, Y.-C., and Muller, R. S., IC processed electrostatic micromotors, in: *IEEE International Electron Devices Meeting (IEDM)*, San Francisco, CA, pp. 666–669, December 11–14, 1988.
16. Sinclair, M. J., A high force low area MEMS thermal actuator, in: *Proceedings of IEEE Intersociety Conference on Thermal and Thermomechanical Phenomena*, Las Vegas, NV, pp. 127–132, May 23–26, 2000.
17. Neukomm, P. A., Bornhauser, H. P., Hochuli, T., Paravicini, R., and Schwarz, G., Characteristics of thin wire shape memory actuators, *Sens. Actuators A: Phys.*, 21(1–3), 247–252, 1990.
18. Wright, J. A., Tai, Y.-C., and Chang, S.-C., A large force fully integrated MEMS magnetic actuator, in: *Proceedings of the International Conference on Solid-State Sensors and Actuators (Transducers '97)*, Vol. 2, Chicago, IL, pp. 793–796, June 16–19, 1997.

19. Niarchos, D., Magnetic MEMS: Key issues and some applications, *Sens. Actuators A*, 106, 255–262, 2003.

20. Daigle, M., Corcos, J., and Wu, K., An analytical solution to circular touch mode capacitor, *Sens. J., IEEE*, 7(4), 502–505, April 2007.

21. Mohamed-el-Hak (Ed.), *MEMS: Applications*, CRC Press, Boca Raton, FL, 2006.

22. Palasagaram, J. N. and Ramadoss, R., MEMS-capacitive pressure sensor fabricated using printed-circuit-processing techniques, *Sens. J., IEEE*, 6, 1374–1375, December 2006.

23. Goldsmith, C. L., Yao, Z., Eshelman, S., and Denniston, D., Performance of low loss RF MEMS capacitive switches, *IEEE Microw. Guided Wave Lett.*, 8(8), 269–271, August 1998.

24. Muldavin, J. B. and Rebeiz, G. M., High-isolation CPW MEMS shunt switches – Part 1: Modeling, *IEEE Trans. Microw. Theory Tech.*, 48(6), 1045–1052, June 2000.

25. Pacheco, S. P. et al., Design of low actuation voltage RF MEMS switch, in: *2000 IEEE MTT-S International Microwave Symposium Digest*, Vol. 1, Boston, MA, pp. 165–168, November 5–8, 2000.

26. Peroulis, D., Pacheco, S., and Kaheti L. P. B., MEMS devices for high isolation switching and tunable filtering, in: *IEEE MTT-S International Microwave Symposium Digest*, Boston, MA, pp. 1217–1220, June 2000.

27. Muldavin, J. B. and Rebeiz G. M., High-isolation inductively-tuned X-band MEMS shunt switches, in: *IEEE MTT-International Microwave Symposium Digest*, Vol. 1, Boston, MA, pp. 169–172, June 11–16, 2000.

28. Hayden, J. and Rebeiz, G. M., Low-loss cascadable MEMS distributed X-band phase shifter, *IEEE Microw. Guided Wave Lett.*, 10(4), 142–144, April 2000.

29. Herfst, R.W, Steeneken, P.G., and Schmitz, J., Identifying degradation mechanisms in RF MEMS capacitive switches, in: *Proceedings of the MEMS 2008*, Tucson, AZ, pp. 168–171, January 13–17, 2008.

30. Blondy, P., Crunteanu, D., Champeaux, C., Catherinot, A., Tristant, P., Vendier, O., Cazaux, J. L., and Marchand, L., Dielectric less capacitive MEMS switches, in: *IEEE MTT-International Microwave Symposium Digest*, Vol. 2, Fort Worth, TX, pp. 573–576, June 6–11, 2004.

31. Fernández-Bolaños, M., Tsamadós, D., Dainesi, P., and Ionescu, A. M., Reliability of RF MEMS capacitive switches and distributed MEMS phase shifters using AlN dielectric, in: *22nd International Conference on Micro Electro Mechanical Systems (MEMS)*, Sorrento, Italy, pp. 638–641, 2009.

32. Lakshminarayanan, B., Mercier, D., and Rebeiz, G. M., High-reliability miniature RF-MEMS switched capacitors, *IEEE Trans. Microw. Theory Tech.*, 56(4), 971–981, April 2008.

33. Barker, N. S. and Rebeiz, G. M, Optimization of distributed MEMS transmission lines phase shifters—U-band and W-band designs, *IEEE Trans. Microw. Theory Tech.*, 48(11), 1957–1966, November 2000.

34. Perruisseau-Carrier, J. et al., Modeling of periodic distributed MEMS, application to the design of variable true-time delay lines, *IEEE Trans. Microw. Theory Tech.*, 54(1), 383–392, 2006.

35. Borgioli, A. et al., Low-loss distributed MEMS phase shifter, *IEEE Microw. Guided Wave Lett.*, 10, 7–9, 2000.

36. Nagra, A. S. and York, R. A., Distributed analog phase shifter with low insertion loss, *IEEE Trans Microw. Theory Tech.*, 47(9), 1705–1711, September 1999.

37. Kim, H.-T. et al., V-band 2-b and 4-b low-loss and low-voltage distributed MEMS digital phase shifter using metal-air-metal capacitors, *IEEE Trans. Microw. Theory Tech.*, 50(12), 2918–2923, 2002.

38. Hayden, J. S. and Rebeiz, G. M., 2-bit MEMS distributed X-band phase shifter, *IEEE Microw. Guided Wave Lett.*, 10(12), 540–542, December 2000.

39. Fernández-Bolaños, M., Lisec, T., Dainesi, P., and Ionescu, A. M., Thermally stable distributed MEMS phase shifter for airborne and space applications, in: *Proceeding of 38th European Microwave Conference*, Amsterdam, the Netherlands, pp. 100–103, October 2008.

40. Barker, N. S. et al., Distributed MEMS true-time delay phase shifters and wide-band switches, *IEEE Trans. Microw. Theory Tech.*, 46(11), 1881–1890, November 1998.
41. Vinoy, K. J. and Varadan, V. K., Design of reconfigurable fractal antennas and RF-MEMS for space-based systems, *Smart Mater. Struct.*, 10, 1211–1223, November 2001.
42. Fernández-Bolaños, M., Lisec, T., Dainesi, P., and Ionescu, A. M., Thermally stable distributed MEMS phase shifter for airborne and space applications, in: *Proceeding of 38th European Microwave Conference*, Amsterdam, the Netherlands, pp. 100–103, October 2008.
43. Koch, M., Evans, A., and Brunnschweiler, A., *Microfluidic Technology and Applications*, Research Studies Press, Baldock, U.K., 2000.
44. Samel, B., Nock, V., Russom, A., Griss, P., and Stemme, G., A disposable lab-on-a-chip platform with embedded fluid actuators for active nanoliter liquid handling, *Biomed. Microdevices*, 9, 61–67, 2007.
45. Braneberg, J. and Gravesen, P., A new electrostatic actuator providing improved stroke length and force, in: *Proceedings of International Conference on Micro Electro Mechanical Systems*, Travenmiinde, Germany, 1992.
46. Zengerle, R., Kluge, S., Richter, M. and Richter, A., A bidirectional silicon micropump, in: *Proceedings of the Microelectromechanical Systems (MEMS'95)*, Amsterdam, the Netherlands, January 29–February 2, 1995.
47. Laser D. J. and Santiago, J. G., A review of micropumps, *J. Micromech. Microeng.*, 14, R35–R64, 2004.
48. Su, Y., Evans, A. G. R., and Brunnschweiler, A., Micromachined silicon cantilever paddles with piezoresistive readout for flow sensing, *J. Micromech. Microeng.*, 6, 69–72, 1996.
49. Asgar, A, Bhagat, S., and Papautsky, I., Enhancing particle dispersion in a passive planar micromixer using rectangular obstacles, *J. Micromech. Microeng.*, 18, 085005, 2008.
50. Suzuki, H., Ho, C.-M., and Kasagi, N., A chaotic mixer for magnetic bead-based micro cell sorter, *J. Microelectromech. Syst.*, 13, 779–790, 2004,.
51. Van Kessel, P. F., Hornbeck, L. J., Meier, R. E., and Douglass, M. R., A MEMS-based projection display, *Proc. IEEE*, 86(8), 1687–1704, August 1998.
52. Maeda, R., Tsaur, J. J., Lee, S. H., and Ichiki, M., Piezoelectric microactuator devices, *J. Electroceram.*, 12(1–2), 89–100, January–March 2004.
53. Davis, W.O., Sprague, R., and Miller, J., MEMS-based pico projector display, in: *2008 IEEE/LEOS International Conference on Optical MEMs and Nanophotonics*, Freiburg, Germany, pp. 31–32, August 11–14, 2008.
54. Doble, N. and Williams, D. R., The application of MEMS technology for adaptive optics in vision science, *IEEE J. Sel. Top. Quantum Electron.*, 10(3), 629–635, May/June 2004.
55. Hokari, R. and Hane, K., A varifocal micromirror with pure parabolic surface using bending moment drive, in: *2008 IEEE/LEOS International Conference on Optical MEMs and Nanophotonics*, Freiburg, Germany, pp. 92–93, August 11–14, 2008.
56. Bierden, P., Bifano, T., and Cornelissen, S., MEMS deformable mirrors for high-performance AO applications, in: *Adaptive Optics for Industry and Medicine: Proceedings of the Sixth International Workshop*, Dainty, C. (Ed.), Imperial College Press, London, U.K., 2008.
57. Sinzinger, S. and Jahns, J., *Microoptics*, Wiley-VCH, Weinheim, Germany, 2003.
58. Werber, A. and Zappe, H., Tunable microfluidic microlenses, *Appl. Opt.*, 44, 3238–3245, June 2005.
59. Mugele, F. and Baret, J.-C., Electrowetting: From basics to applications, *J. Phys.: Condens. Matter*, 17(28), R705–R774, July 2005.
60. Knittel, J., Richter, H., Hain, M., Somalingam, S., and Tschudi, T., A temperature controlled liquid crystal lens for spherical aberration compensation, *Microsyst. Technol.*, 13(2), 161–164, January 2007.
61. Kraft, M., Berge, B., and Peseux, J., Variable focal lens controlled by an external voltage: An application of electrowetting, *Eur. Phys. J. E*, 3, 159–163, 2000.
62. Kraft, M., Kenda, A., Frank, A., Scherf, W., Heberer, A., Sandner, T., Schenk, H., and Zimmer, F., Single-detector micro-electro-mechanical scanning grating spectrometer, *Anal. Bioanal. Chem.*, 385(5), 1259–1266, November 2006.

63. Konidaris, N., Wong, C. W., Jeon, Y., Barbastathis, G., and Kim, S.-G., Analog tunable gratings driven by thin-film piezoelectric microelectromechanical actuators, *Appl. Opt.*, 42(4), 621–626, February 2003.

64. Konidaris, N. P., Kubby, J. A., and Sheinis, A. I., Small solutions to the large telescope problem: A massively replicated MEMS spectrograph, in: *Advanced Optical and Mechanical Technologies in Telescopes and Instrumentation, Proceedings of SPIE*, SPIE, Bellingham, Washington, 7018, 70182I-1–70182I-9, August 2008.

65. Bog, T., Sagberg, H., Bakke, T., Johansen, l.-R., Lacolle, M., and Moe, S. T., Two-state optical filter based on micromechanical diffractive elements, in: *2007 IEEE/LEOS International Conference on Optical MEMs and Nanophotonics*, Freiberg, Germany, pp. 167–168, August 12–July 16, 2007.

66. Bond, T. C., Cole, G. D., Goddard, L. L., and Behymer, E. M., Photonic MEMS for NIR in-situ gas detection and identification, in: *Proceedings of IEEE 2007 Sensors Conference*, pp. 1368–1371, Atlanta, GA, October 28–31, 2007.

67. Musca, C. A., Antoszewski, J., Winchester, K. J., Keating, A. J., Nguyen, T., Silva, K. K. M. B. D., Dell, J. M., et al., Monolithic integration of an infrared photon detector with a MEMS-based tunable filter, *IEEE Electron Device Lett*, 26(12), 888–890, December 2005.

14

MEMS Technologies

Antonio Luque
University of Seville

José M. Quero
University of Seville

Carles Cané
National Microelectronics Center

14.1 Introduction

The term "microelectro-mechanical systems" (MEMS) refers to any device whose characteristic dimension lies in the range of micrometers. This general definition encompasses not only devices whose main effects are electrical or mechanical, but also those that deal with optical, chemical, biological, or others. MEMS are usually referred to as "microsystems" in Europe or "micromachines" in Asia.

Although the term is general, it is usually used to name devices fabricated using micromachining techniques based on those used for microelectronics fabrication. Nevertheless, in recent years other fabrication techniques have been developed exclusively for MEMS, independently from microelectronics industries.

After the inception of the silicon-based microelectronics industry in the 1950s and 1960s, several researchers studied the mechanical and piezoelectric properties of silicon and found that it was also suitable to build mechanical sensors. It was at that time that a now-classical speech was given by Nobel Prize winner Richard Feynman at the 1959 *Annual Meeting of the American Physical Society*. In that speech, called "There's plenty of room at the bottom" [1], Feynman raised the question of whether it would be possible to construct extremely small machines, and explained how physics laws would allow that. It created a lot of enthusiasm in the matter and opened the way for the developments to come in the microscale.

In the following decades, all kinds of MEMS devices were developed, including pressure and acceleration sensors, inkjet printer heads, gyroscopes, RF components, optical elements, and chemical, microfluidic, and biological devices.

The worldwide MEMS market is expected to grow up to 10,000 M\$ in year 2011, and most estimations conclude that it will keep growing at least until 2020 [2].

Although it is a sector that already reached maturity in many aspects, still a lot of research is being directed toward developing new devices and applications or improving fabrication processes, making them less expensive and more reliable.

MEMS and nanotechnology are related fields, but the latter should not be viewed as a mere continuation of the former, going beyond in the miniaturization. MEMS are usually built using a top-down approach (microdevices are constructed from macroscale materials using macroscale equipment), while nanotechnology-based components are usually built bottom-up (making complex systems by putting together molecules or simple elements). Many have come to see MEMS as the natural construction equipment for nanotechnology and nanodevices.

The rest of this chapter contains an introduction to the physical laws governing MEMS behavior, and the description of the fabrication techniques most commonly used in the development of these devices.

14.2 Modeling and Scaling Laws

Traditionally, MEMS devices are considered to have characteristic dimensions in the range of 0.1–$100\,\mu m$. This is a very important miniaturization, relative to the traditional, macroscopic world, that will affect the relative importance of physical phenomena. For instance, in the macro world, at the human scale, the influence of gravitational forces is often considered, but in the micro world, at the micron scale, gravity has a very low incidence when compared with other forces. That is, MEMS lives in a zero-gravity world, and thus the performance of a mechanical MEMS device is independent of its relative position with respect to the Earth. This is just one example of the changes that appear when scaling down the physic laws, and special care should be taken when applying the intuition of our experience in macroscale engineering because many considerations will change.

In order to quantify the scaling effect, we will assume that there is a geometry scaling factor S. Normally, S will be in the range of 10^{-3} to 10^{-5}. This means that a length L_o in the macro world will be scaled down to $L_s = SL_o$ in the micro world. In what follows, we present a brief comparative study of the scaling down of some physical phenomena.

14.2.1 Scale in Size

According to the definition of the scaling factor, a given length L in a device in the macroscale will be scaled by a factor of S, its surface is scaled by S^2, and its volume by S^3. The first conclusion we deduce is that the area to volume ratio will decrease from macro world expectations by a factor of S. Another conclusion is that there is a great reduction of volume with respect to its size. For example, if we have a scaling factor of $S = 0.001$, the volume of the corresponding MEMS device is reduced by 10^{-9}. This fact implies a very large reduction of the volume of the devices compared to its surface or characteristic dimension. Therefore, any physical magnitude related to the volume will suffer a proportional reduction. For this reason, the weight of an object, which is proportional to its volume via the density of the material, will be reduced in the same proportion. This drastic reduction of volumetric properties will have very relevant impact in the behavior of MEMS devices.

14.2.2 Mechanical Properties

The stiffness is the ability of a mechanical system to resist an external force. This is the most important mechanical property because it determines the capability to support an external force. Let us consider a cantilever beam like the one depicted in Figure 14.1a.

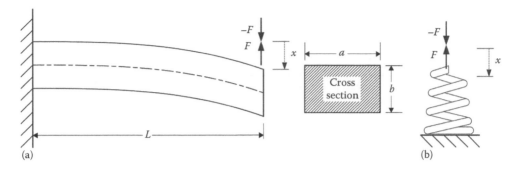

FIGURE 14.1 (a) Deformed cantilever beam when an external force −*F* is applied. (b) An equivalent spring.

The reactive force *F* that appears in its free end when an external force *F* is applied can be expressed as

$$F(x) = \frac{3EI}{L^3} x = \frac{3Eab^3}{12L^3} x \qquad (14.1)$$

where

   E is the Young's modulus
   I is the moment of inertia of the beam's section
   x is the deflection in the free end of the beam

Assuming that *E* is constant during the reduction of size, we find that *F* decreases with S^2. If we compare with an equivalent spring (Figure 14.1b), we have

$$F(x) = K_{elastic}x \quad \Rightarrow \quad K_{elastic} = \frac{3Eab^3}{12L^3} \propto S \qquad (14.2)$$

TABLE 14.1 Scaling Factors of Some Physical Phenomena When a Device Size Is Scaled by a Factor *S*

Physical Property	Scaling Factor
Length	S
Surface	S^2
Volume	S^3
Mass	S^3
Weight	S^3
Elastic forces	S^2
Electrostatic energy	S^3
Electrostatic forces	S^2
Magnetic energy	S^5
Magnetic forces	S^5
Surface tension force	S
Viscous friction	S
Mechanical time constant	S
Natural frequency	S^{-1}
Thermal time constant	S
Electric resistance	S^{-1}

Note: The smaller the *S* exponent, the larger its relative increase with respect to the macro world.

A similar analysis [3] can be generalized to any physical phenomena, and we can obtain the scaling factors given in Table 14.1. There, a large exponent in *S* represents a high decrease of that effect. For this reason, we can deduce that magnetic forces drastically decrease or MEMS structures resonate at much higher frequencies that their counterparts in the macro world.

By comparing these laws in pairs more relevant results are produced. Table 14.2 presents the scaling effect on some ratios. A large *S* exponent means that forces in the numerator are relatively larger with respect to the ones in the denominator. In the above example, if we consider a cantilever beam of 1 m length that is scaled down to 1 mm, its reactive force decreases by a factor of 1,000,000. But what is relevant is the comparison to its weight, which decreases 1,000,000,000. Consequently, the stiffness decreases 1,000 times than its weight when it is scaled down and therefore they have a more significant impact in the mechanical analysis. For this reason in the micro world, we will design our mechanical devices as if there is no gravity.

We can infer other important conclusions from these ratios. The use of very expensive and unusual materials for surface deposition is possible because the mass we need is quite smaller compared to their corresponding macroworld quantities. Conventional

TABLE 14.2 Comparison of Physical Phenomena When a Device Size Is Scaled by a Factor S

Physical Property	Scaling Factor	Comment
Elastic forces/gravitational forces	S^{-1}	Weight is negligible.
Electrostatic forces/magnetic forces	S^{-3}	Depending on the actuators' geometries; but electrostatic forces are generally larger than magnetic ones.
Electrostatic forces/elastic forces	1	Depending on the actuators' geometries and actuation principles.
Inertial forces/viscous friction	S^2	Reynolds number.
Surface tension force/gravity	S^{-2}	Weber number.

Note: The larger the S exponent, the larger the force ratio is with respect to the macro world.

electromagnetic actuators like induction motors quite often used in everyday live are inefficient, and we usually apply electrostatic actuators in the microscale. Furthermore, as electrostatic and elastic forces decrease in the same proportion, they are the most appropriate pair for building mechanical actuators.

But not all ratios are beneficial. For example, viscous friction is a very strong inconvenient when designing movable parts, and mechanical structures can collapse due to surface tension forces when drying after a wet process. All these remarks, and even more, shall be considered when designing MEMS devices.

14.2.3 Modeling

Today, electronic engineers simulate complex VLSI circuits thanks to the availability of many commercial computer-assisted design tools. However, MEMS are much more complex systems for several reasons. First, not only electronic processes are involved, but also mechanical, electrical, or fluidic domains interact in a coupled manner. Second, the reduction in size is very critical, and in many cases the hypothesis of continuous media is no longer valid because we are dealing with such thin layers of material that it is necessary to model them at the molecular level. For these reasons, the analytical description of the behavior of MEMS processes and devices is only possible in specific cases, and the use of simulation tools becomes a real need. But nowadays, MEMS engineers lack of a generic platform to make full-system simulations.

Modeling of system dynamics is normally done by Newtonian formulation, but Hamilton's and Lagrange's equations are two alternatives for the development of the equation of motion. Resonance mode of vibrating devices is quite convenient in mechanical design.

Fluidic modeling is normally performed using Navier–Stokes formulation. But the scaling process originates a dilution of fluids, especially in gases, and a molecular model for processes like mass or thermal transport is necessary.

Another problem is that nonlinear behavior commonly appears when working at small sizes. In these cases, a good alternative for the modeling of these phenomena is the use of neural networks.

In any case, the previous descriptions lead to very complex and coupled set of linear and nonlinear equations. Thus, the use of a numerical software package like MATLAB® or Simulink® is essential for their resolution.

Using analogies with electrical circuits [4] is another way to circumvent the modeling problem. In this way, we can apply the understanding of electrical circuits that electrical engineers have developed for years, and the use of conventional well-known simulators like SPICE becomes possible.

But new problems arise at the microscale level and should be taken into account. For instance, that is the case of the Paschen's effect; this law predicts an increase of the breakdown voltage of gases due to the decrease of the number of molecules when reducing the gap between plates. This fact allows for the use of higher polarization voltages in electrostatic actuators than expected.

14.3 MEMS Materials

Most of the interesting properties of MEMS devices are due to the fact that they can be manufactured using similar, if not the same, facilities than for integrated circuits. The same occurs for the materials, as silicon is the most important material for MEMS because it takes advantage of all the development done for microelectronics. However, in MEMS industry silicon is the first choice not only for its good electrical properties but also thanks to its mechanical properties. Other semiconductor materials, polymers, and ceramics are also being used in new devices. In MEMS, we find not only substrate and thin or thick layer materials that have a function on the device, but also sacrificial materials that are used as intermediate layers for defining the layout of the devices but are not part of the final product. In most cases, the properties of these sacrificial materials have to be rather different than the ones for bulk materials, and not only silicon-based semiconductors and dielectrics are used for this sacrificial purpose. Finally, while the complexity of MEMS devices have evolved to 3-D structures and to smart systems, new materials have appeared and have taken importance as a complement to silicon. In the next paragraphs, the most interesting semiconductor materials will be described along with others like glass, polymers, ceramics, and metals, which allow for low cost, easy manufacturing, and high quality packaging. Figure 14.2 graphically summarizes the most important materials for MEMS and their applications.

When selecting a material, it must be taken into account that properties like uniformity, electrical and mechanical quality, thermal and optical properties, etc., depend not only on the material itself but also on the process for growth or deposition and on if it is in bulk or in film format. In addition, parameters like inherent stress, bucking, roughness, etc., may vary between thick and thin films of the same material. In conclusion, material selection for MEMS highly depends on the requirements of the application.

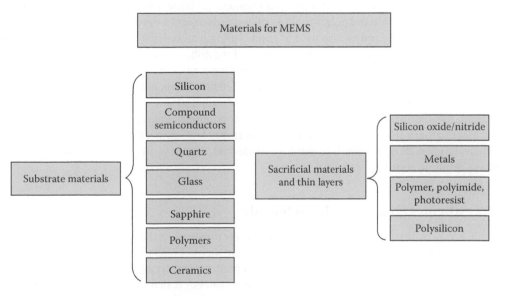

FIGURE 14.2 Materials used in MEMS.

14.3.1 Semiconductor Substrates: Silicon and Other Compound Materials

Silicon is the most important bulk material and the elemental semiconductor for MEMS fabrication because it combines multiple convenient properties: electrical and optical quality; good mechanical properties based on its robustness and elasticity and high stiffness; and easy growth of a silicon oxide layer as a good dielectric material. Because of these characteristics, many MEMS structures and devices can be integrated on silicon alone or combined with other materials (glass, metals, polymers, and so on). Silicon can be chemically and physically etched, and as etch rates depend on the crystal orientation of silicon planes, it is easy to fabricate membranes, beams, and other complex 3-D structures that are the basics of many sensors and actuators.

Thanks to the huge use in microelectronics, silicon is an abundant and inexpensive material and can be obtained in the format of ultrapure crystalline double-side polished wafers for MEMS processing. Both N-type (electron-rich with intentionally added concentration of group V impurities) and P-type (hole-rich with intentionally added concentration of group III impurities) substrates are used depending on the application.

In addition, polysilicon (polycrystalline thin layer of silicon) is also interesting for microelectronics and MEMS and it can be deposited by low-pressure chemical vapor deposition (LPCVD) with very good control. Amorphous silicon can be also deposited as thin or thick film of usually below $5\,\mu m$. Interesting moveable microstructures of polysilicon and amorphous silicon can be fabricated with specific techniques commonly called surface micromachining. Compared to bulk silicon, deposited layers show inherent stress, but this stress can be controlled with the deposition parameters and also with subsequent annealing processes. Porous silicon is another interesting material for some niche applications. Silicon can be also porosified with chemical treatments by HF, and used as sacrificial layer, as a thermal insulator, as mechanical or chemical filter, and for optical devices.

Silicon on insulator (SOI) substrates are also popular for MEMS, despite their higher cost. They are silicon substrates with a buried layer of oxide, which can be used for micromachining silicon as the oxide is a natural etch stop for silicon wet etching. In this way, crystalline bulk silicon of the top layer can be used for creating moveable structures with low stress and uniform and controlled thickness. The main drawback of SOI is the high cost of the wafers, compared to pure silicon.

Silicon is also a good material for intermediate temperature operation, while for high-temperature MEMS, silicon carbide and diamond are more appropriate substrates. In addition, both materials offer advantages of higher hardness and stiffness and resistance to harsh environments, compared to pure silicon. Layers of crystalline and polycrystalline silicon carbide and amorphous synthetic diamond can be deposited by CVD and can be also used as good coating materials for MEMS packaging. The main drawbacks of these materials are the very high cost and the difficulty of growing substrates of a reasonable size that makes it more convenient to work with deposited layers on a low-cost silicon substrate. Another drawback of these materials is the difficulty of etching them, as they are highly inert.

Finally, it has to be mentioned that for specific MEMS applications, such as for optical devices and lasers, high speed or wide band gap devices, the most appropriate materials are some compound semiconductors of groups III–V (GaAs, InP, GaN,...). Controlled layer deposition combined with micromachining techniques allow also for the creation of RF devices, antennas, and other interesting devices for wireless communications.

14.3.2 Silicon Oxide and Silicon Nitride

It is fully accepted that another crucial factor of success of silicon is the simplicity of forming a high quality and stable silicon oxide dielectric from a simple thermal oxidation process. Silicon oxide is a good electrical isolator in microelectronics and also plays an important role in MEMS, as sacrificial layers and also inter-metal films and planarization layers can be created with such a simple material. Its main drawback is that it is difficult to control the inherent stress of the materials when it is used in thin film format. For such a reason, silicon nitride is commonly preferred for membrane and beam fabrication. Nitride can be deposited by CVD and is interesting also as material with good thermal isolation and for being used as etch stop for some silicon chemical reagents.

14.3.3 Insulating Substrates: Quartz, Glass, and Sapphire

Quartz wafers are mostly used as piezoelectric material for some acoustic and RF applications (i.e., SAW, BAW filters and sensors, and so on). Quartz can be found as pure crystalline and amorphous fused quartz wafers depending on the application. For some optical devices that need transparent access, quartz is also a good option.

Glass is a cheaper option that can be used for the same applications, but the potential problem of the impurities of this material has to be taken into account. Glass is also a good option for soft lithography processes, and depending on the material (e.g., Pyrex 7740) it can be strongly attached to silicon wafers via anodic bonding. This makes glass very useful for some microfluidics devices and 3-D integration and as for sensor packaging (e.g., pressure sensors and accelerometers).

Sapphire is a third option if MEMS devices require insulating and/or transparent substrates. Compared to glass and quartz, sapphire is stronger and wear resistant but much more expensive and difficult to etch. Sapphire is also compatible with CMOS manufacturing.

14.3.4 Metals

Compared to microelectronics, a much higher variety of metals is used in MEMS. It is mainly due to the fact that in this case, metals are not only necessary for interconnections but they can also have other important functionalities for the final MEMS device. Thus, the selection of the metal to be used highly depends on the application. Pure aluminum or alloys (Al-Cu, Al-Cu-Si) are the most traditional for low temperature operation devices if only connectivity between devices is required. Aluminum is also a good candidate for optical mirrors in the visible range. However, as its melting point is around 450°C, it cannot be used for some specific devices and processing techniques and when the ambient is corrosive. Thus, for chemical and biochemical applications (i.e., microelectrodes and other chemical sensors), gold, platinum, or iridium are preferred. Platinum and palladium are also used for electrochemical devices because of their high chemical stability. Alloys and pure silver are used for stable counter electrodes. Chromium and titanium are suitable as very thin layers for promoting the adhesion of other metals over dielectric layers and silicon substrates. Finally, other more complex alloys (permalloy, NiCr, ITO, NiCr, and so on) are being integrated on MEMS for some specific magnetic, thermal, or optical devices.

The abovementioned metals can be deposited by sputtering or evaporation techniques. Stresses and other mechanical properties can be tuned with the deposition parameters, which are not only important for thin layer deposition but also for the case of using metals as sacrificial layers.

14.3.5 Ceramics

Ceramics are another family of interesting materials for MEMS. Nowadays different functional ceramics, such as dielectric, piezoelectric, pyroelectric, ferroelectric, and conducting materials, can be very useful when combined with silicon for incorporating some functionalities that could not be reached by silicon alone. Many sensing devices can benefit from such properties.

Ceramics are of lower cost when compared to other already mentioned materials and are good for harsh environment applications as they usually are chemically inert, with good temperature stability and with different electric and magnetic properties. SiCN and PZT are the most common materials used nowadays for MEMS. Ceramics can be made both in bulk format and also as deposited layers.

14.3.6 Polymers

Apart from the already mentioned inorganic materials, another set of organic materials are becoming more and more important in MEMS fabrication. Polymers try to circumvent some of the main drawbacks of the silicon-based structures, which are related to the difficulty of implementing high aspect ratio devices that are of main importance for most mechanical sensors and actuators. Polymers are more elastic, less stiff, and need lower actuating forces to create the same displacements.

On the other hand, polymers of low cost are very appropriate for fluidic applications and single-use medical devices. The good absorption properties of polymers make them interesting for humidity and vapor sensing, too. However, maximum operating temperature (normally below 200°C but up to 400°C for specific resists) of the organic materials is a limitation for some devices.

Polymers can be used in foil format or as thick layers usually deposited by spinning. 3-D structures can be also patterned by hot embossing and molding techniques. Many polymeric materials and composites are being investigated for MEMS applications, and current research on material intends to improve characteristics by changing their chemical composition, looking for the appropriate characteristics on chemical resistance, thermal characteristics, and optical properties from transparent to opaque materials. On the other hand, most polymers can be easily metalized, and thus it is feasible to integrate electrodes on the same structure.

Most well-known materials for MEMS are polyimides, photoresists, and silicones, being SU-8, PMMA, BCB, PDMS, Parylene, and liquid crystal polymers some of the most popular. Thicknesses can range from 1 μm for standard spin-coated photoresists up to 300 μm for thick SU-8 structures. Rigid structures can be obtained after subsequent hardening of the resists. Resistance to harsh environments depends also on the chemical composition of the polymer. Thus, the selection is again based on the application that can range from optical MEMS, BioMEMS, and microfluidics, to packaging and integration of complex sensor and actuator structures. In most cases, the low cost required for the devices of these applications make polymers to be the best solution for high volumes.

14.4 Deposition

As already stated, MEMS fabrication processes mostly rely on already developed processes for microelectronics. The development of planar technologies for integrated circuit fabrication was the breakthrough that allowed the fabrication of huge amounts of integrated circuits at very low unitary cost. Planar technologies consist of a set of processes that allow the fabrication of thousands of chips at the same time on the same substrate following a set of deposition and geometrical definition and etching steps of different materials. In this section, the most interesting materials and deposition techniques for MEMS devices are shortly summarized.

14.4.1 Material Deposition Techniques for MEMS

Deposition techniques are additive processes of materials on substrates for the fabrication of functional electronic devices. For the specific MEMS fabrication, the most interesting materials are silicon as a basic substrate material and polysilicon and other semiconductors, metals, dielectrics, and polymers. These materials can be an active part of MEMS, but can be also isolators and sacrificial layers. Depending on the material and on its use, its addition to the substrate can be made by following physical and chemical techniques that are described below.

14.4.2 Classification of Deposition Techniques

In Figure 14.3, the main techniques are presented and classified according to its nature and characteristics. For their selection for MEMS fabrication, it is important to know that not all of them allow for both front and back side process; electrical and mechanical properties, conformality, and residual stresses

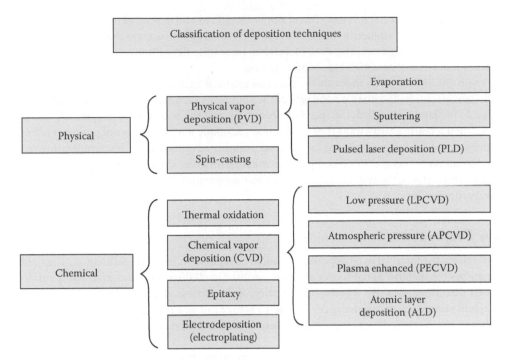

FIGURE 14.3   Classification of deposition techniques.

of the layers deposited also vary among the different techniques. Thicknesses, deposition rates, and cost are also radically different. Thus, it is important to define the specific characteristics of the MEMS device to be fabricated prior to the selection of the deposition technique to be used [5].

14.4.3 Physical Deposition Techniques

The main physical deposition techniques are vapor- and plasma-deposition and spin casting. In these cases, the material to be deposited is transported by physical methods from a source of material to the surface of a substrate where it is deposited without the existence of a chemical reaction on that surface. The main differences among the techniques are the types of materials that can be deposited, the thickness and growth rates that can be achieved, the temperature of operation, and the electrical and mechanical quality of the material.

14.4.4 Physical Vapor Deposition

With these techniques the material to be deposited is removed from a solid source, is transported to the substrate and is finally deposited on it. The main differences among the different physical techniques refer to the process of removing the material from the solid source, the directionality of the deposition and its cleanliness and potential contamination of the substrate with by-products. Compared to most chemical techniques, physical vapor deposition (PVD) is a low-temperature process that is very suitable for back-processes (steps done after metals and other low-melting temperature materials are added to the substrate). However, electrical quality and stresses may be worse because of the low temperature processing. The two main PVD processes for MEMS are sputtering and evaporation, which are described below.

14.4.5 Sputtering

Sputtering is probably the most used technique of deposition of most of the materials used in integrated circuit fabrication and MEMS. With this technique, the desired atoms are physically removed from a target by energized ions Ar^+ (plasma). Then, atoms are ejected and deposited onto the wafer. Energetic ions can be obtained by magnetron or plasma (PECVD) sources in processes that can be also low pressure (LPCVD) but not high vacuum ($P > 0.01$ Torr) for increasing deposition rates. Thicknesses obtained range from hundreds of nanometers to some microns.

The main MEMS materials that can be sputtered are most metals, some dielectrics, and also some piezoelectric ceramics. The main advantages of the sputtering technique are the uniformity and good adhesion of the layers deposited, the speed of deposition, and the wide choice of materials that can be sputtered, being really a well-established production technique with a wide set of parameters to select, depending on the desired properties of the material. Another advantage is the low temperature required for the deposition process (<150°C) compared to other techniques. On the other hand, the main drawbacks are complexity of the system, its cost, and also the potential damage of the substrates because of the bombardment damage.

14.4.6 Evaporation

In evaporation, the material to be deposited is heated in high vacuum until sublimation. Thus, the substrate receives the material as hot source atoms are emitted in all directions and stick in any surface that is on their directional flux. For a better optimization of the process, different substrates are placed in a planetarium around the material source. The energy source can be a resistance (thermal evaporation) or from that of an e-beam (e-beam evaporation). Despite the additional complexity of the e-beam, in both cases the system is simpler than the sputtering.

On the other hand, the main limitation is on the materials that can be evaporated, as only metals can be heated until sublimation with thermal evaporation, although with e-beams some dielectrics can be evaporated. E-beams also allow higher evaporation rates and better material adhesion. Other drawbacks of evaporation compared to sputtering are the poor conformity of the layer if there is no substrate rotation during the evaporation, and the possible substrate contamination from the filament used to heat the material during evaporation. Thicknesses obtained range from hundreds of nanometers to some microns with deposition rates in the range of 5–100 nm/min, which are slightly higher than for sputtering.

14.4.7 Pulsed Laser Deposition

Pulsed laser deposition (PLD) is a much newer technique of deposition compared to the other two techniques already presented. In this technique, a thin film can be deposited on a hot substrate thanks to a high-power pulsed laser beam that is focused on the target under ultrahigh vacuum or a specific gas (i.e., oxygen). The target material is then vaporized in the form of a plume and deposited on the surrounding vacuum. The technique is well used for metals, oxides, and also other complex materials especially for very thin films, as the good control of the laser properties allows to obtain smoother and thinner films compared to sputtering.

14.4.8 Spin Casting

Spin casting or spin coating technique is used for forming thin or thick films of materials that are dissolved in a volatile liquid solvent that are spin coated onto a substrate. Then, the material is cured around 100°C. The main materials that can be spinned are organic photoresists and polyimides in the typical range of 1–20 μm (and up to 200 μm for epoxy like SU-8 resists). Such resists can be used as intermediate materials during processing but also as active materials for MEMS. In addition, spin casting is used for depositing thick (up to 100 μm) inorganic spin-on glass (SoG) layers for uniformly coating surfaces, for depositing thin (up to 0.5 μm) interlayer dielectrics between metal levels for some specific application.

Films are of low quality and amorphous, and the thickness can be controlled with the density of the solution and with the spinner parameters (i.e., speed, acceleration, and time). Thicknesses can reach the order of 100 μm, depending on the material.

14.4.9 Chemical Deposition and Growth Techniques

The addition of materials on a substrate can also be done by chemical techniques, which became very popular at the beginning of the 1960s, when planar techniques were developed for microelectronics fabrication. Compared to physical deposition, now in chemical techniques the material to be deposited is not transported as it is finally deposited, but a chemical reaction occurs on the surface of the substrate and the material is generated. As shown in Figure 14.3, the most important techniques are oxidation, low-pressure and plasma-enhanced CVD, and epitaxy [6]. Fundamentals of these techniques are described in the next paragraphs.

14.4.10 Thermal Oxidation

The main chemical process is the one from which a silicon oxide (SiO_2) is obtained from the high-temperature oxidation of a silicon wafer. Thus, in fact it is not a deposition process but a growth process, as part of the SiO_2 material is the own silicon of the substrate. The importance of this process relies on the fact that the SiO_2 shows very interesting properties, as it is the best natural amorphous insulator of high electrical quality. For that reason, SiO_2 is commonly used as insulating layer between metal lines and within electronic devices. Silicon oxide is also used as a surface passivation layer or as a blocking layer to mask impurity atoms. Oxidation can be masked locally by an oxidation barrier, such as silicon nitride. In MEMS, oxides are also good sacrificial layers for the definition of complex structures. Silicon oxidation followed by etching can be used for sharpening of silicon MEMS features.

Thermal oxidation is based on the transport and reaction of oxidation species on a silicon surface. Once oxidation starts, another process is necessary, which is the diffusion of oxygen through the existing oxide. Growth rate is determined by this diffusion of oxygen. Depending on if the reacting oxygen is pure gas or comes from water vapor, we can differentiate between dry oxide and wet oxide.

$$\text{Dry oxide } Si(s) + O_2(g) \rightarrow SiO_2(s) \quad \text{(dry)}$$

$$\text{Wet oxide } Si(s) + 2H_2O(g) \rightarrow SiO_2(s) + 2H_2(g) \quad \text{(wet)}$$

Oxide properties may be varied by controlling oxygen or water vapor flow and temperature during reaction. The pros and cons of both techniques are as follows: while dry oxide is denser, with higher dielectric strength, the oxidation rate is slow and is mainly dedicated to thin oxides (like for the gate oxide of a MOS transistor). On the contrary, wet oxides show faster growth rates as H_2 speeds the diffusion of oxygen, but the dielectric is of lower quality, being appropriate for field and sacrificial oxides. In both cases, cleanliness is essential as oxide charge is very detrimental for the electrical stability of MOS active devices.

14.4.11 Chemical Vapor Deposition

In CVD, species in vapor phase react at a hot surface to deposit solid films. Thus, unlike oxidation, it does not consume substrate material. CVD is a common MEMS tool for creating thick silicon oxide and silicon nitride films on the wafer surface. In practice, films are typically submicron thick but can be used as conformal coatings. Inherent stress of the deposited material can be reduced by optimizing process parameters or by post annealing.

Many different CVD techniques have been developed over the years, the most important being the low pressure (LPCVD), atmospheric pressure (APCVD), plasma-enhanced (PECVD), metal organic CVD (MOCVD), and more recently atomic layer deposition (ALD) for extremely thin coatings. In all cases the film composition, uniformity electrical and mechanical characteristics, and deposition rates can be varied by controlling the variables of the gas composition, pressure, and temperature of the substrate. CVD films can be polycrystalline or amorphous thin films. In microelectronics and MEMS, principal uses of CVD films are for passivation, insulation, but also for structural and sacrificial layers.

Compared to thermal oxidation, CVD is a faster process that allows not only obtaining layers of silicon oxide but also of other interesting materials such as Si_3N_4, polysilicon, some metals, phosphosilicate glass (PSG), etc., on substrates that do not need to be of silicon, as there is no material consumption. However, as the main drawback, there is the lower electrical and mechanical quality of the layer inherent to the low temperature processing.

14.4.12 Epitaxy

Epitaxy is a chemical deposition technique that allows the deposition of crystalline layers over a crystalline substrate, as deposited atoms move to lattice sites. Thicknesses of epitaxial layers typically range from 1 to 20 μm. For the material to be crystalline it is necessary that the process is at slow (0.2–4 μm/min) and at high temperature (>800°C) in order to have enough time and energy to move atoms to a lattice site.

When the material grown is the same as the substrate (i.e., Si on Si), the process is usually called homoepitaxy, while when the material grown is different (i.e., AlGaAs on GaAs) it is called heteroepitaxy. The most well-known technique for epitaxial growth is the molecular beam epitaxy (MBE).

14.4.13 Electroplating

Electroplating is a well-known technique for depositing thick metals (i.e., Cu, Au, Cr, Ni, permalloy, copper interconnect) that is also used in MEMS fabrication (e.g., magnetic devices and others), when sputtering or evaporation do not reach the desired thickness or step coverage. As an advantage compared to other techniques, electroplating allows the deposition of metal on both vertical and horizontal surfaces at the same time and with the same deposition rate.

The electroplating method is based on the passing of current through an aqueous solution in which the metal to be deposited is dissolved. Positive metal ions from an anode travel to cathode on the substrate and deposit there. The anode is made of the metal to be deposited, while the cathode is the conductive seed of the substrate. If electroplating is not desired on the whole surface of the substrate, the seed or adhesion layer can be previously masked, and if high aspect ration features are to be developed, then it is necessary to create a pattern on thick resist that can be used as a sort of mold for the platted metal.

14.5 Etching

Etching is the selective removal of material from the substrate. Selection is usually made by protecting some areas with a hard mask that has been previously patterned using a photolithographic technique. Etching is wet when this removal is performed in the presence of a liquid element, and dry if otherwise.

14.5.1 Wet Etching

Depending on the shape produced by the etching, it is classified as isotropic or anisotropic. In isotropic etching, all spatial directions are etched at the same speed, while in anisotropic etching one or more directions are etched much more slowly than the rest [7].

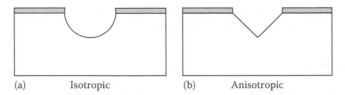

FIGURE 14.4 Etching: (a) isotropic (b) anisotropic.

Anisotropy can be caused by the etched material having an anisotropic internal structure, like in crystalline solids (crystallographic etching), or by a preference in the movement of the reactant ions (this is usually the case of dry etching, see below).

Figure 14.4 shows the difference between isotropic and anisotropic etching, in the same material and using the same mask.

As in an isotropic etching the etching speed is the same in all directions; a phenomenon known as underetching is common in these processes. A part of the material that is supposed to be protected by the mask is also etched away. After enough time, the etching profile will resemble a quarter-cylinder along the mask edge. In isotropic etching, it is not possible to create deep holes, only holes whose depth is equal to one half their width are possible. Usually, this is not a problem for surface micromachined devices (see below), where the thickness of the material to be etched is orders of magnitude smaller than the width of the patterns.

Anisotropy can be measured using a parameter known as anisotropy degree, η_0. It is defined as function of the lateral etching rate v_l and the perpendicular etching rate v_n and it is calculated as follows:

$$\eta_0 = 1 - \frac{v_l}{v_n}$$

When $v_l > v_n$, η_0 is assumed to be 1. Using η_0 the underetching can be calculated as

$$l_u = (1 - \eta_0)d$$

where d is material thickness or the depth to be etched.

In wet etching, atoms or molecules from the solid substrate are transferred to the liquid phase [8]. This way, it is possible to achieve a high precision in the etched depth, but at the same time the process is considerably slow.

A typical distinction in etching is that of surface and bulk micromachining [9,10]. In surface micromachining the material to be etched has a thickness much smaller than the other dimensions. This is the case, for example, of thin films that must be etched to reveal the material beneath, or when a thin layer is removed to leave a structure hanging (see Figure 14.4). In bulk micromachining real 3-D structures are formed, and holes whose depth is comparable to their width are possible. Figure 14.5 shows a bulk micromachining performed anisotropically on a substrate.

Etch rates for a given etchant on a specific material are widely known for most useful combinations [11], so the required time to achieve a desired depth can be calculated. Nevertheless, it is sometimes

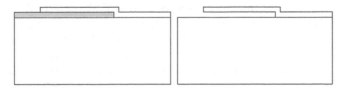

FIGURE 14.5 Surface micromachining.

desirable to have a method for stopping the etch at a predefined depth. Most commonly used etching-stop techniques are electrochemical stopping, p+ doping, and material change detection [12].

14.5.2 Dry Etching

In dry etching the etchant is in vapor phase or, more commonly, in plasma phase. Etching is caused by a chemical reaction on the surface, by ion bombing over the substrate, or by a combination of both [13]. Sometimes plasma resides in the same chamber as the substrate (glow discharge), while at other times, it is generated in a separate chamber, and ions are extracted to make them collide against the substrate (ion beam).

Techniques of dry etching can be classified as chemical or physical, according to the effect that is used to remove the material [14].

In physical dry etching, inert ions are directed toward a surface. Ions do not react with the substrate, but remove material from it by colliding with a high kinetic energy. If ion energy is too low, they will bounce against the surface, and if it is too high, ions will penetrate inside the material, generating ion implantation. Optimal energy for etching lies between 10 and 10,000 eV. The main inconvenient of physical dry etching is its low selectivity between materials, as all of them are etched roughly at the same rate. In addition, it is considerably slower than chemical dry etching.

In chemical dry etching, a chemical reaction is produced between the impacting ions and the surface, so the etch rate is higher. The right combination between physical and chemical etching usually gives the best results in terms of selectivity and etch rate.

Of particular interest is the technique known as deep reactive ion etching or DRIE, which uses a high-density plasma in a low-pressure chamber, and then can be employed to achieve near-vertical walls in the etching [15]. This can be achieved at low temperatures (cryogenic process), or at ambient temperature (Bosch process).

14.6 Molding

By using molding and casting, large quantities of devices can be fabricated at a low cost. Usually, molds are made of a hard material, like silicon or a thick photoresist, while the final devices are made out of a deformable material that can be cured or hardened in some way. Different types of polymers are commonly used for this purpose.

Also, fabrication of microdevices in polymeric materials is accomplished in a shorter time than, e.g., silicon, so they are often used for prototyping or small batches.

Among the polymers commonly used in MEMS, PMMA (poly-methyl-methacrylate) and PDMS (poly-dimethyl-siloxane) are the best known. PMMA is usually available in solid form, and to pattern it, thermal casting or molding is used. After being molded, it exhibits very good structural properties [16].

PDMS is commercially sold as two liquid products: a prepolymer and a curing agent. When mixed in the right proportion, and left to cure, they will form a transparent, elastomeric solid. This process is particularly suitable for microfluidic prototyping, since channels and chambers are easily fabricated, and PDMS-to-PDMS bonding can be made irreversibly and with good hermetic sealing [17].

Typical PDMS processing begins with the fabrication of a mold, which will have the negative shape of the final device to be fabricated. This mold is usually made of SU-8 photoresist on silicon. The mix of PDMS prepolymer and curing agent is poured over the mold and left to cure for 1 h at 80°C. After this time, it solidifies and it can be easily peeled off the wafer. One added advantage of the process is that several layers of PDMS can be peeled before they are completely cured, and then put together to finish the curing. They will be cured together and will form an irreversible bond. This way, hermetically closed cavities can be constructed.

FIGURE 14.6 Bulk micromachining.

The whole process can be finished in less than 4 h, making it extremely convenient for prototyping.

Both PMMA and PDMS are biocompatible and can be used in implants in the human body. Bonding of PMMA and PDMS to other materials is possible, but difficult to achieve after curing. Commonly, the surfaces have to be activated by oxygen plasma. However, the obtained bonds are not permanent, although they are able to withstand moderate pressures.

A different process that makes use of molding is LIGA, an abbreviation of *LIthographie, Galvanoformung und Abformung* (lithography, electroforming and molding) [18]. Using LIGA, high-aspect ratio microstructures can be fabricated. The process starts by coating a substrate with a thick photoresist, which is subsequently exposed to radiation. Depending on the thickness of the photoresist layer, exposing will use traditional UV light, electron beams, or x-ray radiation. After photoresist has been patterned and developed, a thick metal layer is electrodeposited on top of it, filling all the gaps. The remaining of photoresist is then stripped off, leaving just the metal, which will be used as a mold to fabricate the final device (Figure 14.6).

The LIGA process has traditionally been the only possible way to achieve high-aspect ratio microstructures with submicrometer feature size, almost vertical sidewalls, and ample variety of materials [19]. The high cost that was associated with synchrotron light made the process too costly for prototyping or small-volume production, but recently commercial scanners have been made available at lower cost. Nowadays, the major problem related to LIGA is uniformity in the thickness of the electroplated metal layer (usually nickel, gold, or NiFe). Electroplating equipment from DVD industry has been adapted to the LIGA process, providing better uniformity at a lower cost.

LIGA is often employed where high precision is required for the fabrication of mechanical, optical, or fluidic devices, usually to be integrated in more complex systems.

References

1. R. P. Feynman, There's plenty of room at the bottom, *J. Microelectromech. Syst.*, 1 (1992), 60–66.
2. *Status of the MEMS Industry*, Yole Développement Lyon, France, 2003.
3. J. J. Allen, *Micro Electro Mechanical Systems Design*, CRC Press, Berlin, Germany, 2005.
4. S. D. Senturia, *Microsystem Design*, Springer Science, New York, 2001.
5. R. C. Jaeger, Introduction to microelectronic fabrication, in *Modular Series on Solid State Devices*, vol. 5, G. W. Neudeck and R. F. Pierret (Eds.), Addison-Wesley, Reading, MA, 1993.
6. K. F. Jensen, Chemical vapor deposition, in *Microelectronics Processing: Chemical Engineering Aspects*, D. W. Hess and K. F. Jensen (Eds.), American Chemical Society, Washington, DC, 1989, pp. 199–263.
7. K. E. Bean, Anisotropic etching of silicon, *IEEE Trans. Electron Dev.*, 25 (1978), 185–1193.
8. M. Köhler, *Etching in Microsystem Technology*, Wiley-VCH, Weinheim, Germany, 1999.
9. J. M. Bustillo, R. T. Howe, and R. S. Muller, Surface micromachining for microelectromechanical systems, *Proc. IEEE*, 86 (1998), 1552–1574.
10. G. T. A. Kovacs, N. I. Maluf, and K. E. Petersen, Bulk micromachining of silicon, *Proc. IEEE*, 86 (1998), 1536–1551.
11. K. R. Williams and R. S. Muller, Etch rates for micromachining processing, *J. Microelectromech. Syst.*, 5 (1996), 256–269.

12. C. M. A. Ashruf, F. J. French, H. M. Sarro, M. Nagao, and M. Esashi, Fabrication of micromechanical structures with a new electrodeless electrochemical etch stop, in *International Conference on Solid State Sensors and Actuators (Transducers '97)*, June 16–19, 1997, Chicago, IL, pp. 703–706.

13. D. M. Manos and D. L. Flamm, *Plasma Etching: An Introduction*, Academic Press, San Diego, CA, 1989.

14. H. Jansen, H. Gardeniers, M. de Boer, M. Elwenspoek, and J. Fluitman, A survey on the reactive ion etching of silicon on microtechnology, *J. Micromech. Microeng.*, 6 (1996), 14–28.

15. M. Esashi, M. Takanami, Y. Wakabayashi, and K. Minami, High-rate directional deep dry etching for bulk silicon micromachining, *J. Micromech. Microeng.*, 5 (1995), 5–10.

16. A. B. Frazier, Recent applications of polyimide to micromachining technology, *IEEE Trans. Ind. Electron.*, 42 (1995), 442–448.

17. J. R. Anderson, D. T. Chiu, R. J. J. O. Cherniavskaya, J. C. McDonald, H. Wu, S. H. Whitesides, and G. M. Whitesides, Fabrication of topologically complex three-dimensional microfluidic systems in PDMS by rapid prototyping, *Anal. Chem.*, 72 (2000), 3158–3164.

18. W. Menz, LIGA and related technologies for industrial application, *Sens. Actuators A (Phys.)*, 54 (1996), 785–789.

19. B. Loechel, J. Goettert, and Y. M. Desta, Direct LIGA service for prototyping: Status report, *Microsyst. Technol.*, 13 (2007), 327–334.

15

Applications of MEMS

Antonio Luque
University of Seville

José M. Quero
University of Seville

Robert Lempkowski
*Motorola Applied Research
and Technology Center*

Francisco Ibáñez
European Commission

15.1 Introduction

This chapter presents the main sectors where microelectromechanical systems (MEMS) devices are currently being applied. Those sectors include automotive, communications, aerospace, industrial, and energy, among others [1]. MEMS foundries target many of their products to one or more of these fields. It is also expected that the share of each of these sectors in the global MEMS market will change in the coming years, and an estimation on how this may happen is included at the end of chapter.

15.2 Industrial

The ability of MEMS to measure physical parameters makes them particularly attractive to implement monitoring and control functions. In manufacturing environments, MEMS integrated into feedback loop systems contribute to keep process variables within acceptable limits and play a critical role in industrial automation. The monitoring of temperature, pressure, position, vibration, humidity, gas content, presence/absence, angular rate, etc., opens many opportunities for the use of MEMS in industrial contexts. It is relatively frequent to observe the utilization of MEMS developed in a particular application field (communication, consumer, automotive) in industrial systems, once the specific requirements of the application (reliability, performance, cost) are met. This phenomenon is illustrated by MEMS inertial sensors, such as linear accelerometers and angular gyroscopes, initially developed for automotive that at a later stage find their way into industrial applications (motion control and awareness). Reliable, compact, and low-cost MEMS components are gradually replacing traditional bulky and costly sensing approaches for industry.

Inkjet printers are a very relevant case of MEMS "application transfer." Disposable cartridges for inkjet printers have been one of the early market successes of MEMS. The emerging field of printed electronics opens new high-value professional opportunities for the area. Major progress has been made in the use of MEMS for printing and the versatility, resolution, and throughput achieved are expanding the application range well beyond consumer printers.

MEMS accelerometers are increasingly used for vibration detection in industrial applications. They usually serve two types of situations: precision machinery requiring accurate positioning of their moving parts, and vibration analysis of manufacturing equipment to prevent and anticipate failures in the shop floor. Vibration MEMS sensors are a particular type of 3D accelerometers, which can operate as a stand-alone transducers or integrated with a data-capture system. The operating principle varies from piezoelectric, capacitive, resonant frequency to magnetic induction. Recent advances in MEMS technologies allow the measurement of vibrations with high precision over a broad range of frequencies. Of particular interest are MEMS seismic sensors, able to detect vibrations of less than a millionth of the earth's gravitational acceleration and used within seismic data–acquisition systems for oil and gas exploration.

MEMS technologies are entering into industrial application providing higher levels of performance than alternative sensing techniques. Examples that illustrate this trend are gas/air ultrasonic sensors replacing piezoelectric devices. Miniaturized ultrasonic sensors can emit high-frequency ultrasound and thus operate at frequencies ranging from 200 kHz to 5 MHz (most piezoelectric devices are limited to 200 kHz operation). Like other types of MEMS sensors (pressure, accelerometers), ultrasonic sensors make use of a suspended membrane on a silicon substrate. The cavity structure allows both to emit ultrasound waves (a voltage applied to the membrane makes it vibrate at the desired frequency) and to detect them (the received ultrasound affects the output electrode, which makes the membrane vibrate, changing the capacitive characteristics of the sensor). MEMS sensors are used for the nondestructive inspection of materials and structures with the added advantage that no contact with the inspected material is required, providing high versatility in a manufacturing line. Gas/air MEMS sensors are also applied for high accuracy gas flowmeters. Finally, for liquid immersion applications, MEMS sensors provide high-resolution industrial imaging (also relevant in the medical environment) and high-resolution liquid level sensing.

Finally, the combination of MEMS and wireless communications under the WSN (wireless sensor network) concept addresses industrial monitoring and control applications for real-time operation. Air and climate control make use of MEMS to measure the level of critical elements (carbon dioxide, ammonia) and parameters (temperature, light) and take corrective actions to maintain them under the required limits. The monitoring of buildings' and large constructions' "health" is implemented by measuring and analysis of vibrations at key points of the structure. In a longer term, intelligent building systems will be based on the deployment of a large amount of MEMS with the combined abilities to capture information, to communicate wirelessly and to operate from ambient energy.

Reliable MEMS packaging is critical for industrial applications (e.g., see Figure 15.1).

15.3 Automotive

The automotive industry was among the first to make extensive use of MEMS [2] and still nowadays airbag deployment systems are one of the largest applications for which MEMS accelerometers are commercialized. Many MEMS foundries sell capacitive accelerometers specifically designed to detect large decelerations, and their output is used to ignite the explosive device that makes the airbag inflate. Main

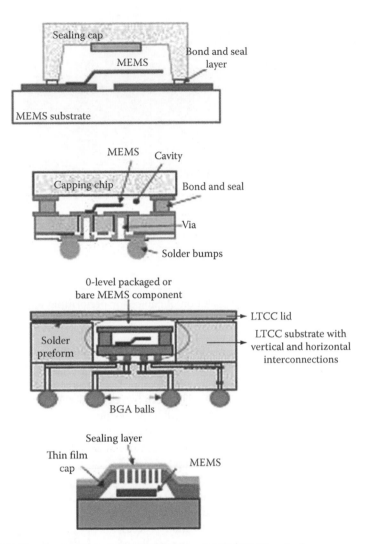

FIGURE 15.1   MEMS packaging for industrial applications (MEMSPACK project).

suppliers of accelerometers for airbag control include Robert Bosch GmbH, Analog Devices, Freescale Semiconductors, and NXP Semiconductors.

But MEMS are also used in occupant-detection systems that make airbags deploy or not depending on whether a person is sitting in the appropriate place. Different sensor configurations are used for this matter, one of the most advanced being an array of hundreds of pressure sensors in the bottom of the seat that can detect if a person is seated and what is their position (this is especially useful for the case of small children). Other systems include infrared detectors, or a single-pressure sensor that monitors the total weight.

Pressure, temperature, and mass-flow sensors are used in different parts of the engine to monitor parameters in the air or gas fed to the engine, or in the engine case itself. By making use of them, intelligent injection systems that adapt to the current conditions can be implemented.

Electronic stability control systems try to avoid the vehicle from overturning, and, if unavoidable, to trigger the deployment of appropriate airbags and/or pretense seat belts. This is accomplished by the use of different sensors, which include accelerometers, gyroscopes, and pressure and speed sensors for tires. Figure 15.2 shows a rollover sensor based on a gyroscope manufactured by Robert Bosch GmbH.

FIGURE 15.2   Rollover sensor by Robert Bosch GmbH.

15.4 Biomedical

In biomedical applications, MEMS are the solution of choice for performing local analysis at a small scale. Disposable analyzers for personal use are starting to be used, thanks to the low cost obtained by batch fabrication [3].

The measurement of glucose, lactose, and other substances in the blood torrent was one of the first applications of miniaturized devices in medicine. Personal glucometers are nowadays readily available. Most of the sensors in the market use an electrochemical method to measure the glucose content. An electrolyte where the glucose will be dissolved is located between two electrodes. When an electrical potential is applied between both electrodes, the current flow will be proportional to the concentration of glucose (or other species) in the electrolyte. In order for the electrical effect to happen, a catalytic enzyme must be present that oxidizes or reduces the analyte molecule. For glucose, the catalyst is usually glucose-oxidase.

Most glucose sensors work by extracting a sample from the human body, and then analyzing it using the amperometric method described above. The fluid sample can come from the blood torrent, or, more commonly, from the interstitial fluid, which is present between live cells. Microfabrication processes present the advantage of being capable of constructing needles of a size, which makes the extraction painless, as their size and length avoids touching the nerves.

More complex systems intended for diabetes patient monitoring also include an insulin pump, which delivers the needed insulin when the measured glucose levels exceed the safety threshold. The glucose monitoring is made continuously at short intervals. Commercial systems include OneTouch Ping from Animas Corporation (see Figure 15.3) and GlucoDay from Menarini, among others.

The final goal of building a lab-on-chip, or micro-total-analysis-system (μTAS), a chip where all the processes of sample extraction, pretreatment, biochemical analysis, and perhaps electronic processing of the results take place at a reasonable cost, is becoming closer, thanks to the low-cost microfabrication processes that allow the use of biocompatible materials to construct disposable components. Most of the research in this direction is based on the use of polydimethylsiloxane (PDMS) for the microfluidic part (see Figure 15.4).

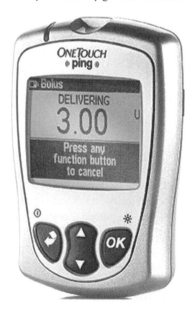

FIGURE 15.3   OneTouch Ping continuous glucose monitor and insulin dispenser. (Courtesy of Animas Corporation.)

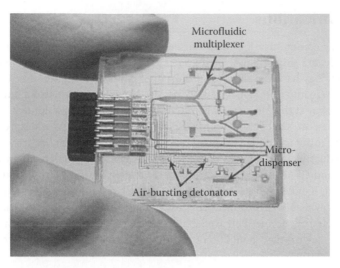

FIGURE 15.4 Microfluidic system built in PDMS with air-bursting detonators for fluidic movement. (Reproduced from Ahn, C.H. et al., *Proc. IEEE*, 92, 154, 2004. With permission.)

Microsystems for controlled drug delivery are also being employed nowadays. Apart from the miniaturized insulin pumps mentioned above, other systems exist. One of them is the smart pill [4], a capsule filled with drug whose surface is perforated with multiple holes. Each hole is sealed by a material that can be dissolved in a controlled way. Some of these smart pills include a wireless interface to trigger the release of drugs into the body.

In other cases, drugs are delivered through transdermal needles [5]. Methods of impulsion to the fluid that contains the drug can include pressurized chambers, microexplosive devices, peristaltic pumps, and thermally expandable materials. An image of microneedles for transdermal drug delivery can be seen in Figure 15.5.

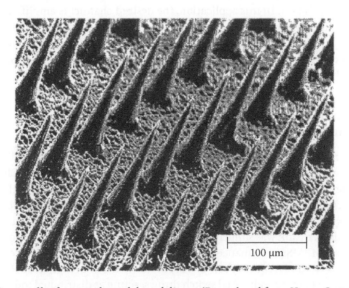

FIGURE 15.5 Microneedles for transdermal drug delivery. (Reproduced from Henry, S. et al., Micromachined needles for the transdermal delivery of drugs, in *Proceedings of the IEEE Eleventh Annual International Workshop on Micro Electro Mechanical Systems*, Heidelberg, Germany, 1998, IEEE, Piscataway, NJ, pp. 494–498. With permission.)

15.5 Communications

MEMS devices in the communications area started out with a promise of becoming resonators, and though it took awhile, are shipping now in volume as replacements for crystals. Television created by light projector technology has been shipping in consumer quantity for quite some time. MEMS speakers have found their way into cell phones as well. MEMS RF switches have had a number of companies developing products, but at this time are only found in military or test-equipment areas, and have not been able to reach the cost, size, or some other aspect that has limited its widespread application. RF MEMS switches have lower insertion loss, good return loss, and very high linearity (lack of harmonics due to non-semiconducting junctions) compared to competitors in complementary metal oxide semiconductor (CMOS) and GaAs. They have reasonable switching speed (microseconds) for many applications, and some can handle high RF power. On the other hand, most designs require higher supply voltages than what is available for handheld devices, and must use converters for their control, though some have been built to function well on available batteries. Here are examples of the communications application of MEMS devices that should become commonplace in the coming years.

15.5.1 MEMS Replacement of Existing System Components

With the cell phone being one of the highest-volume consumer products, MEMS have more than one location in the RF hardware where they can provide a suitable alternative. From the antenna, one SP2T switch is a possible alternative, where its low insertion loss and linearity make it attractive. Since one or more bands of service require frequent and high-speed switching, most MEMS switches today either cannot meet the switching speed, lifetime reliability of 10^{12} switching cycles, or RF power handling during switching. The other applications are very well suited for today's RF MEMS switch, a multi-throw switch in the receiver and transmit chain, and a two-throw switch monitoring transmit power. Other transceiver applications are very similar when it comes to band switching with low loss.

The second application is cable TV function, whereby multi-throw MEMS switches select a cascade of high-pass and low-pass filters, each with differing cutoff frequencies separated through multiple octaves of frequency operation. In this application, the desired channel is amplified through a broadband amplifier, with the high-pass filter selected and lower frequencies rejected. The second amplifier increases the signal and its harmonic, which is selected to reject the harmonic and pass the desired signal. This construction also can create a number of passbands of different widths, depending on the cutoff frequencies chosen.

15.5.2 Antenna, Filters, and Matching Network Step-Tuning Examples in PWB MEMS

Printed Wiring Board (PWB)-based MEMS RF switches were developed using modified processes and similar materials as used for high-density interconnect/embedded passives (HDI/EP) components. By constructing cantilever-beam electrostatic mechanisms, single-pole single throw through single-pole four-throw RF switches were developed with suitable performance in the microwave band (Table 15.1). Combining these switches with embedded lumped element filter components using the same substrate materials, a switched filter bank was simulated, fabricated, and measured results shown.

Typical PWB-based MEMS performance is shown in Table 15.1 for a single switch and multi-throw versions. Additional losses and frequency roll-offs occur due to the line length from the RF input transition (shown in Figure 15.6) to the individual switch throws, acting as distributed inductance and capacitance. Since the same materials and processes are as used for the MEMS device as in PWB manufacture, construction can also be made on the core, with the outer layer of resin acting as the cover for the switch (as depicted in Figure 15.7).

TABLE 15.1 PWB-Based MEMS Performance

	SPST	SP4T
Insertion loss	0.1 dB (2 GHz), 0.3 dB (10 GHz)	0.2 dB (1 GHz), 0.6 dB (2 GHz)
Isolation	>35 dB (2 GHz)	>35 dB (2 GHz)
RF Power handling	5 W (hot switching)	5 W (hot switching)
Speed	<50 μS ON and OFF	<50 μS ON and OFF
Actuation voltage	40–100 VDC	40–100 VDC
Life cycles	$>1 \times 10^7$	$>1 \times 10^7$
Operation frequency	<10 GHz	<6 GHz

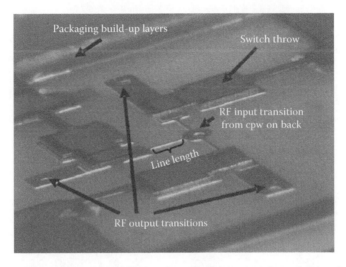

FIGURE 15.6 SP3T MEMS switch on printed wiring board with RF in/out through vias to bottom of package.

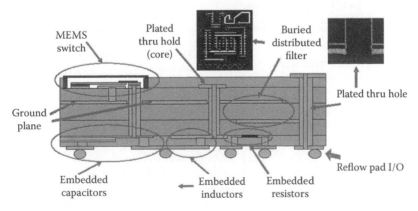

FIGURE 15.7 MEMS and embedded passives integration.

A switch/filter bank has been fabricated with individual packaged and tested switches on a substrate, which contained lumped-element LC filters, as shown in Figure 15.8. After the tests, the package covers were removed to show this early design version.

Similarly, a tunable antenna was demonstrated that used an SP3T switch in a cell phone form factor. A tuning element was added to a PIFA antenna, and the construction according to the scheme shown in Figure 15.9a. The prototype consisted of batteries, a DC-DC converter, control circuits, and the antenna

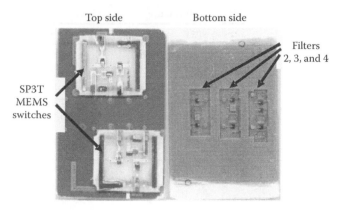

FIGURE 15.8   Module of two MEMS SP3T switches and embedded passives filters.

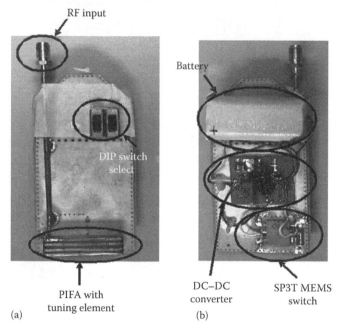

FIGURE 15.9   MEMS switching the same PIFA antenna structure with different shunt components. (a) Front of board showing PIFA and (b) back of board showing MEMS and other components.

and ground plane, as in Figure 15.9b. The dual in-line package (DIP) switch controlled the switches individually, allowing single or multiple shunt elements to be selected at any given time. Shunt capacitors were selected that provided a good match at three cell phone frequencies of interest. In addition, these capacitors were swapped out to determine the potential frequencies that this antenna could possibly be tuned to beyond the normal cell phone bands.

15.5.3 RF Switches

The PWB MEMS SPST (single-pole, single throw) RF switch uses an electrostatically controlled physical structure with a polymer isolation member separating an RF contact that connects the signal path. Multi-throw switches use the same common switch structure, but have RF transmission line structures dividing the input signal to SP2T, SP3T, and SP4T switches [6], typically microstrip lines and coplanar waveguide lines.

The polymer beam extending from the electrostatic actuator top is also stiffened with copper rigidizers, as part of the fabrication process [7].

15.5.4 Meso-MEMS Phase Shifters

Several types of discrete phase shifters are amenable to the topology used in fabricating MEMS structures in PWB technology. Some of the components have been already been developed with suitable performance through X-band, with the remaining portions requiring steps to smaller process rules for operation in the higher frequency bands.

The first type is a switched-line approach, relying on SPMT switch designs. The second type is a hybrid reflective phase shifter, which historically was a phase modulator with shunt p-i-n diodes to ground. That version has two forms, one that relies on SPMT switches, and the other that only needs SPST ones. Lastly, instead of stepped-switch phase shifters with large phase steps, a fine-grain phase shifter is described.

15.5.5 Switched Line Phase Shifters

The most straightforward phase shifter is one that uses back-to-back SPMT switches, with phase shift accomplished via switching in different line lengths corresponding to the desired phase step. This architecture is limited by high frequency performance of the switches. For a four-state switch, SP4T performance for designs on common substrates would have about 2 dB loss at 8 GHz. This would provide a phase shifter with about 5–7 dB loss from 10 to 60 GHz. Alternatively, by using four SP2T switches, each with about 1 dB loss, that construction could yield less losses over the same range since they provide better high-frequency performance due to less parasitic stubs for switches in the open state. It remains to be seen whether the design shrink that is recommended to obtain higher frequency performance will improve the SP4T switches more than using multiple lower throw versions. Both architectures are recommended to determine the trade-offs in loss/frequency performance (see Figure 15.10).

15.5.6 Reflective Hybrid Coupler with Shunt Switches

A common method to provide phase shifting is to have a 90° hybrid coupler with the shunt arms providing differing phase shifts (Figure 15.11). In semiconductor technology, the coupler is typically a Lange

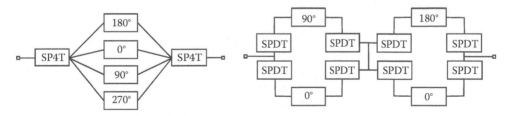

FIGURE 15.10 Switched-line phase shifters.

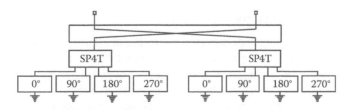

FIGURE 15.11 Branchline hybrid coupler shown with shunt switched-line phase shifters (can substitute reactive elements).

coupler due to its small area, and wire bonding a part of the processes. In PWB technology, since the switches are formed on the top metal layer, the input coupler is better suited to be a branchline hybrid type. SPMT switches are used for the shunt arms, and the desired electrical line lengths from each throw are terminated with a via to the ground on the backside. Once again, this structure relies on high-frequency performance of SPMT switches for the desired number of bits.

An alternative configuration is also available (Figure 15.12), which relies only on SPST switches in a cascade arrangement from the hybrid's shunt arms. This configuration relies on the lowest losses possible, since the reflection mechanism of the open stub is utilized. As each low-loss switch is closed, the line lengths are made electrically longer. At first glance, this approach might be expected to perform better at higher frequencies due to using single-throw switches. The trade-off here is in how well the open-circuit stubs provide a strong reflection considering the fringing of the open stubs and coupling to adjacent structures. This would also be a trade-off in PWB thickness and could be addressed by simulation.

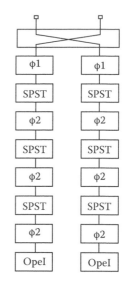

FIGURE 15.12   Branchline hybrid coupler with open-stub switched-line phase shifters. (From Osiander, R. et al., *IEEE Sens. J.*, 4, 525, 2004. With permission.)

15.5.7 Lower Loss Cantilever and Fine Grain Variable MEMS Phase Shifters

The existing meso-MEMS switch design is a dual cantilever beam structure with the electrostatic capacitor having a secondary RF-isolated beam to bridge a gap across the signal trace. The bottom side of the RF beam has a copper contact to perform this function, and is supported by a dielectric material.

The cantilever structure can be modified similar to the bridge designs, having the signal trace complete and the cantilever acting as a reflective shorting bar. Since the RF beam consists of a dielectric sandwiched between two copper layers, if the bottom copper layer that makes the contact is removed, a dielectric layer is applied to the signal trace, providing phase shift proportional to the width of the RF beam.

15.6 Aerospace

MEMS technology is having a significant influence on airspace systems design. The reason for this impact is the drastic reduction in size, mass, power, and cost of individual sensors and actuators used in air and spacecrafts. In space applications, this reduction also affects the launch cost of space systems. But there are some differences when using MEMS devices in aerospace with respect to other fields like industry. First, aerospace specificity demands a strict certification of new components, requiring a sourcing of 20 or 30 years with a rather small production. Second, the MEMS performances should remain stable for its life duration in order to avoid costly periodic ratings. And third, due to safety reasons, the reliability needs to be very high.

15.6.1 Aeroplane Applications

In aircrafts, the most relevant MEMS devices that can be found are inertial sensors (gyros and accelerometers) that combine data from digital magnetometers and GPS in the attitude heading reference system (AHRS). As the specific technical requirements are not so exigent for these applications, it is possible to import these sensors from the automotive market [8] and thus saving manufacturing costs. Another, more sophisticated application is the reduction of fuel consumption in airplanes. Air does not usually glide smoothly around a wing and fuselage; on the contrary, bubbles and roils create a turbulence regime causing friction. To solve this problem, MEMS devices have already been used to cut skin drag by 5%–6% on an F-15 fighter, simply modifying the air flow on the surfaces. Another application field where MEMS are finding a niche is unmanned aerial vehicles (UAV), where the combination of the information generated by commercial off-the-shelf (COTS) MEMS sensors with Kalman filters for the attitude determination is being quite fruitful for the estimation of the state of the continuous-time model of the airplane [9].

15.6.2 Space Applications

It is in space applications where MEMS technology is currently being massively applied because they quite fit the satellite concept. But in this environment, MEMS devices have to operate under very severe conditions such as vacuum, vibrations, shocks, temperature gradients, EMC, etc. Having in mind these difficulties, NASA [10] edited some guidelines in order to facilitate the insertion of this technology into high-reliability applications. In this report, some specific problems in this field are identified, and they are summarized in Table 15.2.

ESA has shown a lot of concern on this technology and it is also interested in MEMS applications at space. In this context, it has started the NEOMEx (near earth object micro explorer) mission, which will provide a focus application for a microsystem-based spacecraft concept.

In the following paragraphs, some examples of MEMS applications in space are presented. They are not intended to be an exhaustive description of the state-of-the-art applications. Many other realizations, like those in communications (like micro-machined filters for RF) or attitude control (like gyros, accelerometers or micro-wheels), are not included due to their similarity with other application fields.

15.6.3 Coarse and Fine Sun Sensors

Solar sensors are used in space for satellite attitude determination. They determine the incident angle of the sun light, thus providing a reference for the position of the satellite. This information is normally complemented with a magnetometer that provides information about the relative position with respect to the earth's magnetic field. An example of a sun sensor manufactured using MEMS technique, which has been used in NANOSAT 1B, is shown in Figure 15.13. This device has been successfully used [11] for the positioning of heliostats in solar power plants and its concept has been adapted for aerospace applications. It consists of a wafer with two photodiodes covered by a cover glass with a metal layer on top. The cover glass acts as a shield for the high energy particle radiation. The metal layer is patterned using lift-off to create an input window. These two wafers are glue-bonded using an optical epoxy resin qualified for space applications.

TABLE 15.2   NASA Classification of MEMS Failure Mechanisms

Mechanical Failure	Environmentally Induced Failure
• Stress-induced failure	• Vibration
• Point defects	• Shock
• Dislocations	• Humidity effects
• Precipitates	• Radiation effects
• Fracture strength	• Particulates
• Fatigue	• Temperature changes
	• Electrostatic discharge
Stiction	Stray stresses
Wear	Parasitic capacitance
Delamination	Dampening effects

Source: Courtesy of NASA/JPL-Caltech; Stark, B., *MEMS Reliability Assurance Guidelines for Space Applications*, JPL Publication 99-1, Jet Propulsion Laboratory, Pasadena, CA, 1999.

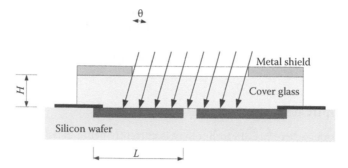

FIGURE 15.13 Cross section of a sun sensor manufactured using MEMS techniques. (From Zhang, K. et al., *J. Microelectromech. Syst.*, 13, 165, 2004. With permission.)

The incident light generates a photocurrent in each photodiode, being its value proportional to the normal irradiance and the photodiode illuminated area. Angle θ can be determined by means of the quotient of the difference and the sum of photocurrents:

$$\frac{I_{f1} - I_{f2}}{I_{f1} + I_{f2}} = \frac{2H \cdot \tan(\theta)}{L} = G \cdot \tan(\theta), \tag{15.1}$$

where
   H is the weight of the cover glass
   L is the length of the photodiode

It is remarkable that G can be regarded as a geometrical gain defined as $2H/L$. So, the sensitivity of the device can be constructively modified by changing the weight of the glass. In Figure 15.14, there is a photograph of the sensor mounted on a PCB with auxiliary electronics and the final aluminum package for its installation.

15.6.4 Shutter for Reflective Control

The temperature stabilization of a satellite is crucial to guarantee its correct operation. For this reason, a balance between the incident radiation and the emitted radiation in the satellite surface should be

(a)   (b)

FIGURE 15.14 Example of a satellite sun sensor that has been used in satellite Nanosat 1B. (a) Sensor with auxiliary electronics and (b) encapsulated sensor.

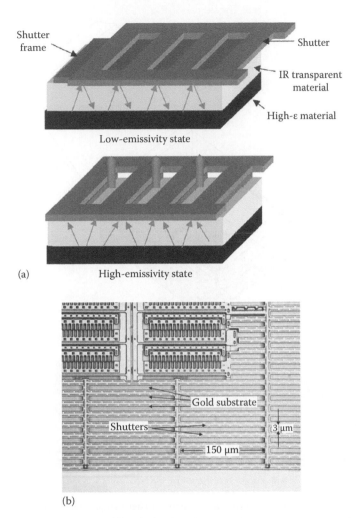

FIGURE 15.15 (a) Shutter concept for emissivity control in satellite. (b) High-resolution optical microscope image of the motors and shutters. (Reproduced from Osiander, R. et al., *IEEE Sens. J.*, 4, 525, 2004.)

achieved. A MEMS device that works as a satellite skin that can vary its emissivity has been designed [12]. It consist of arrays of gold-coated sliding shutters (see Figure 15.15) manufactured via surface micromachining. The first generation of this active thermal management system will be demonstrated on NASA's New Millennium Program ST-5 spacecraft.

15.6.5 Microthrusters

A micropropulsion system will be required in microspacecrafts for attitude control. A correct positioning is required to guarantee station keeping, gravitation compensation, and orbit adjust.

The schematic view of a solid propellant microthruster [13] is depicted in Figure 15.16. Its configuration has no movable parts, like pumps, fuel lines, or valves to avoid leakage. In the design, a silicon layer is wet etched to fabricate the combustion chamber, a convergent-divergent nozzle, and an ignition slot. A specific glass wafer is bonded on top of the silicon layer to form a closed structure. The chamber is then loaded with the solid propellant. Once ignited, the resultant gas expands producing the desired impulse.

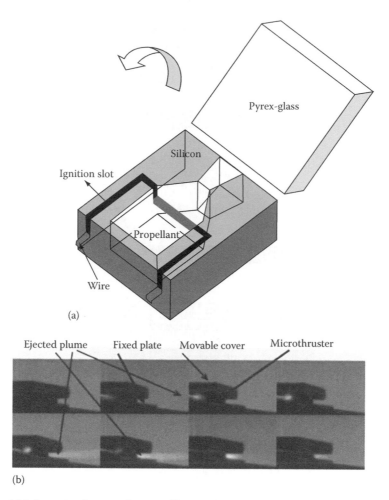

FIGURE 15.16 (a) Schematic of a microthruster. (b) Microthruster firing. Images are acquired at 500 frames/s. (Reproduced from Zhang, K. et al., *J. Microelectromech. Syst.*, 13, 165, 2004.)

15.7 Power and Energy

MEMS technologies are allowing for an increase in the scale of integration of complex systems, including not only sensing and actuation, but also control and communication. Nowadays, there is a tendency to develop the concept of "ambient intelligence," where a network of sensors and actuators hidden in our surroundings will provide ubiquitous and valuable information, computing, and support to end users. But this concept demands the inclusion in microsystems of an energy source to became autonomous and guarantee its functionality for years. There are two possible solutions to supply power to these miniaturized systems: the use of a chemical power source or converting different forms of nonelectrical energy into electrical energy. In the next paragraphs, a review of the most important energy-scavenging alternatives is presented.

15.7.1 Photovoltaic

The most successful example of such energy-scavenging devices is solar cells. These devices takes advantage of MEMS processes as their surfaces are anisotropically etched to create a pyramidal shape that increases light absorption, reducing its reflectivity. But there are a large number of applications where devices receive very limited or null radiation energy. In these cases, new types of scavengers are needed.

15.7.2 Vibration

One potential power source is the vibration of objects. For example, buildings suffer from very small amplitude vibration generated by air-conditioned systems or large machinery movement; human beings also provide a natural source of movement. In these cases, a MEMS device may include a suspended mass that vibrates and a piezoelectric, electromagnetic, or electrostatic generator would generate several microwatts [14]. The main drawback within this approach is that the generated power scales with mass. For this reason, numerous vibration-based generators have been reported in literature, but none of them have yet been commercialized in high-volume applications.

15.7.3 Thermal

There are many heat sources that maintain a difference of temperature with its environment. For instance, a lighting or an air-conditioning system, or even a human body [15], provides a temperature jump respect to its surroundings. The physical principle commonly used to convert heat into electric power is the Seebeck effect, which occurs when you form a circuit with two semiconductor junctions. In the presence of a temperature difference between the junctions, a small current flows around the circuit. The reverse phenomenon is called Peltier effect. Based on this principle, it is possible to fabricate thermoelectric generators capable of producing electric energy using MEMS technologies. But these energy scavengers suffer from the variation of the external environmental conditions and thus additional energy storage is required.

15.7.4 Electromagnetic

Another potential power source is the electromagnetic induction, a physical principle that generates voltage in a conductor by changing the magnetic field around it. Low frequency electromagnetic fields generated by power distribution lines are good candidates to provide energy remotely. However, the energy density is not sufficient for portable applications; also, the necessary antenna that is needed to obtain sufficient radiation would be too large to be used in real applications.

On the contrary, high-frequency electromagnetic fields, radio frequency transmission, for example, have been proved to be very appropriate to transfer energy in short range. That is the case of RFID tags, whose antenna receives sufficient energy to power up a sensor or microcontroller, including a radio communication interface. The main inconvenience of this energy exchange is that the irradiated energy decreases with the square of the distance between antennas, making this solution valid only in short distances.

Finally, it is worth mentioning that another effective way of producing electric energy is harvesting energy with the help of external permanent magnets and a resonating structure in the form of a cantilever or a membrane that contains a coil.

15.7.5 Fuel Cells

There exists a large variety of fuel cells that generate electricity from the chemical oxidation of hydrogen, producing water. Proton-exchange-membrane fuel cell (PEMFC) is one of the most successful alternatives, where a special membrane permeable to protons separates two volumes filled with H_2 and O_2, respectively. For this reason, electrons from ionized hydrogen atoms should flow through an external electrical circuit while their protons can go through the membrane to combine with oxygen molecules. The miniaturization of PEMFC is quite feasible, as the room temperature operation facilitates the simplification or even elimination of a cooling system. Basically, a MEMS PEMFC [16] consists of a sandwich of

TABLE 15.3   Comparison of Power Sources and Energy Scavengers

	Power Density (μW/cm^2)	
	1 Year Lifetime	10 Year Lifetime
Energy scavenging		
Solar (outdoor)	15,000-direct sun	15,000-direct sun
	150-cloudy day	150-cloudy day
Solar (indoor)	6	6
Vibration	100–200	100–200
Acoustic noise	0.003@75 dB	0.003@75 dB
Daily temperature	10	10
Temperature gradient	15@10°C gradient	15@10°C gradient
Power sources		
Batteries (lithium)	89	7
Combustion (μ-engine)	403	40.3
Fuel cells (methanol)	560	56
Nuclear source	850,000 (8% efficiency)	850,000 (8% efficiency)

two electrodes with a membrane electrode assembly (MEA) in between. This MEA includes a thin plastic film permeable to hydrogen ions and two diffusion layers made out of carbon paper. In MEMS technologies, these electrodes are fabricated with a silicon wafer that is etched to create channels and thus facilitate the flow of chemical species. Usually, these cells only precise hydrogen as fuel, as self-breathing structures directly acquire oxygen from air.

But storing hydrogen as compressed gas or liquid presents significant explosion hazards. Nowadays, there is a clear tendency to use a reactor to extract hydrogen from a liquid hydrocarbon such as methanol. An example of such reactor implemented using MEMS technologies can be found in [17].

An interesting alternative to conventional fuel cells are photosynthetic electrochemical ones, which directly harnesses subcellular photosystems [18] isolated from plant cells to perform bioconversion of light energy into electricity.

15.7.6 Microbatteries

Finally, we should not forget microbatteries, and especially rechargeable ones, because they are based on a proven chemical technology for energy storage that will store the energy obtained by the aforementioned scavengers.

As a summary, a comparison of power density of different power sources and energy scavengers [19] is presented in Table 15.3.

15.8 Market Trends

15.8.1 Current Status

For MEMS—like for any other technological domain—the degree of current vitality and future potential is measured to a large extent by the market value generated and the market value projected. In 2007, the estimated size of the MEMS market was $7 billion. The supply chain of MEMS involves equipment and material suppliers, device manufacturers and OEM system integrators, and the $7 billion figure relates exclusively to the device-manufacturing part (packaged MEMS) of the value chain.

The main applications driving current market include pressure sensors, RF-MEMS, microdisplays, and inkjet cartridges.

The MEMS market is well consolidated as it is proven by the high interest with which large semiconductor manufacturers try to capture a part of it. Unlike other markets, the regional distribution of MEMS manufacturing is well balanced across the world. Among the leading suppliers, there are companies headquartered in United States (Hewlett-Packard, Texas Instruments, Freescale), Europe (Bosch, STMicroelectronics), and Asia (Seiko, Canon).

15.8.2 MEMS Market Forecast

The MEMS market is expected to grow from the $7 billion in 2007 to $15 billion in 2012 [20]. This substantial growth would come from new market applications and from the extended market success of products requiring large-scale manufacturing of MEMS (see Figure 15.17).

In this 5 year period, motion sensing for electronics, silicon microphones, and biochips for diagnostic and drug delivery will be within the fast-growing MEMS applications. RF-MEMS will be produced in very large quantities for wireless systems, once the perennial issues of reliability and packaging have found suitable solutions. RF switches for handsets is the application that will generate the largest part of the revenues.

There are MEMS applications under development for which the market acceptance is still uncertain but with a very high volume potential that can materialize in the coming years, for instance, the integration of micromirrors and light engines for microprojection in mobile electronics. Texas Instruments has dominated this segment of MEMS with the DMD (digital micromirror device) technology, but new companies are emerging with alternative approaches, some (Microvision, Miradia) integrate micro/pico projectors in handheld/portable devices, others (Nippon Signal) develop stand-alone miniprojectors that fit in your pocket. MEMS-based displays for portable devices is another innovative approach to visualization.

15.8.3 Future Trends

A trend already perceived in 2007 is the increasing share of the consumer sector in the total MEMS market. Particularly important is the MEMS contribution to new functionalities in video games,

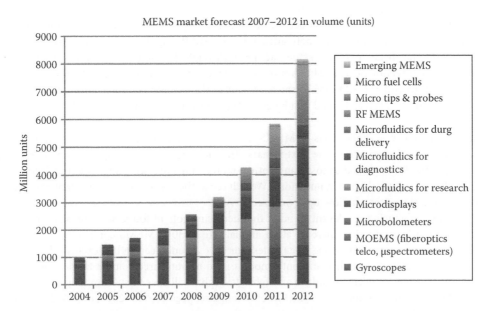

FIGURE 15.17 MEMS market forecast for 2007–2012. (Reproduced from *Status of the MEMS Industry*, 2008 edition, Yole Développement, October 2008. With permission.)

FIGURE 15.18   3D accelerometers from STMicroelectronics.

mobile electronics, lifestyle, and sport applications. This is reflected in the strong growth of MEMS in these areas [20].

Major manufacturers of game consoles (Nintendo, Sony) have introduced in their latest product generation inertial sensors for motion sensing, providing a new experience to users from sport simulation to dance training. New game systems include one or more 3D accelerometers to register changes in direction, speed, and acceleration. The popular Nintendo Wii includes an accelerometer from STMicroelectronics (LIS3L02AL, Figure 15.18) and one from Analog Devices (ADXL330). They are very compact devices featuring low-power operation and high sensitivity and linearity.

Modern laptops and high-end mobile phones include hard disk drive (HDD) protection feature. Accelerometers can detect a drop and "park" the HDD head to avoid data loss. Smart telephones, such as the iPhone, integrate MEMS motion sensors for free-fall detection, plus additional motion-activated functions and gaming input. Image rotation and stabilization are implemented in digital still cameras and camcorders with MEMS accelerometers.

The combination of high product-added value (actual and perceived) and large volume manufacturing makes these applications extremely attractive for MEMS suppliers. Reduced cost to enter the consumer market has been, to a large extent, achieved through the use of larger wafer size (200 mm). More effort in the direction of cost reduction has to be made for MEMS to expand further in this market segment. The growth of MEMS for consumer sector in the coming years will strongly contribute to the total MEMS market. It is estimated that by 2012, the consumer share will exceed 40% of the MEMS market value.

In a more modest way, applications of MEMS in life science and wireless communication will also contribute to the market growth in the next 5 years. The measurement of biological parameters creates many opportunities for medical diagnostic and patient monitoring but also to help individuals to conduct a healthier lifestyle (physical activity, nutrition). We will see in the coming years a substantial increase of MEMS-based smart devices to follow-up sport practice.

In MEMS manufacturing, a relevant trend is the increasing role of foundries. The share of foundry MEMS fabrication is expected to increase at an annual rate of 30%. MEMS companies created in the past years do not own a manufacturing line (fabless) or have an engineering line exclusively for process and product development (fablight), and both models need to outsource volume fabrication to foundries. These new business models (foundries and fables/fablight) will cover a substantial part of the future MEMS market. It is expected that out of the $18 billion market predicted for 2015, over $2 billion will be originated in foundries.

References

1. Microsystems Project Portfolio 2007–2010. *Seventh Research and Development Framework Programme 2007–2013*, June 2010. Directorate-General Information Society and Media. Directorate "Components and Systems". Unit Microsystems. http://cordis.europa.eu/fp7/ict/micro-nanosystems/docs/microsystems-project-portfolio-21-06-2010.pdf.
2. D. S. Eddy and D. R. Sparks, Application of MEMS technology in automotive sensors and actuators, *Proceedings of the IEEE*, 86 (1998), 1747–1755; S. S. Saliterman, *BioMEMS and Medical Microdevices*, Bellingham, WA: The International Society for Optical Engineering, 2006.
3. C. H. Ahn, J.-W. Choi, G. Beaucage, J. H. Nevin, J.-B. Lee, A. Puntambekar, and J. Y. Lee, Disposable smart lab on a chip for point-of-care clinical diagnostics, *Proceedings of the IEEE*, 92 (2004), 154–173.
4. H.-K. A. Tsai, J. Zoval, and M. Madou, Bi-layer artificial muscle valves for drug delivery devices, in *Device Research Conference Digest (DRC '05)*, Vol. 1, Santa Barbara, CA, June 22, 2005, pp. 129–130.
5. S. Henry, D. V. McAllister, M. G. Allen, and M. R. Prausnitz, Micromachined needles for the transdermal delivery of drugs, in *Proceedings of the IEEE Eleventh Annual International Workshop on Micro Electro Mechanical Systems*, Heidelberg, Germany, 1998, Piscataway, NJ: IEEE, pp. 494–498.
6. R. Lempkowski, K. Lian, M. Eliacin, and P. Kulkarni, A PWB-based MEMS switched filter bank using lumped element embedded passives, in *IEEE IECON05*, Raleigh, NC, November 6–10, 2005.
7. R. Ramadoss, S. Lee, K. C. Gupta, Y. C. Lee, and V. M. Bright, RF MEMS capacitive switches for integration with printed circuits and antennas, in *IEEE Antennas and Propagation Symposium Digest*, Vol. 1, Columbus, OH, June 22–27, 2003, pp. 395–398.
8. L. Sherry, C. Brown, B. Motazed, and D. Vos, Automotive-grade MEMS sensors used for general aviation, *Aerospace and Electronic Systems Magazine, IEEE*, 16 (2004), 13–16.
9. J. Jang, and D. Liccardo, Small UAV automation using MEMS, *Aerospace and Electronic Systems Magazine, IEEE*, 22 (2007), 30–34.
10. B. Stark, *MEMS Reliability Assurance Guidelines for Space Applications*. JPL Publication 99-1, Pasadena, CA: Jet Propulsion Laboratory, 1999.
11. J. Quero, C. Aracil, L. Franquelo, J. Ricart, P. Ortega, M. Dominguez et al. Tracking control system using an incident radiation angle microsensor, *IEEE Transactions on Industrial Electronics*, 54 (2007), 1207–1216.
12. R. Osiander, S. Firebaugh, J. Champion, and D. A. Farrar, Microelectromechanical devices for satellite thermal control, *IEEE Sensors Journal*, 4 (2004), 525–531.
13. K. Zhang, S. Chou, and S. Ang, MEMS-based solid propellant microthruster design, simulation, fabrication, and testing, *Journal of Microelectromechanical Systems*, 13 (2004), 165–175.
14. P. Mitcheson, E. Yeatman, G. Rao, A. Holmes, and T. Green, Energy harvesting from human and machine motion for wireless electronic devices, *Proceedings of the IEEE*, 96 (2008), 1457–1486.
15. V. Leonov, P. Fiorini, S. Sedky, T. Torfs, and C. Van Hoof, Thermoelectric MEMS generators as a power supply for a body area network, in *Digest of Technical Papers of the 13th International Conference on Solid-State Sensors, Actuators and Microsystems*, Vol. 1, Seoul, South Korea, 2005, pp. 291–294.
16. X. L. Cong Chen, A self-breathing proton-exchange-membrane fuel-cell pack with optimal design and microfabrication, *Journal of Micromechanical Systems*, 15 (2006), 1088–1097.
17. A. Pattekar and M. Kothare, A microreactor for hydrogen production in micro fuel cell applications, *Journal of Microelectromechanical Systems*, 13 (2004), 7–18.
18. K. Lam, E. Johnson, M. Chiao, and L. Lin, A MEMS photosynthetic electrochemical cell powered by subcellular plant photosystem, *Journal of Microelectromechanical Systems*, 15 (2006), 1243–1250.
19. H. Kulah and K. Najafi, Energy scavenging from low-frequency vibrations by using frequency up-conversion for wireless sensor applications, *IEEE Sensors Journal*, 8 (2008), 261–268.
20. Yole Développement, Status of the MEMS industry, 2008 edition. Yole Développement, Lyon, France, October 2008.

16

Transistors in Switching Circuits

Tina Hudson
*Rose-Hulman Institute
of Technology*

16.1 Large-Signal Models: Use of a Transistor as a Switch

The transistor characteristic equations presented in Chapter 4 constitute the transistors' "large-signal modeling," which includes all modes of operation and a current–voltage relationship for any input. These equations must be used when operating the transistor as a switch [3]. When the transistor is operating as an amplifier, only small input signals are used, which are amplified to large output signals. In this case, simpler linear models can be used for the transistor.

When a transistor operates as a switch, the input changes from a high voltage to a low voltage and back, which causes the output to move from a low voltage to a high voltage and back. When the output moves between a high and low voltage, the transistor changes regions of operation, from saturation to linear, or active to saturation, depending on whether it is a FET or BJT [3]. Therefore, to model the device in this operation, or design a device for this operation, the large-signal characteristic equations must be considered.

16.2 BJTs as Switches

16.2.1 Basic Switch Using an npn

An example of an npn used as a switch is shown in Figure 16.1 [3]. When V_{IN} is at a low voltage (less than V_{BE}), the transistor produces very little current (see Figure 16.1a). With very little current running through R_C, there is very little voltage drop across R_C and V_O is approximately at V_{CC}. This operation requires that V_{IN} be less than V_{BE} in order to turn the transistor off and allow a high output voltage. As a consequence, some consideration should be given to the largest value of V_{IN} that is considered a "low value" and the "turn-on" voltage of the BJT. The low value for V_{IN} should be less than 0.5 V to guarantee that the transistor produces very little current in the "off" state.

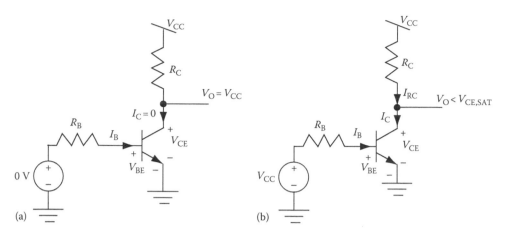

FIGURE 16.1 Basic npn switch circuit. (a) Shows the circuit operation with $V_{IN} = 0\,V$ and (b) shows the circuit operation with $V_{IN} = V_{CC}$.

When V_{IN} is at a high voltage, for example V_{CC}, the transistor sources the maximum amount of current (see Figure 16.1b). This current causes a large voltage drop across the resistor, R_C, which causes the output voltage to drop. Because the output voltage is low, V_{CE} is small, and the npn is placed in the saturation region [3]. In the saturation region, the base current increases, which requires more current to drive the switch [2]. To minimize this extra current, the base resistor, R_B, is added to set the collector current at a value that will place the npn at the edge of the saturation-active region. The resistor, R_C, and the transistor acting in the saturation mode will also further limit the amount of output current that the BJT can draw. The final resting voltage will occur where the current through R_C is equal to the current through the npn transistor.

To find the resulting V_O and I_C values, the analysis proceeds with the assumption that the npn is at the edge of the active region, which places $V_{CE} = V_{Cesat} = 0.2$ to $0.3\,V$ [3]. The analysis proceeds as follows.

The value of I_C that places the transistor in the saturation region must be calculated first as follows:

$$I_{RC} = I_C \tag{16.1}$$

By Kirchoff's voltage law (KVL)

$$V_{CC} - V_{RC} = V_{CE} \tag{16.2}$$

To minimize base current in the low state, set $V_{CE} = V_{CE,SAT} = 0.2\,V$ and by Ohm's law

$$V_{RC} = I_C R_C \tag{16.3}$$

Substitute and rearrange to get

$$I_C = \frac{V_{CC} - V_{CE,SAT}}{R_C} \tag{16.4}$$

Next, find the value of R_B that will produce I_C given the worst-case, or minimum, β value:

$$I_B = \frac{I_C}{\beta} \tag{16.5}$$

KVL at the input loop produces

$$V_{IN} - V_{RB} - V_{BE} = 0 \tag{16.6}$$

Set $V_{RB} = I_B R_B$ and $V_{BE} = V_{BE,ON}$ (the turn-on voltage of the transistor) to get

$$V_{IN} - I_B R_B - V_{BE,ON} = 0 \tag{16.7}$$

Solve for R_B to get

$$R_B = \frac{V_{IN} - V_{BE,ON}}{I_B} \tag{16.8}$$

A numerical example is presented to show how device tolerances affects the analysis. Assume $R_C = 5\,k\Omega$, $100 < \beta < 300$, $V_{CE,SAT} = 0.2\,V$ and $V_{BE,ON} = 0.7\,V$, and $4.6 < V_{INHIGH} < 5$. Find the base resistor needed to guarantee that the transistor is saturated:

$$I_C = \frac{V_{CC} - V_{CE,SAT}}{R_C} = \frac{5\,V - 0.2\,V}{5\,k\Omega} = 960\,\mu A$$

$$I_B = \frac{I_C}{\beta} = \frac{960\,\mu A}{100} = 9.6\,\mu A$$

$$R_B = \frac{V_{IN} - V_{BE,ON}}{I_B} = \frac{4.6\,V - 0.7\,V}{9.6\,\mu A} = 406\,k\Omega$$

Round R_B to a realizable resistor that makes I_B even larger to guarantee that the transistor is in the saturation region. If β is at the high end of the range, I_C will be even larger and push the transistor even further into the saturation region. This will result in a higher current draw on the input, but a lower current draw on the output. However, this solution will guarantee that V_O is at a "low" value.

16.2.2 Switch Circuit with a Resistive Load

The previous analysis assumed that the load was purely capacitive or an open circuit. If the circuit drives a load resistor, shown in Figure 16.2, the analysis changes. When V_{IN} is low (see part (a)), V_O will not reach V_{CC}. If $V_{IN} < V_{BE}$, $I_C \approx 0$; however current still flows through R_C to the load resistor. If $I_C \approx 0$, the circuit acts as if the transistor is not there. Remove the transistor, and the final V_O is the result of a voltage divider between R_L and R_C as follows:

$$V_O = V_{CC} \frac{R_L}{R_L + R_C} \tag{16.9}$$

When V_{IN} is high (see part (b)), the solution for R_B becomes slightly more complex. To find I_C, perform Kirchoff's current law (KCL) at the V_O node as follows:

$$I_{RC} = I_C + I_{RL} \tag{16.10}$$

$$\frac{V_{CC} - V_O}{R_C} = I_C + \frac{V_O}{R_L} \tag{16.11}$$

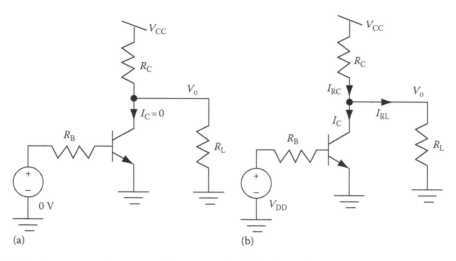

FIGURE 16.2   An npn switch circuit with a resistive load. (a) Shows the circuit operation with $V_{IN} = 0\,V$ and (b) shows the circuit operation with $V_{IN} = V_{CC}$.

To design this circuit at the edge of the saturation-active region, assume $V_O = V_{CEsat} = 0.2\,V$ and solve for I_C:

$$I_C = \frac{V_{CC} - V_{CEsat}}{R_C} - \frac{V_{CEsat}}{R_L} \tag{16.12}$$

When I_C is known, R_B can be solved, as in the previous example, using the worst-case β and the worst-case V_{IN}:

$$I_B = \frac{I_C}{\beta} \tag{16.13}$$

$$R_B = \frac{V_{IN} - V_{BE,ON}}{I_B} \tag{16.14}$$

If the design of this circuit requires a high value for V_{OHIGH} and the load resistor is small, a constraint may be placed on R_C. To keep V_O at a high value, R_C must be less than R_L to minimize the loss of voltage across the voltage divider. However, a small R_C will result in more current. Therefore, if the circuit must drive a small load resistor without an increase in power consumption, the design may require a CMOS implementation (see Section 16.4).

16.2.3 Switch Driving an LED

Another common switch example is shown in Figure 16.3, where the load is an LED (or transformer coil) with a current limiting resistor [3]. This circuit requires a similar analysis as shown in Section 16.2.1, except that the diode drop across the LED must be considered and the transistor must produce enough current to turn the LED on when V_{IN} is high. When V_{IN} is low, I_C is approximately zero, so the voltage across the LED as well as the voltage across R_C are both zero (Figure 16.3a). As a result, V_O is still approximately V_{CC} and the LED will not be lit (or the transformer will not become activated).

When V_{IN} is high, I_C must be chosen to be a value that will turn the LED on to the desired brightness level (Figure 16.3b). If the LED requires 15 mA to emit light, R_C and R_B must be chosen such that the BJT

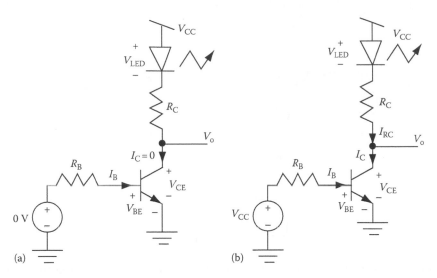

FIGURE 16.3 An npn switch circuit driving an LED in the collector. (a) Shows the circuit operation with $V_{IN} = 0\,V$ and (b) shows the circuit operation with $V_{IN} = V_{CC}$.

produces 15 mA in the saturation region [3]. Without R_C and R_B, the npn is likely to produce too much current and burn out the LED. Therefore, the analysis proceeds as follows:

KVL at the output yields

$$V_{CC} - V_{LED} - I_C R_C = V_{CE,SAT} \tag{16.15}$$

Rearrange and solve for R_C to get

$$R_C = \frac{V_{CC} - V_{LED} - V_{CE,SAT}}{I_C} \tag{16.16}$$

where I_C is the current necessary to illuminate the LED.

To find R_B, follow the same procedure in Section 16.2.1. I_C is the current necessary to illuminate the LED. From I_C, find I_B and by KVL, find R_B.

The following example shows a numerical example to demonstrate how component tolerances should be handled. Find the resistor values to illuminate an LED requiring 15 mA if $100 < \beta < 300$, $V_{LED} = 1.8\,V$, $V_{CC} = 5V$, $V_{CE,SAT} = 0.2\,V$, $V_{BE,ON} = 0.7\,V$, and $4.6 < V_{INHIGH} < 5$:

$$R_C = \frac{V_{CC} - V_{LED} - V_{CE,SAT}}{I_C} = \frac{5V - 1.8V - 0.2V}{15\,mA} = 866\,\Omega$$

$$I_B = \frac{I_C}{\beta} = \frac{15\,mA}{100} = 150\,\mu A$$

$$R_B = \frac{V_{IN} - V_{BE,ON}}{I_B} = \frac{4.6\,V - 0.7\,V}{150\,\mu A} = 26\,k\Omega$$

Round R_B and R_C to realizable resistors that will guarantee at least 15 mA through LED and more than enough current to saturate the transistor. To achieve this, round R_B down so that I_B, and thereby I_C, will be larger than necessary and round R_C up to push the transistor further into the saturation region.

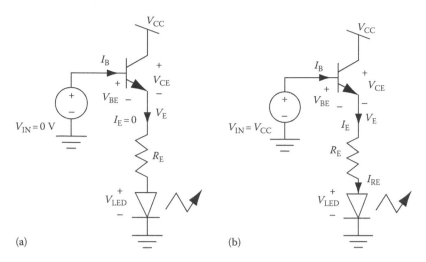

(a)   (b)

FIGURE 16.4 An npn switch circuit driving an LED in the emitter. (a) Shows the circuit operation with $V_{\text{IN}} = 0\,\text{V}$ and (b) shows the circuit operation with $V_{\text{IN}} = V_{\text{CC}}$.

The circuit shown in Figure 16.3 can also be implemented with the LED in the emitter, as shown in Figure 16.4. Since $I_{\text{C}} \approx I_{\text{E}}$, if the collector current can be turned on and off, so also can the emitter current. KVL around the input loop helps to understand how this circuit works (see Figure 16.4a):

$$V_{\text{IN}} - V_{\text{BE,ON}} - I_{\text{E}}R_{\text{E}} - V_{\text{LED}} = 0 \tag{16.17}$$

Solving for V_{IN} produces the following equation:

$$V_{\text{IN}} = V_{\text{BE,ON}} + I_{\text{E}}R_{\text{E}} + V_{\text{LED}} \tag{16.18}$$

To turn the circuit off, V_{IN} should be set low enough to force $I_{\text{E}} = 0$. This will occur when $V_{\text{IN}} < V_{\text{BE,ON}}$, as in the previous examples.

When V_{IN} is high, as in Figure 16.4b, the LED is turned on as follows. By KVL, $V_{\text{E}} = V_{\text{IN}} - V_{\text{BE,ON}}$. Therefore, as V_{IN} rises, so does V_{E}. As V_{E} rises, the LED and R_{E} see a larger voltage across them, resulting in more current draw. R_{E} can be chosen to set the current through the LED at the desired level when V_{IN} is at its lowest turn-on voltage as follows. Rearrange Equation 16.18 to solve for R_{E} and set I_{E} equal to the desired LED current:

$$R_{\text{E}} = \frac{V_{\text{IN}} - V_{\text{BE,ON}} - V_{\text{LED}}}{I_{\text{E}}} \tag{16.19}$$

Notice that there is no base resistor in this implementation. The circuit can turn the LED on and off without pushing the transistor into the saturation region; therefore, the base current limiting resistor is unnecessary. In fact, the only way the transistor can be placed in the saturation region is when $V_{\text{IN}} > V_{\text{CC}}$. If $V_{\text{IN}} = V_{\text{CC}}$,

$$V_{\text{E}} = V_{\text{CC}} - V_{\text{BE,ON}} = V_{\text{CC}} - 0.7 \tag{16.20}$$

Since $V_{\text{CE}} = V_{\text{CC}} - V_{\text{E}}$, substituting Equation 16.20 results in

$$V_{\text{CE}} = V_{\text{CC}} - (V_{\text{CC}} - 0.7) = 0.7 > V_{\text{CE,SAT}} \tag{16.21}$$

This analysis shows the benefit of placing the LED in the emitter rather than the collector. The desired brightness can be achieved with one less resistor and potentially less power. It is important to note that with this implementation, the transistor does not go into the saturation region. If R_C is chosen to set $V_{CE} = V_{CE,SAT}$, V_{BE} will be less than the turn-on voltage, turning the BJT off and thereby turning the LED off.

To provide a final comparison of the two circuits, the specifications for the previous design are implemented using this circuit: illuminate an LED requiring 15 mA if $100 < \beta < 300$, $V_{LED} = 1.8$ V, $V_{CE,SAT} = 0.2$ V, $V_{BE,ON} = 0.7$ V, and $4.6 < V_{INHIGH} < 5$. Using worst-case V_{IN},

$$R_E = \frac{V_{IN} - V_{BE,ON} - V_{LED}}{I_E} = \frac{4.6\,\text{V} - 0.7\,\text{V} - 1.8\,\text{V}}{15\,\text{mA}} = 140\,\Omega \qquad (16.22)$$

Round R_E down to the nearest realizable resistor value to ensure that enough current goes through the LED.

16.2.4 Basic Switch Using a pnp

Any of these circuits can be implemented using a pnp by simply flipping the circuit over as shown in Figure 16.5 and replacing the npn with a pnp [3]. When V_{IN} is high such that $V_{EB} = V_{CC} - V_{IN} < V_{EB,ON}$, the transistor is off and produces very little current (Figure 16.5a). As a result, there is little voltage drop across R_C and V_O goes to approximately 0 V. When V_{IN} is low, R_B sets the base current, which sets the collector current that drives V_O high (Figure 16.5b). The collector current is chosen to drive the transistor into the saturation region to limit the DC power dissipation while the transistor is in the high state. The analysis of this circuit proceeds in the same manner as the npn circuit shown in Section 16.2.1.

KVL at the output yields

$$V_{CC} - V_{EC,SAT} = I_C R_C \qquad (16.22a)$$

Rearrange and solve for the I_C that is necessary to saturate the transistor:

$$I_C = \frac{V_{CC} - V_{EC,SAT}}{R_C} \qquad (16.23)$$

Next, find the value of R_B that will produce I_C given the worst-case, or minimum, β value:

$$I_B = \frac{I_C}{\beta} \qquad (16.24)$$

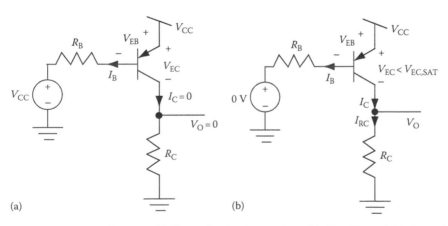

FIGURE 16.5 Basic pnp switch circuit. (a) Shows the circuit operation with $V_{IN} = V_{CC}$ and (b) shows the circuit operation with $V_{IN} = 0$ V.

KVL at the input loop produces

$$V_{CC} - V_{EB} - V_{RB} = V_{IN} \tag{16.25}$$

Note that I_B comes out of the base in a pnp, which changes the polarity of the voltage across R_B.

Substitute $V_{RB} = I_B R_B$ and solve for R_B to get

$$R_B = \frac{V_{DD} - V_{EB} - V_{IN}}{I_B} \tag{16.26}$$

16.3 MOSFETs as Switches

16.3.1 Basic Switch Circuit Using an nFET

FET switch operation is very similar to BJT switch operation. Any of the circuits presented for BJTs can be implemented using FETs. However, there are three primary differences in the analysis of the FET circuits and the BJT circuits: (1) since the gate current for a FET is zero, a base resistor is unnecessary; (2) the saturation voltage depends on the gate-source voltage as shown in Equation 16.27; and (3) because V_{DSsat} and V_{GS} depend on I_D and cannot be approximated as a known value, the characteristic equations must be used.

$$V_{DSsat} = V_{GS} - V_t \tag{16.27}$$

An example of an nFET used as a switch is shown in Figure 16.6 [3]. When V_{IN} is at a low voltage (less than V_t), the transistor produces very little current (see Figure 16.6a). With very little current running through R_D, there is very little voltage drop across R_D and so V_O is approximately at V_{DD}. This operation requires that V_{IN} be less than V_t in order to turn the transistor off and allow a high output voltage. As a consequence, some consideration should be given to the largest value of V_{IN} that is considered a "low value" and the threshold voltage of the FET. If a low value for V_{IN} could be anywhere between 0.7 and 1.2 V, the threshold voltage of the FET should be larger than 1.2 V. Conversely, if the threshold voltage of the FET is 1 V, the maximum "low" value for V_{IN} should be less than 1 V.

When V_{IN} is at a high voltage, for example V_{DD}, the transistor sources the maximum amount of current (see Figure 16.6b). This current causes a large voltage drop across the resistor, R_D, which causes the output voltage to drop. Because the output voltage is low, V_{DS} is small, and the FET is placed in the linear

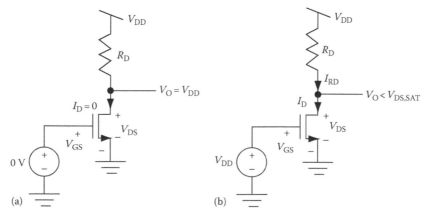

FIGURE 16.6 Basic nFET switch circuit. (a) Shows the circuit operation with $V_{IN} = 0$ V and (b) shows the circuit operation with $V_{IN} = V_{DD}$.

region (*note*: this corresponds to the saturation region of the BJT) [3]. The resistor, R_D, and the transistor acting in the saturation mode will limit the amount of current that the FET can draw. The final resting voltage will occur where the current through R_D is equal to the current through the FET, which is determined by the linear region equation:

$$I_{DS} = K'\left(\frac{W}{L}\right)\left[(V_{GS} - V_t)V_{DS} - \frac{V_{DS}^2}{2}\right] \tag{16.28}$$

Solving the two equations together yields

$$I_{RD} = I_D:$$

$$\frac{V_{DD} - V_{DS}}{R_D} = K'\left(\frac{W}{L}\right)\left[(V_{GS} - V_t)V_{DS} - \frac{V_{DS}^2}{2}\right] \tag{16.29}$$

Substituting in for $V_{GS} = V_{DD} - 0$ and $V_{DS} = V_O - 0$ yields

$$\frac{V_{DD} - V_O}{R_D} = K'\left(\frac{W}{L}\right)\left[(V_{DD} - V_t)V_O - \frac{V_O^2}{2}\right] \tag{16.30}$$

If the circuit has been designed, everything is known except V_O, so the value of V_O can be determined. For the example circuit in Figure 16.6, if $V_t = 1\,V$, $K'(W/L) = 10\,mA/V^2$, $V_{IN} = 5\,V$ and $R_D = 500\,\Omega$, the analysis would proceed as follows:

$$\frac{5 - V_O}{500\,\Omega} = 10\,mA/V^2\left[(5-1)V_O - \frac{V_O^2}{2}\right]$$

Solving for V_O yields

$$V_O = 0.245\,V$$

V_O may be solved using the root function in a calculator or math function program, or solving for the quadratic equation by hand. If using a calculator, be sure to choose a root that makes sense in the circuit.

If the goal is to design the circuit, a desired V_O must be chosen, and then a value for R_D can be determined. A lower V_O will push the transistor deeper into the linear region, further limiting the current through the transistor. If the circuit is driving a purely capacitive load, this condition is advantageous to limit the power consumption. Less current sourced from V_{DD} to ground in the low state will require less power. However, it also requires a larger value for R_D. For the example circuit in Figure 2.6, if $V_t = 2\,V$, $K'(W/L) = 20\,mA/V^2$, and $V_{IN} = 4\,V$, the analysis would proceed as follows:

Choose $V_O = 0.1\,V$. The FET yields the following equation:

$$I_D = K'\left(\frac{W}{L}\right)\left[(V_{DD} - V_t)V_O - \frac{V_O^2}{2}\right] = 20\,mA/V^2\left[(4\,V - 2\,V)0.1\,V - \frac{(0.1\,V)^2}{2}\right] = 3.9\,mA$$

The resistor yields the following equation:

$$I_D = \frac{V_{DD} - V_O}{R_D} = \frac{5\,V - 0.1\,V}{R_D}$$

Setting the two equations equal to one another and solving for R_D yields

$$R_D = \frac{5\,\text{V} - 0.1\,\text{V}}{3.9\,\text{mA}} = 1.25\,\text{k}\Omega$$

Round R_D to the nearest realizable resistor that will place the circuit even further into the linear region. To achieve this, round R_D up to increase the voltage drop across R_D and, thereby, to push the FET further into the saturation region. If the result does not give the necessary power consumption, choose a lower V_O and repeat the process. Alternatively, if a required power consumption is known, find the V_O that will produce the necessary power consumption using the linear equation and then find the value for R_D.

16.3.2 Switch Circuit with a Resistive Load

Recall that the previous analysis assumes that the load is purely capacitive or an open circuit. Like the BJT analysis, if the circuit is driving a load resistor, as shown in Figure 16.7, the value of V_O depends on the voltage divider set-up between R_D and R_L (see Figure 16.7a). If $V_{IN} < V_t$, $I_D = 0$; however, the current still flows through R_D to the load resistor. The final V_O is the result of a voltage divider between R_L and R_D as follows:

$$V_O = V_{DD}\frac{R_L}{R_L + R_D} \tag{16.30a}$$

When V_{IN} is high (see part (b)), the solution for V_O also becomes slightly more complex. To find V_O, perform the KCL at the V_O node as follows:

$$I_{RD} = I_D + I_{RL} \tag{16.31}$$

$$\frac{V_{DD} - V_O}{R_D} = K'\left(\frac{W}{L}\right)\left[(V_{DD} - V_t)V_O - \frac{V_O^2}{2}\right] + \frac{V_O}{R_L} \tag{16.32}$$

Rearranging produces

$$K'\left(\frac{W}{L}\right)\frac{V_O^2}{2} - V_O\left[\frac{1}{R_D} + \frac{1}{R_L} + K'\left(\frac{W}{L}\right)(V_{DD} - V_t)\right] + \frac{V_{DD}}{R_D} = 0 \tag{16.33}$$

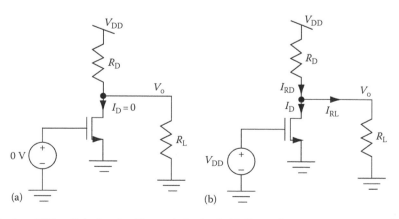

FIGURE 16.7   An nFET switch circuit with a resistive load. (a) Shows the circuit operation with $V_{IN} = 0\,\text{V}$ and (b) shows the circuit operation with $V_{IN} = V_{DD}$.

V_O can again be solved using a root function. If designing this circuit, the requirements on V_{OHIGH} and value of the load resistor may place a constraint on R_D. To keep V_O at a high value, R_D must be less than R_L to minimize the loss of voltage across the voltage divider. However, a small R_D will result in a higher V_{OLOW}. Therefore, if the circuit must drive a small load resistor, the design may require a small transition region between V_{OH} and V_{OL} or a buffer to increase the load resistance seen by the switch.

16.3.3 Switch Circuits Driving an LED

The FET can also be used to drive an LED or transformer in a similar manner as the BJT. However, BJTs typically have stronger drive capability due to the exponential *I–V* relationship rather than the square-law *I–V* relationship [3]. If a high current is necessary, either a BJT or a high power MOSFET may be necessary.

Figure 16.8 shows the FET switch with an LED in the drain [3]. This circuit requires a similar analysis as Section 16.3.1, except the diode drop across the LED must be considered and the transistor must produce enough current to turn the LED on when V_{IN} is high. When V_{IN} is low, I_D is approximately zero, so the voltage across the LED as well as the voltage across R_D are both zero (Figure 16.8a). As a result, V_O is still approximately V_{DD}.

When V_{IN} is high, I_D must be chosen to be a value that will turn the LED on (Figure 16.8b). If the LED requires 15 mA to emit light, V_O must be chosen so that the FET produces 15 mA in the linear region.

$$15\,\text{mA} = K'\left(\frac{W}{L}\right)\left[(V_{DD} - V_t)V_O - \frac{V_O^2}{2}\right] \tag{16.34}$$

Solve for V_O using a root function. Then, using KVL, find an equation for R_D as follows:

$$V_{DD} - V_{LED} - 15\,\text{mA} * R_D = V_O \tag{16.35}$$

Solving for R_D yields

$$R_D = \frac{V_{DD} - V_{LED} - V_O}{15\,\text{mA}} \tag{16.36}$$

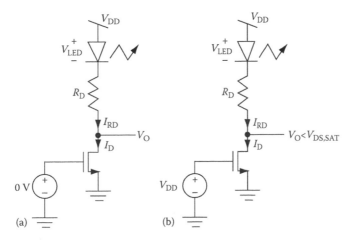

FIGURE 16.8 An nFET switch circuit driving an LED in the drain. (a) Shows the circuit operation with $V_{IN} = 0\,\text{V}$ and (b) shows the circuit operation with $V_{IN} = V_{DD}$.

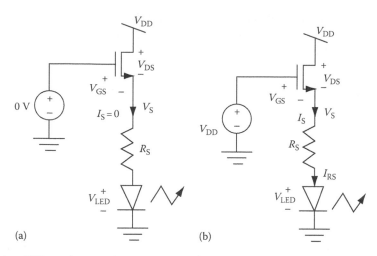

FIGURE 16.9   An nFET switch circuit driving an LED in the emitter. (a) Shows the circuit operation with $V_{IN} = 0\,V$ and (b) shows the circuit operation with $V_{IN} = V_{DD}$.

The LED can be placed in the source just like the BJT circuit (see Figure 16.9). When $V_{IN} < V_{GS}$, as shown in Figure 16.9a, the transistor produces little current making the voltage drop across the LED and the resistor approximately zero and the LED turns off. However, when V_{IN} is high (see Figure 16.9b) the transistor produces maximum current, which turns the LED on. Like the BJT circuit, $V_S = V_{IN} - V_{GS}$. As a result, the FET cannot be placed into the linear region unless V_{IN} goes outside of the rails. The analysis of this fact follows, assuming that V_{IN} is at the highest possible value within the rails: V_{DD}.

$$V_{GS} = V_{IN} - V_S = V_{DD} - V_S \tag{16.37}$$

$$V_{DS} = V_{DD} - V_S \tag{16.38}$$

Therefore,

$$V_{GS} = V_{DS} \tag{16.39}$$

$$V_{DS,sat} = V_{GS} - V_t < V_{GS} \tag{16.40}$$

As a result, the FET will not be in the linear region, so the saturation region equations should be used for this circuit.

In both FET LED circuits, the current through the circuit is limited by a resistor, R_D or R_S. The analysis for the circuit in Figure 16.9 proceeds as follows. Given a desired LED current (I_S), a known voltage drop across the LED (V_{LED}), and a known worst-case input voltage (V_{IN}), the source resistor (R_S) can be determined to set these conditions. The KVL at the input loop yields

$$V_{IN} - V_{GS} - I_S R_S - V_{LED} = 0 \tag{16.37a}$$

All values are known in this equation except V_{GS} and R_S. Since the transistor is in the saturation equation, V_{GS} must be found using the characteristic equation in this region:

$$I_D = \frac{K'}{2}\left(\frac{W}{L}\right)(V_{GS} - V_t)^2 \tag{16.38a}$$

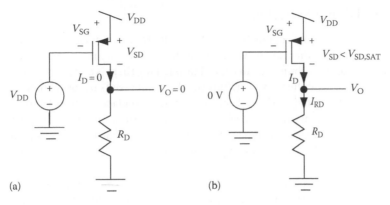

FIGURE 16.10 Basic pFET switch circuit. (a) Shows the circuit operation with $V_{IN} = V_{DD}$ and (b) shows the circuit operation with $V_{IN} = 0\,V$.

Rewriting for V_{GS} yields

$$V_{GS} = \sqrt{\frac{2I_D L}{K'W}} + V_t \tag{16.39a}$$

Once V_{GS} is known, R_S can be found as follows:

$$R_S = \frac{V_{IN} - V_{GS} - V_{LED}}{I_S} \tag{16.40a}$$

16.3.4 Basic Switch Circuit Using a pFET

Like the BJT circuits, any of the nFET circuits can be implemented using a pFET by simply flipping the circuit over, as shown in Figure 16.10 [3]. In this case, a high value of V_{IN} ($V_{DD}-V_{IN} < |V_t|$) will turn the FET off (Figure 16.10a). With no I_D, $V_O = 0$. A low value of V_{IN}, such as 0, will cause the transistor to source maximum current, placing a large voltage across R_D, pushing V_O high and placing the transistor into the linear region (Figure 16.10b). Just like the nFET circuit, V_O settles where $I_D = I_{RD}$.

Setting the FET and resistor currents equal to one another yields

$$K'\left(\frac{W}{L}\right)\left[(V_{SG} - |V_t|)V_{SD} - \frac{V_{SD}{}^2}{2}\right] = \frac{V_S - 0}{R_D} \tag{16.41}$$

Replacing V_{SG} and V_{SD} with known node voltages produces

$$K'\left(\frac{W}{L}\right)\left[(V_{DD} - |V_t|)(V_{DD} - V_O) - \frac{(V_{DD} - V_O)^2}{2}\right] = \frac{V_O}{R_D} \tag{16.42}$$

Using Equation 16.42, V_O may be determined or R_D may be solved in a manner similar to the example shown in Section 16.3.1.

16.4 CMOS Switches

Complementary MOS (CMOS) switches use both n and p devices to produce the switch. Such configurations can improve power consumption in some circuits and prevent signal degradation in others. This section will present the circuits and analysis of some common CMOS switch circuits.

16.4.1 CMOS Digital Switches

The switch circuits described in Sections 16.2 and 16.3 suffer from power consumption when the n-type switches have a high V_{IN} or when the p-type switches have a low V_{IN}. In this state (see Figure 16.11a), $V_{DS} = V_{DSsat}$. Since the drain is connected to the rail through R_D, there is a small current going from V_{DD} to the ground through R_D and the transistor. This small current results in constant power consumption while the circuit is in this state [4]. The resistors are designed to place the transistors in the saturation region to minimize this current draw; however, it is nonzero for these circuits. When the switch is in the opposite state (e.g., V_{IN} is low for the n-type or high for the p-type), the transistor is in off position. As a result, the current through the resistor has no place to go, so $I_D = 0$, unless there is a resistive load. If the load is capacitive, the power consumption in this state is zero.

Using CMOS logic with a capacitive load, shown in Figure 16.11b and c, guarantees that one device or the other is turned off for each state which minimizes power consumption [4]. When V_{IN} is high (Figure 16.11b), $V_{GSp} = 0$ which turns M2 off. M1 is strongly activated and will sink current off the capacitive load until $V_O = 0$, $V_{DS1} = 0$, and $I_{D1} = I_{D2} = 0$. At this point, there is no further power consumption until the switch changes its state.

When V_{IN} is low (Figure 16.11c), $V_{GSn} = 0$, which turns M1 off. M2 is strongly activated and will source current on to the capacitive load until $V_O = V_{DD}$, $V_{SD2} = 0$, and $I_{D1} = I_{D2} = 0$. Once again, at this point there is no further power consumption until the switch changes its state. As a result, the only time this switching circuit draws power is when it switches states.

It is important to keep in mind that if there is a resistive load, even a CMOS circuit will continue to draw power when the circuit is in a steady-state [4]. Additionally, in this case V_O does not necessarily

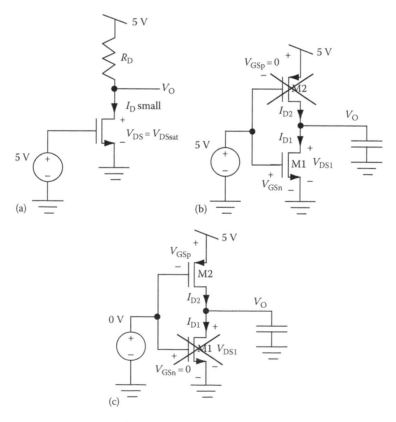

FIGURE 16.11 CMOS switch circuit with capacitive load. (a) Shows the power limitation of an nFET switch circuit, (b) shows the CMOS circuit operation with $V_{IN} = V_{DD} = 5\,V$, and (c) shows the CMOS circuit operation with $V_{IN} = 0\,V$.

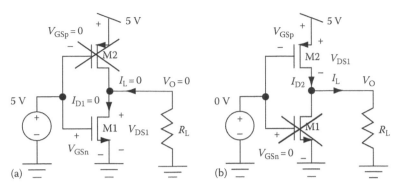

FIGURE 16.12 CMOS switch circuit with resistive load. (a) Shows the circuit operation with $V_{IN} = V_{DD} = 5$ V and (b) shows the circuit operation with $V_{IN} = 0$ V.

reach the rails [4]. An example of a CMOS circuit driving a resistive load is shown in Figure 16.12. When V_{IN} is high (Figure 16.12a), the load resistor, R_L, and the source of M1 are both tied to the ground. In this case, V_O can be pulled all the way to 0 V. If $V_O = 0$ V, both sides of the resistor are effectively grounded so there is no load current to go through M1. As a result, the circuit does not consume power in this state. However, when V_{IN} is low (Figure 13.12b), M2 will try to pull V_O up to V_{DD}. However, as V_O rises, the current through R_L rises as well. The final value for V_O occurs when the current through M2 (I_{D2}) is equal to the current through the load resistor (I_L). Assuming that the transistor is in the linear region at this point, the solution for V_O is found using the following equation:

$$K'\left(\frac{W}{L}\right)\left[(V_{SG} - V_t)V_{SD} - \frac{V_{SD}^2}{2}\right] = \frac{V_O - 0}{R_L} \tag{16.43}$$

Knowing the $V_{SG} = V_{IN}$ and $V_{SD} = V_{DD} - V_O$, Equation 16.43 can be rewritten as follows:

$$K'\left(\frac{W}{L}\right)\left[(V_{IN} - V_t)(V_{DD} - V_O) - \frac{(V_{DD} - V_O)^2}{2}\right] = \frac{V_O - 0}{R_L} \tag{16.44}$$

V_O can be found using a calculator root function. The current draw in this state can be determined by $I_{D2} = I_L = V_O/R_L$.

It is important to note that the complementary logic with BJTs does not have the same power consumption advantage as FETs. Even if the output current draw is reduced, the input current draw is increased, since BJTs absorb more base current when they are in the saturation region. In fact, the most base current draw occurs when V_{CE} or $V_{EC} = 0$, because the collector-base p–n junction diode is most strongly forward biased at this biasing level [2]. Therefore, the net power gain is insignificant. However, if the source can provide the base currents necessary, the advantage of V_O swinging rail-to-rail with a capacitive load may still be a benefit.

16.4.2 CMOS Pass-Gates

The circuits presented so far are used for producing digital output levels (V_O = high or low voltage) or driving current source loads, such as LEDs or transformers. Another common use for switches is to determine whether or not to pass a voltage based on a controlling input. The voltages passed can be digital voltages or analog voltages. The typical circuit used for this operation is called a pass-gate or transmission-gate. A single transistor can be used (a pass-gate), as in Figure 16.13, or a CMOS pair can be used (a transmission gate), as in Figure 16.14 [4]. In these example circuits, two pass-gates are used to select which input sets the value of a common output line.

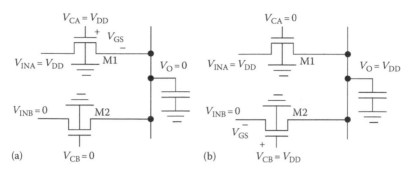

FIGURE 16.13 An nFET pass-gate. (a) Shows the nFET (M1) passing a high voltage (V_{DD}) to V_O and (b) shows the nFET (M2) passing a low voltage (0V) to V_O.

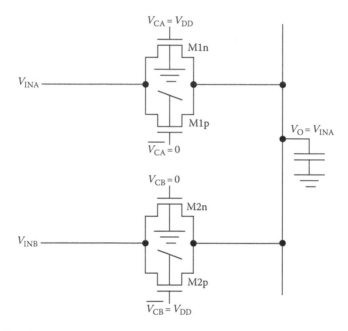

FIGURE 16.14 A CMOS pass-gate.

In the single transistor pass-gate operation (Figure 16.13a), M1 and M2 are the two pass-gates [1,4]. V_{CA} and V_{CB} are typically driven with digital input values equal to the rails of the circuit, but ideally V_{INA} and V_{INB} can be any voltage within the rails. If V_{CA} is set to a high voltage and V_{CB} is set to a low voltage, M1 lets V_{INA} "pass through" to V_O so that $V_O = V_{INA}$. If V_{CA} is set to a low voltage and V_{CB} is set to a high voltage, M2 lets V_{INB} "pass through" to V_O so that $V_O = V_{INB}$. If $V_{CA} = V_{CB} = 0$, V_O floats and typically maintains its value until leakage current through the transistors causes it to change [1,4]. If V_{CA} and V_{CB} are both high, both V_{INA} and V_{INB} try to drive the line to their voltages causing a short between two sources. This condition should be avoided.

In this schematic, the sources are not labeled in these transistors because the source changes according to the voltage level of V_{IN} [1,4]. For an nFET, the source is at a lower potential than the drain. If $V_O = 0$ V and $V_{IN} = 5$ V, the source would be at V_O. If $V_O = 5$ V and $V_{IN} = 0$ V, the source would be at V_{IN}. Since the pass-gate can allow any voltage through, the source location changes for each value that is passed through. If the pass-gates were implemented with pFETs, the source would change locations similarly, except that the source would at a higher potential than the drain. For a FET to be used as a pass-gate, it is essential that the bulk is NOT tied to the source such that the source can change locations.

Otherwise, some conditions will connect the bulk to the wrong potential and forward bias the drain-bulk and source-bulk n–p junctions. Therefore, if a discrete device is being used as a pass-gate, it must be a 4-terminal MOSFET. Since BJTs are not bidirectional devices (the emitter and collector cannot be swapped without seriously degrading the value of β), they cannot be used for pass-gates either.

The limitation of the single-transistor pass-gate is shown in Figure 16.13b. In this example, $V_{INA} = V_{DD}$ and $V_{CA} = V_{DD}$. If initially $V_O = 0\,V$, then the source is at V_O. Therefore, $V_{GS} = V_{CA} - V_O$. V_{INA} charges the capacitor bringing V_O up to V_{DD}. However, as V_O rises, V_{GS} drops. Once V_O tries to go higher than $V_{DD} - V_t$, $V_{GS} < V_t$ and M1 turns off. Therefore, the pass-gate cannot drive $V_O > V_{CA} - V_t$. This limitation does not occur when the nFET is passing a low voltage. If initially $V_{INB} = 0$, $V_{CB} = V_{DD}$, and $V_O = V_{DD}$, the nFET source is at V_{INB}. Since V_{INB} does not change values while V_O is dropping, V_{GS} remains constant and M2 remains on. A similar limitation exists for the pFET pass-gate; however, because the source for a pFET is at a potential higher than the drain, it cannot drive $V_O < |V_t|$.

Since an nFET can pass a low voltage without turning off but not a high voltage, and a pFET can pass a high voltage without turning off but not a low voltage, this limitation can be overcome by using both types of devices in parallel. The same circuit is shown with CMOS transmission-gates in Figure 16.14. In this circuit, both transistors in parallel must be turned on or off simultaneously to allow or block transmission [1,4]. Therefore, assuming digital values drive the gates, the pFET gate is driven with the inverse of the nFET gate. If $V_{CA} = V_{DD}$, $\overline{V_{CA}} = 0$, $V_{CB} = 0$, and $\overline{V_{CB}} = V_{DD}$, then M1n and M1p will be turned on and M2n and M2p will be turned off. As a result, $V_O = V_{INA}$. If $V_{CB} = V_{DD}$, $\overline{V_{CB}} = 0$, $V_{CA} = 0$, and $\overline{V_{CA}} = V_{DD}$, then M2n and M2p will be turned on and M1n and M1p will be turned off. In this case, $V_O = V_{INB}$. When the pass-gate tries to pass a high voltage, the nFET will turn off but the pFET will remain on and pass the value to V_O. Conversely, when the pass-gate tries to pass a low voltage, the pFET will turn off but the nFET will remain on and pass the value to V_O. This solution allows voltages from rail-to-rail to be passed.

References

1. P. E. Allen and D. R. Holberg, *CMOS Analog Circuit Design* (2nd edition). New York: Oxford University Press, 2002.
2. G. W. Neudeck, *Modular Series on Solid State Devices: The Bipolar Junction Transistor* (2nd edition), Vol. 3. Reading, MA: Addison-Wesley Publishing Co., 1989.
3. A. S. Sedra and K. C. Smith, *Microelectronic Circuits* (5th edition). New York: Oxford University Press, 2004.
4. J. P. Uyemura, *Introduction to VLSI Circuits and Systems*. New York: John Wiley & Sons, Inc., 2002.

17

Transistors in Amplifier Circuits

Tina Hudson
*Rose-Hulman Institute
of Technology*

17.1 Using Linear Transistor Models for Amplifiers

When a transistor is being used as an amplifier, the typical operation involves a low-amplitude sinusoidal V_{IN} producing a large-amplitude sinusoidal V_O. Each sinusoidal voltage and current will oscillate around a DC bias point (e.g., V_{IN} may oscillate ±10 mV around a 0 V bias point), while V_O may oscillate ±3 V around a 5 V bias point (see Figure 17.1). The current, I_C, performs similarly, oscillating ±0.3 mA around a 1 mA bias point, but note that I_C does not go negative. To simplify analysis and design, superposition is applied to the amplifier circuit to isolate the DC component of the signal from the AC component. Therefore, a DC analysis, where all AC sources are turned off, is performed independently of the AC analysis, where all DC sources are turned off. The DC analysis will produce the bias points (otherwise known as DC points or quiescent points or Q-points). The AC analysis will provide an expression of the AC gain. The complete operation of the amplifier, as shown in Figure 17.1, is a recombination of the AC and DC components.

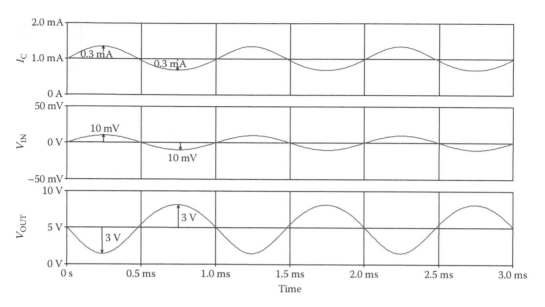

FIGURE 17.1   Decomposition of amplifier signals.

17.2 DC Analysis

To perform the DC analysis, place the circuit in its DC state:

1. Turn off all AC sources. This means shorting AC voltage sources (making them have 0 V across them) and setting all AC current sources to an open circuit (making them have 0 A in them).
2. Capacitors are open at DC, so make all capacitors an open circuit.
3. Using the current-saturating equations (FET = saturation region, BJT = active region), find the key bias points. Common key bias points for a FET are I_{DQ} and V_{OQ}. Similarly, common key bias points for a BJT are I_{CQ} and V_{OQ}. Sometimes, I_{BQ} is also a necessary bias point for BJTs.

For the transistor to be used in an amplifier circuit, it must remain in the current-saturating region. Therefore, when performing the DC analysis, assume that the transistor is operating in this region. If the assumption is wrong, the resulting bias points will indicate that the transistor is not in this region of operation (e.g., $V_{DS} < V_{DS,SAT}$ or $V_{CE} < V_{CE,SAT}$). Important note: since the transistor is operating in the current-saturation region, the DC value for V_{DS} and V_{CE} are not known. As shown in Figures 4.6 and 4.11 in Chapter 4, V_{DS} and V_{CE} can be set to many different values and, approximately, the same output current will be produced.

The following examples will show the DC bias process for some standard amplifier circuits. Each example will be worked for both a FET and a BJT.

17.2.1 High-Gain Amplifier with Input Resistor Biasing

17.2.1.1 MOSFET Implementation

The circuit shown in Figure 17.2a is a common-source amplifier with input resistor biasing. This circuit description means that the source is held constant (common-source) and that the bias points are set by the input resistors, R_{G1} and R_{G2}. In this circuit, the input resistors set the DC value for V_{GS} (called V_{GSQ}), which sets the DC value for I_D (called I_{DQ}). I_{DQ} sets the DC value of V_O (called V_{OQ}).

First, the circuit should be redrawn for DC biasing following the rules presented previously. The DC bias circuit is shown in Figure 17.2b. The AC voltage source is shorted and the gate capacitor is an open

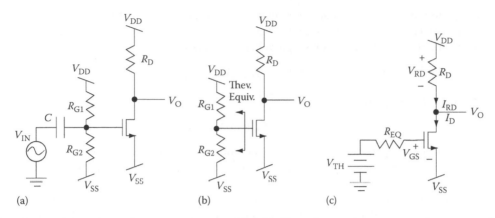

FIGURE 17.2 DC analysis of common-source amplifier. (a) Shows the complete circuit with an input voltage containing both AC and DC components, (b) shows the DC equivalent circuit, and (c) shows the Thevenin equivalent circuit of the circuit in (b).

circuit. It is evident at this point as to why the gate capacitor is necessary. The gate capacitor allows the input voltage to be capacitively coupled onto the gate on top of the DC bias point that is set by the gate resistors. Without the cap, in the DC, the gate would be tied to ground, which would prevent V_{GSQ} from being set to a desired voltage to set a desired I_{DQ}.

In order to obtain a clearer picture of how the gate resistors set V_{GSQ}, the circuit in Figure 17.2b is redrawn in Figure 17.2c using the Thevenin equivalent circuit of the bias resistors. The Thevenin equivalent voltage seen by the gate is just a voltage divider between R_{G1} and R_{G2}, and can be solved as follows using superposition of each DC source:

$$V_{TH} = (V_G \text{ with } V_{SS} \text{ turned off}) + (V_G \text{ with } V_{DD} \text{ turned off})$$

$$V_{TH} = V_{DD} \frac{R_{G2}}{R_{G1} + R_{G2}} + V_{SS} \frac{R_{G1}}{R_{G1} + R_{G2}} \tag{17.1}$$

The Thevenin equivalent resistance is found by turning off both DC sources (V_{DD} and V_{SS}) and finding the equivalent resistance seen by the gate. In this case, if both DC sources are turned off and tied to ground, and the other side of both resistors is connected to the gate, the resistors are in parallel. Therefore, the Thevenin equivalent resistance is

$$R_{EQ} = R_{G1} \,||\, R_{G2} \tag{17.2}$$

With the circuit in Figure 17.2c, it is simple to find V_{GSQ} by using KVL on the input loop:

$$V_{TH} = V_{REQ} + V_{GSQ} + V_{SS} \tag{17.3}$$

Since $I_G = 0$, there is no voltage drop across R_{EQ}, therefore $V_{GSQ} = V_{TH} - V_{SS}$. The calculation of R_{EQ} is unnecessary for the FET circuit since it has no impact on V_{GSQ}. However, R_{G1} and R_{G2} are still required to set V_{TH} and thereby V_{GSQ}. Additionally, the complete Thevenin equivalent transformation process is presented here to show the parallels and differences between this circuit and the BJT equivalent circuit.

Continuing with the analysis, the value of I_{DQ} can be found from V_{GSQ} using the saturation equation:

$$I_{DQ} = \frac{K'}{2}\left(\frac{W}{L}\right)(V_{GSQ} - V_t)^2 \tag{17.4}$$

Once I_{DQ} is found, V_{OQ} can be found using circuit analysis at the output of the circuit. Recall, V_{DS} can be one of a number of different values, so its value cannot be assumed. Therefore, V_{OQ} must be found using Ohm's law and KVL from the top supply voltage as follows:

$$V_{DD} - V_{RD} = V_{OQ} \tag{17.5}$$

By KCL at V_O, $I_{RD} = I_{DQ}$, so

$$V_{DD} - I_{DQ}R_D = V_{OQ} \tag{17.6}$$

The transistor device parameters, $K'/2$ and V_T, depend on the fabrication process and therefore are typically known. W/L is a design parameter for integrated circuit designers. If a discrete transistor is being used, W/L is often included in the K' parameter.

If the circuit has been designed, the bias points can be calculated by Equations 17.2 through 17.6 since R_{G1}, R_{G2}, R_D, and V_{DD} will all be known. If the goal is to design an amplifier, a value for V_{OQ} and I_{DQ} must be chosen first, then the resistors and supplies can be chosen using Equations 17.2 through 17.6. The method to design an amplifier will be covered in Section 17.5.

17.2.1.2 BJT Implementation

The circuit shown in Figure 17.3a is a BJT version of the same circuit, called a common-emitter amplifier with input resistor biasing. In this case, the emitter is held constant (common-emitter) and that the bias points are set by the input resistors, R_{B1} and R_{B2}. Similar to the FET circuit, the input resistors set the DC value for V_{BE} (called V_{BEQ}) that sets the DC value for I_C (called I_{CQ}). I_{CQ} sets the DC value of V_O (called V_{OQ}). However, one difference in the analysis of the BJT circuit and FET circuit is due to the exponential relationship between V_{BE} and I_C. Since I_C changes exponentially with V_{BE}, a 0.1 V change in V_{BE} causes a 100 times change in I_C. As a result, obtaining an accurate calculation of I_{CQ} as a function of V_{BEQ} is difficult. Therefore, the circuit analysis will calculate for I_{BQ} using an assumed value for V_{BE}. Then, I_{CQ} will be found from I_{BQ}.

The basic steps are identical to the FET example. First, the circuit should be redrawn for DC biasing as shown in Figure 17.3b. The AC voltage source is shorted and the base capacitor is open circuited. In order to obtain a clearer picture of how the base resistors set I_{BQ}, the circuit in Figure 17.3b is redrawn in Figure 17.3c using the Thevenin equivalent circuit of the bias resistors. To find the Thevenin equivalent circuit, break the circuit at the base. The Thevenin equivalent calculation is identical to the FET calculation as follows:

$$V_{TH} = (V_B \text{ with } V_{EE} \text{ turned off}) + (V_B \text{ with } V_{CC} \text{ turned off})$$

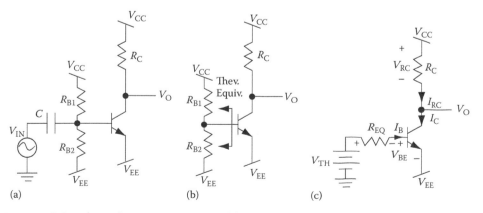

FIGURE 17.3 DC analysis of common-emitter amplifier. (a) Shows the complete circuit with an input voltage containing both AC and DC components, (b) shows the DC equivalent circuit, and (c) shows the Thevenin equivalent circuit of the circuit in (b).

$$V_{TH} = V_{CC} \frac{R_{B2}}{R_{B1} + R_{B2}} + V_{EE} \frac{R_{B1}}{R_{B1} + R_{B2}} \qquad (17.7)$$

The Thevenin equivalent resistance is found in an identical manner as the FET circuit as well:

$$R_{EQ} = R_{B1} \parallel R_{B2} \qquad (17.8)$$

Using the circuit in Figure 17.3c, find I_{BQ} using KVL on the input loop. Since current is going into the base, the voltage across R_{EQ} drops from V_{TH} to V_B, resulting in the following equation:

$$V_{TH} = I_{BQ} R_{EQ} + V_{BEQ} + V_{EE} \qquad (17.9)$$

A value for V_{BEQ} must be assumed. If the circuit is expected to have an I_C on the order of magnitude of 1 mA, $V_{BE} = 0.7\,\mathrm{V}$ is a common assumption. If the circuit is expected to produce an I_C on the order of magnitude of 100 mA, $V_{BE} = 0.8\,\mathrm{V}$ can be used. Rearranging 17.9, I_{BQ} can be solved for as follows:

$$I_{BQ} = \frac{V_{TH} - 0.7 - V_{EE}}{R_{EQ}} \qquad (17.10)$$

Once I_{BQ} has been calculated, I_{CQ} is found from the active region equation:

$$I_{CQ} = \beta \cdot I_{BQ} \qquad (17.11)$$

Once I_{CQ} is found, V_{OQ} can be found in the same manner as the FET circuit. Recall, V_{CE} can be one of a number of different values, so its value cannot be assumed. Therefore, V_{OQ} must be found using Ohm's law and KVL from the top supply voltage as follows:

$$V_{CC} - V_{RC} = V_{OQ} \qquad (17.12)$$

By KCL, $I_{RC} = I_{CQ}$, so

$$V_{CC} - I_{CQ} R_C = V_{OQ} \qquad (17.13)$$

17.2.2 Common-Emitter Amplifier with Emitter Resistor and Load Resistor

17.2.2.1 Implementation Using the npn

The circuit in Figure 17.3 works well when a single amplifier is needed. However, for the design to be replicated, device parameter variation can cause problems with the design. The parameter β can vary significantly from transistor to transistor due to manufacturing variation. If β varies from device to device, I_{CQ} will change from circuit to circuit. The variation in I_{CQ} will cause a variation in V_{OQ}, and in the worst case, push a transistor into the saturation region in the DC condition, making the amplifier useless. In order to fix this problem, a resistor is added between the emitter and the negative supply, shown in Figure 17.4a, which provides negative feedback. If an increase in β causes an increase in I_{CQ}, it also causes the increase in I_{EQ}. The increase in I_{EQ} causes a larger voltage drop across R_E, which by KVL around the input loop, decreases V_{BE} slightly. This decrease in V_{BE} brings I_{CQ} back down closer to the I_{CQ} with the lower β value.

The process for the DC analysis follows the previous example identically, except now the emitter resistor must be accounted for. The DC circuit is drawn and the Thevenin equivalent circuit is found resulting in the circuit in Figure 17.4b. KVL around the input loop produces

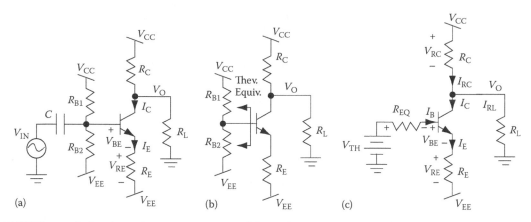

FIGURE 17.4 DC analysis of common-emitter amplifier with emitter resistor. (a) Shows the complete circuit with an input voltage containing both AC and DC components, (b) shows the DC equivalent circuit, and (c) shows the Thevenin equivalent circuit of the circuit in (b).

$$V_{TH} = I_{BQ}R_{EQ} + V_{BEQ} + V_{RE} + V_{EE} \tag{17.14}$$

If β is large, I_{EQ} is approximately equal to I_{CQ}, so Equation 17.14 can be rewritten as

$$V_{TH} = I_{BQ}R_{EQ} + V_{BEQ} + I_{CQ}R_{E} + V_{EE} \tag{17.15}$$

Using the active region equation for I_C and I_B, Equation 17.15 can be rewritten for I_C as

$$V_{TH} = \frac{I_{CQ}R_{EQ}}{\beta} + V_{BEQ} + I_{CQ}R_{E} + V_{EE} \tag{17.16}$$

Using this equation, I_{CQ} can be found and thereby V_{OQ}. With no load resistor, finding V_{OQ} would follow the same procedure shown in Section 17.2.1. In this example, a load resistor was added to examine its impact on the circuit. With the load resistor, V_{OQ} must be found using KCL at the output node as follows:

$$I_{RC} = I_{CQ} + I_{RL} \tag{17.17}$$

Using Ohm's law, the equation can be rewritten as

$$\frac{V_{CC} - V_{OQ}}{R_C} = I_{CQ} + \frac{V_{OQ}}{R_L} \tag{17.18}$$

Rearranging yields

$$\frac{V_{CC}}{R_C} - I_{CQ} = V_{OQ}\left(\frac{1}{R_C} + \frac{1}{R_L}\right) \tag{17.19}$$

Solving for V_{OQ} yields

$$V_{OQ} = \left(\frac{V_{CC}}{R_C} - I_{CQ}\right)\left(\frac{1}{(1/R_C) + (1/R_L)}\right) = \left(\frac{V_{CC}}{R_C} - I_{CQ}\right)(R_C \parallel R_L) \tag{17.20}$$

17.2.2.2 Implementation Using the pnp

Figure 17.5a shows the same circuit using a pnp. Notice, to achieve the pnp version of the same circuit, the npn circuit is flipped vertically and the transistor is replaced with a pnp. Since it is effectively the same circuit, the procedure for the circuit analysis is identical to the npn; however, care must be given to voltage and current directions. To perform the DC analysis, the DC circuit should be drawn (Figure 17.5b), the Thevenin equivalent circuit determined (Figure 17.5c), and then the analysis should proceed as follows:

Recall that a positive V_{EB} turns the transistor on and I_{BQ} comes out of the base of a pnp, resulting in a voltage drop from the base to V_{TH}. KVL around the input loop produces

$$V_{CC} - V_{RE} - V_{EBQ} - I_{BQ}R_{EQ} = V_{TH} \tag{17.21}$$

Using the same large β assumption, Equation 17.21 can be rewritten as

$$V_{CC} - I_{CQ}R_E - V_{EBQ} - \frac{I_{CQ}R_{EQ}}{\beta} = V_{TH} \tag{17.22}$$

Once I_{CQ} is calculated, V_{OQ} can be found using KCL at the output node as follows:

$$I_{CQ} = I_{RQ} + I_{RL} \tag{17.23}$$

Using Ohm's law, the equation can be rewritten as

$$I_{CQ} = \frac{V_{OQ} - V_{EE}}{R_C} + \frac{V_{OQ}}{R_L} \tag{17.24}$$

Solving for V_{OQ} yields

$$V_{OQ} = \left(I_{CQ} + \frac{V_{EE}}{R_C} \right)(R_C \parallel R_L) \tag{17.25}$$

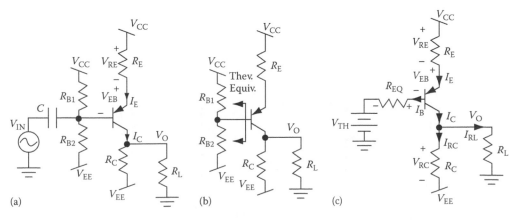

FIGURE 17.5 DC analysis of pnp common-source amplifier with emitter resistor. (a) Shows the complete circuit with an input voltage containing both AC and DC components, (b) shows the DC equivalent circuit, and (c) shows the Thevenin equivalent circuit of the circuit in (b).

17.2.3 High-Gain Amplifier with Current Source Biasing and Capacitively Coupled Load

17.2.3.1 MOSFET Implementation

The circuit shown in Figure 17.6a is called a common-source amplifier with current source biasing. This circuit also solves the parameter variation problem described in Section 17.2.2. In this circuit, the current source draws current, pulling charge off the source capacitor to drop V_{SQ} until $I_{SQ} = I_{BIAS}$. As a result, V_{GSQ} is set by the current source. Since the output current is set directly by the bias current, and not V_{GS}, the output bias points (I_{DQ} and V_{OQ}) do not strongly depend on the transistor parameters. Therefore, the bias point problems associated with parameter variation disappear.

To analyze this circuit, draw the DC circuit as shown in Figure 17.6b. By KCL,

$$I_{SQ} = I_{BIAS}$$

Since $I_{DQ} = I_{SQ}$, for a FET

$$I_{DQ} = I_{BIAS} \tag{17.26}$$

Since the load capacitor is an open circuit in the DC condition, V_{CQ} is found in the same manner as in Section 17.2.1:

$$V_{DQ} = V_{DD} - I_{DQ}R_{D} \tag{17.27}$$

Since there is no current going through the load resistor:

$$V_{OQ} = 0 \tag{17.28}$$

This circuit uses a capacitor to couple V_{DQ} to the load resistor. The advantage of this configuration is that V_{OQ} is equal to zero, which means the output will oscillate around zero. Additionally, the load resistor does not impact the DC biasing of the amplifier. Any of the previous circuits could use a similar load configuration and use the same analysis technique once I_{DQ} has been determined.

17.2.3.2 BJT Implementation

The same circuit can be implemented using a BJT, shown in Figure 17.7a, which is called a common-emitter amplifier with current source biasing. The circuit in DC configuration is shown in Figure 17.7b. The analysis of the circuit is very similar to the FET. By KCL

$$I_{EQ} = I_{BIAS} \tag{17.29}$$

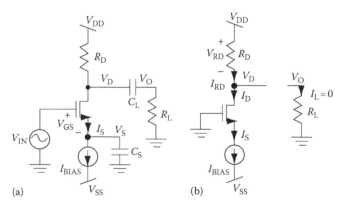

(a)   (b)

FIGURE 17.6 DC analysis of common-source circuit with current source biasing. (a) Shows the complete circuit with an input voltage containing both AC and DC components and (b) shows the DC equivalent circuit.

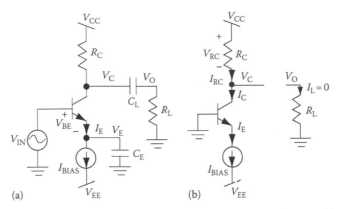

FIGURE 17.7 DC analysis of common-emitter amplifier with current source biasing. (a) Shows the complete circuit with an input voltage containing both AC and DC components and (b) shows the DC equivalent circuit.

Since $I_{CQ} = \alpha I_{EQ} = (I_{EQ}\beta)/(\beta + 1)$, if β is large, then

$$I_{CQ} \cong I_{BIAS} \tag{17.30}$$

The collector voltage, V_{CQ}, and output voltage, V_{OQ}, can be found identically to the FET example:

$$V_{CQ} = V_{CC} - I_{CQ}R_C \tag{17.31}$$

$$V_{OQ} = 0 \tag{17.32}$$

17.2.3.3 Current Source Implementation

The current source can be implemented using a transistor biased with input resistors to produce the desired current, as shown in Figure 17.8a or a current mirror, shown in Figure 17.8b and c. The output voltage of the current mirror, V_Y in each schematic, would be connected to V_S in Figure 17.6 or V_E in Figure 17.7.

The circuit in Figure 17.8a uses bias resistors, R_{G1} and R_{G2}, to set the value of V_{GS} that sets I_{BIAS}. The addition of R_S is to modify V_{GS} to account for process variations, similar to the emitter resistor in Section 17.2.2. These resistors are chosen using the Thevenin equivalent circuit, similar to the analysis in Section 17.2.1, except the voltage across R_S must also be considered in the input loop:

$$V_{TH} - I_G R_{TH} - V_{GS} - I_S R_S = V_{EE} \tag{17.33}$$

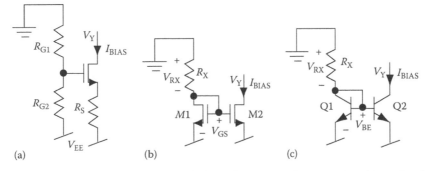

FIGURE 17.8 Current source implementations. (a) Shows the common discrete resistor biasing implementation, (b) shows the integrated circuit current mirror implementation with nFETs, and (c) shows the integrated circuit current mirror implementation with BJTs.

Since $I_G = 0$ and $I_D = I_S = I_{BIAS}$, the equation can be simplified to

$$V_{TH} - V_{GS} - I_{BIAS}R_S = V_{EE} \tag{17.34}$$

Note that one difference between this circuit and the circuits in Sections 17.2.1 and 17.2.2 is that the top supply for the bias resistor–voltage divider is connected to ground, not V_{CC}, to minimize power consumption. This difference will impact the calculation of the Thevenin equivalent voltage as follows:

$$V_{TH} = V_{EE} \frac{R_{G1}}{R_{G1} + R_{G2}} \tag{17.35}$$

The current source in Figure 17.8a could be implemented using a MOSFET, BJT, or JFET. Often JFETs are used to implement the most accurate current mirrors. The process for designing a JFET mirror would be the same as the MOSFET, except the current–voltage relationship for a JFET (see Equation 17.36) would be used to find V_{GS} instead of the current–voltage relationship of the MOSFET:

$$I_D = I_{DSS} \left(1 - \frac{V_{GS}}{V_P}\right)^2 \tag{17.36}$$

where I_{DSS} and V_P are process parameters much like K' and V_t for a MOSFET.

The current mirror in Figure 17.8b uses R_X and a diode-connected transistor (M1) to set the value of I_{BIAS}. Because M1 is diode-connected, the current through M1 sets the input voltage, V_{GS}. Since the gates of M1 and M2 are connected together and the sources of M1 and M2 are connected to the same voltage (V_{EE}), both transistors have the same V_{GS}, which means they will have the same current (e.g., $I_{D1} = I_{D2} = I_{BIAS}$). For the current mirror to work properly, the parameters for M1 must be closely matched to M2 so that I_{D1} will equal I_{D2}.

To design this current mirror, choose a value for R_X that sets the desired value for I_{BIAS} in M1 as follows.

KVL around M1 produces

$$0 - V_{RX} - V_{GS} = V_{EE} \tag{17.37}$$

The value for V_{GS} can be found by knowing I_{D1}, which is equal to I_{BIAS}, and using the MOSFET current–voltage relationship ($I_D = (K'/2)(W/L)(V_{GS} - V_t)^2$) rewritten for V_{GS}:

$$0 - I_{BIAS}R_X - \left(\sqrt{\frac{2I_D}{K'(W/L)}} + V_t\right) = V_{EE} \tag{17.38}$$

Rearranging produces

$$R_X = \frac{-V_{EE} - \left(\sqrt{\left(2I_D/K'(W/L)\right)} + V_t\right)}{I_{BIAS}} \tag{17.39}$$

The MOSFETs in the mirror can also be implemented using BJTs as shown in Figure 17.8c. This current mirror acts similar to the FET mirror. R_X sets the current through the diode-connected transistor (Q1) that sets V_{BE}. Since $V_{BE1} = V_{BE2}$, $I_{C1} \cong I_{C2}$ (if base currents are small).

A similar analysis can be used to find the value for R_X. KVL around Q1 produces

$$0 - V_{RX} - V_{BE} = V_{EE} \tag{17.40}$$

If β is large, base currents are approximately zero and $I_{C1} \cong I_{BIAS}$. Using Ohm's law on R_X results in

$$0 - I_{BIAS}R_X - V_{BE} = V_{EE} \tag{17.41}$$

Rearranging produces

$$R_X = \frac{-V_{EE} - V_{BE}}{I_{BIAS}} \tag{17.42}$$

where V_{BE} is the turn-on voltage of the transistor, typically between 0.5 and 0.8 V.

17.2.4 Alternative High-Gain Amplifier

The next example shows another method of creating a high-gain amplifier.

17.2.4.1 MOSFET Implementation

The circuit shown in Figure 17.9a is called a common-gate amplifier. This circuit is similar to the common-source amplifier with current source biasing, except that gate is held constant while the source oscillates. A coupling capacitor, C_S, is used to isolate the AC input signal from the DC biasing, which is set by the current source. The DC model of this circuit is shown in Figure 17.9b. Notice that the DC circuit looks identical to Figure 17.6b, where $V_D = V_O$. The analysis is identical to the analysis shown in Section 17.2.3. The fundamental differences between the common-source and the common-gate circuits will be discussed in the AC section, 17.3.4.

17.2.4.2 BJT Implementation

The circuit shown in Figure 17.10a is called a common-base amplifier, since the base is held constant. This circuit structure is the BJT equivalent of the common-gate amplifier. Like the common-gate amplifier, the DC model of the circuit (Figure 17.10b) is identical to the common-emitter amplifier shown in Figure 17.7b, except $V_C = V_O$. The analysis follows the analysis shown in Section 17.2.3.

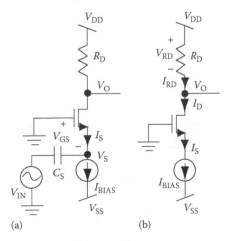

(a)   (b)

FIGURE 17.9 DC analysis of common-gate amplifier. (a) Shows the complete circuit with an input voltage containing both AC and DC components and (b) shows the DC equivalent circuit.

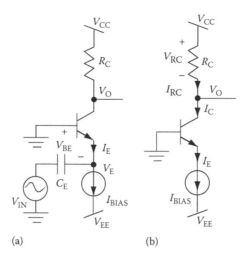

(a)   (b)

FIGURE 17.10 DC analysis of common-base amplifier. (a) Shows the complete circuit with an input voltage containing both AC and DC components and (b) shows the DC equivalent circuit.

17.2.5 Voltage Follower

17.2.5.1 MOSFET Implementation

The final circuit example is a voltage follower, shown in Figure 17.11a. In this circuit, V_O follows V_{IN} with a DC offset. This circuit is also commonly called a common-drain amplifier, since the drain is held constant. The DC configuration of this circuit is shown in Figure 17.11b.

To find the bias points, perform KCL at the output node:

$$I_{DQ} = I_{SQ} = I_{BIAS} + I_L \tag{17.43}$$

Using Ohm's law, substitute for I_L as follows:

$$I_{DQ} = I_{BIAS} + \frac{V_{OQ}}{R_L} \tag{17.44}$$

Find an expression for V_{OQ} using KVL and the FET saturation equation:

$$V_{OQ} = 0 - V_{GS} \tag{17.45}$$

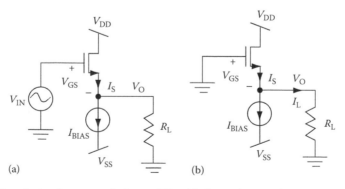

(a)   (b)

FIGURE 17.11 DC analysis of common-drain amplifier. (a) Shows the complete circuit with an input voltage containing both AC and DC components and (b) shows the DC equivalent circuit.

By rewriting the FET saturation equation as follows:

$$V_{GS} = \sqrt{\frac{2I_D}{K'(W/L)}} + V_t \tag{17.46}$$

and substituting into Equation 17.45, an expression for V_{OQ} can be found as follows:

$$V_{OQ} = -\left(\sqrt{\frac{2I_{DQ}}{K'(W/L)}} + V_t\right) \tag{17.47}$$

This expression for V_{OQ} can substituted into Equation 17.44 to find I_{DQ} as follows:

$$I_{DQ} = I_{BIAS} - \frac{\sqrt{\left(2I_{DQ}/K'(W/L)\right)} + V_t}{R_L} \tag{17.48}$$

I_{DQ} can be found using the root function on a calculator or math program or it can be solved iteratively. To perform the iterative solution, guess at a value for I_D, place it into the right-hand side of the equation. Solve for I_D. Use this answer as your next guess. Keep repeating this process until the answer converges. Since I_{DQ} is under the square root, the equation will converge, and generally does so within a few iterations. Once I_{DQ} is determined, V_{OQ} can be found using Equation 17.47.

17.2.5.2 BJT Implementation

The BJT implementation of the voltage follower, called the common-collector amplifier, is shown in Figure 17.12a. Like the FET common-drain amplifier, V_O follows V_{IN} with a DC offset. The DC configuration of this circuit is shown in Figure 17.12b.

Like the FET circuit, the bias points are found by performing KCL at the output node:

$$I_{EQ} = I_{BIAS} + \frac{V_{OQ}}{R_L} \tag{17.49}$$

Find an expression for V_{OQ} using KVL and the V_{BE} assumption of 0.6–0.8 V:

$$V_{OQ} = 0 - V_{BE} \cong -0.7 \text{ V} \tag{17.50}$$

Substitute Equation 17.50 into Equation 17.49 to get

$$I_{EQ} = I_{BIAS} + \frac{-0.7 \text{ V}}{R_L} \tag{17.51}$$

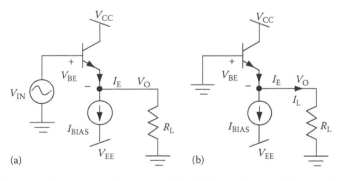

FIGURE 17.12 DC analysis of common-collector amplifier. (a) Shows the complete circuit with an input voltage containing both AC and DC components and (b) shows the DC equivalent circuit.

I_{CQ} can be found from I_{EQ}:

$$I_{CQ} = \frac{\beta}{\beta+1}\left(I_{BIAS} - \frac{0.7}{R_L} \right) \tag{17.52}$$

If β is large, $I_{CQ} \cong I_{EQ}$:

$$I_{CQ} \cong \left(I_{BIAS} - \frac{0.7}{R_L} \right) \tag{17.53}$$

17.3 AC Analysis

17.3.1 Small-Signal Model

Amplification in an amplifier occurs when a small change in the input voltage causes a large change in output current. Because the input voltage (V_{GS} for FETs and V_{BE} for BJTs) changes by a small amount, the nonlinear I–V relationship (FET square-law, BJT exponential) can be approximated by a linear function. As a result, the design and analysis of transistors in amplifier circuits use a linear, "small-signal" model to represent the nonlinear "large-signal" model of the transistor. The linear, small-signal models for a FET and a BJT are shown in Figure 17.13a and b, respectively.

The input resistance of the transistor, r_π for the BJT, is found by approximating the input I–V relationship for the BJT as a linear function around the bias point (e.g., taking the derivative of I_B versus V_{BE} at I_{BQ}) and inverting the result to get a resistance as follows:

$$r_\pi = \frac{1}{\partial I_B / \partial V_{BE}} = \frac{\beta V_T}{I_{CQ}} \tag{17.54}$$

The input resistance of the FET is infinite since the gate current is zero.

The transconductance of the transistor (how well a change in the input voltage turns into a change in output current) is modeled by g_m, which is found by approximating the transconductance of the BJT as a linear function evaluated at I_{DQ}:

$$g_{mFET} = \frac{\partial I_D}{\partial V_{GS}} = \sqrt{2K'\frac{W}{L}I_D} \tag{17.55}$$

$$g_{mBJT} = \frac{\partial I_C}{\partial V_{BE}} = \frac{I_{CQ}}{V_T} \tag{17.56}$$

The output resistance of the transistor, modeled by r_O, is found by approximating the output I–V relationship as a linear function at I_{CQ} and inverting the result to get a resistance. In the ideal case, the output current (I_D or I_C) does not change with output voltage (V_{DS} or V_{CE}). However, due to

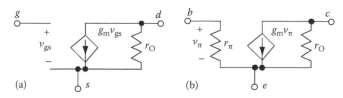

FIGURE 17.13 Small signal model for a FET (a) and a BJT (b).

channel-width and base-width modulation, a small linear change in output current is evident for a change in output voltage. As a result, the output resistance is finite and evaluated as a function of the channel-width/base-width parameters as follows:

$$r_{\text{OFET}} = \frac{1}{\partial I_D / \partial V_{DS}} = \frac{1}{\lambda I_{DQ}} \tag{17.57}$$

$$r_{\text{OBJT}} = \frac{1}{\partial I_C / \partial V_{CE}} = \frac{V_A}{I_{CQ}} \tag{17.58}$$

This small-signal model is used to find an expression for the AC gain, input resistance, and output resistance. The small-signal model of a circuit is drawn using the following rules:

1. Replace the transistor with the small-signal model. Beginners should label the terminal nodes on the small-signal model and on the schematic to keep track of which node goes where in the model.
2. To find the maximum AC gain or resistance, assume that the circuit is operating at a sufficiently high enough frequency to short all capacitors, and a sufficiently low enough frequency that the internal device capacitors are still open circuits. Therefore, short all discrete capacitors.
3. Turn off all DC sources to apply superposition. As stated earlier, voltage sources become short circuits ($V = 0$) and current sources become open circuits ($I = 0$). Note: this means an "AC ground" still has a DC voltage, but the voltage never *changes* from the DC condition.
4. Draw all remaining components connected to each node of the transistor model.

The examples in the DC section will be used again to demonstrate the AC analysis. Each example will start with a general explanation or intuition of how the circuit produces gain using only the most basic circuit analysis and an understanding of how each device operates. Then, the small-signal analysis will provide an expression for the gain, which should reinforce the intuition. Input and output resistance examples will proceed as well.

17.3.2 High-Gain Amplifier with Input Resistor Biasing

17.3.2.1 MOSFET Implementation

The common-source amplifier with input resistor biasing is shown again in Figure 17.14a. In this circuit, the source is held constant by the negative supply. Therefore, any changes to the gate voltage directly impact V_{GS} and therefore I_D. v_{IN} is coupled onto the gate through the gate capacitor. When v_{IN} increases, V_{GS} increases, causing an increase in I_D. When v_{IN} decreases, V_{GS} decreases, causing a decrease in I_D. An increase in I_D causes a larger voltage drop across R_D, which causes v_O to drop. A decrease in I_D causes a smaller voltage drop across R_D, which causes v_O to rise. As a result, an increase in v_{IN} causes a decrease in v_O, resulting in a phase inversion, or negative gain. The final gain depends on three things: (1) how well a change in v_{IN} is translated to a change in V_{GS}, (2) how well a change in V_{GS} is translated into a change in I_D, and (3) how well the change in I_D is translated into a change in v_O.

Figure 17.14b shows the small-signal model of the common-source amplifier. Both supplies are turned off, tying these nodes to ground. As a result, R_{G1} is in parallel with R_{G2} from the gate to ground. The input capacitor is shorted, connecting v_{IN} to the gate. Since the negative supply is turned off, the source is connected to ground. Since the positive supply is turned off, R_D goes from the drain to ground.

Perform circuit analysis to find the desired AC parameter. The AC gain is determined by finding v_O / v_{IN}. To find the gain, start by finding an expression for v_O in terms of known parameters (such as R_D, g_m, r_O, R_{G1}, R_{G2}) and v_{IN}. Once an expression is found where every term is a function of v_O or v_{IN}, algebraic manipulation can provide the gain. The analysis of this circuit proceeds as follows.

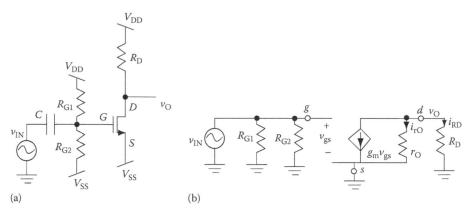

FIGURE 17.14 AC analysis of common-source amplifier. (a) Shows the complete circuit and (b) shows the AC small-signal model.

Find v_O using KCL at v_O to get

$$g_m v_{gs} + i_{RD} + i_{r_O} = 0 \tag{17.59}$$

Using Ohm's law,

$$i_{RD} = \frac{v_O}{RD} \tag{17.60}$$

$$i_{r_O} = \frac{v_O}{r_O} \tag{17.61}$$

Substituting Equations 17.60 and 17.61 into Equation 17.59 produces

$$g_m v_{gs} + v_O \left(\frac{1}{R_D} + \frac{1}{r_O} \right) = 0 \tag{17.62}$$

Rearranging yields

$$v_O = -g_m v_{gs} \frac{1}{(1/R_D) + (1/r_O)} = -g_m v_{gs} (R_D \| r_O) \tag{17.63}$$

The only unknown parameter in this equation is v_{gs}. Next, find an expression for v_{gs}. Since R_{G1} and R_{G2} are in parallel with v_{IN}, these resistors do not have an impact on the input voltage. Therefore,

$$v_{gs} = v_{IN} \tag{17.64}$$

Substituting Equation 17.64 into Equation 17.63 and solving for v_O/v_{IN} provides the gain as follows:

$$A_V = \frac{v_O}{v_{IN}} = -g_m (R_D \| r_O) \tag{17.65}$$

Note that the final expression of the gain supports the intuitive description provided earlier. The gain depended upon three components, all represented in the final gain equation. In this circuit, v_{IN} is directly coupled into the gate, therefore $v_{IN} = v_{gs}$. The parameter g_m represents how well a change in v_{gs}

translates into a change in I_D. R_D translates the change in I_D into a change in v_O. Therefore, the gain is a function of g_m and R_D.

The input resistance can be found using the same circuit shown in Figure 17.14b. However, to find an expression for this specification, replace v_{IN} with a test voltage, V_T, and find the current going through the test voltage, I_T. An expression for the input resistance is found by solving for V_T/I_T. The analysis for r_{in} follows.

First, simplify the circuit showing parallel resistor combinations (R_{G1}/R_{G2} and R_D/r_O), as seen in Figure 17.15a.

Find an expression for I_T using KCL and Ohm's law at the input node:

$$I_T = \frac{V_T}{R_{G1} \| R_{G2}} \tag{17.66}$$

Rearranging and solving for V_T/I_T yields

$$r_{IN} = \frac{V_T}{I_T} = R_{G1} \| R_{G2} \tag{17.67}$$

The output resistance can be found in similar manner as the input resistance, except the test voltage is placed at the output node, as shown in Figure 17.15b, and the input voltage is turned off. The analysis for r_{OUT} follows.

Find an expression for I_T using KCL as the output node:

$$I_T = I_{R_D\|r_O} + g_m v_{gs} = \frac{V_T}{R_D \| r_O} + g_m v_{gs} \tag{17.68}$$

Since the input voltage is turned off, v_g and v_s are both tied to ground. Thereby, $v_{gs} = 0$. As a result, $g_m v_{gs} = 0$. Using this relationship, rearranging, and solving for V_T/I_T yields

$$r_{OUT} = \frac{V_T}{I_T} = R_D \| r_O \tag{17.69}$$

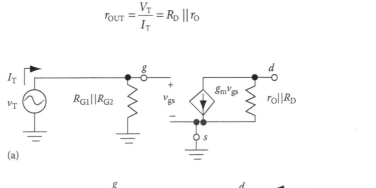

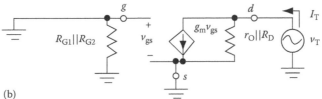

FIGURE 17.15 AC model for input resistance (a) and output resistance (b) of common-source amplifier.

17.3.2.2 BJT Implementation

The equivalent BJT circuit, the common-emitter amplifier with input resistor biasing, is shown again in Figure 17.16a. The basic operation is identical to the FET example. The emitter is held constant by the negative supply allowing any changes to the base voltage to directly impact V_{BE} and therefore I_C. v_{IN} is coupled onto the base through the base capacitor. A change in v_{IN} causes a change in V_{BE} causing a change in I_C. A change in I_C causes a change in voltage across R_C, which causes v_O to change. Similar to the FET circuit, the BJT amplifier will exhibit negative gain; e.g., an increase in v_{IN} will cause an increase in I_C, which causes a decrease in v_O.

Figure 17.16b shows the small-signal model of the common-emitter amplifier. Note that the circuit looks identical to the FET small-signal model circuit, except that the transistor model uses the BJT small-signal model rather than the FET small-signal model. It is also interesting to note that the only significant difference between the FET and BJT small-signal model is the finite input resistance r_π.

Since the FET small-signal model and the BJT small-signal model circuit are so similar, the analysis is similar as well. To find the gain, note that R_C and r_O are in parallel and the current $g_m v_\pi$ runs through them. Given the direction that $g_m v_\pi$ goes, the voltage across R_C is negative to v_O. Evaluate v_O as follows:

$$v_O = -g_m v_\pi \left(R_C \| r_O \right) \tag{17.70}$$

R_{B1}, R_{B2}, and r_π are in parallel with v_{IN} and v_π resulting in

$$v_\pi = v_{IN} \tag{17.71}$$

Substituting Equation 17.71 into Equation 17.70 yields the equivalent gain equation as the FET circuit:

$$A_V = \frac{v_O}{v_{IN}} = -g_m \left(R_C \| r_O \right) \tag{17.72}$$

The final expression of the gain for this circuit also supports the intuitive description. The gain depends upon how well Δv_{IN} translates to a ΔV_{BE}, how well ΔV_{BE} translates to a ΔI_C (represented in the gain equation by g_m), and how well a ΔI_C translates into a Δv_O (represented in the gain equation by R_C).

The simplified small-signal model for the input resistance is shown in Figure 17.17a. The analysis for input resistance is very similar to the FET, except the impact of r_π must be accounted for.

Find an expression for I_T using KCL and Ohm's law at the input node:

$$I_T = \frac{V_T}{R_{B1} \| R_{B2} \| r_\pi} \tag{17.73}$$

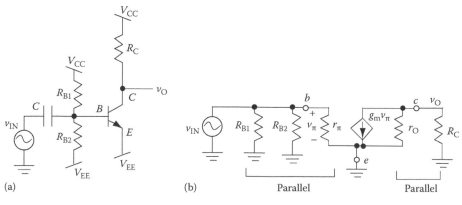

FIGURE 17.16 AC analysis of common-emitter amplifier. (a) Shows the complete circuit and (b) shows the AC small-signal model.

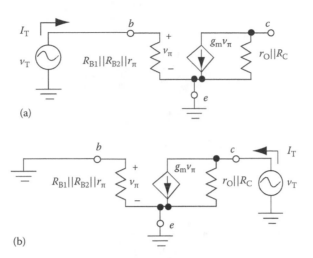

FIGURE 17.17 AC model for input resistance (a) and output resistance (b) of common-emitter amplifier.

Rearranging and solving for V_T/I_T yields the equivalent r_{in} as the FET circuit with the inclusion of r_π:

$$r_{IN} = \frac{V_T}{I_T} = R_{B1} \parallel R_{B2} \parallel r_\pi \tag{17.74}$$

The small-signal model for the output resistance is shown in Figure 17.17b. Find an expression for I_T using KCL and Ohm's law at the output node:

$$I_T = \frac{V_T}{R_C \parallel r_o} + g_m v_\pi \tag{17.75}$$

Since $v_\pi = 0$, $g_m v_\pi = 0$. Rearranging and solving for V_T/I_T yields the same answer found in the FET circuit:

$$r_{OUT} = \frac{V_T}{I_T} = R_C \parallel r_o \tag{17.76}$$

The trade-off between the FET and BJT circuits are now evident. Although the gain equations are identical for the FET and the BJT, g_m for a BJT is typically much larger than the g_m for a FET since this represents how well a change in input voltage translates to a change in output current. The fact that BJTs exhibit an exponential input voltage–output current relationship and FETs exhibit only a square-law input voltage–output current relationship typically makes $g_{mBJT} > g_{mFET}$ for the same bias condition. However, the input resistance of the FET circuit is typically larger than the input resistance of the BJT circuit. R_{B1} and R_{B2} are typically chosen to be large values to minimize the power consumption, making r_π the dominant resistor in the parallel combination for the BJT. As a result, $r_{in,FET} > r_{in,BJT}$.

17.3.3 Common-Emitter Amplifier with Emitter Resistor and Load Resistor

17.3.3.1 Implementation Using a npn

The common-emitter amplifier with emitter resistor is shown again in Figure 17.18a. In this circuit, the emitter is not held constant, which will hurt the gain since v_{IN} is not directly translated to V_{BE}. However, the amount of v_{IN} that is translated to V_{BE} is turned into a voltage gain in the same manner as the common-emitter amplifier previously described: the change in V_{BE} causes a change in I_C. The change in I_C causes a change in the voltage across R_C. The change in voltage across R_C causes a change in v_O in the opposite direction of v_{IN}.

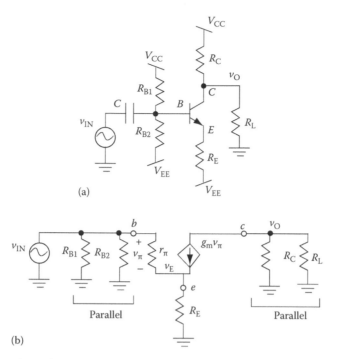

FIGURE 17.18   AC analysis of common-emitter amplifier with emitter resistor. (a) Shows the complete circuit and (b) shows the AC small-signal model.

Figure 17.18b shows the small-signal model of the common-emitter amplifier with an emitter resistor. The circuit looks identical to the previous common-emitter small-signal model circuit with the addition of the emitter resistor, R_E, between the emitter and ground and the load resistor, R_L, between the collector and ground. Because the change in current due to r_O is small compared to the change in current due to $g_m v_\pi$, and the analysis containing r_O is significantly more complex and thereby loses intuition, r_O will be neglected for this analysis.

The output of the circuit is very similar to the previous example. The current $g_m v_\pi$ goes through the parallel resistor combination of R_C and R_L to produce v_O. Using KCL at v_O and Ohm's law, the following expression for v_O can be obtained:

$$v_O = -g_m v_\pi (R_C \parallel R_L) \tag{17.77}$$

To complete the gain equation, an expression for v_π must be found in terms of v_{IN} and v_O. However, the emitter resistor has a significant impact the expression for v_π, since it changes the manner that v_{IN} translates into V_{BE}. The analysis at the input node proceeds as follows.

By KVL at the input loop:

$$v_{IN} = v_\pi + v_E \tag{17.78}$$

An expression for v_E can be found using KCL at v_E and Ohm's law:

$$\frac{v_\pi}{r_\pi} + g_m v_\pi = \frac{v_E}{R_E} \tag{17.79}$$

Rearranging, solve for v_E to get

$$v_E = R_E \left(\frac{v_\pi}{r_\pi} + g_m v_\pi \right) \tag{17.80}$$

Substituting Equation 17.80 into Equation 17.78 yields

$$v_{IN} = v_\pi + R_E \left(\frac{v_\pi}{r_\pi} + g_m v_\pi \right) = v_\pi \left(1 + R_E \left(\frac{1}{r_\pi} + g_m \right) \right) \tag{17.81}$$

Dividing Equation 17.77 by Equation 17.81 produces the final gain equation:

$$A_V = \frac{v_O}{v_{IN}} = \frac{-g_m (R_C \| R_L)}{1 + R_E((1/r_\pi) + g_m)} \tag{17.82}$$

The final gain equation still depends upon g_m and R_C because ΔV_{BE} translates to a ΔI_C (represented in the gain equation by g_m) and ΔI_C translates into a Δv_O (represented in the gain equation by R_C). However, because the emitter is not held constant, $\Delta v_{IN} \neq \Delta V_{BE}$. Since a change in I_C causes a change in V_E through R_E, the gain drops by $(1 + R_E (1/r_\pi + g_m))$.

The emitter resistor also impacts the input resistance, since r_π is no longer in parallel with R_{B1} and R_{B2}. The simplified small-signal model for the input resistance is shown in Figure 17.19a. The analysis proceeds as follows.

Find an expression for I_T using KCL and Ohm's law at the input node:

$$I_T = \frac{V_T}{R_{B1} \| R_{B2}} + \frac{v_\pi}{r_\pi} \tag{17.83}$$

KVL at the input loop provides the exact same expression for v_π that was found in Equation 17.78. Using the same analysis performed previously produces the same expression of V_T as a function of v_π as in Equation 17.81. Rearrange this equation to solve for v_π as a function of V_T:

$$v_\pi = \frac{V_T}{1 + R_E((1/r_\pi) + g_m)} \tag{17.84}$$

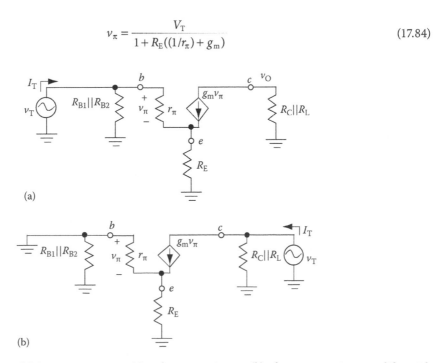

(a)

(b)

FIGURE 17.19 AC model for input resistance (a) and output resistance (b) of common-emitter amplifier with emitter resistance.

Substitute Equation 17.84 into Equation 17.83 and distribute r_π:

$$I_T = \frac{V_T}{R_{B1} \| R_{B2}} + \frac{V_T}{r_\pi + R_E + R_E r_\pi g_m} \tag{17.85}$$

This equation can be further simplified by using the following relationship:

$$r_\pi g_m = \frac{\beta V_t}{I_{CQ}} \cdot \frac{I_{CQ}}{V_t} = \beta \tag{17.86}$$

Substituting Equation 17.86 into Equation 17.85 produces the following equation:

$$I_T = \frac{V_T}{R_{B1} \| R_{B2}} + \frac{V_T}{r_\pi + R_E(1 + \beta)} \tag{17.87}$$

Solving for the input resistance results in the following relationship:

$$r_{IN} = \frac{V_T}{I_T} = \frac{1}{(1/(R_{B1} \| R_{B2})) + (1/(r_\pi + R_E(1 + \beta)))} = R_{B1} \| R_{B2} \| [r_\pi + R_E(1 + \beta)] \tag{17.88}$$

Although the emitter resistor drops the gain, it increases the input resistance. Since R_{B1} and R_{B2} are typically chosen to be large resistors to minimize power consumption, $(r_\pi + R_E (\beta + 1))$ typically dominates the parallel equation. Recall that the input resistance for the common-emitter amplifier without an R_E was $R_{B1} \| R_{B2} \| r_\pi \cong r_\pi$. Therefore, the emitter resistor increases the input resistance by a factor of approximately $R_E (\beta + 1)$.

The small-signal model for the output resistance is shown in Figure 17.19b. R_{B1} and R_{B2} are shorted when the input source is turned off, leaving r_π in parallel with R_E. However, voltage changes at v_C have no impact on v_E through the dependent current source, $g_m v_\pi$, which can have any voltage across it and still produce the same current. Therefore, changes in V_T will not cause any changes in v_π, resulting in $g_m v_\pi = 0$. The circuit analysis proof of this concept can be found by performing KCL at v_E:

$$g_m v_\pi = \frac{v_E}{r_\pi \| R_E} = \frac{-v_\pi}{r_\pi \| R_E} \tag{17.89}$$

Rearranging produces

$$v_\pi \left(g_m + \frac{1}{r_\pi \| R_E} \right) = 0 \tag{17.90}$$

Since $g_m + (1/r_\pi \| R_E)$ represent resistors that cannot be negative, Equation 17.90 can only be true if v_π is equal to 0.

With $g_m v_\pi = 0$, the output resistance follows the following relationship:

$$r_{OUT} = \frac{V_T}{I_T} = R_C \| R_L \tag{17.91}$$

17.3.3.2 Implementation Using a pnp

The common-emitter amplifier with an emitter resistor implemented with a pnp is redrawn in Figure 17.20a. This circuit works identically to the npn circuit with voltages and currents reversed. v_{IN} is coupled onto the base through the base capacitor. When v_{IN} increases, V_{EB} decreases, causing a decrease in I_C.

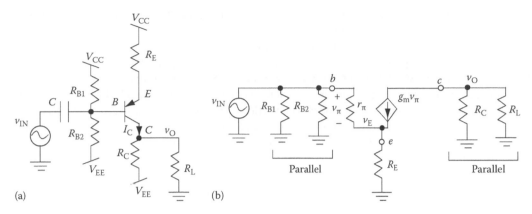

FIGURE 17.20 AC analysis of pnp common-source amplifier with emitter resistor. (a) Shows the complete circuit and (b) shows the AC small-signal model.

When v_{IN} decreases, V_{EB} increases, causing an increase in I_C. An increase in I_C causes a larger voltage drop across R_C, which causes v_O to increase. A decrease in I_C causes a smaller voltage drop across R_C, which causes v_O to drop. As a result, an increase in v_{IN} causes a decrease in v_O, resulting in a phase inversion, or negative gain. Since the emitter is not held constant due to the voltage changes across R_E, v_{IN} is not directly mapped onto V_{EB}, which causes the gain to decline identically to the npn circuit.

To draw the small-signal mode for the pnp common-emitter amplifier, start with the small-signal mode for the BJT (the pnp and the npn small-signal models are identical), as shown in Figure 17.20b. Since V_{CC} and V_{EE} are turned off, R_{B1} and R_{B2} both go from the base to ground. The capacitor is a short circuit connecting v_{IN} to the base. The emitter goes through R_E to ground. The collector goes through R_C to ground and through R_L to ground. Notice, the resulting small-signal circuit is identical to the npn small-signal circuit. As a result, all small-signal model equations, A_V, r_{in}, and r_{OUT}, are identical for npn and pnp circuits.

17.3.4 High-Gain Amplifier with Current Source Biasing and Capacitively Coupled Load

17.3.4.1 MOSFET Implementation

The common-source amplifier with current source biasing is shown again in Figure 17.21a. In this circuit, the source is held constant by the source capacitor, C_S. Therefore, any changes to the gate voltage correspond directly to V_{GS} and therefore I_D. As a result, this circuit operates identically to the common-source amplifier shown in Section 17.3.2. The gate is tied to v_{IN}. Therefore, a Δv_{IN} causes a ΔV_{GS}. A ΔV_{GS}

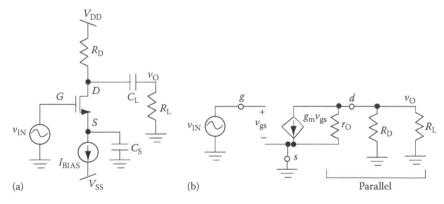

FIGURE 17.21 AC analysis of common-source amplifier with current source biasing. (a) Shows the complete circuit and (b) shows the AC small-signal model.

causes a ΔI_D. A ΔI_D causes a ΔV_{RD}, which causes a Δv_O. This circuit exhibits the same phase inversion as the common-source amplifier with input resistor biasing.

Figure 17.21b shows the small-signal model of this common-source amplifier. v_{IN} is connected to the gate. Since the source capacitor is shorted, the source is connected to ground. Recall that this means V_S does no change in the AC. It does *not* mean $V_S = 0$ in the DC. Since the positive supply is turned off, R_D goes from the drain to ground. Since the load capacitor is shorted, R_L is in parallel with R_D. The resulting circuit is identical to the small-signal model from Section 17.3.2, without R_{G1} and R_{G2}. Since R_{G1} and R_{G2} have no impact on how v_{IN} is translated to V_{GS}, the gain analysis is identical to Section 17.3.2, with the addition of the load resistor that is in parallel with R_D and r_O. As a result, the AC analysis produces the following gain equation:

$$A_V = \frac{v_O}{v_{IN}} = -g_m(R_D \,||\, r_O \,||\, R_L) \tag{17.92}$$

The simplified small-signal model for the input resistance is shown in Figure 17.22a. Without any bias resistors and an infinite gate resistance, the input resistance for this circuit is infinity:

$$r_{IN} = \frac{V_T}{I_T} = \frac{V_T}{0} = \infty \tag{17.93}$$

The small-signal model for the output resistance is shown in Figure 17.22b. This circuit is identical to the common-source example in Section 17.3.2 and, therefore, produces the exact same output resistance, with the addition of the load resistor in parallel with R_D and r_O:

$$r_{OUT} = \frac{V_T}{I_T} = R_D \,||\, r_O \,||\, R_L \tag{17.94}$$

The advantage of this circuit over the circuits in Sections 17.3.2 and 17.3.3 can now be appreciated. The current source biasing eliminates parameter variation while maintaining the largest possible gain and input resistance. As a result, this is the fundamental amplifier configuration that is used in integrated circuit implementations of amplifiers. However, the current source is difficult to make without matching devices. Therefore, discrete implementations may require the circuits from Sections 17.3.2 and 17.3.3.

17.3.4.2 BJT Implementation

The BJT common-emitter amplifier with current source biasing is shown in Figure 17.23a. This circuit operation is identical to the FET operation. A Δv_{IN} causes a ΔV_{BE}, which causes a ΔI_C. A ΔI_C causes a ΔV_{RC}, which causes a Δv_O. The small-signal model for the gain is shown in Figure 17.23b. Since r_π has no impact on the translation of v_{IN} to V_{BE}, the final AC gain is identical to the FET:

$$A_V = \frac{v_O}{v_{IN}} = -g_m(R_C \,||\, r_O \,||\, R_L) \tag{17.95}$$

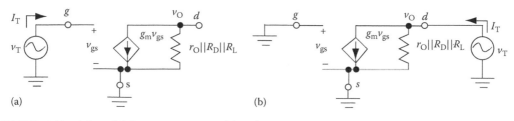

FIGURE 17.22 AC model for input resistance (a) and output resistance (b) of common-source amplifier with current source biasing.

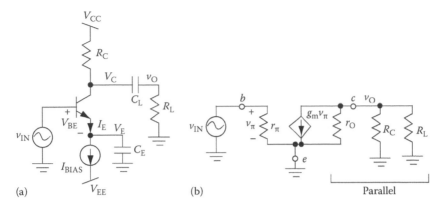

FIGURE 17.23 AC analysis of common-emitter amplifier with current source biasing. (a) Shows the complete circuit and (b) shows the AC small-signal model.

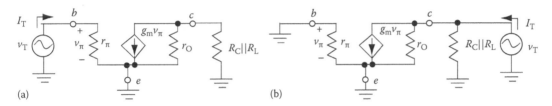

FIGURE 17.24 AC model for input resistance (a) and output resistance (b) of common-emitter amplifier with current source biasing.

The simplified small-signal model for the input resistance is shown in Figure 17.24a. The only resistance seen by the source is r_π. This result is actually similar to the input resistance in the circuit in Section 17.3.2 since r_π generally dominates the input resistance equation:

$$r_{IN} = \frac{V_T}{I_T} = r_\pi \tag{17.96}$$

The small-signal model for the output resistance is shown in Figure 17.24b. Since both sides of r_π are connected to ground, $V_\pi = 0$, which makes $g_m V_\pi = 0$. As a result, this circuit is identical to the FET example and therefore produces the exact same output resistance:

$$r_{OUT} = \frac{V_T}{I_T} = R_C \parallel r_O \parallel R_L \tag{17.97}$$

The trade-off between the FET and BJT circuits is identical to the trade-off presented in Section 17.3.2. Although the FET circuit has a much greater input resistance, the BJT circuit typically has a larger gain due to the larger value for g_m.

17.3.5 Alternative High-Gain Amplifier

17.3.5.1 MOSFET Implementation

Figure 17.25a shows once again the common-gate, high-gain amplifier configuration. In this circuit, changes in v_{IN} are capacitively coupled onto V_S. Since V_G is held constant, changes in V_S are equal and opposite in phase to changes in V_{GS}. As a result, the circuit operates identically to the common-source amplifier shown in Section 17.3.3 without the phase inversion between the input and the output.

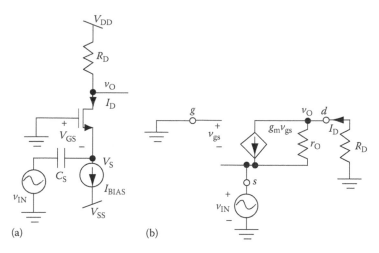

FIGURE 17.25 AC analysis of common-gate amplifier. (a) Shows the complete circuit and (b) shows the AC small-signal model.

An increase in v_{IN} causes a decrease in V_{GS}, which causes a large decrease in I_D. The decrease in I_D causes a decrease in V_{RD}, which causes an increase in v_O. The in-phase operation of this amplifier is evident as a small increase in v_{IN} causes a large increase in v_O.

Figure 17.25b shows the small-signal model of the common-gate amplifier. To find the small-signal voltage gain for this circuit, first find an expression for v_O as follows:

$$v_O = -I_D \cdot R_D \tag{17.98}$$

I_D is found using KCL at the drain as

$$I_D = g_m V_{gs} + \frac{v_O - v_{IN}}{r_0} \tag{17.99}$$

Substituting Equation 17.99 into Equation 17.98 yields

$$v_O = \left(-g_m V_{gs} + \frac{v_{IN} - v_O}{r_0}\right) \cdot R_D \tag{17.100}$$

Inspection of the small-signal model shows that $v_{IN} = -V_{gs}$. Substituting this expression into 17.100 and rearranging produces

$$v_O = v_{IN}\left(g_m R_D + \frac{R_D}{r_0}\right) - \frac{R_D}{r_0} v_O \tag{17.101}$$

Collecting terms yields

$$v_O\left(1 + \frac{R_D}{r_0}\right) = v_{IN}\left(g_m + \frac{1}{r_0}\right)R_D \tag{17.102}$$

Equation 17.102 can be used to solve for the small-signal gain of this amplifier as follows:

$$A_V = \frac{v_O}{v_{IN}} = \frac{(g_m R_D + (R_D/r_0))}{1 + (R_D/r_0)} \tag{17.103}$$

If $r_O \gg R_D$, which is commonly true for good transistors, R_D/r_O will be very small resulting in the typically used approximation for the small-signal gain of the common-gate amplifier:

$$\frac{v_O}{v_{IN}} = g_m R_D \tag{17.104}$$

This expression of the gain follows the behavioral description provided earlier. Since v_{IN} is equal in magnitude to v_{gs}, the only two circuit elements that impact v_O are (1) the amount that I_D changes as a result of changes to v_{gs} (e.g., g_m of the transistor) and (2) the amount that v_O changes as a result of changes to I_D (which occurs due to R_D).

The simplified small-signal model for the input resistance is shown in Figure 17.26a. The change in current due to r_O is small compared to the change in current due to $g_m v_{gs}$, and the analysis containing r_O is significantly more complex and thereby loses intuition, therefore r_O will be neglected for this analysis. Notice in Figure 17.26a that the input connects directly to the transistor's transconductor (voltage-gated current source) rather than the infinite input resistance of the gate. As a result, the input resistance is significantly smaller, as the analysis shows. Using KCL at the source yields

$$I_T = -g_m v_{gs} \tag{17.105}$$

KVL around the input shows

$$v_T = -v_{gs} \tag{17.106}$$

Substituting Equation 17.106 into Equation 17.105 and solving for the V_T/I_T provides the standard equation for the input resistance:

$$r_{in} = \frac{v_T}{I_T} = \frac{1}{g_m} \tag{17.107}$$

This solution assumes that $r_O \gg R_D$.

The small-signal model for the output resistance is shown in Figure 17.26b. This circuit is identical to the common-source example in Section 17.3.2 and therefore produces the exact same output resistance:

$$r_{OUT} = \frac{V_T}{I_T} = R_D \| r_O \tag{17.108}$$

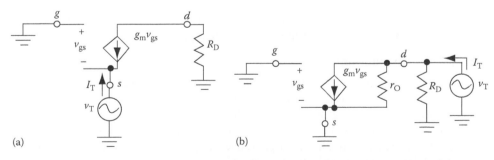

(a)   (b)

FIGURE 17.26   AC model for input resistance (a) and output resistance (b) of common-gate amplifier.

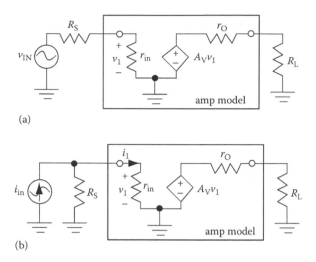

FIGURE 17.27 Impact of small r_{in} on voltage gain (a) versus current gain (b).

Although the magnitude of the small-signal gain is identical for the common-source amplifier with current source biasing and the common-gate amplifier, the input resistances are significantly different, which results in different uses for these two high-gain circuits. To demonstrate the difference in these two circuits, a general amplifier model is used, as shown in Figure 17.27. Once the input resistance, small-signal gain, and output resistance have been determined, any amplifier can be modeled as shown inside the box. Figure 17.27a shows a voltage source driving the amplifier. If the voltage source contains any output resistance (which is typical for sensors, previous stage amplifiers, or even test equipment), the input voltage to the amplifier (v_1) is divided by the source resistor (R_S) and the input resistance (r_{in}), as follows:

$$v_1 = V_{IN} \frac{r_{in}}{R_S + r_{in}} \tag{17.109}$$

If r_{in} is large, as in the case of the common-source circuit, the signal loss is small. However, if r_{in} is small, as in the case of the common-gate circuit, the signal loss can be significant, which can greatly reduce the gain of the entire circuit.

On the other hand, if a current source is driving the amplifier, as in Figure 17.27b, the results are different. In this case, the current divides between the source resistance and the input resistance as follows:

$$I_1 = I_{in} \frac{R_S}{R_S + r_{in}} \tag{17.110}$$

To maximize the amount of current going into the amplifier, r_{in} should be small as in the case of the common-gate circuit. Consequently, common-source amplifiers are typically used with voltage inputs and common-gate amplifiers are typically used with current inputs.

17.3.5.2 BJT Implementation

The equivalent BJT circuit, the common-base amplifier, is shown again Figure 17.28a. The basic operation is identical to the common-gate amplifier. Changes in the input voltage, v_{IN}, is capacitively coupled onto V_E causing V_E to oscillate. Since the base is held constant, changes in V_E result in changes in V_{BE}, equal in magnitude and opposite in phase. If V_E decreases, V_{BE} increases, which causes a dramatic increase in I_C. The increase in I_C results in a larger voltage drop across R_C, which causes v_O to decrease. Therefore, a small decrease in v_{IN} results in a large decrease in v_O, resulting in a positive gain.

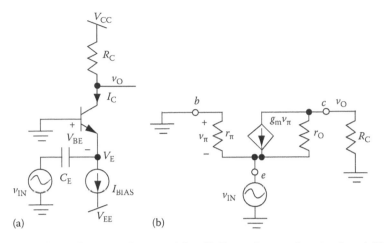

FIGURE 17.28 AC analysis of common-base amplifier. (a) Shows the complete circuit and (b) shows the AC small-signal model.

Figure 17.28b shows the small-signal model for the common-base amplifier. This circuit is identical to the small-signal model of the common-gate amplifier, except the inclusion of r_π. As a result, the analysis is very similar. An alternative method of performing the same analysis follows. First, find an expression for v_O using Ohm's law:

$$v_O = -i_C \cdot R_C \tag{17.111}$$

i_c is found using KCL at the collector as

$$i_c = g_m V_\pi + \frac{v_O - v_{IN}}{r_O} \tag{17.112}$$

Rearranging 17.111 to solve for i_C and substituting that result into 17.112 yields

$$\frac{-v_O}{R_C} = g_m V_\pi + \frac{v_O - v_{IN}}{r_O} \tag{17.113}$$

Inspection of the small-signal model shows that $v_{IN} = -v_\pi$. Substituting this expression into 17.113 and rearranging produces

$$v_O \left(\frac{1}{R_C} + \frac{1}{r_O} \right) = v_{IN} \left(g_m + \frac{1}{r_O} \right) \tag{17.114}$$

Solving for the gain produces

$$A_V = \frac{v_O}{v_{IN}} = \frac{(g_m + 1/r_O)}{(1/R_C) + (1/r_O)} = \left(g_m + \frac{1}{r_O} \right) R_C \| r_O \tag{17.115}$$

If $R_C \ll r_O$, R_C will dominate the parallel connection so that $R_C \| r_O$ is approximately equal to R_C. The gain simplifies to the following expression:

$$\frac{v_O}{v_{IN}} \cong \left(g_m + \frac{1}{r_O} \right) R_C = g_m R_C + \frac{R_C}{r_O} \tag{17.116}$$

Holding the same assumption that $R_C \ll r_O$, R_C/r_O will be very small, resulting in the typically used approximation for the small-signal gain of the common-base amplifier:

$$\frac{v_O}{v_{IN}} = g_m R_C \tag{17.117}$$

This expression of the gain supports the behavioral description, showing how the gain depends on how well the transistor converts a change in voltage into a change in current (g_m) and how well the change in current is converted into a voltage using R_C.

The simplified small-signal model for the input resistance is shown in Figure 17.29a. Like the common-gate circuit, the change in current due to r_O is small compared to the change in current due to $g_m v_\pi$, so r_O will be neglected for this analysis to maintain intuition about the circuit. The input of this circuit connects to the transistor's transconductor like the common-gate circuit rather than r_π, so the input resistance is expected to be small. The analysis proceeds in a similar manner to the common-gate circuit. Using KCL at the source yields

$$I_T = -g_m v_\pi - \frac{v_\pi}{r_\pi} \tag{17.118}$$

KVL around the input shows

$$v_T = -v_\pi \tag{17.119}$$

Substituting Equation 17.119 into Equation 17.118 produces

$$I_T = g_m v_T + \frac{v_{IN}}{r_\pi} = v_T\left(g_m + \frac{1}{r_\pi}\right) = v_T g_m\left(1 + \frac{1}{g_m r_\pi}\right) = v_T g_m\left(1 + \frac{1}{\beta}\right) \tag{17.120}$$

Assuming that β is large so that $1/\beta \approx 0$, the input resistance can be approximated to the same equation found for the common-gate amplifier:

$$r_{in} = \frac{v_T}{I_T} = \frac{1}{g_m} \tag{17.120a}$$

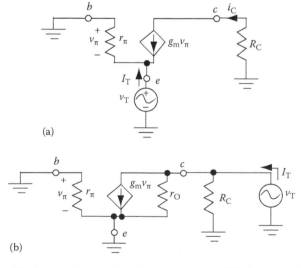

(a)

(b)

FIGURE 17.29   AC model for input resistance (a) and output resistance (b) of common-base amplifier.

The small-signal model for the output resistance is shown in Figure 17.29b. This circuit is identical to the common-emitter example in Section 17.3.2 and therefore produces the exact same output resistance:

$$r_{\text{OUT}} = \frac{V_T}{I_T} = R_C \, || \, r_O \tag{17.121}$$

Just like the common-gate amplifier, the common-base amplifier demonstrates a significantly smaller r_{in} than its high-gain counterpart, the common-emitter amplifier. As a result, this circuit is most commonly used with current source inputs, as is common with many sensors, for the same reason that was presented for the common-gate circuit.

Another use of the common-base amplifier is to obtain higher frequency response. A common-emitter amplifier can be combined with a common-base amplifier as shown in Figure 17.30b. To understand the high-frequency benefit, a discussion of the frequency limitations of the common-emitter circuit is essential. Figure 17.30a shows the common-emitter amplifier with the addition of two parasitic capacitors, C_μ and C_π. These internal capacitors come from the separation of charge due to the depletion region at the p-n junctions of the BJT. At the interface between the n-type emitter and the p-type base, the depletion region separates charge creating a junction capacitor (C_π). Another junction capacitor exists at the interface between the p-type base and the n-type collector (C_μ). In AC operation, the transistor causes a small change in v_{IN} to result in a large change in v_O in the opposite direction. C_μ connects between v_{IN} and v_O. The large gain of the transistor causes a large voltage difference across the capacitor, C_μ, which makes the capacitor act like it has a larger capacitance (called the Miller effect). A larger capacitor results in a smaller pole value, which limits the bandwidth at high frequency.

By combining the common-emitter amplifier with the common-base amplifier as shown in Figure 17.30b, the Miller effect does not happen. This circuit works by a change in v_{in}, resulting in a change in V_{BE1} since V_E is held constant by C_E. This change in V_{BE1} creates a large change in the collector current, I_{C1}. By KCL, $I_{\text{C1}} = I_{\text{E2}}$, so Q2 sees the same change in current. Since Q1 and Q2 see approximately the same change in current, they must also have the same change in V_{BE}. Therefore, $\Delta V_{\text{BE1}} = \Delta V_{\text{BE2}}$. Since V_{C1} changes equal and opposite v_{IN}, the gain $V_{\text{C1}}/v_{\text{IN}} = -1$. However, the large change in current is turned into a voltage by the resistor, R_C. The net result is the same gain as a common emitter:

$$A_V \cong -g_m R_C \tag{17.122}$$

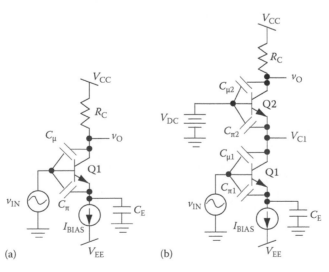

FIGURE 17.30 Use of common-base amplifier for high frequency. (a) Shows intrinsic capacitor components in a standard common-emitter amplifier and (b) shows intrinsic capacitor components in a high-frequency common-emitter amplifier.

The real advantage of this circuit is the improved high-frequency response. The common-emitter transistor (Q1) still has the base-collector capacitor, $C_{\mu 1}$, but the gain across the capacitor is −1. Therefore, the capacitor does not see a large change in voltage across it, so its value is not effectively increased. The junction capacitors around the common-base transistor (Q2) have 1 terminal tied to an AC ground. Therefore, they do not see a large voltage across them due to the gain of a circuit either. As a result, all poles will be at their highest possible values, resulting in the highest possible frequency response.

17.3.6 Voltage Follower

17.3.6.1 MOSFET Implementation

In the voltage-follower circuit, the output voltage follows the input voltage with the DC offset presented in Section 17.2.3, which results in a gain of approximately 1. The purpose of this amplifier is to isolate a high-gain stage from a low load resistance, as shown in Figure 17.31. As shown in Sections 17.3.2 and 17.3.3, a load resistor in the AC model of a high-gain amplifier will be in parallel with the drain or collector resistor (R_D or R_C). If the load resistor is small, the parallel combination will be dominated by R_L, which can decrease the gain significantly. By placing a voltage follower in between the high-gain amplifier and the small load resistor, the high-gain amplifier sees the input resistance of the voltage follower as a load, which is significantly higher than the load resistor. Thereby, the high-gain amplifier retains its high gain. The voltage follower is biased to provide sufficient current to the load.

The MOSFET version of the voltage follower, the common-drain amplifier, is shown in Figure 17.32. The operation of this amplifier can best be understood by looking initially at the circuit with no load resistance, as shown in Figure 17.32a. Given the purpose of the voltage follower, this circuit configuration would never occur in reality; however it is easier to see why the circuit obtains unity gain. KCL at the output voltage of this circuit indicates that I_D must always equal I_{BIAS} even when v_{IN} is oscillating. In order to maintain a constant I_D, a constant V_{GS} must be maintained. By KVL, $v_O = v_{IN} - V_{GS}$. If V_{GS} is constant, v_O will follow v_{IN} with a constant DC offset equal to the value V_{GS}. Removing the DC component of the signal results in unity AC gain.

Figure 17.32b shows the circuit with the load resistor added. In this circuit, as v_O follows v_{IN}, the voltage across the load resistor changes, which causes a change in the load current, I_L. I_S, which equals I_D, is the sum of I_L and I_{BIAS}; therefore, if I_L changes, I_D must change as well. If I_D changes, V_{GS} must also change. However, the change in V_{GS} is small due to the square-root relationship between V_{GS} and I_D. However, since V_{GS} is not perfectly constant, v_O does not follow v_{IN} perfectly. As a result, the gain is a little less than 1.

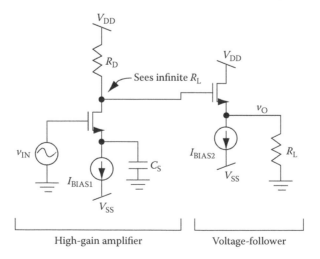

FIGURE 17.31 Behavioral description of the voltage-follower. (a) Shows the behavior without a resistive load, resulting in v_O following v_{IN} and (b) shows the behavior with a resistive load to exhibit the small loss in small-signal gain.

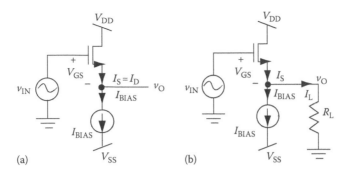

FIGURE 17.32 Behavioral description of the voltage-follower. (a) Shows the behavior without a resistive load, resulting in v_O following v_{IN} and (b) shows the behavior with a resistive load to exhibit the small loss in small-signal gain.

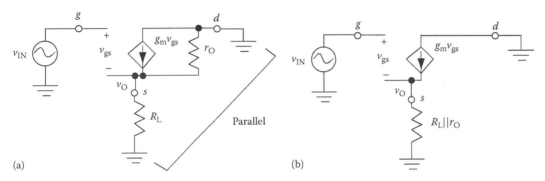

FIGURE 17.33 Small-signal model of common-drain amplifier. (a) Shows the complete small-signal model of the circuit and (b) shows the simplified model.

The small-signal model of the common-drain amplifier is shown in Figure 17.33a. v_{IN} is connected directly to the gate. Since V_{DD} is turned off, the drain is connected directly to ground. The bias current is turned off, resulting in an open circuit. However, the source is connected to ground through R_L. Notice that v_O is at V_S, NOT V_D as in the previous high-gain circuits.

To find the small-signal gain, first notice that r_O and R_L are in parallel and redraw the small-signal model as in Figure 17.33b. The solution for the AC gain follows.

Find an expression for v_O using KCL at v_O and Ohm's law:

$$g_m v_{gs} = \frac{v_O}{R_L \parallel r_O} \tag{17.123}$$

An expression is now necessary for v_{gs} as a function of v_O or v_{IN} or both. By KVL at the input node,

$$v_{IN} - v_O = v_{gs} \tag{17.124}$$

Substituting Equation 17.124 into Equation 17.123 yields

$$g_m(v_{IN} - v_O) = \frac{v_O}{R_L \parallel r_O} \tag{17.125}$$

Collecting terms for v_O and v_{IN} produces

$$g_m v_{IN} = v_O\left(\frac{1}{R_L \parallel r_O} + g_m\right) \tag{17.126}$$

Solving for voltage gain results in

$$\frac{v_O}{v_{IN}} = \frac{g_m}{(1/(R_L \parallel r_O)) + g_m} = \frac{g_m(R_L \parallel r_O)}{1 + g_m(R_L \parallel r_O)} \tag{17.127}$$

If $(R_L \parallel r_O)$ is large, $g_m(R_L \parallel r_O) \gg 1$, and the function approximates to 1. However, as the load resistor decreases, the gain decreases from 1. This follows the intuition described previously, since a smaller load resistance will produce a larger change in I_D as v_O changes, which causes v_O to follow v_{IN} less closely.

The small-signal model for the input resistance is shown in Figure 17.34a. Since there is no gate current, the input resistance is infinite. This result has a significant impact when it is used in conjunction with a high-gain amplifier. Rather than the high-gain amplifier seeing R_L as a load, it sees an open circuit. As a result, the load resistor has no impact on the high-gain amplifier.

The small-signal model for the output resistance is shown in Figure 17.34b. Note that the output of this circuit is at V_S, not V_D, so the test source is placed at V_S (see Figure 17.34b). The circuit analysis proceeds as follows.

R_L is still in parallel with r_O. To find an expression for I_T, use KCL at the output node to get

$$I_T + g_m v_{gs} = \frac{v_T}{R_L \parallel r_O} \tag{17.128}$$

To get an expression for V_{gs}, use KVL at the input node to get

$$v_{gs} = 0 - v_T \tag{17.129}$$

Substituting Equation 17.129 into Equation 17.128 yields

$$I_T - g_m v_T = \frac{v_T}{R_L \parallel r_O} \tag{17.130}$$

Collecting terms yields

$$I_T = v_T \left(g_m + \frac{1}{R_L \parallel r_O} \right) \tag{17.131}$$

Rearranging produces the output resistance

$$r_{OUT} = \frac{V_T}{I_T} = \frac{1}{g_m + (1/(R_L \parallel r_O))} \tag{17.132}$$

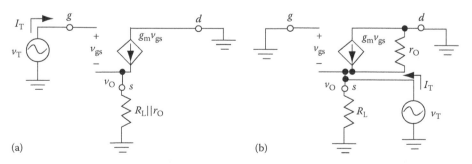

(a)   (b)

FIGURE 17.34 AC model for input resistance (a) and output resistance (b) of common-drain amplifier.

This equivalent resistance can be rewritten to show a group of parallel resistors as follows:

$$r_{OUT} = \frac{1}{(1/(1/g_m)) + (1/(R_L \parallel r_O))} = (1/g_m) \parallel R_L \parallel r_O \tag{17.133}$$

17.3.6.2 BJT Implementation

The BJT version of the voltage follower, the common-collector amplifier, is shown in Figure 17.35a. This circuit operates on the same principle as the common-drain amplifier, except that v_O follows v_{IN} with a DC voltage drop of V_{BE} instead of V_{GS}. Similar to the common-drain amplifier, v_O follows v_{IN} perfectly if I_C is held constant, which is true if R_L is large. If R_L is small, a change in v_O causes a change in I_{RL}, which changes I_C and thereby V_{BE}. However, due to the logarithmic relationship between V_{BE} and I_C, the change in V_{BE} is small, keeping the gain close to 1.

Figure 17.35b shows the small-signal model for voltage gain. In this circuit, r_O is in parallel with R_L, simplifying the small-signal model as shown in Figure 17.35c. The analysis proceeds as follows.

Find an expression for v_O using KCL at v_O:

$$\frac{v_\pi}{r_\pi} + g_m v_\pi = \frac{v_O}{R_L \parallel r_O} \tag{17.134}$$

An expression is now necessary for v_π as a function of v_O or v_{IN} or both. By KVL at the input node:

$$v_{IN} - v_O = v_\pi \tag{17.135}$$

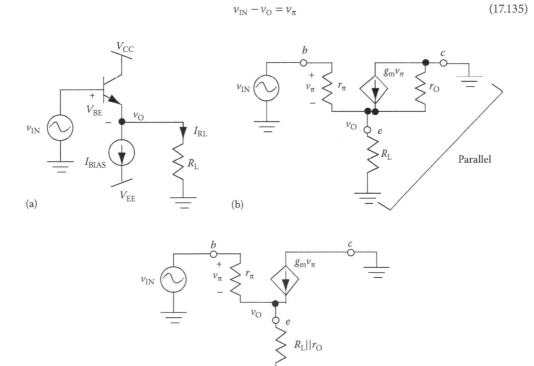

FIGURE 17.35 AC analysis of common-collector amplifier. (a) Shows the complete circuit, (b) shows the AC small-signal model, and (c) shows the simplified small-signal model.

Substituting Equation 17.135 into Equation 17.134 yields

$$\frac{v_{IN} - v_O}{r_\pi} + g_m(v_{IN} - v_O) = \frac{v_O}{R_L \| r_O} \tag{17.136}$$

Collecting terms for v_O and v_{IN} produces

$$v_{IN}\left(g_m + \frac{1}{r_\pi}\right) = v_O\left(\frac{1}{r_\pi} + \frac{1}{R_L \| r_O} + g_m\right) \tag{17.137}$$

To simplify this equation, examine the relationship $(g_m + 1/r_\pi)$ by rewriting $1/r_\pi$ as a function of β:

$$g_m + \frac{1}{r_\pi} = g_m\left(1 + \frac{1}{g_m r_\pi}\right) = g_m\left(1 + \frac{1}{\beta}\right) \cong g_m \tag{17.138}$$

Using this simplification and solving for voltage gain results in

$$\frac{v_O}{v_{IN}} = \frac{g_m}{(1/(R_L \| r_O)) + g_m} = \frac{g_m R_L \| r_O}{1 + g_m R_L \| r_O} \tag{17.139}$$

This gain equation is identical to the common-drain gain equation and follows the intuition similarly: smaller R_L results gain less than unity because larger changes in I_{RL} prevent v_O from following v_{IN} as well.

The small-signal model for the input resistance, shown in Figure 17.36a, is identical to the small-signal model for the gain, with v_{IN} replaced with V_T. To find the input resistance, find an expression for I_T using KCL at the input node and apply Ohm's law:

$$I_T = \frac{v_\pi}{r_\pi} \tag{17.140}$$

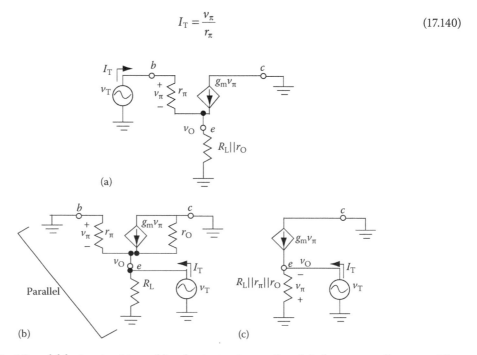

(a)

Parallel

(b)

(c)

FIGURE 17.36 AC model for input resistance (a) and output resistance (b and c) of common-collector amplifier. (c) is a circuit simplification of part (b).

From Equation 17.140, it is evident that an expression for v_π is necessary. Obtain this using KCL at the output node:

$$\frac{v_\pi}{r_\pi} + g_m v_\pi = \frac{v_O}{R_L \parallel r_O} \tag{17.141}$$

While the test current could be written as a function of $V_T - v_O$ or simply as I_T, circuit simplifications will often result if the current is left in terms of v_π. Since v_O is not a parameter in the input resistance, an expression for v_O must be found and substituted into Equation 17.141. The expression for v_O can be found by using KVL across r_π:

$$v_O = V_T - v_\pi \tag{17.142}$$

Substituting Equation 17.142 into Equation 17.141 yields

$$\frac{v_\pi}{r_\pi} + g_m v_\pi = \frac{V_T - v_\pi}{R_L \parallel r_O} \tag{17.143}$$

Rearranging produces

$$v_\pi \left(\frac{1}{r_\pi} + g_m + \frac{1}{R_L \parallel r_O} \right) = \frac{V_T}{R_L \parallel r_O} \tag{17.144}$$

which can be used to solve for V_T as follows:

$$v_\pi \left[(R_L \parallel r_O) \left(\frac{1}{r_\pi} + g_m \right) + 1 \right] = V_T \tag{17.145}$$

Rearranging Equation 17.140 and substituting into Equation 17.145 yields

$$I_T r_\pi \left[(R_L \parallel r_O) \left(\frac{1}{r_\pi} + g_m \right) + 1 \right] = V_T \tag{17.146}$$

Solving for r_{in} results in

$$r_{IN} = \frac{V_T}{I_T} = r_\pi \left[(R_L \parallel r_O) \left(\frac{1}{r_\pi} + g_m \right) + 1 \right] = (R_L \parallel r_O)(1 + r_\pi g_m) + r_\pi = (R_L \parallel r_O)(1 + \beta) + r_\pi \tag{17.147}$$

While this input resistance is not as large as the common-drain circuit, it is approximately β times larger than the load resistance. If β is large, say 100, the common-collector amplifier will make the load seen by the high-gain amplifier appear 100 times larger, preventing the drop in gain due to the small load resistor.

The small-signal model for the output resistance is shown in Figure 17.36b. In this circuit, r_π is in parallel with R_L and r_O, which is redrawn in Figure 17.36c. The analysis proceeds in a similar manner as the common-drain amplifier as follows.

Finding an expression for I_T using KCL at the output node produces

$$I_T + g_m v_\pi = \frac{V_T}{R_L \,||\, r_\pi \,||\, r_O} \tag{17.148}$$

Using KVL and Ohm's law,

$$v_\pi = -v_T \tag{17.149}$$

Substituting Equation 17.149 into Equation 17.148 yields

$$I_T - g_m v_T = \frac{V_T}{R_L \,||\, r_\pi \,||\, r_O} \tag{17.150}$$

Collecting terms results in

$$I_T = v_T \left(g_m + \frac{1}{R_L \,||\, r_\pi \,||\, r_O} \right) \tag{17.151}$$

Rearranging produces the output resistance:

$$r_{OUT} = \frac{V_T}{I_T} = \frac{1}{g_m + (1/(R_L \,||\, r_\pi \,||\, r_O))} = \frac{1}{(1/(1/g_m)) + (1/(R_L \,||\, r_\pi \,||\, r_O))} = (1/g_m) \,||\, R_L \,||\, r_\pi \,||\, r_O \tag{17.152}$$

This result is similar to the common-drain amplifier, with the exception that the equivalent resistance is also in parallel with r_π.

17.4 Swing: Putting AC and DC Together

Another key specification in amplifier operation is the largest output amplitude that the amplifier can produce while still producing a sinusoidal output wave. This parameter is called swing. The maximum swing depends on how far the output can swing while keeping the transistor in the current saturating region (e.g., FET = saturation region, BJT = active region). As a result, neither the AC nor DC components can be ignored in this analysis.

If the input is pushed too high or too low, the output will stop changing as a sinusoid, resulting in some of the sinusoid clipped off. This phenomenon, called "clipping," is shown in Figure 17.37a. If the transistor is pushed out of the current saturating region, clipping will occur. An explanation for clipping in a high-gain amplifier is presented graphically in Figure 17.37b. This figure presents I_C as a function of V_{CE} for different values of V_{BE}. Each curve represents the output current for a different value of V_{BE}. A change in V_{IN} causes a change in V_{BE}, causing the amplifier to move from one curve to the next. However, the resistor transfers the change in current into a change in voltage, which causes the value for V_{CE} to change as demonstrated by the line across the curves. Since the gain is negative, V_{CE} decreases as V_{IN} increases, which is shown by the negative slope of the line. As the input changes sinusoidally, the output moves along this line. If the change in V_{BE} is large enough, the transistor will either (1) be pushed into the saturation region (A), where a change in V_{BE} no longer produces a linear change in current or (2) be turned off (B), where a change in V_{BE} produces zero current. In either case, a further change in V_{BE} does not change V_O, so the output appears clipped.

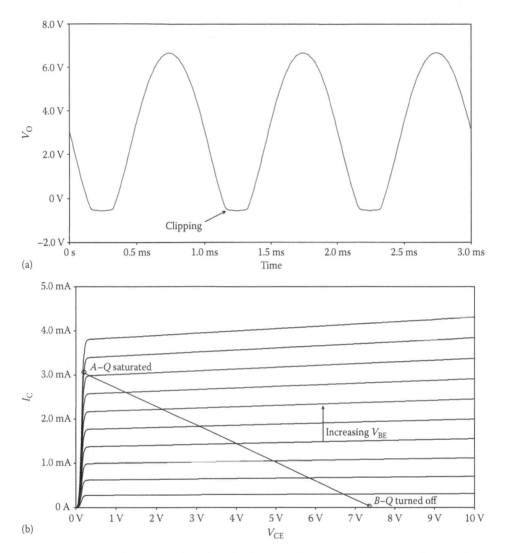

FIGURE 17.37 Amplifier clipping. (a) Shows how the clipping will appear on the output in the time-domain and (b) shows the transistor characteristic curves to demonstrate regions of operation that can cause clipping.

In all cases, swing can be calculated as the worst-case condition between the maximum the output can swing relative to the bias point versus the minimum the output can swing relative to the bias point. This concept is presented graphically in Figure 17.38 and mathematically below:

$$V_{SWING}^+ = V_{OMAX} - V_{OQ} \tag{17.153}$$

$$V_{SWING}^- = V_{OQ} - V_{OMIN} \tag{17.154}$$

The worst-case swing is the minimum between Equations 17.153 and 17.154. Even though the circuit may have the potential to swing higher in one direction (positive in Figure 17.38), the swing will clip on the other side (negative in Figure 17.38) before the potential is met. Therefore, the minimum swing limits the total swing. Although Equations 17.153 and 17.154 are consistent for every circuit, the calculation of V_{OQ}, V_{OMAX}, and V_{OMIN} differ for every circuit. Therefore, several examples will demonstrate the calculation of swing using different transistor types and different load conditions.

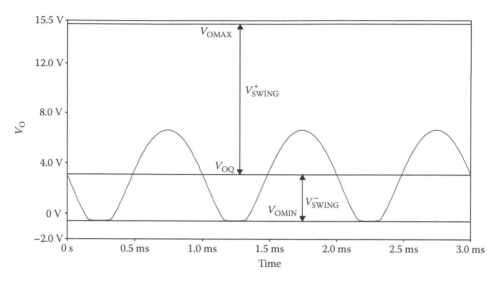

FIGURE 17.38 Amplifier swing.

17.4.1 High-Gain Amplifier with Input Resistor Biasing

17.4.1.1 MOSFET Implementation

The common-source amplifier with input resistor biasing is shown in Figure 17.39. In order to determine the swing, an expression for V_{OQ}, V_{OMAX}, and V_{OMIN} must be obtained. An expression for V_{OQ} was found in Section 17.2.1 and is repeated here for convenience:

$$V_{OQ} = V_{DD} - I_D R_D \qquad (17.155)$$

The value for V_{OMAX} is found by determining which condition, the transistor saturating or the transistor turning off, limits the current (see Figure 17.39a). Due to the negative gain, V_O increases when V_{IN} decreases. Therefore, V_{OMAX} will occur when V_{IN} goes so low that the transistor turns off causing $I_D = 0$. If $I_D = 0$, the voltage across $R_D = 0$, resulting in

$$V_{OMAX} = V_{DD} \qquad (17.156)$$

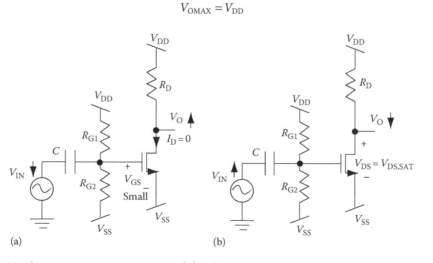

FIGURE 17.39 Swing in a common-source amplifier. (a) Shows the impact on V_O when V_{IN} is decreasing and (b) shows the impact on V_O when V_{IN} is increasing.

The value of V_{OMIN} (see Figure 17.39b) is found by determining which condition limits the current when V_{IN} increases causing V_O to decrease. V_O can decrease until the transistor is pushed into the linear region and starts to clip. The limiting case is at the edge between the linear region and saturation region: $V_{DS,SAT}$. Note: we normally do not know the value for V_{DS}, except in this minimum condition. By KVL, V_{OMIN} can be found relative to the negative supply.

$$V_{OMIN} = V_{SS} + V_{DS,SAT} = V_{SS} + (V_{GSQ} - V_t) \tag{17.157}$$

With Equations 17.155 through 17.157, the maximum swing can be determined. The maximum swing will be

$$V^+_{SWING} = V_{OMAX} - V_{OQ} = V_{DD} - (V_{DD} - I_D R_D) = I_D R_D \tag{17.158}$$

$$V^-_{SWING} = V_{OQ} - V_{OMIN} = (V_{DD} - I_D R_D) - (V_{SS} + V_{DS,SAT}) \tag{17.159}$$

The smallest value calculated from Equations 17.158 and 17.159 will determine the maximum swing possible for the amplifier for a given bias point. The largest possible swing will occur if V_{OQ} is biased exactly in between V_{OMAX} and V_{OMIN}.

17.4.1.2 BJT Implementation

Figure 17.40 demonstrates the BJT implementation of the same circuit, the common-emitter amplifier with input resistor biasing. The output loop of this circuit is identical to the FET equivalent circuit; therefore, the swing analysis is identical, except $V_{CE,SAT}$ is used rather than $V_{DS,SAT}$. As a result, the final swing equations are

$$V^+_{SWING} = V_{OMAX} - V_{OQ} = V_{CC} - (V_{CC} - I_C R_C) = I_C R_C \tag{17.160}$$

$$V^-_{SWING} = V_{OQ} - V_{OMIN} = (V_{CC} - I_C R_C) - (V_{SS} + V_{CE,SAT}) \tag{17.161}$$

The maximum swing is the minimum value of Equations 17.160 and 17.161. Like the FET circuit, maximum swing will occur when V_{OQ} is at the midpoint between V_{OMAX} and V_{OMIN}. However, looking

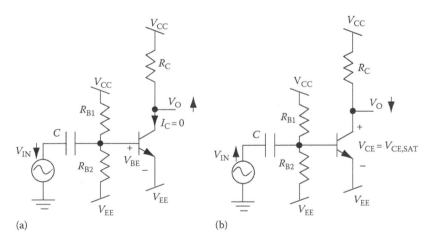

(a)　　　　　　　　　　　　　　　　　　　(b)

FIGURE 17.40 Swing in a common-emitter amplifier. (a) Shows the impact on V_O when V_{IN} is decreasing and (b) shows the impact on V_O when V_{IN} is increasing.

at the gain equation from Section 17.3.2, if $R_C \ll r_O$ (which is typically true), the gain equation can be approximated as follows:

$$A_V = -g_m R_C = \frac{-I_C R_C}{V_T} \tag{17.162}$$

Given this equation, the maximum gain occurs with a maximum value for $I_C R_C$. The bias point equation (17.13) reveals that a larger gain, requiring a larger $I_C R_C$, occurs simultaneously with a lower bias point, V_{OQ}. Once V_{OQ} drops below the midway point between V_{OMAX} and V_{OMIN}, there is a trade-off between gain and swing (e.g., to get a larger gain, maximum swing must drop).

17.4.2 High-Gain Amplifier with Input Resistor Biasing and a Resistive Load

The same circuit in Figure 17.39 is presented in Figure 17.41 using a pFET and a load resistor. Without the load resistor, the analysis would be identical to Section 17.4.1, except V_{OQ} would be relative to the negative supply, and the V_{OMAX} and V_{OMIN} equations would be swapped. However, the load resistor changes V_{OQ} and V_{OMIN}. This example is presented to understand how the p-type device and load resistor impacts the analysis. The analysis proceeds as follows.

As presented in Section 17.2.2, V_{OQ} with a load resistor is determined by KCL at the output node:

$$I_D = \frac{V_{OQ} - V_{SS}}{R_D} + \frac{V_{OQ}}{R_L} \tag{17.163}$$

Rearranging produces

$$V_{OQ} = \left(I_D + \frac{V_{SS}}{R_D} \right) (R_D \parallel R_L) \tag{17.164}$$

V_{OMAX} (see Figure 17.41a) occurs when V_{IN} is decreasing, which increases V_{SG}, so the transistor is not turning off to clip. However, as V_O increases due to negative gain, the transistor may be pushed into the linear region causing the output voltage to clip. Therefore, the limiting case occurs when $V_{SD} = V_{SD,SAT}$ as follows:

$$V_{OMAX} = V_{DD} - V_{SD,SAT} = V_{DD} - (V_{SG} - |V_t|) \tag{17.165}$$

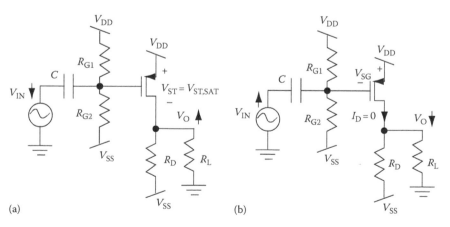

FIGURE 17.41 Swing in a pFET common-source amplifier with a load resistor. (a) Shows the impact on V_O when V_{IN} is decreasing and (b) shows the impact on V_O when V_{IN} is increasing.

V_{OMIN} (see Figure 17.41b) occurs when V_{IN} is increasing, which decreases V_{SG}, potentially causing the transistor to turn off. If the transistor turns off, $I_D = 0$. However, unlike the previous example, even if $I_D = 0$, there is still current flowing through R_D and R_L. If $I_D = 0$, R_D and R_L are left in series, so the expression for V_{OMIN} can be found using a voltage divider between R_D and R_L as follows:

$$V_{OMIN} = V_{SS} \left(\frac{R_L}{R_D + R_L} \right) \tag{17.166}$$

Using the expressions in Equations 17.164 through 17.166, the final swing equations are evaluated as follows:

$$V_{SWING}^{+} = V_{OMAX} - V_{OQ} = (V_{DD} - V_{SD,SAT}) - \left(I_D + \frac{V_{SS}}{R_D} \right)(R_D \| R_L) \tag{17.167}$$

$$V_{SWING}^{-} = V_{OQ} - V_{OMIN} = \left(I_D + \frac{V_{SS}}{R_D} \right)(R_D \| R_L) - \left(V_{SS} \frac{R_L}{R_D + R_L} \right) \tag{17.168}$$

If R_D is significantly larger than R_L, the load resistor can severely limit the allowable swing in addition to dropping the gain. The benefit of placing a voltage follower in between the high-gain stage and a small load resistor, which allows the high-gain stage to see a larger load resistor, is once again evident.

17.4.3 High-Gain Amplifier with Current Source Biasing and Capacitively Coupled Resistive Load

17.4.3.1 MOSFET Implementation

The common-source amplifier with a current source bias is shown in Figure 17.42. In addition to demonstrating the impact of the current source on the swing, this section will also demonstrate the analysis for a capacitively coupled resistive load. Due to the load capacitor, V_{OQ} does not equal V_{DQ}. However, in the AC case, V_D is equal to V_O. If V_D clips, V_O will clip. If V_D does not clip, V_O will not clip. In the previous two sections, this difference was easily extracted apart using superposition. However, in the case of swing, the AC and DC must be considered simultaneously. Clearly, the bias points will be different as demonstrated in Section 17.2.3. However, in the minimum and maximum swing case, the circuit is in AC, so the drain sees the impact of the load resistor. In summary, in the DC case, $V_D \neq V_O$. In the AC case, $V_D = V_O$. In swing, the AC and DC must be added back together, resulting in V_{OQ} *changing* identically to V_{DQ}, but maintaining a DC offset. To include both the AC and DC components in the swing analysis, V_{DQ} and V_{OQ} must be found in the DC case, with the capacitor acting as an open circuit. The difference between V_{DQ} and V_{OQ} is the voltage stored across the capacitor. V_{DMAX} and V_{DMIN} are found in the AC case, assuming that the capacitor is a short circuit. V_{OMAX} and V_{OMIN} will be expressed as V_{DMAX} and V_{DMIN} minus the voltage stored across the capacitor.

The bias points for this circuit were found in Section 17.2.3 with the following result:

$$V_{DQ} = V_{DD} - I_D R_D \tag{17.169}$$

$$V_{OQ} = 0 \text{ V} \tag{17.170}$$

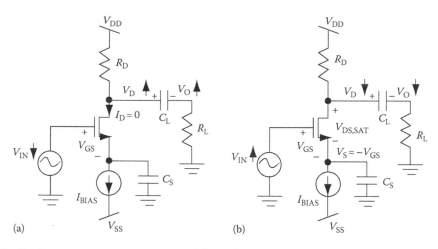

FIGURE 17.42 Swing in a common-source amplifier with current mirror biasing and a capacitively coupled resistive load. (a) Shows the impact on V_O when V_{IN} is decreasing and (b) shows the impact on V_O when V_{IN} is increasing.

The voltage stored across the capacitor is found by

$$V_{CAP} = V_{DQ} - V_{OQ} = V_{DQ} \tag{17.171}$$

Since this amplifier exhibits negative gain, the value for V_{DMAX} is found when V_{IN} is decreasing, turning the transistor off (Figure 17.42a). Therefore, V_{DMAX} is found when $I_D = 0$. At peak swing, V_D sees the impact of the load. Since $I_D = 0$, R_D is in series with R_L and V_{DMAX} is found from a voltage divider:

$$V_{DMAX} = V_{DD}\left(\frac{R_L}{R_D + R_L}\right) \tag{17.172}$$

V_{OMAX} is found by subtracting the voltage stored across the capacitor from V_{DMAX}:

$$V_{OMAX} = V_{DMAX} - V_{CAP} = V_{DMAX} - V_{DQ} \tag{17.173}$$

The value of V_{DMIN} is found when V_{IN} is increasing (Figure 17.42b). V_{DMIN} drops until the transistor is pushed into the linear region, causing the output to clip. Recall that the capacitor holds the source at a constant DC voltage. Therefore, the transistor is pushed into the linear region when V_D is within a $V_{DS,SAT}$ from V_{SQ}:

$$V_{DMIN} = V_{SQ} + V_{DS,SAT} \tag{17.173a}$$

V_{SQ} is found from the DC circuit, where V_{IN} is turned off and $V_G = 0$. KVL around the input loop produces

$$V_{SQ} = V_G - V_{GS} = -V_{GS} \tag{17.174}$$

Recall the definition of $V_{DS,SAT}$ from Chapter F12:

$$V_{DS,SAT} = V_{GS} - V_t \tag{17.175}$$

Substituting Equations 17.174 and 17.175 into Equation 17.173 yields

$$V_{DMIN} = (-V_{GS}) + (V_{GS} - V_t) = -V_t \tag{17.176}$$

To obtain V_{OMIN}, subtract the voltage stored across the load capacitor to get

$$V_{OMIN} = V_{DMIN} - V_{CAP} = V_{DMIN} - V_{DQ} = -V_t - V_{DQ} \tag{17.177}$$

Using the expressions in Equations 17.170, 17.173, and 17.177, the final swing equations are evaluated as follows:

$$V_{SWING}^+ = V_{OMAX} - V_{OQ} = \left(V_{DD} \frac{R_L}{R_D + R_L} - V_{DQ} \right) - V_{OQ} = \left(V_{DD} \frac{R_L}{R_D + R_L} - V_{DQ} \right) \tag{17.178}$$

$$V_{SWING}^- = V_{OQ} - V_{OMIN} = V_{OQ} - (-V_t - V_{DQ}) = 0 + V_t + V_{DQ} \tag{17.179}$$

One of the limitations of the current source biasing becomes evident from this analysis. Although the circuit eliminates the impact of process variation, since V_{SQ} is close to ground, the swing is limited to approximately half the difference in supply voltages.

17.4.3.2 BJT Implementation

The same circuit implemented with a BJT, the common-emitter amplifier with a current source bias, is shown in Figure 17.43. Since the load resistor is attached through a capacitor, the analysis will proceed in a similar manner as the FET example. The minimum and maximum swing will be found for V_C assuming the AC condition and then the minimum and maximum of V_O will be found by subtracting the DC voltage stored across the capacitor.

The bias points for this circuit were found in Section 17.2.3 with the following result:

$$V_{CQ} = V_{CC} - I_C R_C \tag{17.180}$$

$$V_{OQ} = 0 \text{ V} \tag{17.181}$$

The voltage stored across the capacitor is found by

$$V_{CAP} = V_{CQ} - V_{OQ} = V_{CQ} \tag{17.182}$$

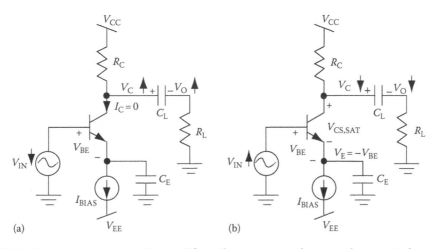

(a) (b)

FIGURE 17.43 Swing in a common-emitter amplifier with current mirror biasing and a capacitively coupled resistive load. (a) Shows the impact on V_O when V_{IN} is decreasing and (b) shows the impact on V_O when V_{IN} is increasing.

The value for V_{CMAX} is found when the transistor is turning off (Figure 17.43a). Therefore, V_{CMAX} is found from a voltage divider:

$$V_{CMAX} = V_{CC}\left(\frac{R_L}{R_D + R_L}\right) \tag{17.183}$$

V_{OMAX} is found by subtracting the voltage stored across the capacitor from V_{CMAX}:

$$V_{OMAX} = V_{CMAX} - V_{CQ} \tag{17.184}$$

The value of V_{CMIN} (Figure 17.43b) is found when the transistor is pushed into the saturation region. Therefore, V_{CMIN} is found when V_C is within a $V_{CE,SAT}$ from V_{EQ}:

$$V_{CMIN} = V_{EQ} + V_{CE,SAT} \tag{17.185}$$

V_{EQ} is found from the DC circuit, where V_{IN} is turned off and $V_B = 0$. KVL around the input loop produces

$$V_{EQ} = V_B - V_{BE} = -V_{BE} \tag{17.186}$$

Substituting Equation 17.186 into Equation 17.185 and using the standard values for V_{BE} and $V_{CE,SAT}$ yields

$$V_{CMIN} = (-V_{BE}) + (V_{CE,SAT}) = -0.7\,\text{V} + 0.2\,\text{V} = -0.5\,\text{V} \tag{17.187}$$

To obtain V_{OMIN}, subtract the voltage stored across the load capacitor to get

$$V_{OMIN} = V_{CMIN} - V_{CQ} = -0.5\,\text{V} - V_{CQ} \tag{17.188}$$

Using the expressions in Equations 17.181, 17.184, and 17.188, the final swing equations are evaluated as follows:

$$V_{SWING}^+ = V_{OMAX} - V_{OQ} = \left(V_{CC}\frac{R_L}{R_D + R_L} - V_{CQ}\right) - V_{OQ} = \left(V_{CC}\frac{R_L}{R_D + R_L} - V_{CQ}\right) \tag{17.189}$$

$$V_{SWING}^- = V_{OQ} - V_{OMIN} = V_{OQ} - (-0.5 - V_{CQ}) = 0 + 0.5 + V_{CQ} \tag{17.190}$$

The maximum swing would be the minimum of Equations 17.189 and 17.190. These results are very similar to the FET circuit.

17.4.4 Swing Nonlinearity

The swing results assume that the transistor is behaving linearly all the way to the point of maximum swing. However, these equations often overestimate the swing since the small-signal model often breaks down at the large input values required to reach maximum swing. The small-signal model assumes that the input voltage (V_{GS} or V_{BE}) is changing with an amplitude that is small enough to approximate the nonlinear input–output characteristic equation ($I_D - V_{GS}$ for FETs and $I_C - V_{BE}$ for BJTs) as a linear line. Maximum output swing requires a larger input voltage, which may cause the input to swing beyond the linear approximation (see Figure 17.44a). As a result, a linear change in V_{IN} causes a nonlinear change in I_D or I_C, resulting in a nonlinear change in V_O. This effect is exaggerated in the BJT. As V_{IN} increases, I_C increases exponentially rather than linearly. This nonlinear relationship results in an expansion of V_O when V_O goes negative (see Figure 17.44b). As V_{IN} decreases by an equal amount, I_C does not decrease linearly, causing a compression of the V_O waveform when V_O goes positive. As a result, the waveform is not exactly sinusoidal. Due to the expansion of V_O on the negative side and the compression of V_O

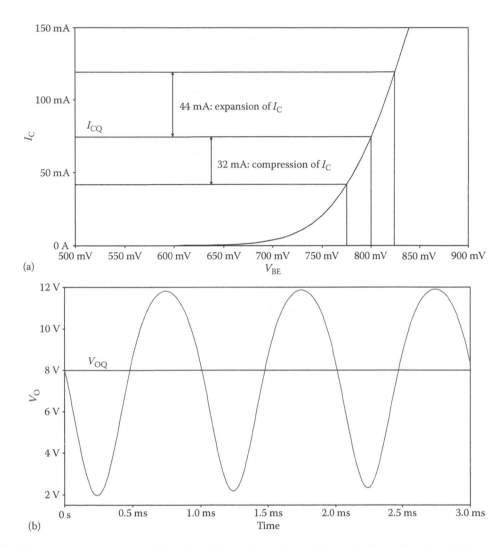

FIGURE 17.44 Swing nonlinearity. (a) Shows the nonlinear change in I_C as V_{BE} changes linearly which causes the output swing characteristic shown in (b).

on the positive side, the output voltage will often clip on the negative side before the positive side can reach the expected swing. Many amplifiers will exhibit nonlinearity before they clip. Although some applications will still work with the harmonic distortion that this nonideal waveform introduces, some applications require a pure sinusoid, which may further limit the maximum swing.

17.5 Design Example

This section will walk through the process of designing an amplifier. The equations derived in the previous sections will be used for the purposes of meeting a design specification.

17.5.1 High-Gain Amplifier with Input Resistor Biasing

Figure 17.45 shows the schematic of a common-source amplifier using a pFET. Design this amplifier to produce $A_V > 150$ and swing >6 V. Assume $K'(W/L) = 10$ mA/V², $|V_t| = 1$ V, and $\lambda = 0.01$.

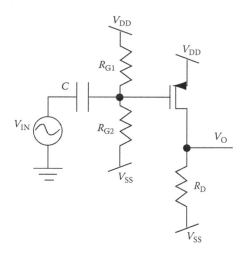

FIGURE 17.45   pFET common-source amplifier design problem.

The gain and the swing determine a range of V_{OQ} values that will allow the amplifier to meet the specifications. Start by using these equations to determine the allowable range of V_{OQ}.

The swing constraints can be determined as follows. First, find a value for V_{OMAX} and V_{OMIN}.

V_{OMAX} occurs when V_O swings high enough to push the transistor into the saturation region:

$$V_{OMAX} = V_{DD} - V_{SD,SAT} = V_{DD} - (V_{SGQ} - |V_t|) \tag{17.191}$$

V_{OMIN} occurs when V_{EB} goes small enough to turn the transistor off and sets $I_D = 0$:

$$V_{OMIN} = V_{SS} \tag{17.192}$$

V_{OQ} can be set to any value that is 6 V less than V_{OMAX} or 6 V greater than V_{OMIN} (see Figure 17.46):

$$V_{SS} + 6\ V < V_{OQ} < [V_{DD} - V_{SGQ} + |V_t|] - 6\ V \tag{17.193}$$

This inequality can be rewritten for $I_D R_D$ using the equation for V_{OQ}. Using Ohm's law across R_D, V_{OQ} is written as

$$V_{OQ} = V_{SS} + I_D R_D \tag{17.194}$$

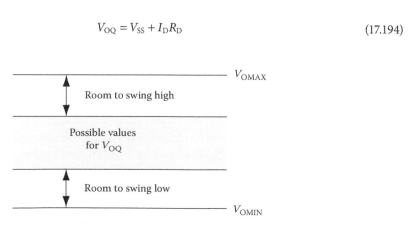

FIGURE 17.46   Swing conditions for design problem.

Substitute Equation 17.194 into Equation 17.193 and rearrange to get an expression for $I_D R_D$:

$$V_{SS} + 6\,V < V_{SS} + I_D R_D < V_{DD} - V_{SGQ} + |V_t| - 6\,V \tag{17.195}$$

$$6\,V < I_D R_D < \left[V_{DD} - \sqrt{\frac{I_D}{2K'W/L}} + |V_t| - 6\,V - V_{SS} \right] \tag{17.196}$$

Choosing supply voltages as $\pm 12\,V$ and substituting in the transistor parameters yields

$$6\,V < I_D R_D < 17 - \sqrt{\frac{I_D}{20\,mA/V^2}} \tag{17.197}$$

Lower supplies will use less power, but limit the gain and/or swing.

The gain equation will set the next constraint on $I_D R_D$, and thereby V_{OQ}. Figure 17.47 shows the small-signal model for the circuit, which looks just like the equivalent nFET circuit shown in Section 17.3.1. Since this is a design problem, simplify the circuit by removing r_O. Typically, $R_D \ll r_O$, so R_D dominates the gain equation. However, if this simplification is made, the design must beat, not barely meet, the gain specification, or r_O will pull the gain below the requirement.

The gain analysis results in the following equation:

$$|A_V| = |-g_m R_D| = \left| -R_D \sqrt{2K' \frac{W}{L} I_D} \right| > 150 \tag{17.198}$$

Rearranging and substituting in the parameters yields

$$R_D \sqrt{I_D} > \frac{150}{\sqrt{2K'W/L}} = 1060 \tag{17.199}$$

By multiplying both sides of the equation by $\sqrt{I_D}$, Equation 17.199 can also be written as

$$R_D I_D > 1060\sqrt{I_D} \tag{17.200}$$

Using Equations 17.197 and 17.200, a value for I_D can be chosen and a value for R_D can be determined that satisfies both equations. Since the circuit is unloaded, the choice of a smaller I_D consumes less power, increases r_O, minimizing its impact on the gain, and provides a higher gain. However, if the circuit has a load resistor, R_D should be chosen to be less than the load resistor, if possible, to minimize the impact of the load resistor on gain and V_{OMIN}. Choose $I_D = 100\,\mu A$ to maximize gain and minimize power. Solving Equation 17.200 yields

$$R_D I_D > 10.6\,V \tag{17.201}$$

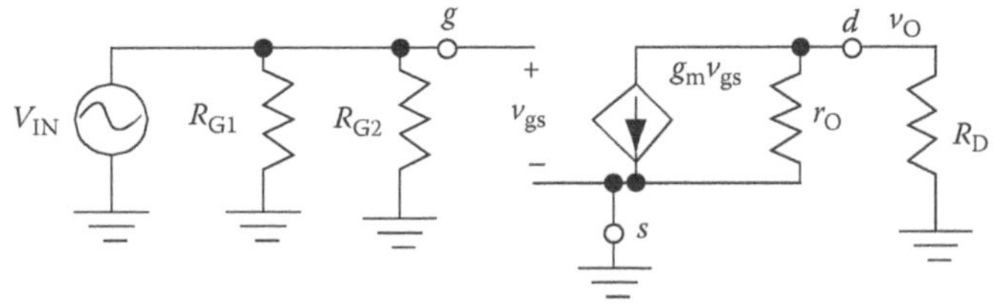

FIGURE 17.47 Small-signal circuit of pFET common-source amplifier.

The gain equation places a tighter constraint on the lower bound of $I_D R_D$. To satisfy both gain and swing, use a combination of Equations 17.197 and 17.201, resulting in the following:

$$10.6 \text{ V} < I_D R_D < 16.9 \text{ V} \tag{17.201a}$$

Choose a value for $I_D R_D$ within the range and solve for R_D. Try not to choose a value too close to either bound. If the choice is too close to the gain boundary (10.6 V), r_O may keep the amplifier from meeting the gain specification. If the choice is too close to the upper boundary, any nonlinearity on the swing may cause clipping. Choose $I_D R_D = 14$ V. Since $I_D = 100\,\mu A$, $R_D = 140\,k\Omega$.

Now that the values for I_D and V_{OQ} have been determined to meet the specifications, the input bias network must be designed to obtain the desired I_D. Performing the analysis presented in Section 17.2.2 yields the following design equation:

$$V_{TH} = V_{DD} - V_{SG} = V_{DD} - \left(\sqrt{\frac{I_D}{K'W/L}} + |V_t| \right) \tag{17.202}$$

Solving for V_{TH} yields

$$V_{TH} = 12 \text{ V} - \left(\sqrt{\frac{100\,\mu A}{10 \text{ mA/V}^2}} + 1 \right) = 10.9 \text{ V} \tag{17.203}$$

The design equation to find the input resistors from the Thevenin equivalent voltage was derived in Section 17.2.2 and shown here:

$$V_{TH} = V_{DD} \frac{R_{G2}}{R_{G1} + R_{G2}} + V_{SS} \frac{R_{G1}}{R_{G1} + R_{G2}} \tag{17.204}$$

Since the gate voltage is close to the top supply, choose a large resistive value for R_{G2}, which also minimizes DC power consumption, and solve for R_{G1}. Choose $R_{G2} = 100\,k\Omega$.

$$11 \text{ V} = 12 \text{ V} \frac{100 \text{ k}\Omega}{R_{G1} + 100 \text{ k}\Omega} - 12 \text{ V} \frac{R_{G2}}{R_{G1} + 100 \text{ k}\Omega} \tag{17.205}$$

Solve for $R_{G1} = 4.8\,k\Omega$.

The design is complete. Using the analysis equations, verify the results meet specification. Eliminate approximations for the verification.

Bias points:

$$V_{TH} = 12 \text{ V} \frac{100 \text{ k}\Omega}{100 \text{ k}\Omega + 4.8 \text{ k}\Omega} - 12 \text{ V} \frac{4.8 \text{ k}\Omega}{100 \text{ k}\Omega + 4.8 \text{ k}\Omega} = 10.9$$

$$I_D = K' \frac{W}{L} (V_{SG} - |V_T|)^2 = 10 \text{ mA/V}^2 (12 \text{ V} - 10.9 \text{ V} - 1)^2 = 100\,\mu A$$

$$V_{OQ} = V_{SS} + I_D R_D = -12 \text{ V} + (100\,\mu A)(140 \text{ k}\Omega) = 2 \text{ V}$$

Gain – be sure to include the impact of r_O to make sure the amplifier will meet the gain specification with this parameter included:

$$|A_V| = \left|-g_m R_D \,\|\, r_O\right| = \left|-\left(140\text{ k}\Omega \,\|\, \frac{1}{(0.01)(100\ \mu\text{A})}\right)\sqrt{2(10\text{ mA/V}^2)(100\ \mu\text{A})}\right| = 173 > 150$$

Swing:

$$V_{\text{OMAX}} = V_{\text{DD}} - V_{\text{SD,SAT}} = 12 - \left(V_{\text{SG}} - |V_t|\right) = 12 - (12\text{ V} - 10.9 - 1) = 10.9\text{ V}$$

$$V_{\text{OMIN}} = V_{\text{SS}} = -12\text{ V}$$

$$V_{\text{SWING}}^{+} = 10.9 - 2 = 8.9\text{ V}$$

$$V_{\text{SWING}}^{-} = 2 - (-12) = 14\text{ V}$$

$$\text{Max swing} = 8.9\text{ V} > 6\text{ V}$$

Design meets all specifications. The design should be simulated, particularly for swing, to make sure that nonlinearity of the transistor does not prevent the design from meeting the specifications.

17.5.2 High-Gain Amplifier with Current Mirror Biasing and Voltage Follower

Figure 17.48 shows the schematic of a common-emitter amplifier driving a voltage follower. Design this amplifier to produce $A_V > 200$ and swing > 5 V. Assume $\beta = 100$, $V_{\text{BE}} = 0.7\text{ V}$, $V_{\text{CE,SAT}} = 0.2\text{ V}$, and $V_A = 75$.

The design process for this circuit is very similar to the previous circuit. One primary difference is the impact of voltage follower on the common-emitter amplifier. Since the voltage follower has a gain of

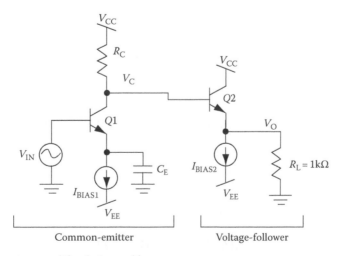

FIGURE 17.48 Two-stage amplifier design problem.

approximately 1, the common-emitter amplifier must meet both the gain and swing specification on its own. The input resistance of the voltage follower acts like a resistive load to the common-emitter amplifier. Section 17.3.5 showed that the input resistance of a common-collector amplifier is approximately βR_L. Therefore, the circuit can be simplified to a common emitter driving a load = $100(1\,k\Omega) = 100\,k\Omega$.

Similar to the previous example, the gain and swing set a range for V_{CQ} that will allow the amplifier to meet the specifications. Both gain and swing must be looked at simultaneously to make sure that both specifications are met.

The swing for this circuit was derived in Section 17.4.3. V_{CMAX} is found when V_{IN} is so low that the transistor is turned off setting $I_{C1} = 0$. The resulting circuit yields a voltage divider between R_C and R_L, where $R_L = r_{in}$ of the voltage follower = $r_{INVF} = 100k\Omega$.

$$V_{CMAX} = V_{CC}\frac{r_{INVF}}{R_C + r_{INVF}} \tag{17.206}$$

V_{CMIN} is found when the transistor is pushed into the saturation region. The emitter capacitor holds the emitter at the bias point, resulting in the following equation for V_{CMIN}:

$$V_{CMIN} = V_{EQ} + V_{CE,SAT} = (0 - V_{BE}) + V_{CE,SAT} = -0.7 + 0.2 = -0.5\text{ V} \tag{17.207}$$

To achieve the 5 V swing, V_{OQ} must be at least 5 V below V_{CMAX} and 5 V above V_{CMIN}, resulting in the following inequality:

$$(-0.5\text{ V} + 5\text{ V}) < V_{CQ} < \left(V_{CC}\frac{r_{INVF}}{R_C + r_{INVF}} - 5\text{ V}\right) \tag{17.208}$$

V_{CQ} was derived in Section 17.2.2 for a load resistor using KCL at the output node:

$$\frac{V_{CC} - V_{CQ}}{R_C} = I_{C1Q} + \frac{V_{CQ}}{r_{INVF}} \tag{17.209}$$

which can be rearranged to produce

$$V_{CQ} = \left(\frac{V_{CC}}{R_C} - I_{C1Q}\right)(R_C \,||\, r_{INVF}) \tag{17.210}$$

The gain for this circuit was derived in Section 17.3.5 and found to be

$$A_V = -g_m R_C \,||\, r_O \,||\, r_{INVF} \tag{17.211}$$

To simplify the design equation, assume that $r_O \gg R_C$ to make the gain equation:

$$|A_V| = |-g_m R_C \,||\, R_L| = \left|-\frac{I_{C1Q}}{V_T}R_C \,||\, r_{INVF}\right| > 200 \tag{17.212}$$

Rearranging and substituting $V_T = 26\,mV$ yields

$$I_{C1Q}(R_C \,||\, r_{INVF}) > 200\,V_T = 5.2\text{ V} \tag{17.213}$$

Using Equations 17.208, 17.210, and 17.213, a range of V_{CQ} values that satisfies both swing and gain can be found. However, these equations cannot be solved simply in their current form. To simplify the equations, a value for R_C is chosen first. To keep the load resistor from dropping the gain and V_{OMAX} high, choose a value of R_C that is less than r_{INVF}. If possible, without producing an R_C value that is too low and thereby an I_C that is too high, choose $R_C = r_{INVF}/10$. Given the value of r_{INVF}, this relationship results in an $R_C = 10\,\text{k}\Omega$. Assume the supplies are connected to ±15 V. If the design results in a wide range for V_{CQ}, a lower rail may be chosen. Using these chosen values in Equations 17.208, 17.210, and 17.213 yields the following equations:

$$4.5\text{ V} < V_{CQ} < 8.6\text{ V} \tag{17.214}$$

$$V_{CQ} = \left(\frac{V_{CC}}{R_C} - I_{CIQ}\right)(R_C\,||\,R_L) = 13.6\text{ V} - 9.1\text{ k}\Omega \cdot I_{CIQ} \tag{17.215}$$

$$I_{CIQ} * 9.1\text{ k}\Omega > 5.2\text{ V} \tag{17.216}$$

Combining Equations 17.214 through 17.216 results in

$$4.5\text{ V} < 13.6\text{ V} - 9.1\text{ k}\Omega * I_{CIQ} < 8.6\text{ V} \tag{17.217}$$

Rearranging provides a limit on I_{CIQ}:

$$1\text{ mA} > I_{CIQ} > 0.549\text{ mA} \tag{17.218}$$

Rearranging Equation 17.216 results in the gain constraint on I_{CIQ}:

$$I_{CIQ} > 0.571\text{ mA} \tag{17.219}$$

The gain further constrains the swing inequality, so the final inequality that satisfies both gain and swing follows:

$$1\text{ mA} > I_{CIQ} > 0.571\text{ mA} \tag{17.220}$$

Choosing a value of I_{CIQ} within the inequality and using $R_C = 10\,\text{k}\Omega$ will set a value for V_{CQ} that meets both swing and gain. To ensure that r_O will not cause the gain to fall under the specification and to ensure that nonlinear swing will not cause V_C to clip, choose a value near the center of the range. Choose $I_{CIQ} = 0.75\,\text{mA}$. To set $I_{CIQ} = 0.75\,\text{mA}$, choose $I_{BIAS1} = 0.75\,\text{mA}$.

To complete the design, I_{BIAS2} must be chosen so that V_O can swing 5 V. When V_O is positive, Q2 provides current to the load. When V_O is negative, the load current is supplied by I_{BIAS2}. The value of I_{BIAS2} must be chosen to bring the output voltage negative. In order to determine the most negative value for V_O, a value for V_{OQ} must be found. The value of V_{OQ} is found by KVL from V_{CQ} to V_{OQ} through the transistor:

$$V_{OQ} = V_{CQ} - V_{BE} \tag{17.221}$$

Using Equation 17.215, V_{CQ} can be determined and then V_{OQ}:

$$V_{CQ} = \left(\frac{V_{CC}}{R_C} - I_{CIQ}\right)(R_C\,||\,R_L) = 6.8\text{ V} \tag{17.222}$$

By Equation 17.221:

$$V_{OQ} = 6.8 - 0.7 = 6.1 \text{ V} \tag{17.223}$$

Given that V_O must swing 5 V, V_{OMIN} can be found by

$$V_{OMIN} = V_{OQ} - 5 \text{ V} = 6.1 \text{ V} - 5 \text{ V} = 1.1 \text{ V}$$

In this case, V_O never goes negative for the 5 V swing, so I_{BIAS2} can be chosen to be any value that will turn Q2 on. I will choose $I_{BIAS2} = 100\,\mu\text{A}$ to save power. However, if V_{OMIN} had been negative, I_{BIAS2} should have been set to a value large enough to sink the current necessary to bring V_O down to the negative value. The minimum current necessary to do this is

$$I_{BIAS2MIN} = \frac{-V_{OMIN}}{R_L} \tag{17.224}$$

Often, the final value is chosen to be a factor of 2 larger than Equation 17.224 to give headroom.

One final consideration is to ensure that I_{CIQ} is large enough that it will not be swamped by I_{B2}. If I_{B2} is approximately the same as I_{CIQ}, then much of the change in I_C due to a change in V_{IN} will be lost to I_{B2}, causing the swing to be nonlinear. A rule of thumb is to ensure that $I_{CIQ} > 10 * I_{B2}$. In this circuit,

$$I_{B2} = \frac{I_{BIAS2}}{\beta} = \frac{100\,\mu\text{A}}{100} = 1\,\mu\text{A}$$

$$I_{CIQ} = 0.75 \text{ mA} > 10\,(\text{IB2}) = 100\,\mu\text{A}$$

The design is complete. Using the circuit components chosen, ensure that the circuit meets the specifications. Minimize the use of assumptions when verifying the answers.

Bias points:

$$I_{CQ} = I_{BIAS1} = 0.75 \text{ mA}$$

$$V_{CQ} = 6.8 \text{ V}$$

Gain:

$A_V = A_{V1} * A_{V2}$; where the input resistance of stage 2 = the load of stage 1

$$|A_{V1}| = g_m(R_C \,||\, R_L \,||\, r_O) = \frac{0.75 \text{ mA}}{26 \text{ mV}}\left(10 \text{ k}\Omega \,||\, 100 \text{ k}\Omega \,||\, \frac{75}{0.75 \text{ mA}}\right) = 240$$

From the gain equation in Section 17.3.4,

$$A_{V2} = \frac{g_m(R_L \,||\, r_O)}{1 + g_m(R_L \,||\, r_O)} = \frac{(100\,\mu\text{A}/26 \text{ mV})(1 \text{ k}\Omega \,||\, (75/100\,\mu\text{A}))}{1 + (100\,\mu\text{A}/26 \text{ mV})(1 \text{ k}\Omega \,||\, (75/100\,\mu\text{A}))} = 0.793$$

$A_V = A_{V1} * A_{V2} = 240 * 0.793 = 190$, which is less than the required gain.

The gain specification was not met due to the impact of r_O and the gain on the voltage follower being a good bit less than 1. To solve this problem, either the value of I_{CIQ} must be raised to increase the gain or the gain of the voltage follower must be improved. Since I_{CIQ} has such a tight constraint, improving the voltage follower gain will be investigated initially.

To improve the gain of the voltage follower, increase the value of I_{BIAS2}. If the load current is too large compared to the bias current, changes in V_O cause significant changes in I_{C2}, which prevents V_{BE} from remaining constant. Increasing I_{BIAS2} will improve this situation. Increase $I_{BIAS2} = 500\,\mu A$ and recalculate:

$$A_{V2} = \frac{g_m(R_L \,||\, r_O)}{1 + g_m(R_L \,||\, r_O)} = \frac{(500\,\mu A/26\,mV)(1\,k\Omega \,||\, (75/500\,\mu A))}{1 + (500\,\mu A/26\,mV)(1\,k\Omega \,||\, (75/500\,\mu A))} = 0.95$$

$A_V = A_{V1} * A_{V2} = 240 * 0.95 = 228$, which is greater than 200

Swing:

$$V_{CMAX} = V_{CC}\left(\frac{R_L}{R_C + R_L}\right) = 15\,V\left(\frac{100\,k\Omega}{10\,k\Omega + 100\,k\Omega}\right) = 13.6\,V$$

$$V_{CMIN} = V_{SS} = -0.5\,V$$

$$V_{SWING}^+ = 13.6\,V - 6.1\,V = 7.5\,V$$

$$V_{SWING}^- = 6.1\,V - (-0.5) = 6.6\,V$$

$$\text{Max swing} = 6.6\,V, \text{ which is} > 5\,V$$

All specifications are now met. Again, the circuit should be simulated to ensure that transistor non-linearity does not compromise the swing specification.

18

A Simplistic Approach to the Analysis of Transistor Amplifiers

Bogdan M.
Wilamowski
Auburn University

J. David Irwin
Auburn University

18.1 Introduction

The traditional approach to the small-signal analysis of transistor amplifiers employs the transistor models with dependent sources, illustrated in Figure 18.1, for both the metal-oxide-silicon (MOS) and bipolar junction transistor (BJT) devices [1,2,4].

In this chapter, techniques for the analysis of transistor circuits will be demonstrated without the use of a small-signal equivalent circuit containing dependent sources [3]. Because of the similarities inherent in the two circuit configurations shown in Figure 18.1, the following analyses will address both MOS and BJT devices in unison.

As a general rule, the small-signal parameters are calculated as a function of the transistor currents. In view of that fact, consider now each type of device.

18.1.1 MOS Transistors

In this case,

$$r_m = \frac{1}{g_m} = \frac{1}{\sqrt{2I_D K}} \tag{18.1}$$

where I_D is the drain biasing current

$$K = \mu C_{ox} \frac{W}{L} = K' \frac{W}{L} \tag{18.2}$$

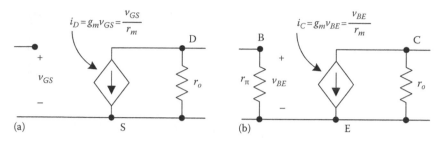

FIGURE 18.1   Small signal equivalent models for: (a) MOS transistor and (b) BJT transistor.

where
  K is the transconductance parameter for a specific transistor with channel length L and channel width W
  K' is a parameter that characterizes the fabrication process and is the same for all transistors in the circuit

The output resistance is given by the expression

$$r_o = \frac{(1/\lambda) + V_{DS}}{I_D} \approx \frac{1}{\lambda I_D} \tag{18.3}$$

where the λ parameter describes the slope of the transistor output characteristics.

18.1.2 Bipolar Junction Transistors

In this case,

$$r_m = \frac{1}{g_m} = \frac{V_T}{I_C} \tag{18.4}$$

and

$$r_\pi = (\beta + 1) r_m \tag{18.5}$$

where
  I_C is the collector biasing current
  V_T is the thermal potential that is equal to about 25 mV at room temperature, as indicated by the expression

$$V_T = \frac{kT}{q} \approx 25 \text{ mV} \tag{18.6}$$

The output resistance is

$$r_o = \frac{V_A + |V_{CE}|}{I_C} \approx \frac{V_A}{I_C} \tag{18.7}$$

where V_A represents the Early voltage that characterizes the slope of the bipolar transistor's output characteristics.

18.2 Calculating Biasing Currents

With MOS transistors, it is assumed that the device operates in the current saturation region. In this region, where $V_{DS} > V_{GS} - V_{th}$, the drain current is given by the expression

$$I_D = \frac{K}{2}\Delta^2 = \frac{K'}{2}\frac{W}{L}\Delta^2 \tag{18.8}$$

where "Δ" = $V_{GS} - V_{th}$ indicates the amount by which the control voltage V_{GS} exceeds the threshold voltage V_{th}.

Example 18.1

Consider the following circuit with the specified parameters (Figure 18.2).

Assuming the gate current is negligible, the gate voltage can be determined from the voltage divider consisting of the resistors R_1 and R_2. Thus, V_G can be calculated as

$$V_G = \frac{R_2}{R_1 + R_2}V_{DD} = 1V \tag{18.9}$$

$$K = K'\frac{W}{L} = 1mA/V^2 \tag{18.10}$$

$$\Delta = V_{GS} - V_{th} = V_G - V_S - V_{th} = 0.4\ V \tag{18.11}$$

Therefore, the biasing drain current is

$$I_D = \frac{K}{2}\Delta^2 = 80\ \mu A \tag{18.12}$$

When the MOS transistor has a series resistor, R_S, connected to the source, determination of the biasing drain current is slightly more complicated. Therefore, this situation is analyzed in Example 18.2.

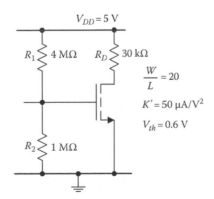

FIGURE 18.2 Circuit for Example 18.1.

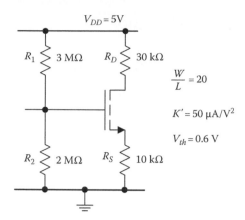

FIGURE 18.3   Circuit for Example 18.2.

Example 18.2

The primary difference between this case and the previous one is the presence of R_S (Figure 18.3).

Once again, using the voltage divider consisting of R_1 and R_2, the gate voltage V_G can be found from the expression

$$V_G = \frac{R_2}{R_1 + R_2} V_{DD} = 2\,\text{V} \tag{18.13}$$

$$K = K' \frac{W}{L} = 1\text{mA/V}^2 \tag{18.14}$$

and by definition

$$\Delta = V_{GS} - V_{th} = V_G - V_S - V_{th} \tag{18.15}$$

Employing Ohm's law and using (18.8) yields

$$V_S = I_D R_S = \frac{K}{2} \Delta^2 R_S \tag{18.16}$$

Combining Equations 18.15 and 18.16 produces

$$\Delta = V_G - \frac{K}{2} \Delta^2 R_S - V_{th} \tag{18.17}$$

which can be written as

$$\frac{KR_S}{2} \Delta^2 + \Delta - (V_G - V_{th}) = 0 \tag{18.18}$$

This is a quadratic equation, the solution of which is

$$\Delta = \frac{-1 \pm \sqrt{1 + 2KR_S(V_G - V_{th})}}{KR_S} \tag{18.19}$$

Since Δ must be positive, then

$$\Delta = \frac{1}{KR_S}\left(\sqrt{1+2KR_S(V_G - V_{th})}-1\right) = \frac{1}{10}\left(\sqrt{1+2\cdot10(1.4)}-1\right) = 0.43385\,\text{V} \qquad (18.20)$$

and thus the drain current is

$$I_D = \frac{K}{2}\Delta^2 = 96.615\,\mu\text{A} \quad \text{or} \quad I_D = \frac{V_G - \Delta - V_{th}}{R_S} = 96.615\,\mu\text{A} \qquad (18.21)$$

Other methods could be used to solve the quadratic equation; e.g., I_D or V_S could serve as unknowns. However, in these cases, it would be difficult to decide which root is the correct answer. This problem does not arise when Δ is selected as the unknown.

Consider now determining the biasing currents for BJTs. In this case, a different tack is needed and will be demonstrated in Example 18.3. The following assumptions are usually made:

$$V_{BE} \approx 0.7\,\text{V} \qquad (18.22)$$

$$I_C = \beta I_B \qquad (18.23)$$

$$I_E \approx I_C \qquad (18.24)$$

Example 18.3

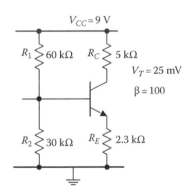

$V_{CC} = 9\,\text{V}$

R_1 60 kΩ R_C 5 kΩ

$V_T = 25\,\text{mV}$

$\beta = 100$

R_2 30 kΩ R_E 2.3 kΩ

FIGURE 18.4 Circuit for Example 18.3.

A BJT common-emitter circuit is shown in Figure 18.4, together with the transistor parameters.

First of all, it is assumed that the base current is negligible and the base voltage V_B can be determined using the voltage divider consisting of R_1 and R_2. This assumption yields the following voltages and currents:

$$V_B = \frac{R_2}{R_1 + R_2}V_{CC} = 3\,\text{V} \qquad (18.25)$$

$$V_E = V_B - V_{BE} = 3 - 0.7 = 2.3 \qquad (18.26)$$

$$I_C \approx I_E = \frac{V_E}{R_E} = \frac{V_B - 0.7}{R_E} = 1\text{mA} \qquad (18.27)$$

It is important to note that with this approximate approach, the current gain β is not needed. In situations where this approximation does not hold, i.e., the base current cannot be neglected, then the resistor divider of Figure 18.4 must be replaced by the Thevenin equivalent circuit, shown in Figure 18.5, where

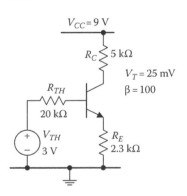

$V_{CC} = 9\,\text{V}$

R_C 5 kΩ

$V_T = 25\,\text{mV}$

R_{TH} $\beta = 100$

20 kΩ

V_{TH} R_E

3 V   2.3 kΩ

FIGURE 18.5 Equivalent circuit for Example 18.3.

$$R_{TH} = R_1 \| R = \frac{R_1 R_2}{R_1 + R_2} \quad \text{and} \quad V_{TH} = \frac{R_2}{R_1 + R_2}V_{CC} \qquad (18.28)$$

Applying Kirchhoff's voltage law to the circuit in Figure 18.5 yields the equation

$$V_{TH} = R_{TH}I_B + V_{BE} + I_E R_E \qquad (18.29)$$

Using Equations 18.22 through 18.24, Equation 18.29 can be rewritten as

$$I_C = \frac{V_{TH} - 0.7}{(R_{TH}/\beta) + R_E} = 0.92\,\text{mA} \tag{18.30}$$

18.3 Small-Signal Analysis

Given the transistor currents, the small-signal parameters r_m and r_o can be determined using Equations 18.1 and 18.3 for MOS transistors and Equations 18.4 and 18.7 for bipolar transistors. MOS transistors can operate in one of three configurations: CS—common source, CD—common drain, and CG—common gate. In a similar manner, bipolar transistors operate in one of the following three configurations: CE—common emitter, CC—common collector, and CB—common base. These different configurations will now be analyzed.

18.3.1 Common-Source and Common-Emitter Configurations

Figure 18.6 illustrates a transistor in the common-emitter configuration. The biasing current for this structure was determined in Example 18.3. An inspection of the circuit indicates that the incremental collector/emitter current Δi_C can be found from Ohm's law as

$$\Delta i_C = \frac{\Delta v_{in}}{r_m + R_E} \tag{18.31}$$

Furthermore, this incremental current, Δi_C, will create an incremental output voltage of

$$\Delta v_{out} = -\Delta i_C R_C \tag{18.32}$$

Substituting Equation 18.30 into Equation 18.32 yields

$$\Delta v_{out} = -\Delta v_{in} \frac{R_C}{r_m + R_E} \tag{18.33}$$

Therefore, the voltage gain of this single stage amplifier is

$$A_V = \frac{\Delta v_{out}}{\Delta v_{in}} = -\frac{R_C}{r_m + R_E} \tag{18.34}$$

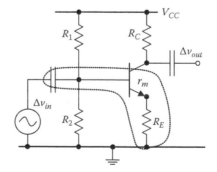

FIGURE 18.6 Small signal analysis of common emitter amplifier.

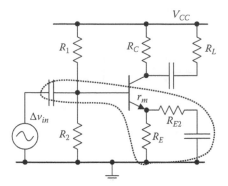

FIGURE 18.7 Small signal analysis of the modified common emitter amplifier.

If the transistor circuit of Figure 18.6 is modified to contain the additional elements shown in Figure 18.7, then Equation 18.31 must be modified as follows:

$$\Delta i_C = \frac{\Delta v_{in}}{r_m + R_E \parallel R_{E2}} \tag{18.35}$$

and the effective load resistance would be the parallel combination of R_C and R_L. As a consequence, the new form for Equation 18.32 is

$$\Delta v_{out} = -\Delta i_C \left(R_C \parallel R_L \right) \tag{18.36}$$

and the transistor voltage gain is

$$A_V = \frac{\Delta v_{out}}{\Delta v_{in}} = -\frac{R_C \parallel R_L}{r_m + R_E \parallel R_{E2}} \tag{18.37}$$

At this point, it is important to note that the traditional lengthy derivation of the gain, which employs the transistor model with a dependent current source as shown in Figure 18.1b, yields the voltage gain A_V:

$$A_V = -\frac{g_m(R_C R_L / R_C + R_L)}{1 + g_m(R_E R_{E2} / R_E + R_{E2})} \tag{18.38}$$

which is, of course, the same as Equation 18.36. If circuit configuration has an additional series base resistance, R_B, then Equation 18.35 should be rewritten as

$$\Delta i_C = \frac{\Delta v_{in}}{(R_B / \beta) + r_m + R_E \parallel R_{E2}} \tag{18.39}$$

and then

$$A_V = \frac{\Delta v_{out}}{\Delta v_{in}} = -\frac{R_C \parallel R_L}{(R_B / \beta) + r_m + R_E \parallel R_{E2}} \tag{18.40}$$

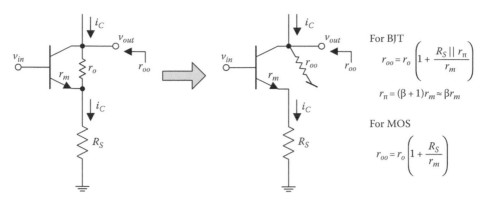

FIGURE 18.8   Evaluation of the output resistance of a transistor.

In the foregoing analysis, the transistor output resistance r_{oo} was ignored. If this output resistance should be included in the calculations, then Equation 18.36 must be slightly modified to include it as follows:

$$A_V = \frac{\Delta v_{out}}{\Delta v_{in}} = -\frac{R_C \,||\, R_L \,||\, r_{oo}}{r_m + R_E \,||\, R_{E2}} \tag{18.41}$$

The calculation of r_{oo} is relatively complicated and requires the use of the traditional small-signal analysis using dependent sources. Figure 18.8 provides the results of this analysis in which slightly different equations must be used for bipolar and MOS transistors. The output resistance of the BJT with R_S connected to its emitter is

$$r_{oo} = r_o + \frac{r_o \left(R_S \,||\, r_\pi \right)}{r_m} + R_S \,||\, r_\pi \approx r_o \left(1 + \frac{R_S \,||\, r_\pi}{r_m} \right) \tag{18.42}$$

where $r_\pi = (\beta + 1)r_m$. In this case, the MOS transistor can be considered as a BJT with $r_\pi = \infty$ and thus,

$$r_{oo} = r_o + \frac{r_o R_S}{r_m} + R_S \approx r_o \left(1 + \frac{R_S}{r_m} \right) \tag{18.43}$$

Calculation of an amplifier's input and output resistances are also an important part of small-signal analysis. The circuit in Figure 18.6 indicates that the input resistance is

$$r_{in} = R_1 \,||\, R_2 \,||\, \left(\beta(r_m + R_E) \right) \tag{18.44}$$

Note that the assumption that $\beta + 1 \approx \beta$ is consistently employed to simplify the equations. This assumption is based upon the fact that there is no good reason to use $\beta + 1$ because the actual value of β is never really known, and furthermore, β fluctuates with temperature (about 1% per °C). Thus, an analysis of the circuit in Figure 18.7 indicates that

$$r_{in} = R_1 \,||\, R_2 \,||\, \left(\beta(r_m + R_E \,||\, R_{E2}) \right) \tag{18.45}$$

Clearly, if the BJT in Figure 18.6 or Figure 18.7 is replaced by a MOS device ($\beta = \infty$), then the input resistance would simply be

$$r_{in} = R_1 \,||\, R_2 \tag{18.46}$$

18.3.2 Common-Drain and Common-Collector Configurations

Figure 18.9 illustrates a BJT in the common-collector configuration. An inspection of this circuit indicates that the incremental collector/emitter current Δi_C is

$$\Delta i_C = \frac{\Delta v_{in}}{r_m + R_E} \tag{18.47}$$

and the output voltage is

$$\Delta v_{out} = \Delta i_C R_E \tag{18.48}$$

Therefore, the voltage gain is

$$A_V = \frac{\Delta v_{out}}{\Delta v_{in}} = \frac{R_E}{R_E + r_m} \tag{18.49}$$

Note carefully that Equation 18.48 is simply the equation for a voltage divider with resistors R_E and r_m:
In the event that a base resistance is present, as shown in Figure 18.10, then using Kirchhoff voltage law:

$$\Delta v_{in} = R_B \Delta i_B + (r_m + R_E)\Delta i_C = \left[\frac{R_B}{\beta} + (r_m + R_E)\right]\Delta i_C \tag{18.50}$$

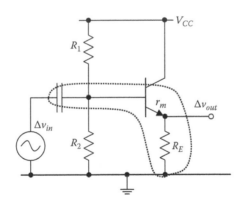

FIGURE 18.9 Transistor circuit in the common collector configuration.

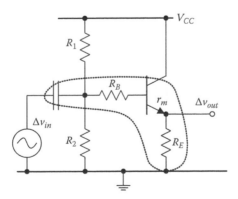

FIGURE 18.10 Transistor circuit in the common collector configuration.

and thus

$$\Delta i_C = \frac{\Delta v_{in}}{(R_B/\beta) + r_m + R_E} \tag{18.51}$$

$$\Delta v_{out} = \Delta i_C R_E \tag{18.52}$$

and once again, the voltage gain is determined from the resistor divider equation

$$A_V = \frac{\Delta v_{out}}{\Delta v_{in}} = \frac{R_E}{(R_B/\beta) + r_m + R_E} \tag{18.53}$$

18.3.3 Common-Gate and Common-Base Configurations

Figure 18.11 is an illustration of a transistor in the common-collector configuration. Note that in this case, the incremental collector/emitter current Δi_C is

$$\Delta i_C = \frac{\Delta v_{in}}{r_m} \tag{18.54}$$

The small-signal analysis for this situation ignores the resistor R_E since it is connected in parallel with the ideal voltage source. The voltage drop across resistor R_C is

$$\Delta v_{out} = \Delta i_C R_C \tag{18.55}$$

and thus the voltage gain is the ratio of two resistors:

$$A_V = \frac{\Delta v_{out}}{\Delta v_{in}} = \frac{R_C}{r_m} \tag{18.56}$$

The addition of a base resistor, as indicated in Figure 18.12, makes the circuit slightly more complicated, and in this case, Equation 18.56 must be modified to include this R_B resistor. The resulting equation is

$$A_V = \frac{\Delta v_{out}}{\Delta v_{in}} = \frac{R_C}{r_m + (R_B/\beta)} \tag{18.57}$$

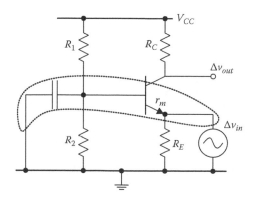

FIGURE 18.11   Transistor circuit in the common base configuration.

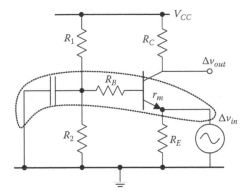

FIGURE 18.12   Transistor circuit in common base configuration with series resistance in the base.

Note that voltage drop across R_B is $\Delta v_{RB} = i_B R_B$ and since $i_C = \beta i_B$ then $\Delta v_{RB} = i_C R_B/\beta$. The voltage drop across r_m is $i_C r_m$. Since the denominator in Equation 18.57 represents the sum of these two voltage drops, Equation 18.57 indicates that the voltage gain is also equal to the ratio of two resistors.

In a small-signal analysis, MOS transistors and BJTs are treated basically the same, even though their small-signal parameters r_m and r_o are calculated differently. One may also treat a MOS transistor as a BJT with a current gain $\beta = \infty$

Example 18.4

Consider now the use of the proposed method for the same circuits that were analyzed in a classical way in Chapter 17 of this book. Figure 17.14a of the previous chapter is repeated in Figure 18.13.

An inspection of the circuit indicates that

$$r_{in} = R_1 \| R_2 = \frac{R_1 R_2}{R_1 + R_2} \qquad (18.58)$$

where r_{in} is the parallel combination of the two biasing resistors R_1 and R_2. In the MOS transistor case, its input resistance can be neglected. The output resistance is again a parallel combination of R_D and r_o

$$r_{out} = R_D \| r_o = \frac{R_D r_o}{R_D + r_o} \qquad (18.59)$$

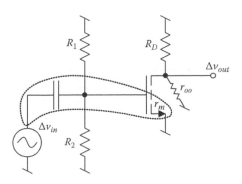

FIGURE 18.13   MOS transistor circuit in the common source configuration.

The gain of the amplifier is then the ratio of the output resistance to r_m, provided that there is no series resistance connected to the source:

$$A_V = -\frac{R_D \| r_o}{r_m} \tag{18.60}$$

This same result would require several pages of analysis using the traditional approach.

Example 18.5

Consider now the circuit shown in Figure 18.14a, which is identical to the one in Figure 17.18a in Chapter 17. The amplifier gain, and the input and output resistances can be found using the approach presented here. In this case, the amplifier gain is

$$A_V = \frac{-R_C \| R_L \| r_{oo}}{r_m + R_E} \tag{18.61}$$

where r_{oo} is given by Equation 18.43, in which R_s is replaced with R_E.

The traditional approach, outlined in the previous chapter where the effect of r_o was ignored, yielded the gain equation

$$A_V = \frac{-g_m(R_C \| R_L)}{1 + R_E((1/r_\pi) + g_m)} \tag{18.62}$$

With the substitutions $g_m = 1/r_m$ and $r_\pi = \beta\, r_m$ the gain becomes

$$A_V = \frac{-R_C \| R_L}{r_m(1 + R_E((1/\beta r_m) + (1/r_m)))} = \frac{-R_C \| R_L}{(r_m + R_E((1/\beta) + 1))} \approx \frac{-R_C \| R_L}{r_m + R_E} \tag{18.63}$$

The input resistance of r_{in} of the circuit shown in Figure 18.14a represents the parallel combination of the two biasing resistors R_1, R_2 and $\beta(r_m + R_E)$

$$r_{in} = R_1 \| R_2 \| \beta(r_m + R_E) \tag{18.64}$$

The output resistance is the parallel combination of R_C, R_L, and r_{oo}

$$r_{out} = R_C \| R_L \| r_{oo} = R_C \| R_L \| \left(r_o \left(1 + \frac{R_E \|(r_m \beta)}{r_m}\right)\right) \tag{18.65}$$

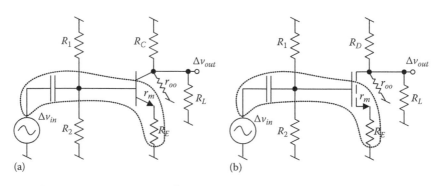

FIGURE 18.14 Common emitter/source configurations (a) Bipolar transistor and (b) MOS transistor.

In the MOS transistor case (Figure 18.14b), the input resistance of the transistor can be neglected, and the output resistance is once again a parallel combination of R_D, R_L, and r_{oo}, so

$$r_{in} = R_1 \| R_2 \tag{18.66}$$

The output resistance is the parallel combination of R_C, R_L, and r_{oo}

$$r_{out} = R_D \| R_L \| r_{oo} = R_D \| R_L \| \left(r_o \left(1 + \frac{R_S}{r_m} \right) \right) \tag{18.67}$$

18.4 Circuits with PNP and PMOS Transistors

Circuits with PNP and PMOS transistors can be handled in a manner similar to that employed for circuits with NPN and NMOS transistors. In order to avoid confusion, the best approach is the use of mirrored circuits, as shown in Figure 18.15. Note that the mirror image circuit is exactly the same as the original one, but simply drawn differently. In the mirror image circuit, the reference voltage is changed so that the bottom node is at 0 V potential and the top node is at −8 V potential. With this configuration, the dc analysis for the mirror image circuit is now essentially identical to that of the NPN transistor in Example 18.3. The approximate solution for this case yields

$$V_B = \frac{4 \text{ k}\Omega}{4 \text{ k}\Omega + 14 \text{ k}\Omega} (-9 \text{ V}) = 2 \text{ V} \tag{18.68}$$

$$V_E = V_B - V_{BE} = -2 - (-0.7) = -1.3 \text{ V} \tag{18.69}$$

$$I_C \approx I_E = \frac{V_E}{R_E} = \frac{-1.3 \text{ V}}{1.3 \text{ k}\Omega} = -1 \text{ mA} \tag{18.70}$$

With the more accurate approach, the resistor divider consisting of the 4 and 12 kΩ resistors must be replaced by a Thevenin equivalent circuit in which

$$R_{TH} = 4 \text{ k}\Omega \| 14 \text{ k}\Omega = \frac{4 \text{ k}\Omega \cdot 14 \text{ k}\Omega}{4 \text{ k}\Omega + 14 \text{ k}\Omega} = 3.111 \text{ k}\Omega \quad \text{and} \quad V_{TH} = \frac{4 \text{ k}\Omega}{4 \text{ k}\Omega + 14 \text{ k}\Omega} (-9 \text{ V}) = 2 \text{ V} \tag{18.71}$$

Now the accurate value of current can be determined as

$$I_C = \frac{V_{TH} - 0.7}{(R_{TH}/\beta) + R_E} = \frac{2 - 0.7}{(3.111 \text{ k}\Omega/100) + 1.3 \text{ k}\Omega} = 0.977 \text{ mA} \tag{18.72}$$

If the mirror image circuit, shown in Figure 18.16, is used, the small-signal analysis can also be used in a manner similar to that employed for NPN transistors. First, the small-signal parameters are calculated as

$$r_m = \frac{V_T}{I_C} = \frac{25 \text{ mV}}{1 \text{ mA}} = 25 \text{ }\Omega \tag{18.73}$$

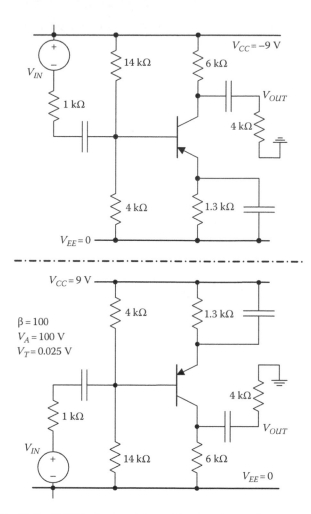

FIGURE 18.15   Circuit with the PNP bipolar transistor and it mirror image above.

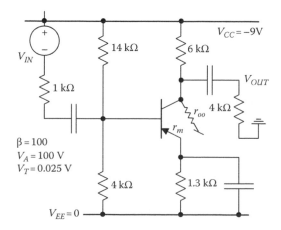

FIGURE 18.16   Mirrored circuit from Figure 18.15 with the inclusion of the small signal parameters.

$$r_o = \frac{V_A}{I_C} = \frac{100 \text{ V}}{1 \text{ mA}} = 100 \text{ k}\Omega \tag{18.74}$$

Then, the input and load resistances are computed. The input resistance, as seen by input capacitor, is

$$r_{in} = 14 \text{ k}\Omega \,\|\, 4 \text{ k}\Omega \,\|\, (\beta \cdot r_m) = 14 \text{ k}\Omega \,\|\, 4 \text{ k}\Omega \,\|\, 2.5 \text{ k}\Omega = 1.386 \text{ k}\Omega \tag{18.75}$$

Because the emitter of the PNP transistor is grounded, $r_{oo} = r_o$. The load resistance, i.e., the resistance between the collector and ground, is

$$r_L = 6 \text{ k}\Omega \,\|\, 4 \text{ k}\Omega \,\|\, r_{oo} = 6 \text{ k}\Omega \,\|\, 4 \text{ k}\Omega \,\|\, 100 \text{ k}\Omega = 2.344 \text{ k}\Omega \tag{18.76}$$

Because of the presence of the $1 \text{ k}\Omega$ series source resistance, the gain calculation must be done in two steps. First, the gain (loss) from the signal source to the transistor base is calculated using the resistor divider:

$$A_{V1} = \frac{r_{in}}{r_{in} + 1 \text{ k}\Omega} = \frac{1.386 \text{ k}\Omega}{1.386 \text{ k}\Omega + 1 \text{ k}\Omega} = 0.581 \tag{18.77}$$

The voltage gain of the transistor from base to collector is

$$A_{V2} = \frac{r_L}{r_m} = \frac{2.344 \text{ k}\Omega}{25 \, \Omega} = 93.76 \tag{18.78}$$

Then, the total circuit gain is

$$A_V = A_{V1} A_{V2} = 0.581 \cdot 93.76 = 54.5 \tag{18.79}$$

18.5 Analysis of Circuits with Multiple Transistors

In integrated circuits, especially MOS technology, it is much easier, and cheaper, to fabricate transistors than resistors. Therefore, a new concept in circuit design was developed resulting in a circuit in which the number of transistors is typically larger than the number of resistors. In addition, in integrated circuits, it is not possible to use large values of capacitance. Figure 18.17 shows how the blocking capacitor can be replaced by an additional transistor. The circuit in Figure 18.17a has the voltage gain of

$$A_V = -\frac{R_D \,\|\, r_{oo}}{r_m} \approx -\frac{R_D}{r_m} \tag{18.80}$$

In this case, it is assumed that the signal frequency is large enough so that the capacitor C shorts the resistor R_S. This assumption is, of course, invalid if the signal frequency is lower than $1/R_S C$. In the modified circuit, i.e., Figure 18.17b, the resistance seen by the source is equal to parallel combination of R_S and the r_m of the M2 transistor. Note that circuit of Figure 18.17b also works well at low frequencies. The voltage gain is then

$$A_V = -\frac{R_D \,\|\, r_{oo}}{r_{m1} + r_{m2} \,\|\, R_S} \approx -\frac{R_D}{2 r_m} \tag{18.81}$$

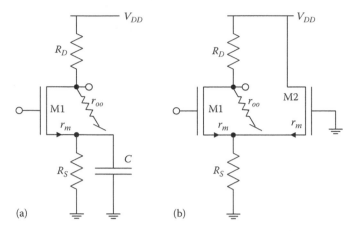

FIGURE 18.17 Replacement of the blocking capacitor in the traditional circuit (a) by a transistor in (b) which is the configuration for integrated circuit design.

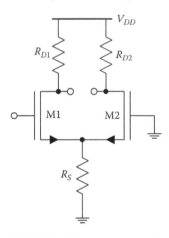

FIGURE 18.18 A simple differential pair.

Proceeding up a notch, consider the circuit shown in Figure 18.18. The voltage gain from the input to the drain of M1 is given by Equation 18.81. The gain calculation from the input to the drain of M2 should be done in two steps. First, the common-drain approach is used to calculate the gain from the input to the node associated with the transistor sources:

$$A_{V1} = -\frac{r_{m2} \parallel R_S}{r_{m1} + r_{m2} \parallel R_S} \approx 0.5 \tag{18.82}$$

Then, the gain from the source to the drain of M2 is determined using the common gate configuration, which yields

$$A_{V2} = +\frac{R_{D2} \parallel r_{oo}}{r_{m2}} \approx \frac{R_{D2}}{r_m} \tag{18.83}$$

Finally, the total voltage gain is

$$A_V = A_{V1} A_{V2} = \frac{R_{D2}}{2r_m} \tag{18.84}$$

The next example is an analysis of a simple amplifier with a current mirror as shown in Figure 18.19. The current mirror is composed of the transistors M3 and M4, and it is assumed that current i_3 is equal to current i_1. A popular technique used to analyze the differential amplifier employs the assumption that half of the input voltage is applied at input 1 and another half, with opposite sign, is applied at input 2. This voltage on the nodes associated with the sources of M1 and M2 is not changing and is considered to be a virtual ground. Therefore, the input 1 signal, which is equal to $0.5v_{in}$, drives r_{m1}:

$$\Delta i_1 = \frac{0.5\Delta v_{in}}{r_{m1}} \tag{18.85}$$

and the other half of the input signal, with opposite sign, drives r_{m2}:

$$\Delta i_2 = -\frac{0.5\Delta v_{in}}{r_{m2}} \tag{18.86}$$

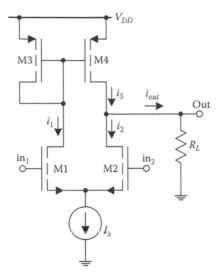

FIGURE 18.19 Amplifier with a current mirror.

Since $i_3 = i_1$, then

$$\Delta i_{out} = \Delta i_3 - \Delta i_2 = \Delta i_1 - \Delta i_2 = 0.5\Delta v_{in}\left(\frac{1}{r_{m1}} + \frac{1}{r_{m2}}\right) = \frac{\Delta v_{in}}{r_m} \tag{18.87}$$

and the incremental output voltage is

$$\Delta v_{out} = \Delta i_{out} R_L = \frac{R_L}{r_m}\Delta v_{in} \tag{18.88}$$

Hence, the voltage gain is

$$A_v = \frac{\Delta v_{out}}{\Delta v_{in}} = \frac{R_L}{r_m} \tag{18.89}$$

If the output resistance, r_o, of each transistor must be considered, then these resistances should be connected in parallel with the loading resistor R_L. Under this condition,

$$A_v = \frac{\Delta v_{out}}{\Delta v_{in}} = \frac{R_L \,||\, r_{oP} \,||\, r_{oN}}{r_m} \tag{18.90}$$

The next example considers the analysis of the two-stage amplifier shown in Figure 18.20. Assuming all transistors are the same size, one finds that the drain currents of transistors M5, M6, M7, and M8 are equal to I_S. In addition, the drain currents of transistors M1, M2, M3, and M4 are equal to $0.5I_S$, assuming an equal division of biasing currents between M1 and M2. From this knowledge of the transistor currents, the small signal parameters, r_m and r_o, can be found using Equations 18.1 and 18.3. An inspection of this circuit, and use of Equation 18.88, yields the voltage gain of the first stage as

$$A_{v1} = \frac{r_{oP} \,||\, r_{oN}}{r_m} = \frac{r_{o4} \,||\, r_{o2}}{r_m} \tag{18.91}$$

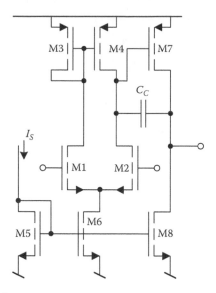

FIGURE 18.20 Two stage amplifier.

Note that in this stage there is no R_L. In addition, it is assumed that the source current was equally divided and $r_{m1} = r_{m2} = r_m$.

The second stage of the circuit in Figure 18.20 is the common-source amplifier consisting of transistor M7 with transistor M8 acting as the load. Therefore, the voltage gain of the second stage is

$$A_{v2} = \frac{r_{o7} \| r_{o8}}{r_{m7}} \tag{18.92}$$

The total voltage gain is

$$A_v = A_{v1} A_{v2} = \frac{r_{o4} \| r_{o2}}{r_{m1}} \frac{r_{o7} \| r_{o8}}{r_{m7}} \tag{18.93}$$

If this circuit is considered to be a transconductance amplifier, then the transconductance parameter, G_m, is

$$G_m = \frac{r_{o4} \| r_{o2}}{r_{m1}} \frac{1}{r_{m7}} \tag{18.94}$$

and the output resistance of the circuit is

$$r_{out} = r_{o7} \| r_{o8} \tag{18.95}$$

Obviously, for MOS input stages, the input resistance is $r_{in} = \infty$.

It can be shown that other key parameters of the amplifier of Figure 18.20 can also be found using the simplistic analysis method. For example, the location of the first pole, i.e., the first-corner frequency shown in Figure 18.21, is given by the expression

$$\omega_0 = \frac{1}{A_v r_{m1} C_C} = \frac{1}{A_{v1} A_{v2} r_{m1} C_C} \tag{18.96}$$

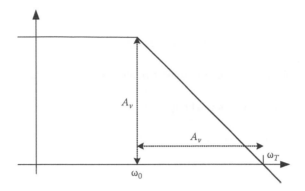

FIGURE 18.21 Frequency response of the amplifier of Figure 18.20.

Then the Gain-Bandwidth product, GB, for the amplifier is equal to the cutoff frequency defined by the time constant $r_{m1}C_C$

$$GB = A_{v1}A_{v2}\omega_0 = \omega_T = \frac{1}{C_C r_{m1}} \tag{18.97}$$

The second pole is located in the high frequency range above ω_T, and it can usually be ignored because the amplifier gain at this frequency is smaller than one. If this is not the case, then the value of C_C should be chosen large enough that the ratio between the second and first pole frequencies is larger than A_V.

The Slew Rate, SR, is dependent upon the speed with which the capacitor C_C can be charged. With the largest possible input voltage difference, the maximum charging current would be the source current of M6 that is approximately equal to I_S. Therefore,

$$SR = \frac{I_S}{C_C} \tag{18.98}$$

The circuit in Figure 18.20 has a relatively large output resistance, given by Equation 18.94, and should be considered an Operational Transconductance Amplifier (OTA), rather than an Operational Amplifier (OPAMP). An OTA can be converted into an OPAMP by adding a unity gain buffer to the circuit. This additional element could be a simple voltage follower, as shown in Figure 18.22a, or more advanced push–pull amplifier, as shown in Figure 18.22b.

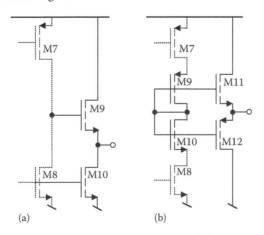

(a)   (b)

FIGURE 18.22 Output buffers used to reduce the output resistance of the amplifier in Figure 18.20.

References

1. Allen, P. and D. Holberg, *CMOS Analog Circuit Design*, Oxford University Press, Oxford, U.K., 2002, ISBN 0-19-511644-5.
2. Jaeger, R. C. and T. N. Blalock, *Microelectronic Circuit Design*, 3rd edition, McGraw-Hill, New York, 2008.
3. Wilamowski, B. M., Simple way of teaching transistor amplifiers, in *ASEE 2000 Annual Conference*, St. Louis, MO, June 18–21, 2000, CD-ROM session 2793.
4. Wilamowski, B. M. and R. C. Jaeger, *Computerized Circuit Analysis Using SPICE Programs*, McGraw-Hill, New York, 1997.

19

Analog and Digital VLSI Design

Vishal Saxena
Boise State University

R. Jacob Baker
Boise State University

19.1 Introduction

Very-large-scale integration (VLSI) is a term used to describe integrated circuit designs using thousands, or more, of field effect transistors (FETs) [Mead79]. VLSI chips can be implemented using several methods, including gate arrays, standard cells, and full-custom design [Baker08]. In this short tutorial, we will focus on full-custom design or design at the transistor level. We will use complementary metal-oxide semiconductor (CMOS) technology that includes an n-type metal-oxide FET, n-type MOSFET or NMOS for short, and a p-type MOSFET or PMOS.

19.2 CMOS Devices and Layout

Figure 19.1 shows the schematic symbol and layout of NMOS and PMOS devices. When the body of the MOSFET is not drawn, it is assumed to be connected to ground, for the NMOS device, and to *VDD*, for the PMOS device. For digital design, we think of the MOSFET as a switch. When the gate of the NMOS is driven to *VDD*, the switch is closed. When the gate of the NMOS is at ground, the switch is open. When the gate of the PMOS is connected to ground, the switch turns on, hence the reason for the bubble on the MOSFET's gate, and when the PMOS gate is high (at *VDD*), the switch is open. This complementary behavior is why the technology is called CMOS.

Figure 19.2 shows the layout of the PMOS and NMOS devices. The NMOS device sits in a p-type material called a p-well while the PMOS device sits in an n-well. The active areas for NMOS and PMOS devices are heavily doped n+ and p+, respectively. Connections to the wells are made from metal through a contact to n+, for the n-well, or p+, for the p-well. These connections are called the body, B, of the MOSFET, see Figure 19.3. For digital design, in most situations, again, the body of the NMOS is connected to ground while the body of the PMOS is connected to *VDD*. The gate of the MOSFET is made using polysilicon (poly). Poly crossing active forms a MOSFET.

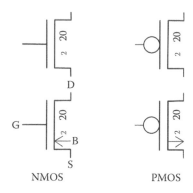

FIGURE 19.1 Schematic symbols for NMOS and PMOS devices.

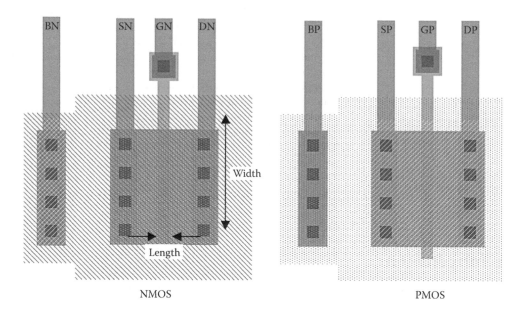

FIGURE 19.2 Layout of NMOS and PMOS devices.

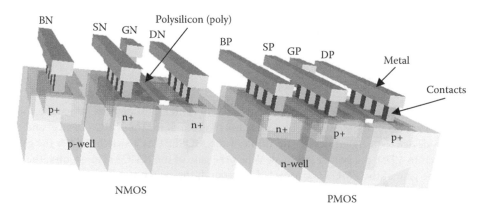

FIGURE 19.3 Three-dimensional views of NMOS and PMOS devices.

19.3 Electrical Behavior for Digital Design

The current that flows in the NMOS device when the source is connected to ground and the drain/gate are connected to *VDD* is the MOSFET's on current, $I_{on,n}$. The resistance between the source and drain can be estimated using

$$R_n = \frac{VDD}{I_{on,n}} \cdot \frac{L}{W} \tag{19.1}$$

where
    L is the length of the MOSFET
    W is the MOSFET's width, see Figure 19.2

Generally, in digital design, the MOSFET's length is set to the minimum while the width is increased to reduce the MOSFET's switching resistance. Figure 19.4 shows how the NMOS is thought of as a resistor for digital design when its gate is driven high to *VDD*.

For the PMOS device, Figure 19.5, the switch turns on when the gate is connected to ground. The effective switching resistance is determined by measuring the PMOS on current, $I_{on,p}$, with the gate and drain at ground with the source connected to *VDD* and is given by

$$R_p = \frac{VDD}{I_{on,p}} \cdot \frac{L}{W} \tag{19.2}$$

Again, notice the complementary behavior of the NMOS and PMOS devices.

When gate, *G*, is high the MOSFET is a resistor. $R_n = \frac{VDD}{I_{on,n}} \cdot \frac{L}{W}$

FIGURE 19.4 Digital model of an NMOS device.

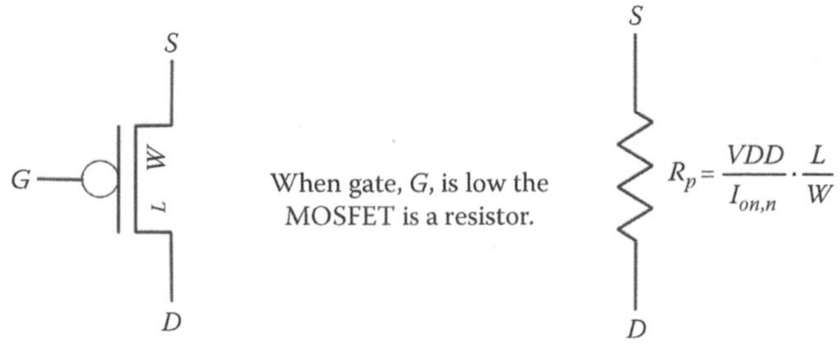

When gate, *G*, is low the MOSFET is a resistor. $R_p = \frac{VDD}{I_{on,n}} \cdot \frac{L}{W}$

FIGURE 19.5 Digital model of a PMOS device.

19.4 Electrical Behavior for Analog Design

Characterizing the electrical behavior of the MOSFET for analog design is considerably more complicated than it is for digital design. Figure 19.6 shows a schematic for extracting the drain current variation with changes in drain-source voltage, V_{DS}, and gate-source voltage, V_{GS}. Typical current–voltage
curves are seen in Figure 19.7. The gate-source voltage is held constant and greater than the threshold
voltage, V_{THN} (here 800 mV), while the drain source is varied from 0 to 5 V. The change in drain current
is then plotted.

When the drain current varies linearly with V_{DS}, the reciprocal of the slope of the curve is the resistance between the MOSFET's drain and source. In this region, the MOSFET is said to be operating in
the triode region (aka linear or ohmic region). When the drain current curves flatten out, or saturate,

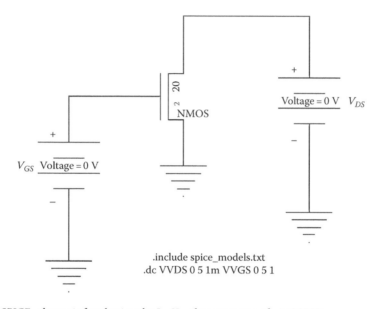

FIGURE 19.6   SPICE schematic for plotting the I_D–V_{DS} characteristics of a MOSFET.

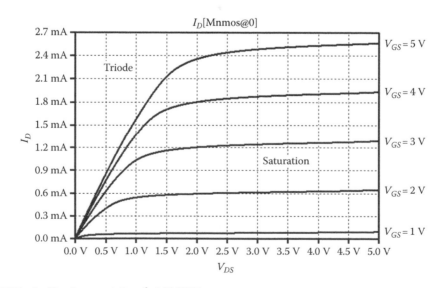

FIGURE 19.7 I_D–V_{DS} characteristics of a MOSFET.

the MOSFET is operating in the saturation region. The drain current at the border between saturation and triode is called $I_{D,sat}$. The drain-source voltage at this border point is called $V_{DS,sat}$. When the MOSFET is operating in the saturation region, it behaves like a current source of $I_{D,sat}$ in parallel with a resistor, r_o. Again, the reciprocal of the slope of the drain current in the saturation region tells us the resistance between the drain and source, r_o, or

$$r_o = \frac{1}{\lambda \cdot I_{D,sat}} \tag{19.3}$$

where λ has units of V^{-1} and generally ranges from 0.01 to 0.2 V^{-1}. The behavior of the NMOS device operating in the saturation region can be described using

$$I_D = \frac{\beta_n}{2} \cdot (V_{GS} - V_{THN})^2 \cdot (1 - \lambda(V_{DS} - V_{DS,sat})) \tag{19.4}$$

for

$$V_{GS} > V_{THN} \quad \text{and} \quad V_{DS} \geq V_{GS} - V_{THN} = V_{DS,sat} \tag{19.5}$$

Note also that

$$I_{D,sat} = \frac{\beta_n}{2} \cdot (V_{GS} - V_{THN})^2 \tag{19.6}$$

While these equations were specified for the NMOS device, by swapping V_{GS} with V_{SG}, V_{DS} with V_{SD}, and V_{THN} with V_{THP} they can also characterize the behavior of the PMOS device (all quantities are positive).

While r_o is used to characterize how I_D changes with changes in V_{DS}, we also need to characterize how I_D varies with AC changes, v_{gs}, around a fixed DC V_{GS} (the device's transconductance, g_m). Using

$$\frac{d(i_d + I_D)}{dv_{gs}} = \frac{d}{dv_{gs}} \left[\frac{\beta_n}{2} (v_{gs} + V_{GS} - V_{THN})^2 \right] \tag{19.7}$$

or, assuming $v_{gs} \ll V_{GS}$ (small-signal approximation)

$$g_m = \beta_n \cdot (V_{GS} - V_{THN}) = \sqrt{2 I_D \beta_n} \tag{19.8}$$

This equation is useful because it tells us that if we turn the MOSFET on with a $V_{GS} > V_{THN}$ and apply a small AC signal on top of this, v_{gs}, then the current that flows, i_d is $g_m \cdot v_{gs}$.

19.5 Digital VLSI Design

The most basic element used in digital VLSI design is the inverter, Figures 19.8 and 19.9. When the input is a logic "1" or *VDD*, the PMOS device is off while the NMOS device is on (behaves like a resistor). When the input goes low, to ground or logic "0," the PMOS turns on, while the NMOS shuts off. The delay associated with driving a capacitive load, C_{load}, is

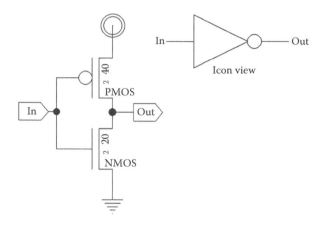

FIGURE 19.8 A CMOS inverter and icon.

$$t_{PHL} = 0.7 \cdot R_n C_{load} \tag{19.9}$$

for a high-to-low delay, and

$$t_{PLH} = 0.7 \cdot R_p C_{load} \tag{19.10}$$

for a low-to-high delay.

19.5.1 Logic Gate Design

The power of CMOS digital VLSI is the ease with which logic functions can be implemented. Figure 19.10 shows how NMOS devices are used to implement NAND and NOR functions. For the NAND function when all inputs, A AND B AND C, are high, the output is pulled to ground. In the NOR implementation, any of the inputs A OR B OR C pull the output to ground. Figure 19.11 shows the PMOS implementation of NAND and NOR functions. For the NAND function, the output goes high when the inputs are all low. For the NOR function, only one input need be high to pull the output high. Notice how the NMOS and PMOS circuits and inputs are complementary (hence, again, the name CMOS).

Figure 19.12 shows the schematic of a 3-input NAND gate. The NAND gate is the preferred topology for VLSI design using CMOS technology because the switching resistance of the PMOS devices is larger than the switching resistance of the NMOS. Putting the NMOS devices in series is thus more desirable than putting the PMOS devices in series for fast switching speeds. For a 3-input NAND gate, the delay associated with driving a load capacitance from *VDD* to ground (high-to-low propagation delay) is

$$t_{PHL} = 0.7 \cdot 3R_n C_{load} \tag{19.11}$$

while the delay associated charging a capacitive load is

$$t_{PLH} = 0.7 \cdot 3R_p C_{load} \tag{19.12}$$

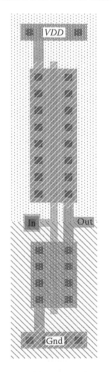

FIGURE 19.9 Layout of a CMOS inverter.

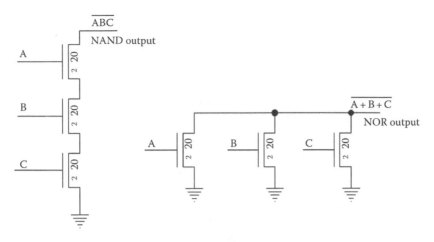

FIGURE 19.10 NAND and NOR implementation using NMOS devices.

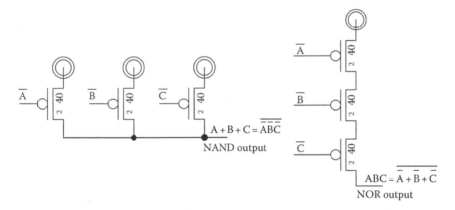

FIGURE 19.11 NAND and NOR implementation using PMOS devices.

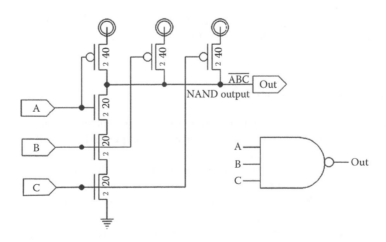

FIGURE 19.12 A CMOS NAND gate.

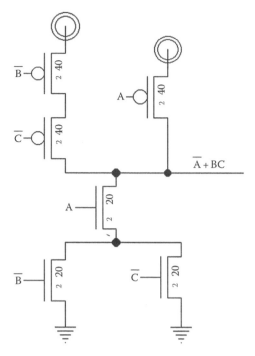

FIGURE 19.13   A complex CMOS logic gate.

19.6 Complex Logic Gate Design

More complex logic functions can easily be implemented using CMOS technology. Consider a single logic gate to implement the logic function:

$$Z = \overline{A} + BC \tag{19.13}$$

The schematic to implement this logic function is seen in Figure 19.13. Notice how the parallel connection of NMOS devices, those connected to $\overline{B}$ and $\overline{C}$, for example are implemented with a series combination of PMOS devices. The series combination of NMOS, the NMOS controlled by A in series with the NMOS connected to $\overline{B}$ and $\overline{C}$, is implemented using a parallel combination of PMOS. The implementation of the logic function is complementary. Note that the more inputs to the gate, the larger the delay associated with driving a load capacitance. This limits the complexity of a single gate in a practical situation.

19.7 Latches and Flip-Flops

The CMOS logic gates presented in the last section are used to implement combinatorial logic. On the other hand, latches and flip-flops are used for realizing sequential logic with storage elements. A latch is a storage circuit whose output changes when the clock (CLK) signal is high. Latch implementation in CMOS is based on the cross-coupled inverter circuit shown in Figure 19.14a. The positive feedback in the cross-coupled latch recirculates and stores the input signal until it is forced to change again. Figure 19.14b shows a level-sensitive latch. When the clock is high, the data input D drives the inverters and the input data is registered. When the clock goes low, the input is disconnected and the inverters are connected in the cross-coupled manner. At this instance, the value of the input D is stored in the latch.

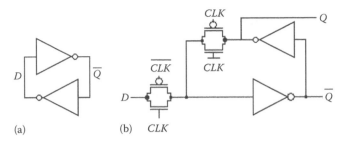

FIGURE 19.14 (a) A basic latch formed using cross-coupled inverters and (b) a CMOS level-sensitive latch.

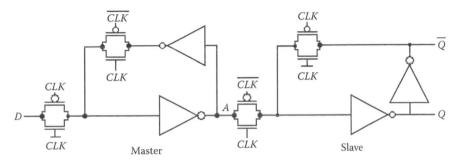

FIGURE 19.15 A positive edge-triggered D-Flip-Flop.

In practical digital systems, it is desirable for perfect synchronization that system state changes exactly on the clock edges instead of depending on the clock levels. This is achieved by employing edge-triggered D-flip-flops (D-FF) for storage. A D-FF is realized by cascading two level-sensitive latches as shown in Figure 19.15. When the clock is low, the first "master" stage tracks the D input whiles the second "slave" stage holds the previous input. At the instance, when the clock goes high, the master captures the input and transfers it to the slave. This causes the output Q to change with the input D only when the clock transitions high. When the clock goes back low again, the value of Q recirculates around the cross-coupled inverters in the slave. The output state Q can change again, following the input D, when the clock goes back high again and thus forming a *positive* edge-triggered flip-flop. Similarly, a *negative* edge-triggered flip-flop can be realized by swapping CLK and CLK signals in the positive edge-triggered D-FF.

19.8 Analog VLSI Design

When we do digital VLSI design, we treat the MOSFET as a switch. When we do analog design, we treat the MOSFET, as mentioned in Section 19.4, as a transconductance amplifier (voltage input, v_{gs}, and current output, i_d). In other words, we can think of the gain of the NMOS device as

$$g_m = \frac{i_d}{v_{gs}} \tag{19.14}$$

This equation assumes that the drain is connected to a low resistance, a voltage source to keep the MOSFET in saturation, Figure 19.6. If the MOSFET drives its own resistance, r_o, so the output voltage is $v_{out} = i_d \cdot r_o$ we can re-write Equation 19.14 as

$$\frac{v_{out}}{v_{in}} = g_m \cdot r_o = \frac{i_d \cdot r_o}{v_{gs}} \tag{19.15}$$

The term $g_m \cdot r_o$ is often called the MOSFET's open-circuit gain.

19.9 Biasing for Analog Design

For analog design, the drain current, I_D, the gate-source voltage, V_{GS}, and, in some cases, the drain-source voltage, V_{DS}, must be set. The proper selection of these electrical characteristics is said to bias the MOSFETs. The equations governing MOSFET operation were given earlier (19.4) through (19.6). Figure 19.16 shows a biasing circuit that can be used in analog design. The PMOS device is a long-length device to drop a significant voltage. The gate-source voltage of the NMOS is, for general design, set so that it is the threshold voltage plus 5% of *VDD*.

The output of the circuit, bias_N, is used in a current mirror, Figure 19.17. Note that the voltage bias_N is also the V_{GS} of the NMOS device. This controls the current in the device noting that the device must be operating in the saturation region, Equation 19.5.

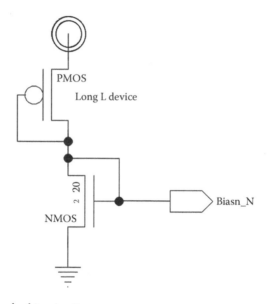

FIGURE 19.16   A simple analog bias circuit.

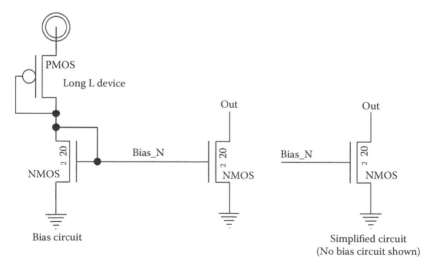

FIGURE 19.17   Biasing a current mirror.

19.10 Differential Amplifier

Figure 19.18 shows the schematic of a differential amplifier, or diff-amp for short. The inputs to the diff-amp are applied to the gates of M1 and M2, the gates of the diff-pair. The input common-mode range of the diff-amp is set by the minimum and maximum input voltages required to keep all MOSFETs operating in saturation. We will call the difference in the input signals, v_{in}, or

$$v_{in} = v_{ip} - v_{im} \tag{19.16}$$

The average of the inputs is the common-mode voltage, v_{cm}:

$$v_{cm} = \frac{v_{ip} + v_{im}}{2} \tag{19.17}$$

The minimum input v_{cm} is limited by the minimum voltage across the current sink, $V_{DS,sat}$ added to M1/M2's V_{GS} or

$$v_{cm,min} = V_{GS} + V_{DS,sat} = 2V_{DS,sat} + V_{THN} \tag{19.18}$$

When the common-mode voltage starts to fall below this minimum voltage, the devices start to shut off. For the maximum input common-mode voltage, M1 and M2 must be kept in the saturation region. Their drains are held at $VDD - V_{SG}$ where V_{SG} is the source-gate voltage of the PMOS current mirror load. Using Equation 19.5,

$$v_{cm,max} = VDD - V_{SG} + V_{THP} \approx VDD - V_{SD,sat} \tag{19.19}$$

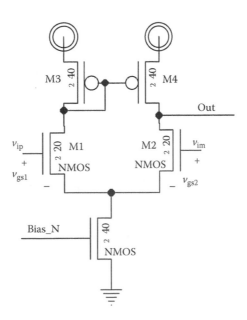

FIGURE 19.18 Schematic of a differential-amplifier (diff-amp).

The AC performance of the diff-amp can be characterized by applying an AC voltage, v_{ip}, between the gates of M1 and M2:

$$v_{in} = v_{gs1} + v_{gs2} \tag{19.20}$$

The AC currents in M1 and M2 must be equal but opposite since ideally, no AC current will flow in the current sink used to bias the diff-pair. The current in the NMOS current sink is fixed (DC) and thus does not vary. Therefore, we can write

$$i_{d1} = -i_{d2} \rightarrow v_{gs1} = -v_{gs2} = \frac{v_{in}}{2} \tag{19.21}$$

The current i_{d1} is mirrored via M3 over to M4, so the total current driven into the output node is $2i_{d1}$. The AC output voltage is thus

$$v_{out} = 2i_{d1} \cdot r_{on} \parallel r_{op} \tag{19.22}$$

or knowing $v_{in} = v_{gs1}/2 = i_{d1} \cdot 2/g_m$

$$\frac{v_{out}}{v_{in}} = g_m \cdot r_{on} \parallel r_{op} \tag{19.23}$$

19.11 Op-Amp

Figure 19.19 shows the basic two-stage op-amp formed with a diff-amp and a common-source output stage. The DC gain of this op-amp is

$$A_v = -g_{m1}R_1 \cdot g_{m2}R_2 = -g_{mn1}\left(r_{op1} \parallel r_{on1}\right) \cdot g_{mp2}\left(r_{op2} \parallel r_{on2}\right) \tag{19.24}$$

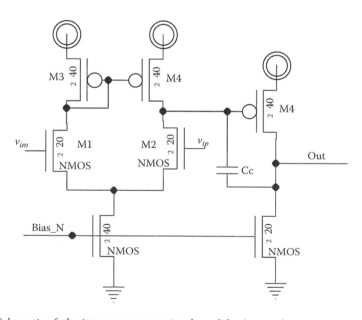

FIGURE 19.19   Schematic of a basic two-stage operational-amplifier (op-amp).

where

   R_1 is the net resistance attached to the *node*-1 (output of the first stage)
   R_2 is the net resistance attached to *node*-2 (output of the second stage)

   The transconductances for the first- and second-stage amplifiers are g_{m1} and g_{m2}, respectively. In Equation 24, g_{m1} is equal to the transconductance of the NMOS M1 and M2 (i.e., g_{mn1}) in Figure 19.19, while g_{m2} is the transconductance of the PMOS M4 (i.e., g_{mp2}). Since the op-amp has two prominent nodes with large capacitive loading, the op-amp has two poles associated with its frequency response. A pole in frequency domain causes the magnitude response to decrease by 20 dB per decade in frequency, and the net phase decreases by 90°. Now, if the op-amp has two poles, located close to each other on the frequency axis, a net phase change of 180° occurs. If for any frequency, the gain is greater than unity and the net phase difference across the op-amp is close to 180°, the op-amp will act like an oscillator and display oscillations in the time-domain transient response. Thus, to stabilize the op-amp, the two poles are separated from each other by employing *frequency compensation*. This is achieved by employing Miller's effect, where a large capacitance, called the compensation capacitor, is connected between the outputs of the first and second stages of the op-amp. This leads to *pole-splitting* whereby the dominant pole associated with the output node is pushed close to DC, and the second pole is pushed out to a higher frequency. Also at higher frequencies, the compensation capacitor shorts the input and the output of the op-amp, leading to a zero in the frequency response. Analytically, it can be shown that the zero is located on the right half s-plane (RHP) and leads to an increase in the magnitude response by +20 dB/decade and causes a decrease in the phase response by 90°. The frequency location of the RHP zero is given by

$$f_z = \frac{g_{m2}}{2\pi C_c} \tag{19.25}$$

The dominant pole that also sets the 3 dB bandwidth of the open-loop response is located at

$$f_1 \approx \frac{1}{2\pi g_{m2} R_2 C_c} \tag{19.26}$$

while the nondominant pole location is given as

$$f_2 \approx \frac{g_{m2} C_c}{2\pi \left(C_c C_1 + C_1 C_2 + C_c C_2 \right)} \tag{19.27}$$

The resulting AC transfer function of the op-amp is

$$A_v(f) = \frac{v_{out}(f)}{v_{in}(f)} = g_{m1} R_1 \cdot g_{m2} R_2 \frac{\left(1 - jf/f_z\right)}{\left(1 + jf/f_1\right)\left(1 + jf/f_2\right)} \tag{19.28}$$

Figure 19.20 shows the pole-splitting in the s-plane when Miller compensation is employed in the op-amp. The resulting frequency response after compensation is shown in Figure 19.21.

   An important parameter for an op-amp is its unity-gain frequency or gain bandwidth. It is defined as the frequency at which the op-amp's open-loop magnitude response has unity gain. The location of unity-gain frequency is estimated as

$$f_{un} = \frac{g_{m1}}{2\pi C_c} \tag{19.29}$$

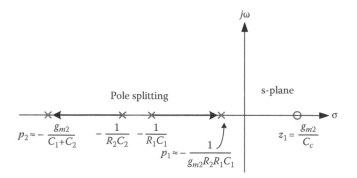

FIGURE 19.20 Pole-zero plot showing pole-splitting in a Miller-compensated two-stage op-amp.

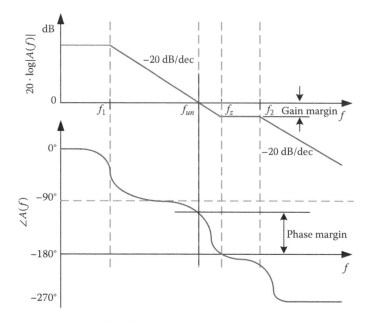

FIGURE 19.21 Frequency response of a Miller-compensated two-stage op-amp.

The stability of the op-amp is characterized by two parameters, namely *phase margin* (PM) and *gain margin* (GM). To determine the PM, we look at the net phase shift when the gain of the op-amp is unity. The amount of phase shift away from 180° is defined as the PM. GM is the difference (in dB) between the open-loop gain and the unity gain, when the phase is equal to 180°. Ideally, the op-amp should look like a single-pole response with a PM of 90°. A PM of 90° signifies an R-C circuit like first-order transient response for the op-amp. As the RHP zero and the second pole come closer to the unity-gain frequency, the PM decreases from the ideal value of 90°. A PM of 60° or higher is considered reasonable for a stable op-amp. Op-amp design involves multiple trade-offs between the op-amp gain, unity-gain frequency, PM, power consumption, and layout area for given specifications. Additional information on op-amp compensation methods and design trade-offs is found in [Baker08].

19.12 Comparator

A comparator is a decision-making circuit that compares two analog voltages and generates a corresponding digital output. The output is high if the analog input at the positive terminal (v_{in+}) is larger than the signal at the negative input terminal (v_{in-}) and low for the opposite case. Figure 19.22 shows

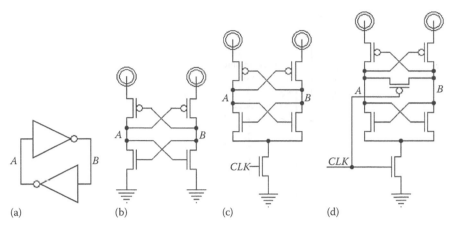

FIGURE 19.22 Implementation of a clocked comparator: (a) latch using cross-coupled inverters, (b) transistor implementation of (a), (c) addition of a clocked switch, and (d) using a PMOS switch to erase latch memory.

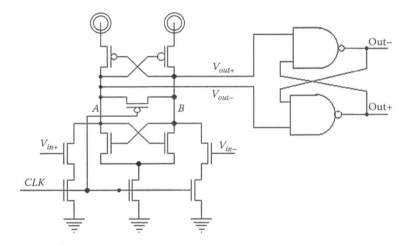

FIGURE 19.23 A complete clocked comparator.

the implementation of a clocked comparator realized from a basic cross-coupled latch. If an imbalance is created between the two nodes, *A* and *B*, the voltage difference between the two is amplified by the inverters. Thus, if initially node *A* is at higher voltage than node *B*, the voltage at node *A* will continue to increase until it settles close to *VDD* and node *B* settles near the ground. The latch is modified to make decision when the clock (*CLK*) goes high by adding a clocked switch to turn the latch on when the clock is high. Another switch is employed to erase the comparator memory from the previous decision by equilibrating the nodes *A* and *B* when *CLK* is low.

A complete clocked comparator is shown in Figure 19.23. Here, a pair of common-source amplifiers is used to amplify the input signal and create an imbalance in the latch. The digital output of the comparator is further amplified and stored by an SR (set-reset) flip-flop that is constructed by two cross-coupled NAND gates.

19.13 Data Converters

Data converters form the interface boundary between the analog and digital domains. An analog-to-digital converter (ADC) samples an analog signal at a given clock rate, quantizes it, and represents it in its equivalent digital code. On the other hand, a digital-to-analog converter (DAC) converts a digital

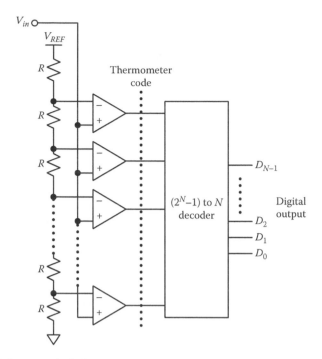

FIGURE 19.24 Block diagram of a flash ADC.

input code into its analog representation. ADCs can be categorized into two broad categories, namely Nyquist rate and oversampling data converters. Nyquist-rate ADCs operate at a sampling frequency, set by Nyquist's sampling theorem, which is double the bandwidth of the input analog signal. A wide variety of Nyquist-rate ADC architectures are available spanning the spectrum of speed and resolution: flash, pipelined, subranging, folding, integrating, and successive approximation.

Flash or parallel ADCs offer the highest sampling speed at low resolution. A Flash ADC employs 2^{N-1} comparators in parallel for N-bit resolution. The comparators compare the input signal with equally spaced voltage references generated by a string of resistors as shown in Figure 19.24. The resulting output code of the comparators, called a thermometer code, is then converted to its digital equivalent output. Flash converters have been typically limited to a maximum of 8 bit resolution and operate at speeds up to 1–4 GHz in CMOS technology.

For moderate resolution of 10–13 bits and moderate sampling speeds (100–400 MHz in CMOS), pipelined ADCs are utilized. A pipelined ADC employs N digitizing stages connected in series. Each of the pipelined stages carries out 1 bit conversion using a comparator and passes on the unconverted residue on to the next stages. Each stage of the pipelined converter follows the following algorithm, as illustrated in Figure 19.25:

1. Sample the input and compare it with the voltage reference $V_{REF}/2$, where V_{REF} is the reference voltage for data converter typically equal to the supply voltage range. The output of the comparator is the resulting output bit for that stage.
2. If the input v_{in} is greater than $V_{REF}/2$ (i.e., comparator output is one), subtract $V_{REF}/2$ from v_{in} and pass the result to the amplifier. If v_{in} is smaller than $V_{REF}/2$, pass the original input to the amplifier. The output after the each stage is called the *residue* and is passed on to the next stage for subsequent conversion.
3. Multiply the summed output by a fixed gain of two and pass the result to the sample and hold of the next stage.

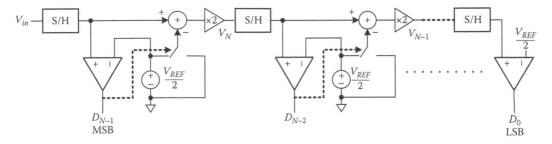

FIGURE 19.25 Block diagram of a pipelined ADC.

The successive approximation ADC searches through all the possible digital codes by successively comparing them with the sampled analog input. A DAC is employed to cycle through the digital codes using binary search. Each time, the DAC output is compared with the sampled input and the result stored in a successive approximation register (SAR). The content of the SAR represents the digital output code corresponding to the analog input. The successive approximation ADCs are simple to implement and is the architecture of choice for high-speed and high-resolution data conversion. However, the resolution of these ADCs is limited by the accuracy of the DAC employed.

The oversampling ADCs employ digital signal processing (DSP) to achieve much higher resolution (up to 24 bits) than the Nyquist-rate data converters. A basic oversampling ADC block diagram is shown in Figure 19.26.

The oversampling ADC operates at a sampling frequency that is a large multiple of the input signal bandwidth and constitutes of a delta-sigma ($\Delta\Sigma$) modulator followed by a digital filter. The $\Delta\Sigma$ modulator incorporates a loop filter, $H(z)$, a pair of lower-resolution ADC and DAC connected in a feedback loop as seen in Figure 19.26. The $\Delta\Sigma$ modulator operating at a higher oversampling frequency effectively pushes (or shapes) the quantization noise of the low-resolution ADC to the higher frequencies beyond the input signal bandwidth. This shaped-out out-of-band quantization noise is filtered out by the digital filter, leading to a higher signal-to-noise ratio (SNR) and hence, higher bit resolution. The ratio of the sampling frequency to the signal bandwidth is defined as the oversampling ratio (OSR). If the order of the $\Delta\Sigma$ loop filter, $H(z)$, is M, the effective increase in bit resolution is roughly given as

$$N_{inc} = \left(M + \frac{1}{2} \right) \cdot \log_2 (OSR) \tag{19.30}$$

Thus, the maximum possible signal bandwidth is traded with higher resolution using DSP techniques. The $\Delta\Sigma$ ADCs are gaining prominence with further advancement in the speed of the CMOS technologies. The feedback structure of the $\Delta\Sigma$ modulator desensitizes the SNR to the errors arising due to manufacturing mismatches across the transistors. Also, the $\Delta\Sigma$ modulation techniques push much of the signal processing to the digital domain from the analog domain, which is favorable for manufacturability in nano-scale CMOS technologies.

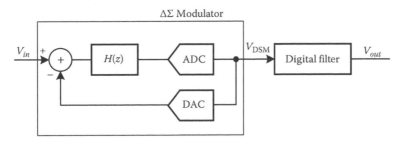

FIGURE 19.26 Block diagram of an oversampling ADC.

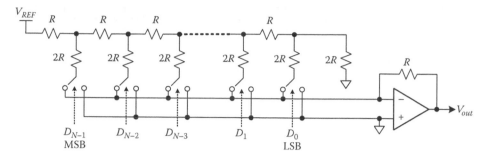

FIGURE 19.27   A DAC using *R*-2*R* ladder network.

The DAC architectures are simpler when compared to the ADCs. Here, the digital code representing the signal is interpolated to its equivalent analog representation. Similar to the ADCs, a wide variety of DAC architectures are available. An example of DAC employing *R*-2*R* ladder is shown in Figure 19.27. The *R*-2*R* ladder is a compact structure that generates the binary-weighted references ($=V_{REF}/2^i$) by voltage division. The input digital bits control whether the resistors in the ladder are connected to ground or to the summing amplifier. The amplifier sums up the individual binary-weighted digital input bits and its output (v_{out}) represents the analog equivalent of the digital input code. Other DAC architectures include current steering, charge scaling, cyclic, pipelined and delta-sigma topologies. The reader is referred to [Baker08,Baker09] for a detailed exposition on data converters.

19.14 Conclusion

An introduction to transistor-level Analog and Digital VLSI design has been presented. The CMOS devices are first characterized using the models presented in the chapter and then they are used to design circuits with the desired performance specifications. Digital circuits like logic gates and clocked storage elements are utilized to build complex data paths. On the other hand, analog circuit blocks like op-amps and comparators are used to construct signal processing systems including the data converters. ADCs are categorized into various architectures that offer a trade-off between the conversion bandwidth and the resolution. A brief tutorial on ADC architectures has been presented in this chapter along with an example of DAC.

References

[Baker08] R.J. Baker, *CMOS Circuit Design, Layout and Simulation*, revised 2nd edition, Wiley-IEEE, New York, 1039pp., 2008.

[Baker09] R.J. Baker, *CMOS Mixed-Signal Circuit Design*, 2nd edition, Wiley-IEEE, Hoboken, NJ, 329pp., 2009.

[Mead79] C. Mead and L. Conway, *Introduction to VLSI Systems*, Addison-Wesley Pub, Reading, MA, 396pp., 1979.

III

Digital Circuits

20

Digital Design— Combinational Logic

Buren Earl Wells
The University of Alabama in Huntsville

Sin Ming Loo
Boise State University

20.1 Introduction

Digital design has become the most modern and efficient way of implementing a wide variety of electronic systems. These systems are the heart of today's computer, wireless, and multimedia technology and are becoming more ubiquitous in nature where they are often embedded in an ever-increasing range of appliances from toasters to automobiles. A major property of digital systems is that they utilize finite size number representation and discrete state behavior to implement the desired functionality to the needed degree of precision. To be useful, such systems must be interfaced through analog-to-digital (A/D) and digital-to-analog (D/A) converters, because the real world is actually analog in nature (i.e., real-world parameters such as time, voltage, current, temperature, etc., have an infinite number of possible values). The focus of the this chapter and the next (sequential logic) are to introduce the basic concepts associated with digital logic and describe how the basic digital logic building blocks can be designed and combined together to perform the required task. Digital logic is most effectively classified as being composed of two main types of logic circuitry: combinational and sequential. In combinational sections of a digital design, the values of the outputs are a function of the current values associated with the inputs, whereas in sequential logic, the outputs will not only depend upon the current set of inputs but on the past values of the inputs. The focus of this chapter is to introduce the basic concepts associated with combinational digital logic circuitry with the important aspects associated with sequential logic being explored in Chapter 21.

20.2 Number Representation

Digital circuitry is implemented using semiconductor technology that utilizes a very large number of transistors that are placed in silicon integrated circuits, ICs. The most natural way for us humans to represent numbers is decimal, but most digital circuitry represents numbers in binary. This is because transistor-level designers have been more successful in implementing reliable and area efficient two-state (on/off) circuitry than in implementing multistate circuitry. The most common signed number representation in binary is called two's complement. Two's complement is an expanded version of the

Positional notation $B_{n-1}B_{n-2}\ldots B_1 B_0$

where $B_i = \{0,1\}$ and n = number of bits used in representation

$$\text{Value} = B_{n-1}(-2^{n-1}) + B_{n-2}2^{n-2} + \cdots + B_1 2^1 + B_0 2^0$$

For $n = 4$ bit representation

$$5_{10} = 0(-2^3) + 1(2^2) + 0(2^1) + 1(2^0) \;\Rightarrow\; 0101_2$$

$$-5_{10} = 1(-2^3) + 0(2^2) + 1(2^1) + 1(2^0) \;\Rightarrow\; 1011_2$$

FIGURE 20.1   An example of positive and negative representation in 2's complement.

standard binary positional notation in which the position of each bit is related to the position of the next bit by the multiplier of 2. In two's complement notation, the multiplier for the most significant bit is negative while the multiplier of all other bits is positive. Figure 20.1 gives an example of basic positive and negative integer representation in two's complement binary.

The standard two's complement notation is easily made to account for fractions by extending the positional notation to allow for integer powers less than 0. This is called fixed point. Fixed point often suffers from the finite number of bit positions reserved for the integer or fraction (or both). This has lead to the development of floating point formats that allow the exponent to be adjusted in a manner that allows the most significant number of bits to always be maintained. It should also be noted that humans rarely interface with digital hardware using binary. Rather users interact with the system using standard decimal (which is usually transmitted in ASCII character format), which is then converted into binary before being processed by the digital system with the output being conversely converted back into human-readable format. The use of such data and number formats of ASCII and Binary Coded Decimal (BCD) facilitate this process.

20.3 Two-Valued Boolean Logic

Boolean algebra is a branch of mathematics that adapts many of the concepts used in traditional philosophical propositional logic to a form that can be easily implemented in digital hardware. It utilizes a set of statements that are made up of variables and constants, which are acted upon by a set of operators to produce the outputs that can serve as an input variable to other Boolean statements. The values of the Boolean variables and constants are restricted to one of two possible values (True or False, or 0 or 1), which corresponds to two specified voltage ranges.

It can be shown that any combinational function can be implemented using a combination of the basic AND, OR, and NOT operators. The corresponding set of gates is easily implemented in digital hardware in a manner that is dependent upon the underlying device technology. In Boolean algebra, a truth table can be used to show the values of the output for all possible values of the set of inputs. An example of the truth tables for all three of these main types of gates is shown in Figure 20.2. An AND gate is characterized by the fact that all inputs must be at a logic 1 before the output is a 1. An OR gate is characterized by the fact that if any of the inputs are at a logic 1, then the output is also at a logic 1, and a NOT gate simply inverts (switches) the value of its logic input. The functionality of applying the logic operations on the set of inputs to produce an output can also be expressed using Boolean algebra notation. In this notation, the input and outputs of a logical network are represented as variables or constants in much the same way as conventional algebraic statements. The operations performed by these three gates are expressed using standard algebraic symbols, where the AND operation represents Boolean multiplication, and the OR operation represents Boolean addition. By definition, the NOT operation is a unary operator. It is expressed by putting a line over the variable of interest.

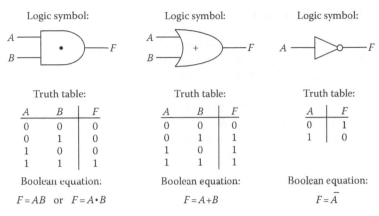

FIGURE 20.2 Digital logic basic operators (AND, OR, and NOT). The truth table and Boolean equations are shown in the figure.

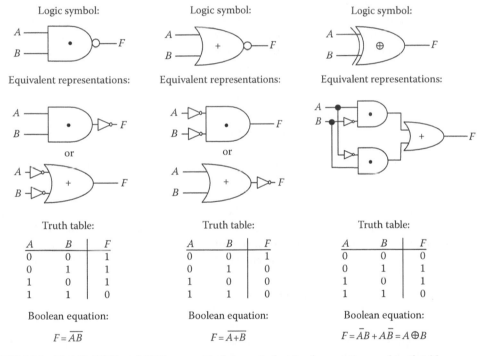

FIGURE 20.3 NAND, NOR, and XOR gates with their equivalent implementations and truth tables.

The set of three, AND, OR, and NOT gates is logically complete, meaning that all combinational Boolean expressions can be expressed using some combination of these gates. It is often more convenient to use other Boolean logic gates and operators that are derived from these basic gates. Figure 20.3 illustrates the NAND, NOR, and XOR (exclusive OR) gate symbols, equivalent representations, truth tables, and Boolean expressions.

Boolean algebraic expressions exhibit standard associative and commutative properties. There are many laws, theorems, and definitions that are present in Boolean algebra some of which are shown in Figure 20.4.

20.4 Logic Minimization

A Boolean expression can be placed directly in a canonical form where it can then be minimized in terms of the number of gates and the number of inputs required for each gate. Such minimization often results in an easier-to-understand logical expression that can result in less underlying digital circuitry being needed to implement the desired logic function.

There are two main conical forms for combinational Boolean functions. The disjunctive form is the Boolean sum (OR-ing) of the set of minterms present in the function whereas the conjunctive form is the Boolean product (AND-ing) of the set of maxterms present in the function. A minterm is a product term that contains a single copy of all the function's input variables in their true or complemented form. Each minterm is a logic 1 for one distinct set of input combinations, which forces the overall disjunctive Boolean Function to logic 1 at these input values. Minterms are constructed by determining what combination of inputs will result in the function being a logic 1. Each of these input combinations will generate a separate minterm. Each minterm then is simply the product of all the true or complemented input variables, with the true form being present whenever that variable is a logic 1 when the minterm is present, and the complemented form being present whenever the variable is a logic 0. Conversely, maxterms are constructed by determining what combination of input will result in the overall logic function being a logic 0. Each maxterm is simply the sum of all the true or complemented input variables, with the true form being present whenever the variable is a logic 0 when the maxterm is present, and the complemented form being present whenever the variable is a logic 1. Minterms and maxterms are given numbers based upon the row of the corresponding truth table where they occur. This of course assumes that the set of input variables obtain a specified ordering from most significant to least significant bit.

The canonical forms of Boolean expressions most often can be simplified. This requires that one consider the set of minterms that differ from one another only at a single variable position. This variable can then be eliminated from the expression using the simplification theorem shown above and the two terms replaced by a single reduced term. The process can then be continued until there are no more possible groupings. A grouping that is composed of the largest possible number of minterms is called a prime implicant. If a prime implicant contains a midterm that is not present in any other prime implicant, it is considered to be *essential* and must be part of any valid solution. If not, it is considered to be nonessential and a minimal solution may be possible that does not include this implicant.

The Quine–McCluskey algorithm [1] represents an ordered/tabular way of accomplishing this process that is amenable to automation. The Karnaugh map method [1] is an equivalent graphical method that is easier to apply to smaller four to six variable Boolean functions. Both methods employ two main steps as shown in Figure 20.5 where an unoptimized function is simplified. The first step is to find the set of prime implicants that make up the function by applying the simplification theorem to product terms that differ from one another at only one variable position. Prime implicants that were uniquely formed from one or more minterms are then declared to be essential. The second step is utilizing a minimum number of the nonessential prime implicants in the final solution. From this example, one can note that the complexity of the canonical form expression is significantly reduced and that there can be more than one equivalent minimal expression.

Original function in disjunctive conical (sum of products) form:

$$F = \bar{A}\bar{B}C\bar{D} + \bar{A}\bar{B}CD + \bar{A}BC\bar{D} + \bar{A}BCD + A\bar{B}\bar{C}\bar{D} + A\bar{B}\bar{C}D + ABC\bar{D} + AB\bar{C}D + ABC\bar{D}$$

$\Rightarrow$ Nine 4-input AND gates, and One 9-input OR gate

DeMorgan's laws :

$$\overline{(AB)} + \bar{A} + \bar{B} \qquad \overline{(A+B)} = \bar{A}\bar{B}$$

Simplification theorem:

$$XY + X\bar{Y} = X \qquad (\overline{\bar{X}}) = X$$

Exclusive OR definition:

$$A \oplus B = \bar{A}B + A\bar{B}$$

FIGURE 20.4 DeMorgan's law, simplification theorem, and exclusive OR definition.

Truth table for
original function

Step 1: Finding prime implicants:
repeated application of $xy + xy' = x$
to combine terms and reduce size
of operators

A	B	C	D	F	
0	0	0	0	0	
0	0	0	1	0	
0	0	1	0	1	m_2
0	0	1	1	1	m_3
0	1	0	0	0	
0	1	0	1	0	
0	1	1	0	1	m_6
0	1	1	1	1	m_7
1	0	0	0	1	m_8
1	0	0	1	1	m_9
1	0	1	0	0	
1	0	1	1	0	
1	1	0	0	1	m_{12}
1	1	0	1	1	m_{13}
1	1	1	0	1	m_{14}
1	1	1	1	0	

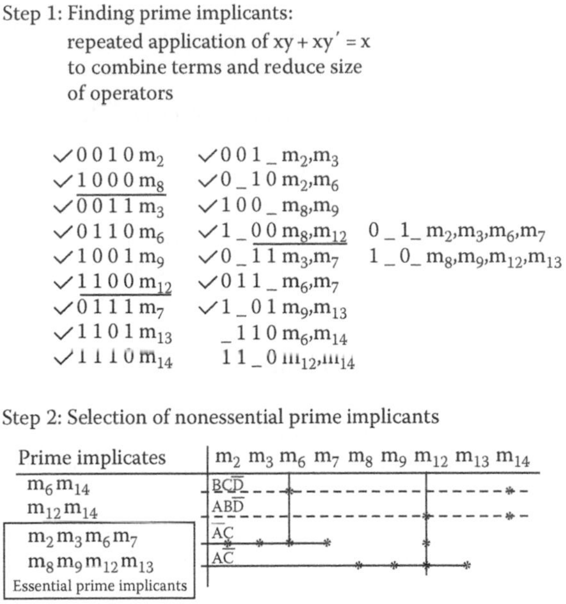

(a)

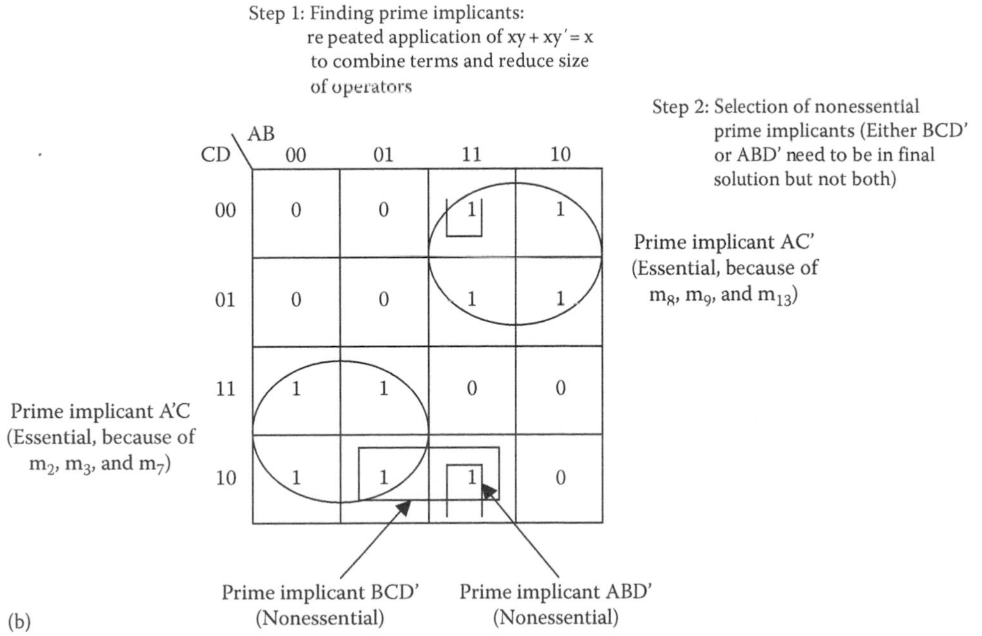

✓ 0 0 1 0 m_2 ✓ 0 0 1 _ m_2,m_3
✓ 1 0 0 0 m_8 ✓ 0 _ 1 0 m_2,m_6
✓ 0 0 1 1 m_3 ✓ 1 0 0 _ m_8,m_9
✓ 0 1 1 0 m_6 ✓ 1 _ 0 0 m_8,m_{12} 0 _ 1 _ m_2,m_3,m_6,m_7
✓ 1 0 0 1 m_9 ✓ 0 _ 1 1 m_3,m_7 1 _ 0 _ m_8,m_9,m_{12},m_{13}
✓ 1 1 0 0 m_{12} ✓ 0 1 1 _ m_6,m_7
✓ 0 1 1 1 m_7 ✓ 1 _ 0 1 m_9,m_{13}
✓ 1 1 0 1 m_{13} _ 1 1 0 m_6,m_{14}
✓ 1 1 1 0 m_{14} 1 1 _ 0 m_{12},m_{14}

Step 2: Selection of nonessential prime implicants

Prime implicates	m_2 m_3 m_6 m_7 m_8 m_9 m_{12} m_{13} m_{14}
$m_6 m_{14}$	$BC\overline{D}$ ----
$m_{12} m_{14}$	$AB\overline{D}$ ----
$m_2 m_3 m_6 m_7$	$\overline{A}C$
$m_8 m_9 m_{12} m_{13}$	$A\overline{C}$
Essential prime implicants	

Step 1: Finding prime implicants:
re peated application of $xy + xy' = x$
to combine terms and reduce size
of operators

Step 2: Selection of nonessential
prime implicants (Either BCD'
or ABD' need to be in final
solution but not both)

Prime implicant AC'
(Essential, because of
m_8, m_9, and m_{13})

Prime implicant A'C
(Essential, because of
m_2, m_3, and m_7)

Prime implicant BCD'
(Nonessential)

Prime implicant ABD'
(Nonessential)

(b)

FIGURE 20.5 Logic simplification methods: (a) a logic minimization example using Quine–McCluskey technique and (b) Karnaugh map method.

Reduced function:

$$F = \bar{A}C + A\bar{C} + BC\bar{D} \quad \text{or} \quad F = \bar{A}C + A\bar{C} + AB\bar{D}$$

$\Rightarrow$ Two 2-input AND gates, One 3-input AND gate, and One 3-input OR gate

20.5 Common Combinational Elements

In addition to the derived gates (XOR, NAND, etc.), it is often useful to utilize larger combinational devices. As with the derived gates, each of these devices can be created from the basic set of logic operators. One such combination element is the multiplexer, which can be used to choose from a set of inputs, one input to connect to the output. Such multiplexers often have Z^N inputs that are controlled by N select lines. In addition to their use as data selectors, multiplexers can be used to implement arbitrary Boolean functions (see Figure 20.6).

Another higher-level combinational device is the decoder. Decoders often are used to produce a one-hot encoded output that responds to the input in a manner where only one of the output lines is a logic 1 at a time and the particular line that is active corresponds to the output whose index value is the same and that placed on the select line. Figure 20.7 illustrates the functionality of a 3-to-8 decoder.

An encoder is in many ways the reverse of a decoder. It converts the one-hot encoding that occurs when one of the inputs is active (logic 1) into a binary number that corresponds to the index of the input where this has occurred. The case where more than one input is active at a time can be handled by giving the higher-indexed inputs priority over the lower-indexed inputs as shown in the truth table of Figure 20.7. Of course, these higher-level objects are actually constructed using the lower-level logic as also shown in the figure.

Adders are examples of other combinational modules that can be constructed using logical gates. These can be defined hierarchically by first developing single-bit half and full adder modules. A half adder is one that adds 2 bits to produce a sum and a carry. A full adder adds two bits together along with a carry to produce a sum and a carry. The truth table and its logic implementation for these 1 bit adders are shown in Figure 20.8.

Once the half and full adders have been constructed, they can be combined and replicated in a hierarchical manner to form a multibit (ripple) adder by connecting the carry outs to carry in of subsequent bit stages, as shown in Figure 20.9.

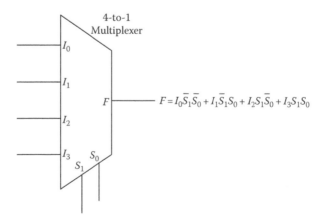

FIGURE 20.6 4-to-1 Multiplexer. The F function specifies which data input will be selected with the use of selector inputs.

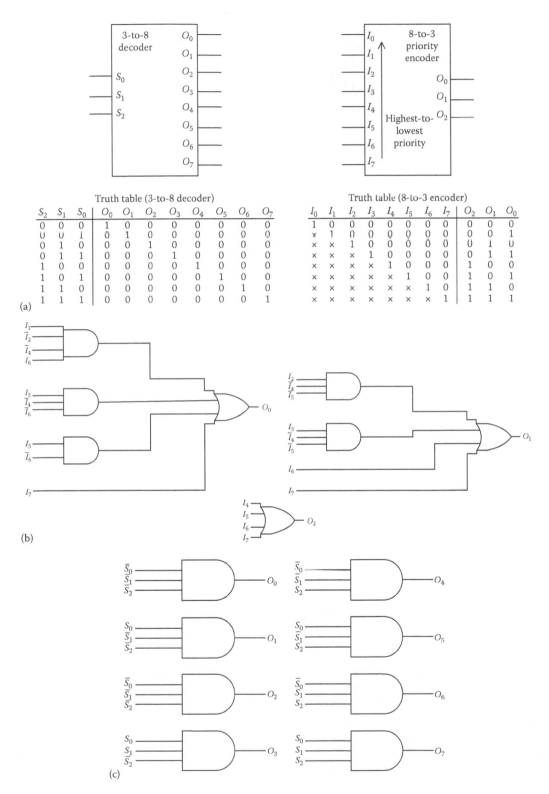

FIGURE 20.7 Decoder and encoder: (a) decoder and encoder block diagrams, (b) encoder implementation, and (c) decoder implementation.

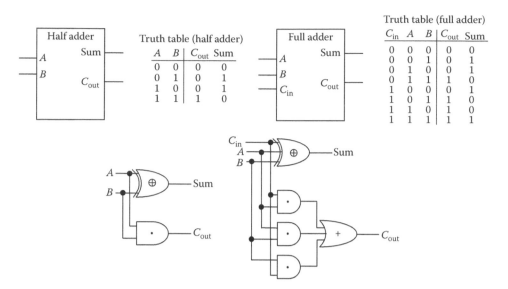

Truth table (half adder)

A	B	C_{out}	Sum
0	0	0	0
0	1	0	1
1	0	0	1
1	1	1	0

Truth table (full adder)

C_{in}	A	B	C_{out}	Sum
0	0	0	0	0
0	0	1	0	1
0	1	0	0	1
0	1	1	1	0
1	0	0	0	1
1	0	1	1	0
1	1	0	1	0
1	1	1	1	1

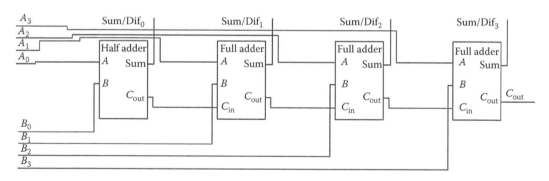

FIGURE 20.8   Half adder and full adder.

FIGURE 20.9   A 4 bit adder design using the 1 bit adders shown in Figure 20.8.

20.6 Modern Combinational Design Practices

A basic knowledge of Boolean algebra is very helpful to digital design engineers [1], but most of the low-level design optimization is now handled using complex software tools that synthesize the design into logic circuitry and allow one to simulate the functionality of the design before it is implemented. These tools often support design entry using modeling languages where the functionality of circuitry is specified, not at the gate level but at a very high level of abstraction. These languages are called hardware description languages (HDLs) and modern HDLs include VHDL [2], Verilog [3], System C [4], and Impulse C [5]. These tools allow for the design engineer to effectively manage the complexity of designs that contain logic circuitry that often contain millions of gates. It is also common for design engineers to prototype their designs using reconfigurable hardware before implementing their designs in application-specific hardware.

References

1. C. H. Roth, *Fundamentals of Logic Design*, 5th edition, ISBN: 0534378048, Thomson Learning, Belmont, CA, 2001.
2. S. Brown and Z. Vranesic, *Fundamentals of Digital Logic with VHDL Design*, 2nd edition, McGraw Hill, New York, 2005.
3. S. Brown and Z. Vranesic, *Fundamentals of Digital Logic with Verilog Design*, 2nd edition, McGraw Hill, New York, 2003.
4. T. Grötker, S. Liao, G. Martin, and S. Swan, *System Design with SystemC*, Kluwer Academic Publishers, Norwell, MA, 2004.
5. D. Pellerin and S. Thibault, *Practical FPGA Programming in C*, Prentice Hall, Englewood Cliffs, NJ, 2005.

References

21

Digital Design—Sequential Logic

Sin Ming Loo
Boise State University

Arlen Planting
Boise State University

21.1 Combinational and Sequential Logic

Memory is the difference between combinational and sequential logic. With combinational logic, the output is generated from inputs (after some gate propagation delays through the logic gates). Propagation delay is the length of time from when the input to a logic gate is valid, to the time when the output of that logic gate is valid. The propagation time will determine how soon a new set of inputs can be processed with the output correctly generated. A sequential logic design has *memory* elements. The output is dependent on not only the current inputs, but also on the state. This means that sequential logic needs to know where it was and where it is now in order to determine where it is going with the current set of inputs. For example, a full adder is a combinational logic component as it adds input A and B to produce the sum. However, an up-down counter is a sequential logic component because the counter knows the current value; depending on the up-down control input, it will determine the next count value. Figure 21.1 shows an illustration of the differences between combinational and sequential logic.

In addition to memory, sequential logic requires a clock. This clock controls the operation of the sequential circuit. Since all operations' progress is based on a common clock, such a circuit is called a synchronous circuit.

21.2 Memory Elements

A memory element can maintain a binary state (0 or 1) as long as the power to the circuit is maintained. There are two basic types of memory elements: latch and flip-flop [1,2,3]. As described in the definitions below, latches are level sensitive and flip-flops are edge sensitive. In this write-up, D latches and D flip-flops are described and are shown in Figure 21.2. The information at data input D is the next state value. The value at output Q is the current state value. Knowing the locations of the current and next state values is important in wiring up a state machine.

Latch—For latches, the information presented at data input D is transferred to output Q when the enable (*en*) is asserted (active-low or active-high). This is termed level sensitive.

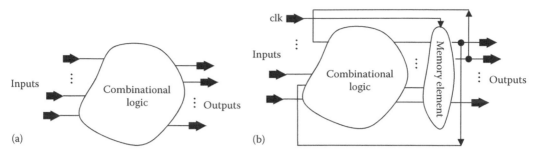

FIGURE 21.1   (a) Combination and (b) sequential logic examples.

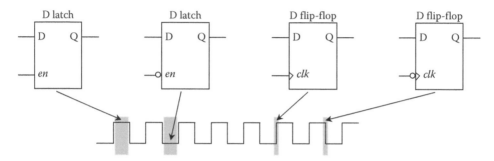

FIGURE 21.2   D latch versus D flip-flop. The (gray) highlighted areas show when the latches or flip-flops are active.

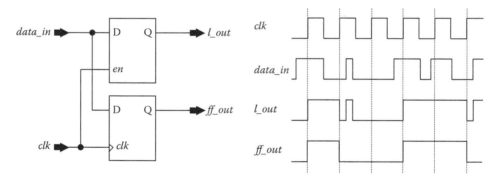

FIGURE 21.3   Timing diagrams showing the difference between latch and flip-flop.

Flip-flop—For flip-flops, the information presented at the data input D is transferred to output Q when the clock (*clk*) is transitioning from low-to-high (positive-edge) or high-to-low (negative-edge). This is termed edge sensitive.

The difference between a latch and flip-flop can be demonstrated using a simple timing diagram (see Figure 21.3). As shown in the timing diagram, the latch (*l_out*) output continues to change as the data changes when *clk* = "1." This is not the case for flip-flop (*ff_out*) output, which only changes during positive clock edge and holds the value until the next edge.

21.3 Designing an Up Counter

The simplest example of a sequential circuit is a counter design [3]. In this section, counter design using state machine method is demonstrated (see Figure 21.4). For this state machine design, the example shows the steps of using the state diagram, state table, and state encoding to design a 2 bit up counter

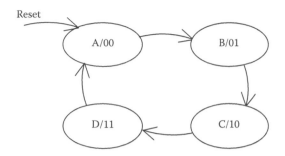

FIGURE 21.4 State diagram of an up counter. Each node indicates the state name and the output produced by the state.

$(00 \rightarrow 01 \rightarrow 10 \rightarrow 11 \rightarrow 00,...)$ that will reset to 00 when 11 is reached. The state diagram of this counter is shown in Figure 21.4. It has four states: A, B, C, and D. The output of each state is indicated inside the oval. States A, B, C, and D have outputs of "00," "01," "10," and "11," respectively. It is important to know the status of the state machine when the reset button is pressed, which is why the reset is in the state diagram. In this state machine design, whenever it is reset, it will always go to State A. There is no special condition for transitioning between states; only a clock edge (in this case, positive edge with the use of positive-edge D flip-flop) is necessary.

From the state diagram, a state table can be generated. This state table (see Table 21.2) presents the information in the state diagram in a simple table format.

With the above state table, an encoding scheme is needed for the state variables A, B, C, and D. Since we have four states, two bits will be enough for the representation. This is called minimized bit encoding. With this encoding scheme, A will be encoded as "00," B as "01," C as "10," and D as "11." A new table (Table 21.2) is generated for this encoding scheme. The output column is the same as in Table 21.1.

Now, the question is how are the present state $(y_1 y_0)$, next state $(Y_1 Y_0)$, and output $(z_1 z_0)$ interpreted in hardware (Figure 21.5). These can be determined using the values from Table 21.2.

Using a logic minimization technique, one can determine the next-state equations $(Y_1 Y_0)$ feeding into the D input of flip-flops, and the output equations $(z_1 z_0)$. This can be determined by using y_1 and y_0 as the variables or as an index in the K-map method. The logic minimizing technique will have the following equations:

$$Y_1 = y_1 \bar{y}_0 + \bar{y}_1 y_0 = y_1 \oplus y_0, \quad Y_0 = \bar{y}_0$$

$$z_1 = y_1, \quad z_0 = y_0$$

TABLE 21.1 State Table for Up Counter in Figure 21.4

Present State	Next State	Output
A	B	00
B	C	01
C	D	10
D	A	11

TABLE 21.2 State-Assigned Table with Minimized Bit Encoding

Present State $y_1 y_0$	Next State $Y_1 Y_0$	Output $z_1 z_0$
00	01	00
01	10	01
10	11	10
11	00	11

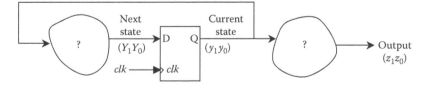

FIGURE 21.5 D flip-flop.

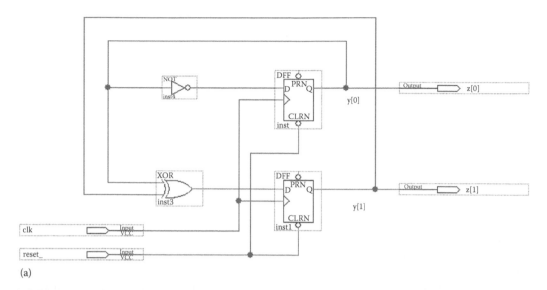

(a)

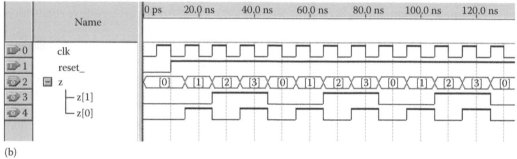

(b)

FIGURE 21.6 The up counter state machine as designed and simulated in Altera Quartus II. (a) Counter logic circuit implementation. (b) Simulation results. The counter is reset at the beginning (using reset low signal). The output counts up to "11" and rolls over "00" to start over.

TABLE 21.3 State-Assigned Table with One-Hot Encoding

Present State $y_3y_2y_1y_0$	Next State $Y_3Y_2Y_1Y_0$	Output z_1z_0
0001	0010	00
0010	0100	01
0100	1000	10
1000	0001	11

Since we used two bits to encode the state machine (state machine representation), we will use exactly two D flip-flops. Using the above equations, we can obtain this counter design (see Figure 21.6). The simulation results for this counter are shown in the same figure.

Now, let's try another state machine encoding technique. This encoding technique—called one-hot coding—will use more resources but has the potential to be faster. This encoding technique is called one-hot encoding because for X number of states, you will need X flip-flops with one active at a time. With one-hot encoding, Table 21.1 can be encoded as shown in Table 21.3, with A as "0001," B as "0010," C as "0100," and D as "1000."

From Table 21.3, one can tell that there are many don't-care states, which is how one-hot coding takes advantage in order to obtain an implementation with simpler next-state logic gates. The logic circuit implementation and simulation results are shown in Figure 21.7. Note that for this simulation to function correctly, it is necessary to preset the first flip-flop for the first clock cycle (using *start*_signal):

$$Y_3 = y_2, \quad Y_2 = y_1, \quad Y_1 = y_0, \quad Y_0 = y_3$$

$$z_1 = y_3 + y_2, \quad z_0 = y_3 + y_1$$

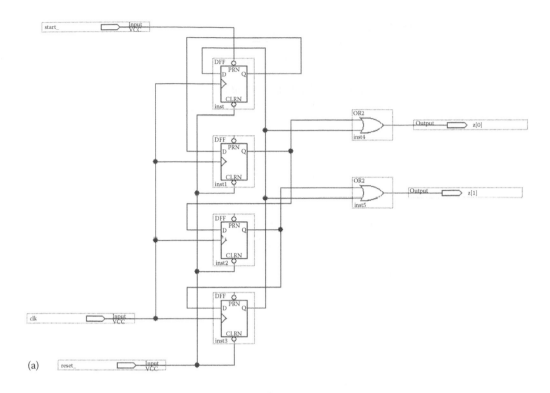

(a)

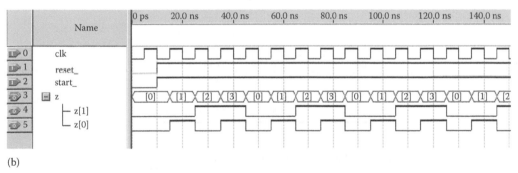

(b)

FIGURE 21.7 One-hot implementation of Figure 21.4. (a) Counter logic circuit implementation, one-hot. (b) Simulation results. The counter is reset (using reset_, an active low signal) and preset (using start_, an active low signal) at the start of simulation. The output counts up to "11" and rolls over "00" to start over.

Let's get back to the original counter (Figure 21.4), but with an up/down (ud) control bit added. When $ud = 0$, the counter counts up. When $ud = 1$, the counter counts down. This newly modified state diagram is shown in Figure 21.8. The state table for Figure 21.8 is shown in Table 21.4. For this state table, the Next State portion has two columns, with ud to differentiate the state transition. The state table with minimized bit encoding (with A encoded as "00," B encoded as "01," C encoded as "10," and D encoded as "11") is shown in Table 21.5.

The next state and output equations can be determined by using y_1, y_0, and ud as the variables or as an index in the K-map method. The logic minimizing technique will have the following equations:

$$Y_1 = y_1 \oplus y_0 \oplus ud, \quad Y_0 = \bar{y}_0$$

$$z_1 = y_1, \quad z_0 = y_0$$

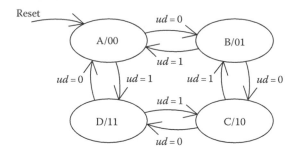

FIGURE 21.8 Up/down counter (Moore implementation).

TABLE 21.4 State Table for Moore Implementation of Up/Down Counter (Figure 21.8)

| Present State | Next State | | Output |
	$ud = 0$	$ud = 1$	
A	B	D	00
B	C	A	01
C	D	B	10
D	A	C	11

TABLE 21.5 State Table with Minimized Bit Encoding

| Present State ($y_1 y_0$) | Next State | | Output ($z_1 z_0$) |
	$ud = 0$ ($Y_1 Y_0$)	$ud = 1$ ($Y_1 Y_0$)	
00	01	11	00
01	10	00	01
10	11	01	10
11	00	10	11

TABLE 21.6 State Table with One-Hot Encoding

| Present State $y_3 y_2 y_1 y_0$ | Next State | | Output |
	$ud = 0$ $Y_3 Y_2 Y_1 Y_0$	$ud = 1$ $Y_3 Y_2 Y_1 Y_0$	
0001	0010	1000	00
0010	0100	0001	01
0100	1000	0010	10
1000	0001	0100	11

Let's try the one-hot encoding. With one-hot encoding, Table 21.6 can be encoded as follows, with A as "0001," B as "0010," C as "0100," and D as "1000."

The next state and output equations are

$$Y_3 = \overline{ud}y_2 + udy_0, \quad Y_2 = \overline{ud}y_1 + udy_3, \quad Y_1 = \overline{ud}y_0 + udy_2, \quad Y_0 = \overline{ud}y_3 + udy_1$$

$$z_1 = y_2 + y_3, \quad z_0 = y_1 + y_3$$

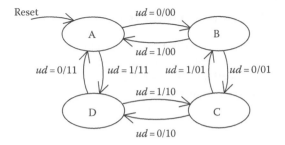

FIGURE 21.9 Up/down counter (Mealy implementation).

The above is a Moore state machine; we can modify the same counter design so that it will work as a Mealy state machine. With a Moore state machine, the output depends only on the state. However, a Mealy state machine depends on not only the state, but also on the present value of inputs. Thus, we modify the counter to the following. The output is at the edge, not at the node (see Figure 21.9)! Note that the output of a Mealy state machine should be registered so that the output will change synchronously.

The state table and the encoded state table (with A encoded as "00," B encoded as "01," C encoded as "10," and D encoded as "11") are shown in Tables 21.7 and 21.8, respectively. One can also see clearly the difference between a Moore and a Mealy state machine implementation. The output column depends on *ud*. The Moore implementation as shown in Table 21.5 doesn't depend on *ud*.

The next state and output equations are

$$Y_1 = y_1 \oplus y_0 \oplus ud, \quad Y_0 = \overline{y_0}$$

$$z_1 = \overline{ud}y_1 + y_1y_0 + \overline{y_1}\,\overline{y_0}ud, \quad z_0 = \overline{ud}y_0 + ud\overline{y_0}$$

TABLE 21.7 State Table for Mealy Implementation of Up/Down Counter (Figure 21.9)

Present State	Next State		Output	
	$ud = 0$	$ud = 1$	$ud = 0$	$ud = 1$
A	B	D	00	11
B	C	A	01	00
C	D	B	10	01
D	A	C	11	10

TABLE 21.8 State Table with Minimized Bit Encoding for Mealy Implementation of Up/Down Counter

Present State (y_1y_0)	Next State		Output	
	$ud = 0$ (Y_1Y_0)	$ud = 1$ (Y_1Y_0)	$ud = 0$ (z_1z_0)	$ud = 1$ (z_1z_0)
00	01	11	00	11
01	10	00	01	00
10	11	01	10	01
11	00	10	11	10

21.4 Designing a Sequence Detector

What if you want to watch for a 3 bit sequence in a serial input data stream? How does one go about designing a sequential logic circuit to watch for the expected sequence? Can a state machine be used to design such a logic circuit for detecting a sequence? Let's show an example that will detect "101" in a serial input data stream. It is important to note that an overlapping sequence is allowed, meaning that "10101" has two sequences. If it is a fixed windowing sequence, there is only one sequence in "10101." Figure 21.10 describes the state machine we are about to design.

Since the sequence to be detected is "101," three states will be necessary. Figure 21.11 shows the state diagram. This state diagram may look like scratches without any order. In fact, there is a systematic way to complete such a state diagram. First, one needs to determine the number of states necessary. In this example, we need a reset state (*Init* state) so that all reset events including initial startup will be directed to this state. The design will be a Moore state machine. This means that the outputs are determined by just the state. In addition to the *Init* state, we need a state when 1 is detected (*Got1*), one when 10 is detected (*Got10*), and one for when 101 is detected (*Got101*).

What about the transition edges—how would one go about adding those edges? The strategy is to add those transition edges that will get to the desired output—in this case, when 101 is detected. These edges are shown with dashed lines—there are three in this case. To complete the design, all the other edges will need to be added. Note that since we are dealing with binary logic, each state will have two output transitions (0, 1).

Table 21.9 shows the state table for the sequence detector in Figure 21.11. Table 21.10 shows the encoded state table, with *Init* encoded as "00," *Got1* encoded as "01," *Got10* encoded as "10," and *Got101* encoded as "11."

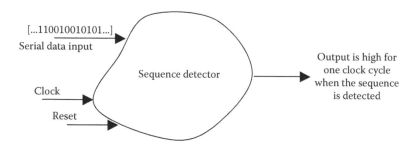

FIGURE 21.10   Sequence detector for detecting sequence "101."

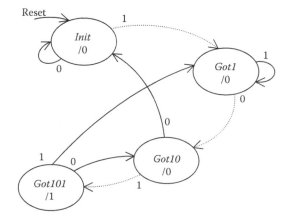

FIGURE 21.11   State diagram of sequence detector shown in Figure 21.10.

TABLE 21.9   State Table for Sequence Detector

	Next State		
Present State	$SI = 0$	$SI = 1$	Output
Init	Init	Got1	0
Got1	Got10	Got1	0
Got10	Init	Got101	0
Got101	Got10	Got1	1

TABLE 21.10   State Table with Minimized Bit Encoding for Sequence Detector

	Next State		
Present State $(y_1 y_0)$	$SI = 0$ $(Y_1 Y_0)$	$SI = 1$ $(Y_1 Y_0)$	Output (z)
00	00	01	0
01	10	01	0
10	00	11	0
11	10	01	1

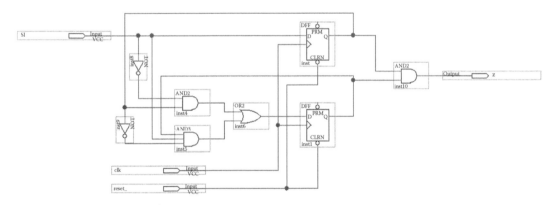

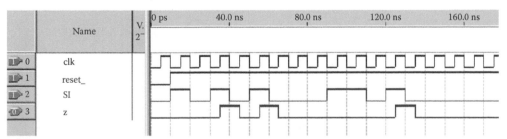

FIGURE 21.12   A sequence detector implementation with simulation results.

Using SI and $y_1 y_0$ as indexes or variables, the next state and output equations are as follows (the logic circuit and simulation results are shown in Figure 21.12):

$$Y_1 = y_0 \overline{SI} + SI y_1 \overline{y}_0, \quad Y_0 = SI$$

$$z = y_1 y_0$$

21.5 Summary

This chapter presents an overview of sequential logic design. Different methods to design state machines were demonstrated. In this chapter, the state machine's next state and output equations were determined manually. As of the writing of this chapter, there are tools, such as Xilinx State Diagram Editor, that will translate a state diagram to hardware with the help of synthesis tools. The use of hardware description language (Verilog or VHDL) can also simplify the implementation of sequential logic. Instead of implementing the design manually, the design can be put together by describing it using HDL constructs such as the *case* statement. This chapter does not attempt to be complete in presenting sequential logic design, but it makes an effort to present state machine design that is one of the most important concepts in sequential logic design.

References

1. C. H. Roth, *Fundamentals of Logic Design*, 3rd edition, ISBN: 0-314-85292-1, Nest Publishing Co., St. Paul, MN, 1985.
2. M. M. Mano and M. D. Ciletti, *Digital Design*, 4th edition, ISBN: 0131989243, Prentice Hall, Upper Saddle River, NJ, 2007.
3. S. Brown and Z. Vranesic, *Fundamental of Digital Logic with Verilog Design*, 2nd edition, ISBN: 9780073380339, McGraw Hill, Boston, MA, 2008.

22

Soft-Core Processors

Arlen Planting
Boise State University

Sin Ming Loo
Boise State University

22.1 Introduction [1–6]

A soft-core processor is a microprocessor core that can be implemented entirely by using digital logic synthesis. It is typically instantiated on programmable hardware such as a field programmable gate array (FPGA). (The main feature of an FPGA is that it is reconfigurable with different digital designs, allowing functionality to be changed an unlimited number of times by downloading a new file without physically changing the chip.)

In order to better understand what a soft-core processor is, it might be beneficial to consider the basic elements of construction and usage. Since a typical environment for creation of soft-core processors is on an FPGA, let's review the development cycle of digital designs with FPGAs. First, a digital design is encoded via schematic capture or a hardware description language. Then, the design is synthesized and transformed into a form that can be used to configure the FPGA. This form is usually called the configuration file. The configuration file is then used to configure the FPGA, which transforms the FPGA into the required digital solution (Figure 22.1a).

If the required digital design for the first input file above is that of a microprocessor, then the resulting transformation of the FPGA yields a microprocessor (Figure 22.1b). This type of microprocessor is called a soft-core processor.

Once an FPGA has been transformed into a microprocessor, its usage patterns mirror those of a traditional (discrete) microprocessor. The processor must still be programmed using traditional programming languages and methods (Figure 22.2).

Many traditional devices that are interfaced to discrete microprocessors (such as memory, switches, buttons, LEDs, etc.) can also be connected to soft-core processors. The interface logic that is typically produced in the form of discrete components can be realized as logic instantiated in the FPGA along with the soft-core processor (Figure 22.3).

The interface logic typically supplied by vendors of commercial soft-core processors includes gpio (general purpose input/output), Universal Asynchronous Receiver/Transmitter (UART), Inter-Integrated Circuit (I²C), Serial Peripheral Interface (SPI), and others. These packages of device logic are also referred

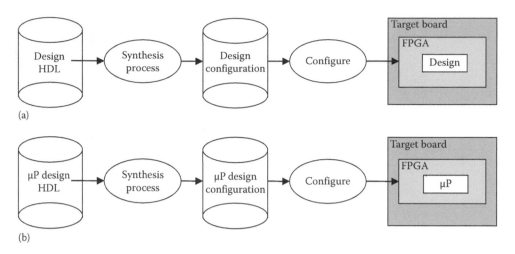

FIGURE 22.1 (a) Typical FPGA design flow. (b) Soft-core processor design flow for FPGA.

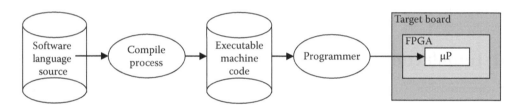

FIGURE 22.2 Programming of soft-core processor.

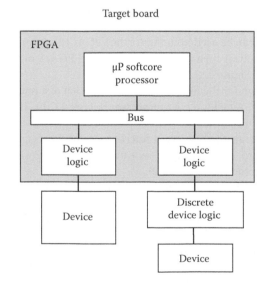

FIGURE 22.3 Device interfacing to soft-core processor.

to as intellectual property or IP cores. Open-source IP cores can be found and custom in-house solutions can be developed. This interface logic is sometimes referred to as glue logic, which, on an FPGA, is often in the form of a hardware definition language (HDL).

22.2 Processor Core Options

There are many soft microprocessor cores available, both closed and open source. Xilinx and Altera are the two major vendors; other vendors include Lattice, Actel, and Quicklogic. The leading 32 bit soft-core processors include Altera Nios II [7] and Xilinx Microblaze [8]. The LatticeMico32 [9] is a 32 bit soft-core microprocessor available for free with an open IP core licensing agreement. Many of the open cores are porting older processor designs to FPGA technology, and are good sources for studying soft-core processor design and architecture. Commercially available soft-core processors are particularly useful for integration into existing designs, plus projects and courses that focus on the use of processors rather than processor architecture.

22.3 Processor Definition Process

In the previous discussion of soft-core processor design flow (as shown in Figure 22.1a), it was assumed that a microprocessor (μP) definition was available. This μP Design HDL file can be developed in-house from scratch (custom), or obtained from open core sources. The custom method could take many man-years of effort. In many cases, these methods for the creation of the processor description are not cost-effective. The vendors of FPGA technology have invested in the development of software tools to streamline the process of soft-core definition. As is common in the software industry, the tools hide the proprietary aspects (source) of the processor but provide the capability to create one or more instances of the processor.

Just as the basic logic gate building blocks such as AND, OR, NOT, NAND, NOR, etc. are provided by the vendor for inclusion in digital designs, the vendors also provide a variety of processor models to be instantiated into projects. The ability to create a custom processor configuration design element for inclusion is usually done with a separate tool (e.g., EDK for Xilinx, SOPC for Altera). The processor design tool allows for drag-and-drop inclusion of the processor model and glue logic for specific attached devices. For each component, optional parameters may be specified for custom situations. Features such as a device memory address (memory mapped devices) and IRQ numbering may be specified. Cache sizes, clock sources, and bus connections are also available for inclusion.

The components available for inclusion can come from a variety of sources. Many times, these components are the device logic referred to in Figure 22.3. The soft-core vendor supplies many of the basic components such as the processor and its primary bus, but many of the other components can come from third-party vendors, open-source entities, or from in-house development.

The result of the processor design tool is a module that is created by compiling the selected items into a form that can be instantiated into a standard HDL design (Figure 22.4). (The highlighted portion of Figure 22.4 shows the processor definition process; the grayed-out portion is the standard hardware synthesis process.)

22.4 Software Development Aspects

Hardware components selected as part of the processor definition also have software aspects, in that the components often integrate software device drivers to support the selected devices (Figure 22.5).

For custom in-house developed soft cores, an accompanying assembler/compiler will need to be developed (unless the user likes to create machine code by hand). The creation of an assembler/compiler is not for the faint-hearted! For commercially provided soft cores, the vendors provide ready-made software development tools, including assemblers, C compilers, and debuggers, usually in an integrated development environment (IDE). These tools are often gcc-based and are generally free.

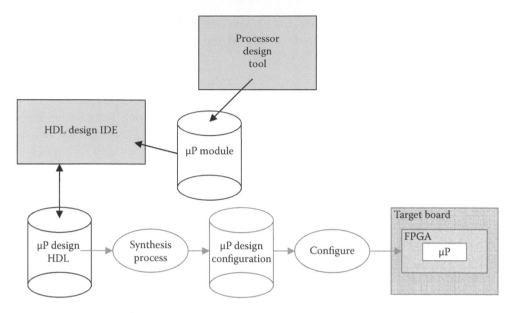

FIGURE 22.4   Processor definition process.

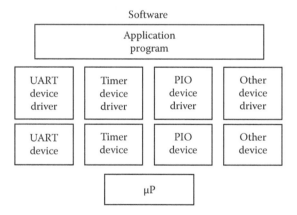

FIGURE 22.5   Integrated device drivers.

The basic difference between software development for a soft-core processor and for a discrete processor arises from the fact that the soft-core processor can (and does) change. With such a moving target, the work involved with translating each hardware change into an appropriate change of software to accommodate the hardware can be daunting. To reduce this chore, vendors have provided processes that allow the software to track with the hardware changes.

22.5 Utilization of Soft-Core Processors

Up to this point, our discussion has been from the perspective of the development cycles. During those cycles, the target system is connected to a desktop workstation running development software to configure the FPGA with a soft-core processor and then download and run or debug the software. Interface between the target device and the development workstation is generally via the JTAG connection. Once the development has been completed, the target device must be untethered from the development workstation and made to work independently. Without the connection to the development workstation, another way must be found to place executable code into memory at the system reset address/program start location and transfer control to that code (boot the system).

To boot the system, traditional computer systems typically have a flash-based (random access flash) device containing executable code at the program start location. Since flash is too slow for large systems, the code in the flash device is used only to retrieve the remaining executable system code and place it in the much faster RAM for execution. This process is termed boot strapping or initial program load (IPL). Microcontroller systems often have all of the executable code contained in random flash memory, and are essentially instant-on as system reset immediately starts executing the code directly from the flash memory. The practice of running from flash restricts the speed at which these systems run, but also contributes to their lower power requirements.

If the target device for the soft-core processor is an FPGA (usually RAM-based), both the processor and the programming are lost when power is removed. This means that the soft-core processor must first be instantiated on the FPGA before the boot process can be carried out. Sequential flash memory is usually used to store the configuration file (as shown in Figure 22.1b), which is used at power on to configure the FPGA and re-establish the soft-core processor when power is applied.

Once the soft-core processor has been re-established, it must address the same issue that other microprocessors face; i.e., how to load the executable code into memory and transfer control to that code. A variety of solutions exist for soft-core processors in addition to those available for traditional processors, and the appropriate solution may depend on which devices are already attached to the system. Binding a memory image of the executable code to the processor configuration is one solution. Partitioning a flash device with one section for the processor configuration and another for executable code is another.

22.6 Custom Instructions

With a hardware-based microprocessor, the only possible change is the software that the processor executes. Providing custom instructions for a discrete processor requires a large investment in hardware. The capability to add custom instructions to a soft-core processor allows the designer to provide new functionality without having to replace the underlying hardware. This provides a means for moving many software solutions into hardware on the FPGA platform, which may actually have better performance than a discrete processor with the solution in software.

22.7 Soft-Core Processor on an ASIC vs. FPGA

A soft-core processor can be implemented via different semiconductor devices, such as an application-specific integrated circuit (ASIC) or FPGA. An ASIC is customized for a specific application, while the FPGA's programmable logic blocks and interconnects allow a single FPGA to be used and reused in many different applications.

On an ASIC, the soft core and other design logic can be synthesized into a gate-level netlist that describes connectivity of the electronic design. Logic gates are placed and routed according to the netlist, and turned into photomasks to create the chip. The final product is hardwired logic gates formed from transistors and their interconnections. For an FPGA, the netlist is used to generate a configuration file that will be used to program lookup tables and configurable logic blocks within the FPGA.

22.8 Design Issues

Design considerations when choosing a processor include power usage, speed (performance), flexibility, device support, nonrecurring engineering (NRE) costs, and development time.

Traditional discrete microprocessors—such as x86 families, ARM, and PowerPC—are usually optimized for a wide range of applications, and the hardware features are fixed at the foundry during fabrication. The functionality of a discrete processor is changed by changing the software. In a sense, a discrete processor is software programmable. Dedicated boards require that a connection resource be permanently allocated for a specific purpose.

Processor cores can be classified as hard or soft. A hard-core processor is a processor embedded into a programmable device. It is called "hard" because it is in the silicon. Since it is part of the silicon, it offers better predictability than a soft core in terms of timing performance and area. Hard-core processors added to an FPGA offer performance trade-offs somewhere between an ASIC and an FPGA. A hard-core processor is more flexible than a discrete processor because it has access to the surrounding programmable logic resources, but less configurable than a soft-core processor because the silicon area taken up by the core cannot be reprogrammed.

Soft-core processors can be customized to the particular application in the field. In conjunction with an FPGA, the soft-core processor is constructed with look-up tables, multipliers, dividers, and on-chip memory. Due to direct access to the on-chip programmable logic resources and a customizable general-purpose bus, a soft-core processor is easily extendable.

Though soft-core processors implemented in an FPGA generally have higher power consumption and lower performance characteristics than discrete microprocessors, they provide much greater flexibility and reconfigurable device support. Since the design is in digital form, the processor characteristics can be altered simply by changing the code and reprogramming the device. The designer can specify memory width, ALU functionality, number and types of peripherals, and memory address space parameters at compile time. For projects requiring a hardware implementation, the soft-core processor implemented on an FPGA is faster and more economical than an ASIC in low to medium quantity production.

22.9 Applications for Soft-Core Processors

Soft-core processors are often used as building blocks in a larger digital design, with the goal of simplifying data processing of the overall system (i.e., hardware/software codesign). In a sense, the design is not fixed. That's why a soft-core processor is very well suited for use in FPGA designs.

Soft-core processors are suitable for educational purposes (such as courses on microprocessors, hardware/software codesign, and rapid prototyping). They are also valuable for commercial prototyping and testing uses. The integration of soft-core processors into existing digital designs provides the capability to rapidly add functionality.

Some vendors are also adding memory management units (MMU) that allow for the full-blown implementation of linux (most soft-core processors can handle uClinux, which doesn't require MMU).

References

1. D. Arbinger and J. Erdmann, Designing with an embedded soft-core processor, *Embedded Systems Conference Silicon Valley 2006*. Available: http://www.embedded.com.
2. Soft Microprocessor, http://en.wikipedia.org/wiki/Soft_processor, Date visited: December 11, 2008.
3. Listing of Soft Core Microprocessors at Opencores.com, http://www.opencores.com/projects?cat=10& lang=0&stage=0&lic=0, Date visited: December 11, 2008 and June 23, 2010.
4. Field-Programmable Gate Array, http://en.wikipedia.org/wiki/Field-programmable_gate_array, Date visited: December 11, 2008.
5. Soft Core Processor, http://encyclopedia2.thefreedictionary.com/Softcore, Date visited: December 11, 2008.
6. J.O. Hamblen, T.S. Hall, and M.D. Furman, 2008. *Rapid Prototyping of Digital Systems, SOPC Edition*. Springer Massachusetts.

7. Altera Nios II Soft-core Processor Website, http://www.altera.com/products/ip/processors/nios2/ni2-index.html, Date visited: December 14, 2008.
8. Xilinx Microblaze Soft-core Processor Website, http://www.xilinx.com/products/design_resources/proc_central/microblaze.htm, Date visited: December 14, 2008.
9. Lattice Semiconductor LatticeMico32 Open Source Soft-core Processor Website, http://www.latticesemi.com/products/intellectualproperty/ipcores/mico32/index.cfm?source=topnav, Date visited: December 14, 2008.

23

Computer Architecture

Victor P. Nelson
Auburn University

A digital computer is a device capable of solving problems and manipulating information under the direction of a given program of instructions. The hardware of a digital computer is a set of digital logic circuits that receives information from one or more sources, processes this information, and sends the results to one or more destinations. Digital computers allow the automation of arithmetic operations and provide an inexpensive way to solve complex numeric problems; store, retrieve, and communicate information; and control robots, appliances, automobiles, games, manufacturing plants, and a variety of other processes and machines.

In this chapter, we introduce the basic hardware and software elements of a digital computer and examine different computer architectures constructed from these elements. As examples, we will consider the architectures of two general-purpose microprocessors, the Intel Pentium and ARM7DMI, and two microcontrollers, from the Freescale HCS12 device family and the Intel 8051-compatible device families.

23.1 Hardware Organization

The primary hardware elements of a digital computer are a central processing unit (CPU), memory, and assorted input and output (I/O) devices, as illustrated in Figure 23.1. The CPU comprises a control unit, which coordinates the actions of the other elements in the computer; one or more arithmetic and logic units (ALUs), which are digital logic circuits that manipulate data as instructed by the control unit; and a set of registers, which are high-speed storage locations used to temporarily store data, addresses, and

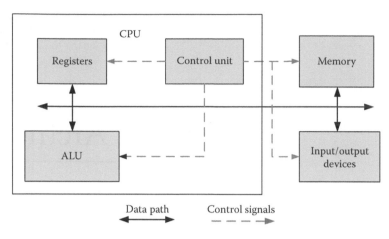

FIGURE 23.1   Computer hardware organization.

other information within the CPU. The ALU, registers, memory, and I/O devices make up the data path of the computer.

Each ALU is unique in the types of data that it can manipulate and the set of operations that it can perform on these data. Most ALUs support operations on binary integers of various sizes. Some also include operations to manipulate floating-point numbers, decimal numbers, and various nonnumeric data. Typical ALU operations include

- Arithmetic (add, subtract, multiply, divide, compare)
- Logical (AND, OR, Exclusive-OR [XOR], complement, bit test/set/clear)
- Shift and rotate data

The control unit of a CPU is responsible for fetching program instructions from memory, interpreting or decoding the instruction codes, and executing instructions by issuing control signals to the elements of the data path. The control unit coordinates all operations of the ALU, memory, and I/O devices by continuously cycling through a set of operations that cause instructions to be fetched from the memory and executed. This sequence of events is called the instruction cycle of the computer, and is illustrated in Figure 23.2. An instruction cycle includes five basic steps:

1. An instruction is fetched from the memory into the control unit of the CPU.
2. The control unit decodes the instruction, i.e., determines from the instruction code what operations to perform.
3. Any data, called operands, needed to perform these operations are accessed from CPU storage registers, retrieved from memory, or read from input devices.
4. The operation is performed on these operands.
5. The result is saved in a register, written to a memory location, or sent to an output device.

Program instructions and data are stored and retrieved from the memory of the computer. If a single memory is used for both, as is the case in most general-purpose computers and illustrated in Figure 23.3a, the computer is said to have a Von Neumann architecture, after Jon Von Neumann who is credited

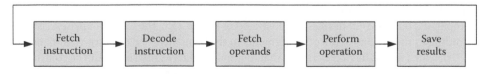

FIGURE 23.2   Instruction fetch and execute cycle.

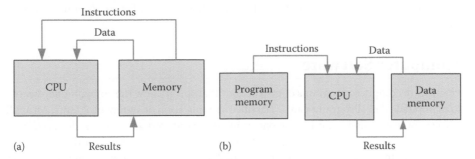

FIGURE 23.3 Computer system memory architectures: (a) Von Neumann architecture and (b) Harvard architecture.

with developing the first stored program computer. A computer that uses one memory for instructions and a separate memory for data, as shown in Figure 23.3b, is referred to as having a Harvard architecture. Many microcontrollers fall into this category. In addition, a number of high-performance CPUs use Harvard architectures to enable instruction and data memories to be accessed concurrently.

The instruction set architecture of a computer refers to the organization of a computer instruction set as seen from a programmer's point of view. Every instruction set architecture is unique in how it supports different data types; operations on data; and access to information in registers, memories, and I/O devices.

Information is transferred between a computer and the outside world through various I/O devices. Programs are usually transferred into the memory of a computer from such peripheral equipment as magnetic, optical, or flash memory peripheral storage devices. Data to be used by a program can likewise be transferred into the memory from keyboards, scanners, magnetic disks, analog-to-digital converters, communication channels, and other input devices. A program may output data to several types of peripherals. Cathode-ray tubes (CRTs), liquid crystal display (LCD) panels, and light-emitting diodes (LEDs) are often used to display the results of a program's calculations, and various types of printers are used to produce permanent results. Digital-to-analog converters, motor drivers, communication channels, plotters, magnetic disks, and other recording equipment are a few commonly used output devices.

Computers are often classified according to levels of integration. A mainframe computer is a large machine whose circuitry is typically contained on several circuit boards or cabinets of circuit boards. A microprocessor is an integrated circuit (IC) chip containing a complete CPU. Personal computers (PCs) are typically built around microprocessors and include a video display system, a keyboard for data entry, and disk drives for information storage. Common PC add-ons include printers; pointing devices, such as mice, track balls, and joysticks; CDROMs (compact disk read-only memory) and tape drives for mass storage of information; sound generators for multimedia applications; and modems and network interface hardware for communication with other computers. Engineering workstations are similar to PCs, although workstations are oriented more toward intensive graphics applications and networking.

To improve performance, some processors incorporate high-speed memory, called cache memory, within the CPU itself. Superscalar processors further improve performance by integrating multiple ALUs and other functional units within a single CPU to allow multiple instructions to be executed concurrently.

A microcontroller is a complete computer on a single IC chip, comprising CPU, memory, and various peripheral functions, such as timers, digital-to-analog converters, display drivers, communication functions, and interfaces to external sensors, actuators, and other devices. To make room for these elements on the chip and keep the cost low, microcontroller CPUs often have fewer capabilities than general-purpose microprocessors. Microcontrollers are primarily used in embedded control systems, in which the computer is embedded within the hardware of such products as automobile engines, kitchen appliances, cell phones, electronic games, music players, communication equipment, and industrial control systems.

23.2 Computer Software

Computer software comprises programs of instructions that specify how the computer hardware is to be utilized to manipulate data. Programs can be classified as application or system programs. An application program is a set of instructions, specified in one or more programming languages, designed to perform a given task according to a specified algorithm. System programs comprise all of the support software provided on a computer system to aid programmers in developing and executing application programs.

23.2.1 Programming Languages

The individual steps of an algorithm or task to be performed by a program must be expressed using the statements of a programming language. Every CPU has a unique native machine language that is recognized by its hardware. Machine language is a set of binary codes that inform the CPU of the operations to be performed and operands (data items) to be used in these operations. All digital computer instructions must be represented by these binary codes before the computer can interpret them.

Rather than writing programs directly as sequences of binary codes, a symbolic representation of a CPU's machine language, called assembly language, is often used to develop programs, especially for applications that require very small or very efficient programs, such as control systems embedded into such products as home appliances and automobiles. The assembly language allows a programmer to symbolically specify operations to be performed on the data stored in the internal registers and in the memory of a processor without becoming bogged down in creating binary code sequences. A system program called an assembler translates the symbolic assembly language instructions into a machine language so that the program code can be interpreted by the CPU.

The following is an 8051 assembly language instruction and its equivalent machine language:

```
Assembly language:   ADD A,#25
Machine language:    00100100 00011001
```

This instruction tells the 8051 CPU to add 25 to the contents of its accumulator (A register) and replace the contents of the accumulator with the computed sum. The first 8 bits of the machine language code indicate that the operation is ADD, that register A is to be used, and that an 8-bit data value follows as the second byte of the instruction code.

An assembly language programmer must be intimately familiar with the architecture of the specific CPU being programmed to efficiently design algorithms for a given application in the assembly language of that CPU. In contrast, high-level languages allow a programmer to express an algorithm for a given application in a more natural way, independent of any particular computer architecture. Hundreds of different high-level languages have been developed, many tailored to specific applications. Among the first were FORmula TRANslation (FORTRAN), developed for numeric applications, COBOL (COmputer Business Oriented Language) for business applications, and PROLOG and LISP to support artificial intelligence and expert system applications. Some languages, such as BASIC, Perl, PHP, Java, C, and C++, are more general purpose, supporting a wide variety of different applications.

Before a computer can execute a program written in a compiled high-level language, such as C and C++, the program must first be translated to the native machine language of that computer by a system program called a compiler. One of the primary benefits of using a high-level language to write a program is that the program can be targeted at different computer architectures by simply recompiling it into each architecture's machine language, as illustrated in Figure 23.4. Thus, a high-level language

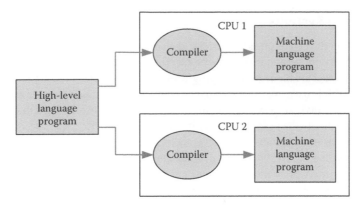

FIGURE 23.4 Compilation of a high-level language program for two different CPU architectures.

program is portable across different computer architectures. The process of recompiling a program for a different computer architecture is referred to as porting the program to the new architecture.

The following example of a C language statement is a natural way to indicate that the value of variable g is to be set to the sum of the values of variables b and c:

$$g = b + c;$$

This statement might be translated by a C compiler to the following sequence of assembly language instructions for a Freescale HCS12 microcontroller:

```
LDD    b    ;load variable b into accumulator D (ACCD)
ADDD   c    ;add variable c to the value in ACCD
STD    g    ;store the value in ACCD into variable g
```

The same C instruction might be translated to the following for an ARM processor:

```
LDR    r0, = b      ;point to variable b
LDR    r1, [r0]     ;load variable b into register r1
LDR    r0, = c      ;point to variable c
LDR    r2, [r0]     ;load variable c into register r2
ADD    r3, r1, r2   ;write the sum r1 + r2 into register r3
LDR    r0, = g      ;point to variable g
STR    r2, [r0]     ;store the sum in variable g
```

In addition to assemblers and compilers, other system programs assist in the development of application programs. These include text editors to create and alter the text of a program, linkers to link together multiple program segments, including subprograms from software libraries, and program debugging tools. An integrated development environment (IDE) incorporates all of these tools within a single user interface, including editing, assembly/compilation, and debugging for a family of CPUs.

Some programming languages, such as BASIC, Perl, PHP, and Javascript, are interpreted rather than compiled. Instead of compiling the entire program into the machine language prior to execution, each instruction of an interpreted language is individually read, translated to machine language instructions, and executed at run time. This occurs each time the instruction is executed, making it slower than a compiled program. However, since a program written in an interpreted language does not need to be compiled in advance, programs can be written, changed, and tested quickly.

23.2.2 Operating Systems

When a computer is dedicated to one specific task, the application program can be stored permanently in the memory of the computer and executed with no other support software. General-purpose computing systems, however, execute many different programs, which change from day to day and from user to user. In addition, multiple users may need to share a single system, or a single user may wish to concurrently perform multiple tasks on a single system. In such cases, a control program called an operating system is used to coordinate the usage of the facilities of the computer.

An operating system is a program that interprets user commands typed at a keyboard, selected by clicking on an icon with a mouse, or read from a file. Some commands are executed by programs built into the operating system, while other commands correspond to programs that reside on a disk or are otherwise supplied by a user. An operating system also manages the file system on the computer, which comprises a directory of files stored on a disk or tape and programs that locate and access these files when requested by a program. Operating systems also coordinate access to printers, networks, and other I/O devices, and manage CPU time and memory space.

The first PCs were controlled by single-user operating systems, such as MSDOS. A single-user operating system interprets one command at a time from the user, executes the corresponding program, and then waits for another command to be entered. No other program may be executed until the current one is finished. Once a user has control of the computer, the user may modify and re-execute programs, or execute several different programs before turning the machine over to the next user.

In the single-user environment, much time can be spent idling while waiting for user inputs or data transfers involving slow I/O devices. To exploit this idle time, multitasking and multiuser operating systems allow CPU time to be shared by multiple programs. The operating system passes control of the CPU from program to program, with each program allowed to execute for a small allotment of time or until it becomes stalled waiting for input/output. In this manner, the execution of a program is interleaved with the execution of other programs until it has completed. The end result is that programs execute concurrently with each program appearing to have exclusive control of the CPU. Linux [1] is an example of a multiuser, multitasking operating system used in a wide variety of embedded systems, PCs, workstations, and larger systems. Multiple users can issue commands to the operating system of one computer from different terminals, with each user running several programs at the same time.

Process control and many other embedded system applications require real-time operation, in which the computer must respond to various events and perform designated actions within the given time constraints as the events occur. In such applications, a real-time operating system (RTOS) can be used to coordinate the execution of processes in response to these events, including scheduling and synchronization of processes, controlling access to resources, and managing communication between processes.

23.3 Information Representation in Digital Computers

All information in a computer must be represented in patterns of 1's and 0's. The assignment of a meaning to a bit pattern is called coding. In general, an n-bit pattern of information can represent 2^n unique items. CPUs generally support a limited number of pattern lengths. In most general-purpose computers, the smallest pattern size is 8 bits, referred to as a byte, although some of the early microcontrollers worked primarily with 4-bit nibbles of information, and some application-specific CPUs (such as digital signal processors, or DSPs) work exclusively with larger pattern sizes. In general, the primary pattern length supported by a CPU is referred to as its word size, which is usually an integral number of bytes. For example, the word size is 32 bits in the Pentium, PowerPC, and ARM CPUs; 16 bits in the Freescale HCS12 and Microchip PIC24F; and 8 bits in the Atmel AVR and 8051-compatible microcontrollers. Some advanced processors support 64-bit and larger information patterns.

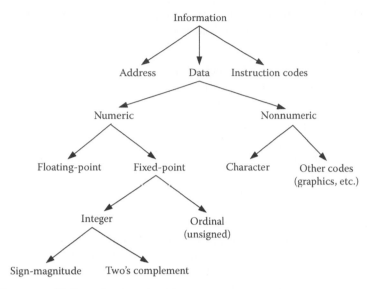

FIGURE 23.5 Taxonomy of information representation.

Figure 23.5 shows the taxonomy of different information types found in a computer. Addresses are pointers to storage locations in memory or I/O device interfaces. Instruction codes tell the CPU what operation to perform in a given program step. Data are items to be manipulated by computer instructions.

An n-bit data pattern has 2^n unique codes, allowing up to 2^n items to be represented. Data types can be classified as numeric or nonnumeric. Numeric data formats can vary widely from application to application, although, in many situations, simple binary integers are sufficient. Binary integers can be unsigned or signed. In an n-bit unsigned (ordinal) number, $b_{n-1}b_{n-2}...b_1b_0$, each bit is weighted by a power of 2, representing the number $b_{n-1} \times 2^{n-1} + b_{n-2} \times 2^{n-2} + ... + b_1 \times 2^1 + b_0$ with values in the range $[0...(2^n - 1)]$ (see Table 23.1).

Signed integers can be represented in several ways. An n-bit sign-magnitude number format uses the leftmost bit to represent the sign of the number and the remaining $n - 1$ bits to represent the magnitude or absolute value of the number. A sign bit of 0 indicates a positive value and 1 indicates a negative value. An n-bit sign-magnitude number can represent numbers in the range $[-(2^{n-1} - 1)... + (2^{n-1} - 1)]$. Note that there are two representations for the number 0, namely, +0 and −0 (see Table 23.1).

Digital logic circuits that add and subtract numbers in the sign-magnitude format are inefficient. Therefore, digital computers normally use the two's complement number system to represent signed numbers. In an n-bit two's complement number system, positive numbers are represented as they are in the sign-magnitude format; the most significant bit is 0 and the remaining $n - 1$ bits indicate the magnitude. The negative of a value A is represented by its two's complement, defined as $2^n - A$. This simplifies arithmetic hardware by allowing one to compute $A - B$ by forming the two's complement of B and then adding it to A, i.e., $A - B = A + (-B)$, thereby eliminating the need for special subtraction circuits.

TABLE 23.1 Integer Number Ranges for Different Coding Methods

Bits	Ordinal	Sign-Magnitude	Two's Complement
4	0...15	−7...+7	−8...+7
8	0...255	−127...+127	−128...+127
16	0...65,535	−32,767...+32,767	−32,768...+32,767
32	0...4,294,967,295	−2,147,483,647...+2,147,483,647	−2,147,483,648...+2,147,483,647

The two's complement is fairly easy to compute; one method is to complement all of the bits and then add 1 to the result. For example, the 8-bit two's complement number system representation of the value -5_{10} would be computed as follows:

1. Write the 8-bit binary code for $+5_{10} = 00000101_2$.
2. Complement the bits: $\overline{00000101} = 11111010$.
3. Add one to the result: $11111010 + 1 = 11111011$.

Therefore, the 8-bit code 11111011 represents the value -5_{10} in the two's complement number system. In contrast, the sign-magnitude code for the value -5_{10} is 10000101, which is the code for $+5_{10}$ with the leftmost bit set to 1 to indicate a negative value. The reader is referred to [2] for additional algorithms and examples involving signed numbers.

Financial and other applications may require the manipulation of decimal, rather than binary, numbers. In such cases, a binary-coded decimal (BCD) coding is used, in which each decimal digit is represented independently by its 4-bit binary equivalent. Generally, two BCD digits are packed into an 8-bit byte; this is referred to as packed BCD. For example, the packed BCD representation of 25_{10} is coded by packing into 8 bits the binary code for 2 (0010) and the binary code for 5 (0101). The result is the packed BCD code 00100101. Many general-purpose CPUs provide special arithmetic hardware and instructions to assist in manipulating BCD numbers without requiring conversions to and from binary.

If both integer and fractional numbers are needed, a fixed-point or floating-point number format is used. The fixed-point notation partitions the n bits used to represent a number into two fixed-length parts: k bits to represent the fraction part and $n - k$ bits to represent the integer part, as illustrated in Figure 23.6. A binary point is assumed to be at a fixed position between the two parts. An n-bit fixed-point number is said to have n bits of precision, and can accurately represent a value to within 2^{-k}, the value of the least significant (leftmost) bit.

The range of a set of numbers is the span between the largest and smallest magnitudes. For the number format in Figure 23.6, the largest value is approximately 2^{n-k}, as determined by the number of integer bits, and the smallest value is 2^{-k}, which is a function of the number of fraction bits. Integers are special cases of fixed-point numbers, with the binary point to the immediate right of the least significant bit.

Scientific and many other applications require very wide ranges of numeric values. In these cases, floating-point number formats are used. A floating-point number is generally written in the form $\pm m \times r^e$, where m is the mantissa, r the radix, and e the exponent of the number. Floating-point numbers are represented in a computer system by packing the codes for e and m into a single storage location, as illustrated in Figure 23.7, with the sign of the mantissa usually stored in the leftmost bit. The code for radix r need not be stored, since it is always known. The most common radix value is $r = 2$, corresponding to binary numbers. The IBM 360 mainframe architecture used a floating-point format based on $r = 16$.

Prior to the mid-1980s, each computer manufacturer developed its own scheme for encoding floating-point numbers, making it difficult to transfer data from one computer architecture to another. In 1985, American National Standards Institute (ANSI) and the IEEE developed the *IEEE Standard for Binary Floating-Point Arithmetic*, ANSI/IEEE Standard 754-1985 [3], which defines standard single-precision (32-bit) and double-precision (64-bit) floating-point formats, shown in Figure 23.7, and related operations that are now used in most computer systems.

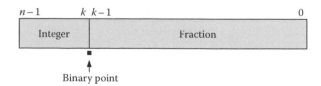

FIGURE 23.6   Fixed-point number format.

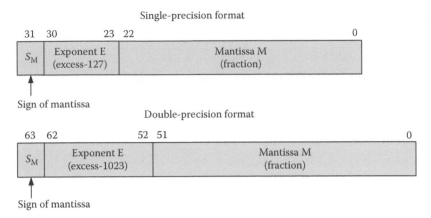

FIGURE 23.7 IEEE Standard 754-1985 floating-point formats. (Reproduced from *IEEE Standard for Binary Floating-Point Arithmetic*, ANSI/IEEE Std. 754-1985, IEEE, Inc., New York, 1985. With permission.)

The mantissa is stored in a normalized sign-magnitude format. The sign bit is placed in the leftmost bit of the 32-bit code, allowing a number to be easily identified as positive or negative. The magnitude of the mantissa is of the form *1.F*, where *F* is a 23-bit binary fraction stored in bits 22-0 for the single-precision format, and a 53-bit fraction stored in bits 52-0 for the double-precision format. Normalized forms are characterized by their most significant digits being nonzero, and are used to ensure a single unique representation for each floating-point number. For example, the following are a few of the many possible representations of the number 1408.

$$10.11 \times 2^9$$

$$1.011 \times 2^{10}$$

$$0.1011 \times 2^{11}$$

$$0.01011 \times 2^{12}$$

Requiring the mantissa to be normalized to the form *1.F* forces the unique code 1.011×2^{10} to be used to represent this number.

The 8-bit exponent is stored in the biased, excess-127 format. In the excess-*N* number format, a bias of *N* is added to each value to force numbers in the range $[-N... +(N-1)]$ to be represented by a linearly increasing number sequence $[0...(N-1)]$. In the IEEE Standard 754-1985 format, a bias of 127 is added to each exponent value, with the result stored in bits 30–23. Exponents in the range $[-127... +126]$ are therefore represented by increasing binary codes from $[00000001...11111110]$. The exponent codes 00000000 and 11111111 are reserved to indicate special conditions. For example, the constant zero is represented by the all-0's word, and +/− infinity by an exponent of all 1's and a mantissa of all 0's.

Nonnumeric data is represented in a computer by designing a coding scheme that assigns a unique binary code to each data item. Alphanumeric and special character information is commonly represented by the American Standard Code for Information Interchange (ASCII) code, listed in Table 23.2. Each printable character has a unique 7-bit code recognized by most devices that send or receive alphanumeric information, such as printers and terminals. For example, suppose that we want to encode the message "ADD 1." This message has five characters, the fourth being a space or blank. In the ASCII code, our message becomes

```
       A         D         D       space        1
    1000001   1000100   1000100   0100000    0110001
```

TABLE 23.2   Seven-Bit ASCII Character Codes ($c_6c_5c_4c_3c_2c_1c_0$)

$c_3c_2c_1c_0$	*000	001	010	011	100	101	110	111
				$c_6\,c_5\,c_4$				
0000	NUL	DLE	SP	0	@	P	'	p
0001	SOH	DC1	!	1	A	Q	a	q
0010	STX	DC2	"	2	B	R	b	r
0011	ETX	DC3	#	3	C	S	c	s
0100	EOT	DC4	$	4	D	T	d	t
0101	ENQ	NAK	%	5	E	U	e	u
0110	ACK	SYN	&	6	F	V	f	v
0111	BEL	ETB	'	7	G	W	g	w
1000	BS	CAN	(	8	H	X	h	x
1001	HT	EM	)	9	I	Y	i	y
1010	LF	SUB	*	;	J	Z	j	z
1011	VT	ESC	+	;	K	[	k	{
1100	FF	FS	'	<	L	\	l	|
1101	CR	GS	-	=	M	]	m	}
1110	S0	RS	.	>	N	^	n	~
1111	S1	US	/	?	O	_	o	DEL

ASCII characters are often padded with an extra zero on the left to allow each code to fit exactly into one 8-bit byte of memory, or the 8th bit might be used to extend the code to 128 additional symbols.

Many other codes have been developed to represent graphical information, special symbols, audio signals, and a wide variety of other information.

23.4 Computer Programming Model

The programming model of a computer system refers to the assembly language programmer's view of the system. From the programmer's perspective, a computer is a collection of registers and memory for storing information, a set of instructions to manipulate data and control program flow, and various types of data that can be manipulated by the instruction set.

A computer instruction specifies to the control unit what operation is to be performed, where to obtain operands for the operation, and where to store the result of the operation, if a result is produced. As with other information, an instruction must be encoded into patterns of ones and zeros. Instruction codes are generally subdivided into separately coded fields, as illustrated in Figure 23.8. These fields include the operation code (opcode), which specifies what the instruction is to do, and one or more operand specifiers, which indicate the operands to be used for the instruction.

The number of operand specifiers encoded in an instruction differs between CPUs. Most reduced instruction set computer (RISC) processors, such as ARM [4], Sparc [5], PowerPC [6], and MIPS [7], use a three-operand format for all ALU instructions, specifying two source operands and a destination. For example, the following ARM instruction tells the CPU to read two operands from registers R2 and R3, add the operands, and store the sum in register R1:

ADD   R1,R2,R3   ;(R2)+(R3)→R1

Operation code	Operand specifier 1	Operand specifier 2	...

FIGURE 23.8   Instruction code format.

Pentium ALU instructions use a two-operand format [8]. Both operand sources are specified, and the left-hand operand is also the destination for the result. The following instruction adds the operands in registers AX and BX and stores the sum in AX:

$$\text{ADD} \quad \text{AX,BX} \quad ;(\text{AX})+(\text{BX}) \to \text{AX}$$

The source operand that was originally in AX is replaced with the computed sum. Therefore, if the programmer wishes to retain the original source operand, it must be copied from AX to some other place before executing the above instruction. This provides less flexibility for the programmer than the three-operand format, but instructions can be more compact, since they require only two operand specifiers.

The HCS12 [9] is one of a number of CPUs that use a one-operand format. The programmer specifies one of the source operands. The other source operand comes from a default location, with that location also receiving the result. The following instruction adds the constant 15 to the accumulator register A, with the sum stored in A, replacing the source operand:

$$\text{ADDA} \quad \#15 \quad ;(\text{A})+15 \to \text{A}$$

The opcode ADDA tells the CPU that accumulator A provides the second source operand and is the destination. The operand must be moved into accumulator A by a previous instruction, and the result may need to be stored elsewhere to allow A to be used for a subsequent instruction. This provides less flexibility for the programmer, but instructions can be shorter since there is only one operand specifier. Note that the 8051 and Pentium specify the source of the data as the right-hand operand and the destination as the left-hand operand, while the HCS12 specifies one operand as part of the instruction mnemonic.

Instruction operands may be embedded within the instruction code, retrieved from CPU registers, or read from external memory locations. Some CPUs also recognize special I/O device addresses; others access information from I/O devices via their memory address space. RISC architectures restrict ALU operands to register immediate values only; memory may only be accessed by special load and store instructions that move data between registers and memory.

23.4.1 CPU Registers

A register is simply an n-bit binary storage element. CPU registers are used to temporarily hold data and memory address values that might be needed in the near future. Being located within the CPU, registers can be accessed more quickly and more efficiently than external memory. In addition, since the number of registers is relatively small, operand specifiers need only be a few bits, as compared to memory addresses that require considerably more bits to represent. For example, a 32-bit ARM instruction code contains two or three 4-bit operand specifiers, each of which identifies one of 16 registers (R0–R15) to be used as operands. In contrast, ARM memory addresses are 32-bits each, so it is not possible to include a memory address within a 32-bit instruction code.

Every instruction set architecture has its own distinctive set of program-accessible registers that may be used to store data, addresses, and control or status information. Figure 23.9 shows the register sets of four popular microprocessors.

Every CPU contains one register called a program counter (PC) or instruction pointer (IP) that always contains the address in the memory of the next program instruction to be executed. The PC/IP register is updated automatically as each instruction is executed, so that it points to the next instruction to be fetched from memory.

One or more CPU registers are available to the programmer to hold memory addresses or information used to compute addresses. All 16 ARM registers (R0–R15) can hold either data or addresses. Likewise, Pentium registers EBX, ESI, EDI, and EBP can be used for both addresses and data. In contrast, 8- and 16-bit microcontrollers usually have very limited memory-addressing capabilities. For example, the

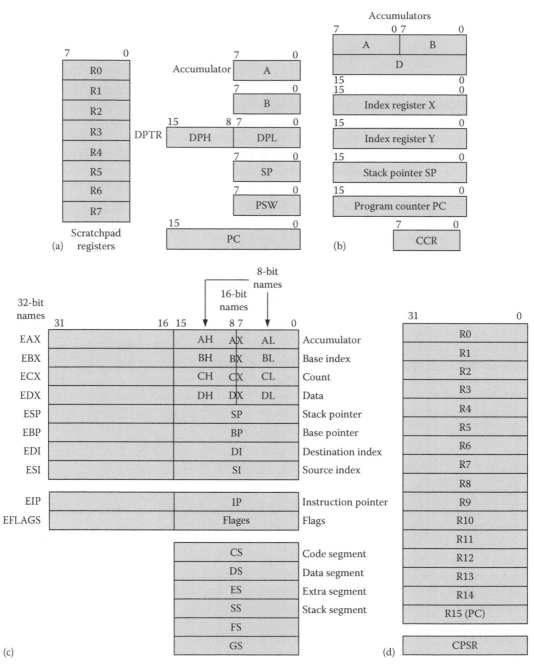

FIGURE 23.9 Register sets of four common CPUs: (a) 8051/8052, (b) Freescale HCS12 (From Cady, F.M., *Software and Hardware Engineering: Assembly and C Programming for the Freescale HCS12 Microcontroller*, 2nd edn., Oxford University Press, New York, 2008. With permission.), (c) Intel Pentium (From Brey, B.B., *The Intel Microprocessors: 8086/8088, 80286, 80386, 80486, Pentium, Pentium Pro Processor, Pentium II, Pentium III, Pentium 4: Architecture, Programming and Interfacing*, 7th edn., Pearson-Prentice Hall, Upper Saddle River, NJ, 2006. With permission.), and (d) ARM. (From Hohl, W., *ARM Assembly Language: Fundamentals and Techniques*, CRC Press, Boca Raton, FL, 2009. With permission.)

HCS12 has only two registers, index registers X and Y, that can be used in memory addressing. Methods for addressing memory will be discussed in Section 23.4.4.

The 8051 and HCS12 microcontrollers are similar to some older computers in that arithmetic and logic operations utilize accumulator registers, i.e., registers in which results are "accumulated." The 8051 has a single accumulator, labeled A in Figure 23.9a, and the HCS12 has two accumulators, labeled A and B in Figure 23.9b. In the HCS12, the two 8-bit accumulators can be concatenated into a double accumulator, labeled D in Figure 23.9b, for 16-bit operations. Arithmetic operations such as addition and subtraction combine the number in the accumulator with a second operand and write the result back to the accumulator, overwriting the original accumulator contents. The second operand can be an immediate value or the contents of another register or memory location. To combine two data values from memory, one of them must be moved to the accumulator prior to the operation. After the operation, the result can be moved to a memory location if desired. In the HCS12, most instructions can use either the A or the B accumulator, and the A and B accumulators can be used together as a single 16-bit accumulator, referred to as register D.

The Pentium and ARM CPUs give more flexibility to a programmer by providing a number of general-purpose registers, any of which can supply operands for, or receive the results of, arithmetic and logical operations. In the ARM, any of its 16 registers, 32 bits each, R0–R15, may be a source and/or a destination in any instruction. Likewise, the Pentium provides eight general-purpose registers that can be used as operand sources/destinations. Pentium supports 8-bit, 16-bit, and 32-bit ALU operations, using the corresponding register names shown in Figure 23.9c. For example, an 8-bit operation might use the AL or AH register, a 16-bit operation the AX register, and a 32-bit operation the EAX register. All four are located within the same 32-bit register. AL is the low 8 bits, AH the next 8 bits, and AX the low 16 bits of the EAX register. The following illustrate 8-, 16-, and 32-bit addition operations in the Pentium; the data size is implied by the register names:

```
ADD   AL,BH      ;AL + BH → AL (8-bit bytes)
ADD   AX,BX      ;AX + BX → AX (16-bit words)
ADD   EAX,EBX    ;EAX + EBX → EAX (32-bit long words)
```

While the eight Pentium general registers can be used in all arithmetic and logical instructions, most of these registers also have additional special functions and are named accordingly. The four general-purpose registers EAX, EBX, ECX, and EDX are used by some instructions as an accumulator, a base address register, a count register, and an I/O-addressing register, respectively. The four index and pointer registers ESI, EDI, ESP, and EBP are used in memory addressing. Each half of the four general registers can be used independently in 8-bit operations, and hence the high and low parts of these registers are labeled AH/AL, BH/BL, CH/CL, and DH/DL.

Most CPUs contain a processor status register (PSR), sometimes called a condition code register, that contains information about internal CPU conditions and about ALU operations that have been performed. PSRs usually contain one or more condition code bits, or flags, which characterize the result of a previous arithmetic or logical operation performed in the ALU. These allow decisions to be made based on the outcomes of these operations. Table 23.3 lists the condition code flags of the HCS12, which

TABLE 23.3 HCS12 Condition Code Flags

Flag	Status of Last Result if Flag = 1
Z (Zero)	Result zero
N (Sign)	Result negative
C (Carry)	Carry out of the most significant bit of the result
V (Overflow)	Result out of range for the given number of bits
H (Half Carry)	Carry from bit 3 to bit 4 of result

Source: Cady, F.M., *Software and Hardware Engineering: Assembly and C Programming for the Freescale HCS12 Microcontroller*, 2nd edn., Oxford University Press, New York, 2008.

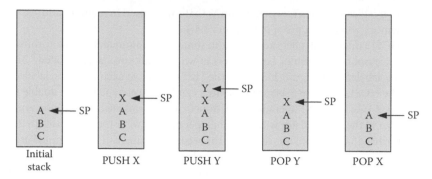

FIGURE 23.10 Push-down stack PUSH and POP operations.

are typical of those found in most CPUs. The half carry flag is used to support operations on BCD values, representing a carry from one decimal digit to the next within a byte.

In addition to regular memory addressing, many CPUs support a special last-in/first-out data structure in memory called a push-down stack. A stack is a convenient place to temporarily save information and subsequently restore it. A CPU running a program may be temporarily interrupted to execute some other program. The state of the running program is saved temporarily on a stack while the other program is executed, and then restored from the stack when that program is finished, allowing the original program to continue where it left off.

A dedicated stack pointer (SP) register contains the address of the top element on the stack. An operation called PUSH adds an element to the stack, and an operation called POP or PULL removes an element from the stack, as illustrated in Figure 23.10. The SP automatically increments and decrements as elements are added to and removed from the stack. ARM does not have dedicated PUSH/PULL instructions, although the memory-addressing capabilities of the memory load and store instructions can perform basic stack operations, using any of the 16 registers as an SP. In lieu of a stack, ARM saves its program counter and the current PSR in registers R14 and SPSR (saved processor status register), respectively, for subroutine calls and interrupts.

23.4.2 Immediate Operands

A constant to be used as an operand for an instruction may be encoded directly within the instruction as the operand specifier. Such a data value is referred to as an immediate operand, because it is immediately available to the CPU from the fetched instruction, without having to access additional storage locations. The following instructions add the immediate operand 5 to a designated CPU register:

```
8051:      ADD    A,#5        ;A + 5 → A
HCS12:     ADDA   #5          ;ACCA + 5 → ACCA
Pentium:   ADD    EAX,5       ;EAX + 5 → EAX
ARM:       ADD    R1,R2,#5    ;R2 + 5 → R1
```

In general, immediate values are encoded with the same number of bits as the other operand(s) of the instruction, and appended to the instruction code. In ARM and other RISC processors, however, all instructions are encoded in 32-bits, which must include the opcode and all operand specifiers. Hence, ARM limits an immediate operand to an 8-bit value, with an optional left or right shift. This value will be extended by the CPU to 32 bits before performing the operation.

23.4.3 Memory Organization

A computer system utilizes memory elements for storing program instructions, data, and other information. From the viewpoint of the instruction set, memory is an array of words, each identified by a unique address that indicates its location within the memory. The concept of a memory address is equivalent to that of a telephone number. Every telephone is assigned a unique number comprising an area code, an exchange, and a number within that exchange. Similarly, each memory location is assigned a unique address that identifies a memory module and a specific storage location within that module.

Each memory word contains one or more addressable bytes, as illustrated in Figure 23.11. The number of bits in the data path of the CPU determines the number of bytes per word of the memory. For example, the 8051 has an 8-bit data path and, thus, has a byte-wide memory organization, as in Figure 23.11a. The HCS12 has a 16-bit data path and supports operations on both bytes and words. Therefore, memory must be byte-addressable, i.e., each byte of memory must have a unique address. Figure 23.11b shows two 16-bit memory formats. The HCS12 uses the little endian format, i.e., the least significant byte of data is placed in the lower numbered address. The Motorola 68000 uses the big endian format, in which the least significant byte of data is placed in the higher numbered address.

The Pentium has a 32-bit data path and supports operations on 32-bit words, 16-bit half-words, and 8-bit bytes. The ARM likewise has a 32-bit data path, but all arithmetic and logical operations are performed exclusively on 32-bit data. Bytes, half-words, and words can be transferred between the ARM CPU registers and memory, but a byte or half-word read from memory is converted into 32 bits before being placed in the target register. Memory for these CPUs is organized as 4 bytes per word, as shown in Figure 23.11c. Both CPUs can access one, two, or all 4 bytes of a memory word with a single read or write operation. While the Pentium uses the little endian format, the ARM processor is unique in that it can be configured at power-up for either big or little endian format, as desired by the system designer.

The number of addressable memory locations in a computer is a function of the number of bits used by the CPU to represent memory addresses. An N-bit address can address 2^N locations. For example, the 8051 and HCS12 use a 16-bit address, allowing them to address 2^{16} = 64 kB of memory. The ARM uses 32 address bits, and can address 2^{32} = 4 GB of memory, organized as 4 bytes per word for a total of 2^{30} = 1 G words.

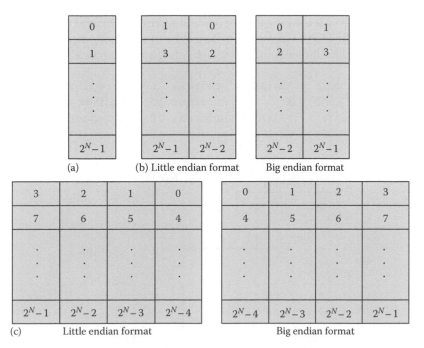

FIGURE 23.11 Byte-addressable memory organizations (2^N bytes): (a) byte-wide, (b) word-wide, and (c) double-word-wide organizations.

23.4.4 Memory Addressing

To retrieve an operand from memory or write a result to memory, the address of that memory location must be specified by the instruction being executed. In most CPUs, memory addresses may be specified directly or indirectly.

23.4.4.1 Direct Addressing

The location of the operand is explicitly specified either numerically or symbolically when writing the instruction. The operand address is embedded within the assembled instruction code, as illustrated in Figure 23.12. The following are examples of direct addresses specified by assembly language instructions. (The ARM does not support direct addressing.)

```
HCS12:      STAA   103      ;contents register A to M[103]
8051:       LD     R0,103   ;contents of M[103] to register R0
Pentium:    MOV    AL,BOB   ;BOB represents memory address 103
```

The last example uses the symbolic label BOB to represent the memory address 103. The actual address of BOB is determined by the assembler and inserted into the instruction code when the instruction is assembled.

23.4.4.2 Indirect Addressing

It is often the case that a programmer does not know in advance where a particular operand will be located in memory at the time an instruction is executed; in such cases, the operand address is determined while the program is running. Examples include arrays of numbers that are accessed using a starting address of the array and an index into the array, and data accessed using pointers. In these cases, the address is specified indirectly. The CPU is either told where to find the address, in a register or memory location, or how to calculate the address.

Figure 23.13 illustrates register indirect addressing of the data in memory location 103. The operand specifier in the instruction code indicates that the operand address is contained in register R. Since there are relatively few registers in a CPU, as compared to the number of addressable memory locations, the operand specifier is only a few bits. The following are examples of register indirect addressing for several CPUs:

```
Pentium:    MOV    AL,[BX]   ;address 103 in register BX
HCS12:      LDAA   X         ;address 103 in index register X
8051:       MOV    @R1,A     ;address 103 in register R1
ARM:        LDRB   R2,[R3]   ;address 103 in register R3
```

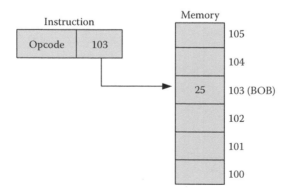

FIGURE 23.12   Direct memory addressing.

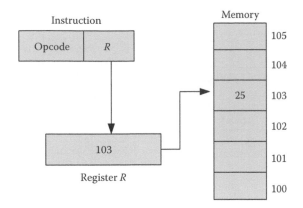

FIGURE 23.13   Indirect memory addressing.

The ARM and HCS12 support special auto-increment and auto-decrement modes of register indirect addressing, in which the address register is automatically incremented after, or decremented before, a memory access. This simplifies access to tables of data in which the elements are to be accessed sequentially, as illustrated in the following example, which computes the sum of a list of 32-bit numbers in a table:

```
        LDR    R0,#100      ;Load size of TABLE into R0
        LDR    R1,=TABLE    ;Load address of TABLE into R1
        LDR    R2,#0        ;Initialize SUM to 0 in R2
Loop:   LDR    R3,[R1],#4   ;Read element TABLE(I) and increment pointer
        ADD    R2,R2,R3     ;SUM = SUM + TABLE(I)
        DBEQ   R0,Loop      ;Decrement R0 and repeat above until R0 = 0
```

The LDR instruction at location *Loop* adds 4 to the address in register R1 (called post-increment) after reading the 4-byte (32-bit) data word, leaving R1 pointing to the next element of the table.

The HCS12 and a few other CPUs also support memory indirect addressing modes, in which the address of an operand is contained in a designated memory location. This requires two memory accesses: one to fetch the operand address and another to fetch the operand itself.

23.4.4.3 Base/Indexed Addressing

Data are often stored in tables, lists, records, or other data structures for which addresses are specified in two parts: a base (beginning) address of the data structure and an offset, or index, from the beginning. The operand specifier either directly or indirectly identifies the base address and the index.

The base address, the index, or both are usually retrieved from registers. The following example instructions, illustrated in Figure 23.14, designate an address register for the base address and specify a constant index of 3. This form of base-indexed addressing is useful for accessing records and similar data structures in which each element is at a known offset from the beginning of the structure.

```
Pentium:   MOV    AX,3[BX]     ;address = (BX) + 3
ARM:       LDRB   R0,[R1,#3]   ;address = (R1) + 3
HCS12:     LDAA   3,X          ;address = (X) + 3
```

For accessing linear arrays of data, it may be more convenient to explicitly specify the base address as a constant, and then designate a register containing an index as shown in Figure 23.15 and the following

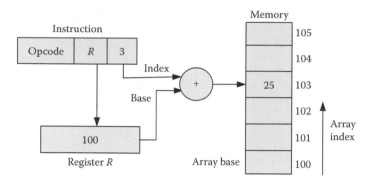

FIGURE 23.14 Addressing a record element with variable base and constant offset.

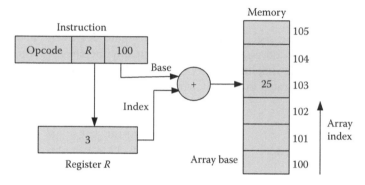

FIGURE 23.15 Addressing an array with constant base and variable offset.

instructions. ARM limits the constant to a 12-bit number, to fit within the 32-bit instruction code, so not all base addresses can be written as constants.

```
Pentium:   MOV    AX,100[BX]    ;address = 100 + (BX)
ARM:       LDRB   R0,[R1,#100]  ;address = 100 + (R1)
6805:      LDAA   100,X         ;address = 100 + (X)
```

Alternatively, some CPUs allow a base address to be in one register and an index in a second register, as shown in Figure 23.16 and in the following examples:

```
Pentium:   MOV    AX,[BX][SI]   ;address = (BX) + (SI)
ARM:       LDR    R1,[R2,R3]    ;address = (R2) + (R3)
HCS12:     LDY    D,X           ;address = (D) + (X)
```

The following program shows how to store a data word in an indexed array variable, TABLE(I), in the ARM, where I is a variable to be used as the array index:

```
LDR    R0,=TABLE    ;R0 points to TABLE
LDR    R1,=I        ;R1 points to index I
LDR    R2,[R1]      ;Load index I into R2
STRB   R3,[R0,R1]   ;Store R3 at TABLE(I)
```

ARM and Pentium support a scaled index addressing mode, in which the index is multiplied by a scale factor corresponding to the number of bytes in the accessed data item. This allows a simple index

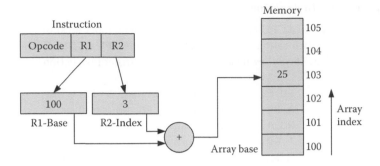

FIGURE 23.16 Addressing an array using both base and index registers.

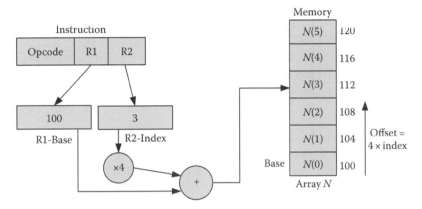

FIGURE 23.17 Scaled index for an array of 4-byte data values.

to be automatically converted into a displacement from the beginning of a table of values, as illustrated in Figure 23.17 and in the following examples, where ARM and Pentium registers R2 and ECX, respectively, contain index value 3. Note that the array index 3 corresponds to an offset of 12 bytes from the base of the array. In the ARM instruction, the designation SHL #2 tells the CPU to shift the number in R2 to the left by 2 bit positions, effectively multiplying that value by 4.

```
ARM:        LDR    R0,[R1,R2,SHL #2]    ;Memory address = (R1) + (4*R2)
Pentium:    MOV    EAX,[EBX + 4*ECX]    ;Memory address = (EBX) + (4*ECX)
```

23.4.4.4 Program-Counter-Relative Addressing

It is often necessary to jump or branch from one point in a program to another. Instead of specifying the target address within the branch instruction, most CPUs compute the target address by adding to the program counter a displacement from the current instruction to the target instruction. This is referred to as program-counter-relative addressing. The advantage of doing this is that the code can be made position independent and relocatable. This means that the program can be loaded in any location in memory without reassembling or recompiling it, since branch instructions are only dependent on the distance to each target address and not on the absolute value of the address.

23.4.5 Computer Instruction Types

Digital computer instructions can be organized into seven basic categories: data transfer, arithmetic, logical, shift/rotate, control transfer, input/output, and processor control.

23.4.5.1 Data Transfer Instructions

Data transfer instructions load a CPU register with data from a memory location, store the contents of a CPU register into a memory location, and move data from one CPU register to another, or from one memory location to another. The following are example data transfer instructions:

```
8051:     MOV    R0,#5         ;move constant 5 into R0
          MOV    A,R0          ;register R0 to accumulator A
          MOV    R0,A          ;accumulator A to register R0
          XCH    A,R0          ;exchange/swap contents of A and R0
HCS12:    LDAA   MEMY          ;memory to accumulator A
          STAA   MEMY          ;accumulator A to memory
          TBA                  ;accumulator B to accumulator A
Pentium:  MOV    AX,MEMY       ;memory to register AX
          MOV    MEMY,AX       ;register AX to memory
          MOV    AX,BX         ;register BX to register AX
          XCHG   AX,BX         ;exchange/swap contents of AX and BX
ARM:      LDR    R0,[R1]       ;load R0 from memory
          STR    R0,[R1]       ;store R0 into memory
          SWP    R0,[R1]       ;swap words between R0 and memory
```

Other examples of data transfer instructions are PUSH and POP operations using the SP register, as described earlier, and instructions to load address registers with the computed operand addresses. The following instructions compute the sum of a base address and an index, and place the result into an address register, to be used as a pointer:

```
Pentium:  LEA    DI,TABLE[SI]   ;DI points to TABLE + (SI)
HCS12:    LEAX   TABLE,X        ;X points to TABLE + (X)
```

23.4.5.2 Arithmetic Instructions

Instructions such as add, subtract, multiply, and divide perform binary arithmetic on integer operands. Not all CPUs provide all four functions. Some simple microcontrollers provide only add and subtract instructions and leave it to the programmer to write short programs to perform multiplication or division. Some CPUs include additional instructions to increment and decrement binary numbers to facilitate counting operations and the modification of memory addresses.

CPUs that support decimal number formats provide special instructions to perform binary arithmetic on BCD numbers, either directly or indirectly. The Motorola 68000 and the IBM 360 had special BCD add and subtract instructions, while most other CPUs, including Pentium, HCS12, and 8051, support the addition of BCD numbers by using the normal binary ADD instruction followed by a special *decimal adjust* instruction to correct the result. The following example illustrates the addition of two packed BCD numbers:

```
HCS12:    ADDA   N    ;binary sum of A + N in A
          DAA         ;decimal adjust result in AL
```

One must often perform arithmetic on multi-precision numbers, which have more bits than the word size of the CPU. This is done as with pencil and paper, where one adds the two least significant digits of a pair of numbers, producing a digit and possibly a carry. If there is a carry, 1 is added to the next pair of digits and so on. CPU add and subtract operations set the carry flag in the PSR to 0 or 1 to indicate that a carry was generated, or a borrow was required, to complete the operation. An add-with-carry or subtract-with-borrow instruction uses the carry flag to adjust the next pair of digits accordingly. The following example illustrates the computation of $H = H + G$ on a Pentium, where H and G are 32-bit numbers and CF is the carry flag:

$$CF$$

$$G_1 G_0$$

$$\pm \frac{H_1 H_0}{H_1^* H_1^*}$$

```
MOV   AX,G       ;Get low word of G
ADD   H,AX       ;Add to low word of H and set CF (CF, H₀* = G₀ + H₀)
MOV   AX,G + 2   ;Get high word of G
ADC   H + 2,AX   ;Add with carry to high word of H (H₁* = G₁ + H₁ + CF)
```

23.4.5.3 Logical Instructions

Logical instructions apply the Boolean AND, OR, and XOR operators to the corresponding bits of two operands. This gives the computer the ability to selectively set, clear, complement, or test individual bits, or groups of bits, within a memory location or I/O device register. Table 23.4 summarizes the three Boolean operators applied to a 1-bit Boolean variable.

The AND operator can be used to force selected bits of a word to 0, as illustrated in Figure 23.18a. The second operand is a bit pattern called a mask that contains a 0 in each bit position that is to be forced to 0, and a 1 in each bit position that is to be left unchanged. Similar masks can be created for the OR operator to force selected bits to 1, and for the XOR operator to force selected bits to be complemented. These are illustrated in Figure 23.18b and c, respectively. The corresponding HCS12 logical instructions are also shown in Figure 23.18 for each case.

Many I/O devices contain a status register whose bits reflect the readiness of the device to perform an operation. The AND operator can be used to isolate a selected bit of a byte read from a status register to determine if that bit is 0 or 1. This is illustrated in Figure 23.19, which shows the HCS12 logical AND instruction ANDA #%00000010.

The mask 00000010 forces all bits to 0 except for bit b_1. If the result is 00000000, then the zero flag is set and it follows that $b_1 = 0$; otherwise, the zero flag is cleared, indicating a nonzero result, from which it can be implied that $b_1 = 1$. The result of a logical AND instruction can be tested by a conditional jump instruction, such as the following HCS12 instruction, which jumps to location B1_ZERO if the zero flag is set, or otherwise continues with the next sequential instruction:

TABLE 23.4 Boolean Operations on Bit Variable b

AND	OR	XOR
$b \cdot 0 = 0$	$b + 0 = b$	$b \oplus 0 = b$
$b \cdot 1 = b$	$b + 1 = 1$	$b \oplus 1 = \bar{b}$

```
BEQ   B1 _ ZERO   ;Jump if zero flag set (b₁ = 0)
```

$$b_7\, b_6\, b_5\, b_4\, b_3\, b_2\, b_1\, b_0 \qquad b_7\, b_6\, b_5\, b_4\, b_3\, b_2\, b_1\, b_0 \qquad b_7\, b_6\, b_5\, b_4\, b_3\, b_2\, b_1\, b_0$$

$$\underline{\wedge\ 1\ 1\ 1\ 1\ 1\ 1\ 0\ 1} \qquad \underline{\vee\ 0\ 0\ 0\ 0\ 0\ 0\ 1\ 0} \qquad \underline{\oplus\ 0\ 0\ 0\ 0\ 0\ 0\ 1\ 0}$$

$$b_7\, b_6\, b_5\, b_4\, b_3\, b_2\, 0\, b_0 \qquad b_7\, b_6\, b_5\, b_4\, b_3\, b_2\, 1\, b_0 \qquad b_7\, b_6\, b_5\, b_4\, b_3\, b_2\, \bar{b_1}\, b_0$$

(a) ANDB #%11111101   (b) ORAB #%00000010   (c) XORB #%00000010

FIGURE 23.18 Logical operations used to alter a selected bit: (a) Clear b_1, (b) Set b_1, and (c) Toggle b_1.

For example, assume that a printer interface contains a status register in which the rightmost bit indicates whether the printer is ready to accept another character to print. The following Pentium program loop will repeat continuously as long as the "printer ready" bit is 0. The CPU will exit the loop and continue as soon as the ready bit becomes 1.

$$b_7\ b_6\ b_5\ b_4\ b_3\ b_2\ b_1\ b_0$$

$$\begin{array}{c} \land \end{array} \frac{0\ 0\ 0\ 0\ 0\ 0\ 1\ 0}{0\ 0\ 0\ 0\ 0\ 0\ b_1\ 0}$$

FIGURE 23.19   Using logical AND to isolate 1 bit.

```
Check:   IN    AL,PrintStatus    ;read printer status register
         AND   AL,00000010       ;isolate "printer ready" bit
         JZ    Check             ;go back to check if printer not ready
```

The HCS12 and some other microcontrollers provide extensions of the logical operators to explicitly set, clear, and test bits of a byte. These are illustrated in the following examples:

```
         BCLR   PortA,%00000101        ;clear bits 2 and 0 of the byte at PortA
         BSET   PortA,%10000000        ;set bit 7 of the byte at PortA
Wait:    BRCLR  PortA,%00000010,Wait   ;branch to Wait if bit 1 of PortA is 0
```

Note that the third instruction effectively performs the same function as the printer status check loop above. It reads the byte at PortA, masks all but bit 1, and branches to address Wait if the bit is clear (0). This instruction would therefore repeat until the bit is eventually not clear (1).

23.4.5.4 Shift and Rotate Instructions

Shift and rotate instructions slide bits right or left within a register or memory location, as illustrated in Figure 23.20. These can be used for extracting or combining bit fields within an operand, to convert data between parallel and serial forms, and to perform multiplication and division by powers of 2.

A logical shift operation shifts the bits right or left by one bit position, with the vacated bit replaced by a 0. For unsigned numbers, this is equivalent to dividing or multiplying the number by 2. An arithmetic right shift implements a divide-by-2 operation on a two's complement number by preserving the sign bit as the operand is shifted. Some CPUs allow an operand to be shifted by more than 1 bit position with a single instruction. The following HCS12 example packs two BCD digits into a single byte by shifting one digit 4 bits to the left, and then combining the two digits:

```
         LSLA              ;shift BCD digit to upper nibble of A
         LSLA
         LSLA
         LSLA
         ORAA   Memy       ;combine two BCD digits from A and Memy
```

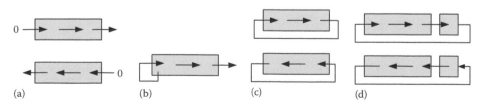

       (a)   (b)   (c)   (d)

FIGURE 23.20   Shift and rotate operations: (a) logical shift, (b) arithmetic shift right, (c) circular rotate, and (d) circular rotate through carry.

A *circular rotate* instruction performs a shift operation while replacing the vacated bit with the bit shifted out of the other end of the operand. An alternate rotate operation is often provided that rotates the number through the carry flag of the PSR. In most CPUs, the bit shifted out of an operand is copied to the carry flag of the PSR, where it can be tested or used to support multi-precision shift operations. A multi-precision number can be shifted by using the carry flag as a link between parts of the number, allowing a bit shifted out of one part to be shifted into the other using a rotate-through-carry operation. The following Pentium example multiplies a 32-bit number by 2, by shifting 1 byte at a time 1 bit to the left:

```
SHL   NUMBER       ;shift memory byte 1 bit left
RLC   NUMBER + 1   ;shift carry and 2nd byte 1 bit left
RLC   NUMBER + 2   ;shift carry and 3rd byte 1 bit left
RLC   NUMBER + 3   ;shift carry and 4th byte 1 bit left
```

23.4.5.5 Control Transfer Instructions

The normal flow of a program is to execute instructions in the order listed in the program, fetched from sequential memory addresses. To control this flow, the program counter increments automatically after each instruction. Jump, branch, and subroutine call instructions interrupt the normal flow by transferring the control of the program to some instruction other than the next one in sequence. This supports the design of looping and decision-making programs, as well as supporting procedure and function calls. The following are examples of instructions that unconditionally transfer the control of a program to location X within the current program:

```
8051/Pentium:   JMP Target
HCS12:          JMP Target or BRA Target
ARM:            B Target
```

Decision making and looping require conditional branch instructions that jump only if a given condition is true, but continue with the next sequential instruction if the condition is false. Conditional branch instructions test selected bits of the PSR, which reflect the result of a previous arithmetic or logical operation. The following Pentium program adds a list of four numbers in memory, decrementing the SI register at the end of each iteration and repeating the ADD, DEC, and JGE instructions as long as SI is greater than or equal to 0:

```
        MOV   SI,3            ;set counter to 3
        MOV   AL,0            ;clear accumulator
Start:  ADD   AL,TABLE[SI]    ;add next element of TABLE
        DEC   SI             ;subtract 1 from SI
        JGE   Start           ;repeat if SI ≥ 0
```

The relationship between two operands can be tested by subtracting them and then testing the resulting condition codes according to Table 23.5. Most CPUs provide a compare instruction (CMP) that performs the subtraction and sets the condition code flags, but does not save the actual result in a destination. The following HCS12 program branches to location RICK if the unsigned number in accumulator A is less than or equal to 10, using the "branch if less or same" instruction to test the result of the CMP.

```
Check:  CMP   #10    ;subtract 10 from A
        BLS   RICK   ;go to RICK if A lower than or same as 10
```

TABLE 23.5   Condition Codes for Relational Operators

Condition	Symbol	Relation	Number Type	Boolean Condition
Zero	Z	A = B	Both	Z
Not zero	NZ	A ≠ B	Both	$\bar{Z}$
Greater than	G	A > B	Signed	$\overline{(N \oplus V) + Z}$
Greater than or equal	GE	A ≥ B	Signed	$\overline{N \oplus V}$
Less than	L	A < B	Signed	$N \oplus V$
Less than or equal	LE	A ≤ B	Signed	$(N \oplus V) + Z$
Above	A	A > B	Unsigned	$\overline{C + Z}$
Above or equal	AE	A ≥ B	Unsigned	$\bar{C}$
Below	B	A < B	Unsigned	C
Below or equal	BE	A ≤ B	Unsigned	$C + Z$

Modular programming requires the ability to partition software into subroutines, such as procedures and functions, which can be invoked as needed. This is supported by special subroutine call instructions that jump from a main program to the start of a subroutine after saving a pointer to the next instruction in the main program, allowing a return to the main program after completing the subroutine.

A subroutine call (CALL) or jump to subroutine (JSR) instruction typically pushes the current program counter onto the system stack to save the address of the next instruction in the main program, and then loads the program counter with the address of the subroutine. A return (RET) or return from subroutine (RTS) is executed as the last instruction of the subroutine to pop the program counter from the stack and, thus, return to the main program. The ARM does not support a system stack; a subroutine is called with a jump and link (JAL) instruction, which saves the program counter in register R14 before loading the PC with the target address. The subroutine returns to the main program by moving the saved return address from register R14 to the program counter as in the following instructions:

```
MOV   PC,R14   ;copy R14 to PC
BX    R14      ;branch (indirect) to the address in R14
```

23.4.5.6 Input/Output Instructions

A few CPUs utilize separate address spaces for memory and I/O devices. Hence, there is both a memory address 0 and an I/O address 0. In these cases, special instructions are provided to read information into the CPU from an input device and to write information from the CPU to an output device. The Intel CPUs support an isolated I/O address space that can be accessed only by the two special instructions IN and OUT as follows:

```
IN    AL,25   ;data from IO address 25 to AL register
OUT   25,AL   ;data from AL register to IO address 25
```

23.4.5.7 Processor Control Instructions

These instructions manipulate various hardware elements within the CPU, and are therefore CPU specific. The reader is referred to [4–10] for descriptions of processor control instructions for specific CPUs.

23.4.6 Interrupts and Exceptions

Events often occur that require the interruption of normal instruction processing to perform some special action. Such exceptional events, or exceptions, can be triggered by signals from devices external to the CPU, or by conditions detected within the CPU.

For example, a cell phone microcontroller might use a timer to interrupt the CPU once per second to make it update an image of a clock displayed on the screen. Microcontrollers used in process control are typically interrupted by sensors that detect various conditions in the plant that require immediate attention. An example of an internally detected condition is an attempt to divide a number by 0, which cannot produce a valid result. This type of exceptional condition should suspend normal processing to abort the operation and send a warning message to the user.

A primary advantage of an external interrupt is that a CPU may work in parallel with one or more external processes, such as printing a document, and be interrupted only when the process requires attention. The alternative is to continuously monitor the process by checking a status register in the device to detect when the device requires attention. Such monitoring would prevent the CPU from doing other work while waiting for the device to be ready, whereas with interrupts the CPU can do other jobs until interrupted.

When an exception condition is detected, a CPU typically responds as follows:

1. Complete the instruction currently in progress to reach a convenient stopping point.
2. Save the current program counter on the system stack or in a designated register, preserving a pointer to the next instruction that would have been executed had the program not been interrupted.
3. Determine the condition that requested the interrupt. Many CPUs execute a special interrupt acknowledgment operation to allow an external interrupting device to identify itself with a unique number called an interrupt vector.
4. Load the program counter with the starting address of a program to perform the service requested by the interrupting device. This program is called an interrupt service routine (ISR). Where used, an interrupt vector points to a memory location containing the starting address of the ISR. Some CPUs restrict the ISR starting address to a fixed address in memory, while others require that the ISR address be stored in one specific memory location.
5. Fetch and execute the instructions of the ISR.
6. Upon completion of the ISR, execute an interrupt return instruction to restore the original program counter from the stack, allowing the CPU to return to the point in the original program at which it was interrupted.

Since a running program may be interrupted at any time, the ISR should begin by saving any registers that will be modified within the ISR, and then restore them before returning to the main program. This will allow the main program to continue as if the interrupt had not occurred.

23.5 Evaluating Instruction Set Architectures

Many different metrics are used as indicators of computer performance. Perhaps the most commonly cited is MIPS, which stands for millions of instructions per second. Unfortunately, MIPS figures do not indicate how much work is done by each CPU instruction. In some CPUs, instructions perform very primitive operations, while in others, each instruction may do a considerable amount of work. Therefore, simply knowing how many instructions a CPU can execute per second provides only a partial picture of how fast a computer can perform.

As the computer architecture evolved from the first computers in the 1940s through the machines of the 1970s and 1980s, the sizes of the instruction sets grew as designers became able to incorporate more circuit devices on a single IC chip. High-level languages became widespread, and emphasis was placed on making compiled programs as efficient as possible. For this reason, CPU instruction sets were expanded to include any instruction that might be needed to implement a high-level language statement. Thus, instructions became more powerful, with compilers producing fewer instructions per program. However, this growth in the number and power of instructions required larger and more complex CPU control units, slowing down the entire processor.

In 1970, researchers at IBM designed the model 801 computer, in which the instruction set was reduced to only those that were used frequently. Later, researchers at Stanford and Berkeley observed that only a small core of computer instruction sets were executed for the majority of the time. They developed RISCs that could execute programs faster than complex instruction set computers (CISC) by streamlining CPU designs. These efforts evolved to the Sparc, MIPS, PowerPC, ARM, and other RISC processors.

The true performance of a CPU can be measured by the time, T_{exec}, that it takes to execute a given program:

$$T_{exec} = \left(\frac{I}{P}\right) \times \left(\frac{C}{I}\right) \times \left(\frac{T}{C}\right) = (\# \text{Instructions}) \times CPI \times T_{CLK}$$

where
   I/P is the number of instructions in the program
   $C/I = CPI$ is the average number of CPU clock cycles required to execute an instruction
   $T/C = T_{CLK}$ is the CPU clock period (time per clock cycle), which is the inverse of the clock frequency

CISC designs strive to minimize T_{exec} by reducing the number of machine instructions needed to execute a high-level language program, minimizing I/P. Designers of CPUs like the Digital Equipment Corporation's VAX minicomputer, and the Intel 80x86 and Motorola MC680x0 microprocessors attempted to anticipate the needs of compiler writers and provided as many instructions as might be needed. However, the cost of providing a large number of instructions is increased hardware complexity, which increases the factors CPI and T_{CLK}.

RISC designs target the CPI and T_{CLK} factors at the expense of increased numbers of instructions (I/P). Simplifying the instruction set can reduce the number of clock cycles required to execute each instruction. The target of most RISC processors is a single clock cycle per instruction. In addition, simplifying the hardware often enables the clock period to be shortened. However, more instructions are required to perform a given task than for an equivalent CISC machine.

With transistor-switching-time improvements slowing down, computer architects have turned to parallelism to obtain additional improvements. One method of utilizing parallelism is a superscalar design, in which multiple functional units are contained in the CPU and multiple instructions fetched and executed in parallel. For example, the Intel Pentium includes two integer and one floating-point ALU that can be used concurrently. The Pentium control unit fetches multiple instruction codes at one time from memory, and can simultaneously initiate operation in one, two, or all three of these units, instead of executing the instructions sequentially. Likewise, the SUN Super Sparc and the Motorola PowerPC 601 CPUs each contain an integer unit, a floating-point unit, and a branch processing unit that can be used to execute multiple instructions concurrently.

References [7,11,12] provide thorough discussions of computer performance and the effects of architectural features on program execution times.

23.6 Computer System Design

Figure 23.21 illustrates the basic interconnection of memory and I/O devices to a CPU. Information is transferred between a CPU and selected memory or I/O devices via a data bus, which is a set of parallel signal lines, each carrying one data bit. The CPU selects one memory location or I/O device to receive or provide data by broadcasting its address to all devices in the system over an address bus. Logic circuits in each device interface compare the address on the bus to its assigned value to determine if it is the one being addressed. This is called decoding the address.

Data transfers are coordinated by the CPU, using one or more control signals to indicate the type, direction, and timing of each data transfer. The direction of a data transfer is either device-to-CPU (a CPU read cycle) or CPU-to-device (a CPU write cycle). Freescale processors and many other CPUs issue a

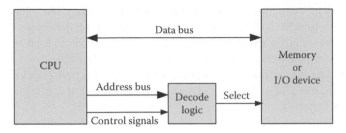

FIGURE 23.21 Computer system address and data buses.

control signal $R/\overline{W}$ at the beginning of the cycle, setting it high to designate a read cycle and low to indicate a write cycle. Some CPUs, such as the Intel processors, use two control lines to indicate the type of cycle: $\overline{RD}$ to signify a read cycle and $\overline{WR}$ to signify a write cycle. (The overbar indicates that these signals are active-low, i.e., a logic 0 level signals that the indicated operation is active.)

The timing of a data transfer is critical. The CPU must signal a device when it is time to begin and when it is time to end each data transfer. The Freescale processors issue an enable (E) or a data strobe signal ($\overline{DS}$) that activates when the transfer is to begin and returns to the inactive level when the transfer is finished. In the Intel processors, $\overline{RD}$ and $\overline{WR}$ act as data strobes for read and write cycles, respectively.

Figure 23.22 illustrates the read and write cycle bus timing for CPUs that use control signals compatible with those of the Freescale CPUs, Figure 23.22a, and for CPUs that use Intel-compatible control signals, Figure 23.22b.

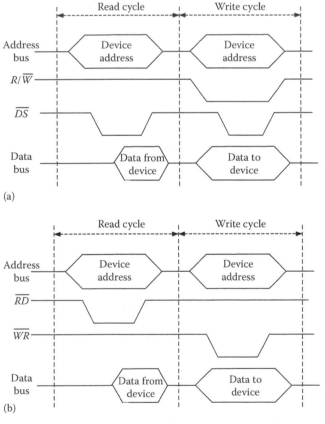

FIGURE 23.22 Data transfer timing between CPU and memory for: (a) Freescale-compatible control signals and (b) Intel-compatible control signals.

The number of bits in a data bus may or may not match the width of the internal data path of the CPU. For example, the Intel 8086 CPU has internal and external 16-bit data buses. The Intel 8088 CPU is identical to the 8086, including the 16-bit internal data path, but it has an 8-bit external data bus. The Intel Pentium has a 32-bit internal data path and a 64-bit external data bus.

The rate at which data can be transferred between a CPU and an external device is referred to as the bandwidth (BW) of the bus, and is a function of the bus speed (number of data transfers per second) and the number of bits per transfer:

$$BW = (\text{transfers/second}) \times (\text{bits/transfer})$$

Given identical bus speeds, the external bus BW of an 8088 CPU would be half of that of an 8086, even though they have identical internal data paths.

For CPUs that support byte-addressable memory with 16-bit and wider buses, control signals are provided to indicate which byte or bytes of the bus are to be used to transfer data. The memory controller of the NXP LPC2292 microcontroller [13], based on an ARM CPU, generates four "byte lane" select lines, $BLS[3]$, $BLS[2]$, $BLS[1]$, and $BLS[0]$, corresponding to sets of eight data lines (byte lanes), D_{31-24}, D_{23-16}, D_{15-8}, and D_{7-0}, respectively. Each byte lane select signal activates if that "byte lane" is to be used for the data transfer. So, all four would activate during a 32-bit data transfer, two during a 16-bit transfer, and one during an 8-bit transfer. Likewise, the 32-bit data buses of the Intel 80386 and 80486 have four byte-enabled control lines: $\overline{BE_3}$, $\overline{BE_2}$, $\overline{BE_1}$, and $\overline{BE_0}$, corresponding to data bus lines D_{31-24}, D_{23-16}, D_{15-8}, and D_{7-0}, respectively. The Pentium has eight such control lines for its 64-bit bus.

HCS12 microcontrollers, which have 16-bit external buses, activate a special control line, $\overline{LSTRB}$ (lower byte strobe), when the low byte of the data bus is to be used for a data transfer. The address line A_0 indicates a byte within a 16-bit word. $A_0 = 1$ indicates that the low byte is to be used and $A_0 = 0$ indicates that the high byte is to be used. However, for a 16-bit data transfer, both $A_0 = 0$ and $\overline{LSTRB} = 0$, to access both bytes. This is summarized in Table 23.6, which compares the HCS12 to the low 16 bits of the Pentium data bus. Note that the HCS12 uses the big endian format, while the Pentium uses the little endian format.

Reliable data transfers require synchronization of the CPU and the device being accessed. A synchronous bus implies that the data transfer is synchronized to the clock signal that drives the CPU, and must be completed within a designated number of clock periods. A designer must select memory and I/O devices whose access times are short enough to fit within this constraint.

In contrast, an asynchronous bus does not synchronize data transfers to a reference clock. Instead, the CPU signals the device to begin the transfer and the device signals the CPU when it has completed the data transfer. This allows each data transfer to take as much or as little time as required for the device being accessed, and therefore allows slower devices to be used. The Motorola 68000 has an input signal called $\overline{DTACK}$ (data transfer acknowledge) that must be activated by each accessed device to signal the CPU that the transfer has been completed.

A semisynchronous bus synchronizes data transfers to the system clock, as in the synchronous case, but allows a slow device to signal the CPU that it needs more time to complete a data transfer. For example, the 8086 has an input called *READY* that would normally be held high, but can be pulled low by a device until it is ready to complete a data transfer. When the CPU sees *READY* = 0, it enters a special wait state for one clock cycle and checks the *READY* signal again, repeating this wait state indefinitely until *READY* has been activated by the device.

On most microcontrollers and some older microprocessors, chip package sizes have a limited number of signal pins. In these cases, it is impractical to allocate a separate pin for each address,

TABLE 23.6 Byte and Word Transfers on a 16-Bit Bus

Data Transfer Type	Pentium		HCS12	
	$\overline{BE_1}$	$\overline{BE_0}$	$\overline{LSTRB}$	A_0
Byte: D_{7-0}	1	0	0	1
Byte: D_{15-8}	0	1	1	0
Word: D_{15-0}	0	0	0	0

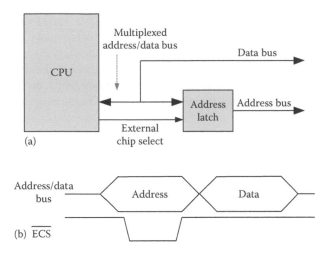

FIGURE 23.23 Multiplexed address and data buses: (a) Address is latched to demultiplex the bus and (b) Timing of the multiplexed bus.

data, and control signal. To reduce the pin count, a single pin may be used for two or more different functions, i.e., two or more signals are time-multiplexed over one signal line. It is common for expandable microcontrollers to use a set of signal pins to transmit an address to external memory, and then use the same pins to transfer data. The CPU provides a control signal to indicate when each type of information is on the pins, allowing the bus to be demultiplexed.

For example, the HCS12 multiplexes its 16-bit data bus DB_{15-0} and address bits ADR_{15-0} on signal pins ADR/DB_{15-0}, as illustrated in Figure 23.23a. For each data transfer, the CPU first broadcasts an address on ADR_{15-0} and then uses the same pins to transfer the data, as illustrated in the timing diagram of Figure 23.23b. A signal called $\overline{ECS}$ (external chip select) is set activated by the CPU whenever an address is present on ADR_{15-0}, and low otherwise. Since the CPU removes the address before the data is transferred, a latch must be provided external to the CPU chip, as shown in Figure 23.23a, to save the address while $\overline{ECS} = 0$ and hold it for the duration of the data transfer.

23.7 Hierarchical Memory Systems

General-purpose computer systems often utilize a hierarchy of memory devices, as shown in Figure 23.24. Memory devices differ in how they are accessed, their storage capacity, volatility, cost per bit, and access time.

Computer memory is classified as primary memory if any storage location within the memory can be accessed directly by the CPU; otherwise, it is classified as secondary memory. Primary or direct-access memories comprise a set of numbered storage locations that are accessed by supplying the address of the location to be accessed, along with one or more control signals to indicate whether information is to be read from memory or written to memory.

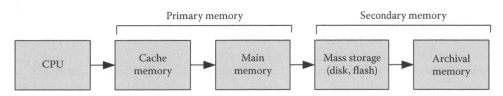

FIGURE 23.24 Computer system memory hierarchy.

Secondary memory devices are used for bulk or mass storage of programs and data, and include rotating magnetic devices, such as disks and magnetic tapes, flash memory devices, optical devices such as CDROMs, and a variety of other devices. In contrast to primary memory, the information in secondary memory devices is not accessed directly. The CPU accesses the device through a special controller that searches the device to locate the desired item. When found, an entire block of information is usually transferred into primary memory, where the desired items can be accessed in a more convenient fashion.

23.7.1 Memory Characteristics

Volatility refers to the permanence of data stored in a memory. A read-only memory (ROM) preserves information permanently and cannot be rewritten. This is useful for storing programs and data values that will not change. Read-write memories can be erased and/or rewritten. Some read-write memories retain information only while powered up, while others, like disk or flash, can retain data indefinitely until erased or rewritten.

The capacity of a memory refers to the total number of bits of information that can be stored. Capacity is a function of the mechanism used to access the memory. A direct access memory is limited in size by the number of address bits provided by the CPU. The capacity of a disk drive is determined partially by its physical characteristics and control circuitry, and the ability of the operating system software to keep track of the information on the disk. Archival secondary storage devices with removable media provide virtually unlimited capacity.

The cost per bit of storage decreases as one moves farther from the CPU in Figure 23.24. High-speed cache memories are more expensive than slower main memories, while storage on most disk drives is often orders of magnitude less expensive per bit. Archival storage allows large quantities of information to be saved on inexpensive tapes or diskettes with minimal cost.

As the capacity increases and the cost per bit goes down in the memory hierarchy, the performance parameters, or access and cycle times, of the memory become longer. Memory access time is the length of time required to retrieve (read) a word from the memory, and memory cycle time is the minimum interval of time required between successive memory operations. The access time of a memory determines how quickly information can be obtained by the CPU, while the cycle time determines the rate at which successive memory accesses may be made. In general, direct access devices can be accessed more quickly and more often than secondary devices.

23.7.2 Semiconductor Memory Technologies

Most computers built prior to 1970, some of which are still in operation today, utilized arrays of magnetic cores as their primary memory elements, while a few specialized systems, particularly in space vehicles, utilized plated wire as a replacement for magnetic core in applications where radiation hardness was required. In today's digital computers, primary memories are constructed of semiconductor IC chips.

Semiconductor memories are available as read-only or read-write devices. ROM is used to store programs, tables, and other data that will not be modified while the computer is operating. Table 23.7 lists several ROM technologies, which differ in how the devices are programmed and/or erased. ROMs are programmed when they are manufactured, while PROMs are field-programmable, i.e., they are programmed by the system designers that use them. Neither ROMs nor PROMs are alterable; they must be discarded when their contents are no longer valid.

EPROM, EEPROM, and FLASH memories are field-programmable devices that can also be erased and reprogrammed. EPROMs are erased by exposing the storage cells to ultraviolet light to free electrical charge trapped in the memory cells, while EEPROM and FLASH memories are erased electrically. Selected locations can be erased and reprogrammed in an EEPROM, while with a FLASH memory, an

TABLE 23.7 Nonvolatile Memory Technologies

Acronym	Device Type
ROM	Read-only memory
PROM	Programmable read-only memory
EPROM	Erasable programmable read-only memory
EEPROM	Electrically erasable programmable read-only memory
FLASH	Flash memory—EEPROM that is erased by erasing the entire memory array
FeRAM	Ferroelectric RAM—ferroelectric material enables device to retain data while power is removed

entire block of bits in the device must be erased before a location in that block can be reprogrammed. Ferroelectric RAM (FeRAM) technology utilizes ferromagnetic material to retain information when power is removed. While not as dense as FLASH memory, FeRAM devices are faster, consume less power, and allow more read/write cycles.

Read-write memory is commonly referred to as RAM, which is an acronym for random access memory. In reality, both ROM and RAM are random access memories since their contents can be accessed in any random order. Nonetheless, the term RAM has become associated with read and write memory.

RAM chips are further classified as static or dynamic. Once written, a static RAM retains information until its power is removed; the RAM cells will be in random states when power is restored. Each memory cell contains the equivalent of a digital flip-flop circuit to achieve this static storage. In contrast, a dynamic RAM retains information for only a short time as charge stored on a capacitor associated with a single transistor. This charge slowly leaks away over a short period of time. Therefore, to retain information, the contents of a dynamic RAM must be periodically refreshed, i.e., read and rewritten.

Because more transistors are used for each storage cell, static RAM devices have a higher cost per bit than dynamic RAMs, and can hold fewer bits than a dynamic RAM chip fabricated with a comparable technology. However, extra control circuitry is required for dynamic RAMs to refresh the memory, and dynamic RAMs are typically slower than comparable static devices. For small systems, the complexity and cost of the extra control circuitry for dynamic RAMs make them less attractive than static RAMs. However, for systems that use large amounts of RAM, the extra cost of the refresh circuitry is negligible compared to the lower cost per bit of dynamic RAM devices. Therefore, most PCs, workstations, and larger computers use dynamic RAMs for their main memories.

23.7.3 Memory System Organization

The organization of a primary memory system is determined primarily by the address and data buses of the host CPU. A memory chip is organized as $2^K \times N$, which means that there are K address input lines and N data I/O lines, and thus a total of 2^K addressable N-bit locations. An address decoder within the chip selects one location corresponding to each K-bit address and either sends the information from that location off the chip or writes new information into that location.

Using memory devices organized as $2^K \times N$ in a system that has J total address lines, where $J \geq K$, means that the system can accommodate $2^J/2^K = 2^{J-K}$ devices. These devices are usually organized hierarchically into banks, as illustrated in Figure 23.25, with the memory address partitioned into a bank number, a chip number, and an on-chip address, as shown. Each part of the address is decoded at a different level of the hierarchy. One level of address decoding selects one of the banks, another selects a chip within the bank, and the on-chip decoder selects one location within the selected chip.

If the system data bus width B is wider than the number of data pins D on the memory chip, then B/D chips must be used to create one addressable block of memory, with each chip connected to a different D-bit section of the data bus.

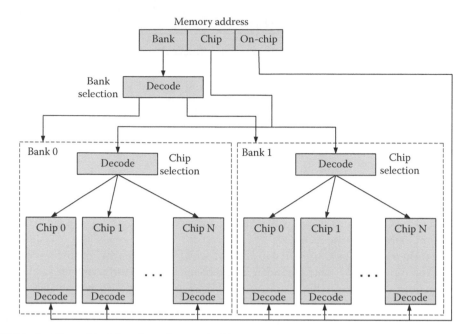

FIGURE 23.25   Hierarchical decoding of memory addresses.

For example, consider a CPU bus interface that comprises 16 data lines and 24 address lines, using a big endian format. If a 16 MB memory system is to be constructed of 1 MB RAM chips ($2^{20} \times 8$), then chips must be connected to the data bus in pairs, one connected to data bus lines D_{7-0}, containing odd-numbered bytes, and one to data bus lines D_{15-8}, containing even-numbered bytes, for a total of 2 MB of RAM. 16 MB of RAM requires $16/2 = 8$ pairs of chips. Three address bits A_{23-21} must be decoded to select a pair of chips, with 20 address bits A_{20-1} decoded within each chip. The remaining address bit A_0 is used within the CPU to determine whether the even- or odd-numbered byte or both are to take part in the data transfer.

If only four pairs of chips fit on one circuit board, then two boards are used, with address bit A_{23} selecting one of the boards and A_{22-21} selecting a pair of chips within the board, as illustrated in Figure 23.25, where the boards are labeled "banks."

23.7.4 Cache Memory

CPU speeds have increased at dramatic rates from year to year. As a result, the design of memory systems that can keep information moving in and out of the CPU, without making it wait, is becoming more difficult. Memory chips with short access times are expensive. An alternative to building a large memory out of expensive high-speed RAM chips is to use a small high-speed memory to hold the information most likely to be used by a CPU, with the remaining information kept in the main memory. The main memory can be built with lower-speed and, therefore, less expensive RAM chips.

Memory accesses often exhibit a property called locality of reference. Spatial locality of reference means that if a memory location is accessed, then there is a high probability of accessing the next sequential location in memory. Programs tend to exhibit high degrees of spatial locality of reference, because instructions are fetched from sequential memory locations until a jump or branch is executed. Temporal locality of reference means that if an address has been accessed, then there is a high likelihood that it will be accessed again in the near future. Data variables often exhibit good temporal locality of reference, as do instructions that are contained in loops.

A cache memory exploits locality of reference by holding, in high-speed memory, a subset of the main memory locations most likely to be needed by the CPU. Each memory reference is sent first to the cache.

If the requested information is there, a *cache hit* is said to occur and the information is passed quickly to the CPU. If a *cache miss* occurs, i.e., if the information is not found in the cache, then the slower main memory must be accessed. The average access time is given by

$$T_{access} = HT_{cache} + (1 - H)(T_{cache} + T_{main})$$

where

    H is the hit ratio of the cache, i.e., the percentage of cache accesses that result in hits

    T_{cache} and T_{main} are the access times of the cache and main memories, respectively

For example, a memory system with a 90% hit ratio with T_{cache} = 20 ns and T_{main} = 100 ns would have an average access time of T_{access} = 30 ns, which is much closer to that of the higher-speed cache memory than the slower main memory.

Writes to memory can be handled in two different ways in systems with cache memory. In a *write through* strategy, each item is written directly to the main memory on every write cycle, with the cache updated concurrently. In a *write back* strategy, all writes are done only to the cache, leaving the cache and main memory contents temporarily inconsistent. Later, when information must be replaced in the cache, all modified cache entries are copied back to the main memory. This reduces the total number of main memory accesses when multiple updates are made to a single data item.

Cache memory designs and performance are examined extensively in [7,11,12].

23.7.5 Virtual Memory Management

Increasing program sizes and workload demands for PCs, workstations, handheld computers, and larger systems have made it impractical to provide enough primary memory to store a user's entire program and data. This is especially true in multiuser and multitasking environments in which memory and CPU time are shared by multiple programs. Referring to the locality of reference principle described in Section 23.7.4, it is usually the case that a CPU does not need immediate access to all instructions and data of a program at any given time. Therefore, it is sufficient to make a subset of this information available in memory, with the rest held on a disk or other secondary storage device until needed. This allows a program's address space to be much larger than the available or allocated physical memory.

This is achieved by using two different address spaces: a virtual or logical address space that specifies locations within a program, and a physical address space that identifies physical storage locations in the main memory. Since different portions of a program are placed in memory at different times, each location from the logical address space must be mapped to the current physical address containing that information, as illustrated in Figure 23.26. This mapping is handled by a memory management unit (MMU), which is a hardware element that translates each logical address to the corresponding physical address.

A common method for translating logical to physical addresses is paging, in which a user's program is partitioned into fixed-length blocks called pages. Physical memory is likewise partitioned into fixed-length blocks, usually of the same length as the logical pages, so that each logical page fits exactly into one physical page of memory. Each logical address is divided into an *n*-bit page number and a *k*-bit offset within the page, as illustrated in Figure 23.27. The *n*-bit logical page number is mapped to an *m*-bit page number in the physical address space. Since logical and physical page sizes are the same, the offset portion of the logical address is used directly as the offset portion of the physical address without going through the translation process. Only the address bits corresponding to the page number need to be translated.

As logical pages are placed in the physical memory, a page table is created and updated. As illustrated in Figure 23.27, the page table contains a descriptor for each logical page containing a present bit (P) to

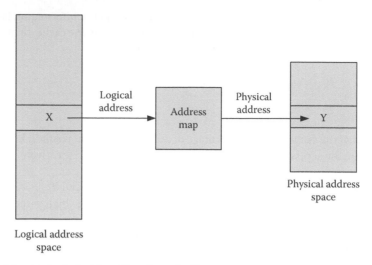

FIGURE 23.26   Mapping logical address X to physical address Y.

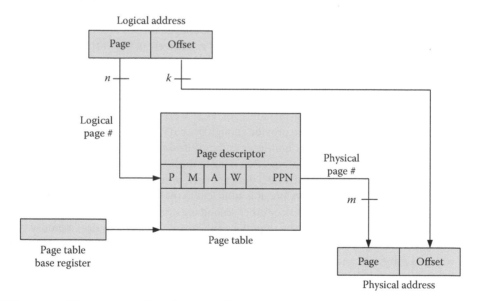

FIGURE 23.27   Address mapping through a page table.

indicate whether that page is currently resident in memory and, if so, the corresponding physical page number (PPN). The page descriptor may also contain a modified bit (M) to indicate whether the page has been modified since being loaded into memory; an accessed bit (A) to indicate if the page has been accessed; and possible restrictions on how the page may be used, such as a writeable bit (W) to indicate that the page may be altered. These bits are used by the operating system to help manage the pages in memory.

To perform an address mapping, the MMU uses the logical page number as an index into the page table, as shown in Figure 23.27, to fetch the page descriptor. If the page is present in memory, the physical page number is retrieved from the page descriptor and concatenated with the page offset to produce the physical address. If the page is not present in memory, a page fault is said to occur. The CPU is interrupted and the operating system called to find the requested page on the disk and load it into memory. Then the page table is updated and the original memory request is repeated.

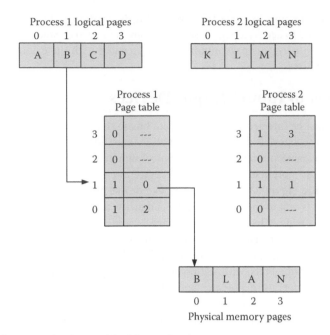

FIGURE 23.28   Address mapping in a multitasking environment.

In a multiuser or multitasking system, each process has its own page table. This is illustrated in Figure 23.28, which shows two processes sharing CPU time. Pages A and B of Process 1 are currently in memory along with pages L and N of Process 2. The page table of each process keeps track of which pages are in memory. When Process 1 is running, logical address 1, corresponding to page B, is mapped to page 0 of physical memory. When Process 2 is running, page L is accessed by mapping logical address 1 into page 1 of physical memory. If Process 1 attempts to access page D, which is not currently in memory, it will be suspended to allow the operating system to retrieve page D from the disk and replace one of the current pages in memory. Then Process 1 will be allowed to repeat the access to page D and continue processing.

An alternative method of address translation is segmentation, in which information is organized by the programmer into segments of items that share common characteristics. Each segment has a segment number and an offset within the segment. Segment numbers are translated in the same manner as page numbers, indexing into a segment table to retrieve a segment descriptor that points to the beginning of a physical memory area. Unlike pages, segments can be of arbitrary size, and can be loaded at any memory address.

The segment size is stored in the segment descriptor and compared to each segment offset to ensure that the requested information is not outside the bounds of the segment. The segment offset is then added to the starting memory address to form the complete physical address. The use of arbitrary segment sizes allows a programmer to use only as much memory as needed, although memory allocation is difficult, since the operating system is not working with fixed-sized blocks, and therefore free memory may become fragmented as segments are moved in and out of memory. There may be a sufficient amount of free memory to accommodate a new segment, but if the free memory is not contiguous, the segment cannot be loaded.

Occasionally, a combination of segmentation and paging is used. In this case, information is organized by the programmer into segments, with protection information placed in the segment descriptor. Then each segment is partitioned, transparent to the programmer, into fixed-sized pages, simplifying the memory allocation process, since logical and physical pages will be the same size, preventing fragmentation. Each logical address comprises a segment number, a page number, and a page offset. The segment number is used to access a segment descriptor in a segment table.

The segment descriptor points to a page table containing descriptors of the pages of that segment. The page number is then used to access a page descriptor from the selected page table, which provides the physical page number. Finally, this page number is concatenated with the original page offset to form the physical memory address.

23.8 Interfaces to Input/Output Devices

An external device must be interfaced to a CPU's data bus so that data can be transferred by program instructions between the external device and the CPU. As is the case with memory, each I/O device interface must be addressable and respond to bus control signals issued by the CPU. The most common approach is to use one or more registers in the I/O device interface as buffers between the device and the CPU, as illustrated in Figure 23.29. Data is sent to a device by writing it to a register in the device interface, from where it is sent to the external device. Likewise, data from an external device is accessed by the CPU by reading a register in the device interface.

An external device may require data in some format other than the parallel binary digits provided by the CPU. For example, the transmission of data over a telephone line via a modem requires that each byte be converted into a serial stream of bits, with additional control bits prepended and appended to each transmitted byte. Data from temperature, pressure, and other sensors is often produced as continuous analog voltages or currents. To be processed by a digital computer, analog values must be sampled and represented by digital values that can be read by the CPU. A similar function must be provided by digital-to-analog converters to send information from a CPU to an analog device. There are many different device interface chips that automatically convert data from one format to another. In most of these, the CPU simply transfers parallel digital data to and from registers in the device interface. The conversion and transmission of data between these registers and the external devices is handled automatically by circuits in the interface.

Interfaces to disk drives, graphics display terminals, communication links, and other complex I/O devices operate in a similar manner. Despite the significant differences between the characteristics of different types of peripheral devices, most of them are viewed by the CPU as a set of addressable registers. These device interfaces incorporate intelligent control circuits that respond to commands sent by a CPU to special command registers within the device interface. Parameters and data are likewise transferred to and from registers in these interface circuits.

The operation of these circuits is monitored by reading the status information from registers in the device interfaces. For example, the following HCS12 and Pentium program segments read and test the rightmost bit of a device status register. If that bit is 1, the device is known to be busy and the program repeats the test. If the bit is 0, the device is not busy, and new data can be sent to a data register in the device interface.

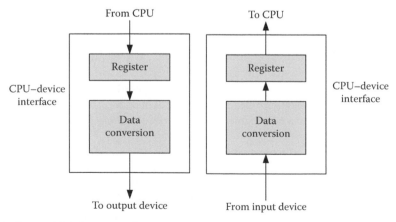

FIGURE 23.29 External I/O device interfaces.

```
HCS12:
          Check:    BRSET    Status,$01,Check    ;test bit 0 of device status
                                                  register
                    STAA     Data                ;Send new data to device
                                                  when not busy
Pentium:
          Check:    IN       AL,Status           ;read device status register
                    TEST     AL,01H              ;mask all but busy bit
                    JNZ      Check               ;repeat if busy = 1
                    MOV      AL,BL               ;get byte to send
                    OUT      Data,AL             ;send data to device
```

23.9 Microcontroller Architectures

Microcontrollers are used in applications that require low cost and chip count, combining on a single chip a CPU, a ROM, a RAM, and various peripheral functions and I/O interfaces. Typical applications include embedded controllers for kitchen appliances, automotive electronics, cellular phones, home electronics (TVs, VCRs, etc.), process control, and many other applications. In general, it has proven cost effective to use a single microcontroller chip to replace circuitry that would otherwise require several digital and/or analog ICs. In addition, the programmability of microcontrollers allows features to be changed or added with very little extra cost.

Microcontroller CPUs are available with data path sizes from 4 to 32 bits. On-chip RAM is generally small compared to on-chip ROM, usually 64-4 Kbytes. In embedded applications, RAM is used primarily to store a few temporary variables and perform "scratch" work; hence, on-chip RAM is commonly referred to as scratchpad memory. Microcontrollers generally provide a number of parallel I/O pins. These are used to send control signals to LEDs, actuators, displays, etc., to receive signals from switches, sensors, and other devices, and to pass data to/from other external devices. Most microcontrollers also offer programmable serial communication interfaces. Other special I/O interfaces may be found in selected chips. I/O interfaces and other special functions are accessed by the instruction set via control, status, and data registers with permanently assigned addresses.

Programmable timers/counters are among the most common functions included in microcontrollers. These are used to provide timing for digital alarm clocks, cooking cycle times in microwave ovens, automobile engine control timing, generation of periodic waveforms, timing of bit transmissions for serial I/O, measurement of time periods between signal changes detected on external I/O line events, and counting signal changes.

Figure 23.30 shows a block diagram of an 8051-compatible microcontroller [10,15]. In addition to the CPU, 128 bytes of RAM, 4 Kbytes of ROM, and four 8-bit parallel I/O ports, the 8051 includes two programmable timers, a serial communication port, interrupt control logic, and an external bus interface. When activated, parallel ports P0 and P2 are disabled and these pins become the external data and address buses. Data is multiplexed with the low address byte on the 8 pins of port P0, with the remaining address bus bits on the eight pins of port P2. Additional memory and I/O function interface chips can be accessed via this external bus to expand the capabilities of the microcontroller. The Freescale HCS12 microcontrollers can be expanded off-chip in a similar manner.

The basic architecture of the Freescale MC9S08 family of devices is shown in Figure 23.31 [14]. As in the 8051, devices in the MC9S08 family include CPU, RAM, Flash, general-purpose I/O pins, a programmable timer, serial communication interfaces, and interrupt support logic. Unlike the 8051 and the HCS12, the MC9S08 is not expandable; the data bus cannot be accessed external to the chip. Therefore, no additional memory may be used other than what is provided on chip, and all signals to and from the

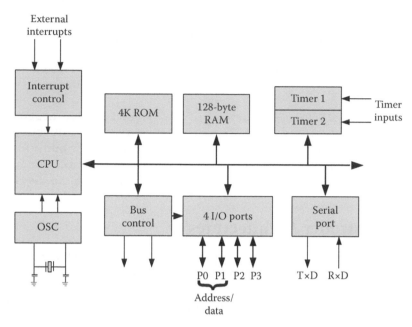

FIGURE 23.30 8051-compatible microcontroller block diagram. (Reprinted from Intel Corporation, *Embedded Microcontrollers*, Chapter 2, pp. 2–3, Figure 1, Intel order number 270646. With permission.)

outside world must go through the provided parallel, serial, or other special I/O pins. Because applications differ in their memory and I/O requirements, dozens of different configurations of the MC9S08 are offered. Table 23.8 summarizes eight members of the MC9S08 family. As can be seen in this table, members of the MC9S08 family differ in the amounts of on-chip RAM and FLASH memory, number and types of communication modules, analog-to-digital converters, timer channels, I/O pins, package sizes, and other special functions. In addition, versions are available that operate at different power supply voltages and clock speeds, and most devices are available in several packages. A system designer must select a device from the MC9S08 family that most closely matches the needs of his or her application. Many other manufactures also supply single-chip microcontrollers as families of nonexpandable devices.

Microcontroller instruction sets generally support smaller memory spaces and provide fewer arithmetic and other high-level functions than general-purpose CPUs. However, many microcontroller CPUs include special instructions to facilitate I/O operations and to manipulate individual bits of I/O data, since these operations typically dominate microcontroller applications more so than arithmetic computations.

For designers needing more computation power in their embedded applications, or who wish to take advantage of the vast quantity of system and application software available for popular general-purpose CPUs, several IC manufacturers have developed microcontrollers around general-purpose 16- and 32-bit CPUs. One such device is the NXP LPC2000 microcontroller family [13], which is built around the basic 32-bit ARM7 CPU, with typical microcontroller memory and I/O device interfaces.

A number of microcontrollers are also available with instruction sets tailored to specific applications. For example, microcontrollers are available whose instructions sets have been designed specifically for digital signal processing (DSP), which relies heavily on multiply-accumulate and other mathematical computations.

A recent trend in microcontroller design is to create custom system-on-chip (SoC) devices for customers from libraries of intellectual property (IP) "cores." IC cores range from predesigned and characterized layouts of such functions as CPUs, RAMs, ROMs, I/O interface ports, and other special

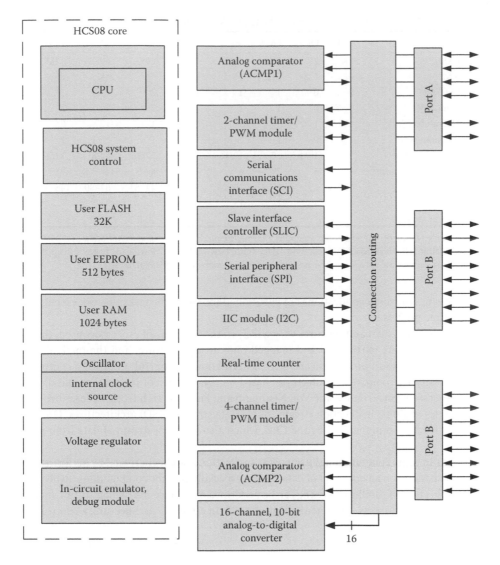

FIGURE 23.31 Freescale MC9S08EL32 block diagram. (Modified from *MC9S08EL3, MC9S08EL16, MC9S08SL16, MC9S08SL8, Data Sheet*, Rev. 3, Freescale Semiconductors, Austin, TX, July 2008.)

functions, to register-transfer-level models of these functions that can be synthesized into a target technology. A designer creates an SoC by specifying the desired CPU(s) and exactly the types and amounts of memory I/O devices needed for the intended application. The chip layout is then created by the designer or the manufacturer by placing and interconnecting the IP cores, and the chip is fabricated for the customer in a relatively short time.

23.10 Multiple Processor Architectures

System designers have long been faced with applications requiring more computing power than can be provided by a single computer. Computing throughput requirements can be orders of magnitude greater than can be provided by even the fastest computers, despite the fact that CPU performances have increased dramatically over the years. Such high-throughput requirements can only be met by the use of

TABLE 23.8 Sample Freescale Semiconductor MC9S08 Microcontroller Configurations

Family[a]	KA1	QD2	QG8	EL32	SH32	JM60	LC60	QE128
FLASH	1K	2K	8K	32K	32K	60K	60K	128K
RAM	62	128	512	1K	1K	4K	4K	8K
SCI (UART)	0	0	1	1	1	2	1	2
SPI modules	0	0	1	1	1	2	2	2
I²C modules	0	1	1	1	1	1	1	2
Comparators	1	0	1	1	1	1	1	2
ADC bits/channels	0	10/4	10/8	10/16	10/16	12/12	12/8	12/24
16-bit timer channels	1	5	2	6	4	8	4	9
Package pins	6–8	8	8–24	20–28	16–28	44–64	64–80	44–80
Max GPIO pins	4	4	13	13	26		26	
Other modules						USB	LCD	

Source: PowerPC™ Microprocessor Family: The Programmer's Reference Guide, Motorola Inc., 1995, International Business Machines Inc., 1991–1995. (http://www.cebix.net/downloads/bebox/PRG.pdf)

[a] Part name = MC9S08 + family + package option (for example, MC9S08QD2CSC).

parallel processing, in which applications are partitioned into multiple tasks that are executed concurrently by multiple processors.

Multiple processor system architectures vary widely in the numbers of processors used, the methods in which applications are partitioned and mapped onto an architecture, and the methods for interconnecting processors to communicate and share information. Flynn [16] proposed a commonly used method for classifying computer architectures, based on the number of instruction and data streams that can be processed concurrently. The Von Neumann and Harvard architectures described in Section 23.1 are examples of single-instruction stream, single-data stream (SISD) architectures. One stream of instructions is fetched from memory by the CPU, which also fetches a stream of data from memory as needed.

Throughputs in some computationally intensive applications can be increased by performing a single operation concurrently on an entire set of data. This is especially true for computations involving vectors and matrices. This type of parallel processing can be performed on a single instruction stream, multiple data stream (SIMD) architecture, as illustrated in Figure 23.32. In an SIMD architecture, a control processor

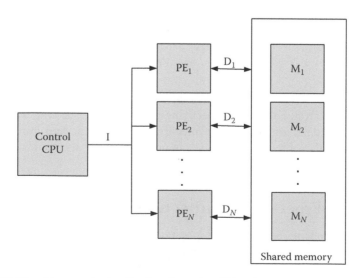

FIGURE 23.32 SIMD architecture: instruction I performed on *N* data items.

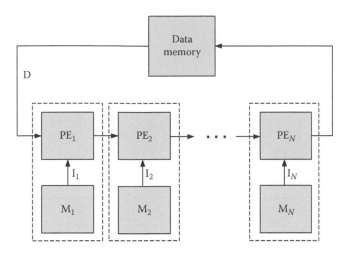

FIGURE 23.33   MISD architecture: each PE performs a different operation on one data item.

fetches program instructions and identifies those instructions that involve computations on sets of numbers. As shown in Figure 23.32, each such instruction, I, is broadcast to N processing elements (PE_1, PE_2, ..., PE_N), which perform that operation concurrently on N data items (D_1, D_2, ..., D_N) accessed from a shared memory. SIMD architectures are commonly referred to as array processors. The reader is referred to Hwang's textbook [17], which discusses the architectural features of a number of SIMD machines.

A multiple instruction stream, single data stream (MISD) architecture comprises N processing elements, each of which performs a different operation (I_1, I_2, ..., I_N) on a single data item, D, in an assembly-line fashion, as shown in Figure 23.33. MISD principles have found their way into single processor architectures in the form of pipelining. In a pipelined CPU, the steps needed to process each instruction are performed by separate hardware modules. As each module completes its part, the module passes the instruction on to the next module and begins processing the next instruction. If there are N such modules, then up to N instructions can be processed at one time within a CPU, producing one new result every clock period rather than one every N clock periods. Pipelined CPU designs are discussed extensively in [9,11,12].

MISD principles have also been applied to experimental data flow computers, in which data is passed from processor to processor. Whenever a processor receives all the data items required for an operation, it performs that operation and passes the results on to other processors. In this manner, computations are triggered by the flow of data through the system.

The most widely used multiple processor architectures are multiple instruction stream, multiple data stream (MIMD) configurations. In an MIMD system, each processor performs its own assigned tasks, and accesses its own stream of data. Consequently, MIMD architectures can be applied to a much broader range of problems than the more specialized SIMD and MISD configurations.

There are numerous ways to configure an MIMD system. The number of processors can range from as few as two to hundreds or even thousands of elements. The most significant differences between MIMD architectures are related to the manner in which processors cooperate in solving problems. In general, MIMD architectures can be classified according to the degree of coupling between processors. In a loosely coupled system, the processors are autonomous and communicate primarily by exchanging messages through a communication network. In a tightly coupled system, the processors are closely synchronized and work closely with each other in solving problems. The required degree of coupling will determine how the processors should be interconnected.

The simplest MIMD interconnection method is the shared bus, as illustrated in Figure 23.34. Multiple CPUs can cooperate in solving problems by sharing information in a global memory that is accessed via

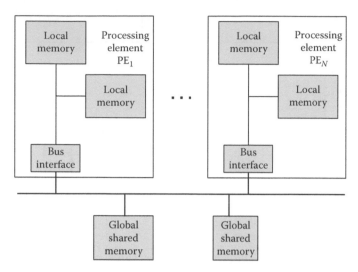

FIGURE 23.34 Shared-bus, shared-memory multiprocessor.

a shared bus. Because the available bus BW is limited, each processing element typically uses a private local memory as a cache for non-shared programs and data. The majority of each processor's memory accesses are to this local memory, with traffic on the system bus, limited to accessing data that must be shared between processes. Many commercial microprocessors contain special bus interface signals and functions to support connections to shared buses. Therefore, many multiprocessor systems have been developed using off-the-shelf CPU and memory boards.

A number of shared-bus standards have been developed to support networks of low-cost microcontrollers in automobiles and other embedded applications. Two of the more widely used are the I²C bus (Inter-Integrated Circuit) and the CAN bus (Controller Area Network) [18], both of which are serial buses, and supported by modules built into many microcontrollers. A physical I²C bus comprises a serial data line (SDL) and a serial clock line (SCL), supporting data rates of 100 to 400 Kb/s. The bus master is responsible for transmitting both data bits and clock. The bus is a multimaster bus, which means that any module connected to the bus may act as the master at any time. A bus transaction may begin any time the bus is idle (when SDL and SCL both are high), with the master sending a start bit, a 7-bit slave address, and then one or more data bytes. If a master detects that the level of SDL is not what was transmitted, it stops transmitting. This ensures that if multiple devices start transmitting simultaneously, all but one will stop, thereby allowing the highest priority message to be completed. The CAN bus uses a similar protocol to support multiple masters, but with more options and higher potential data rates.

If a single bus is unable to provide sufficient BW to handle all of the shared-memory accesses, processors will be forced to wait for access to the bus, degrading system performance. In such cases, the BW can be improved by using multiple buses. In the extreme, maximum memory BW can be achieved by interconnecting processors and memories with a crossbar switch, as shown in Figure 23.35. A separate switch and bus are provided between each processing element and each shared-memory module. Therefore, any permutation of N processing elements concurrently accessing N shared memories can be achieved. Conflicts occur only when two processors must access the same shared memory.

To reduce the number of switches and buses, many other connection networks have been proposed for shared-memory multiprocessors [16,17]. Many of these networks allow a limited number of permutations of concurrent connections between N processing elements and N shared-memory modules. However, since there are fewer switches, conflicts often occur in the connection network that block the progress of one processor while the other accesses memory, even though the processors may be attempting to access different memory modules. The tutorial by Wu and Feng [19] provides an excellent overview of interconnection networks for multiprocessor systems.

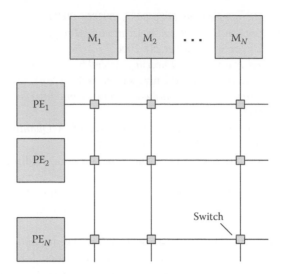

FIGURE 23.35 MIMD architecture based on a crossbar switch.

Networks of PCs, workstations, and industrial computers are typical examples of loosely coupled systems. Loosely coupled systems contain autonomous processing elements with no shared resources. The processors share information by exchanging messages through a communication network. The most common interconnection methods for computer networks are buses and rings.

Ethernet is one of the most widely used interconnection buses for computer networks. Similar to the shared-bus configuration of Figure 23.34, the Ethernet bus is accessed using a carrier-sense, multiple-access with collision detection (CSMA/CD) protocol. All processors continuously "listen" to the bus. When any processor wishing to transmit a message senses that the bus is free, it may begin transmitting preliminary information, followed by the message. If multiple processors begin transmitting at the same time, a collision is said to occur and the information on the bus becomes garbled, allowing the collision to be detected by all processors on the bus. When a collision is detected, all processors back off and wait a random amount of time before attempting to transmit again. Each processor makes its own decisions regarding access to the bus. Consequently, processors can be added to and removed from the bus without disturbing the rest of the network.

Many other communication network topologies and protocols have been developed for multiple processor systems. The interested reader is referred to [16,17] for further information.

References

1. *Linux Online*, http://www.linux.org/info
2. Nelson, V.P. et al., *Digital Logic Circuit Analysis & Design*, Prentice-Hall, Englewood Cliffs, NJ, 1995.
3. *IEEE Standard for Binary Floating-Point Arithmetic*, ANSI/IEEE Std. 754-1985, IEEE, Inc., New York, 1985.
4. Hohl, W. *ARM Assembly Language: Fundamentals and Techniques*, CRC Press, Boca Raton, FL, 2009.
5. *The SPARC Architecture Manual*, Version 8, SPARC International Inc., Menlo Park, CA, 1992 (http://www.sparc.com/standards/V8.pdf).
6. *PowerPC™ Microprocessor Family: The Programmer's Reference Guide*, Motorola Inc., 1995, International Business Machines Inc., 1991–1995 (http://www.cebix.net/downloads/bebox/PRG.pdf).
7. Patterson, D.A. and Hennessy, J.L., *Computer Organization & Design: The Hardware/Software Interface*, 3rd edn, Morgan Kaufmann Publishers, Borlington, MA, 2005.

8. Brey, B.B., *The Intel Microprocessors: 8086/8088, 80286, 80386, 80486, Pentium, Pentium Pro Processor, Pentium II, Pentium III, Pentium 4: Architecture, Programming, and Interfacing*, 7th edn, Pearson-Prentice Hall, Upper Saddle River, NJ, 2006.

9. Cady, F.M., *Software and Hardware Engineering: Assembly and C Programming for the Freescale HCS12 Microcontroller*, 2nd edn., Oxford University Press, New York, 2008.

10. Stewart, J.W., *The 8051 Microcontroller: Hardware, Software and Interfacing*, Regents/Prentice-Hall, Englewood Cliffs, NJ, 1993.

11. Hennessy, J.L. and Patterson, D.A., *Computer Architecture: A Quantitative Approach*, 3rd edn., Morgan Kaufmann Publishers, San Francisco, CA, 2002.

12. Stone, H.S., *High-Performance Computer Architecture*, 3rd edn., Addison-Wesley Publishing Co., Boston, MA, 1993.

13. *LPC21xx and LPC22xx User Manual*, Rev. 03, NXP Semiconductors, Netherlands B.V., April 2008.

14. *MC9S08EL3, MC9S08EL16, MC9S08SL16, MC9S08SL8, Data Sheet*, Rev. 3, Freescale Semiconductors, Austin, TX, July 2008.

15. Atmel, *Low Pin Count 8-bit Microcontroller with A/D Converter and 16 KBytes Flash Memory, T89C5115, AT89C5115*, Rev. 4128G-8051-02/08, Atmel Corporation, San Jose, CA, September 2008.

16. Flynn, M.J., Very high-speed computing systems, *Proc. IEEE*, 54(12), December 1966, 1901–1909.

17. Hwang, K., *Advanced Computer Architecture: Parallelism, Scalability, Programmability*, McGraw-Hill, Inc., New York, 1993.

18. Wolf, W. *Computers as Components: Principles of Embedded Computing System Design*, 2nd edn., Chapter 8, Morgan Kaufman Publishers, San Francisco, CA, 2008.

19. Wu, C. and Feng, T., *Tutorial: Interconnection Networks for Parallel and Distributed Processing*, IEEE Computer Society Press, Silver Spring, MD, 1984.

20. Intel Corporation, *Embedded Microcontrollers*, Chapter 2, pp. 2–3, Figure 1, Intel order number 270646, ISBN1-55512-203-5.

24

FPGAs and Reconfigurable Systems

Juan J.
Rodriguez-Andina
University of Vigo

Eduardo de la Torre
*Polytechnic University
of Madrid*

24.1 Introduction

Since their advent, microprocessors were for years the only efficient way to provide electronic systems with programmable functionality. Although the hardware structure of microprocessors is fixed, they are capable of executing different sequences of basic operations (instructions). The programming process mainly consists of choosing the right instructions and sequences for the target application.

Another way of achieving programmable functionality is to use devices whose internal hardware resources and interconnects are not totally configured by default. In this case, the programming process (generically called configuration) consists of choosing, configuring, and interconnecting the resources to be used. Programmable logic matrices, whose basic structure is shown in Figure 24.1, were the first attempts to implement this second approach for programmability. However, for a number of years their use was quite limited, mainly due to technological reasons.

Currently, programmable matrices can be found in programmable logic devices (PLDs), whose main application is in glue logic and finite state machines. The basic structure of PLDs is shown in Figure 24.2. Configuration consists of creating the connections between rows and columns of the programmable matrices, and of configuring the macrocells.

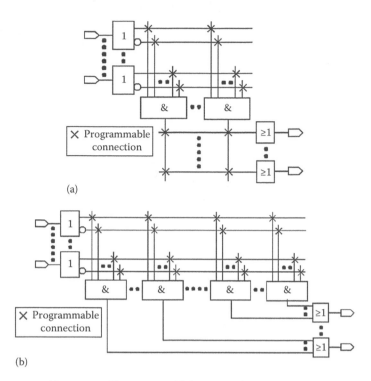

FIGURE 24.1   Programmable matrices: (a) programmable logic array (PLA) and (b) programmable array logic (PAL).

The extension of the gate array technique to post-manufacturing customization, based on the idea of using arrays of custom logic blocks (LBs) surrounded by a perimeter of I/O blocks (IOBs), all of which could be assembled arbitrarily (Xilinx 2004), gave rise to the field programmable gate array (FPGA) concept, depicted in Figure 24.3a. LBs are relatively low-complexity elements, capable of providing basic combinational and sequential functionality, as shown by the example in Figure 24.3b, which corresponds to the most usual, SRAM-based, FPGAs. On them, logic functions are mainly implemented using look-up tables (LUTs), which are $2^n \times 1$ SRAM blocks, and the configuration bits (those that define the configuration of the device) are also stored in SRAM cells.

The original application of FPGAs in rapid prototyping has been complemented with new applications that take advantage of the characteristics of current devices, such as high speed, very large number of components and supported protocols, or the availability of intellectual property (IP) cores. While application-specific integrated circuits (ASICs) are usually the preferred implementation platform for final production, mainly because of their high-performance and low-power possibilities, as well as their low cost for high volume production, there are some problems associated with them: longer development times and increasing non-recurring costs. In this context, FPGAs have become a main player even for deployment in mass production quantities, particularly when time-to-market constraints are high, or some reconfiguration capabilities in the hardware are needed.

In this chapter, the main characteristics and structure of modern FPGAs are described, together with the software resources that enable the efficient exploitation of their many capabilities. It is not intended to provide a comprehensive list of resources, tools, and their corresponding devices and vendors [readers can refer to specialized literature for this (Rodriguez-Andina et al. 2007)]. After that, the main concepts related to reconfigurable systems are presented, with emphasis on their FPGA implementation. Finally, the most significant current application domains of FPGAs are discussed.

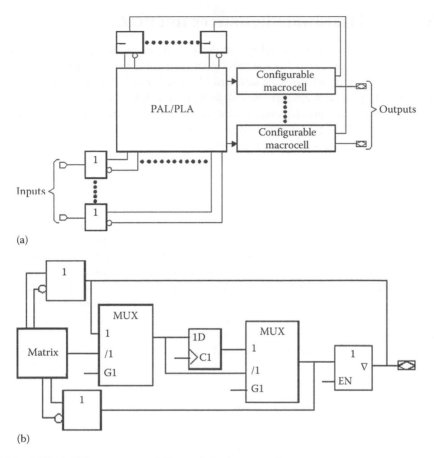

(a)

(b)

FIGURE 24.2   (a) Basic PLD structure and (b) sample basic macrocell.

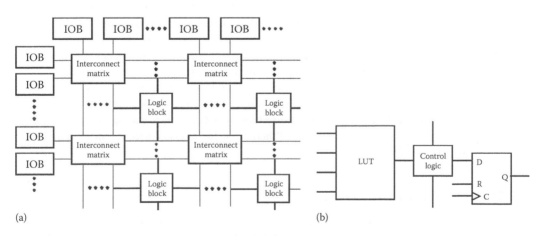

(a)   (b)

FIGURE 24.3   (a) FPGA concept and (b) sample basic LB.

24.2 Advanced Hardware Resources in FPGAs

FPGA devices have reached a level of development that puts them on the edge of microelectronics fabrication technology advancements. This has enabled an impressive amount of resources to be readily available on a single chip. Apart from the standard LBs, hardware resources implemented in FPGAs greatly differ depending on the manufacturer and, for a given manufacturer, on the architecture/family of devices. The resources that provide most advantages to designers can be classified as follows: integrated functional blocks, I/O signal conditioning and special (e.g., radiation-tolerant) devices. Due to its significance, reconfiguration is addressed in Section 24.5.1.

24.2.1 Integrated Functional Blocks

The most usual functional blocks that can be found in current FPGAs are as follows:

1. *Memories*, which allow processing speed to be increased, I/O pins to be more efficiently used, and the design of the system at the board level to be simplified. Large amounts of internal memory are available in current FPGAs, in many different arrangements: RAM (single port, simple dual-port, true dual-port and bidirectional dual-port), ROM, or shift registers. From these structures, it is also possible to implement others such as FIFOs or content addressable memories (CAMs). Advantage can also be taken of internal memory blocks to build complex functions, such as arbitrary waveform generators, whose values can be obtained from a table stored in memory.
2. *Clock managers.* Phase-locked loops (PLLs) and delay-locked loops (DLLs), can be used to compensate clock propagation delays throughout the FPGA, to correct clock duty cycle or phase shifts, or to multiply/divide clock frequency.
3. *Arithmetic circuits.* Some devices incorporate a relatively large amount of low-complexity arithmetic blocks. Complex arithmetic blocks (DSP blocks) are found in advanced FPGAs. Multipliers with non-standard 9, 18, or 36 bit data inputs (available in some devices) allow performance to be improved, because DSP functions not often require exact 16, 24, or 32 bit precision.
4. *Transceivers.* In some devices, standard as well as user-defined communication protocols can be easily implemented by means of blocks that encode/decode and serialize/deserialize data, include transmission and reception buffers, and can perform clock synchronization.
5. *Integrated processors.* Embedded processors and peripherals are available in complex FPGA architectures, in order to support system-on-chip (SoC) solutions. Processors include RISC or 8051-based microcontroller cores. Soft processors (i.e., processing cores built from standard FPGA resources) are available from Xilinx (PicoBlaze, MicroBlaze) and Altera (Nios, Nios II). The PicoBlaze core is a basic 8 bit microcontroller implemented using a reduced number of logic blocks. The MicroBlaze core corresponds to a 32 bit RISC microprocessor including a standard set of peripherals. The Nios processor provides a more complex but flexible solution. For example, register bank, instruction, and data sizes can be configured and the user can add custom instructions to the CPU. Peripherals include universal asynchronous receivers/transmitters (UARTs), timers, compare and capture units, multiply–divide units, memory controllers, Ethernet media access controllers, or even analog-to-digital converters (ADCs).

24.2.2 I/O Signal Conditioning

Specific resources associated to the I/O pins allow FPGAs to be connected to other devices operating at different voltages, without the need for additional interface circuitry and, therefore, significantly simplify PCBs and reduce costs. I/O pins are usually grouped in banks, each one of which can be used with a different I/O standard. Some architectures also include internal terminating resistors, which allow signal integrity (and therefore bandwidth) to be improved and ease the design of the PCBs.

24.2.3 Special Devices

24.2.3.1 Nonvolatile FPGAs

One-time programmable (OTP) devices are available, providing advantages in some applications, mainly because they do not need auxiliary external resources for configuration at power-on, their switch resistance and capacitance (and usually their power consumption) are lower, and their noise immunity larger than those of their SRAM-based competitors. Antifuse is the main technology used in nonvolatile FPGAs, but there also exist SRAM-based devices with internal nonvolatile configuration memory, as well as flash-based devices.

24.2.3.2 Low-Power Devices

Current handheld and portable electronics demand low-power devices and small, cheap, efficient power converters, and batteries. In addition, the reduction in power consumption simplifies thermal management and improves reliability. FPGAs are usually not power-efficient due to the overhead required to provide configurability. Moreover, technology scaling down leads to increasing leakage current, therefore increasing static power consumption. To overcome these problems, low-power FPGAs have been developed. Low-power operating modes can be efficiently implemented in antifuse devices, which, in addition, usually dissipate less dynamic power than SRAM-based ones, as mentioned above. Several techniques can be used to reduce power consumption in SRAM-based FPGAs (Khan 2006). On one hand, transistors with different sizes can be distributed throughout the devices in order to have them used for high- or low-speed sub circuits as needed, thus optimizing dynamic power consumption. Thick oxide can be used in "slow" transistors, therefore reducing static consumption. Further reductions can be achieved by using low-k dielectrics and increasing the amount of local interconnecting resources. The amount of interconnects can be reduced by using increasingly complex logic blocks, with which more complex functions can be implemented, and embedded specialized blocks (as described in Section 24.2.1).

24.2.3.3 Radiation-Tolerant Devices

FPGAs are being increasingly used in aerospace applications. Antifuse radiation-hardened FPGAs, are inherently more radiation-tolerant than SRAM-devices. However, radiation-hardened versions of the latter are also available, providing significant advantages for aerospace applications, because of their higher performance and reconfiguration capabilities, with respect to antifuse devices.

24.2.3.4 Mixed-Signal FPGAs

Analog I/O blocks and configurable analog blocks for signal conditioning are available in some devices, including analog multiplexers, gate drivers, and ADCs.

24.3 IP Cores

These are pre-designed (in many cases parameterizable) and verified functions that allow design-synthesis-verification time to be reduced. They should (unfortunately not all do) provide predefined and highly optimized performance. They cover application domains such as communications, multimedia, signal processing, and automotive.

24.4 Software Tools for FPGAs

Specific powerful software tools are needed to take advantage of the many different complex hardware resources included in current FPGAs. In fact, the lack of suitable design tools can make advanced features useless. Although many efforts are still needed in this area, some resources are already available, which are described in Sections 24.4.1 through 24.4.5.

24.4.1 SoC Design Tools

The implementation of FPGA-based SoCs or embedded systems results in very complex tasks involving software and hardware developers. From the hardware point of view, the main problem is efficient IP integration, including design, synthesis, simulation, and verification. From the software point of view, the main problem is to debug the software system (made up of a real-time operating system (RTOS), drivers, and custom software), in real hardware.

There are tools that support the design based on soft or hard embedded processor cores. In general, they include peripheral IPs, soft processor cores customization tools, software development tools, hardware/software debuggers, hardware verification tools, libraries, and software and hardware code downloads to development boards. On the other hand, traditional tools used for ASIC implementations have been adapted for FPGA.

Finally, some existing tools support FPGA-based SoC design from C language, by translating the C code to synthesizable hardware descriptions.

24.4.2 DSP Design Tools

There are software design tools to implement DSP applications on FPGAs, which accelerate the migration from traditional software algorithms to faster hardware implementations. Existing tools automatically translate Simulink® models to synthesizable hardware descriptions that are to be used with FPGA implementation tools. Those tools provide Simulink libraries including common DSP, arithmetic, bus manipulation, control logic, storage, imaging, and communication functions. Advanced options like HDL co-simulation and hardware-in-the-loop are also supported.

24.4.3 Software and Hardware Debugging Tools

Debugging complex real-time embedded designs that take advantage of the many different resources available in today's FPGAs can be a non-trivial, highly time-consuming task. As a consequence, efficient in-circuit FPGA verification tools are needed. These tools can be either implemented in hardware (using the device's logic resources) or software (using either hard or soft embedded processors). Debugging capabilities are needed for both hardware- and software-implemented subsystems.

The main component of any FPGA debugging tool is the measurement instrumentation, which must be capable of capturing logic values of signals while the application is running in-system at operational speed. The basic operation is as follows: the internal nodes to be observed are defined, their activity is monitored, and the sequence of values is transmitted in real time to a host.

Most debugging tools are based on IP blocks (which could therefore be instantiated several times) implementing logic analyzer functionalities.

24.4.4 Power Management Tools

In addition to the manufacturing techniques and special operation modes to reduce power consumption discussed in Section 24.2.3.2, tools are needed to manage power in two basic directions. On one hand, design tools (i.e., synthesis, place and route, etc.) must include strategies to minimize consumption. On the other hand, it is necessary to estimate the power that a given design will dissipate in practice.

In order to be feasible, power-aware design strategies need to be supported by tools that provide power consumption estimates as accurate as possible, which could be used to support design decisions. Currently, power calculator tools are available from FPGA vendors.

As many complex designs are based on IP blocks, these can be responsible for a significant part of the power consumption. IP power modeling is currently an important research area, in which promising results are being obtained. In some cases, different versions of a given IP functionality are available (Khan 2006), each one optimized for different requirements (e.g., performance, logic usage, balanced, and others) and therefore exhibiting different power consumption, which should be carefully examined.

24.4.5 Signal Integrity and Mixed-Signal Design Tools

The use of FPGAs in high-speed and/or mixed-signal applications is increasing. Because of the complexity of such systems, modeling languages and tools are often used to accelerate behavioral simulation and PCB implementation tasks.

Most FPGA vendors provide IBIS (I/O buffer information specification) and HSPICE models of their devices, in order to allow board-level signal integrity analysis within EMC guidelines, crosstalk, and timing analysis.

Recently, other modeling languages oriented to the design and simulation of mixed-signal systems (like VHDL-AMS, Verilog-AMS, and MAST) have been introduced in applications where FPGAs are used as digital support. The advantage of mixed-signal languages is that designers can easily design and simulate with both analog and digital components from the same environment, enabling a top-down mixed-signal design methodology.

24.5 Role of FPGAs in Reconfigurable Systems

When hardware reconfiguration capabilities are required for a given application, using FPGAs offers many advantages and opportunities. Although reconfigurable systems are not limited to FPGAs, these are the most significant devices at the commercial level. Other possibilities exist, based on custom devices with specific reconfiguration features, mainly oriented towards hardware reconfigurable computing systems. However, these devices are intended to overcome some problems of FPGAs in very specific areas, like ultra-fast reconfiguration time (one clock cycle to reconfigure a complete device). Therefore, the focus herein is on the use of FPGAs, the advantages of applying their reconfiguration capabilities in-field, the different reconfiguration alternatives, commercial and industrial approaches, and, finally, on what the future role of FPGAs in reconfigurable systems will be. It should be noted that not all FPGAs are reconfigurable, but some are OTP devices. Obviously, this section deals with reconfigurable FPGAs.

Configurable devices are configured, with sequences of bits arranged in configuration files called bitstreams. Some years ago, FPGAs could only be reconfigured by using a full bitstream that programmed all configuration bits in the device. This configuration had to be static, typically immediately after system power-up. However, an increasing number of FPGA technologies are permitting the reconfiguration of portions of the FPGA, even while the rest of the device is kept working normally. This possibility is referred to as partial reconfiguration, whereas if the device can be reconfigured at run-time, it is referred to as a run-time reconfigurable system (RTRS). A special subset within RTRSs is composed of systems that can reconfigure themselves, which are referred to as self-reconfigurable systems (SRSs).

Granularity is another important aspect in reconfiguration. It is defined as the size of the functional elements that are reconfigured at the same time. Large granularity reconfiguration applies to systems where complex IP cores are replaced in the logic and, in this case, size is related to large portions of the FPGA and referred to as a percentage of their area. This way, there are reconfigurable systems that reconfigure a significant part (e.g., one half) of the FPGA, or use a slot-based approach, where the FPGA is divided into several (normally equal) slots, each of which can be reconfigured separately. Medium granularity applies to a portion of an IP core, and it is typically used to reconfigure functionality at register level. Finally, small granularity refers to the reconfiguration of a small number of configuration bits, applying typically to the values in a LUT, to the content of a flip-flop, or to an interconnection.

One of the major problems derived from the use of reconfiguration capabilities in FPGAs is the matching between the reconfiguration granularity that is desired and the granularity supported by the specific FPGA technology. The smallest reconfigurable area in FPGAs is a variable factor that depends on the manufacturer technology and family. Some FPGA families support column reconfiguration, that is, the minimum reconfigurable unit is a column of LBs. Other (newer) families support rectangular reconfiguration (the portion being reconfigured does not necessarily have to span a whole column), but the rectangle size is not restricted to a single element of the FPGA. Even though some reconfiguration techniques in commercial FPGAs claim that run-time reconfiguring an area with the same programming than it had (before reprogramming) is a glitch-less operation, there are many restrictions derived from the atomic reconfiguration unit that can be handled. For instance, if the content of a flip-flop has to be modified, all flip-flop contents in the same column for a column-based reconfigurable FPGA have to be modified. In order to do this, system execution has to be stopped, a read-back operation has to be produced to read the content of all flip-flops in the same column, then the desired flip-flop value has to be modified, and all flip-flops reprogrammed back. This does not allow real run-time reconfiguration but, if the atomic reconfiguration unit had been a single flip-flop, operation at run-time would have been possible. This, together with faster reconfiguration times are the main claims of the research community.

There are many reconfigurable system models, although most of them rely on the use of microprocessors and reconfigurable fabric. According to Al-Hashimi (2006) and Compton and Hauck (2002), there are several types of coupling between both parts, as shown in Figure 24.4:

- *External standalone*: The reconfigurable hardware is a fully independent device connected to the inputs and outputs of the microprocessor.
- *Coprocessor unit or attached processor unit*: In both cases, the reconfigurable hardware is closer to the microprocessor than in the previous case. In the coprocessor case, the reconfigurable hardware is closest to the microprocessor, and it can operate as a functional resource of the microprocessor itself. In the second case, the reconfigurable hardware is accessed after the cache memories, that is, in the secondary bus.
- *Reconfigurable functional unit*: The reconfigurable hardware is embedded into the microprocessor. This structure is the easiest to define custom instructions in the processor with its associated functional hardware.
- *Microprocessor embedded in the reconfigurable hardware*: In this case, the microprocessor could be a hard core, embedded in the reconfigurable fabric, or a soft core, placed using a part of the reconfigurable fabric itself.

The partial reconfiguration topic has started to produce applications which benefit from it. The main benefits of disruptive (non real-time) partial reconfiguration are reduced programming time and the

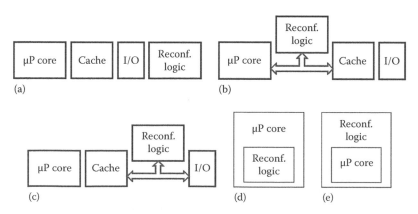

FIGURE 24.4   Different coupling alternatives: (a) external standalone, (b) coprocessor unit, (c) attached processor unit, (d) reconfigurable functional unit, and (e) embedded microprocessor.

possibility of silicon reuse, since the same device can be used for different tasks. In addition to these, the added benefits of using PRTRS, instead of conventional reconfigurable devices, are briefly summarized as follows:

- *Increased system performance*: The system stays operative, while a portion is being reconfigured. Therefore, as the system does not need to be stopped, there is no theoretical loss of performance.
- *Enhanced system updates*: By modifying only a portion of the device, system updates do not affect the remaining devices, connected to the FPGA, which can keep functioning normally.
- *Hardware sharing*: Partial reconfiguration permits not only having several applications sharing the same FPGA, allowing a smaller and cheaper device to be selected (like full reconfiguration), but also running them in parallel. This characteristic is gaining importance with the increasing FPGA integration level.
- *Shorter reconfiguration energy*: Partial reconfiguration deals with smaller configuration times compared to full device updates. This results in reduced configuration times and therefore in reduced power consumption during reconfiguration.
- *Less storage resources*: Partial configuration files require smaller storage sizes for programming, and this is important for resource-restricted devices.

Partial reconfiguration is valuable for devices that operate in environments where applications cannot be interrupted while the system is being adapted. It is also suitable for highly parallel systems that can time-share the same piece of the FPGA resources. Without this capability, it would be necessary to stop the system during device updates and to reconfigure the entire FPGA to support a different application, losing all previous configurations.

24.5.1 FPGAs as Reconfigurable Elements

The design of an FPGA-based PRTRS system involves several issues that need to be analyzed. The first one is to select a device that supports these reconfiguration techniques. Related to this, the partial reconfiguration possibilities of some commercial FPGAs are shown in Section 24.5.1.1.

The device needs to be logically partitioned into fixed and reconfigurable areas, so that reconfigurable cores, stored as partial bitstreams, can safely fit into the target position. The definition of this reconfiguration architecture is important, because it is necessary to match the size of reconfigurable areas and the targeted reconfiguration granularity, a process which requires good knowledge of the internal FPGA architecture. Hardware partitioning issues will be reviewed in Section 24.5.1.2.

Partial reconfiguration, especially at run-time, also requires tool support. In some cases, these tools have to run in restricted devices that need to autonomously handle reconfiguration themselves. Tools for reprogramming, adding, deleting, or relocating pieces of hardware into FPGA silicon areas have to be considered. This issue will be covered in Section 24.5.1.3.

Finally, it is also important to take into account that the communication between the microprocessor and the reconfigurable element, or among several reconfigurable elements in case there is more than one, could be a bottleneck, and the coupling between all software and hardware tasks must fulfill communication requirements. For an ASIC approach, the problem of choosing a suitable communication scheme is a challenge, but this problem is even harder when considering a reconfigurable environment, since communication requirements may be unknown before deciding the communication infrastructure. In this sense, reconfigurable communications may be a solution to solve the problem, and this topic will be presented in Section 24.5.1.4. Special attention is given to networks on chip (NoCs), more specifically to reconfigurable NoC approaches.

24.5.1.1 Commercial FPGAs with Partial Reconfiguration Support

The Altera Excalibur devices allow dynamic configuration of the FPGA while the on-chip processor is active for other tasks, such as running an operating system. The user can modify the device functionality by storing the configurable system function in an external nonvolatile memory, and use the processor to configure the hardware without the need to reboot. This results in systems with a single reconfigurable coprocessor that is composed of the entire reconfigurable array. Additionally, more recently, some Altera's FPGAs permit the reconfiguration of some elements, like serializers/deserializers or PLLs.

Differently, Atmel's FPGAs have the ability to implement cache logic designs, where part of the FPGA can be reprogrammed without loss of register data, while the remainder of the FPGA continues to operate without disruption, which is real-time partial reconfiguration. The main drawback of these FPGAs is their small size and the bit-based reconfiguration access method, which permits the highest flexibility, but requires very low-level reconfiguration control.

Xilinx' FPGAs can also be partially reconfigured. The configuration bitstream format allows a designer to modify one or more configuration packets and perform partial reconfiguration by accessing some portions of the FPGA reconfiguration memory. However, each device family has different reconfiguration features: the low-cost Spartan 3 series permit the reconfiguration of entire device columns that span from the FPGA top side to the bottom side, including top and bottom I/O blocks. When a single device row has to be changed, the information for the entire column has to be sent. The first Spartan 3 family does not include an internal configuration access port (ICAP, a mechanism to internally access the configuration memory), and thus it is not well suited for designing self-reconfigurable systems. Some subfamilies like the Spartan 3A, 3AN, and 3ADSP one, include such a port. On the other side, all Xilinx high-performance FPGA families permit glitch-less reconfiguration and include an ICAP configuration port. Virtex II and Virtex II Pro families require column-based reconfiguration, while in the newer families, Virtex 4 and Virtex 5, reconfiguration frames do not span the entire FPGA height, but several FPGA rows (16 in Virtex 4 and 10 in Virtex 5) instead. They also have double ICAP support. These improvements in new families are probably a result of the push of the research community that has put efforts in designing architectures, tools, and applications. Nevertheless, migration to the new platforms is being relatively slow, although new research is being based on these devices.

24.5.1.2 Virtual Architecture FPGA Partitions for Partial Reconfiguration

The selection of a suitable reconfigurable device is conditioned not only by the aforementioned reconfiguration features and restrictions, but also by the internal architecture. The internal topology of the FPGA has to be analyzed, and a partition into several areas has to be done. Basically, these are fixed area, reconfigurable area, and communications infrastructure.

The fixed area of the FPGA is the portion of the logic that does not change in any configuration. It is normally devoted to external off-chip communications, internal communications management, and self-reconfiguration. It is typically placed in FPGA regions where irregularities prevent them from being mapped as reconfigurable areas. For column-based FPGAs, these blocks are placed in the leftmost or rightmost sides of the FPGAs, and only I/O blocks close to these areas are used for off-chip interconnects.

The reconfigurable area has to have a fixed position for the connections with the fixed area, but the rest of the logic inside can be freely reconfigured. Several architectures have been proposed with different numbers of reconfigurable areas, with different sizes, but for most column-based reconfigurable FPGAs, column-based reconfigurable areas are defined.

Some approaches define just two regions (fixed and reconfigurable), with different geometries and proportions. There are also quite a few slot-based approaches, where the reconfigurable area is divided into equal-size portions of logic, with the possibility of programming individual IP cores in each of these slots. In many of these approaches most of the slots are column-based, following a 1D organization, although there are a more reduced number of approaches that follow a 2D organization. Figure 24.5 shows examples of 1D and 2D partitions.

Washington University has created a platform for partial run-time reconfigurable systems oriented to telecommunications, called field programmable port eXtender (FPX) (Horta et al. 2002), with applications in rapid prototyping of telecommunication routers and firewalls. It uses a two-slot partition, which is connected through a ring network.

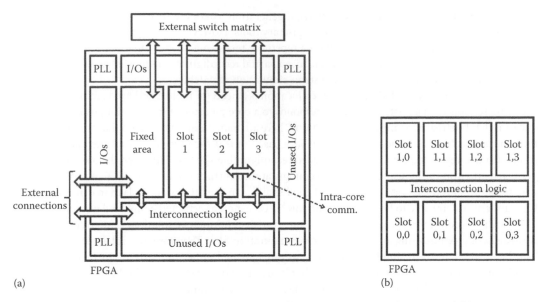

FIGURE 24.5 Examples of (a) 1D partition showing possible interconnection mechanisms and (b) 2D partition.

The term slot for general 1D partitions was first used in (Ullmann et al. 2004). A 1D multi-slot architecture has been proposed that uses a 1 bit serial bus for communications in the first versions (Palma et al. 2002), and a NoC in the most recent ones (Moller et al. 2006). Another 1D approach is described by Walder and Platzner (2004), where partial reconfiguration is performed in a fully transparent way by the use of a so-called hardware operating system. The influence on hardware reconfiguration as an operating system task has also been addressed in (Becker et al. 2007), where hardware–software multitasking needs are analyzed. It is also important to mention the Erlangen slot machine (Bobda et al. 2005), where local and shared memory access and off-chip interconnections by using top and bottom I/O blocks of a 1D column-based partition approach are presented.

More recent developments follow 2D partition approaches, like the dynamically reconfigurable NoC (DRNoC) approach (Krasteva et al. 2008) where a method to implement partial reconfigurable partitions is presented, and the communication infrastructure is a reconfigurable NoC.

A sensitive slot-based partition should take into account the possibility of reallocating a core into any slot position, being capable of using IP cores that could span through more than one slot, while still achieving real-time reconfiguration, if required.

It is common that FPGAs have some non-regular regions dedicated to memories, DSP blocks, embedded hard microprocessor cores, etc. In this case, the slot partition is not trivial, and the compromise between reconfiguration granularity and number of slots has to be solved.

The communication infrastructure is in charge of linking slots between them, as well as slots with the fixed area. As for the ASIC intra-communication problem, bus approaches and NoC approaches are the solutions more frequently implemented.

24.5.1.3 Tool Support for Partial Reconfiguration

The complexity of partially reconfigurable systems requires the support of tools to automate several design and in-field operation tasks. They can be classified into the following categories:

- Tools to support design flows for the creation of partial reconfiguration bitstreams.
- Tools to manipulate partial reconfiguration bitstreams, so that a core can be placed into any slot position in the FPGA.
- Toolsets, which may also be embedded and packed as hardware-aware operating systems, to support reconfiguration during device operation.

Regarding the design flows for partial bitstream generation, there are several commercially available solutions integrated in the manufacturers' proprietary tools. However, these approaches are not flexible enough, in the sense that they do not handle repetitive tasks in a friendly manner, they do not perform any kind of bitstream manipulation, and deep partial reconfiguration knowledge is required. Also, these tools do not help in reconfigurable systems simulation and debug.

Partial bitstream manipulation for core reallocation is a need for multi-slot-based architecture partitions. These tools read a bitstream that corresponds to a core placed in a specific slot position, and produce another bitstream for another slot position. There are many tools of this kind that are derived from the JBits application, but since this library is Java based, it is difficult to have implementations able to run on feature-restricted devices like embedded reconfigurable systems. However, the possibility of fine grain reconfiguration by using low-level reconfiguration routines (mainly for LUT modification and wire rerouting) produces very complete and good results. On the other side, tools than can run without JBits underneath are able to be executed with low CPU cost on a restricted embedded microprocessor.

Regarding reconfiguration support at run-time, available commercial solutions seldom cover the most basic tasks of programming and reading back configuration files. Therefore, several academic solutions have been adopted, from simple control systems implemented either in hardware or in software running in the embedded processor, to complete hardware operating system-based solutions. Furthermore, some solutions extend already existing operating systems, like Linux.

24.5.1.4 On-Chip Communications in Reconfigurable Systems

On-chip communications are an important challenge for all SoC designs, but this problem is even more important for reconfigurable systems, since the communication needs may change, and even may be unknown for future hardware configurations. Therefore, scalable and flexible communication structures are needed.

Hardware tasks need to exchange data between them, as well as with off-chip components. Normally, the fixed area of the internal FPGA architecture partition is used for this last purpose.

Hardware tasks, placed in slots or whatever arrangement is used, need to be designed with fixed position connections. These connections are generally called bus macros, following the name defined by Xilinx in their design flow. However, these macros are not the only solution, and many alternatives have been proposed in order to increase the connectivity of modules. They allow either bus or point-to-point connections, which are useful for either real point-to-point links or to provide access to a NoC infrastructure.

The move from the bus to the NoC approach has been followed also in the reconfigurable systems area. The number of slots for past FPGA technologies was not high enough to justify, in most cases, the need of a NoC, because buses are a simple and flexible solution for a low number of cores. However, as FPGA capacity is increasing, the number of cores allowed in a simple FPGA is higher, and NoCs are being considered as a promising solution for these larger systems.

NoC approaches in reconfigurable systems are typically associated with 2D FPGA partition approaches, and regular meshes and heterogeneous networks are being analyzed. Work is being conducted on the possibility of reconfiguring the communication infrastructure, apart from the cores themselves, so that the available communication resources can be fitted to the variable communication needs for a given configuration with a set of hardware and software tasks.

The solution in Ullmann et al. (2004) has switch matrixes that can be reconfigured. The latest works on the Erlangen machine (Bobda et al. 2005) show the use of a reconfigurable NoC, called DyNoC, which allows core grouping and the network reconfiguration to bypass the portions of the NoC that are used by the merging of two adjacent cores. The CoNoChi (Pionteck et al. 2006) and DRNoC (Krasteva et al. 2008) NoCs are more flexible solutions since network interfaces and some parts of the routers can be modified. DRNoC may reconfigure switch matrixes, network interfaces and routers' parameters, allowing not only NoC communications but a combination of these ones with point-to-point connections and bus-based solutions.

24.6 Applications

The way in which digital systems have been implemented over the years has been strongly influenced by the evolution of the programmability capabilities described in Section 24.1. For many years, complex digital processing and control systems have been based on architectures whose processing cores are microprocessors or, more recently, microcontrollers or DSPs. Currently, FPGAs have become an interesting alternative to these structures, because of their flexibility, abundance of resources (which, as described above, include embedded and soft processors), advanced features, and low cost. The systematic use of HDLs and high-level synthesis tools dramatically simplifies the design process and reduces time to market. Besides, the possibility to integrate in a single device several functionalities that previously required dedicated ICs led to the SoC concept. Although microprocessors, microcontrollers, and, to a lesser extent, DSPs keep being devices of paramount importance for industrial electronics, there are three main reasons that boost the widespread use of FPGAs:

- To take advantage of the possibility of readily and efficiently implementing complex functionalities in hardware, outperforming their software counterparts.
- To take advantage of the reconfiguration capabilities of FPGAs.
- To take advantage of SoCs for saving area and power, as well as for simplifying board design.

In Sections 24.6.1 through 24.6.6, the apparently limitless applications of FPGAs are grouped in generic fields, which are associated with the resources that justify the use of FPGAs on them.

24.6.1 Configurable Computing

In many applications, general-purpose microprocessor hosts need to be complemented with specialized coprocessors. The host supports system-level operation as well as the low-level tasks it is able to perform efficiently, whereas specific algorithms are run on the coprocessor. Configurable computing can help area and power consumption to be reduced, for instance by correctly choosing the right operator formats for each operating mode or application.

In this area, FPGAs provide the possibility for the coprocessor hardware to be adapted to changes affecting the algorithms or the data flow. The use of integrated processors as hosts, in a SoC structure, greatly simplifies the implementation of such solutions, particularly because it simplifies the communication between the host and the coprocessor. The availability of embedded memory also contributes to achieve efficient implementations in this area. In addition, partial reconfiguration allows resource usage to be optimized. The main issue in the development of these kinds of systems is the efficient partition of tasks in hardware and software, which involves co-design, co-simulation, and co-debug.

A hardware accelerator for computation of shortest paths for mobile robots is presented in Sridharan and Priya (2007), where an efficient FPGA implementation is achieved by taking advantage of embedded memory blocks. In Kung et al. (2009), a Nios II-embedded soft processor is combined with a hardware implementation of current vector controllers for permanent-magnet synchronous motor drives, providing an advantageous solution in terms of performance, complexity, and cost. By using an FPGA instead of a computer with a RTOS, the sampling period of force measurements in a bilateral teleoperation system can be reduced one order of magnitude (Ishii et al. 2007), improving performance accordingly. Advantages of FPGA hardware acceleration can also be obtained in the control of power converters.

FPGA reconfiguration and hardware acceleration capabilities are particularly useful in applications such as neural and fuzzy controllers. Another interesting application field is that of configurable data acquisition and signal generation systems. Advantage can be taken for their design of embedded processors and memory, transceivers, and configurable I/O interfaces. Educational platforms can also take advantage of configurability, allowing students to use them in different courses and experiments in the same platform, whose structure and operation is adapted accordingly.

24.6.2 Rapid System Prototyping

In FPGA-based prototyping systems, a digital circuit is mapped onto one or more programmable devices. This hardware prototype can be connected to an external system for in-circuit functional verification. This approach has two main advantages:

- Low cost and easy modification and/or bug fixing, because no prototype is actually fabricated.
- Short design cycle through relatively simple simulation and verification.

On the other hand, there are also some minor drawbacks as follows:

- The operation is slower due to the programmable interconnects. However, given the high performance of current devices, this is not a limiting factor for many applications.
- For complex prototypes, several interconnected FPGAs are needed. In this case, prototyping platforms are very complex not only from the hardware but also from the software point of view.

Many FPGA-based flexible prototyping solutions are increasingly available. FPGA prototyping allows control algorithms to be validated before being applied to physical systems, by taking advantage of the possibility of efficiently carrying out computation-intensive tasks in hardware through the use of FPGAs. In a study by Bieser and Muller-Glaser (2005), an FPGA is used as part of an automotive prototyping system, taking advantage of its reconfigurability and parallel processing capabilities. The FPGA is used to support the output signal generation and input signal processing at the logic level, allowing hardware modules to be reused according to the specifications of each automotive system. An experience on testing complex pulse patterns for power switches is presented in Fernandez et al. (2006), where complex real-time calculations, for instance those of switching angles in a three-phase inverter, are carried out from values previously calculated offline and stored in RAM blocks.

Rapid prototyping capabilities are significantly enhanced by the use of hardware-in-the-loop platforms, which allow designs to be verified directly in hardware. For instance, the FPGA implementation of a control system can be combined with a simulated plant.

24.6.3 Communication Processors and Interfaces

Communications applications are characterized by strong requirements regarding speed, power consumption, quality, and adaptability to new emerging standards or rapidly evolving protocols. Using FPGAs in this area allows time to market to be reduced and systems to be updated according to standards evolution, without modifying PCBs, and also to be remotely configured. Embedded processors and Ethernet controllers allow remote upgrades or maintenance to be carried out. IP protection (discussed in Section 24.6.5) is a very important capability in this case.

FPGAs are widely used for connecting complex, high-speed peripherals to general-purpose processors in embedded real-time systems. In this case, the FPGA acts as an interface, providing a standard communication among many different components and a processor. The availability of integrated RAM and the I/O compatibility of electrical levels are key features in these applications.

An application domain in which the use of FPGAs is significantly advantageous is that of fieldbuses. Being dominated by proprietary systems, it is difficult to guarantee compatibility between devices from different manufacturers. In addition, standards (and, in particular, protocols) in this area are continuously evolving. Reconfiguration becomes a suitable solution to adapt these systems to this diversity of approaches without the need for changes in the hardware to be made. Another interesting application field is that of FPGA-based upconverters/downconverters for digital transmitters/receivers. Integrated PLLs/DLLs ease the design of these systems.

24.6.4 Digital Signal Processing and Digital Control

Despite its easy high-level programming, the application of DSPs in systems with high-speed and processing power requirements is limited by their fixed architecture, due to the following reasons:

- Fixed memory and data sizes
- Low number of multiply-accumulate (MAC) units
- Low number and limited performance of buses
- Limited I/O resources

The use of several DSPs increases cost and complexity and also reduces performance. On the other hand, FPGAs provide a much more flexible and configurable architecture. Even if more difficult to program, some advanced features are making them a viable alternative to DSPs:

- Embedded multipliers and DSP blocks, which allow complex arithmetic operations to be performed. Currently, DSPs containing several MAC units are available, but are still far from what can be achieved using FPGAs, and at a very high cost.
- Parallel processing capabilities.
- Integrated memory blocks, which ease data storage.
- Processing support for higher bandwidth data streams.
- Large number of I/O pins and compatibility of electrical levels.

DSP blocks are extensively used in Mato et al. (2008) which, together with the implementation of predistortion functions in internal RAM blocks, allow the efficient correction of nonlinear effects in power amplifiers for digital television broadcasting to be achieved in real time. In Salcic et al. (2006), the FPGA implementation of an adaptive digital controller allows the performance and accuracy problems associated with DSP implementations (due to sequential program execution and limited computing power) to be overcome. In that work, advantage is taken of internal RAM blocks to store intermediate results, and complex functions (floating-point divisions by a constant, responsible for the maximum delay in the circuit) are implemented in LUTs. Specific algorithm functionalities can be developed and stored in libraries of IP blocks for reuse in different applications. An example of such an approach applied to sensorless control in induction motor drives is reported in Charaabi et al. (2005). A three-phase synchronous PWM generator, in which frequency division and arithmetic operations (e.g., 4 bit multiplications) are efficiently performed in an FPGA is presented in (Omar et al. 2004). High-precision, high-speed motor control is addressed in Wen et al. (2006), where sine/cosine waveform generation, fault protection, and other control-related functions are implemented in an FPGA and the configuration capabilities of I/O blocks allow a direct actuation on a driver IC from FPGA pins to be performed. The online measurement and estimation of electrical magnitudes in power converters can be efficiently moved to the digital domain by using FPGAs, as demonstrated in Acero et al. (2007), in which mixed-signal designs are supported by VHDL-AMS-based tools. In Chan et al. (2007), a MATLAB®/Simulink modular approach is used for the design of embedded feedback controllers. It is shown that the implementation of distributed arithmetic, taking advantage of the parallel computing capabilities of FPGAs, results in a reduction of both complexity and power consumption.

A subject of increasing interest is the digital generation of control signals for electronic switches in power converters. While traditional power conversion systems use analog control schemes, the increasing complexity of power systems, motivated by the growing number of subsystems to supply energy, the growth in number of supply outputs, the existence of hybrid systems which imply energy being taken from several alternative points (hybrid cars, for instance), and the use of new topologies with complex control (for instance, multistage power converters for microprocessors or multilevel converters), is producing a rapid change from analog to digital control. Digital control has the possibility of exchanging information with the load, in case this one is digital, allowing complex voltage regulations as well as increasing energy efficiency awareness. Additionally, digital control is more immune to noise than their analog counterparts. Possibilities of loop programming, monitoring, logging and remote operation, soft start/stops, programmable control algorithms, fault detection, output voltage trimming, multiphase synchronization, are all new fields in the power conversion area because of the use of digital control. The main drawbacks of digital control applied to power conversion are the need of (fast) ADCs for control algorithms and regulation, and the associated problems of limited signal resolution, which is common to many other digital processing applications. The use of DSPs is a solution for a reduced number of applications in this area, whereas the most speed demanding ones rely on the use of ASICs or FPGAs. There is a continuous growth in the number of companies which offer integrated products for digital control applications, but FPGAs are also a good opportunity for custom control solutions. The benefits are the possibility of implementing complex algorithms, with a high number of inputs and outputs, with increased flexibility and reprogramming capabilities.

Finally, it is worth mentioning other applications like active filters and the control of AC machine drives where FPGAs are being successfully and advantageously used. In the latter, the use of mixed-signal FPGAs provides significant advantages.

24.6.5 IP Protection

FPGAs can be used to preserve IP, providing solutions that are not only effective but also cheaper than ASIC ones. In nonvolatile FPGAs, security bits are available to prevent users from reading the configuration of the devices back, therefore avoiding the details of the system becoming accessible. In very complex systems that do not fit, or are not intended to be implemented, on FPGAs, a nonvolatile FPGA could be added to provide IP protection.

Encryption is available in some volatile devices, which have to be configured at power-on, as well as at any time a reconfiguration is needed, in order to prevent the configuration bitstream from being read by third parties. The security key is stored in the FPGA and used during configuration to decrypt the bitstream.

In low-cost FPGAs, where bitstream encryption is not available, the handshaking token technique (Feng and Seely 2005) can be used. It works by disabling the functionality of a design within the FPGA, until handshaking tokens are passed to it from a secure external device, in a similar way software licensing works.

24.6.6 Fault Tolerance

As can be clearly noticed in Figure 24.3a, the structure of FPGAs is inherently redundant, particularly taking into account that in many applications a non-negligible amount of resources remains unused. Advantage can be taken of these facts by using spare logic and interconnect resources to replace those affected by faults, through a reconfiguration process. In this way, FPGA-based designs can be provided with fault tolerance capabilities. Obviously, the applicability of fault tolerance approaches is limited by the reconfiguration capabilities of the devices, discussed in a Section 24.5.

Fault tolerance is particularly important in some application domains of electronic systems, for example, whenever safety of people or the environment can be compromised by the occurrence of faults in the system. Other examples are in space or automotive applications, where transient faults like single-event upsets (SEUs) are very prone to occur. Due to this, much research is being conducted for the characterization of the effects of SEUs, particularly in SRAM-based FPGAs.

24.7 Conclusions

Although it is clear that programmable logic cannot compete with the higher performance of other hardware-based solutions, like ASICs, the added flexibility of FPGAs or its lower cost for low volume can be a primary decision factor. Also, it is a real competitor sector, with undisputed benefits, when it is compared with software-based approaches. In some application areas like high-end DSP processing, the natural concurrent execution of hardware offers highest results. However, the need for many of these applications to adapt to time-varying high-level conditions (let us think, for instance, about a filter for a receiver in a communication channel with variable conditions) may make the software-based solutions more attractive because of its flexibility in being reprogrammed. But, when both requirements coexist, that is, high-performance execution and adaptability or flexibility are required together, programmable hardware solutions have a significant niche market.

If microprocessors have been a key player for control systems for years, and many applications are based on this technology due to its flexibility, FPGAs are a key piece in the reconfigurable, high-performance world. FPGA vendors, aware of this tendency and conscious about the higher integration obtained because of the microelectronic technology improvements, are integrating more and more custom blocks, like memories, microprocessor cores, DSP blocks, or custom communication peripherals. So, a new concept is appearing, which is being well accepted by industry, that deals with integrated systems on a programmable chip. There are tools that try to match with this concept, capable of integrating microprocessor cores with a high variety of peripherals, with reduced design efforts. There are also remarkable advances in debugging tools for this type of systems, which also reduce time to market. Nevertheless, the immature status of the tools intended to support these new paradigms is probably the main limiting factor for a faster evolution of hardware reconfiguration technologies.

The area of reconfigurable computing devices, which are slowly being introduced in the supercomputing facilities, is covered with two different approaches: either by using custom, reconfiguration-specific devices, or commercial FPGAs. Within this last approach, dynamic and partial reconfiguration capabilities are being explored by research centers to offer even higher flexibility for reprogramming these devices, although this path has lower penetration for industry.

There are still some other areas where opportunities for improvement exist. Power consumption, or even more important, computing performance per energy consumed, needs to be reduced so that the flexibility of reprogramming does not include this penalty. Tools need to improve in both the design of integrated complex SoCs, as well as for run-time reconfigurable systems management. Even though these problems are still far from being solved, the perception of the benefits of programmable devices, as well as the increasing costs competitiveness, is producing a good momentum for the use of this technology, which can be measured, for instance, in terms of new designs per year or new products adopting it.

References

Acero, J., Navarro, D., Barragan, L.A., Garde, I., Artigas, J.I., and Burdio, J.M. 2007. FPGA-based power measuring for induction heating appliances using sigma–delta A/D conversion. *IEEE Transactions on Industrial Electronics* 54: 1843–1852.

Al-Hashimi, B.M. (ed.) 2006. *System-on-Chip: Next Generation Electronics*. IET Press, London, U.K.

Becker, J., Donlin, A., and Hubner, M. 2007. New tool support and architectures in adaptive reconfigurable computing. In *Proceedings of the IFIP International Conference on Very Large Scale Integration 2007*, October 15–17, Atlanta, GA.

Bieser, C. and Muller-Glaser, K.D. 2005. COMPASS—A novel concept of a reconfigurable platform for automotive system development and test. In *Proceedings of the 16th IEEE International Workshop on Rapid System Prototyping*, June 8–10, Montreal, Canada.

Bobda, C., Majer, M., Ahmadinia, A., Haller, T., Linarth, A., and Teich, J. 2005. The Erlangen slot machine: Increasing flexibility in FPGA-based reconfigurable platforms. In *Proceedings of the 2005 IEEE International Conference on Field-Programmable Technology*, December 11–14, Singapore.

Chan, Y.F., Moallem, M., and Wang, W. 2007. Design and implementation of modular FPGA-based PID controllers. *IEEE Transactions on Industrial Electronics* 54: 1898–1906.

Charaabi, L., Monmasson, E., Naassani, A., and Slama-Belkhodja, I. 2005. FPGA-based implementation of DTSFC and DTRFC algorithms. In *Proceedings of the 31st Annual Conference of the IEEE Industrial Electronics Society*, November 6–10, Raleigh, NC.

Compton, K. and Hauck, S. 2002. Reconfigurable computing: A survey of systems and software. *ACM Computing Surveys* 34:171–210.

Feng, J. and Seely, J.A. 2005. Design security with waveforms. In *Proceedings of the 2005 Software Defined Radio Technical Conference and Product Exposition*, November 14–18, Orange County, CA.

Fernandez, C., Zumel, P., Sanz, M., Lazaro, A., Lopez, C., and Barrado, A. 2006. Fast prototyping of control circuits for power electronics, based on FPGA. In *Proceedings of the 32nd Annual Conference of the IEEE Industrial Electronics Society*, November 6–10, Paris, France.

Horta, E.L., Lockwood, J.W., Taylor, D.E., and Parlour, D. 2002. Dynamic hardware plugins in an FPGA with partial run-time reconfiguration. In *Proceedings of the 39th Design Automation Conference*, June 10–14, New Orleans, LA.

Ishii, E., Nishi, H., and Ohnishi, K. 2007. Improvement of performances in bilateral teleoperation by using FPGA. *IEEE Transactions on Industrial Electronics* 54: 1876–1884.

Khan, M. 2006. Power optimization in FPGA designs. In *Proceedings of the Synopsys Users Group—SNUG San Jose*, March, San Jose, CA.

Krasteva, Y.E., de la Torre, E., and Riesgo, T. 2008. Virtual architectures for partial runtime reconfigurable systems. Application to Network on Chip based SoC emulation. In *Proceedings of the 34th Annual Conference of the IEEE Industrial Electronics Society*, November 10–13, Orlando, FL.

Kung, Y.-S., Fung, R.-F., and Tai, T.-Y. 2009. Realization of a motion control IC for X–Y table based on novel FPGA technology. *IEEE Transactions on Industrial Electronics* 56: 43–53.

Mato, J.L., Pereira, M., Rodriguez-Andina, J.J., Farina, J., Soto, E., and Perez, R. 2008. Distortion mitigation in RF power amplifiers through FPGA-based amplitude and phase predistortion. *IEEE Transactions on Industrial Electronics* 55: 4085–4093.

Moller, L., Soares, R., Carvalho, E., Grehs, I., Calazans, N., and Moraes, F. 2006. Infrastructure for dynamic reconfigurable systems: choices and trade-offs. In *Proceedings of the 19th Annual Symposium on Integrated Circuits and Systems Design*, August 28–September 1, Ouro Preto, Brazil.

Omar, A.M., Rahim, N.A., and Mekhilef, S. 2004. Three-phase synchronous PWM for flyback converter with power-factor correction using FPGA ASIC design. *IEEE Transactions on Industrial Electronics* 51: 96–106.

Palma, J.C., de Mello, A.V., Moller, L., Moraes, F., and Calazans, N. 2002. Core communication interface for FPGAs. In *Proceedings of the 15th Annual Symposium on Integrated Circuits and Systems Design*, September 9–14, Porto Alegre, Brazil.

Pionteck, T., Koch, R., and Albrecht, C. 2006. Applying partial reconfiguration to Networks-on-Chips. In *Proceedings of the 16th International Conference on Field-Programmable Logic and Applications*, August 28–30, Madrid, Spain.

Rodriguez-Andina, J.J., Moure, M.J., and Valdes, M.D. 2007. Features, design tools, and application domains of FPGAs. *IEEE Transactions on Industrial Electronics* 54: 1810–1823.

Salcic, Z., Cao, J., and Nguang, S.K.. 2006. A floating-point FPGA-based self-tuning regulator. *IEEE Transactions on Industrial Electronics* 53: 693–704.

Sridharan, K. and Priya, T.K. 2007. A hardware accelerator and FPGA realization for reduced visibility graph construction using efficient bit representation. *IEEE Transactions on Industrial Electronics* 54: 1800–1804.

Ullmann, M., Hubner, M., Grimm, B., and Becker, J. 2004. An FPGA run-time system for dynamical on-demand reconfiguration. In *Proceedings of the 18th International Parallel and Distributed Processing Symposium*, April 26–30, Santa Fe, NM.

Walder, H. and Platzner, M. 2004. A runtime environment for reconfigurable hardware operating systems. In *Proceedings of the 14th International Conference on Field-Programmable Logic and Applications*, August 30–September 1, Leuven, Belgium.

Wen, Z., Chen, W., Xu, Z., and Wang, J. 2006. Analysis of two-phase stepper motor driver based on FPGA. In *Proceedings of the 2006 IEEE International Conference on Industrial Informatics*, August 16–18, Singapore.

Xilinx Staff. 2004. Celebrating 20 years of innovation. *Xcell Journal* 48: 14–16.

IV

Digital and Analog Signal Processing

25

Signal Processing

James A. Heinen
Marquette University

Russell J.
Niederjohn
Marquette University

25.1 Introduction

Signal processing is a very broad field. Examples include filtering of electrical signals to remove 60 Hz interference, equalization of audio signals, processing of electrocardiograms to determine features of the heartbeat, enhancement of noisy speech signals to improve intelligibility, and enhancement of images to emphasize edges.

A signal is a function of one or more independent variables (usually time in the case of one-dimensional signals). Only real-valued, one-dimensional signals will be considered here. This rules out, e.g., images, which are two dimensional. Furthermore, only deterministic signals will be considered, ruling out random signals, which may be described only in a probabilistic sense. Both continuous-time ("analog") and discrete-time ("digital") signals will be described, as will the sampling (analog-to-digital conversion) process. These signals will be studied directly in the time domain and indirectly in the frequency (transform) domain. Both continuous-time and discrete-time systems ("signal processors") will be discussed as well.

Because of the vast amount of information available on signal processing, only that deemed most fundamental will be discussed here. Several excellent texts on signal processing (Ambardar, 2007; Baher, 2001; Ingle and Proakis, 2007; Kuo and Gan, 2005; McClellan et al., 2003; Mitra, 2006; Oppenheim and Schafer, 1989, 2010; Proakis and Manolakis, 2007; Stearns, 2003), linear systems (Chen, 2004; ElAli, 2004; Gajić, 2003; Lathi, 2005; Sherrick, 2005), and filters (Antoniou, 1993; Lam, 1979; Paarmann, 2001; Schlichthärle, 2000) are listed in the references at the end of this discussion.

25.2 Continuous-Time Signals

Continuous-time signals are functions of the form $x(t)$, $-\infty < t < \infty$, where t is an independent variable, normally time. Only physically meaningful signals, or those reasonably approximated as such, will be considered in this discussion. Signals with infinite discontinuities will not be allowed. Also disallowed will be signals with an infinite number of finite discontinuities and/or maxima and minima in a finite time range. Impulses (defined below) and signals with finite discontinuities will be considered, even though they are not physically meaningful.

25.2.1 Common Signals

Some continuous-time signals often encountered in practice, and/or which may be easily described mathematically, are described below and illustrated in Figure 25.1.

The unit step is the signal:

$$u(t) = \begin{cases} 1, & t \geq 0 \\ 0, & t < 0 \end{cases}.$$

Its integral is the unit ramp:

$$r(t) = \begin{cases} t, & t \geq 0 \\ 0, & t < 0 \end{cases} = \int_{-\infty}^{t} u(\tau) d\tau.$$

The derivative (appropriately defined) of the unit step is the unit impulse:

$$\delta(t) = \begin{cases} \infty, & t = 0 \\ 0, & t \neq 0 \end{cases} = \frac{du(t)}{dt}.$$

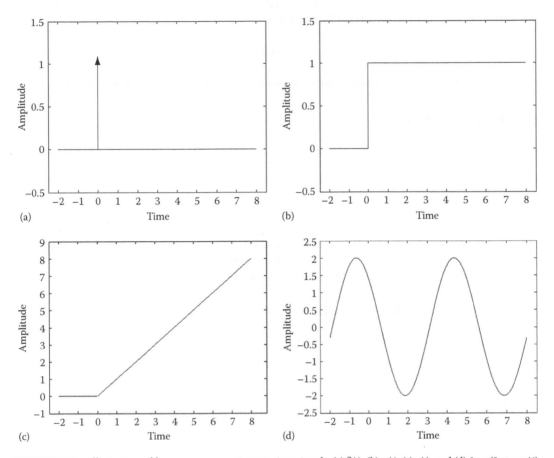

FIGURE 25.1 Illustration of four common continuous-time signals: (a) $\delta(t)$, (b) $u(t)$, (c) $r(t)$, and (d) $2\cos(2\pi t + \pi/4)$.

To completely define the unit impulse, it is necessary to impose the condition that

$$\int_{-\infty}^{\infty} \delta(t)dt = 1.$$

This leads to the so-called "sampling property" of the impulse, namely,

$$\int_{-\infty}^{\infty} x(t)\delta(t - t_o)dt = x(t_o),$$

provided that $x(t)$ is continuous at t_o.

Sinusoids may be written in the form

$$x(t) = A\cos(\omega t + \theta),$$

where

    A is the amplitude

    $\omega = 2\pi f$ is the angular frequency (f being the frequency)

    θ is the phase

A very general signal that is important physically is

$$x(t) = At^p e^{-at} \cos(\omega t + \theta) u(t),$$

where $p \geq 0$. This signal includes steps, ramps, sinusoids, and damped sinusoids as special cases.

25.2.2 Periodic Signals

Periodic continuous-time signals are those satisfying the condition

$$x(t + T) = x(t), \quad -\infty < t < \infty,$$

for some fixed $T > 0$. It is noted that if this condition is satisfied for a given T, it is also satisfied for all integer multiples of T. The smallest T for which it is satisfied is called the period. The frequency of a periodic signal with period T is given by $f = 1/T$. Its angular frequency is $\omega = 2\pi f = 2\pi/T$. Normally, T is specified in units of seconds, ω in radians/second, and f in Hertz. Sinusoids are simple but fundamental examples of periodic signals. Signals that are not periodic are said to be aperiodic.

25.3 Time-Domain Analysis of Continuous-Time Signals

25.3.1 Basic Operations on Signals

Various mathematical operations may be performed on a signal or a combination of signals. Obvious ones include differentiating or integrating a signal, and combining two or more signals using addition, subtraction, multiplication, or division. Other operations are also important in the context of signal analysis. Amplitude scaling a signal $x(t)$ produces a modified signal

$$y(t) = Ax(t),$$

where $A \neq 0$ is a constant. Time shifting produces

$$y(t) = x(t - t_o),$$

where t_o is fixed. Time reversal results in

$$y(t) = x(-t),$$

and time scaling leads to

$$y(t) = x(at),$$

where $a > 0$ is a constant.

25.3.2 Convolution

As will be seen later, a very important operation used to combine two continuous-time signals is convolution. The convolution of the signals $x_1(t)$ and $x_2(t)$ is defined as

$$y(t) = x_1(t) * x_2(t) = \int_{-\infty}^{\infty} x_1(\tau) x_2(t - \tau) d\tau, \quad -\infty < t < \infty.$$

This involves the time reversal and time shifting of the signal $x_2(\tau)$ to produce $x_2(t - \tau)$. (It is noted that unless interpreted carefully, the notation $y(t) = x_1(t) * x_2(t)$ can lead to confusion. For instance, $y(t - t_o) = x_1(t) * x_2(t - t_o)$, not $x_1(t - t_o) * x_2(t - t_o)$, as might be expected.

A positive-time signal is one satisfying the condition

$$x(t) = 0, \quad t < 0.$$

Steps and ramps are clearly examples of positive-time signals. If the signals $x_1(t)$ and $x_2(t)$ are positive time, their convolution is also positive time and may be simplified to the form

$$y(t) = x_1(t) * x_2(t) = \begin{cases} \int_0^t x_1(\tau) x_2(t - \tau) d\tau, & t \geq 0 \\ 0, & t < 0 \end{cases}.$$

25.4 Frequency-Domain Analysis of Continuous-Time Signals

It is often both convenient and informative to study continuous-time signals in the frequency (or transform) domain. The original signal is transformed by some appropriate mathematical operation to result in a frequency-domain quantity (often complex valued in nature). This frequency-domain quantity allows for the interpretation of the signal in terms of the basic constituent components of various frequencies. The particular frequency-domain representation used will, at least in part, depend on the characteristics of the signal being studied.

25.4.1 Fourier Series

Suppose $x(t)$ is a periodic signal with period T_o and angular frequency $\omega_o = 2\pi/T_o$. Then $x(t)$ may be considered as consisting of a (possibly infinite) linear combination of sinusoids with angular frequencies at integer multiples of ω_o. Specifically

$$x(t) = a_o + \sum_{k=1}^{\infty} (a_k \cos(k\omega_o t) + b_k \sin(k\omega_o t)).$$

Written this way, $x(t)$ is called a (trigonometric) Fourier series with coefficients, a_o, a_k, b_k, $k = 1, 2, \ldots$. These coefficients may be calculated from the original signal using the formulas

$$a_o = \frac{1}{T_o} \int_0^{T_o} x(t) dt,$$

$$a_k = \frac{2}{T_o} \int_0^{T_o} x(t) \cos(k\omega_o t) dt, \quad k = 1, 2, \ldots,$$

$$b_k = \frac{2}{T_o} \int_0^{T_o} x(t) \sin(k\omega_o t) dt, \quad k = 1, 2, \ldots.$$

It is noted that, because the integrands are periodic, these integrals may be calculated over any range of t values of length T_o, e.g., $-T_o/2$ to $T_o/2$. It is also noted that a_o is the average (or "dc") value of the signal $x(t)$, as may be seen from its definition. The kth term in the series is referred to as the kth harmonic, with the first harmonic also called the fundamental.

If $x(t)$ is an even signal (i.e., one satisfying $x(-t) = x(t)$, $-\infty < t < \infty$), then $b_k = 0$, $k = 1, 2, ...$ On the other hand, if $x(t)$ is an odd signal (i.e., one satisfying $x(-t) = -x(t)$, $-\infty < t < \infty$), then $a_k = 0$, $k = 0, 1, 2, ...$ Thus, an even signal consists only of cosine components (the cosine is even) and an odd signal consists only of sine components (the sine is odd).

Using trigonometric identities, the trigonometric Fourier series may be rewritten as an exponential (or complex) Fourier series. This takes the form, using j to represent $\sqrt{-1}$,

$$x(t) = \sum_{k=-\infty}^{\infty} X_k e^{jk\omega_o t},$$

where

$$X_k = \frac{1}{T_o} \int_0^{T_o} x(t) e^{-jk\omega_o t} dt, \quad k = ..., -1, 0, 1, 2, ...$$

It is observed that X_k is complex-valued. The coefficients of these two Fourier series representations are related by the formulas

$$a_o = X_o,$$

$$a_k = 2Re\{X_k\}, \quad k = 1, 2, ...,$$

$$b_k = -2Im\{X_k\}, \quad k = 1, 2, ...$$

While it is not always expressed this way, it is instructive to think of X_k and $x(t)$ as forming a transform pair, i.e., a representation in which X_k is uniquely determined by $x(t)$ and vice versa. Thus,

$$X_k = FS\{x(t)\}$$

and

$$x(t) = FS^{-1}\{X_k\}.$$

In this context, it is clear that X_k and $x(t)$ contain exactly the same information, but in a different form.

As an example, consider the signal $x(t)$, which is assumed to be periodic with period T_o, and which is defined over one period by

$$x(t) = \begin{cases} 1, & 0 \leq t < \dfrac{T_o}{2} \\ 0, & \dfrac{T_o}{2} \leq t < T_o. \end{cases}$$

Using the defining relation, it is easily determined that the exponential Fourier series coefficients are

$$X_o = \frac{1}{2}; \quad X_k = j\frac{(-1)^k - 1}{2\pi k}, \quad k \neq 0.$$

The trigonometric Fourier series coefficients are likewise easily found to be

$$a_o = \frac{1}{2}; \quad a_k = 0, \quad k > 0$$

$$b_k = \frac{1 - (-1)^k}{\pi k}, \quad k > 0.$$

Being complex valued, X_k may be written as

$$X_k = |X_k| \angle \varphi_k = |X_k| e^{j\varphi_k},$$

where $\varphi_k = \arg\{X_k\}$. Under our assumption that $x(t)$ is real-valued, $|X_k|$ is an even function of k and φ_k is an odd function of k, i.e., respectively,

$$|X_{-k}| = |X_k|, \quad \varphi_{-k} = -\varphi_k.$$

Thus, the information contained in X_k, $k < 0$ is redundant. When plotted versus k (the frequency index), $|X_k|$ is called the magnitude spectrum and φ_k the phase spectrum of $x(t)$, illustrated in Figure 25.2 for the example under consideration. These spectra aid in visualizing the frequency-domain characteristics of the periodic signal under consideration.

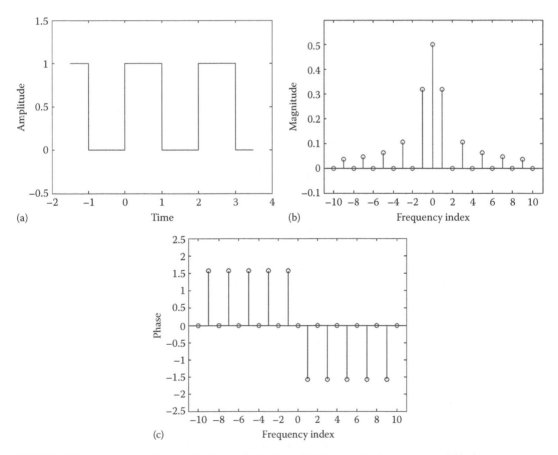

FIGURE 25.2 Illustration of (a) a periodic signal ($T_o = 2$), and (b) its magnitude spectrum and (c) phase spectrum.

Some important properties of the exponential Fourier series are summarized in the following table.

Periodic Signals	Fourier Series Coefficients
$Ax_1(t) + Bx_2(t)$	$AX_{1,k} + BX_{2,k}$
$x(t - t_o)$	$X_k e^{-jk\omega_o t_o}$
$x(-t)$	X_{-k}
$x(at)^+, a > 0$	X_k
$\dfrac{dx(t)}{dt}$	$jk\omega_o X_k$

$x(t)$, $x_1(t)$, $x_2(t)$ are arbitrary periodic signals with period T_o and angular frequency ω_o, and with Fourier series coefficients X_k, $X_{1,k}$, $X_{2,k}$, respectively. A, B, a, t_o are arbitrary constants.

Note that $x(at)$ has period T_o/a and angular frequency $a\omega_o$.

The Fourier series of a signal having finite discontinuities will necessarily contain an infinite number of terms. It is often necessary to approximate such a Fourier series by a truncated series containing only a finite number of terms. Of course, this truncation may take place at any point in the series. Fortunately, including an additional term will never increase the mean-squared error between the truncated series and the true signal. On the other hand, it is very difficult for a Fourier series to represent a discontinuity. Regardless of the number of terms included, the truncated Fourier series exhibits oscillations near each discontinuity. As additional terms are included, these oscillations move closer to the discontinuity, but they never disappear. The existence of these oscillations is known as Gibb's phenomenon.

25.4.2 Fourier Transforms

Fourier series are useful only for analyzing periodic signals. In addition to periodic signals, certain aperiodic signals may be studied using the Fourier transform. This representation may loosely be considered as arising from the Fourier series representation when the fundamental period is allowed to approach infinity.

A continuous-time signal $x(t)$ will have a Fourier transform if it is absolutely integrable, i.e., if

$$\int_{-\infty}^{\infty} |x(t)| \, dt < \infty.$$

With proper interpretation, certain signals not satisfying this condition will also have Fourier transforms. These include steps, constants, and all periodic signals.

The Fourier transform of $x(t)$ is defined as

$$X(j\omega) = FT\{x(t)\} = \int_{-\infty}^{\infty} x(t) e^{-j\omega t} \, dt, \quad -\infty < \omega < \infty.$$

$x(t)$ may be recovered uniquely from $X(j\omega)$ using the inverse Fourier transform given by

$$x(t) = FT^{-1}\{X(j\omega)\} = \frac{1}{2\pi} \int_{\infty}^{\infty} X(j\omega) e^{j\omega t} \, d\omega.$$

Note that while $X(j\omega)$ is considered as a function of ω, for convenience it is often written, as we have done, as a function of $j\omega$. While periodic signals have all their energy concentrated at only certain values of frequency (multiples of the fundamental frequency), other signals allowed by the Fourier transform have their energy spread out over, in general, all frequencies. Some common Fourier transforms are listed in the following table:

$x(t)$	$X(j\omega)$
$\delta(t)$	1
1	$2\pi\delta(\omega)$
$u(t)$	$\pi\delta(\omega) + \dfrac{1}{j\omega}$
$e^{-at}u(t), \quad a > 0$	$\dfrac{1}{a + j\omega}$
$te^{-at}u(t), \quad a > 0$	$\dfrac{1}{(a + j\omega)^2}$
$\sin(\omega_o t)$	$j\pi[\delta(\omega + \omega_o) - \delta(\omega - \omega_o)]$
$\cos(\omega_o t)$	$\pi[\delta(\omega + \omega_o) + \delta(\omega - \omega_o)]$

As is true for the exponential Fourier series coefficients X_k, the Fourier transform $X(j\omega)$ is, in general, complex-valued, i.e.,

$$X(j\omega) = \left|X(j\omega)\right| \angle\varphi(j\omega) = \left|X(j\omega)\right|e^{j\varphi(j\omega)},$$

where $\varphi(j\omega) = \arg\{X(j\omega)\}$. For our assumed real-valued $x(t)$, $\left|X(j\omega)\right|$ is an even function of ω and $\varphi(j\omega)$ is an odd function of ω, i.e., respectively

$$\left|X(-j\omega)\right| = \left|X(j\omega)\right|, \quad \varphi(-j\omega) = -\varphi(j\omega).$$

Thus, the information contained in $X(j\omega)$, $\omega < 0$, is redundant. When plotted versus ω, $\left|X(j\omega)\right|$ is called the magnitude spectrum and $\varphi(j\omega)$ the phase spectrum of $x(t)$. An example is shown in Figure 25.3. While the spectra based on the Fourier series exist only at integer values of the frequency index k, the Fourier transform spectra exist, in general, at all frequencies.

Signals not satisfying the absolute integrability condition stated above, will, if they are Fourier transformable, normally contain frequency-domain impulses in their Fourier transforms. As a specific case, if $x(t)$ is a periodic signal with fundamental angular frequency ω_o and with Fourier series $X_k = FS\{x(t)\}$, then its Fourier transform is given by

$$X(j\omega) = \sum_{k=-\infty}^{\infty} X_k \delta(\omega - k\omega_o),$$

with all energy clearly still concentrated (now, mathematically, by means of impulse functions) at integer multiples of ω_o.

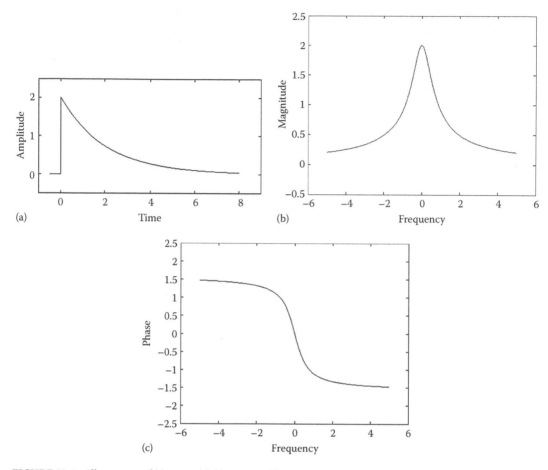

FIGURE 25.3 Illustration of (a) a signal $(x(t) = 2e^{-0.5t}u(t))$, and (b) its magnitude spectrum and (c) phase spectrum.

Some important properties of the Fourier transform are summarized in the following table:

Signals	Fourier Transforms
$Ax_1(t) + Bx_2(t)$	$AX_1(j\omega) + BX_2(j\omega)$
$x(t - t_o)$	$X(j\omega)e^{-j\omega t_o}$
$x(-t)$	$X(-j\omega)$
$x(at), \quad a > 0$	$\dfrac{1}{a}X\left(j\dfrac{\omega}{a}\right)$
$\dfrac{dx(t)}{dt}$	$j\omega X(j\omega)$
$\displaystyle\int_{-\infty}^{t} x(\tau)d\tau$	$\dfrac{X(j\omega)}{j\omega} + \pi X(0)\delta(\omega)$
$x_1(t) * x_2(t)$	$X_1(j\omega)X_2(j\omega)$

$x(t)$, $x_1(t)$, $x_2(t)$ are arbitrary signals with Fourier transforms $X(j\omega)$, $X_1(j\omega)$, $X_2(j\omega)$, respectively. A, B, a, t_o are arbitrary constants.

25.4.3 Laplace Transforms

As seen above, the Fourier transform is a generalization of the Fourier series, in that it allows for the analysis of a broader range of continuous-time signals than does the Fourier series. The (single-sided) Laplace transform is likewise a generalization of the Fourier transform, but with the restriction that the signal must be positive time, i.e., zero for $t < 0$.

A positive-time signal $x(t)$ will have a Laplace transform if

$$\int_{0-}^{\infty} |x(t)| e^{-\sigma t} dt < \infty$$

for some real σ. (The notation 0− is used to indicate that the lower limit of the integral is taken "just to the left of 0." In practice, this means that impulses at $t = 0$ are to be included entirely in the integral.) Since $\sigma = 0$ is allowed in this condition, if $x(t)$ is a positive-time signal with a Fourier transform, it will also have a Laplace transform, but the converse is not true. For example, a unit ramp has a Laplace transform but not a Fourier transform. On the other hand, nontrivial periodic signals and constants (which are necessarily nonzero for $t < 0$) do not have Laplace transforms even though they have Fourier transforms.

The Laplace transform of the positive-time signal $x(t)$ is defined as

$$X(s) = LT\{x(t)\} = \int_{0-}^{\infty} x(t) e^{-st} dt,$$

where $s = \sigma + j\omega$ is a complex variable. In general, this integral will converge only for certain choices of σ. $x(t)$ may be recovered uniquely from $X(s)$ using the inverse Laplace transform given by

$$x(t) = LT^{-1}\{X(s)\} = \frac{1}{2\pi j} \int_{\sigma_1 - j\infty}^{\sigma_1 + j\infty} X(s) e^{st} ds,$$

where σ_1 is appropriately chosen. (This formula is rarely directly used in practice because of its complexity. Instead, tables of Laplace transform pairs are used, with complicated transforms first being decomposed into simpler ones using partial fraction expansion techniques.) Some common Laplace transforms are listed in following table:

$x(t)$	$X(s)$
$\delta(t)$	1
$u(t)$	$\dfrac{1}{s}, \quad Re\{s\} > 0$
$e^{-at}u(t)$	$\dfrac{1}{s+a}, \quad Re\{s\} > -a$
$te^{-at}u(t)$	$\dfrac{1}{(s+a)^2}, \quad Re\{s\} > -a$
$\sin(\omega_o t)u(t)$	$\dfrac{\omega_o}{s^2 + \omega_o^2}, \quad Re\{s\} > 0$
$\cos(\omega_o t)u(t)$	$\dfrac{s}{s^2 + \omega_o^2}, \quad Re\{s\} > 0$

For a positive-time signal $x(t)$ having a Fourier transform $X(j\omega)$, it is the case that

$$FT\{x(t)\} = LT\{x(t)\}|_{\sigma=0}$$

or

$$X(j\omega) = X(s)|_{s=j\omega},$$

where $X(s)$ is the Laplace transform of $x(t)$. (This explains our choice of notation for the Fourier transform.) Thus, in this case, the Fourier transform is simply the Laplace transform evaluated on the $j\omega$ (imaginary) axis in the s (complex) plane.

Some important properties of the Laplace transform are summarized in the following table:

Signals	Laplace Transforms
$Ax_1(t) + Bx_2(t)$	$AX_1(s) + BX_2(s)$
$x(t - t_o),\quad t_o \geq 0$	$X(s)e^{-st_o}$
$x(at), a > 0$	$\dfrac{1}{a}X\left(\dfrac{s}{a}\right)$
$\dfrac{dx(t)}{dt}$	$sX(s) - x(0-)$
$\displaystyle\int_{-\infty}^{t} x(\tau)d\tau$	$\dfrac{X(s)}{s}$
$x_1(t) * x_2(t)$	$X_1(s)X_2(s)$

$x(t), x_1(t), x_2(t)$ are arbitrary signals with Laplace transforms $X(s), X_1(s), X_2(s)$, respectively. A, B, a, t_o are arbitrary constants.

25.5 Continuous-Time Signal Processors

A continuous-time signal processor (system) is a device (see Figure 25.4) that acts on an input signal $x(t)$, modifying it in some manner to produce an output signal $y(t)$. This may be represented abstractly as

$$y(t) = \mathcal{H}\{x(t)\}.$$

(Here, $x(t)$ and $y(t)$ should be thought of in their totality, rather than at specific instants t. Just as in the case of convolution, this notation can lead to confusion. It does not imply that $y(t)$ is a function of $x(t)$ at only the same instant t. $y(t)$ could, e.g., depend on all past values of $x(\tau)$, $\tau \leq t$.)

This abstract definition is very broad. In practice, because of mathematical tractability, only certain relatively simple classes of signal processors are considered. In this treatment, we will discuss only those continuous-time signal processors that can be described by linear constant-coefficient ordinary differential equations. Such signal processors are of necessity linear, time-invariant, and casual. (In the familiar case of

FIGURE 25.4 Continuous-time signal processing system.

electrical circuits, they can be built using resistors, capacitors, inductors, and operational amplifiers.) A system is linear if

$$\mathcal{H}\{Ax_1(t) + Bx_2(t)\} = A\mathcal{H}\{x_1(t)\} + B\mathcal{H}\{x_2(t)\}, \quad -\infty < t < \infty$$

for all signals $x_1(t)$, $x_2(t)$, and all constants A, B. It is time-invariant if

$$y(t - t_o) = \mathcal{H}\{x(t - t_o)\}, \quad -\infty < t < \infty,$$

for all input signals $x(t)$ and corresponding output signals $y(t)$ and all constants t_o. A system is casual if $\mathcal{H}\{x(t)\}$ depends only on $x(\tau)$, $\tau \leq t$ for all signals $x(t)$.

25.6 Time-Domain Analysis of Continuous-Time Signal Processors

Under our assumptions, a continuous-time signal processor may be represented by a linear constant-coefficient ordinary differential equation. That is,

$$\sum_{r=0}^{N} a_r \frac{d^r y(t)}{dt^r} = \sum_{r=0}^{M} b_r \frac{d^r x(t)}{dt^r}.$$

Ordinarily, $M \leq N$ and, often, this equation is normalized so that $a_N = 1$. This describes an implicit relationship between the input $x(t)$ and output $y(t)$. In specific cases, for a known input $x(t)$ and initial conditions on $y(t)$ and its derivatives at $t = 0$, it may be solved for $x(t)$, $t \geq 0$. Laplace transforms may be used for this purpose, as may other techniques.

An alternative means for describing our signal processor, in the form of an explicit relationship between input and output, is based on the use of its impulse response. The impulse response is defined as

$$h(t) = \mathcal{H}\{\delta(t)\},$$

i.e., the output when the input is a unit impulse. Since we have assumed causality, it may be shown that $h(t)$ is a positive-time signal, i.e., that $h(t) = 0$, $t < 0$. Because the signal processor is linear and time invariant, it may additionally be shown that for any input $x(t)$ the corresponding output is given by the convolution of $h(t)$ and $x(t)$, namely,

$$y(t) = \mathcal{H}\{x(t)\} = h(t) * x(t).$$

This is a very significant result, since it shows that all information regarding the signal processor (at least from an input–output viewpoint) is contained in its impulse response.

A very desirable property of signal processors is stability. A system is said to be (bounded-input bounded-output) stable if whenever $x(t)$ is bounded (i.e., $|x(t)| \leq K_x < \infty$, $-\infty < t < \infty$ for some K_x) then the corresponding $y(t)$ is also bounded (i.e., $|y(t)| \leq K_y < \infty$, $-\infty < t < \infty$ for some K_y). (The alternative would normally be undesirable.) It is possible to easily determine stability directly from $h(t)$. Specifically, the signal processor is stable if and only if $h(t)$ is absolutely integrable, i.e., if and only if

$$\int_{-\infty}^{\infty} |h(t)| dt < \infty.$$

25.7 Frequency-Domain Analysis of Continuous-Time Signal Processors

In the frequency (transform) domain, a signal processor is characterized by its transfer function $H(s)$. $H(s)$ may be determined by Laplace transforming the convolution relationship to obtain

$$Y(s) = H(s)X(s),$$

where

$$H(s) = LT\{h(t)\}.$$

Thus, the transfer function is the Laplace transform of the impulse response, well defined since $h(t)$ is a positive-time signal.

Alternatively, the input–output differential equation may be Laplace transformed (assuming zero initial conditions) to obtain

$$\sum_{r=0}^{N} a_r s^r Y(s) = \sum_{r=0}^{M} b_r s^r X(s),$$

from which

$$H(s) = \frac{\displaystyle\sum_{r=0}^{M} b_r s^r}{\displaystyle\sum_{r=0}^{N} a_r s^r}.$$

Thus, $H(s)$ is a rational function (ratio of polynomials) in s. The impulse response $h(t)$, being the inverse Laplace transform of a rational function, will necessarily consist only of terms of the form

$$At^p e^{-at} \cos(\omega t + \theta)u(t),$$

where $p \geq 0$.

The roots of the numerator polynomial are called the zeros of $H(s)$ and the roots of the denominator polynomial are called the poles of $H(s)$. It may be shown that a signal processor is stable if and only if $M \leq N$ and all poles lie strictly in the left half of the complex s plane (i.e., have strictly negative real parts).

Fourier transforming the convolution relationship leads to

$$Y(j\omega) = H(j\omega)X(j\omega),$$

where

$$H(j\omega) = FT\{h(t)\}.$$

$H(j\omega)$ is called the frequency response of the signal processor, and is the Fourier transform of the impulse response. The magnitude of $H(j\omega)$ is the magnitude response and its angle is the phase response. It is seen that

$$H(j\omega) = H(s)\big|_{s=j\omega},$$

where $H(s)$ is the transfer function. That is, the frequency response is the transfer function evaluated on the $j\omega$ (imaginary) axis in the complex s plane. Thus, we may also write

$$H(j\omega) = \frac{\sum_{r=0}^{M} b_r(j\omega)^r}{\sum_{r=0}^{N} a_r(j\omega)^r}.$$

The frequency response of a signal processor is also useful for determining its output in the case of a periodic input. If $x(t)$ is a periodic input signal with Fourier series coefficients X_k, then the output signal $y(t)$ will also be periodic (with the same fundamental angular frequency ω_o) and will have Fourier series coefficients given by

$$Y_k = H(jk\omega_o)X_k.$$

This relationship may also be used for the steady-state analysis of systems, i.e., systems that have been operating for a sufficiently long time that all transient terms may be neglected.

It is thus clear that the way a signal processor processes an input signal (either aperiodic or periodic) is determined by its frequency response. It is thus very informative to have this quantity available, particularly in visual form as embodied in its magnitude and phase spectra.

25.8 Continuous-Time (Analog) Filters

A continuous-time (analog) filter is a signal processor designed to allow input signal components of certain frequencies to pass through to the output while preventing input signal components of other frequencies from doing so. Ideally, the frequency response $H(j\omega)$ of a filter should have magnitude one at those frequencies we wish to allow to pass through to the output and magnitude zero at those frequencies disallowed.

25.8.1 Common Filter Types

The most common types of filters are lowpass, highpass, bandpass, and bandstop. In the ideal case, these are described, respectively, by the following relationships (see Figure 25.5):

$$H_{lp}(j\omega) = \begin{cases} 1, & |\omega| \leq \omega_c \\ 0, & \text{otherwise} \end{cases},$$

$$H_{hp}(j\omega) = \begin{cases} 0, & |\omega| < \omega_c \\ 1, & \text{otherwise} \end{cases},$$

$$H_{bp}(j\omega) = \begin{cases} 1, & \omega_l \leq |\omega| \leq \omega_u \\ 0, & \text{otherwise} \end{cases},$$

$$H_{bs}(j\omega) = \begin{cases} 0, & \omega_l < |\omega| < \omega_u \\ 1, & \text{otherwise} \end{cases}.$$

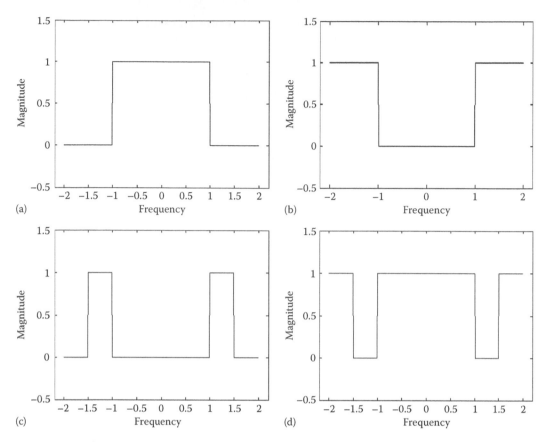

FIGURE 25.5 Illustration of four ideal filter types: (a) lowpass filter, (b) highpass filter, (c) bandpass filter, and (d) bandstop filter.

The values ω_c, ω_l, and ω_u are the cutoff frequencies. Those ranges of frequencies for which the frequency response is 1 are the passbands, and those for which it is 0 are the stopbands. The bandwidth is defined as ω_c for a lowpass filter and $\omega_u - \omega_l$ for a bandpass filter.

Unfortunately, ideal filters are noncausal and are thus not physically realizable. They may, however, be quite closely approximated in practice. In the nonideal (practical) case, the frequency response values are not identically 1 in the passbands and not identically 0 in the stopbands. Furthermore, the response does not abruptly change from 1 to 0 between passbands and stopbands. Instead, it gradually changes in transition bands, which separate the passbands and stopbands. While cutoff frequencies are uniquely defined for ideal filters, in the case of practical filters, any frequency within a transition band may appropriately be called a cutoff frequency. Often, those frequencies at which the magnitude response is $1/\sqrt{2}$ (approximately $-3\,\mathrm{dB}$) are chosen as the cutoff frequencies.

In many instances, the phase response of a filter is of little consequence. In certain applications, however, it is required that the phase response be linear, i.e., that the phase be linearly related to frequency in the passbands. Again, this may only be approximated in practice. Linear phase ensures that the shape of a signal within the passband of a filter, rather than simply its frequency content, is preserved in the filtering process.

25.8.2 Filter Design

The design of a filter normally begins by specifying the locations of the passbands and stopbands, the allowable variations from the ideal values in these bands (tolerances), and the locations and widths of the transition bands. Generally, the tighter the tolerances and the narrower the transition bands, the higher the filter order will have to be to meet the specifications. Normally, our goal will be to meet the specifications with the lowest-order filter possible.

Despite the type of filter desired, an appropriate prototype lowpass filter $H_p(s)$, with a cutoff frequency of one, is often designed first. Design formulas exist for translating the original specifications to the prototype filter. Various standard lowpass prototypes are available. The simplest of these is perhaps the Butterworth filter. The transfer function of an Nth-order Butterworth filter is given by

$$H_{B_N}(s) = \frac{1}{B_N(s)},$$

where $B_N(s)$ is a Butterworth polynomial. The following table lists expressions for the first four of these:

N	$B_N(s)$
1	$s + 1$
2	$s^2 + 1.414s + 1$
3	$s^3 + 2s^2 + 2s + 1$
4	$s^4 + 2.613s^3 + 3.414s^2 + 2.613s + 1$

The poles of $H_{B_N}(s)$ are evenly spaced on the unit circle in the left half of the complex plane. The magnitude response of a Butterworth filter monotonically decreases from one to zero and has a value of $1/\sqrt{2}$ (−3 dB) at $\omega = 1$. Figure 25.6 shows the magnitude and phase responses of a typical lowpass Butterworth filter.

Chebyshev filters are equiripple in the passband (i.e., their magnitude response oscillates between certain tolerance limits) and monotonic in the stopband. These characteristics are interchanged in the case of inverse Chebyshev filters. Elliptic (Cauer) filters are equiripple in both the passband and the stopband. Bessel filters achieve very nearly linear phase in the passband. In each case, various design formulas exist for choosing the appropriate filter order and other filter parameters.

Once the prototype lowpass filter $H_p(s)$ (with cutoff frequency of one) is determined, an appropriate frequency transformation is employed to result in the final filter design $H(s)$ as follows:

$$H(s) = H_p(s)\big|_{s=T(s)}.$$

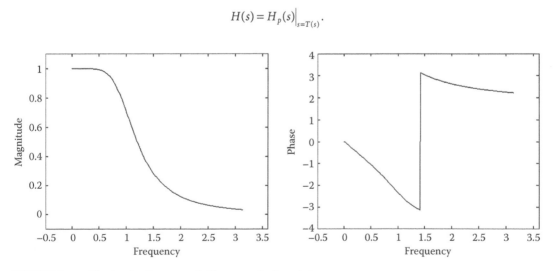

FIGURE 25.6 Third-order Butterworth filter magnitude and phase responses.

That is, s is replaced by $T(s)$ in the expression for $H_p(s)$. These transformations are summarized in the following table for cutoff frequencies of ω_c, ω_l, and ω_u:

Filter Type	$T(s)$
Lowpass	$\dfrac{s}{\omega_c}$
Highpass	$\dfrac{\omega_c}{s}$
Bandpass	$\dfrac{\left(\omega_u - \omega_l\right)s}{s^2 + \omega_u\omega_l}$
Bandstop	$\dfrac{s^2 + \omega_u\omega_l}{\left(\omega_u - \omega_l\right)s}$

25.9 Sampling

Sampling is the process of converting a continuous-time signal into a discrete-time signal. This may be represented abstractly by

$$x[n] = S\{x(t)\},$$

where $x(t)$, $-\infty < t < \infty$, is the original continuous-time signal and $x[n]$, $-\infty < n < \infty$, n being integer-valued, is a discrete-time signal consisting of samples of $x(t)$ (see Figure 25.7). More precisely,

$$x[n] = x(nT_s) = x(t)\big|_{t=nT_s},$$

where $T_s > 0$ is called the sampling period, assumed here to be fixed. The sampling frequency is $f_s = 1/T_s$ and the angular sampling frequency is $\omega_s = 2\pi f_s = 2\pi/T_s$. It is noted that n is really a time index, but will usually be referred to as time. $x[n]$ is thus seen to be simply a sequence of numbers. (The square-bracket notation, which will be used throughout to distinguish discrete-time from continuous-time quantities, follows that used by Oppenheim and Schafer (1989).)

The actual process of sampling a physical signal is considerably more complicated than what is implied by this abstract description. The final result, namely, a sequence of numbers, is however quite meaningful if, say, these numbers are fed to a digital computer for further processing. The physical sampling (or analog-to-digital conversion) process may be modeled by a switch that closes instantaneously, or more realistically, very briefly, at instants separated by T_s units of time. This produces narrow pulses of height $x(nT_s)$, which may further be considered as impulses. This leads to the "impulse-sampled" signal

$$x_s(t) = \sum_{n=-\infty}^{\infty} x(nT_s)\delta(t - nT_s) = \sum_{n=-\infty}^{\infty} x[n]\delta(t - nT_s).$$

Oddly enough, this impulse-sampled signal $x_s(t)$ is in fact a continuous-time signal. Clearly, from an information content viewpoint, the signal $x_s(t)$ and the sequence $x[n]$ are equivalent. Depending on the task at hand, one or the other of these representations may be the most convenient to consider.

In practice, when we sample a signal we wish to do it in such a manner that the samples comprise a reasonably accurate representation of the original continuous-time signal. Intuitively this suggests that we take a great many samples (T_s small, f_s large), especially if the signal is changing rapidly, so as not to lose any significant information. On the other hand, efficiency (both in

FIGURE 25.7 Sampling a continuous-time signal to produce a discrete-time signal.

how fast we must sample and perhaps how many samples we must store) dictates that we not take too large a number of samples. Trade-offs must therefore be made in choosing an appropriate sampling frequency for a given signal.

It is instructive to consider the sampling process in the frequency domain. If $x(t)$ is a continuous-time signal with Fourier transform $X(j\omega)$, then its impulse-sampled counterpart $x_s(t)$ will have Fourier transform

$$X_s(j\omega) = \frac{1}{T_s} \sum_{m=-\infty}^{\infty} X(j(\omega - m\omega_s)).$$

That is, $X_s(j\omega)$ consists of copies of $X(j\omega)$, shifted by all possible multiples of the angular sampling frequency ω_s, added together and scaled by $1/T_s$. $X_s(j\omega)$ is thus seen to be periodic with period ω_s.

In general, the various copies of $X(j\omega)$ will overlap each other, a phenomenon called aliasing. This prevents the recovery of $X(j\omega)$ from $X_s(j\omega)$, implying that information is lost in the sampling process. On the other hand, if $x(t)$ is bandlimited to half the sampling frequency, i.e., if

$$X(j\omega) = 0, \quad |\omega| \geq \frac{\omega_s}{2},$$

then no aliasing or overlap occurs, and the original signal may be recovered from its samples. This may be summarized in the form of a theorem, the "sampling theorem," which states that no information is lost in the sampling process if the sampling frequency is at least twice that of the highest frequency component present in the original signal. This minimum sampling frequency is called the Nyquist rate. The sampling theorem provides a quantitative basis for the choice of sampling frequency, and validates our intuition that rapidly varying signals must be sampled more frequently than slowly varying ones.

25.10 Discrete-Time Signals

Discrete-time signals are functions (sequences) of the form $x[n]$, $-\infty < n < \infty$, where n is an independent integer-valued variable, normally a time index (usually, simply referred to as "time"). For our purposes, we will assume $x[n]$ to be real valued.

25.10.1 Common Signals

Some discrete-time signals often encountered in practice, and/or which may be easily described mathematically, are described below and illustrated in Figure 25.8.

The unit step is the signal

$$u[n] = \begin{cases} 1, & n \geq 0 \\ 0, & n < 0 \end{cases}.$$

Its first difference is the unit impulse

$$\delta[n] = \begin{cases} 1, & n = 0 \\ 0, & n \neq 0 \end{cases} = u[n] - u[n-1].$$

The unit ramp is given by

$$r[n] = \begin{cases} n, & n \geq 0 \\ 0, & n < 0 \end{cases}.$$

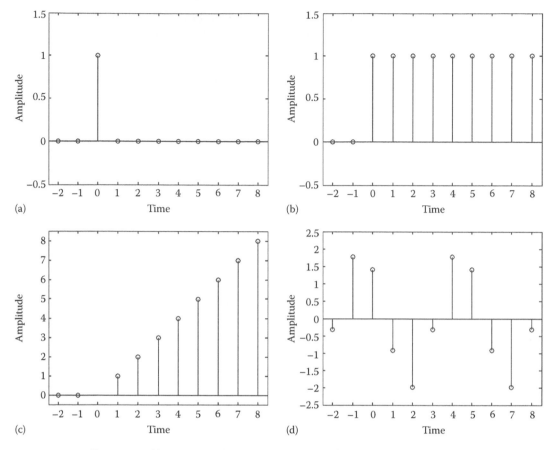

FIGURE 25.8   Illustration of four common discrete-time signals: (a) $\delta[n]$, (b) $u[n]$, (c) $r[n]$, and (d) $2\cos(2\pi n + \pi/4)$.

Discrete-time sinusoids may be written in the form

$$x[n] = A\cos(\omega n + \theta).$$

A very general signal that is important physically is

$$x[n] = An^p a^n \cos(\omega n + \theta)u[n],$$

where $p \geq 0$.

25.10.2 Periodic Signals

Periodic discrete-time signals are those satisfying the condition

$$x[n + N] = x[n], \quad -\infty < n < \infty,$$

for some fixed integer $N > 0$. As is the case for continuous-time periodic signals, if this condition is satisfied for a given N, it is also satisfied for all integer multiples of N. The smallest such N is called the period. Its frequency is $1/N$ and its angular frequency is $2\pi/N$.

Oddly enough, not all discrete-time sinusoids are periodic. The sinusoid $A\cos(\omega n + \theta)$ is periodic if and only if $\omega = 2\pi f$, where f is a rational number (ratio of integers). In this case, the period N is the smallest integer that is an integer multiple of $1/f$.

25.10.3 Finite-Duration Signals

As the name implies, finite-duration signals are signals that are nonzero for only a finite range of time values. We will be somewhat restrictive and define $x[n]$ to be a finite-duration signal if it satisfies the condition

$$x[n] = 0, \ n < 0 \quad \text{and} \quad n > N - 1.$$

This signal thus has, at most, N nonzero values and will be said to be of length N.

There is a close relationship between finite-duration signals and periodic signals. Specifically, if $x[n]$ is periodic with period N, then

$$y[n] = \begin{cases} x[n], & 0 \leq n \leq N - 1 \\ 0, & \text{otherwise} \end{cases}$$

is a finite-duration signal of length N. On the other hand, if $x[n]$ is a length-N finite-duration signal, then

$$y[n] = \sum_{m=-\infty}^{\infty} x[n - mN]$$

is periodic with period N. Each therefore contains the same information, but in a different form.

25.11 Time-Domain Analysis of Discrete-Time Signals

25.11.1 Basic Operations on Signals

Just as in the case of continuous-time signals, various mathematical operations may be performed on a discrete-time signal or a combination of signals. Basic ones include combining two or more signals using addition, subtraction, multiplication, or division.

The (backward) difference of a signal $x[n]$ is given by

$$y[n] = \Delta x[n] = x[n] - x[n - 1].$$

The accumulation of a signal $x[n]$ is given by

$$y[n] = \sum_{m=-\infty}^{n} x[m].$$

The difference and accumulation operations are inverses of each other, and may be thought of as discrete-time analogs of differentiation and integration, respectively.

Other operations important in signal analysis include amplitude scaling a signal $x[n]$ to produce

$$y[n] = Ax[n],$$

where $A \neq 0$ is a constant. Time shifting produces

$$y[n] = x[n - n_o],$$

where n_o is a fixed integer. Finally, time reversal results in

$$y[n] = x[-n].$$

25.11.2 Convolution

The discrete-time convolution of the two signals $x_1[n]$ and $x_2[n]$ is defined as

$$y[n] = x_1[n] * x_2[n] = \sum_{m=-\infty}^{\infty} x_1[m]x_2[n-m], \quad -\infty < n < \infty.$$

This involves time reversal and time shifting of the signal $x_2[m]$ to produce $x_2[n-m]$. (It is once again noted that caution must be exercised in interpreting the notation $y[n] = x_1[n] * x_2[n]$. For instance, $y[n - n_o] = x_1[n] * x_2[n - n_o]$.)

A positive-time signal is one satisfying the condition

$$x[n] = 0, \quad n < 0.$$

Unit steps, ramps, and impulses are positive time. If the signals $x_1[n]$ and $x_2[n]$ are both positive time, then their convolution is also positive time and may be written in the simplified form

$$y[n] = x_1[n] * x_2[n] = \begin{cases} \sum_{m=0}^{n} x_1[m]x_2[n-m], & n \geq 0 \\ 0, & n < 0 \end{cases}.$$

25.11.3 Periodic Convolution

If $x_1[n]$ and $x_2[n]$ are two periodic signals with period N, their convolution would clearly not be meaningful. Such signals may be combined, instead, using the operation of periodic convolution, defined by

$$y[n] = x_1[n] \circledast x_2[n] = \sum_{m=0}^{N-1} x_1[m]x_2[n-m], \quad -\infty < n < \infty.$$

The resulting signal $y[n]$ is also periodic with period N. It should be noted that periodic convolution is not ordinarily something we would wish to do—rather, it is an operation forced upon us by the mathematics.

25.12 Frequency-Domain Analysis of Discrete-Time Signals

As is also true for continuous-time signals, much can be gained by studying discrete-time signals in the frequency (or transform) domain. Again, various transforms are used, depending on the nature of the signal and the task at hand.

25.12.1 Discrete Fourier Series

Suppose $x[n]$ is a periodic discrete-time signal with period N. Then $x[n]$ may be considered as consisting of a finite linear combination of discrete-time complex exponentials. Specifically, it takes N such signals to represent $x[n]$, which can be written as

$$x[n] = \frac{1}{N} \sum_{k=0}^{N-1} X[k]e^{j(2\pi/N)kn}.$$

The coefficients $X[k]$ of the discrete Fourier series may be obtained from the original signal using the formula

$$X[k] = \sum_{n=0}^{N-1} x[n]e^{-j(2\pi/N)kn}, \quad k = 0, 1, 2, \ldots, N-1.$$

It is noted that $X[k]$ is itself a (generally complex) periodic signal with period N, and as such is defined for all integers k. However, only N values of $X[k]$ are required to determine (or represent) $x[n]$.

$X[k]$ and $x[n]$ form a transform pair with $X[k]$ uniquely determined by $x[n]$ and vice versa. Thus, we write

$$X[k] = DFS\{x[n]\}$$

and

$$x[n] = DFS^{-1}\{X[k]\}.$$

$X[k]$ and $x[n]$ thus contain exactly the same information, but in a different form.

As an example, consider the signal $x[n]$, which is assumed to be periodic with period N (where N is even), and which is defined over one period by

$$x[n] = \begin{cases} 1, & 0 \leq n < \dfrac{N}{2} \\ 0, & \dfrac{N}{2} \leq n < N \end{cases}.$$

Using the defining relation, it is easily determined that the discrete Fourier series coefficients are

$$X[k] = \frac{1 - (-1)^k}{1 - e^{-j(2\pi/N)k}}$$

for all integers k.

Being complex valued, $X[k]$ may be written as

$$X[k] = |X[k]| \angle \varphi[k] = |X[k]| e^{j\varphi[k]},$$

where $\varphi[k] = \arg\{X[k]\}$. Under our assumption that $x[n]$ is real valued, $|X[k]|$ is an even function of k and $\varphi[k]$ is an odd function of k, i.e., respectively,

$$|X[-k]| = |X[k]|, \quad \varphi[-k] = -\varphi[k].$$

Thus, the information contained in $X[k]$, $k < 0$, is redundant. In addition, since $X[k]$ is periodic, all but N consecutive values of $X[k]$ are redundant. Combining these two results leads to the conclusion that $X[k]$ is completely determined by $X[k]$, $0 \leq k \leq M$, where $M = N/2$ for N even and $M = (N-1)/2$ for N odd. When plotted versus k (the frequency index), $|X[k]|$ is called the magnitude spectrum and $\varphi[k]$ the phase spectrum. The spectra for this example are illustrated in Figure 25.9.

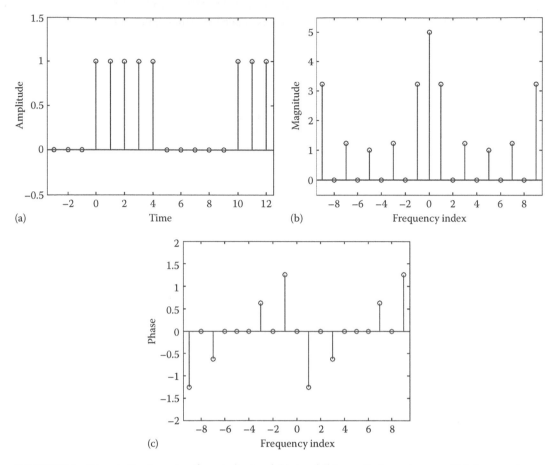

FIGURE 25.9 Discrete Fourier series of a periodic signal: (a) signal, (b) magnitude spectrum, and (c) phase spectrum.

Some important properties of the discrete Fourier series are summarized in the following table:

Discrete-Time Periodic Signals	Discrete Fourier Series Coefficients
$Ax_1[n] + Bx_2[n]$	$AX_1[k] + BX_2[k]$
$x[n - n_o]$	$X[k]e^{-j(2\pi/N)kn_o}$
$x[-n]$	$X[-k]$
$x_1[n] \circledast x_2[n]$	$X_1[k]X_2[k]$

$x[n], x_1[n], x_2[n]$ are arbitrary discrete-time periodic signals with period N and with discrete Fourier series coefficients $X[k], X_1[k], X_2[k]$, respectively. A, B are arbitrary constants and n_o is an arbitrary integer constant.

25.12.2 Discrete Fourier Transforms

As seen earlier, from an information content viewpoint, there is no essential difference between a periodic signal and a finite duration signal. The discrete Fourier series is the frequency-domain representation of a periodic signal. The frequency-domain representation of a length-N finite-duration signal $x[n]$ is the discrete Fourier transform given by

$$X[k] = DFT\{x[n]\} = \begin{cases} \sum_{n=0}^{N-1} x[n]e^{-j(2\pi/N)kn}, & k = 0, 1, 2, \ldots, N-1 \\ 0, & \text{otherwise} \end{cases}.$$

The inverse discrete Fourier transform is given by

$$x[n] = DFT^{-1}\{X[k]\} = \begin{cases} \dfrac{1}{N}\sum_{k=0}^{N-1} X[k]e^{j\frac{2\pi}{N}kn}, & n = 0, 1, 2, \ldots, N-1 \\ 0, & \text{otherwise} \end{cases}.$$

As an example, consider the finite-duration signal $x[n]$, which is defined by

$$x[n] = \begin{cases} 1, & 0 \leq n < \dfrac{N}{2} \\ 0, & \text{otherwise} \end{cases},$$

where N is even. Using the defining relation, it is easily determined that the discrete Fourier transform is

$$X[k] = \begin{cases} \dfrac{1-(-1)^k}{1-e^{-j(2\pi/N)k}}, & k = 0, 1, 2, \ldots, N-1 \\ 0, & \text{otherwise} \end{cases}.$$

This is shown in Figure 25.10.

It is seen that there is no fundamental difference between the discrete Fourier series and the discrete Fourier transform (justifying the use of the same notation for both). Whether one is dealing with the discrete Fourier series or the discrete Fourier transform is thus a matter of interpretation. If $x[n]$ and $X[k]$ are both assumed to be periodic, then the $X[k]$ are the discrete Fourier series coefficients of $x[n]$. On the other hand, if $x[n]$ and $X[k]$ are both assumed to be finite duration, then $X[k]$ is the discrete Fourier transform of $x[n]$.

Since from a formal viewpoint the quantities involved are periodic, the discrete Fourier series is perhaps the more fundamental interpretation. The properties of the discrete Fourier transform thus follow from those of the discrete Fourier series, and in a given situation are best interpreted by making all quantities periodic, invoking the corresponding property of the discrete Fourier series, and then making all quantities finite duration again.

25.12.3 Fast Fourier Transforms

The calculation of the discrete Fourier transform would appear to be very straightforward and, indeed, it is. However, the number of arithmetic operations (additions and multiplications) necessary to compute even a moderately large discrete Fourier transform can be quite excessive. To overcome this difficulty, various clever algorithms have been developed for the efficient calculation of the discrete Fourier transform. These algorithms are collectively known as fast Fourier transforms (FFTs).

The simplest FFT is probably the decimation-in-time algorithm. In this algorithm N is required to be a power of two. In deriving this FFT algorithm, the length-N transform is decomposed into two length-$N/2$ transforms. Each of these is decomposed into two length-$N/4$ transforms. This process is continued until one obtains all length-2 transforms. Because of their shape when represented in flow graph form, these simple length-2 calculations are known as butterflies. For the decimation-in-time FFT algorithm, the number of arithmetic calculations is proportional to $N \log_2 N$. This compares with N^2 in the case of

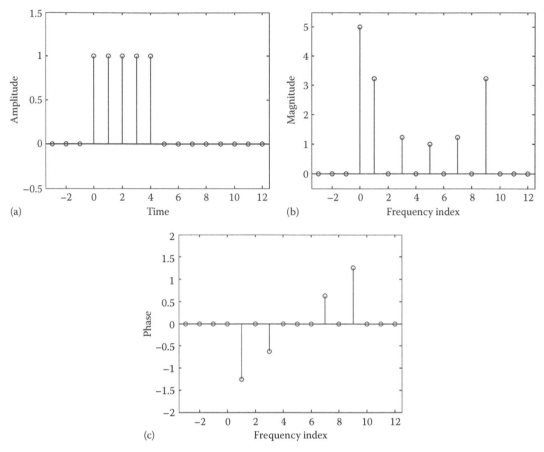

FIGURE 25.10 Discrete Fourier transform of a finite-duration signal: (a) signal, (b) magnitude spectrum, and (c) phase spectrum.

the direct evaluation of the discrete Fourier transform. For large N, the savings in computation can be very significant. For instance, for $N = 2^{10} = 1024$, the number of computations is reduced by a factor of more than 100.

25.12.4 Discrete-Time Fourier Transforms

Periodic discrete-time signals may be represented by the discrete Fourier series and finite-duration discrete-time signals by the discrete Fourier transform. A larger class of discrete-time signals may be represented by the discrete-time Fourier transform.

A discrete-time signal $x[n]$ will have a discrete-time Fourier transform if it is absolutely summable, i.e., if

$$\sum_{n=-\infty}^{\infty} |x[n]| < \infty.$$

With proper interpretation, certain signals not satisfying this property will also have discrete-time Fourier transforms. These include steps, constants, and all periodic signals.

The discrete-time Fourier transform of $x[n]$ is defined as

$$X(e^{j\omega}) = DTFT\{x[n]\} = \sum_{n=-\infty}^{\infty} x[n]e^{-j\omega n}, \quad -\infty < \omega < \infty.$$

It is noted that ω is a continuously varying real quantity, not an integer, despite the fact that $x[n]$ is a discrete-time signal. $x[n]$ may be recovered uniquely from $X(e^{j\omega})$ using the inverse discrete-time Fourier transform given by

$$x[n] = DTFT^{-1}\{X(e^{j\omega})\} = \frac{1}{2\pi} \int_{-\pi}^{\pi} X(e^{j\omega})e^{j\omega n} d\omega.$$

Note that we have written $X(e^{j\omega})$ as a function of $e^{j\omega}$ for convenience, rather than as a function of ω. Some common discrete-time Fourier transforms are listed in the following table.

$x[n]$	$X(e^{j\omega})$
$\delta[n]$	1
1	$2\pi \sum_{m=-\infty}^{\infty} \delta(\omega - 2\pi m)$
$u[n]$	$\dfrac{1}{1-e^{-j\omega}} + \pi \sum_{m=-\infty}^{\infty} \delta(\omega - 2\pi m)$
$a^n u[n],\ |a|<1$	$\dfrac{1}{1-ae^{-j\omega}}$
$na^n u[n],\ |a| < 1$	$\dfrac{ae^{-j\omega}}{(1-ae^{-j\omega})^2}$
$\sin(\omega_o n)$	$j\pi \sum_{m=-\infty}^{\infty} [\delta(\omega + \omega_o - 2\pi m) - \delta(\omega - \omega_o - 2\pi m)]$
$\cos(\omega_o n)$	$\pi \sum_{m=-\infty}^{\infty} [\delta(\omega + \omega_o - 2\pi m) + \delta(\omega - \omega_o - 2\pi m)]$

The quantity $X(e^{j\omega})$ is periodic in ω with a period of 2π. In fact, the expression for $X(e^{j\omega})$ is seen to be an exponential Fourier series with coefficients $x[-n]$ and with ω interpreted as the independent variable. While periodic and finite-duration signals have all their energy concentrated at N distinct frequencies, other signals allowed by the discrete-time Fourier transform have their energy spread out over, in general, all angular frequencies in the range $-\pi$ to π (or any range of width 2π).

The discrete-time Fourier transform is, in general, complex valued and may, thus, be written as

$$X(e^{j\omega}) = \left|X(e^{j\omega})\right| \angle\varphi(e^{j\omega}) = \left|X(e^{j\omega})\right| e^{j\varphi(e^{j\omega})},$$

where $\varphi(e^{j\omega}) = \arg\{X(e^{j\omega})\}$. For our assumed real-valued $x[n]$, $|X(e^{j\omega})|$ is an even function of ω and $\varphi(e^{j\omega})$ is an odd function of ω, i.e., respectively,

$$\left|X(e^{-j\omega})\right| = \left|X(e^{j\omega})\right|, \quad \varphi(e^{-j\omega}) = -\varphi(e^{j\omega}).$$

Thus, the information contained in $X(e^{j\omega})$, $\omega < 0$, is redundant. In addition, since $X(e^{j\omega})$ is periodic with period 2π, it may be concluded that $X(e^{j\omega})$ is completely determined by $X(e^{j\omega})$, $0 \le \omega \le \pi$. When plotted

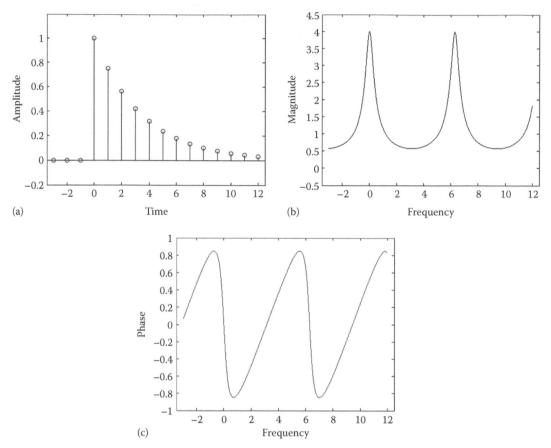

FIGURE 25.11 Discrete-time Fourier transform of $x[n] = (0.75)^n u[n]$: (a) signal, (b) magnitude spectrum, and (c) phase spectrum.

versus ω, $\left| X(e^{j\omega}) \right|$ is called the magnitude spectrum and $\varphi(e^{j\omega})$ the phase spectrum. An example is shown in Figure 25.11.

Signals not satisfying the absolute summability condition stated above will, if they are discrete-time Fourier transformable, normally contain frequency-domain impulses in their transforms. As a specific case, if $x[n]$ is a periodic signal with period N and with discrete Fourier series coefficients $X[k] = DFS\{x[n]\}$, then its discrete-time Fourier transform is given by

$$X(e^{j\omega}) = \frac{2\pi}{N} \sum_{m=-\infty}^{\infty} \sum_{k=0}^{N-1} X[k]\delta\left(\omega - \frac{2\pi}{N}k + 2\pi m \right).$$

For ω in the range $0 \le \omega < 2\pi$, this reduces to

$$X(e^{j\omega}) = \frac{2\pi}{N} \sum_{k=0}^{N-1} X[k]\delta\left(\omega - \frac{2\pi}{N}k \right),$$

with all energy clearly still concentrated (now, mathematically, by means of impulse functions) at the N frequencies $2\pi k/N$, $k = 0, 1, 2, \ldots, N - 1$.

Some important properties of the discrete-time Fourier transform are summarized in the following table:

Discrete-Time Signals	Discrete-Time Fourier Transforms
$Ax_1[n] + Bx_2[n]$	$AX_1(e^{j\omega}) + BX_2(e^{j\omega})$
$x[n - n_o]$	$X(e^{j\omega})e^{-j\omega n_o}$
$x[-n]$	$X(e^{-j\omega})$
$x_1[n] * x_2[n]$	$X_1(e^{j\omega})X_2(e^{j\omega})$

$x[n]$, $x_1[n]$, $x_2[n]$ are arbitrary discrete-time signals with discrete Fourier transforms $X(e^{j\omega})$, $X_1(e^{j\omega})$, $X_2(e^{j\omega})$, respectively. A, B are arbitrary constants and n_o is an arbitrary integer constant.

25.12.5 z-Transforms

As seen above, the discrete-time Fourier transform is a generalization of the discrete Fourier series, in that it allows for the analysis of a broader range of discrete-time signals than does the discrete Fourier series. The z-transform is likewise a generalization of the discrete-time Fourier transform.

A signal $x[n]$ will have a z-transform if

$$\sum_{n=-\infty}^{\infty} \left| x[n]r^{-n} \right| < \infty$$

for some real $r > 0$. Since $r = 1$ is allowed in this condition, if $x[n]$ has a discrete-time Fourier transform, it will also have a z-transform, but the converse is not true.

The z-transform of a signal $x[n]$ is defined as

$$X(z) = ZTx[n] = \sum_{n=-\infty}^{\infty} x[n]z^{-n},$$

where $z = re^{j\omega}$ is a complex variable. In general, this sum will converge only for certain choices of r. $x[n]$ may be recovered uniquely from $X(z)$ using the inverse z-transform given by

$$x[n] = ZT^{-1}X(z) = \frac{1}{2\pi j} \oint_C X(z)z^{n-1}dz,$$

where C is an appropriately chosen contour in the complex z plane. (This formula is rarely directly used in practice because of its complexity. Instead, tables of z-transform pairs are used, with complicated transforms first being decomposed into simpler ones using partial fraction expansion techniques.) Some common z-transforms are listed in the following table:

$x[n]$	$X(z)$
$\delta[n]$	1
$u[n]$	$\dfrac{z}{z-1}, \quad |z|>1$
$a^n u[n]$	$\dfrac{z}{z-a}, \quad |z|>|a|$
$na^n u[n]$	$\dfrac{az}{(z-a)^2}, \quad |z|>|a|$
$\sin(\omega_o n)u[n]$	$\dfrac{z\sin(\omega_o)}{z^2-2z\cos(\omega_o)+1}, \quad |z|>1$
$\cos(\omega_o n)u[n]$	$\dfrac{z^2-z\cos(\omega_o)}{z^2-2z\cos(\omega_o)+1}, \quad |z|>1$

For a signal $x[n]$ having a discrete-time Fourier transform $X(e^{j\omega})$, it is the case that

$$DTFT\{x[n]\} = ZT\{x[n]\}\big|_{r=1}$$

or

$$X(e^{j\omega}) = X(z)\big|_{z=e^{j\omega}},$$

where $X(z)$ is the z-transform of $x[n]$. (This explains our choice of notation for the discrete-time Fourier transform.) Thus, the discrete-time Fourier transform is simply the z-transform evaluated on the unit (radius one) circle in the z (complex) plane.

If the signal $x[n]$ is additionally assumed to be finite duration with length N, then

$$DFT\{x[n]\} = DTFT\{x[n]\}\big|_{\omega=(2\pi/N)k}$$

or

$$X[k] = X(e^{j\omega})\big|_{\omega=(2\pi/N)k},$$

where $X[k]$ is the discrete Fourier transform of $x[n]$. That is, the discrete Fourier transform is the discrete-time Fourier transform evaluated at equally spaced frequency values (or the z-transform evaluated at equally spaced points around the unit circle).

Some important properties of the z-transform are summarized in the following table:

Discrete-Time Signals	z-Transforms
$Ax_1[n] + Bx_2[n]$	$AX_1(z) + BX_2(z)$
$x[n - n_o]$	$X(z)z^{-n_o}$
$x[-n]$	$X(z^{-1})$
$x_1[n] * x_2[n]$	$X_1(z)X_2(z)$

$x[n]$, $x_1[n]$, $x_2[n]$ are arbitrary discrete-time signals with z-transforms $X(z)$, $X_1(z)$, $X_2(z)$, respectively. A, B are arbitrary constants and n_o is an arbitrary integer constant.

25.13 Discrete-Time Signal Processors

FIGURE 25.12 Discrete-time signal processing system.

A discrete-time signal processor (system) is a device or algorithm (see Figure 25.12) that acts on an input signal $x[n]$, modifying it in some manner to produce an output signal $y[n]$. This may be represented abstractly as

$$y[n] = \mathcal{H}\{x[n]\}.$$

(Here, $x[n]$ and $y[n]$ should be thought of in their totality, rather than at specific instants n. Despite appearances, this notation does not imply that $y[n]$ is a function of $x[n]$ at only the same instant n. $y[n]$ could, e.g., depend on all past values of $x[m]$, $m \leq n$.)

Only certain relatively simple classes of discrete-time signal processors are considered in practice. In this treatment, we will discuss only those that can be described by linear constant-coefficient difference equations. Such signal processors are of necessity linear, time invariant, and causal. They may be implemented, in hardware or software, using only summers, constant multipliers, and delay (or memory) elements. A system is linear if

$$\mathcal{H}\{Ax_1[n] + Bx_2[n]\} = A\mathcal{H}\{x_1[n]\} + B\mathcal{H}\{x_2[n]\}, \quad -\infty < n < \infty,$$

for all signals $x_1[n]$, $x_2[n]$ and all constants A, B. It is time invariant if

$$y[n - n_o] = \mathcal{H}\{x[n - n_o]\}, \quad -\infty < n < \infty,$$

for all input signals $x[n]$ and corresponding output signals $y[n]$ and all integer constants n_o. A system is casual if $\mathcal{H}\{x[n]\}$ depends only on $x[m]$, $m \leq n$, for all signals $x[n]$.

25.14 Time-Domain Analysis of Discrete-Time Signal Processors

Under our assumptions, a discrete-time signal processor may be represented by a linear constant-coefficient difference equation. It may seem natural to write such an equation in terms of the differences in the input and output, and this may be done. However, it is customary, and more convenient, to simply use delayed versions of the input and output. The equation then takes the form

$$\sum_{r=0}^{N} a_r y[n - r] = \sum_{r=0}^{M} b_r x[n - r].$$

Often, this equation is normalized so that $a_0 = 1$. The equation may thus be solved for $y[n]$ in terms of current and past values of $x[n]$ and past values of $y[n]$. As such, it may be implemented directly and solved for $y[n]$ recursively, provided that $x[n]$ and a sufficient number of initial values of $y[n]$ are known. This equation may also be solved using z-transforms and other techniques.

An alternative means for describing our signal processor is based on the use of its impulse response. The impulse response is defined as

$$h[n] = \mathcal{H}\{\delta[n]\},$$

i.e., the output when the input is a unit impulse. Since we have assumed causality, it may be shown that $h[n]$ is a positive-time signal, i.e., that $h[n] = 0$, $n < 0$. Because the signal processor is linear and time invariant, it may additionally be shown that for any input $x[n]$, the corresponding output $y[n]$ is given by the convolution of $h[n]$ and $x[n]$, namely,

$$y[n] = \mathcal{H}\{x[n]\} = h[n] * x[n].$$

This is a very significant result, since it shows that all information regarding the signal processor (at least, from an input–output viewpoint) is contained in its impulse response.

A very desirable property of signal processors is stability. A system is said to be (bounded-input bounded-output) stable if whenever $x[n]$ is bounded (i.e., $|x[n]| \leq K_x < \infty$, $-\infty < n < \infty$, for some K_x), then the corresponding $y[n]$ is also bounded (i.e., $|y[n]| \leq K_y < \infty$, $-\infty < n < \infty$, for some K_y). (The alternative would normally be undesirable.) It is possible to easily determine stability directly from $h[n]$. Specifically, the signal processor is stable if and only if $h[n]$ is absolutely summable, i.e., if and only if

$$\sum_{n=-\infty}^{\infty} |h[n]| < \infty.$$

Unlike continuous-time systems, there are two distinctly different types of discrete-time systems, depending on the nature of $h[n]$. If $h[n]$ is a finite-duration signal, then the system is said to be FIR (i.e., to have a finite impulse response). If this is not the case, then the system is IIR (i.e., having an infinite impulse response). For an FIR system, the difference equation reduces to

$$y[n] = \sum_{r=0}^{M} b_r x[n-r],$$

and $h[n]$ is a finite-duration signal of length $M + 1$ with

$$h[n] = \begin{cases} b_n, & 0 \leq n \leq M \\ 0, & \text{otherwise} \end{cases}.$$

Since their impulse responses are clearly always absolutely summable, FIR signal processors are always stable.

25.15 Frequency-Domain Analysis of Discrete-Time Signal Processors

In the frequency (transform) domain, a signal processor is characterized by its transfer function $H(z)$. $H(z)$ may be determined by z-transforming the convolution relationship to obtain

$$Y(z) = H(z)X(z),$$

where

$$H(z) = ZT\{h[n]\}.$$

Thus, the transfer function is the z-transform of the impulse response.

Alternatively, the input–output difference equation may be z-transformed to obtain

$$\sum_{r=0}^{N} a_r z^{-r} Y(z) = \sum_{r=0}^{M} b_r z^{-r} X(z),$$

from which

$$H(z) = \frac{\sum_{r=0}^{M} b_r z^{-r}}{\sum_{r=0}^{N} a_r z^{-r}}.$$

Thus, $H(z)$ is a rational function in z^{-1}, which, if preferred, may also be written as a rational function in z. As such, the impulse response $h[n]$ will necessarily consist only of terms of the form

$$An^p a^n \cos(\omega n + \theta)u[n],$$

where $p \geq 0$.

When written as polynomials in z, the roots of the numerator are the zeros of $H(z)$ and those of the denominator are the poles of $H(z)$. It may be shown that a signal processor is stable if and only if all of its poles lie strictly inside the unit circle in the complex z plane (i.e., have magnitudes less than one). It is noted that all poles of an FIR signal processor are at the origin, which is thus seen to be consistent with our previous comment that such systems are always stable.

Discrete-time Fourier transforming the convolution relationship leads to

$$Y(e^{j\omega}) = H(e^{j\omega})X(e^{j\omega}),$$

where

$$H(e^{j\omega}) = DTFT\{h[n]\}.$$

$H(e^{j\omega})$ is called the frequency response of the signal processor, and is the discrete-time Fourier transform of the impulse response. The magnitude of $H(e^{j\omega})$ is the magnitude response and its angle is the phase response. It is seen that

$$H(e^{j\omega}) = H(z)\big|_{z=e^{j\omega}},$$

where $H(z)$ is the transfer function. That is, the frequency response is the transfer function evaluated on the unit circle in the complex z plane. Thus, we may also write

$$H(e^{j\omega}) = \frac{\sum_{r=0}^{M} b_r e^{-j\omega r}}{\sum_{r=0}^{N} a_r e^{-j\omega r}}.$$

The frequency response of a signal processor is also useful for determining its output in the case of a periodic input. If $x[n]$ is a periodic input signal with period N (not the same N used in the difference equation) and with discrete Fourier series coefficients $X[k]$, then the output signal will also be periodic with period N. Its discrete Fourier series coefficients will be given by

$$Y[k] = H(e^{j(2\pi/N)k})X[k].$$

This relationship may also be used for the steady-state analysis of systems, i.e., systems that have been operating for a sufficiently long time that all transient terms may be neglected.

It is thus clear that the way a signal processor processes an input signal (either aperiodic or periodic) is determined by its frequency response. It is thus very informative to have this quantity available, particularly in visual form as embodied in its magnitude and phase spectra.

25.16 Discrete-Time (Digital) Filters

Just as in the case of continuous-time (analog) filters, a discrete-time (digital) filter is a signal processor designed to allow signal components of certain frequencies to pass through to the output while preventing input signal components of other frequencies from doing so. The frequency response $H(e^{j\omega})$ should ideally have magnitude 1 in the passbands and magnitude 0 in the stopbands.

25.16.1 Common Filter Types

The common types of discrete-time filters are the same as those for continuous-time filters, the only difference being that the frequency response $H(e^{j\omega})$ is completely specified by its values in the frequency range $0 \le \omega \le \pi$ as opposed to $0 \le \omega < \infty$. The comments made regarding ideal and nonideal (practical) continuous-time filters apply directly to discrete-time filters as well.

Some significant differences with continuous-time filters exist regarding linear phase. Again, linear phase ensures that the shape of a signal is preserved in the filtering process. In discrete-time filters, however, true linear phase may in fact be achieved, but only if the filter is FIR. The FIR filter

$$H(z) = \sum_{r=0}^{M} b_r z^{-r}$$

will have linear phase if and only if

$$b_r = b_{M-r}, \quad 0 \le r \le M,$$

i.e., if and only if the coefficients b_r are symmetric. It is important to point out that if M is odd, then $H(z)$ will have a zero at $z = -1$, or equivalently $H(e^{j\omega})$ will be zero at $\omega = \pi$. This rules out the use of odd values of M in this situation if it is desired to construct a highpass or bandstop filter.

25.16.2 FIR Filter Design

Different techniques are employed for the design of FIR and IIR discrete-time filters, so these will be discussed separately. For simplicity, in the case of FIR filters, we will only consider linear-phase filters with M even.

One method for designing FIR filters is known as windowing. This procedure begins by specifying the impulse response $h_i[n]$ of the desired ideal filter, assumed to have zero phase. Anticipating an even M, this quantity is given in the following table for the common filter types, with cutoff frequencies ω_c, ω_l, and ω_u.

Filter Type	$h_i[0]$	$h_i[n], n \neq 0$
Lowpass	$\dfrac{\omega_c}{\pi}$	$\dfrac{\sin(n\omega_c)}{n\pi}$
Highpass	$1 - \dfrac{\omega_c}{\pi}$	$\dfrac{-\sin(n\omega_c)}{n\pi}$
Bandpass	$\dfrac{\omega_u - \omega_l}{\pi}$	$\dfrac{\sin(n\omega_u) - \sin(n\omega_l)}{n\pi}$
Bandstop	$1 - \dfrac{\omega_u - \omega_l}{\pi}$	$\dfrac{\sin(n\omega_l) - \sin(n\omega_u)}{n\pi}$

A simple way to obtain an FIR filter from $h_i[n]$ is to truncate it at $n = \pm M$ and then shift the response to the right by $M/2$ to result in a casual linear-phase FIR filter. However, since it is recalled that the $h_i[n]$ (actually $h_i[-n]$) are the Fourier series coefficients of $H_i(e^{j\omega})$ (with ω interpreted as the independent variable), this process amounts to Fourier series truncation, with its attendant Gibb's phenomenon oscillations.

Since these oscillations are normally unacceptable, a better approach is to reduce the effects of Gibb's phenomenon by multiplying the ideal impulse response by a tapered function known as a window, after first shifting the response to the right to produce causality. The impulse response $h[n]$ of the resulting filter is given by

$$h[n] = h_i\left[n - \frac{M}{2}\right]w[n], \quad -\infty < n < \infty,$$

where $w[n]$ is a symmetric window satisfying

$$w[n] = \begin{cases} w[M-n], & 0 \le n \le M \\ 0, & \text{otherwise} \end{cases}.$$

The resulting filter is then

$$H(z) = \sum_{r=0}^{M} b_r z^{-r},$$

where $b_r = h[r], 0 \le r \le M$.

Numerous choices for windows exist. Some commonly used windows are listed in the following table and shown in Figure 25.13:

Window	$w[n], \ 0 \le n \le M$
Rectangular	1
Bartlett	$1 - \dfrac{2}{M}\left\|n - \dfrac{M}{2}\right\|$
Hanning	$\dfrac{1}{2} - \dfrac{1}{2}\cos\left(\dfrac{2\pi n}{M}\right)$
Hamming	$0.54 - 0.46\cos\left(\dfrac{2\pi n}{M}\right)$

Note that $w[n] = 0$ outside of this range.

It is noted that the use of a rectangular window is equivalent to simply truncating the Fourier series. As we move down in the table, the windows are increasingly better at reducing the oscillations caused by Gibb's phenomenon. Generally, windows that are good in reducing these oscillations do so at the expense of increased transition bandwidths.

As an example of using windowing to design a linear-phase FIR filter, suppose we wish to construct a highpass filter with a cutoff frequency $\omega_c = \pi/4$ and a length $M = 30$. Choosing a Hanning window and following the procedure just discussed results in the filter

$$H(z) = \sum_{r=0}^{30} b_r z^{-r},$$

where

$$b_r = \frac{\sin\left((r-15)(\pi/4)\right)}{2(r-15)\pi}\left[\cos\left(\frac{\pi r}{15}\right) - 1\right], \quad r = 0,1, \ldots, 30, r \ne 15,$$

and $b_{15} = 3/4$. Figure 25.14 illustrates the results of this design procedure.

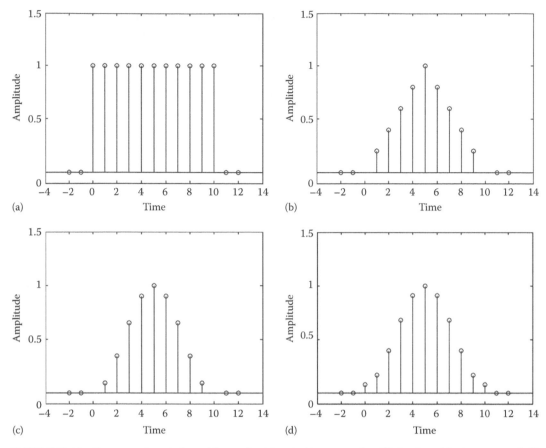

FIGURE 25.13 Four popular windows (shown for $M = 10$): (a) rectangular, (b) Bartlett, (c) Hanning, and (d) Hamming.

While windowing has the advantage of being very simple, it has some significant drawbacks. There is the fact that there is no direct link with the tolerances specified by the designer and the fact that the tolerances in the various filter bands are not independently controllable.

A very popular method for FIR filter design, which does not suffer from these drawbacks, is the Parks-McClellan algorithm. Readily available and easy-to-use computer implementations of this algorithm exist. The Parks-McClellan algorithm results in FIR filters that are equiripple in all passbands and stopbands. The designer simply specifies the passband and stopband edges, the desired magnitude response in each band, the relative tolerances, and the filter order. The algorithm produces the resulting filter coefficients. Since the magnitude response in the transition bands is not the concern of the algorithm, the magnitude response in these regions must be separately checked to ensure acceptability.

25.16.3 IIR Filter Design

The design of an IIR discrete-time filter usually involves the design of a continuous-time lowpass prototype filter, its transformation to a lowpass IIR discrete-time filter, and, if necessary, a frequency transformation to produce the desired filter type. When simultaneously discussing continuous-time and discrete-time quantities, we will use Ω to represent continuous-time angular frequency while retaining ω for discrete-time angular frequency.

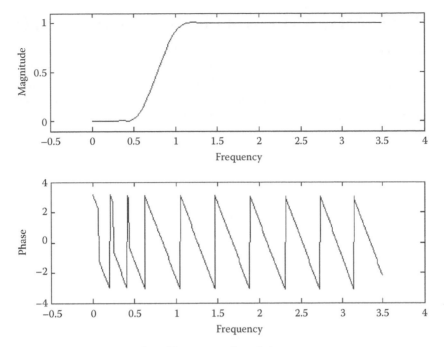

FIGURE 25.14 Highpass FIR linear-phase filter magnitude and phase responses.

The design process begins by establishing the discrete-time filter specifications in precisely the same manner as in the case of continuous-time filters. These specifications are then translated to equivalent specifications on the continuous-time filter using the formula

$$\Omega = \tan\left(\frac{\omega}{2}\right)$$

to translate specific frequency values such as cutoff frequencies and band edges. Using techniques discussed earlier, a continuous-time lowpass prototype filter $H_p(s)$ is designed to meet the translated specifications. This filter is then transformed to a lowpass IIR discrete-time filter using the formula

$$H_{lp}(z) = H_p(s)\big|_{s=B(z)},$$

where

$$B(z) = \frac{z-1}{z+1}.$$

$B(z)$ is known as a bilinear transformation.

If a lowpass filter is desired, the design process is complete. If a highpass discrete-time filter $H_{hp}(z)$ is desired, it may easily be determined using the frequency transformation formula

$$H_{hp}(z) = H_{lp}(-z).$$

Assuming that the cutoff frequency of $H_{lp}(z)$ is ω_c, this results in $H_{hp}(z)$ having a cutoff frequency of $\pi - \omega_c$. Bandpass and bandstop filters may also be obtained in this manner, but these require considerably more complicated frequency transformations that double the order of the filter.

25.17 Discrete-Time Analysis of Continuous-Time Signals

Because of their convenience and efficiency, discrete-time techniques are often employed in the analysis of continuous-time signals. In particular, an FFT is often used as a spectral analysis tool to determine the frequency content of a continuous-time signal. It is thus important to consider the relationship between continuous-time and discrete-time Fourier transforms. As in the previous section, Ω will be used for continuous-time angular frequency and ω for discrete-time angular frequency.

As discussed earlier, if the continuous-time signal $x(t)$, with Fourier transform $X(j\Omega)$, is impulse sampled with an angular sampling frequency Ω_s, a continuous-time signal $x_s(t)$ is produced with Fourier transform

$$X_s(j\Omega) = \frac{1}{T_s} \sum_{m=-\infty}^{\infty} X(j(\Omega - m\Omega_s)),$$

where
$\Omega_s = 2\pi / T_s$
T_s is the sampling period

Now, $x[n] = S\{x(t)\} = x(nT_s)$, i.e., the sampled version of $x(t)$, is a discrete-time signal. As such it will have a discrete-time Fourier transform $X(e^{j\omega})$, which may be shown to be given by

$$X(e^{j\omega}) = X_s\left(j\frac{\omega}{T_s}\right) = \frac{1}{T_s} \sum_{m=-\infty}^{\infty} X\left(j\left(\frac{\omega}{T_s} - m\Omega_s\right)\right).$$

As is true of all discrete-time Fourier transforms, $X(e^{j\omega})$ is periodic with period 2π, and is hence completely specified by its values in the range $-\pi \le \omega \le \pi$. Thus, if the original signal is bandlimited and if sampling is carried out at the Nyquist rate or greater, no aliasing will occur and this expression reduces to

$$X(e^{j\omega}) = X_s\left(j\frac{\omega}{T_s}\right) = \frac{1}{T_s} X\left(j\frac{\omega}{T_s}\right), \quad -\pi \le \omega \le \pi.$$

This is justified since, as a result of the absence of overlap, only the $m = 0$ term in the infinite sum will be present in this frequency range. This shows that, under the circumstances stated (namely, no aliasing), the discrete-time Fourier transform $X(e^{j\omega})$ of the discrete-time signal $x[n] = x(nT_s)$ is simply an amplitude- and a frequency-scaled version of the Fourier transform $X(j\Omega)$ of the continuous-time signal $x(t)$, with the frequencies related by $\Omega = \omega / T_s$.

Now, if the discrete-time signal $x[n] = x(nT_s)$ is additionally assumed to be of finite duration with length N, then, as stated earlier, the discrete Fourier transform $X[k]$ of $x[n]$ is given by

$$X[k] = X(e^{j(2\pi/N)k}), \quad k = 0,1,2, \ldots, N-1,$$

and hence

$$X[k] = \frac{1}{T_s} X\left(j\frac{2\pi}{NT_s}k\right), \quad k = 0,1,2, \ldots, N-1.$$

That is, the discrete Fourier transform values are amplitude-scaled samples of the continuous-time Fourier transform $X(j\Omega)$ at the frequency values

$$\Omega = \frac{2\pi}{NT_s}k, \quad k = 0,1,2, ..., N-1.$$

(Because of redundancy, only the first half of these values, for $k = 0,1,2,...,(N/2)-1$, are useful. Here N is assumed to be even.) This relationship clearly justifies the utility of using an FFT (to evaluate the discrete Fourier transform) in performing the spectral analysis of a continuous-time signal.

Unfortunately, the assumption that $x[n]$ is of finite duration is at odds with the assumption that $x(t)$ is bandlimited. If $x(t)$ is bandlimited, it, and hence $x[n]$, must be of infinite duration. Thus, in reality, the expression for $X[k]$ is an approximation. However, this approximation may still be quite useful if applied carefully.

Several comments are in order regarding the use of an FFT for spectral analysis. Prior to sampling, it is often beneficial to pass the original signal through a continuous-time lowpass filter. Such an "anti-aliasing" filter helps to ensure that the signal to be sampled is bandlimited and free of high-frequency noise.

Additional zero-valued samples are sometimes appended to the end of the discrete-time signal obtained by the sampling process. This "zero padding" may be done to provide a signal length equal to a power of two for use with an FFT. It may also be done to increase the resolution of the Fourier transform. This may be seen by observing that, in the expression for $X[k]$, increasing N both provides more frequency samples and more closely spaced frequency samples.

Prior to taking an FFT, the discrete-time signal $x[n]$ may first be multiplied by a window, $w[n]$, of the type discussed earlier. While this certainly alters the frequency content of the signal, this effect may be outweighed by the benefit gained due to the tapered nature of the window, which reduces unnatural discontinuities introduced by artificially time limiting the signal.

Because of the frequency sampling that occurs when using an FFT, a phenomenon known as "leakage" may sometimes occur. This happens when a frequency component of the original signal falls between two frequencies of the FFT. The "energy" in this component is then distributed ("leaks") to nearly frequencies, thereby somewhat obscuring the true frequency component.

Finally, it should be noted that other more sophisticated and generally better methods for spectral analysis exist, but these are beyond the scope of this discussion.

25.18 Discrete-Time Processing of Continuous-Time Signals

Because of the desirable properties of discrete-time systems, they are often used in circumstances in which the goal is to process (filter) a continuous-time input to produce a continuous-time output. This involves sampling to produce a discrete-time signal, discrete-time filtering, and, finally, the conversion of the resulting discrete-time signal to a continuous-time signal. Since the input and output signals are continuous time, the process is, in fact, equivalent to continuous-time filtering (even though it is implemented in discrete time).

Ordinarily, such a process begins by passing the original signal through a continuous-time anti-aliasing filter to ensure that the signal $x(t)$ to be processed is bandlimited (or, at least, nearly so). The signal $x(t)$ is then sampled at the Nyquist rate (or greater) to produce the discrete-time signal $x[n] = x(nT_s)$. This signal is then passed through the discrete-time filter $H(e^{j\omega})$ to produce an output $y[n]$. The signal $y[n]$ is then converted to a continuous-time signal $y(t)$ in such a manner that $y(nT_s) = y[n]$. This process (of digital-to-analog conversion) may be viewed as consisting of first generating $y_s(t)$, an impulse-sampled version of $y[n]$. The frequency response of $y_s(t)$ is given by

$$Y_s(j\Omega) = \frac{1}{T_s} \sum_{m=-\infty}^{\infty} Y(j(\Omega - m\Omega_s)).$$

To complete the process, the signal $y_s(t)$ is passed through a "reconstruction" filter (a continuous-time lowpass filter with a gain of T_s in the passband) to remove all but the $m = 0$ term in the above expression.

Under these circumstances, it may be shown that the equivalent continuous-time filter $H_e(j\Omega)$ (with input $x(t)$ and output $y(t)$) has frequency response

$$H_e(j\Omega) = \begin{cases} H(e^{j\Omega T_s}), & -\dfrac{\pi}{T_s} < \Omega < \dfrac{\pi}{T_s} \\ 0, & \text{otherwise} \end{cases}.$$

That is, in the range $-(\pi/T_s) < \Omega < (\pi/T_s)$, $H_e(j\Omega)$ is simply a frequency-scaled version of the discrete-time frequency response $H(e^{j\omega})$, where the frequencies are related by $\omega = \Omega T_s$ or $\Omega = \omega/T_s$. It is noted that because of the requirement for bandlimiting, true highpass and bandstop filters cannot be constructed in this manner.

References

Ambardar, A. 2007. *Digital Signal Processing: A Modern Introduction*, Nelson-Thomson Canada, Toronto, Canada.

Antoniou, A. 1993. *Digital Filters: Analysis, Design, and Applications*, 2nd edn., McGraw-Hill, New York.

Baher, H. 2001. *Analog and Digital Signal Processing*, 2nd edn., Wiley, Hoboken, NJ.

Chen, C.-T. 2004. *Signals and Systems*, 3rd edn., Oxford University Press, New York.

ElAli, T. S. 2004. *Discrete Systems and Digital Signal Processing with MATLAB®*, CRC Press, Boca Raton, FL.

Gajić, Z. 2003. *Linear Dynamic Systems and Signals*, Pearson Higher Education, Upper Saddle River, NJ.

Ingle, V. J. and Proakis, J. G. 2007. *Digital Signal Processing Using MATLAB®*, 2nd edn., Nelson-Thomson Canada, Toronto, Canada.

Kuo, S. M. and Gan, W.-S. 2005. *Digital Signal Processors: Architectures, Implementations, and Applications*, Pearson Higher Education, Upper Saddle River, NJ.

Lam, H. Y.-F. 1979. *Analog and Digital Filters: Design and Realization*, Prentice-Hall, Englewood Cliffs, NJ.

Lathi, B. P. 2005. *Linear Systems and Signals*, 2nd edn., Oxford University Press, New York.

McClellan, J. H., Schafer, R. W., and Yoder, M. A. 2003. *Signal Processing First*, Pearson Higher Education, Upper Saddle River, NJ.

Mitra, S. J. 2006. *Digital Signal Processing: A Computer Based Approach*, 3rd edn., McGraw-Hill, New York.

Oppenheim, A. V. and Schafer, R. W. 1989. *Discrete-Time Signal Processing*, Prentice-Hall, Englewood Cliffs, NJ.

Oppenheim, A. V. and Schafer, R. W. 2010. *Discrete-Time Signal Processing*, 3rd edn., Pearson Higher Education, Upper Saddle River, NJ.

Paarmann, L. D. 2001. *Design and Analysis of Analog Filters: A Signal Processing Perspective*, Springer, Berlin, Germany.

Proakis, J. G. and Manolakis, D. G. 2007. *Digital Signal Processing: Principles, Algorithms, and Applications*, 4th edn., Pearson Higher Education, Upper Saddle River, NJ.

Schlichthärle, D. 2000. *Digital Filters: Basics and Design*, Springer, Berlin, Germany.

Sherrick, J. D. 2005. *Concepts in Systems and Signals*, 2nd edn., Prentice-Hall, Englewood Cliffs, NJ.

Stearns, S. D. 2003. *Digital Signal Processing with Examples in MATLAB®*, CRC Press, Boca Raton, FL.

26

Analog Filter Synthesis

Nam Pham
Auburn University

Bogdan M.
Wilamowski
Auburn University

26.1 Introduction

Analog filters are essential in many different systems that electrical engineers are required to design in their engineering career. Filters are widely used in communication technology as well as in other applications. Although we discuss and talk a lot about digital systems nowadays, these systems always contain one or more analog filters internally or an interface with the analog world [SV01].

There are many different types of filters such as Butterworth filter, Chebyshev filter, inverse Chebyshev filter, Cauer elliptic filter, etc. The characteristic responses of these filters are different. The Butterworth filter is flat in the stop-band but does not have a sharp transition from the pass-band to the stop-band while the Chebyshev filter has a sharp transition from the pass-band to the stop-band but it has the ripples in the pass-band. Oppositely, the inverse Chebyshev filter works almost the same way as the Chebyshev filter, but it does have the ripples in the stop-band rather than the pass-band. The Cauer filter has ripples in both pass-band and stop-band; however, it has lower order [W02,KAS89]. The analog filter is a broad topic, and this chapter will focus more on the methodology of synthesizing analog filters only (Figures 26.1 and 26.2).

Section 26.2 will present methods to synthesize four different types of these low-pass filters. Then we will go through design example of a low-pass filter that has 3 dB attenuation in the pass-band, 30 dB attenuation in the stop-band, the pass-band frequency at 1000 rad/s and the stop-band frequency at 3000 rad/s to see four different results corresponding to four different synthesizing methods.

26.2 Methods to Synthesize Low-Pass Filter

26.2.1 Butterworth Low-Pass Filter

ω_p—pass-band frequency
ω_s—stop-band frequency
α_p—attenuation in pass-band
α_s—attenuation in stop-band

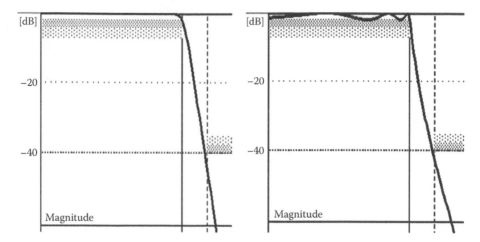

FIGURE 26.1   Butterworth filter (left), Chebyshev filter (right).

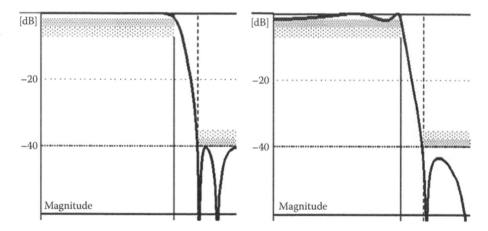

FIGURE 26.2   Inverse Chebyshev filter (left), Cauer elliptic filter (right).

Butterworth response (Figure 26.3):

$$\left|T(j\omega)\right|^2 = \frac{1}{1+\left(\omega^{2n}/\omega_0^{2n}\right)}$$

There are three basic steps to synthesize any type of low-pass filters. The first step is calculating the order of a low-pass filter. The second step is calculating poles and zeros of a low-pass filter. The third step is designing circuits to meet pole and zero locations; however, this part is another topic of analog filters, so it will be not be covered in this work [W90,WG05,WLS92].

Following are steps to design Butterworth low-pass filter:

Step 1—Calculate order of filter:

$$n = \frac{\log[(10^{\alpha_s/10}-1)(10^{\alpha_p/10}-1)]^{1/2}}{\log(\omega_s/\omega_p)} \quad (n \text{ needs to be rounded up to integer value})$$

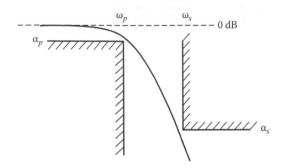

FIGURE 26.3 Butterworth filter characteristic.

Step 2—Calculate pole and zero locations:
Angle if n is odd:

$$\Omega = \pm \frac{k180°}{n}; \quad k = 0, 1, \ldots, \frac{n-1}{2}$$

Angle if n is even:

$$\Omega = \pm \left(0.5 + \frac{k}{n} \right) 180°; \quad k = 0, 1, \ldots, \frac{n-2}{2}$$

Normalized pole locations:

$$a_k = -\cos(\Omega); \quad b_k = \pm \sin(\Omega); \quad (\omega_0 = 1)$$

$$\omega_0 = \frac{(\omega_p \omega_s)^{1/2}}{[(10^{\alpha_s/10} - 1)/(10^{\alpha_p/10} - 1)]^{1/(4n)}}; \quad Q_k = \left| \frac{1}{2a_k} \right|$$

Step 3—Design circuits to meet pole and zero locations (not covered in this work) (Figure 26.4).

Example 26.1: Design the Low-Pass Butterworth Filter Assuming $\alpha_p = 3$ dB, $\alpha_s = 30$ dB, $\omega_p = 1000$ rad/s, $\omega_s = 3000$ rad/s

Step 1—Calculate order of filter:

$$n = \frac{\log[(10^{30/10} - 1)(10^{3/10} - 1)]^{1/2}}{\log(3000/1000)} = 3.1456 \Rightarrow n = 4$$

Step 2—Calculate pole and zero locations.
Normalized values of poles and ω_0 and Q:

$-0.38291 + 0.92443i$	1.00059	1.30656
$-0.38291 - 0.92443i$	1.00059	1.30656
$-0.92443 + 0.38291i$	1.00059	0.54120
$-0.92443 - 0.38291i$	1.00059	0.54120

Normalized values of zeros $\Rightarrow$ none.

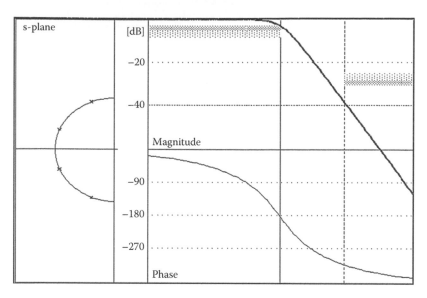

FIGURE 26.4 Pole-zero locations, magnitude response, and phase of Butterworth filter.

26.2.2 Chebyshev Low-Pass Filter

ω_p—pass-band frequency
ω_s—stop-band frequency
α_p—attenuation in pass-band
α_s—attenuation in stop-band

Chebyshev response (Figure 26.5):

$$|T(j\omega)|^2 = 1/(1+\varepsilon^2 C_n^2(\omega))$$

Step 1—Calculate order of filter:

$$n = \frac{\ln[4*(10^{\alpha_s/10}-1)/(10^{\alpha_p/10}-1)]^{1/2}}{\log[(\omega_s/\omega_p)+((\omega_s^2/\omega_p^2)-1)^{1/2}]} \quad (n \text{ needs to be rounded up to integer value})$$

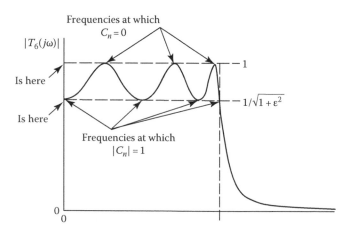

FIGURE 26.5 Chebyshev filter characteristic.

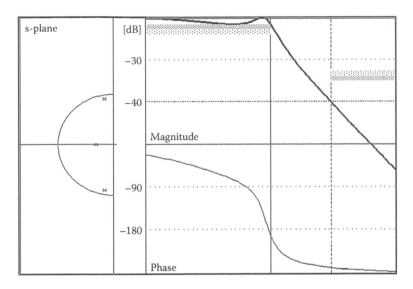

FIGURE 26.6 Pole-zero locations, magnitude response, and phase of Chebyshev filter.

Step 2—Calculate pole and zero locations:

$$\Omega = 90° + \frac{90°}{n} + \frac{(k-1)180°}{n}$$

$$\varepsilon = \left[10^{\alpha_p/10} - 1 \right]^{1/2}; \quad \gamma = \frac{\sinh^{-1}(1/\varepsilon)}{n}$$

$$a_k = \sinh(\gamma)\cos(\Omega); \quad b_k = \cosh(\gamma)\sin(\Omega); \quad \omega_k = a_k^2 + b_k^2; \quad Q_K = \left| \frac{\omega_k}{2a_k} \right|$$

Step 3—Design circuits to meet pole and zero locations (not covered in this work) (Figure 26.6).

Example 26.2: Design the Low-Pass Chebyshev Filter Assuming $\alpha_p = 3\,dB$, $\alpha_s = 20\,dB$, $\omega_p = 1000\,rad/s$, $\omega_s = 3000\,rad/s$

Step 1—Calculate order of filter:

$$n = \frac{\ln[4 * (10^{30/10} - 1)/(10^{3/10} - 1)]^{1/2}}{\log[(3000/1000) + ((3000^2/1000^2) - 1)^{1/2}]} = 2.3535 \Rightarrow n = 3$$

Step 2—Calculate pole and zero locations.
   Normalized values of poles and ω_0 and Q:

−0.14931 + 0.90381*i*	0.91606	3.06766
−0.14931 − 0.90381*i*	0.91606	3.06766
−0.29862		

Normalized values of zeros $\Rightarrow$ none.

26.2.3 Inverse Chebyshev Low-Pass Filter

ω_p—pass-band frequency
ω_s—stop-band frequency
α_p—attenuation in pass-band
α_s—attenuation in stop-band

Inverse Chebyshev response (Figure 26.7):

$$\left|T_{IC}(j\omega)\right|^2 = \frac{\varepsilon^2 C_n^2(1/\omega)}{1+\varepsilon^2 C_n^2(1/\omega)}$$

The method to design the inverse Chebyshev low-pass filter is almost the same as the Chebyshev low-pass filter. It is just slightly different.

Step 1—Calculate order of filter.
n = order of the Chebyshev filter

Step 2—Calculate pole and zero locations:

$$P_{ic} = \frac{1}{a_k + b_k}, \quad \text{find zeros} \quad \omega_i = \frac{1}{\cos[\Pi * i/(2n)]}; \quad i = 2k-1:1,3,5\ldots < np$$

Note: two conjugate poles on the imaginary axis.

Step 3—Design circuits to meet pole and zero locations (not covered in this work) (Figure 26.8).

Example 26.3: Design the Low-Pass Inverse Chebyshev Filter Assuming $\alpha_p = 3\,dB$, $\alpha_s = 30\,dB$, $\omega_p = 1000\,rad/s$, $\omega_s = 3000\,rad/s$

Step 1—Calculate order of filter:

$$n = \frac{\ln[4*(10^{30/10}-1)/(10^{3/10}-1)]^{1/2}}{\log[(3000/1000)+((3000^2/1000^2)-1)^{1/2}]} = 2.3535 \implies n = 3$$

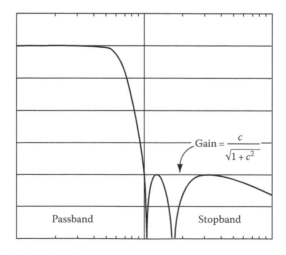

FIGURE 26.7 Inverse Chebyshev filter characteristic.

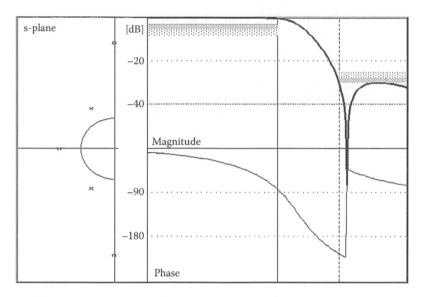

FIGURE 26.8   Pole-zero locations, magnitude response, and phase of inverse Chebyshev filter.

Step 2—Calculate pole and zero locations.
   Normalized values of poles and ω_0 and Q:

−0.6613 + 1.29944i	1.45803	1.10240
−0.6613 − 1.29944i	1.45803	1.10240
−1.60734		

Normalized values of zeros:

3.4641i	3.4641i	3.4641
−3.4641i	−3.4641i	3.4641

26.2.4 Cauer Elliptic Low-Pass Filter

Cauer elliptic response (Figure 26.9):

$$|T(jw)|^2 = \frac{1}{1+\varepsilon^2 R_n^2(w,L)}$$

Designing the Cauer elliptic filter is more complicated than designing the three previous filters. In order to calculate the transfer function of this filter, a mathematical process is summarized as below. Although the low-pass Cauer elliptic filter has ripples in both stop-band and pass-band, it has lower order than the three previous filters (Figure 26.10), that is, the advantage of the Cauer elliptic filter:

$$k = \frac{\omega_p}{\omega_s} \tag{26.1}$$

$$k' = \sqrt{1-k^2} \tag{26.2}$$

$$q_0 = \frac{0.5(1-\sqrt{k'})}{(1+\sqrt{k'})} \tag{26.3}$$

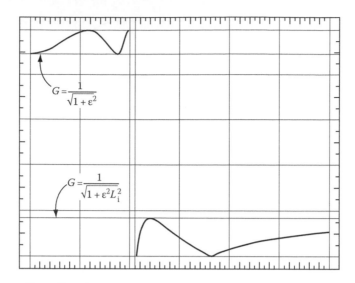

FIGURE 26.9 Cauer elliptic filter characteristic.

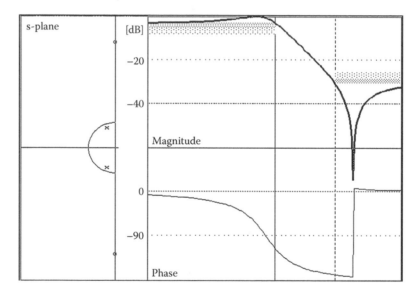

FIGURE 26.10 Pole-zero locations, magnitude response, and phase of Cauer elliptic filter.

$$q = q_0 + 2q_0^5 + 15q_0^9 + 150q_0^{13} \tag{26.4}$$

$$D = \frac{10^{0.1\alpha_s} - 1}{10^{0.1\alpha_p} - 1} \tag{26.5}$$

$$n \geq \frac{\log(16D)}{\log(1/q)} \tag{26.6}$$

$$\Lambda = \frac{1}{2n} \ln \frac{10^{0.05\alpha_p} + 1}{10^{0.05\alpha_p} - 1} \tag{26.7}$$

$$\sigma_0 = \left| \frac{2q^{1/4} \sum_{m=0}^{\infty} (-1)^m q^{m(m+1)} \sinh[(2m+1)\Lambda]}{1 + 2\sum_{m=1}^{\infty} (-1)^m q^{m^2} \cosh(2m\Lambda)} \right| \tag{26.8}$$

$$\omega = \sqrt{\left(1 + k\sigma_0^2\right)\left(1 + \frac{\sigma_0^2}{k}\right)} \tag{26.9}$$

$$\Omega_i = \left| \frac{2q^{1/4} \sum_{m=0}^{\infty} (-1)^m q^{m(m+1)} \sinh((2m+1)\pi\mu/n)}{1 + 2\sum_{m=1}^{\infty} (-1)^m q^{m^2} \cosh\left(\frac{2m\pi\mu}{n}\right)} \right| \tag{26.10}$$

$$\mu = \begin{cases} i & \text{for odd } n \\ i - \dfrac{1}{2} & \text{for even } n \end{cases} \quad i = 1, 2, \ldots, r \tag{26.11}$$

$$V_i = \sqrt{\left(1 - k\Omega_i^2\right)\left(1 - \frac{\Omega_i^2}{k}\right)} \tag{26.12}$$

$$A_{01} = \frac{1}{\Omega_i^2} \tag{26.13}$$

$$B_{0i} = \frac{(\sigma_0 V_i)^2 + (\Omega_i \omega)^2}{(1 + \sigma_0^2 \Omega_i^2)^2} \tag{26.14}$$

$$B_{1i} = \frac{2\sigma_0 V_i}{1 + \sigma_0^2 \Omega_i^2} \tag{26.15}$$

$$H_0 = \begin{cases} \sigma_0 \displaystyle\prod_{i=1}^{r} \frac{B_{0i}}{A_{0i}} & \text{for odd } n \\ 10^{-0.05\alpha_p} \displaystyle\prod_{i=1}^{r} \frac{B_{0i}}{A_{0i}} & \text{for even } n \end{cases} \tag{26.16}$$

Example 26.4: Design the Low-Pass Cauer Elliptic Filter Assuming $\alpha_p = 3\,\text{dB}$, $\alpha_s = 30\,\text{dB}$, $\omega_p = 1000\,\text{rad/s}$, $\alpha_s = 3000\,\text{rad/s}$

$n = 1.9713 \Rightarrow n = 2$. This filter is the second order low pass filter.
Normalized values of poles and ω_0 and Q:

$-0.31554 + 0.97313i$	0.85360	1.35259
$-0.31554 + 0.97313i$	0.85360	1.35259

Normalized values of zeros:

$4.18154i$	$4.18154i$	4.18154
$-4.18154i$	$-4.18154i$	4.18154

26.3 Frequency Transformations

Four typical methods of deriving a low-pass transfer function that satisfies a set of given specifications are presented. However, there are a lot of applications in the real world of designing, which require not only the low-pass filters but also the band-pass filters, the high-pass filters, and the band-rejection filters. A designer can design any type of filter by designing a low-pass filter first. When a low-pass filter is achieved, the desired filter can be derived by "frequency transformation." In other words, the understanding of methods to design a low-pass filter is the basic but not the trivial task.

26.3.1 Frequency Transformations Low-Pass to High-Pass

$$Z(s) = S = \frac{1}{s}; \quad j\Omega = \frac{1}{j\omega} \quad \Rightarrow \quad \Omega\frac{1}{\omega}; \quad \begin{array}{l} -1 \leq \Omega \leq 1 \quad \text{frequency of low-pass passband} \\ -1 \leq \omega \leq 1 \quad \text{frequency of high-pass passband} \end{array}$$

Frequency transformation transforms the pass-band of the low-pass, centered around $\Omega = 0$, into that of the high-pass, centered around $\omega = \infty$ (Figure 26.11). Similarly, it transforms the low-pass stop-band that is centered around $\Omega = \infty$ into that of the high pass, centered around $\omega = 0$. Consequently, the frequency transformation function $Z(s)$ has a zero in the center of the pass-band of the high-pass (at $\omega = \infty$) and a pole in the center of the high-pass, stop-band (at $\omega = 0$) [SV01]:

$$T(S) = \frac{\omega_0^2}{S^2 + (\omega_0 S/Q) + \omega_0^2} \quad \Rightarrow \quad T(s) = \frac{\omega_0^2}{(1/s^2) + (\omega_0/Qs) + \omega_0^2} = \frac{s^2}{(1/\omega_0^2) + (s/\omega_0^2 Q) + s^2}$$

$T(S)$: low-pass transfer function; $T(s)$: high-pass transfer function.

26.3.2 Frequency Transformations Low-Pass to Band-Pass

$$Z(s) = S = \frac{s^2 + \omega_c}{Bs} = \frac{\omega_c(s^2 + \omega_c^2)}{B\omega_c s} \quad \Rightarrow \quad \Omega = \frac{\omega^2 - \omega_c}{B\omega}; \quad \omega_c^2 = \omega_1\omega_2; \quad B = \omega_2 - \omega_1$$

Frequency transformation transforms the pass-band of the low-pass, centered around $\Omega = 0$, into that of the band-pass, centered around $\omega = \omega_c$. Similarly, it transforms the low-pass stop-band that is

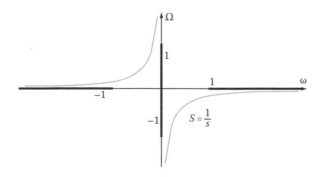

FIGURE 26.11   Frequency transformations low-pass to high-pass.

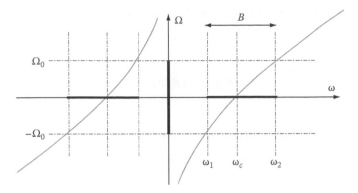

FIGURE 26.12   Frequency transformations low-pass to band-pass.

centered around $\Omega = \infty$ into that of the band-pass, centered around $\omega = 0$ (Figure 26.12). Consequently, the frequency transformation function $Z(s)$ has zeros in the center of the pass-band of the band-pass (at $\omega = \pm\, \omega_c$) and poles in the center of the band-pass, stop-band (at $\omega = 0$ and $\omega = \infty$) [SV01]:

$$T(S) = \frac{\omega_0^2}{S^2 + (\omega_0 S/Q) + \omega_0^2} \quad \Rightarrow \quad T(s) = \frac{s^2 B^2 \omega_0^2}{s^4 + (\omega_0 B s^3/Q) + (2\omega_c^2 + B^2 \omega_0^2)s^2 + (\omega_0 B \omega_c^2 s/Q) + \omega_c^4}$$

$T(S)$: low-pass transfer function; $T(s)$: band-pass transfer function.

26.3.3 Frequency Transformations Low-Pass to Band-Stop

$$Z(s) = S = \frac{Bs}{s^2 + \omega_c^2} \quad \Rightarrow \quad \Omega = \frac{-B\omega}{\omega^2 - \omega_c^2}; \quad \omega_c^2 = \omega_1 \omega_2; \quad B = \omega_2 - \omega_1$$

Frequency transformation transforms the pass-band of the low-pass, centered around $\Omega = 0$, into that of the band-stop, centered around $\omega = 0$ and $\omega = \infty$ (Figure 26.13). Similarly, it transforms the low-pass, stop-band that is centered around $\Omega = \infty$ into that of the band-stop, centered around $\omega = \omega_c$.

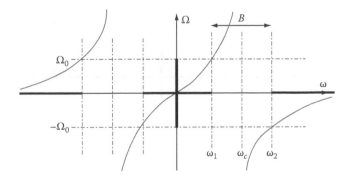

FIGURE 26.13   Frequency transformations low-pass to band-stop.

Consequently, the frequency transformation function $Z(s)$ has zeros in the center of the pass-band of the band-stop (at $\omega = 0$ and $\omega = \infty$) and poles in the center of the band-stop, stop-band (at $\omega = \pm\,\omega_c$) [SV01]:

$$T(S) = \frac{\omega_0^2}{S^2 + (\omega_0 S/Q) + \omega_0^2} \quad \Rightarrow \quad T(s) = \frac{\omega_0^2 s^4 + 2\omega_0^2\omega_c^2 s^2 + \omega_0^2\omega_c^4}{\omega_0^2 s^4 + (\omega_0 B s^3/Q) + (2\omega_c^2\omega_0^2 + B^2)s^2 + (\omega_0 B\omega_c^2 s/Q) + \omega_c^4\omega_0^2}$$

26.3.4 Frequency Transformation Low-Pass to Multiple Band-Pass

Frequency transformation transforms the pass-band of the low-pass, centered around $\Omega = 0$, into that of the multiple band-pass, centered around $\omega = 0$ and $\omega = \omega_{z1}$. Similarly, it transforms the low-pass, stop-band that is centered around $\Omega = \infty$ into that of the multiple band-pass, centered around $\omega = \omega_{p1}$ and $\omega = \infty$. Consequently, the frequency transformation function $Z(s)$ has zeros in the center of the pass-band of multiple band-pass and at ω_{z1} (at $\omega = 0$ and $\omega = \pm\omega_{z1}$) and poles in the center of the band-stop of multiple pass-band (at $\omega = \pm\omega_c$ and $\omega = \infty$) [SV01] (Figure 26.14):

$$Z(s) = S = \frac{s(s^2 + \omega_{z1}^2)}{B(s^2 + \omega_{P1}^2)} \quad \Rightarrow \quad \Omega = \frac{\omega(\omega^2 - \omega_{z1}^2)}{B(\omega^2 - \omega_{P1}^2)}$$

Transfer functions from the low-pass frequency S to the frequency s of other types of filters are recognized and can be written under the following form:

$$Z(s) = \frac{H(s^2 + \omega_{z1}^2)(s^2 + \omega_{z2}^2)\dots(s^2 + \omega_{zn}^2)}{(s^2 + \omega_{p1}^2)(s^2 + \omega_{p2}^2)\dots(s^2 + \omega_{pn}^2)}$$

or

$$\Omega(\omega) = \frac{H(\omega^2 - \omega_{z1}^2)(\omega^2 - \omega_{z2}^2)\dots(\omega^2 - \omega_{zn}^2)}{(\omega^2 - \omega_{p1}^2)(\omega^2 - \omega_{p2}^2)\dots(\omega^2 - \omega_{pn}^2)}$$

$Z(s)$ has zeros where the desired filter has pass-bands and poles where it has stop-bands. The function $Z(s)$ is called Foster Reactance function.

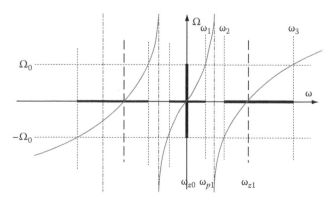

FIGURE 26.14   Frequency transformation low-pass to multiple band-pass.

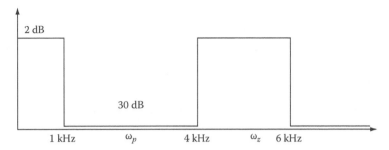

FIGURE 26.15 Frequency transformation by foster reactance function.

Example 26.5: Writing Transfer Function of the Filter in Figure 26.15 Assuming $\alpha_p = 2\,dB$, $\alpha_s = 30\,dB$, $\omega_1 = 1\,KHz$, $\omega_2 = 4\,KHz$, $\omega_3 = 6\,KHz$

We can write the transfer function of the filter Figure 26.15 as following

$$Z(s) = \frac{Hs(s^2 + \omega_z^2)}{(s^2 + \omega_p^2)} \quad \text{or} \quad \Omega(\omega) = \frac{H\omega(\omega^2 - \omega_z^2)}{(\omega^2 - \omega_p^2)}$$

The transfer function has zeros at $\omega = 0$, $\omega = \omega_z$, and poles at $\omega = \omega_p$ and $\omega = \infty$.

At corner frequencies $\omega_1 = 1\,kHz$, $\omega_2 = 4\,kHz$, and $\omega_3 = 6\,kHz$, the values of $\Omega(\omega)$ are equal to 1, −1, and 1, respectively. Therefore, the transformation $\Omega(\omega)$ can be rewritten into multi-equations corresponding to $\omega = \omega_1, \omega_2, \omega_3$. Three equations with three unknowns always have solutions

$$1 = \frac{H\omega_1(\omega_1^2 - \omega_z^2)}{\omega_1^2 - \omega_p^2}$$

$$-1 = \frac{H\omega_2(\omega_2^2 - \omega_z^2)}{\omega_2^2 - \omega_p^2}$$

$$1 = \frac{H\omega_3(\omega_3^2 - \omega_z^2)}{\omega_3^2 - \omega_p^2}$$

$$\begin{cases} \omega_z^2 = 22 \\ \omega_p^2 = 8; \text{ so the Foster Transfer Function is } S = \frac{(1/3)s^3 + (22/3)s}{s^2 + 8} \\ H = \frac{1}{3} \end{cases}$$

26.4 Summary and Conclusion

Analog filters have been used broadly in communication. Understanding the methods to synthesizing analog filters is extremely important and is the basic step in designing analog filters. Four different synthesizing methods were presented, each method will result in different characteristics of filters. This chapter also presented steps to design other types of filters from the low-pass filter by writing the frequency transfer function.

References

[KAS89] M.R. Kobe, J. Ramirez-Angulo, and E. Sanchez-Sinencio, FIESTA—A filter educational synthesis teaching aid, *IEEE Trans. Educ.*, 32(3), 280–286, August 1989.

[SV01] R. Schaumann and M.E. Van Valkenburg, *Analog Filter Design*, Oxford University Press, Oxford, U.K., 2001.

[W02] S. Winder, *Analog and Digital Filter Design*, Newnes, Woburn, MA, 2002.

[W90] B.M. Wilamowski, A filter synthesis teaching-aid, in: *Proceedings of the Rocky Mountain ASEE Section Meeting*, Golden, CO, April 6, 1990.

[WG05] B.M. Wilamowski and R. Gottiparthy, Active and passive filter design with MATLAB, *Int. J. Eng. Educ.*, 21(4), 561–571, 2005.

[WLS92] B.M. Wilamowski, S.F. Legowski, and J.W. Steadman, Personal computer support for teaching analog filter analysis and design courses, *IEEE Trans. Educ.*, E-35(4), 351–361, 1992.

27

Active Filter Implementation

Nam Pham
Auburn University

Bogdan M.
Wilamowski
Auburn University

John W. Steadman
*University of South
Alabama*

27.1 Introduction

There are many ways to implement filters. These could be passive LC filters [JW09], cascade active filters using operational amplifiers (op-amps), active filters using operational transconductance amplifiers (OTAs) [AWD07,TDU03], or switched capacitor or switched current filters [J76,P08]. As a part of the design process, the desired high-pass, band-pass, or band-stop prototype must be often converted to the low-pass filter prototype [B74].

Many different circuits can be used to realize any given transfer function. For purposes of this chapter, several of the most popular types of realizations are presented. Much more detailed information on various circuit realizations and the advantages of each may be found in the literature, in particular Van Valkenburg [V82], Huelseman and Allen [HA80], and Chen [C86]. Generally, the design trade-offs in making the choice of the circuit to be used for the realization involve considerations of the number of elements required, the sensitivity of the circuit to changes in component values, and the ease of tuning the circuit to given specifications. Accordingly, limited information is included about these characteristics of the example circuits in this chapter.

Each of the circuits described here is commonly used in the realization of active filters. When implemented as shown and used within the appropriate gain and bandwidth specifications of the amplifier, they will provide excellent performance. Computer-aided filter design programs are available that simplify the process of obtaining proper element values and simulation of the resulting circuits [KAS89,MLS92].

27.2 Circuit Realization

Various electronic circuits can be found to implement any given transfer function. Cascade filters and ladder filters are two of the basic approaches for obtaining a practical circuit. Cascade realizations are much easier to design and to tune, but ladder filters are less sensitive to element variations. In cascade realizations, the transfer function is simply factored into first- and second-order parts. Circuits are built for the individual parts and then cascaded to produce the overall filter. For simple to moderately complex filter designs, this is the most common method, and the remainder of this section is devoted to several examples of the circuits used to obtain the first- and second-order filters. For very high-order transfer functions, ladder filters should be considered, and further information can be obtained by consulting the literature.

In order to simplify the circuit synthesis procedure, very often ω_0 is assumed to be equal to one, and then after a circuit is found, the values of all capacitances in the circuit are divided by ω_0. In general, the following magnitude and frequency transformations are allowed:

$$R_{new} = K_M R_{old} \quad \text{or} \quad C_{new} = \frac{C_{old}}{K_F K_M} \quad \text{or} \quad L_{new} = K_M \frac{L_{old}}{K_F} \tag{27.1}$$

where K_M and K_F are magnitude and frequency scaling factors, respectively.

Cascade filter designs require the transfer function to be expressed as a product of first- and second-order terms. For each of these terms, a practical circuit can be implemented. Examples of these circuits are presented. In general, the following first- and second-order terms can be distinguished.

27.2.1 First-Order Low-Pass (Figure 27.1)

27.2.2 First-Order High-Pass

While several passive realizations of first-order filters are possible (low-pass, high-pass, and lead-lag), the active circuits shown here are inexpensive and avoid any loading of the other filter sections when the individual circuits are cascaded. Consequently, these circuits are preferred unless there is some reason to avoid the use of the additional op-amp. Note that a second-order filter can be realized using one op-amp as shown in the following paragraphs; so, it is common practice to choose even-order transfer functions, thus avoiding the use of any first-order filters (Figure 27.2).

27.2.3 Two Popular Second-Order Low-Pass Circuits

This filter is non-inverting, and unity gain, i.e., H must be one, and the scaling factors shown in Equation 27.1 should be used to obtain reasonable element values. This is a very popular filter for realizing second-order functions because it uses a minimum number of components, and since the operation amplifier is in the unity gain configuration, it has very good bandwidth (Figure 27.3).

Another useful configuration for second-order low-pass filters uses the op-amp in its inverting "infinite gain" mode, as shown in Figure 27.4.

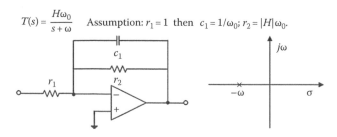

$$T(s) = \frac{H\omega_0}{s + \omega} \quad \text{Assumption: } r_1 = 1 \text{ then } c_1 = 1/\omega_0; r_2 = |H|\omega_0.$$

FIGURE 27.1 Low-pass circuit with one real pole. (This filter is inverting, i.e., H must be negative, and the scaling factors shown in Equation 27.1 should be used to obtain reasonable values for the components.)

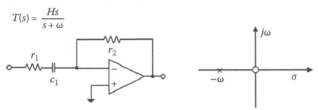

$$\text{Assumption: } r_1 = 1 \text{ then } c_1 = 1/\omega_0; \; r_2 = |H|\omega_0.$$

$$T(s) = \frac{Hs}{s + \omega}$$

FIGURE 27.2   High-pass circuit with one real pole and one zero at the origin. (This filter is inverting, i.e., H must be negative, and the scaling factors shown in Equation 27.1 should be used to obtain reasonable values for the components.)

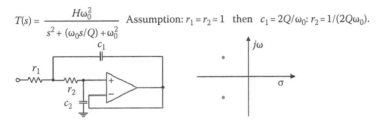

$$T(s) = \frac{H\omega_0^2}{s^2 + (\omega_0 s/Q) + \omega_0^2}$$

$$\text{Assumption: } r_1 = r_2 = 1 \text{ then } c_1 = 2Q/\omega_0; \; r_2 = 1/(2Q\omega_0).$$

FIGURE 27.3   Low-pass circuit with two conjugate poles.

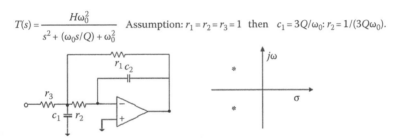

$$T(s) = \frac{H\omega_0^2}{s^2 + (\omega_0 s/Q) + \omega_0^2}$$

$$\text{Assumption: } r_1 = r_2 = r_3 = 1 \text{ then } c_1 = 3Q/\omega_0; \; r_2 = 1/(3Q\omega_0).$$

FIGURE 27.4   Low-pass circuit with two conjugate poles.

This circuit has the advantage of relatively low sensitivity of ω_0 and Q to variations in component values. In this configuration, the op-amp's gain-bandwidth product may become a limitation for high-Q and high-frequency applications [B74]. There are several other circuit configurations for low-pass filters. The references given at the end of the chapter will guide the designer to alternatives and the advantages of each.

27.2.4 Second-Order High-Pass Filter

Second-order high-pass filters may be designed using circuits very much like those shown for the low-pass realizations. For example, the Sallen–Key high-pass filter is shown (Figures 27.5 through 27.7).

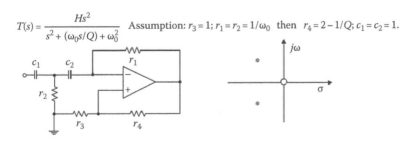

$$T(s) = \frac{Hs^2}{s^2 + (\omega_0 s/Q) + \omega_0^2}$$

$$\text{Assumption: } r_3 = 1; \; r_1 = r_2 = 1/\omega_0 \text{ then } r_4 = 2 - 1/Q; \; c_1 = c_2 = 1.$$

FIGURE 27.5   High-pass circuit with two conjugate poles and two zeros at the origin.

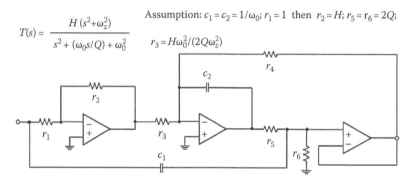

$$T(s) = \frac{H(s^2 + \omega_z^2)}{s^2 + (\omega_0 s/Q) + \omega_0^2}$$

Assumption: $c_1 = c_2 = 1$; $r_1 = 1/2Q\omega_0$) then $r_3 = 1/Q\omega_0$);

$r_2 = r_4 = 2Q/\omega_0$.

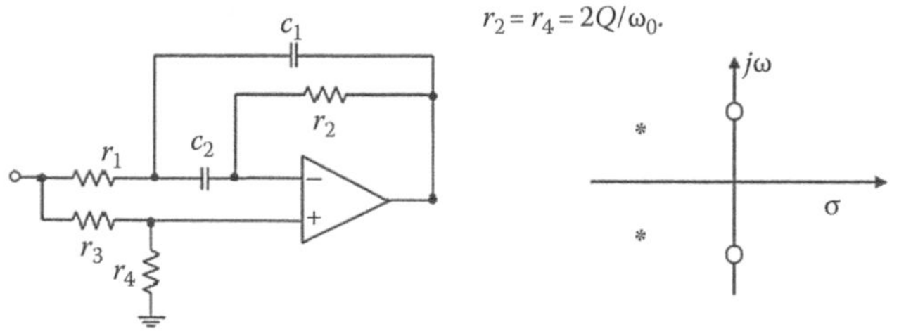

FIGURE 27.6 Band-stop circuit with two conjugate poles and two conjugate zeros. (The primary advantage of this circuit is that it requires a minimum number of components. For applications where no tuning is required and the Q is low, this circuit works very well. When the band-stop filter must be tuned, the three-operational-amplifier circuit is preferable.)

$$T(s) = \frac{H(s^2 + \omega_z^2)}{s^2 + (\omega_0 s/Q) + \omega_0^2}$$

Assumption: $c_1 = c_2 = 1/\omega_0$; $r_1 = 1$ then $r_2 = H$; $r_5 = r_6 = 2Q$;

$r_3 = H\omega_0^2/(2Q\omega_z^2)$

FIGURE 27.7 Band-stop circuit with two conjugate poles and two conjugate zeros.

The foregoing circuits provide a variety of useful first- and second-order filters. For higher order filters, these sections are simply cascaded to realize the overall transfer function desired. Additional detail about these circuits as well as other circuits used for active filters may be found in the references.

27.3 Circuits with Placement of Poles and Zeros

Assume that the low-pass filter function was already found. The next step to design filters is how circuits with different locations of pole and zeros are recognized. The list below are some typical circuits with their poles and zero locations.

27.3.1 High-Pass Filter (Figure 27.8)

$$T(s) = \frac{s^2 + \omega_z^2}{s^2 + (\omega_0 s/Q) + \omega_0^2}$$

$$a = \frac{1 + \sqrt{1 + 4Q^2}}{2Q^2}; \quad r = \sqrt{a + 1}; \quad \omega = \frac{\omega_z}{\omega_0};$$

$$R_1 = \frac{r}{a}; \quad R_2 = \frac{r}{2}; \quad R_3 = \frac{r}{a + 1}; \quad R_4 = r$$

$$C_1 = \frac{1}{\omega^2}; \quad C_2 = \left(1 - \frac{1}{\omega^2}\right); \quad C_3 = 2; \quad C_4 = 1$$

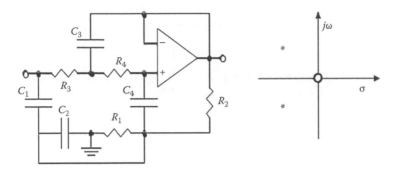

FIGURE 27.8 Circuit with two conjugate poles and two zeros at the origin.

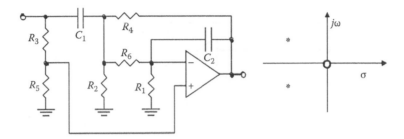

FIGURE 27.9 Circuit with two conjugate poles and two zeros at the origin.

27.3.2 High-Pass Filter (Figure 27.9)

$$T(s) = \frac{K(s^2 + \omega_z^2)}{s^2 + (\omega_0 s/Q) + \omega_0^2}$$

$$\omega_z > \omega_0$$

$$a = \sqrt{\frac{\omega_z}{\omega_0}}; \quad b = -1.5 + \sqrt{1.25 + a}; \quad e = \sqrt{Q(b+2)}$$

$$C_1 = 1; \quad C_2 = eC_1$$

$$r_a = \frac{1}{\omega_0 \sqrt{C_1 C_2}}; \quad r_b = \frac{r_a}{b}; \quad d = \frac{r_a(1 + 1/e)}{r_b} + \frac{2}{e}$$

$$R_1 = r_b; \quad R_2 = r_b; \quad R_3 = 1; \quad R_4 = r_a; \quad R_5 = \frac{R_3}{d}; \quad R_6 = r_a$$

27.3.3 Low-Pass Filter (Figure 27.10)

$$T(s) = \frac{\omega_0^2}{s^2 + (\omega_0 s/Q) + \omega_0^2}$$

$$R_1 = 1; \quad R_2 = 1; \quad R_3 = 10^6; \quad R_4 = 1$$

$$C_1 = \frac{3Q}{\omega_0}; \quad C_2 = \frac{1}{3Q\omega_0}$$

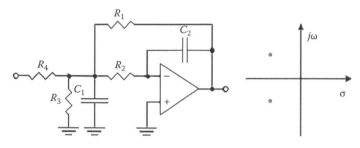

FIGURE 27.10   Circuit with two conjugate poles.

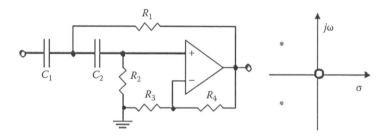

FIGURE 27.11   Circuit with two conjugate poles and two zeros at the origin.

27.3.4 High-Pass Filter (Figure 27.11)

$$T(s) = \frac{s^2}{s^2 + (\omega_0 s/Q) + \omega_0^2}$$

$$a = 3 - \frac{1}{Q}$$

$$R_1 = R_2 = \frac{1}{\omega_0}; \quad R_3 = 1; \quad R_4 = a - 1$$

$$C_1 = C_2 = 1$$

27.3.5 Band-Pass Filter (Figure 27.12)

$$T(s) = \frac{K\omega_0^2 s}{s^2 + (\omega_0 s/Q) + \omega_0^2}$$

Assumption : $r = 1$ and $c = \dfrac{1}{\omega_0}$ then

$$R_1 = Q; \quad R_2 = r; \quad R_3 = \left| \frac{Q}{K} \right|; \quad R_4 = R_5 = R_6 = r$$

$$C_1 = C_2 = c$$

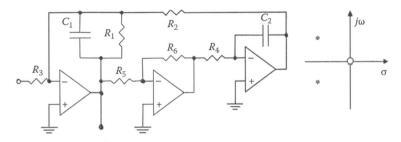

FIGURE 27.12   Circuit with two conjugate poles and one zero at the origin.

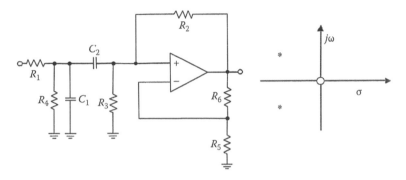

FIGURE 27.13   Circuit with two conjugate poles and one zero at the origin.

27.3.6 Band-Pass Filter (Figure 27.13)

$$T(s) = \frac{H(\omega_0 s/Q)}{s^2 + (\omega_0 s/Q) + \omega_0}$$

Assumption: $C_1 = C_2 = R_5 = 1$ then

$$R_1 = \frac{(4Q/\sqrt{2}) - 1}{H}; \quad R_2 = R_3 = \frac{\sqrt{2}}{\omega_0}; \quad R_4 = \frac{(4Q/\sqrt{2}) - 1}{(4Q/\sqrt{2}) - 1 - H}; \quad R_6 = 3 - \frac{\sqrt{2}}{\omega_0}$$

27.3.7 Band-Stop Filter (Figure 27.14)

$$T(s) = \frac{s + \omega_z^2}{s^2 + (\omega_0 s/Q) + \omega_0^2}$$

Assumption: $C_1 = C_2 = 1;$ $C_3 = 2$ then

$$a = \frac{1}{2Q^2} + \sqrt{\frac{1 + 4Q^2}{4Q^4}}; \quad r_a = \frac{\sqrt{a + 1}}{\omega_0}; \quad r_b = \frac{1}{\omega_z * r_a}; \quad r_a > \frac{1}{\omega_0}$$

$$R_2 = \frac{r_a * r_b}{r_b(1 + a) - r_a}; \quad R_3 = r_a; \quad R_4 = \frac{r_a}{2}; \quad R_5 = \frac{r_a}{a}$$

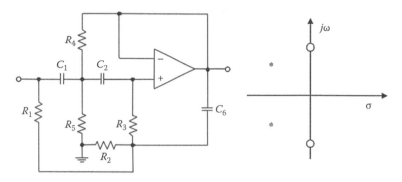

FIGURE 27.14   Circuit with two conjugate poles and two zeros.

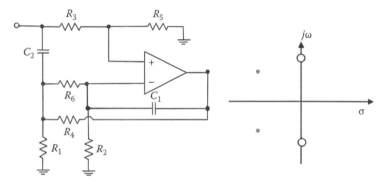

FIGURE 27.15   Circuit with two conjugate poles and two zeros.

27.3.8 Band-Stop Filter (Figure 27.15)

$$T(s) = \frac{s + w_z^2}{s^2 + (\omega_0 s/Q) + \omega_0^2}$$

$$a = \frac{\omega_z^2}{\omega_p^2}; \quad b = \sqrt{\frac{\omega_z^2}{\omega_p^2} + 1.25} - 1.5; \quad c = Q^2(b+2)^2; \quad C_1 = 1; \quad C_2 = c * C_1$$

$$R_4 = \frac{1}{\omega_p \sqrt{C_1 C_2}}; \quad R_1 = \frac{R_4}{b}; \quad R_2 = R_1; \quad R_3 = 1; \quad R_5 = \frac{R_3}{(R_4/R_2)(1 + (1/c)) + (2/c)}; \quad R_6 = R_4$$

27.4 Design Example

Example 27.1

Design a low-pass Chebyshev filter that satisfies the following attenuation and gain specification $\alpha_p = 0.5\,\text{dB}$, $\alpha_s = 40\,\text{dB}$, $f_p = 1000\,\text{Hz}$, $f_s = 2000\,\text{Hz}$ and implement this design using second-order low-pass Sallen–Key non-inverting filters.

Step 1:

$$n = \frac{\ln[4 * (10^{\alpha_s/10} - 1)/(10^{\alpha_p/10} - 1)]^{1/2}}{\log\left[(\omega_s/\omega_p) + \left((\omega_s^2/\omega_p^2) - 1\right)^{1/2}\right]} = \frac{\ln[4 * (10^{0.5/10} - 1)/(10^{40/10} - 1)]^{1/2}}{\log[(2000/1000) + ((2000^2/1000^2) - 1)^{1/2}]} = 4.82$$

$n = 5 =>$. This is the fifth-order filter.

Step 2:

$$\varepsilon = [10^{\alpha p/10} - 1]^{1/2} = [10^{0.5/10} - 1]^{1/2} = 0.3493;$$

$$\gamma = \sinh^{-1}\frac{1/\varepsilon}{n} = \sinh^{-1}\frac{1/3.493}{5} = 0.3548$$

$$\Omega = 90° + \frac{90°}{n} + \frac{(k-1)180°}{n} = 1.885, 2.5133, 3.1416, 3.7699, 4.3982 \,(\text{rad})$$

$$a_k = \sinh(\gamma)\cos(\Omega); \quad b_k = \cosh(\gamma)\sin(\Omega); \quad \omega_k = a_k^2 + b_k^2; \quad Q_k = \left|\frac{\omega_k}{2a_k}\right|$$

Pole locations: $-0.11196 \pm j1.0116$, $-0.29312 \pm j0.62518$, -0.36232.

$$Q_k = 4.5450, 1.1778, 1.1778, 4.5450, 0.5; \quad \omega_k = 1.0177, 0.69048, 0.69048, 1.0177, 0.36232.$$

This filter has two pairs of conjugate poles and one real pole. To design this filter, two circuits that have conjugate poles and another circuit having only one pole are needed. By picking up the proper circuits from the table above and cascading them together, the desired filter will be designed (Figures 27.16 and 27.17).

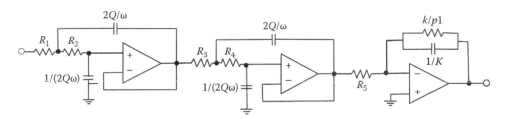

FIGURE 27.16 Cascading low-pass filter.

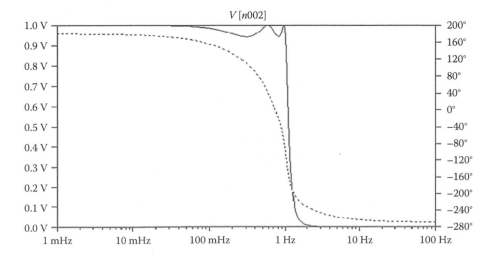

FIGURE 27.17 Output without scaling.

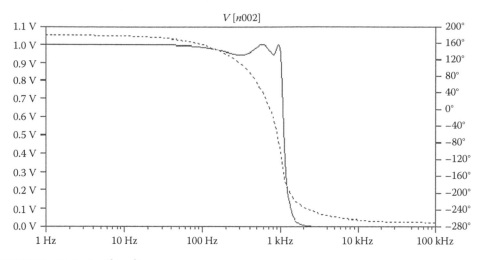

FIGURE 27.18   Output with scaling.

As a result from the p-spice simulation, the cut-off frequency is 1 Hz that was not scaled yet. Therefore, in order to get the cut-off frequency at 1 kHz, the values of capacitors and/or resistors have to be scaled by dividing and/or multiplying by a factor that is equal to the value of cut-off frequency (Figure 27.18).

Example 27.2

Two fundamental methods are the cascade approach and the simulation of lossless ladders. The fundamental reasons behind cascade realization are lower sensitivities to component tolerances, and the simplicity and flexibility of the design of low-order modules rather than of the complete high-order filter. The only condition to be satisfied for the approach to be valid is that the cascaded stages do not load each other so that they do not interact. However, the use of op-amps generally limits the frequency range over which our designs could be employed with predictable results. A problem arises with low-frequency op-amp-based active RC filters because integrated capacitors are limited to around 30 pF. Much larger values consume too much silicon area on the integrated circuit chip. As a consequence, the design calls for huge resistors that also consume too much area and cause considerable noise. And another method that is also usable from the low audio range, but extends to applications at hundreds of megahertz, avoids op-amps altogether and instead obtains the required gain from *transconductance amplifiers* by using the concept of state variables.

Design the given Chebyshev low-pass filter and implement this filter by using transconductance amplifiers (Figures 27.19 and 27.20).

Step 1: To implement this circuit by the transconductance amplifier, the state equations need to be written and all current variables have to be converted into voltage variables by using $R_0 = 1$ and $I_x = V_x/R_0$.

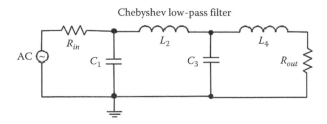

FIGURE 27.19   Low-pass filter.

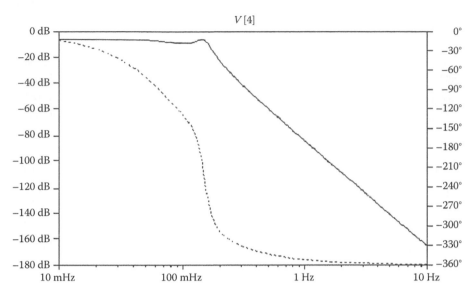

FIGURE 27.20 Output of LC low-pass filter.

$$V_1 = \left(\frac{V_{in} - V_1}{R_{in}} - I_2\right)\frac{1}{SC_1} \quad \Rightarrow \quad V_1 = \frac{1}{SC_1 + (1/R_{in})}\left(\frac{V_{in}}{R_{in}} - I_2\right) \quad \Rightarrow \quad V_1 = \frac{1}{SC_1 + (1/R_{in})}\left(\frac{V_{in}}{R_{in}} - \frac{V_2}{R_{02}}\right)$$

$$I_2 = \frac{1}{SL_2}(V_1 - V_3) \quad \Rightarrow \quad \frac{V_2}{R_{02}} = \frac{1}{SL_2}(V_1 - V_3) \quad \Rightarrow \quad V_2 = \frac{R_{02}}{SL_2}(V_1 - V_3)$$

$$V_3 = \frac{1}{SC_3}(I_2 - I_4) \quad \Rightarrow \quad V_3 = \frac{1}{SC_3}\left(\frac{V_2}{R_{02}} - \frac{V_4}{R_{04}}\right)$$

$$I_4 = \frac{1}{SL_4}(V_3 - I_4R_{out}) \quad \Rightarrow \quad I_4 = \frac{1}{SL_4 + R_{out}}V_3 \quad \Rightarrow \quad V_4 = \frac{R_{04}}{SL_4 + R_{out}}V_3$$

Step 2: Draw the circuit (Figure 27.21 and 27.22).

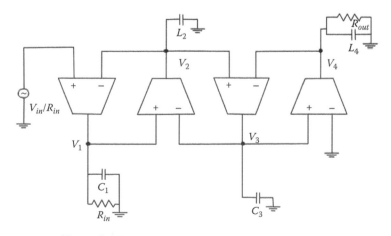

FIGURE 27.21 Low-pass filter with OTAs.

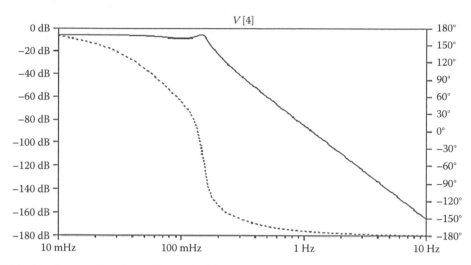

FIGURE 27.22   Output of band-pass filter with OTAs.

Example 27.3

Design the band-pass filter from the given Chebyshev low-pass filter and implement this filter by using transconductance amplifiers.

Step 1: To transfer a low-pass filter to a band-pass filter, each passive element has to be converted. An inductor in a low-pass filter will become an inductor in series with a capacitor in a band-pass filter. And a capacitor in a low-pass filter will become an inductor in parallel with a capacitor in a band-pass filter (Figures 27.23 and 27.24).

Step 2: To implement this circuit by the transconductance amplifiers the state equations need to be written and all current variables have to be converted into voltage variables by using $R_0 = 1$ and $I_x = V_x/R_0$.

$$I_1 = \frac{V_2}{SL_1} \quad \Rightarrow \quad I_1 R_{01} = \frac{V_2 R_{01}}{SL_1} \quad \Rightarrow \quad V_1 = \frac{V_2}{S(L_1/R_{01})}$$

$$V_2 = \frac{1}{SC_2}\left(\frac{V_{in} - V_2}{R_{in}} - I_1 - I_3\right) \quad \Rightarrow \quad V_2 = \frac{1}{SC_2}\left(\frac{V_{in} - V_2}{R_{in}} - \frac{V_1}{R_{01}} - \frac{V_3}{R_{03}}\right) \quad \Rightarrow \quad V_2 = \frac{R_{in}}{1 + SC_2 R_{in}}\left(\frac{V_{in}}{R_{in}} - \frac{V_1}{R_{01}} - \frac{V_3}{R_{03}}\right)$$

$$I_3 = \frac{1}{SL_3}(V_2 - V_4 - V_6) \quad \Rightarrow \quad I_3 R_{03} = \frac{1}{SL_3}(V_2 - V_4 - V_6)R_{03} \quad \Rightarrow \quad V_3 = \frac{1}{S(L_3/R_{03})}(V_2 - V_4 - V_6)$$

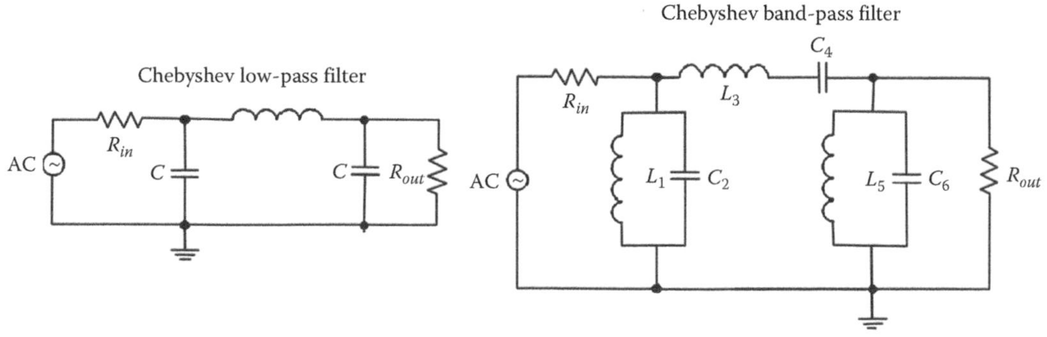

Chebyshev low-pass filter

Chebyshev band-pass filter

FIGURE 27.23   Low-pass filter and band-pass filter.

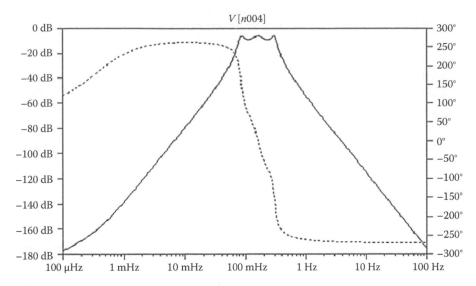

FIGURE 27.24   Output of LC band-pass filter.

$$V_4 = \frac{I_3}{SC_4} \quad \Rightarrow \quad V_4 = \frac{V_3}{S(R_{03}C_4)}$$

$$I_5 = \frac{V_6}{SL_5} \quad \Rightarrow \quad I_5R_{05} = \frac{V_6R_{05}}{SL_5} \quad \Rightarrow \quad V_5 = \frac{V_6}{S(L_5/R_{05})}$$

$$V_6 = \frac{R_{out}}{1+R_{out}SC_6}(I_3-I_5) \quad \Rightarrow \quad V_6 = \frac{R_{out}}{1+SR_{out}C_6}\left(\frac{V_3}{R_{03}} - \frac{V_5}{R_{05}}\right)$$

Step 3: Draw schematic (Figures 27.25 and 27.26).
As discussed in this example, the ladder circuits can be implemented by the transconductance ampli-fiers. In the same manner, the ladder circuits can also be implemented by the op-amps. However, the circuits with op-amps normally require more components, so it is not the optimum design.

   Besides some fundamental methods to implement filters by using op-amps, transconductance amplifiers, and LC ladders, there is another way to implement filters by using switched capacitors to

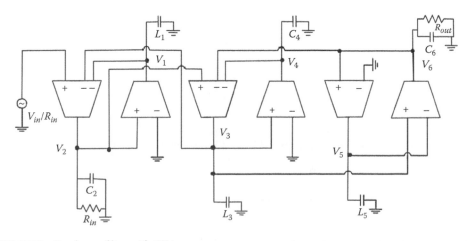

FIGURE 27.25   Band-pass filter with OTAs.

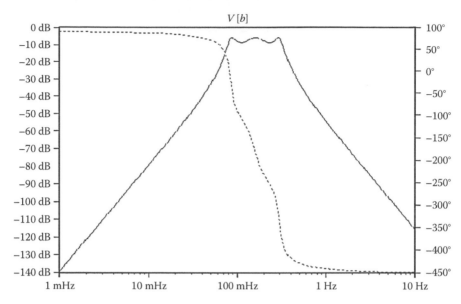

FIGURE 27.26 Output of band-pass filter with OTAs.

simulate resistance. Switched-capacitor filters were developed to be able to design accurate analog filters at voice-band frequencies economically in fully integrated form. Furthermore, these filters provide reduced power, compact design, and compatibility with digital systems. The advantage of this technology is that it allows us to realize accurate RC time constants without the use of resistors by using switches. Note that it is not easy to implement good resistors in integrated circuit technologies.

The capacitor is connected to the voltage v_1 during the phase φ_1, it stores the charge $q_1 = Cv_1$. Thereafter, in the phase φ_2, the capacitor is connected to the voltage v_2, the capacitor recharges to $q_2 = Cv_2$. Therefore, the charge transferred from v_1 to v_2 is

$$\Delta q = q_1 - q_2 = C(v_1 - v_2)$$

The switch is flipped periodically with a clock frequency $f_c = 1/T$ that is very large compared to the signal frequency $\omega = 2\pi f$ of the two voltage sources v_1 and v_2. Then the average transferred charge can be considered as a current

$$i \approx \frac{\Delta q}{T} = \Delta q f_c = f_c C(v_1 - v_2)$$

As long as the condition $f_c \gg f$ is valid, the switched capacitor behaves approximately like an equivalent resistor (Figure 27.27):

$$R_{equ} \approx \frac{v_1 - v_2}{i} = \frac{1}{f_c C}$$

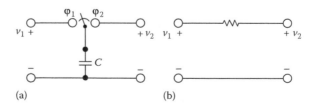

FIGURE 27.27 (a) A switch capacitor and (b) its resistor equivalent.

27.5 Summary and Conclusion

There are many methods to implement filters. Filters can be implemented by using op-amps, LC ladder circuits, transconductance amplifiers, as well as switched capacitors. Method of cascade realization can lower the sensitivity of the circuit on account of the component tolerance. This method is also simple and flexible. LC ladder circuits are also recognized to implement filters. The limit of this type of circuit is the loss of component because inductors and capacitors are not ideal. Transconductance amplifiers are also used to implement the prototype of LC ladder circuit by writing state variables. With this type of concept, LC ladder circuit can be implemented in integrated circuits. Generally, to implement filters, we have to consider all factors: sensitivity, power, compatibility, accuracy, etc.

References

[AWD07] W.M. Anderson, B.M. Wilamowski, and G. Dundar, Wide band tunable filter design implemented in CMOS, in: *Proceedings of the 11th INES 2007—International Conference on Intelligent Engineering Systems*, Budapest, Hungary, pp. 219–223, June 29–July 1, 2007.

[B74] Budak, *Passive and Active Network Analysis and Synthesis*, Boston, MA: Houghton Mifflin, 1974.

[C86] W.K. Chen, *Passive and Active Filters, Theory and implementations*, New York: Wiley, 1986.

[HA80] L.P. Huelsernan and P.E. Allen, *Introduction to the Theory and Design of Active Filters*, New York: McGraw-Hill, 1980.

[J76] D.E. Johnson, *Introduction to Filter Theory*, Englewood Cliffs, NJ: Prentice Hall, 1976.

[JW09] M. Jagiela and B. M. Wilamowski, A methodology of synthesis of lossy ladder filters, in: *Proceedings of the 13th IEEE Intelligent Engineering Systems Conference, INES 2009*, Barbados, April 16–18, 2009.

[KAS89] M.R. Krobe, J. Ramirez-Angulo, and E. Sanchez-Sinencio, FIESTA-A filter educational synthesis teaching aid, *IEEE Trans. Educ.* 12(3), 280–286, August 1989.

[MLS92] B.M. Wilamowski, S.P. Legowski, and J.W. Steadman, Personal computer support for teaching analog filter analysis and design, *IEEE Trans. Educ.* 35(4), 144–152, November 1992.

[P08] S.A. Pactitis, *Active Filters Theory and Design*, Boca Raton, FL: CRC Press, 2008.

[TDU03] W. Tangsrirat, T. Dumawipata, and S. Unhavanich, Realization of lowpass and bandpass leapfrog filters using OAs and OTAs, in: *SICE 2003 Annual Conference*, Vol. 3, Fukui University, Fukui City, Japan, pp. 4–6, 2003.

[V82] M.E. Van Valkenburg, *Analog Filter Design*, New York: Holt, Rinehart & Winston, 1982.

22.5 Summary and Conclusion

28

Designing Passive Filters with Lossy Elements

Marcin Jagiela
University of Information Technology and Management in Rzeszów

Bogdan M. Wilamowski
Auburn University

28.1 Introduction

The role of filters in many electrical and microwave systems cannot be overestimated. The frequency response of a filter can impact significantly on the performance of a whole system. The continuously evolving communication systems demand high-performance filters with the characteristics as close to the ideal as possible. The lack of ideal reactive elements makes all the classical design methods insufficient in practical application. High-Q components are often large in size, which results in larger and heavier filters. The most advanced high-Q technologies are expensive (e.g., multimode dielectric resonator filters) or require cooling systems (superconducting filters). On the other hand, using low-Q reactive elements, particularly inductors, affects the response of the filter in a variety of ways. Apart from increasing passband insertion loss and decreasing stopband attenuation, passband edges of the filter response become rounded which strongly affects the selectivity of the filter. The deterioration in the example of Chebychev filter characteristic due to element losses is shown in Figure 28.1.

Since a certain amount of additional passband loss can often be compensated for by the amplifiers existing in the system, the main objective when designing a lossy filter is to maintain a prescribed selectivity of the filter. Thus, a method of design is needed where losses are taken into account and compensated for [1,2].

28.2 Background

The general structure of a lossless ladder filter is shown in Figure 28.2.

Every branch of the ladder consists of an inductor, a capacitor, or a resonant circuit. Hence, admittances Y_1 and impedances Z_k are purely imaginary.

The transfer function of the passive network of Figure 28.2 built with lossless reactive elements is the ratio of two polynomials (28.1):

$$H(s) = \frac{N(s)}{D(s)} = \frac{a_m s^m + a_{m-1} s^{m-1} + \cdots + a_1 s + a_0}{b_n s^n + b_{n-1} s^{n-1} + \cdots + b_1 s + b_0} \tag{28.1}$$

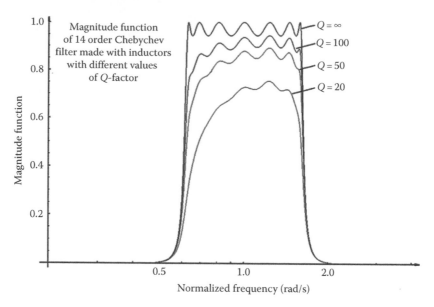

FIGURE 28.1 Magnitude functions of 14-order Chebychev filter made with inductors with different values of Q-factors.

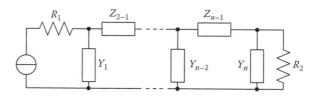

FIGURE 28.2 The general structure of ladder filter.

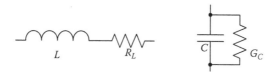

FIGURE 28.3 Models of lossy inductor and lossy capacitor.

where constant coefficients a_i and b_k are real positive numbers. Now, consider the filter of Figure 28.2 built with lossy elements. Each lossy inductor can be represented by an ideal inductor L in series with resistance R_L and each lossy capacitor can be represented by an ideal capacitor C in parallel with a conductance G_s, as shown in Figure 28.3.

In order to use the models of Figure 28.3, it is assumed that the resistance R_L and conductance G_C are constants, that is, they do not depend on the frequency.

The transfer function (28.2) of the lossy filter differs from the transfer function of its lossless prototype:

$$H'(s) = \frac{a'_m s^m + a'_{m-1} s^{m-1} + \cdots + a'_1 s + a'_0}{b'_n s^n + b'_{n-1} s^{n-1} + \cdots + b'_1 s + b'_0} \tag{28.2}$$

Its poles and zeros are shifted to the left on the complex plane, as shown in Figure 28.4, and the magnitude function is deformed, particularly near the passband edges.

Complex plane

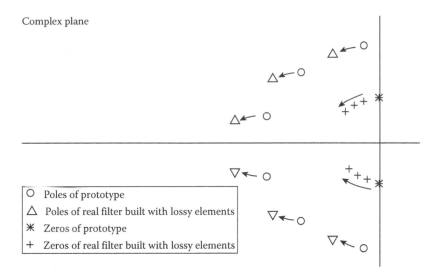

O	Poles of prototype
△	Poles of real filter built with lossy elements
✳	Zeros of prototype
+	Zeros of real filter built with lossy elements

FIGURE 28.4 Movement of poles and zeros due to losses of reactive elements.

Coefficients b'_k in (28.2) are functions of filter elements and of added losses. With fixed values of losses, the coefficients b'_k are multilinear functions of filter elements. Provided the added losses are small enough, it is possible to perturb the element values of the lossy filter so as to move the poles of (28.2) to their original places. It is done by comparing coefficients b'_k to b_k and solving the system of equations (28.3) obtained in this way:

$$\begin{cases} b'_0 = b_0 \\ b'_1 = b_1 \\ \quad \vdots \\ b'_n = b_n \end{cases} \tag{28.3}$$

Despite the fact that the position of zeros cannot be restored, the magnitude function of the lossy filter with the element values modified in this way can be almost ideal.

To express losses of reactive elements in terms of Q-factor rather than in terms of resistance and conductance, one can use the lossy models shown in Figure 28.5.

To use models shown in Figure 28.5, it is assumed that Q varies linearly with the frequency. If more detailed models have to be used, for example, with distributed capacitance for inductors and/or with frequency-dependant resistance, optimization methods are recommended to improve the characteristics of a lossy filter [3].

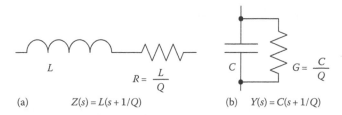

(a) $Z(s) = L(s + 1/Q)$ (b) $Y(s) = C(s + 1/Q)$

FIGURE 28.5 Losses of inductor (a) and capacitor (b) expressed in terms of Q-factor ($\omega = 1$ for prototype filters).

28.3 Method

In the first step of designing the lossy filter, requirements for the filter are determined, such as type of filter (lowpass, highpass, etc.), edge frequencies, necessary attenuations, and the filter order. A good starting point is a prototype filter, normalized to $\omega_0 = 1$, obtained by the traditional methods [4–6]. Thus, with the aid of the existing formulas or tables, based on the classical approach, filter circuit, its transfer function, and element values are computed. Example of prototype filter is shown in Figure 28.6.

The obtained filter is not yet feasible, since the ideal reactive elements were assumed during the process of synthesis.

Now, the prototype circuit has to be redrawn in order to include losses. Each inductor in the circuit is replaced by its lossy model shown in Figure 28.5a and, if it is necessary, each capacitor is replaced by its lossy model shown in Figure 28.5b. As an example, filter of Figure 28.6 with the losses added is shown in Figure 28.7.

After introducing losses, the transfer function of the lossy filter has to be calculated. The simplest algorithm to determine the transfer function of a ladder is to start from the right side of a ladder and work toward the left, determining the currents and the voltages in the series and shunt branches alternately. The impedances and admittances shown in Table 28.1 are useful when obtaining the transfer function.

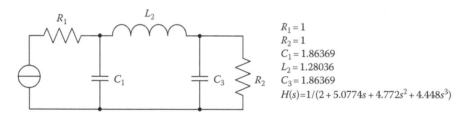

$R_1 = 1$
$R_2 = 1$
$C_1 = 1.86369$
$L_2 = 1.28036$
$C_3 = 1.86369$
$H(s) = 1/(2 + 5.0774s + 4.772s^2 + 4.448s^3)$

FIGURE 28.6   Example of prototype filter with element values and transfer function.

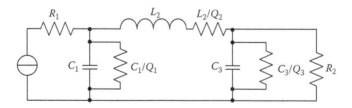

FIGURE 28.7   Filter of Figure 28.6 with losses added.

TABLE 28.1   Impedances and Admittances of Lossy Elements and Resonant Circuits

Lossy inductor	$Z(s) = L(s + 1/Q_L)$
Lossy capacitor	$Y(s) = C(s + 1/Q_C)$
Lossy parallel resonant circuit	$Z(s) = \dfrac{LQ_C(1 + Q_L s)}{CL + Q_C Q_L + CL(Q_C + Q_L)s + CLQ_C Q_L s^2}$
Lossy series resonant circuit	$Y(s) = \dfrac{CQ_L(1 + Q_C s)}{CL + Q_C Q_L + CL(Q_C + Q_L)s + CLQ_C Q_L s^2}$

TABLE 28.2 Impedances and Admittances of Lossy Elements and Resonant Circuits ($Q_C = \infty$)

Lossy inductor	$Z(s) = L(s + 1/Q_L)$
Lossy parallel resonant circuit	$Z(s) = \dfrac{L(1 + Q_L s)}{Q_L + CLs + CLQ_L s^2}$
Lossy series resonant circuit	$Y(s) = \dfrac{CQ_L s}{Q_L + CLs + CLQ_L s^2}$

Values of Q_C and Q_L can be different for each reactive element. Often it is possible to assume infinite values of Q-factor for capacitors. Then, the formulas shown in Table 28.1 become a little bit simpler, and are shown in Table 28.2.

As an example of a process of introducing losses to the ladder and recalculating its transfer function, consider a three-order lowpass filter shown in Figure 28.6.

The circuit obtained after replacing the reactive elements with their lossy models is shown in Figure 28.7. Now, let us determine its transfer function, assuming $Q1 = Q2 = Q3 = Q$ and denoting $P = (1/Q)$. This process is very awkward and performing it by hand is very complicated, so one of the available tools (e.g., Mathematica, or freeware such as Maxima, etc.) is recommended to obtain the results. Below, the consecutive Mathematica commands are presented with output to show the whole process of calculating the transfer function of lossy filter from Figure 28.7:

Input:
```
Iout=Vout/R2
VC3=Vout
IC3=VC3*C3*(s+P)
IL2=IC3+Iout
VL2=IL2*L2*(s+P)
VC1=VL2+VC3
IC1=VC1*C1*(s+P)
IR1=IC1+IL2
VR1=IR1*R1
Vin=VR1+VC1
Tf=Vout/Vin
```

Output (Transfer function of the lossy filter)

$$
\begin{aligned}
&R2/\ (L2P+R1+C1L2R1P^2+R2+C3L2R2P^2+C1R1R2P+C3R1R2P+C1C3L2R1R2P^3 \\
&\quad + (L2+2C1L2R1P+2C3L2R2P+C1R1R2+C3R1R2+3C1C3L2R1R2P^2)\,s \\
&\quad + (C1L2R1+C3L2R2+3C1C3L2R1R2P)\,s^2+C1C3L2R1R2s^3\,) \tag{28.4}
\end{aligned}
$$

After obtaining the transfer function, the system of equations is created. It is done by comparing denominator coefficients from the transfer function of the lossy filter and from the transfer function of the prototype. Before solving the system, all the values of Q-factors have to be established. For a singly terminated structure, the system has then $n + 1$ unknowns and $n + 1$ equations, where n is the filter order. For a doubly terminated ladder, the system has one degree of freedom, since the number of unknowns is $n + 2$ (n reactive elements and two terminating resistors).

In order to solve the obtained system of equations, numerical methods are recommended, since an analytical approach is complicated even for low filter orders and for higher orders it is simply impossible.

For every numerical method, an initial point is needed to start iterations. For all the lossy filters designed with the presented method, a satisfactory initial point is made of element values of a prototype filter.

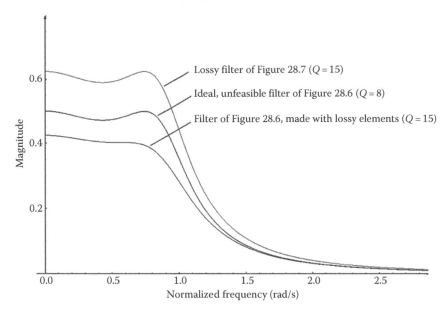

FIGURE 28.8 Magnitude functions for different configurations of filter of Figure 28.6.

As an example, consider a system of equations created by comparing the coefficients of transfer function of the lossy filter (28.4) to the transfer function of the prototype filter shown in Figure 28.5. Setting values of $P = 0.067$ ($Q = 15$) and $R_2 = 1.25$, we get a system

$$
\begin{cases}
0.45 + 0.067\,L1 + 0.002\,C1\,L1 + 0.067\,L2 + 0.0003\,C1\,L1\,L2 + R1 + 0.03\,C1\,R1 + 0.004\,C1\,L2\,R1 \\[4pt]
\quad = 2L1 + 0.06\,C1\,L1 + L2 + 0.0133333\,C1\,L1\,L2 + 0.45\,C1\,R1 + 0.133333\,C1\,L2\,R1 \\[4pt]
\quad = 5.0074\,0.45\,C1\,L1 + 0.2\,C1\,L1\,L2 + C1\,L2\,R1 = 4.77239\,C1\,L1\,L2 = 4.44713 \\[4pt]
Initial\ point\,(L1 = 1.86369, C1 = 1.28036, L2 = 1.86369, R1 = 1)
\end{cases}
\tag{28.5}
$$

After solving system (28.4) using prototype elements as the initial point, solution (28.5) determines the element values of the lossy filter

$$
\begin{cases}
L1 = 1.7, \\
C1 = 1.29, \\
L2 = 2.0267, \\
R1 = 0.436
\end{cases}
\tag{28.6}
$$

Each reactive element of the filter has Q-factor 15. Magnitude functions of the prototype filter and lossy filter are shown in Figure 28.8.

28.4 Solving the System of Equations

When solving the system (28.3) to find element values of a lossy filter, several problems can occur. When using a Newton method, it may happen that the Jacobian matrix is singular. This problem can be easily fixed by perturbing slightly the initial point by trial and error.

When using Mathematica, the commands shown below can be used to solve system (28.5) and to obtain solution (28.6) using the Newton method with command *FindRoot*. It is assumed that commands (28.4) have been already performed.

Input:
```
P=1./15; R2=1.25;
InitialPoint={{L1,1.86369},{C1,1.28036},{L2,1.86369},{R1,1}}
LeftSide=CoefficientList[Denominator[Tf],s]
RightSide={2,5.00774,4.77239,4.44713}
FindRoot[LeftSide==RightSide,InitialPoint]
```

Output:
$$\{L1 \rightarrow 1.70086, C1 \rightarrow 1.29011, L2 \rightarrow 2.02668, R1 \rightarrow 0.436044\} \tag{28.7}$$

For the singly terminated filter, the system has no degrees of freedom, that is, the number of unknowns matches the number of equations. If the introduced losses are not too large, the solution can always be found. For the lossy filter, the obtained solution has to be real and positive. As it was mentioned, the right sides of the system of equations (28.3) are coefficients of prototype transfer function (28.1). This function may be written in several equivalent ways, since one can always multiply or divide the numerator and the denominator by the same factor. The question that appears now is: does the solution depend on the chosen factor? The answer is positive, which means that with the factor chosen improperly, a positive solution may not be found. For example, it may happen that $b_0' = 1 + R1 + L2C3$ and $b_0 = 0.5$ which obviously will not lead us to the positive solution. Instead of looking for the appropriate factor to reduce the transfer function (28.1), one can transform the system of equations (28.3) in the following way:

$$\begin{cases} b_0' b_0 = b_0 b_0' \\ b_1' b_0 = b_1 b_0' \\ \quad \vdots \\ b_n' b_0 = b_n b_0' \end{cases} \tag{28.8}$$

In system (28.8), all the left sides are multiplied by b_0 and all the right sides are multiplied by b_0'. In this way, the first equation is always true and can be omitted. The solution of (28.8) obtained numerically will not necessarily satisfy system (28.3), but it will always satisfy system (28.9):

$$\begin{cases} b_0' = kb_0 \\ b_1' = kb_1 \\ \quad \vdots \\ b_n' = kb_n \end{cases} \tag{28.9}$$

where k is a constant. Such a solution, if it is positive, is sufficient to build a lossy filter. This method can also be used for the doubly terminated network, where the system of equations has one degree of freedom.

28.5 Limitations for Losses

From the filter designer's point of view, some of the most important issues that have to be considered when designing a lossy filter are how the introduced losses affect the filter response and what are the limitations for losses. Unfortunately, there is no simple answer for these questions and every type of filter requires special consideration. When designing a singly terminated lossy filter, the minimal value of reactive elements Q-factor can be determined numerically. It is done by setting Q-factors big enough to find the solution of system (28.3) and next by lowering the Q-factors as long as system (28.3) has real positive solution. There is a general rule concerning the limit for Q. The higher the order of the filter,

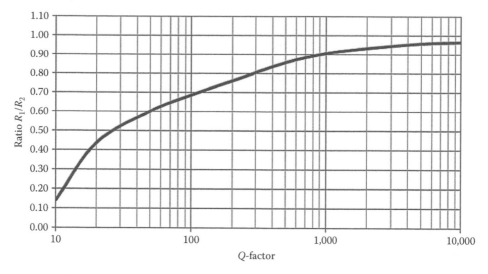

FIGURE 28.9 Relationship between ratio R_1/R_2 and Q-factor of reactive elements for filter of Figure 28.7.

the higher the Q demanded to find the solution. Moreover, if Q reaches its limit, it may happen that the attenuation of the filter is not acceptable or the shape of the magnitude function is much too degraded due to shifting of zeros. In that case, it is enough to introduce a slightly bigger Q than minimal needed to obtain a solution.

As was mentioned, for the doubly terminated filters, system (28.3) has one degree of freedom, which means that one of the unknown lossy filter elements may be fixed. The choice is arbitrary, but if one of the terminating resistors is chosen to establish, one can try to introduce the losses, and watch their influence on the ratio of these terminating resistors. Lower the values of Q-factor, the more asymmetric the filter becomes.

As an example, consider the filter of Figure 28.7. The relationship between the introduced Q-factor and the ratio R_1/R_2 is shown in Figure 28.9.

This relationship may be considered as the trade-off between the level of the introduced losses and the symmetry of the doubly terminated filter.

28.6 Conclusion

The method described above is based on poles restoring approach. Along with its simplicity, it gives good results and the obtained filters can be implemented as traditional ladders as well as into microelectronic form, where inductors are simulated by active circuits. Other issues concerning designing lossy filters, particularly focused on microwave filters, are widely described in the literature [7]. The coupling matrix synthesis method based on a lossy transversal network model is presented in [8]. An effective optimization method for lossy ladders is described in detail in the classical paper [9].

References

1. M. Jagiela and B. Wilamowski, A methodology of synthesis of lossy ladder filters, in: *INES 2009, International Conference on Intelligent Engineering Systems*, April 16–18, 2009, Barbados, pp. 45–50.
2. B.S. Senior, I.C. Hunter, and J.D. Rhodes, Synthesis of lossy filters, in: *32nd European Microwave Conference*, October 2002, Milan, Italy, pp. 1–4.
3. C. Fu and H. Wang, An efficient optimization for passive filter design, in: *INDIN 2008, 6th IEEE International Conference on Industrial Informatics*, July 13–16, Daejeon, Korea, 2008, pp. 631–634.

4. B.M. Wilamowski and R. Gottiparthy, Active and passive filter design with MATLAB, *International Journal on Engineering Educations*, 21(4), 561–571, 2005.
5. R. Koller and B. Wilamowski, LADDER—A microcomputer tool for advanced analog filter design and simulation, *IEEE Transactions on Education*, 39(4), 478–487, November 1996.
6. B.M. Wilamowski, S.F. Legowski, and J.W. Steadman, Personal computer support for teaching analog filter analysis and design courses, *IEEE Transactions on Education*, E-35(4), 351–361, 1992.
7. R. Levy, R.V. Snyder, and G. Matthaei, Design of microwave filters, *IEEE Transactions on Microwave Theory and Techniques*, 50(3), 783–793, March 2002.
8. V. Miraftab and M. Yu, Generalized lossy microwave coupling matrix synthesis and design using mixed technologies, *IEEE Transactions on Microwave Theory and Techniques*, 56(12, Part 2), 3016–3027, December 2008.
9. C.A. Desoer and S.K. Mitra, Design of lossy ladder filters by digital computer, *IRE Transactions on Circuit Theory*, 8(3), 192–201, September 1961.

V

Electromagnetics

29

Electromagnetic Fields I

Sadasiva M. Rao
Auburn University

Tyler N. Killian
Auburn University

Michael E. Baginski
Auburn University

29.1 Introduction

In this chapter, we shall study two important phenomena, electricity and magnetism, and show how these two phenomena combine to give rise to electromagnetism. Electricity and magnetism have been known to mankind since ancient times. Thales of Miletus, a Greek mathematician who lived almost 600 years before Christ, reported that amber produces sparks when rubbed with silk cloth and attracts fluff. He also noted that a certain natural material, called landstone, has attractive powers. The names *electricity* and *magnetism* owe their existence to these discoveries. The word electricity was derived after *electron*, the Greek word for amber, and *magnetism* was derived from *Magnesia*, the place where landstone was found. Over the next twenty-five centuries, scientists, as well as ordinary folk, observed electricity and magnetism in a variety of situations. However, in 1831, Michael Faraday, a British scientist, experimentally demonstrated that these two seemingly different phenomena, in fact, originate from the same source, that is, from charge. Thus, our study of electricity and magnetism begins with charge, and current, which is yet another important quantity that gets generated when the charge is varied with time.

29.2 Charge, Current, and Continuity Equation

In this section, we study two fundamental quantities, charge and current, and the relationship between these two quantities.

29.2.1 Electric Charge

There are four fundamental quantities in nature: mass, length, time, and charge. Like other quantities, charge also preserves the principle of conservation which implies that charge can neither be created nor destroyed. The unit for charge is the *coulomb*, named after the scientist who quantified the force between charges.

We know from basic high school physics that all matter is composed of molecules and each molecule consists of several atoms. Furthermore, we know that each atom, no matter what material it constitutes, consists of electrons, protons, and neutrons. Electrons are negatively charged particles while protons are positively charged. Finally, neutrons, as the name indicates, are electrically neutral. Thus, we know that charge can be either positive or negative.

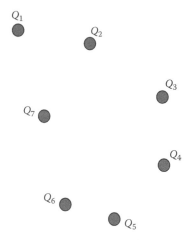

FIGURE 29.1 Discrete charges in free space.

Since electrons and protons are so tiny in size, we refer them as point charges. Mathematically, any charge with zero mass and zero volume may be called a point charge. Usually, the point charge is denoted by the Q.

Next, we take a look at various charge distributions. If we arrange several point charges, both positive and negative, over a certain region of space, as shown in Figure 29.1, then we have a *discrete charge distribution*. The total charge (Q_T) in this charge distribution is given by the summation of all the individual charges. Thus, we can write

$$Q_T = \sum_{i=1}^{N} Q_i \quad (C) \tag{29.1}$$

However, if the charge is distributed over a region in a *continuous* manner, then we can have a linear charge distribution (q_l), a surface charge distribution (q_s), or a volume charge distribution (q_v). For the linear charge distribution, the charge is spread over a line such as a wire. For the surface charge distribution, the charge is painted over a surface such as a plate. Finally, for the volume charge distribution, the charge is assembled in a finite volume such as a container. All these cases are depicted in Figure 29.2.

For the continuous case, the total charge (Q_T) for a given charge distribution is given by

$$Q_T = \int_l q_l dl = \int_S q_s ds = \int_v q_v dv$$

Next, we will look at the concept of current. Generally, while studying circuit theory, electrical engineering students are introduced to current as the rate of change of charge. In the following section, we present a more general definition of current applicable for field theory as well as circuit theory.

q_l q_s q_v

FIGURE 29.2 Charge distributions in free space.

29.2.2 Electric Current

Generally speaking, current is the result of charges in motion. Whenever charges move from one position to another, we have current. However, the charge transport can happen in a variety of ways and as a result we have different types of current.

First, consider the situation as shown in Figure 29.3. Here, we have a charge cloud moving in space. Since the charges are moving, we have current. Also, note that the charges are moving in empty space. Such a current is known as *convection current*.

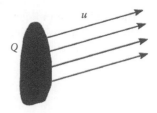

FIGURE 29.3 Charge cloud moving in free space.

Next, we can have charges moving in a material medium and causes current flow. In the material medium, charge transport takes place due to the external forces. This type of current is known as *conduction current*. In circuit theory, we mostly deal with conduction currents since the current is mainly flowing in the wires.

There is also yet another of type of current known as *displacement current* which occurs due to a displacement of charges. We shall study this phenomenon a little later.

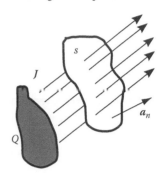

FIGURE 29.4 Charge cloud passing through a surface.

Regardless of the nature of current, let us consider a charge distribution moving at a velocity U, as shown in Figure 29.3. The charge density is equal to q_v C/m^3. The current density J may be defined as

$$J = q_v U. \tag{29.2}$$

Next, we allow this charge cloud to pass through a surface S, as shown in Figure 29.4. Then, we can calculate the total current I passing through the surface S as

$$I = \int_S J \cdot a_n ds \tag{29.3}$$

where a_n is the unit vector normal to the surface. Here, we note that the current density J is a vector quantity, whereas the current I is a scalar quantity. Also note that the current I is maximum when J and a_n are pointing in same direction. Further, we know that the current is measured in *amperes*, which implies that J is measured in A/m^2.

It is clear from the previous discussion that charges and current are interrelated. This is an important observation that has profound implications as we shall see in the following sections.

29.2.3 Continuity Equation

We have stated earlier that charge can neither be created nor destroyed. So, let us consider the following situation as shown in Figure 29.5.

Here, we have a certain amount of charge Q enclosed in a container of volume V. Let S be the surface bounding the volume. The amount of charge in the container may be increased or decreased by making the charge flow into or out of the container, respectively. Since the charge flow results in current, from the principle of conservation, we have

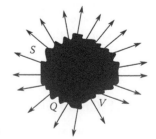

FIGURE 29.5 Charge conservation and continuity equation.

$$I = -\frac{dQ}{dt} \tag{29.4}$$

The minus sign in Equation 29.4 signifies that the level of charge inside the container is decreasing when there is a positive outward current flow. Furthermore, using Equations 29.2 and 29.3, Equation 29.4 may be written as

$$\int_S J \cdot a_n ds = -\int_V \frac{\partial q_v}{\partial t} dv \tag{29.5}$$

Finally, using the vector identity given by

$$\int_V \nabla \cdot A \, dv = \oint_S A \cdot a_n ds \tag{29.6}$$

where
    A is any arbitrary vector in a volume *V*
    a_n represents the unit outward normal vector to the surface *S* bounding the volume*

We can write Equation 29.5 as

$$\nabla \cdot J = -\frac{\partial q_v}{\partial t} \tag{29.7}$$

Equations 29.5 and 29.7 represent the *continuity equation* in integral and differential form, respectively.

29.3 Electrostatic and Magnetostatic Fields

In this section, our study is focused on the concept of static fields, *viz.*, electrostatic and magnetostatic fields. Before we undertake such a task, however, we first need to know the meaning of the word *static*. We define a quantity as static quantity when it does *not* change with respect to time, for example, static charge. Obviously, the static charge has no velocity and, consequently, does not generate current. On the other hand, if the charge varies with time, then it is no longer a static quantity. Plus, we also have current in this situation.

The current generated by a time-varying charge can be static, however. This is due to the fact that current flow can be steady, which implies that the rate of change of charge is constant. Of course, in a more general situation, neither charge nor current is static.

The moral of this story is that we can have static charges and static currents, although, not at the same time. In the following subsection, we discuss what the static charges do when situated in an open empty space.

29.3.1 Coulomb's Law and Electric Field

Let us consider an infinite space filled with an only one type of material (or even an empty space like a vacuum). Such a space may be defined as a *homogeneous* space. Next, let us consider a simple situation where two point charges Q_1 and Q_2 are located in this infinite, homogeneous space, as shown in Figure 29.6. We assume that there are no other charges in the whole space. Then, Coulomb's law states that *The electric force F_{21} exerted on Q_2 by Q_1 is proportional to the magnitude of the charges, Q_1 and Q_2, and inversely proportional to the square of the distance R separating them.*

---

* Equation 29.6 is, in fact, the well-known divergence theorem in vector calculus.

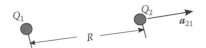

FIGURE 29.6 Two point charges in free space.

Mathematically, referring to Figure 29.6, Coulomb's law may be given by

$$F_{21} \propto \frac{Q_1 Q_2}{R^2} a_{21} \quad \Rightarrow \quad F_{21} = k \frac{Q_1 Q_2}{R^2} a_{21} \qquad (29.8)$$

where

k is a proportionality constant which depends on the surrounding medium

a_{21} is the unit vector along the line joining Q_1 and Q_2 and pointing away from Q_2

It is very important to note the direction of a_{21} as one cannot afford to make a mistake at this stage. In the rationalized meters-kilogram-seconds (RMKS) system, the constant k for air (or vacuum) is given by

$$k = \frac{1.0}{4\pi\varepsilon_0} \quad \text{and} \quad \varepsilon_0 = 8.854 \times 10^{-12} \qquad (29.9)$$

Coulomb's law may be interpreted in two ways. One way of understanding the law is to imagine that the charge Q_1 is directly exerting the force on Q_2 given by Equation 29.8. However, we can also interpret the law in another way. Here, we say that the charge Q_1 merely creates a field, known as *electric field* (*E*) around it and this field, in turn, causes the force on Q_2. Thus, we can see, according the second interpretation, the charge Q_2 is merely required to detect the presence of the electric field E. In fact, the presence of charge Q_1 is sufficient for the creation of an electric field.

Although both arguments are valid, the second argument is more appealing and accepted by the scientific community at large. Thus, following the second line of argument and referring to Figure 29.7, we define the electric field E, measured in volts per meter (V/m), at a point P as

$$E = \frac{Q}{4\pi\varepsilon_0 R^2} a_R \qquad (29.10)$$

In Equation 29.10, E is the electric field at P due to the point charge located at a distance R meters. Further, a_R is the unit vector along the line joining the location of the point charge and the point of observation and is pointing away from the point charge. It is very important to note all these quantities since they are fundamental to what we will learn later.

Note that in Equation 29.10, we dropped all the subscripts and refer to the point charge Q developing a field E surrounding it. However, we can always calculate the electrostatic force using the relation

$$F = qE \qquad (29.11)$$

where q is the test charge required to detect the electric field.

Next, we consider the electric field for a given charge distribution. In Figures 29.1 and 29.2, we have shown discrete charges and other different types of charge distributions. In all these cases, the total electric field is obtained using the principle of superposition, that is, calculating the field from individual charges and performing vector addition. Mathematically, we have

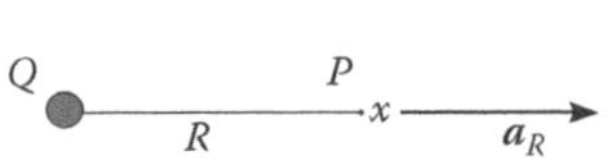

FIGURE 29.7 Definition of electric field.

$$E_T = \sum_{i=1}^{N} \frac{Q_i}{4\pi\varepsilon_0 R_i^2} a_{R_i}$$

$$= \int_l \frac{q_l dl}{4\pi\varepsilon_0 R^2} a_R = \int_S \frac{q_s ds}{4\pi\varepsilon_0 R^2} a_R = \int_V \frac{q_v dv}{4\pi\varepsilon_0 R^2} a_R \qquad (29.12)$$

In the following, we consider some important facts about the electric field:

- The sources for the electric field are charges.
- The charges can be both positive and negative.
- The charges of opposite polarities attract each other. Similarly, charges of same polarity repel each other.
- The electric field emanates from the positive charge and terminates on the negative charge. If we have positive (negative) charge at some place and no negative (positive) charge in the vicinity, then we assume that the negative (positive) charge is at infinity.

29.3.2 Biot–Savart Law and Magnetic Flux Density

First, let us consider the concept of a *current element*. Mathematically, we may define a current element $I\,\boldsymbol{dl}$ as a piece of wire of length dl carrying current I and pointing in the direction of the arrow. According to this definition, we may consider a current loop as a summation of current elements as shown in Figure 29.8. Thus, the current element forms the basic building block for a current distribution just like a point charge forms the building block for a charge distribution. In fact, we may consider any current distribution as a collection of current elements.

However, a word of caution is necessary. Unfortunately, the current element is *not* a physical quantity, that is, we cannot have an isolated current element sitting in an empty space like point charges do. Current elements must form a closed loop so that there can be a current flow. We all know that if the loop is broken, the current through the wire is zero.

Next, we consider two current elements $I_1\boldsymbol{dl}_1$ and $I_2\boldsymbol{dl}_2$ located in infinite homogenous space, as shown in Figure 29.9. Then, Biot–Savart law states that the magnetic force $d\boldsymbol{F}_{21}$ exerted on current element $I_2\boldsymbol{dl}_2$ by $I_1\boldsymbol{dl}_1$ is proportional to the magnitude of the currents, I_1 and I_2, and inversely proportional to the square of the distance R separating them.

Mathematically, referring to Figure 29.9, Biot–Savart law may be given by

$$d\boldsymbol{F}_{21} \propto \frac{I_1\boldsymbol{dl}_1 \times (I_2\boldsymbol{dl}_2 \times \boldsymbol{a}_{21})}{R^2} \quad \Rightarrow \quad d\boldsymbol{F}_{21} = k\frac{I_1\boldsymbol{dl}_1 \times (I_2\boldsymbol{dl}_2 \times \boldsymbol{a}_{21})}{R^2} \tag{29.13}$$

where

k is a proportionality constant that depends on the surrounding medium

$\boldsymbol{a}_{21}$ is the unit vector along the line joining the centers of $I_1\boldsymbol{dl}_1$ and $I_2\boldsymbol{dl}_2$ and pointing away from $I_2\boldsymbol{dl}_2$

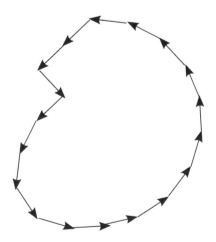

FIGURE 29.8　An arbitrary loop formed by current elements.

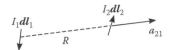

FIGURE 29.9 Two current elements in free space.

Further, note that the sign "×" in Equation 29.13 represents vector cross product *not* multiplication.

In the RMKS system, the constant k for air (or vacuum) is given by

$$k = \frac{\mu_0}{4\pi} \quad \text{and} \quad \mu_0 = 4\pi \times 10^{-7} \tag{29.14}$$

Again, Biot–Savart law may be interpreted in two ways just like Coulombs' law. One way of understanding the law is to imagine that the current element $I_1 dl_1$ is directly inducing the force on $I_2 dl_2$ given by Equation 29.13. However, we can also interpret the law in another way. Here, we say that the current element $I_1 dl_1$ merely creates a field, known as *magnetic flux density* around it and this field, in turn, causes the force on $I_2 dl_2$. Thus, we can see, according to the second interpretation, that the current element $I_2 dl_2$ is only required to detect the presence of the magnetic flux density B.

Following the second line of argument, we define the magnetic flux density B, measured in Webers per square meter (Wb/m^2), at a point P as

$$B = \frac{\mu_0}{4\pi} \int_l \frac{I dl \times a_R}{R^2} \tag{29.15}$$

Note that, in Equation 29.15, we dropped all the subscripts and introduced an integral. The integration appears due to the fact that the current element must belong to a closed loop.

In the following, we list some important facts about the magnetic flux density:

- The sources for the magnetic flux density are currents that flow in a closed path.
- The current element is a vector quantity.
- The magnetic flux lines always form closed loops implying that the flux lines have no beginning or end.

Again, it is very important to study Figure 29.9 and note all the quantities. Particularly, note the direction of the unit vectors. In order to illustrate the importance of these quantities, we present a few examples in the following.

29.3.3 Illustrative Examples

Example 29.1

Consider three point charges $Q_1 = Q_2 = Q_3 = Q$ located at $(a, 0, 0)$, $(0, a, 0)$, and $(0, 0, a)$, as shown in Figure 29.10a. (1) Calculate the total electric field E_T at the origin $(0, 0, 0)$. (2) If the charges are replaced by current elements, as shown in Figure 29.10b, calculate the magnetic flux density B_T at the origin.

Solution

The total electric field E_T at $(0, 0, 0)$ is the sum of the electric fields from each charge. Notice that the distance from each charge to the point of observation is a. Thus, using Equation 29.10,

$$E_T = E_1 + E_2 + E_3$$

$$= \frac{Q}{4\pi\varepsilon_0 a^2}(-a_x) + \frac{Q}{4\pi\varepsilon_0 a^2}(-a_y) + \frac{Q}{4\pi\varepsilon_0 a^2}(-a_z)$$

$$= \frac{-Q}{4\pi\varepsilon_0 a^2}\left[a_x + a_y + a_z\right] V/m$$

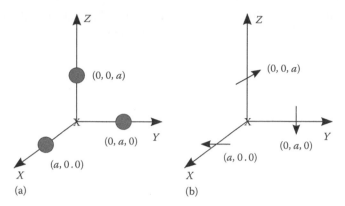

FIGURE 29.10   Three point charges and three current elements in free space.

Next, let us calculate the magnetic flux density. Using Equation 29.15, we have

$$B_1 = \frac{\mu_0}{4\pi} \frac{Idl}{a^2}(-\boldsymbol{a}_y) \times (-\boldsymbol{a}_x) = \frac{\mu_0 Idl}{4\pi a^2}(-\boldsymbol{a}_z)$$

$$B_2 = \frac{\mu_0}{4\pi} \frac{Idl}{a^2}(-\boldsymbol{a}_z) \times (-\boldsymbol{a}_y) = \frac{\mu_0 Idl}{4\pi a^2}(-\boldsymbol{a}_x)$$

$$B_3 = \frac{\mu_0}{4\pi} \frac{Idl}{a^2}(-\boldsymbol{a}_x) \times (-\boldsymbol{a}_z) = \frac{\mu_0 Idl}{4\pi a^2}(-\boldsymbol{a}_y)$$

Thus, we have

$$B_T = -\frac{\mu_0 Idl}{4\pi a^2}\left[\boldsymbol{a}_x + \boldsymbol{a}_y + \boldsymbol{a}_z\right] \quad (\text{Wb/m}^2)$$

Example 29.2

Consider a straight wire of length $2l$ meters carrying a uniform charge density q_l C/m, as shown in Figure 29.11a. (1) Calculate the electric field $\boldsymbol{E}$ at a point P, located d meters from the axis of the wire. (2) If the wire is carrying uniform current, as shown in Figure 29.11b, then calculate the magnetic flux density at the point P.

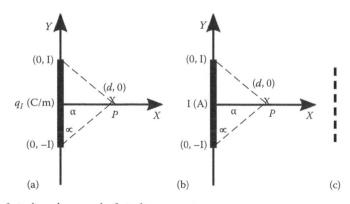

FIGURE 29.11   A finite line charge and a finite line current.

Solution

In this example, we again use the principle of superposition as applied in Example 29.1. In order to do this, we subdivide the wire into very small segments. Note that, however small the segment may be, it still has some finite length given by δl. We attack this problem by viewing each small segment as a point charge. Then we know how to compute the electric field due to this approximated point charge. Finally, the total electric field is the summation of fields from the charge located on each segment.

However, there is one important difference. Since we do not have isolated charges, the final step in the solution procedure is not *summation* but *integration*.

Referring to Figure 29.11c, the charge **dQ** on each segment is given by

$$dQ = q_l dl = q_l dy$$

$$R = \sqrt{d^2 + y^2}$$

$$\boldsymbol{a}_R = \frac{(d-0)\boldsymbol{a}_x + (0-y)\boldsymbol{a}_y}{R}$$

Thus, we have

$$\boldsymbol{E} = \int_{-l}^{l} \frac{q_l dy}{4\pi\varepsilon_0} \frac{(d\boldsymbol{a}_x - y\boldsymbol{a}_y)}{(d^2 + y^2)^{3/2}}$$

$$= \frac{q_l}{4\pi\varepsilon_0} \left[d\boldsymbol{a}_x \int_{-l}^{l} \frac{dy}{(d^2 + y^2)^{3/2}} - \boldsymbol{a}_y \int_{-l}^{l} \frac{y dy}{(d^2 + y^2)^{3/2}} \right] \quad \text{(V/m)}$$

The two integrals in the previous step may be evaluated using simple substitutions or, alternatively, using Tables of Integration. Notice that the second integral, after substituting the limits, is equal to zero. Thus, we have

$$\boldsymbol{E} = \boldsymbol{a}_x \frac{q_l}{4\pi\varepsilon_0 d} \left[\frac{2\ell}{\sqrt{d^2 + \ell^2}} \right] = \boldsymbol{a}_x \frac{q_l}{2\pi\varepsilon_0 d} \cos\alpha$$

where α is the angle as shown in Figure 29.11a.

For part (b), referring to Figure 29.11b, we define each current element as

$$I\,\boldsymbol{dl} = I\,dy\boldsymbol{a}_y$$

$$R = \sqrt{d^2 + y^2}$$

$$\boldsymbol{a}_R = \frac{(d-0)\boldsymbol{a}_x + (0-y)\boldsymbol{a}_y}{R}$$

Thus, we have

$$d\boldsymbol{B} = \frac{\mu_0}{4\pi} \frac{I\,dy\boldsymbol{a}_y \times (d\boldsymbol{a}_x - y\boldsymbol{a}_y)}{(d^2 + y^2)^{3/2}}$$

$$= (-\boldsymbol{a}_z) \frac{\mu_0 I d}{4\pi} \frac{dy}{(d^2 + y^2)^{3/2}}$$

and

$$B = (-a_z)\frac{\mu_0 I d}{4\pi}\int_{-l}^{l}\frac{dy}{(d^2+y^2)^{3/2}}$$

$$= (-a_z)\frac{\mu_0 I d}{4\pi}\left[\frac{2\cos\alpha}{d^2}\right]$$

$$= (-a_z)\frac{\mu_0 I}{2\pi d}\cos\alpha \ (\text{W/m}^2)$$

Now, we use the result of the previous example to derive an important result. Let us consider the situation where the finite length is changed to infinite length, that is, $l \rightarrow \infty$. For this case, $\alpha = 0$ and $\cos\alpha = 1$. Thus, for an infinite line along the y-axis, the fields at a point d meters away along the x-axis are given by

$$E = a_x\frac{q_l}{2\pi\varepsilon_0 d} \ (\text{V/m}) \quad \text{and} \quad B = (-a_z)\frac{\mu_0 I}{2\pi d} \ (\text{W/m}^2) \tag{29.16}$$

Further, we note the following:

- a_x is the radial vector from the axis of the wire to the point of observation, which may be referred to, in general, as a_d.
- $-a_z$ is actually the vector obtained by taking the cross product of the current vector, a_y, with the radial vector a_x. For a general case, this may be written as $a_l \times a_z$.

Thus, in general, for an infinite wire, the electric field and the magnetic flux density are given by

$$E = \frac{q_l}{2\pi\varepsilon_0 d}a_d \ (\text{V/m}) \tag{29.17}$$

$$B = \frac{\mu_0 I}{2\pi d}a_l \times a_d \ (\text{W/m}^2) \tag{29.18}$$

Example 29.3

Consider a circular plate of radius a meters carrying a uniform surface charge density q_s C/m^2, as shown in Figure 29.12a.

1. Calculate the electric field E at a point $P(0,0,d)$.
2. If the plate is carrying uniform current with surface current density $J_s = J_0 a_y$ A/m, as shown in Figure 29.12b, then calculate the magnetic flux density at the point P.

Solution
Again, following the procedures of previous examples, let us divide the circular plate into elemental pieces as shown in the figure. The area of each piece is $dS = (d\rho)(\rho \, d\phi)$. Thus, we have

$$dQ = q_s dS = q_s(d\rho)(\rho \, d\phi)$$

$$R = \sqrt{d^2+\rho^2}$$

$$a_R = \frac{d a_z - \rho a_\rho}{R} = \frac{d a_z - \rho\cos\phi a_x - \rho\sin\phi a_y}{R}$$

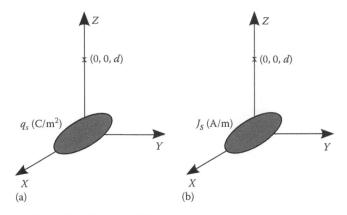

FIGURE 29.12 A disk of (a) surface charge and (b) ϕ-directed current density.

Therefore,

$$d\boldsymbol{E} = \frac{q_s \rho d\rho d\phi}{4\pi\varepsilon_0} \frac{d\boldsymbol{a}_z - \rho\cos\phi\boldsymbol{a}_x - \rho\sin\phi\boldsymbol{a}_y}{(d^2 + \rho^2)^{3/2}}$$

Now, to calculate the electric field $\boldsymbol{E}$, we need to integrate the above expression. Note that we need to perform a double integration with the limits $\rho \to 0$ to a and $\phi \to 0$ to 2π. It is easy to prove that the integrals along $\boldsymbol{a}_x$ and $\boldsymbol{a}_y$ directions are zero. Thus, we have

$$\boldsymbol{E} = \boldsymbol{a}_z \frac{q_s d}{4\pi\varepsilon_0} \int\limits_{\rho=0}^{a}\int\limits_{\phi=0}^{2\pi} \frac{\rho d\rho d\phi}{(d^2 + \rho^2)^{3/2}}$$

$$= \boldsymbol{a}_z \frac{q_s d}{2\varepsilon_0} \int\limits_{\rho=0}^{a} \frac{\rho d\rho}{(d^2 + \rho^2)^{3/2}}$$

$$= \boldsymbol{a}_z \frac{q_s d}{2\varepsilon_0} \left[\frac{-1}{\sqrt{d^2 + \rho^2}} \right]_0^a$$

$$= \boldsymbol{a}_z \frac{q_s d}{2\varepsilon_0} \left[\frac{1}{d} - \frac{1}{\sqrt{d^2 + a^2}} \right] \ \ (\text{V/m})$$

In the above expression, the integral on ρ is evaluated using tables of integration.

For part (b), referring to Figure 29.12b, we have

$$I d\boldsymbol{l} = J_0(d\rho)(\rho d\phi)\boldsymbol{a}_y$$

$$R = \sqrt{d^2 + \rho^2}$$

$$\boldsymbol{a}_R = \frac{d\boldsymbol{a}_z - \rho\boldsymbol{a}_\rho}{R} = \frac{d\boldsymbol{a}_z - \rho\cos\phi\boldsymbol{a}_x - \rho\sin\phi\boldsymbol{a}_y}{R}$$

Thus,

$$dB = \frac{\mu_0}{4\pi} \frac{J_0(d\rho)(\rho\, d\phi)(\boldsymbol{a}_y) \times (d\boldsymbol{a}_z - \rho\cos\phi\boldsymbol{a}_x - \rho\sin\phi\boldsymbol{a}_y)}{(d^2 + \rho^2)^{3/2}}$$

$$= \frac{\mu_0 J_0 \rho\, d\rho\, d\phi}{4\pi(d^2 + \rho^2)^{3/2}} [d\boldsymbol{a}_x + \rho\cos\phi\boldsymbol{a}_z]$$

To obtain the total magnetic flux density, we integrate the previous expression between the limits ϕ from 0 to 2π and ρ from 0 to a. Note that the first part of the expression, that is, the x-component, is same as that of part (a). Also, the z-component vanishes since the integration of $\cos\phi$ over 0 to 2π is zero. Thus, we have

$$B = (\boldsymbol{a}_x)\frac{\mu_0 J_0 d}{2}\left[\frac{1}{d} - \frac{1}{\sqrt{d^2 + a^2}}\right] (\text{W/m}^2)$$

Now, we use the result of the previous example to derive another important result. Let us consider the situation where the radius of the plate is extended to infinity, that is, $a \to \infty$. Thus, for an infinite plate situated at $z = 0$, the fields at a point d meters away from the plate along the z-axis are given by

$$E = \boldsymbol{a}_z \frac{q_s}{2\varepsilon_0} (\text{V/m}) \quad \text{and} \quad B = (\boldsymbol{a}_x)\frac{\mu_0 J_0}{2} (\text{W/m}^2)$$

Further, we note the following:

- $\boldsymbol{a}_z$ is the normal vector from the plane of the plate to the point of observation, which may be referred to, in general, as $\boldsymbol{a}_n$.
- $\boldsymbol{a}_x$ is actually the vector obtained by taking a cross product of the current direction vector, $\boldsymbol{a}_y$, with the normal vector $\boldsymbol{a}_n = \boldsymbol{a}_z$.

Thus, in general, for an infinite plate, the electric field and the magnetic flux density are given by

$$E = \frac{q_s}{2\varepsilon_0}\boldsymbol{a}_n (\text{V/m}) \quad \text{and} \quad B = \frac{\mu_0 \boldsymbol{J}_s \times \boldsymbol{a}_n}{2} (\text{W/m}^2) \qquad (29.19)$$

Also, note that in Equation 29.19, the normal vector $\boldsymbol{a}_n$ is pointing into the region where the fields are observed.

Before closing this section, we note the following important observations:

1. The sources for the electric field and magnetic flux density are charges and currents, respectively.
2. The basic building blocks for the case of electric field and magnetic flux density are point charges and elemental currents, respectively.
3. The total fields for discrete or continuous source distributions are obtained using the principle of superposition.
4. For elemental sources (i.e., point charges and elemental currents), the fields vary as the inverse of the square of the distance.
5. For infinitely long line sources, the fields vary as the inverse of the distance.
6. For infinite surface source distributions, the fields are constant no matter how far the observation point is located.

29.4 Potential Theory

In the previous section, we have learned a general technique to calculate E and B at any observation point once the source distribution is given. The technique presented is applicable to all situations, at least in theory. However, the method is cumbersome when the source distribution is complicated. This is because the procedure involves developing three integrals, one for each component, and evaluating them. Furthermore, all operations are vector operations, which are somewhat tricky. Thus, there is a need to develop alternate methods that may provide mathematically easier solution procedures. It is widely believed that *potential theory*, as discussed in the following sections, provides such a method.

29.4.1 Concept of Potential Energy

We begin our study by taking a closer look at the concept of potential energy. From fundamental physics, we are all familiar with the concept of objects being moved by a force field, *viz.*, the gravitational field. The fundamental concept that should be borne in mind is that whenever there is work done on the system, such as moving an object in a force field, the amount of work done is transferred to the system as potential energy.

This concept is also valid for force fields E and B, which we are presently studying. Thus, if we move a point charge either in an electric field or a flux field, work must be done to counter the force exerted by these fields. Obviously, the work done is stored in the system as the potential energy for the point charge in the presence of the field. Now this system is capable of performing some useful functions. In other words, the system acquires certain potential or capability to perform given tasks.

Mathematically, we may express incremental work done or the potential energy stored as

$$dW = -F \cdot dl \qquad (29.20)$$

where dl is the incremental displacement along the direction of the movement. It is obvious that since the force fields are vector fields, the amount of work done depends on the direction of movement. Also, the negative sign in Equation 29.20 implies that work must be done to counter the force exerted by the fields.

Next, the total work done when a point charge moved from point A to point B is given by

$$W_{AB} = -\int_A^B F \cdot dl \text{ (J)} \qquad (29.21)$$

29.4.2 Electric Potential

We define electric potential, or simply *potential*, as the work done per unit charge and it is given by

$$V_{AB} = \frac{W_{AB}}{Q} = \frac{-1}{Q}\int_A^B F \cdot dl = \frac{-1}{Q}\int_A^B QE \cdot dl$$

$$= -\int_A^B E \cdot dl \qquad (29.22)$$

Notice that potential is, in reality, the potential difference between two points, that is, A and B. Thus, we always measure potential as a difference between two given points, that is, the initial point and the final point. The initial point is also known as *reference point*. If the reference point is at *infinity*, then the potential is known as *absolute potential*. The units of potential are volts (V).

If we carry out the integral in Equation 29.22 over a closed loop, we notice that the initial and final points are same. Obviously, the result is then zero. This implies that the potential difference between two points that are identical to each other is zero. Thus, we have

$$\oint_l \mathbf{E} \cdot d\mathbf{l} = 0 \tag{29.23}$$

Equation 29.23 is also telling us that the electrostatic field is *conservative*, which implies that the work done in this force field is *independent* of the path. Note that, in general, all force fields are *not* conservative. We just happened to get lucky.

Using Stoke's theorem, we can rewrite Equation 29.23 as

$$\nabla \times \mathbf{E} = 0 \tag{29.24}$$

which says that the electrostatic field is *curl-free* or *irrotational*.

29.4.3 Relationship between *E* and *V*

We have already seen in the previous section the relationship between the electric field *E* and the potential *V*, which is given by Equation 29.22. This equation is useful when we need to calculate the potential difference between two points in a given electric field. However, in many situations, we need to compute the electric field once the potential is known. Thus, in the following, we derive another relationship between *E* and *V*, which enables us to do just that.

From Equation 29.22, we can write the following equation as

$$dV = -\mathbf{E} \cdot d\mathbf{l} \tag{29.25}$$

Now, consider the left hand side of Equation 29.25. Using basic calculus involving partial derivatives, we have

$$dV = \frac{\partial V}{\partial x} dx + \frac{\partial V}{\partial y} dy + \frac{\partial V}{\partial z} dz$$

$$= \nabla V \cdot d\mathbf{l} \tag{29.26}$$

Using Equation 29.26, we can write Equation 29.25 as

$$\mathbf{E} \cdot d\mathbf{l} = -\nabla V \cdot d\mathbf{l}$$

$$\Rightarrow \mathbf{E} = -\nabla V \tag{29.27}$$

Thus, we see that we can calculate the electric field by taking the gradient of the potential function. This is an important result and provides us the new method for which we are looking. In this new method, we compute the potential function for a given charge distribution and then obtain the electric field by taking the gradient. Although this is an indirect method of calculating the electric field, it is definitely an easier one because of the following reasons:

- Computing V is straightforward and relatively easy because we are dealing with scalar quantities.
- For a given charge distribution, we only need to evaluate one integral, that is, potential integral.
- The electric field is obtained by differentiating the potential function, which is comparatively easy.
- No computation of unit vectors.
- Method is general and applicable to all situations.

29.4.4 Potential Function for a Given Charge Distribution

Before we develop the potential function for a given charge distribution, we calculate the potential function for a point charge located at the origin.

Consider a point charge Q located at the origin, as shown in Figure 29.13.

The absolute potential at R meters from the point charge is given by

$$V = -\int_{\infty}^{R} \mathbf{E} \cdot \mathbf{dl}$$

$$= -\int_{\infty}^{R} \frac{Q}{4\pi\varepsilon_0 r^2} \mathbf{a}_r \cdot \mathbf{a}_r \, dr = -\frac{Q}{4\pi\varepsilon_0} \int_{\infty}^{R} \frac{dr}{r^2} = \frac{Q}{4\pi\varepsilon_0 R} \tag{29.28}$$

Thus, the potential function of a point charge varies as the inverse of the distance from its location.

Next, we consider the potential function for a given charge distribution. Referring to Figures 29.1 and 29.2, the potential function is obtained using the principle of superposition, that is, calculating the potential from individual charges and obtaining the sum. Mathematically, we have

$$V = \sum_{i=1}^{N} \frac{Q_i}{4\pi\varepsilon_0 R_i} = \int_l \frac{q_l dl}{4\pi\varepsilon_0 R} = \int_S \frac{q_s ds}{4\pi\varepsilon_0 R} = \int_v \frac{q_v dv}{4\pi\varepsilon_0 R} \tag{29.29}$$

29.4.5 Vector Potential

Now let us take a look at the calculation of $\mathbf{B}$. Unfortunately, for this case we do not have a physical quantity, such as potential, which is a scalar. The primary reason for this difficulty is our sources, which, for this case, are vectors. Since the sources (currents) are vectors, all the quantities directly related to these sources must necessarily be vectors. Thus, we define a quantity known as *Vector potential A*, with the following characteristics:

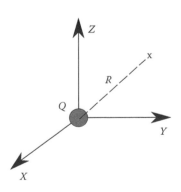

FIGURE 29.13 Potential of a point charge.

- The quantity, vector potential $\mathbf{A}$, is analogous to the potential function V with one difference, that is, it is a vector function. Also, the direction of $\mathbf{A}$ is simply the direction of the current source.
- The magnetic flux density $\mathbf{B}$ is obtained by the relation

$$\mathbf{B} = \nabla \times \mathbf{A} \tag{29.30}$$

- The units for $\mathbf{A}$ are Wb/m.
- The most important concept to notice is, unlike V, $\mathbf{A}$ is *not* a physical quantity. It is only a mathematical tool to compute $\mathbf{B}$.

Keeping these facts in mind, we write A for current distributions as follows:

- For a current element $I \, dl$, we have

$$A = \frac{\mu_o}{4\pi} \frac{I \, dl}{R} \tag{29.31}$$

- For a linear current distribution, we have

$$A = \frac{\mu_o}{4\pi} \int_l \frac{I \, dl}{R} \tag{29.32}$$

- For a surface current distribution J_s, we have

$$A = \frac{\mu_o}{4\pi} \int_s \frac{J_s \, ds}{R} \tag{29.33}$$

- For a volume current distribution J_v, we have

$$A = \frac{\mu_o}{4\pi} \int_v \frac{J_v \, dv}{R} \tag{29.34}$$

29.4.6 Illustrative Examples

Example 29.4

Consider three point charges $Q_1 = Q_2 = Q_3 = Q$ located at $(a, 0, 0)$, $(0, a, 0)$, and $(0, 0, a)$, as shown in Figure 29.10a.

1. Calculate the total potential V_T at the origin.
2. If the charges are replaced by current elements, as shown in Figure 29.10b, calculate the vector potential A_T at the origin.

Solution

The total potential $V_T(0, 0, 0)$ is the sum of the potentials from each charge. Notice that the distance from each charge to the point of observation is a. Thus, using Equation 29.28,

$$V_T = V_1 + V_2 + V_3$$

$$= \frac{Q}{4\pi\varepsilon_0 a} + \frac{Q}{4\pi\varepsilon_0 a} + \frac{Q}{4\pi\varepsilon_0 a}$$

$$= \frac{3Q}{4\pi\varepsilon_0 a} \ (\text{V})$$

Using Equation 29.30, we have

$$A_1 = \frac{\mu_0}{4\pi} \frac{I dl}{a} (-\boldsymbol{a}_y)$$

$$A_2 = \frac{\mu_0}{4\pi} \frac{Idl}{a} (-\boldsymbol{a}_z)$$

$$A_3 = \frac{\mu_0}{4\pi} \frac{Idl}{a} (-\boldsymbol{a}_x)$$

Thus, we have

$$A_T = -\frac{\mu_0 Idl}{4\pi a} [\boldsymbol{a}_y + \boldsymbol{a}_z + \boldsymbol{a}_x] \text{ (Wb/m)}$$

Example 29.5

Consider a straight wire of length $2l$ meters carrying a uniform charge density q_1 C/m, as shown in Figure 29.11a. (1) Calculate the potential V at a point P, located $\boldsymbol{x}$ meters from the axis of the wire. Then, calculate the electric field $\boldsymbol{E}$ using this result. (2) If the wire is carrying uniform current, as shown in Figure 29.11b, then calculate $\boldsymbol{A}$ and $\boldsymbol{B}$ at the point P.

Solution
In this example also, we use the principle of superposition as applied in Example 29.2.

Referring to Figure 29.11c, the charge dQ on each segment is given by

$$dQ = q_l dl = q_l dy$$

$$R = \sqrt{x^2 + y^2}$$

and

$$dV = \frac{q_l dy}{4\pi\varepsilon_0 \sqrt{x^2 + y^2}}$$

The integration is straightforward and we have

$$V = \frac{q_l}{4\pi\varepsilon_0} \int_{-l}^{l} \frac{dy}{\sqrt{x^2 + y^2}}$$

The integral in the previous step may be evaluated using simple substitutions or tables of integration. Thus, we have

$$V = \frac{q_l}{4\pi\varepsilon_0} \left[\ln\left(\sqrt{x^2 + y^2} + y\right) \right]_{-l}^{l}$$

$$= \frac{q_l}{4\pi\varepsilon_0} \left[\ln\left(\sqrt{x^2 + l^2} + l\right) - \ln\left(\sqrt{x^2 + l^2} - l\right) \right]$$

Next, the electric field **E** may be calculated as

$$E = -\nabla V = -\frac{\partial V}{\partial x}a_x = a_x\frac{q_l}{4\pi\varepsilon_0 x}\left[\frac{2l}{\sqrt{x^2+l^2}}\right] = a_x\frac{q_l}{2\pi\varepsilon_0 x}\cos\alpha$$

For part (b), referring to Figure 29.11b, we define each current element as

$$Idl = Idya_y$$

$$R = \sqrt{x^2+y^2}$$

$$dA = \frac{\mu_0}{4\pi}\frac{Idya_y}{\sqrt{x^2+y^2}}$$

and

$$A = a_y\frac{\mu_0 I}{4\pi}\int_{-l}^{l}\frac{dy}{\sqrt{x^2+y^2}}$$

$$= a_y\frac{\mu_0 I}{4\pi}\left[\ln\left(\sqrt{x^2+y^2}+y\right)\right]_{-l}^{l}$$

$$= a_y\frac{\mu_0 I}{4\pi}\left[\ln\left(\sqrt{x^2+l^2}+l\right)-\ln\left(\sqrt{x^2+l^2}-l\right)\right]$$

To obtain **B**, we carry out the following steps:

$$B = \nabla\times A = a_z\frac{\partial A_y}{\partial x} = -a_z\frac{\mu_0 I}{4\pi x}\left[\frac{2l}{\sqrt{x^2+l^2}}\right] = -a_z\frac{\mu_0 I}{2\pi x}\cos\alpha$$

29.5 Gauss' Law and Ampere's Law

In this section, we shall study yet another method to calculate the electric field and magnetic flux density from a given source distribution. Admittedly, this new method is *not* as general as the previous two techniques and is applicable to some selected problems only. Although the method is a restricted one, it is very easy to understand and also simple to evaluate. The procedure is based on simple intuitive concepts and provides a lot of insight in understanding the field concepts. We begin the study by looking at the electrostatic case.

29.5.1 Gauss' Law

Gauss' law may be stated as follows:

The charge residing inside a closed surface is proportional to the electric field emanating from the surface.

The Gauss' law may be explained as follows:

Consider a lump of positive charge placed inside a box, as shown in Figure 29.14. The box can be of any shape. However, we assume

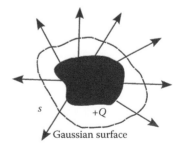

FIGURE 29.14 Positive charge inside a box made of electrically transparent material.

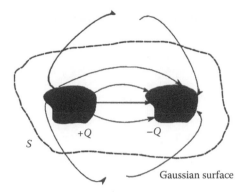

FIGURE 29.15 Positive and negative charges inside a box made of electrically transparent material.

that the box is made of an electrically transparent material. Because the box is transparent, we have an electric field coming out of the box. Remember that positive charge produces an electric field that will terminate somewhere on the negative charge. If we collect all the electric field coming out of the box and sum it up, then Gauss' law states that the result is proportional to the charge inside the box.

Now, consider another situation, as shown in Figure 29.15. Here we have both positive charge and negative charge. Note that, for this case, the electric field coming out of the positive charge is terminating on the negative charge. Now let us consider a box that encloses both charges. If we sum up all the electric field emanating from and entering into this surface, Gauss' law states that the summation would be zero. This is because same amount of electric field emanating from one side of the box is entering from other side and summation always turns out to be zero.

Mathematically, we can write Gauss' law as

$$\oint_S \boldsymbol{E} \cdot \boldsymbol{a}_n ds \propto Q_{enclosed}$$

where

$\oint_S \boldsymbol{E} \cdot \boldsymbol{a}_n ds$ gives us the summation of the electric field over the closed surface S

$Q_{enclosed}$ is the net charge inside the box

Note that the net charge is actually the sum of all positive and negative charges residing in the box. The variational relation may be converted into an equation by introducing a constant. Thus, we have

$$\oint_S \boldsymbol{E} \cdot \boldsymbol{a}_n ds = \frac{Q_{enclosed}}{\varepsilon_0} \tag{29.35}$$

Notice that the surface S must necessarily be a closed surface.

One of the simple applications of Gauss' law is if we know the electric field, even if we do not know the actual charge producing it, we can enclose the electric field in a closed box and replace the actual source by an equivalent charge. We may consider this application as somewhat analogous to Thevenin's theorem in circuit theory.

Next, we look at an alternate form of Gauss' law. We know that $Q_{enclosed}$ may be written as

$$Q_{enclosed} = \int_v q_v dv$$

Then, using the Divergence theorem, we can write Equation 29.35 as

$$\nabla \cdot E = \frac{q_v}{\varepsilon_0} \qquad (29.36)$$

which is known as the differential form of Gauss' law.

29.5.2 Ampere's Circutal Law

Just like Gauss' law for the electric field, we also have Ampere's law for the magnetic flux density, which may be stated as follows:

The total current crossing a surface is proportional to the line integral of the magnetic flux density enclosing the surface.

Mathematically, Ampere's law may be written as

$$\oint_C B \cdot dl \propto I_{crossing}$$

where C is the contour enclosing the surface S.

The variational relation may be changed into an equation by introducing a constant. Thus, we have

$$\oint_C B \cdot dl = \mu_0 I_{crossing} \qquad (29.37)$$

Next, we derive an alternate form of Ampere's law. We know that $I_{crossing}$ may be written as

$$I_{crossing} = \int_S J \cdot a_n ds$$

Then, using Stoke's theorem, we can write Equation 29.37 as

$$\nabla \times B = \mu_0 J \qquad (29.38)$$

which is known as the differential form of Ampere's law.

29.5.3 Applications

Gauss' law and Ampere's law may be used to calculate the electric field and magnetic flux density, respectively, from a given source distribution. However, the method we are going to discuss is *not* a general method and is applicable only for specific situations. In order to use either Gauss' law or Ampere's law to calculate the field distribution, the source distribution must have the following property:

The source distribution should be such that the field component of interest can be a function of one variable only.

If the field varies as a function of more than one variable, this method cannot be applied and we have to use the methods discussed in the previous section. Further, we remark that for a given problem, it is not always obvious whether this method is applicable or not. A good amount of experience is called for before we start seeing the light. Despite all these limitations, the method turns out to be very simple when applicable. In the following, we consider some illustrative examples to understand the method.

Example 29.6

Consider an infinite length of wire carrying a uniform charge density q_l C/m, as shown in Figure 29.16a.

1. Calculate the electric field E at a point P, located ρ meters from the axis of the wire.
2. If the infinite wire is carrying current along the z-direction, as shown in Figure 29.16b, then calculate B using Ampere's law.

Solution

Since the wire is infinitely long, we can safely divide the wire into two semi-infinite halves along the point of observation, as shown in Figure 29.16a. If we consider two elemental charges, one above the xy-plane and one below at equal distances, it is very easy to prove that the axial components cancel and only the radial component survives. Since we can apply this procedure for any point of observation, we say that, in general, only the radial component is non-zero. Further, we note that, due to symmetry, the radial component is only a function of the radial distance from the axis of the wire. Thus, we have $E = E_\rho a_\rho$, and E_ρ is a function of ρ only for this problem. We will calculate E_ρ using Gauss' law.

1. In order to use Gauss' law, we must have a closed surface (box). Since we do not have any surface around the line charge, we simply draw a cylindrical surface, closed at both ends, around the line charge, as shown in Figure 29.17a. Now, we apply Gauss' law using this surface.

Note that we can draw a surface of any shape. However, the solution will be easier if we draw a surface that fits into the coordinate system. The cylindrical surface, which we imagined, is known as a *Gaussian surface*, in general.

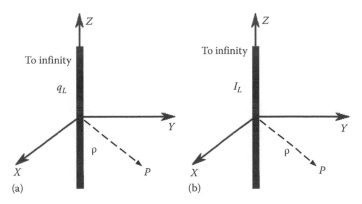

FIGURE 29.16 (a) Infinite line charge and (b) infinite line current.

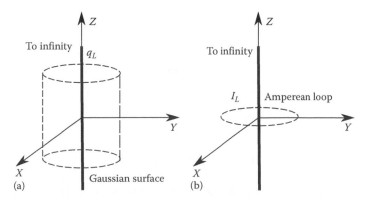

FIGURE 29.17 (a) Gaussian surface and infinite line charge and (b) Amperean loop and infinite line current.

Referring to Equation 29.35, we have

$$\oint_{S} \boldsymbol{E} \cdot \boldsymbol{a}_n ds = \oint_{S} E_\rho \boldsymbol{a}_\rho \cdot \boldsymbol{a}_\rho ds = E_\rho 2\pi l$$

and

$$Q_{enclosed} = q_l l$$

Thus, we have

$$E_\rho 2\pi\rho l = \frac{q_l l}{\varepsilon_0}$$

which implies

$$E_\rho = \frac{q_l}{2\pi\varepsilon_0\rho}$$

2. The method using Ampere's law has the same restrictions as the method using Gauss' law. Here, we define an Amperean loop around the wire, as shown in Figure 29.17b, and apply Ampere's law. For this case, we have $\boldsymbol{B} = B_\phi \boldsymbol{a}_\phi$ only. Referring to Equation 29.37, we have

$$\oint_{C} \boldsymbol{B} \cdot \boldsymbol{dl} = \oint_{C} B_\phi \boldsymbol{a}_\phi \cdot \boldsymbol{a}_\phi dl = B_\phi 2\pi\rho$$

and

$$I_{crossing} = I$$

Thus, we have

$$B_\phi 2\pi\rho = \mu_0 I$$

which implies

$$B_\phi = \frac{\mu_0 I}{2\pi\rho}$$

Example 29.7

Consider an infinite plate carrying a uniform surface charge density q_s C/m², as shown in Figure 29.18a. (1) Calculate the electric field $\boldsymbol{E}$ at a point P, located h meters from the plane of the plate. (2) If the infinite plate is carrying current density $\boldsymbol{J}_s = J_0 \boldsymbol{a}_y$, as shown in Figure 29.18b, then calculate $\boldsymbol{B}$ using Ampere's law.

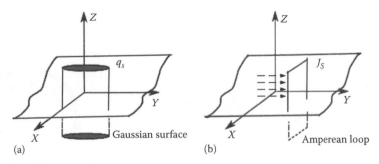

FIGURE 29.18 (a) Gaussian surface and infinite plane charge. (b) Amperean loop and infinite plane current.

Solution

(1) For this case, it is easy to prove that the electric field has only z-component and it is independent of x and y. Drawing a cylindrical box, closed at both ends, as in the Gaussian surface, we have the following:

Referring to Equation 29.35, we have

$$\oint_S E \cdot a_n ds = \oint E_z a_z \cdot a_z = E_z A + E_z A = 2 E_z A$$

and

$$Q_{enclosed} = q_s A$$

In the previous equation, A is the area of the cross section of the cylindrical box. Note that we need to evaluate the closed surface integral on both ends of the box and hence we have the factor 2 in the equation. Thus, we have

$$E_z(2A) = \frac{q_s A}{\varepsilon_0}$$

which implies,

$$E_z = \frac{q_s}{2\varepsilon_0}$$

(2) For this case, it is easy to prove that $B = B_x a_x$ above the plate and $B = B_x(-a_x)$ below the plate. This can be done by either using the right hand rule or considering two infinitely long filament currents at equal distance on either side of the x-axis. Considering an Amperean loop, as shown in the figure and referring to Equation 29.37, we have

$$\oint_C B \cdot dl = \oint_C B_x a_x \cdot dl = B_x l + B_x l = 2 B_x l$$

and

$$I_{crossing} = J_0 l$$

Thus, we have

$$B_x(2l) = \mu_0 J_0 l$$

which implies,

$$B_x = \frac{\mu_0 J_0}{2}$$

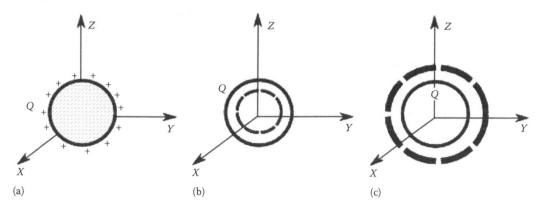

FIGURE 29.19   (a) Charge on a spherical surface. (b) Internal problem. (c) External problem.

Example 29.8

Consider a charge Q C distributed uniformly over the surface of a sphere of radius a, as shown in Figure 29.19a. Calculate the electric field E at a point P, located r meters from the center of the sphere. Consider both cases, that is, (1) $r < a$, and (2) $r > a$.

Solution

For this case, it is easy to prove that the electric field has only a r-component and it is independent of θ and ϕ due to the symmetrical nature of the charge distribution.

Case (1): Let us consider the situation $r < a$. Drawing a sphere with radius r as the Gaussian surface, shown with dotted lines in Figure 29.19b, we note that the charge enclosed by the Gaussian surface is zero. Hence, we have $E_r = 0$ for $r < a$.

Case (2): Again, let us consider a Gaussian sphere of radius r and $r > a$, shown with dotted lines in Figure 29.19c. Now, we see that all the charge is enclosed by the Gaussian sphere. Thus, we have

$$Q_{enclosed} = Q$$

Next, we have

$$\int_S E \cdot a_n ds = \oint_S E_r a_r \cdot a_r ds = E_r 4\pi r^2$$

Then, using Equation 29.35, we have

$$E_r = \frac{Q}{4\pi\varepsilon_0 r^2}$$

Example 29.9

Consider a current I A flowing over the surface of an infinitely long circular cylinder of radius a, as shown in Figure 29.20a. Calculate the magnetic flux density B at a point P, located ρ meters from the axis of the cylinder. Consider both cases, that is, (1) $\rho < a$ and (2) $\rho > a$.

Solution

For this case, it is easy to prove that B has only ρ-component and it is independent of z and ϕ due to the symmetrical nature of the current distribution.

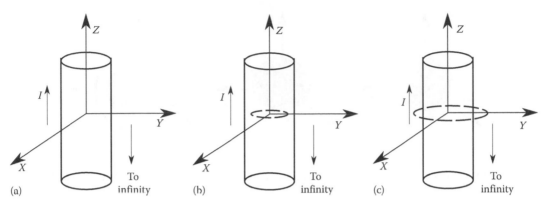

FIGURE 29.20 (a) Infinitely long circular cylinder with radius *a* carrying current *I*. (b) Amperean loop inside the cylinder. (c) Amperean loop outside the cylinder.

Case (1): Let us consider the situation $\rho < a$. Drawing an Amperean loop of radius ρ, as shown in Figure 29.20b, we note that the current crossing the surface formed by the Amperean loop is zero. Hence, we have $B_\rho = 0$ for $\rho < a$.

Case (2): Again, let us consider an Amperean loop of radius $\rho > a$, as shown in Figure 29.20c. Now, we see that all the current is crossing the surface enclosed by the Amperean loop. Thus, we have

$$I_{crossing} = I$$

Next, we have

$$\oint_C \boldsymbol{B}d\boldsymbol{l} = \oint B_\phi \boldsymbol{a}_\phi \cdot \boldsymbol{a}_\phi dl = B_\phi 2\pi\rho$$

Then, using Equation 29.37, we have

$$B_\phi = \frac{\mu_0 I}{2\pi\rho}$$

Example 29.10

Consider a charge Q C distributed uniformly over the volume of a sphere, of radius *a*, as shown in Figure 29.21a. Calculate the electric field $\boldsymbol{E}$ at a point *P*, located *r* meters from the center of the sphere. Consider both cases, that is, (1) $r > a$ and (2) $r > a$

Solution

Case (1): Let us consider the situation $r < a$. Drawing a sphere with radius *r* as the Gaussian surface, shown with dotted lines in Figure 29.21b, we note that the charge enclosed by the Gaussian surface is proportional to the ratio of volumes of the Gaussian sphere and the charged sphere. Thus, we have

$$Q_{enclosed} = Q\frac{r^3}{a^3}$$

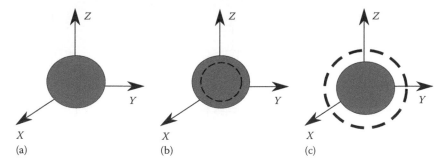

FIGURE 29.21   (a) Charge distributed inside a spherical volume. (b) Internal problem. (c) External problem.

Next, we have

$$\oint_S \mathbf{E} \cdot \mathbf{a}_n ds = \oint_S E_r \mathbf{a}_r \cdot \mathbf{a}_r ds = E_r 4\pi r^2$$

Then, using Equation 29.35, we have

$$E_r = \frac{Qr}{4\pi\varepsilon_0 a^3}$$

Case (2): Again, let consider a Gaussian sphere of radius r and $r > a$. Now, referring to Figure 29.21c, we see that all the charge is enclosed by the Gaussian sphere. Thus, we have

$$Q_{enclosed} = Q$$

Next, we have

$$\oint_S \mathbf{E} \cdot \mathbf{a}_n ds = \oint_S E_r \mathbf{a}_r \cdot \mathbf{a}_r ds = E_r 4\pi r^2$$

Then, using Equation 29.35, we have

$$E_r = \frac{Q}{4\pi\varepsilon_0 r^2}$$

Example 29.11

Consider a current I A flowing along the volume of an infinitely long circular cylinder of radius a, as shown in Figure 29.22a. Calculate the magnetic flux density $\mathbf{B}$ at a point P, located ρ meters from the axis of the cylinder. Consider both cases, that is, (1) $\rho < a$ and (2) $\rho > a$.

Solution

Case (1): Let us consider the situation $\rho < a$. Drawing an Amperean loop of radius ρ, we note that the current crossing the surface formed by the Amperean loop is given by

$$I_{crossing} = I\frac{\rho^2}{a^2}$$

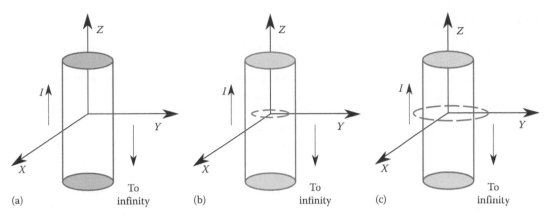

FIGURE 29.22 (a) Current flowing inside a cylindrical volume. (b) Internal problem. (c) External problem.

Next, we have,

$$\oint_C \boldsymbol{B}\,dl = \oint B_\phi \boldsymbol{a}_\phi \cdot \boldsymbol{a}_\phi dl = B_\phi\, 2\pi\rho$$

Then, using Equation 29.37, we have

$$B_\phi = \frac{\mu_0 I \rho}{2\pi a^2}$$

Case (2): Again, let us consider an Amperean loop of radius $\rho > a$. Now, referring to the figure, we see that all the current is crossing the surface enclosed by the Amperean loop. Thus, we have

$$I_{crossing} = I$$

Next, we have

$$\oint_C \boldsymbol{B}\,dl = \oint B_\phi \boldsymbol{a}_\phi \cdot \boldsymbol{a}_\phi dl = B_\phi\, 2\pi\rho$$

Then, using Equation 29.37, we have

$$B_\phi = \frac{\mu_0 I}{2\pi\rho}$$

29.6 Material Interaction

In this section, we study the interaction of materials with electric field and magnetic flux density. Recall that so far we have assumed the sources are located in a homogeneous medium. However, in a practical situation, the electric field or the magnetic flux cross from one medium to other and we need to examine this phenomenon in more detail.

29.6.1 Polarization

From the electric field point of view, one may classify all the materials into two categories: conductors and dielectrics. For both materials, the material parameter, conductivity (σ), which depends on the number of electrons situated in the outermost orbit of respective atoms, plays a significant role. If σ is very large ($>10^5$), the material is classified as a conductor, otherwise as a dielectric.

For conductors, we note that the field penetrating the material is zero. However, for a dielectric, the field penetration is significant given by a phenomenon called *polarization* (*P*).

It is noted that when a dielectric material is subjected to an electric field, each atom behaves like a small dipole (defined as two equal and opposite charges separated by a small distance) and net effect of all these dipoles result in polarization.

Mathematically, the total field is the applied field plus the field induced due to polarization. It is customary to define a new quantity, in the presence of dielectrics, to take care of the polarization phenomenon, known as electric flux density (*D*), measured in C/m², given by

$$D = \varepsilon_o E + P \tag{29.39}$$

Another way of taking care of polarization is to define a quantity known as relative permittivity (ε_r), which is equal to unity for free-space (air) and greater than 1 for all other dielectric materials. Using this quantity, we define the following relationships, given by

$$P = (\varepsilon_r - 1)\varepsilon_o E \quad \text{and} \quad D = \varepsilon_r \varepsilon_o E \tag{29.40}$$

29.6.2 Magnetization

In a similar way, from the magnetic flux point of view, it is noted that when a magnetic material is subjected to magnetic flux, each atom behaves like small magnet and the net effect of all these dipoles result in *magnetization* (*M*).

Mathematically, the total magnetic flux is the applied flux plus the flux induced due to magnetization. It is customary to define a new quantity, in the presence of magnetic materials to take care of the magnetization phenomenon, known as magnetic field (*H*), measured in A/m, and is given by

$$H = \frac{B}{\mu_o} - M \tag{29.41}$$

Another way of taking care of magnetization is to define a quantity known as relative permeability (μ_r), which is equal to unity for free-space (air) and for most materials except ferromagnetic materials ($\sim 10^6$). Using this quantity, we define the following relationships, given by

$$M = (\mu_r - 1)H \quad \text{and} \quad B = \mu_r \mu_o H \tag{29.42}$$

29.6.3 Boundary Conditions

We note that the electric and magnetic fields must satisfy certain boundary conditions when propagating from one material medium to another material medium. Assuming the material interface is locally planar, it is customary to define the boundary conditions in terms of the tangential and normal components with respect to the interface. Assuming there are no sources on the interface, we have

$$E_{t1} = E_{t2} \quad \text{and} \quad H_{t1} = H_{t2} \tag{29.43}$$

$$D_{n1} = D_{n2} \quad \text{and} \quad B_{n1} = B_{n2} \tag{29.44}$$

where the subscripts "*t*" and "*n*" refer to tangential and normal components, respectively.

However, if the interface contains sources, then we have

$$D_{n2} - D_{n1} = q_s \quad \text{and} \quad H_{t2} - H_{t1} = J_s \tag{29.45}$$

where q_s and J_s represent the surface charge density and surface current density of the sources located on the interface, respectively, and the normal points into medium #2 from medium #1.

The derivation of Equations 29.43 through 29.45 is trivial and may be found in any elementary textbook on electromagnetic theory.

29.7 Faraday's Law, Displacement Current, and Maxwell's Equations

In 1820, Hans Christian Oersted demonstrated experimentally that electric current can produce magnetic fields. Soon after, Michael Faraday devised experiments to produce electric current from magnetic field. His initial experiments were not successful because he used steady magnetic fields that produced no electrical phenomenon. Then, almost by accident, he discovered that if a device generates a time-changing magnetic field, then it can produce current in a neighboring circuit even when there is no physical contact between the two. His experimental observations have become famous and known today as Faraday's law.

29.7.1 Faraday's Law

The Faraday's law may be stated as any time-varying magnetic flux linking a neighboring loop would cause currents in the loop. Mathematically, the law can be stated as

$$\oint_C E \cdot dl = -\frac{d}{dt} \oint_S B \cdot a_n ds \tag{29.46}$$

where S represents any arbitrary, open surface enclosed by the contour C.

In Equation 29.46, the negative sign eliminates the impossible situation of developing infinite fields, introduced by Henri Frederic Emile Lenz.

Using the Stoke's theorem, Faraday's law can be transformed into an alternate form, given by

$$\nabla \times E = -\frac{\partial B}{\partial t} \tag{29.47}$$

29.7.2 Displacement Current

Soon after the discovery of Faraday's law, Maxwell proved mathematically that Ampere's law is inconsistent when the time-varying situation is considered. The proof is as follows.

Since $\nabla \cdot \nabla \times A = 0$ for any vector A, using Ampere's law given by Equation 29.38, it can be shown that

$$\nabla \cdot \nabla \times H = \nabla \cdot J = 0 \quad \Rightarrow \quad \nabla \cdot J = 0 \tag{29.48}$$

However, the continuity equation given by Equation 29.7 shows that $\nabla \cdot J$ need not be zero always. Hence, to remove this inconsistency, Maxwell modified Ampere's law by adding another term to Equation 29.38 as

$$\nabla \times H = J + \frac{\partial D}{\partial t} \tag{29.49}$$

where $\partial D/\partial t$ is referred to as the displacement current density.

29.7.3 Maxwell's Equations

Once the inconsistency in Ampere's law was removed, Maxwell wrote down his famous four equations as

$$\nabla \times E = -\frac{\partial B}{\partial t} \Rightarrow \oint_C E \cdot dl = -\frac{d}{dt} \int_S B \cdot a_n ds \tag{29.50}$$

$$\nabla \times H = J + \frac{\partial D}{\partial t} \Rightarrow \int_C H \cdot dl = \int_S J \cdot a_n ds + \frac{d}{dt} \int_S D \cdot a_n ds \tag{29.51}$$

$$\nabla \cdot D = q_v \Rightarrow \oint_S D \cdot a_n ds = Q \tag{29.52}$$

$$\nabla \cdot B = 0 \Rightarrow \oint_S B \cdot a_n ds = 0 \tag{29.53}$$

and claimed that all the electromagnetic phenomenon must satisfy these four equations along with the continuity equation (Equation 29.7).

29.8 Summary

In this chapter, we defined electric field and magnetic flux density using Couloumb's law and Biot–Savart law. We have also developed three methods to compute the E and B fields for a given source distribution.

The first method is based on the definition of fields from elemental sources and uses superposition to obtain the fields at any point of observation. Although this method is general, unfortunately, it is cumbersome.

The second method is based on the potential theory. This method is general and applicable to all situations. Actually, this is the preferred method in most situations.

Lastly, we developed yet another method based on Gauss/Ampere's law. This method is easy and conceptually simple to apply. Unfortunately, this method is applicable to very few situations only.

We have described Faraday's law, which establishes the connection between electric and magnetic fields and briefly described the contribution of Maxwell to the electromagnetic phenomenon.

30

Propagating
Electromagnetic Fields

Michael E. Baginski
Auburn University

Sadasiva M. Rao
Auburn University

Tyler N. Killian
Auburn University

30.1 Maxwell's Equations in the Time Domain and Phasor Domain

In this chapter, we will begin by investigating electromagnetic wave propagation in complex media via a solution of the Maxwell's equations. This will be followed by a section that focuses on analyzing wave propagation on terminated transmission lines and typical impedance-matching techniques. Next, electromagnetic wave propagation inside rectangular waveguides, dielectric waveguides, and optical fiber is characterized. The final section is an overview of the transmission and reception characteristics of antennas.

30.1.1 Maxwell's Equations in Complex (Phasor) Domain

Maxwell's Equations are the governing equations that describe all macroscopic electromagnetic field behavior. They may be stated in either the differential or integral form and are associated with earlier work by Faraday, Ampere, and Gauss, as shown below.

Faraday's law of induction (Maxwell's first equation):

$$\nabla \times \mathbf{E} = -\frac{\partial \mathbf{B}}{\partial \mathbf{t}} \tag{30.1}$$

$$\oint_C E \cdot dl = -\oint_S \frac{\partial B}{\partial t} \cdot dS \qquad (30.2)$$

Ampere's law (Maxwell's second equation):

$$\nabla \times \mathbf{H} = \mathbf{J} + \frac{\partial \mathbf{D}}{\partial \mathbf{t}} \qquad (30.3)$$

$$\oint_C H \cdot dl = \int_S \left(J + \frac{\partial D}{\partial t} \right) \cdot dS \qquad (30.4)$$

Gauss' law (Maxwell's third equation):

$$\nabla \cdot \mathbf{D} = \mathbf{q_v} \qquad (30.5)$$

$$\oint_S D \cdot dS = \int_v q_v \, dv \qquad (30.6)$$

Conservation of magnetic flux (Maxwell's fourth equation):

$$\nabla \cdot \mathbf{B} = 0 \qquad (30.7)$$

$$\oint_S B \cdot dS = 0 \qquad (30.8)$$

Current continuity equation can then be obtained by taking the divergence of Equation 30.3 and substituting $q_v = \nabla \cdot \mathbf{D}$:

$$\nabla \cdot \mathbf{J} + \frac{\partial \mathbf{q_v}}{\partial \mathbf{t}} = 0 \qquad (30.9)$$

$$\oint_S J \cdot dS = -\int_v \frac{\partial q_v}{\partial t} \, dv \qquad (30.10)$$

The equations indicate that a time-varying magnetic field gives rise to a time-varying electric field (Equations 30.1 and 30.2), and time-varying electric fields or current densities induce time-varying magnetic fields (Equations 30.3 and 30.4). Equations 30.5 and 30.6 show the relationship between charge density and the total electric flux. Finally, Equations 30.7 and 30.8 mathematically state that magnetic charges (monopoles) do not exist.

30.1.2 Maxwell's Equations in Complex (Phasor) Domain

The differential form of Maxwell's Equations depends on three spatial coordinates and time. However, in many circumstances, the source of the electromagnetic phenomena is sinusoidal or time harmonic at a frequency of f or radian frequency $\omega = 2\pi f$. If the medium is linear, then all field quantities vary sinusoidally, and we may express the behavior in terms of complex or phasor quantities using the definitions shown below:

$$\mathbf{A} = \mathbf{A}(x, y, z)\cos(\omega t + \phi)$$

$$\mathbf{A} = \mathrm{Re}\{\tilde{\mathbf{A}}e^{j\omega t}\}$$

$$\tilde{\mathbf{A}} = \mathbf{A}(x, y, z)e^{j\phi}$$

Here, **A** and **Ã** represent vector quantities, but the relationship holds for scalar terms as well. All temporal derivatives in Maxwell's equations can therefore be expressed in the phasor domain by simply multiplying the quantity by $j\omega$, as shown below:

$$\frac{\partial \tilde{\mathbf{A}}}{\partial t} \Leftrightarrow j\omega \tilde{\mathbf{A}}$$

The differential form of Maxwell's equations may then be expressed in the phasor domain, as shown in Equations 30.11 through 30.15:

$$\nabla \times \tilde{\mathbf{E}} = -j\omega \tilde{\mathbf{B}} \tag{30.11}$$

$$\nabla \times \tilde{\mathbf{H}} = \tilde{J} + j\omega \tilde{\mathbf{D}} \tag{30.12}$$

$$\nabla \cdot \tilde{\mathbf{D}} = q_v \tag{30.13}$$

$$\nabla \cdot \tilde{\mathbf{B}} = 0 \tag{30.14}$$

$$\nabla \cdot \tilde{\mathbf{J}} = -j\omega q_v \tag{30.15}$$

Equation 30.11 can be defined in terms of $\tilde{\mathbf{E}}$ and $\tilde{\mathbf{H}}$ using the relationships $\tilde{\mathbf{J}} = \sigma \tilde{\mathbf{E}}$ and $\tilde{\mathbf{D}} = \varepsilon \tilde{\mathbf{E}}$, and expressed as

$$\nabla \times \tilde{\mathbf{H}} = (\sigma + j\omega\varepsilon)\tilde{\mathbf{E}} \tag{30.16}$$

At a given frequency f, a material can be classified as a **good conductor** if the conduction current density is much larger than the displacement current density ($\sigma \gg \omega\varepsilon$) or a **lossy dielectric** if the opposite is true ($\sigma \ll \omega\varepsilon$).

30.1.3 Uniform Plane Waves

Maxwell demonstrated that electric and magnetic fields travel through space in the form of waves at the velocity of light in the medium. This is mathematically demonstrated by developing the wave equation for a linear medium.

By assuming that an electromagnetic field is present in a linear, charge-free medium, we can derive the wave equation in the phasor domain using the two curl Equations 30.11 and 30.12. We begin by taking the curl of Equation 30.17 and use the identity $\tilde{\mathbf{B}} = \mu\tilde{\mathbf{H}}$, as shown below:

$$\nabla \times \tilde{\mathbf{E}} = -j\omega \tilde{\mathbf{B}} \tag{30.17}$$

$$\nabla \times \nabla \times \tilde{\mathbf{E}} = -j\omega \nabla \times \tilde{\mathbf{B}} \tag{30.18}$$

$$\nabla \times \nabla \times \tilde{\mathbf{E}} = -j\omega\mu \nabla \times \tilde{\mathbf{H}} \tag{30.19}$$

By using substituting Equation 30.16 into Equation 30.19, the equation becomes

$$\nabla \times \nabla \times \tilde{\mathbf{E}} = -j\omega u \sigma \tilde{\mathbf{E}} + \omega^2 \mu \varepsilon \tilde{\mathbf{E}} \tag{30.20}$$

A vector identify is then used to reduce the $\nabla \times \nabla \times \tilde{\mathbf{E}}$ into two parts, as shown below:

$$\nabla \times \nabla \times \tilde{\mathbf{E}} = \nabla \nabla \cdot \tilde{\mathbf{E}} - \nabla^2 \tilde{\mathbf{E}} \tag{30.21}$$

Since the region is assumed to be homogenous and charge free ($q_v = 0$), a simplification can be made ($\nabla \cdot \tilde{\mathbf{E}} = 0$) resulting in the final form of the wave equation in the phasor domain, as shown below:

$$\nabla^2 \tilde{\mathbf{E}} + (\omega^2 \mu \varepsilon - j\omega u \sigma)\tilde{\mathbf{E}} = 0 \tag{30.22}$$

An identical equation can be developed for the $\tilde{\mathbf{H}}$ field using the same mathematical operations. This is achieved by taking the curl of Equation 30.16 $\nabla \times \nabla \times \tilde{\mathbf{H}} = (\sigma + j\omega \varepsilon) \nabla \times \tilde{\mathbf{E}}$ and substituting (30.11) resulting in

$$\nabla^2 \tilde{\mathbf{H}} + (\omega^2 \mu \varepsilon - j\omega u \sigma)\tilde{\mathbf{H}} = 0 \tag{30.23}$$

These equations are more commonly expressed in a more compact form known as the Helmholtz equations (Equations 30.24 and 30.25), as shown below:

$$\nabla^2 \tilde{\mathbf{E}} - \gamma^2 \tilde{\mathbf{E}} = 0 \tag{30.24}$$

$$\nabla^2 \tilde{\mathbf{H}} - \gamma^2 \tilde{\mathbf{H}} = 0 \tag{30.25}$$

where γ is the propagation constant, defined as

$$\gamma = \sqrt{j\omega\mu(\sigma + j\omega\varepsilon)}.$$

The real part of γ is referred to as the attenuation constant α and the imaginary part the phase constant β:

$$\gamma = \sqrt{j\omega\mu(\sigma + j\omega\varepsilon)} = \alpha + j\beta \tag{30.26}$$

If we assume that propagation is limited to waves traveling in the $\mathbf{a}_z$ direction having only an $\mathbf{a}_x$-oriented electric field (E_x) and an $\mathbf{a}_y$-oriented magnetic field H_y, Equations 30.24 and 30.25 become

$$\frac{\partial^2 E_x}{\partial z^2} - \gamma^2 E_x = 0 \tag{30.27}$$

$$\frac{\partial^2 H_y}{\partial z^2} - \gamma^2 H_y = 0 \tag{30.28}$$

Solutions to Equations 30.27 and 30.28 are given in (30.29) and (30.30):

$$E_x = (E_0^+ e^{-\gamma z} + E_0^- e^{\gamma z})\boldsymbol{a}_x \tag{30.29}$$

$$H_y = (H_0^+ e^{-\gamma z} + H_0^- e^{\gamma z})\boldsymbol{a}_y \tag{30.30}$$

The first terms of each equation describe a plane wave propagating in the $\boldsymbol{a}_z$ direction having magnitudes of E_0^+, H_0^+, respectively, with the second term depicting a plane wave propagating in the $\boldsymbol{a}_z$ direction having magnitudes of E_0^-, H_0^-, respectively.

The above equations can be expressed in the time domain using the relationship given in Equations 30.11 and 30.26 and is shown below:

$$E(z,t) = (E_0^+ e^{-\alpha z} \cos(\omega t - \beta z + \theta_{E^+}) + E_0^- e^{\alpha z} \cos(\omega t + \beta z + \theta_{E^-}))\boldsymbol{a}_x$$

$$H(z,t) = (H_y^+ e^{-\alpha z} \cos(\omega t - \beta z + \theta_{H^+}) + H_y^- e^{\alpha z} \cos(\omega t + \beta z + \theta_{H^-}))\boldsymbol{a}_y$$

where θ_E and θ_H are the angles associated with the $\tilde{E}_x$ and $\tilde{E}_y$, respectively.

Both electric and associated magnetic fields propagate with a phase velocity v_p that is given as

$$v_p = \frac{\omega}{\beta} \tag{30.31}$$

The characteristic impedance of the region (η) is defined as the ratio of the electric field to the magnetic field, as shown below:

$$\eta = \frac{\tilde{E}}{\tilde{H}} \tag{30.32}$$

For a wave traveling in the positive $\boldsymbol{a}_z$ direction, the simplest method to determine η is by taking the curl of the electric field and using Equation 30.17 and the identify $\tilde{B} = \mu\tilde{H}$. This results in an expression for η that is dependent on only the constitutive parameters of the media (30.35):

$$\nabla \times \tilde{E} = -j\omega\mu\tilde{H} = -\gamma E_x^+ e^{-\gamma z}\boldsymbol{a}_y \tag{30.33}$$

$$\text{or} \quad \tilde{H} = \frac{j\nabla \times \tilde{E}}{\omega\mu} = \frac{-j\gamma E_x^+ e^{-\gamma z}}{\omega\mu} \tag{30.34}$$

$$\eta = \frac{j\omega\mu}{\gamma} = \sqrt{\frac{j\omega\mu}{\sigma + j\omega\varepsilon}} \tag{30.35}$$

Using the same mathematical procedure, the impedance of waves traveling in the $-\boldsymbol{a}_z$ direction is given as

$$\frac{E_0^-}{H_0^-} = -\frac{j\omega\mu}{\gamma} = -\eta \tag{30.36}$$

Two very useful relationships that relate the η and the direction of propagation $\mathbf{a}_R$ to $\tilde{E}$ and $\tilde{H}$ are shown below:

$$\tilde{H} = \frac{1}{\eta}\mathbf{a}_R \times \tilde{E} \tag{30.37}$$

$$\tilde{E} = -\eta\,\mathbf{a}_R \times \tilde{H} \tag{30.38}$$

where $\mathbf{a}_R$ is a unit vector in the direction of propagation. It should be noted that the above relationships (30.37) and (30.38) are only valid for propagating waves.

30.1.4 Propagation in a Lossless Media

In lossless ($\sigma = 0$), charge-free ($\nabla \cdot \tilde{E} = 0$) media, γ is equal to the propagation constant $j\beta$ and Equations 30.27 and 30.28 become

$$\frac{\partial^2 E_x}{\partial z^2} + \beta^2 E_x = 0 \tag{30.39}$$

$$\frac{\partial^2 H_y}{\partial z^2} + \beta^2 H_y = 0 \tag{30.40}$$

The solutions to these equations are

$$E_x = (E_0^+ e^{-j\beta z} + E_0^- e^{j\beta z})\,\boldsymbol{a_x} \tag{30.41}$$

$$H_y = (H_0^+ e^{-j\beta z} + H_0^- e^{j\beta z})\,\boldsymbol{a_y} \tag{30.42}$$

and the phase velocity v_p and intrinsic impedance η of the media are given as

$$v_p = \frac{\omega}{\beta} = \frac{1}{\mu\varepsilon} \tag{30.43}$$

$$\eta = \sqrt{\frac{\mu}{\varepsilon}} \tag{30.44}$$

30.1.5 Propagation in a Low-Loss Dielectric

In the electrostatics section, we discussed how power losses are associated with both conduction and polarization currents. Waves propagating in a lossy medium induce conduction current $\tilde{J}_{cond} = \sigma\tilde{E}$ and a displacement (polarization) current $\tilde{J}_{disp} = j\omega\tilde{D} = j\omega\varepsilon\tilde{E}$. The total current density is the sum of each of these and sometimes referred to as the **Maxwell current density**:

$$\tilde{J}_{total} = \nabla \times \tilde{H} = \tilde{J}_{cond} + \tilde{J}_{disp} \tag{30.45}$$

The quantities σ and ε may be frequency dependent. However, over a relatively small frequency band, we may assume the conductivity σ and permittivity ε are constant. The permittivity of the dielectrics is complex and given by

$$\varepsilon_c = \varepsilon' - j\varepsilon'' = \varepsilon_r - j\left(\varepsilon'' + \frac{\sigma}{\omega}\right) \tag{30.46}$$

Therefore, (30.45) may be expressed simply as

$$\nabla \times \tilde{\mathbf{H}} = \tilde{\mathbf{J}}_{total} = j\omega\varepsilon_r \tilde{\mathbf{E}} \tag{30.47}$$

30.1.6 Propagation in a Good Conductor

If the conduction current is much greater than the displacement current ($\sigma \gg \omega\varepsilon$), then the material is considered a good conductor and several simplifying assumptions can be made. Since $\sigma \gg \omega\varepsilon$, the propagation constant (γ) becomes

$$\gamma = \sqrt{j\omega\mu\sigma} = \frac{(1+j)}{\sqrt{2}}(\omega\mu\sigma)$$

and therefore α and β are equal and given as

$$\alpha = \frac{\omega\mu\sigma}{\sqrt{2}} \quad \text{and} \quad \beta = \frac{\omega\mu\sigma}{\sqrt{2}} \tag{30.48}$$

The total current density is approximately equal to the conduction current density $\tilde{\mathbf{J}}_{total} = \sigma\tilde{\mathbf{E}}$. For a wave propagating in the $+\mathbf{a}_z$ direction, the magnitude of the $\tilde{\mathbf{E}}$ and $\tilde{\mathbf{J}}_{cond}$ rapidly decay and are given as

$$\tilde{\mathbf{E}} = E_0^+ \varepsilon^{-az}$$

$$\tilde{\mathbf{J}} = \sigma E_0^+ \varepsilon^{-az}$$

The point at which the magnitude of the electric field has decreased to e^{-1} is commonly referred to as the skin depth (δ) expressed as

$$\delta = \sqrt{\frac{1}{\pi f \mu \sigma}} \tag{30.49}$$

30.1.7 Reflection and Transmission of Plane Waves

When electromagnetic waves are normally incident on a planar region ($\tilde{\mathbf{E}}$, $\tilde{\mathbf{H}}$ are perpendicular to each other and parallel to the planar region) having different constitutive parameters (ε, μ, σ), a transmitted wave and a reflected wave are created. The analysis of this behavior begins by assuming a normally incident wave ($\tilde{\mathbf{E}}^i$, $\tilde{\mathbf{H}}^i$) in region (1) propagating in the $\mathbf{a}_t$ direction resulting in a transmitted wave ($\mathbf{E}^t$, $\tilde{\mathbf{H}}^t$) in region (2) and reflected wave ($\tilde{\mathbf{E}}^r$, $\tilde{\mathbf{H}}^r$) wave in region (1), as shown in Figure 30.1 and given by

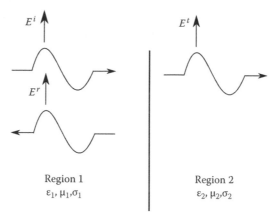

FIGURE 30.1 Incident, transmitted, and reflected electric fields at a boundary.

$$\tilde{\mathbf{E}}^i = E_0^i e^{-\gamma_1 z}\mathbf{a}_x \quad \text{and} \quad \tilde{\mathbf{H}}^i = H_0^i e^{-\gamma_1 z}\mathbf{a}_y \tag{30.50}$$

$$\tilde{\mathbf{E}}^r = E_0^r e^{\gamma_1 z}\mathbf{a}_x \quad \text{and} \quad \tilde{\mathbf{H}}^r = H_0^r e^{\gamma_1 z}\mathbf{a}_y \tag{30.51}$$

$$\tilde{\mathbf{E}}^t = E_0^t e^{-\gamma_2 z}\mathbf{a}_x \quad \text{and} \quad \tilde{\mathbf{H}}^t = H_0^t e^{-\gamma_2 z}\mathbf{a}_y \tag{30.52}$$

The differential voltage (dV) and current drop that occurs over the length dz are given by

$$\frac{dV}{dz} = -IR - L\frac{dI}{dt} \tag{30.53}$$

$$\frac{dI}{dz} = -VG - C\frac{dV}{dt}. \tag{30.54}$$

Equations 30.50 through 30.52 can be stated in terms of the electric field by making use of Equations 30.35 and 30.36, as shown below:

$$\tilde{\mathbf{E}}^i = E_0^i e^{-\gamma_1 z} \quad \text{and} \quad \tilde{\mathbf{H}}^i = \frac{E_0^i}{\eta}e^{-\gamma_1 z} \tag{30.55}$$

$$\tilde{\mathbf{E}}^r = E_0^r e^{\gamma_1 z} \quad \text{and} \quad \tilde{\mathbf{H}}^r = \frac{-E_0^r}{\eta}e^{\gamma_1 z} \tag{30.56}$$

$$\tilde{\mathbf{E}}^t = E_0^t e^{-\gamma_2 z} \quad \text{and} \quad \tilde{\mathbf{H}}^t = \frac{E_0^t}{\eta}e^{-\gamma_2 z}. \tag{30.57}$$

The total field of each region is given as

$$\tilde{\mathbf{E}}_1 = E_0^i e^{-\gamma_1 z} + E_0^r e^{\gamma_1 z}\mathbf{a}_y \quad \text{and} \quad \tilde{\mathbf{H}}_1 = H_0^i e^{-\gamma_1 z} + H_0^r e^{\gamma_1 z}\mathbf{a}_y \tag{30.58}$$

$$\tilde{\mathbf{E}}_2 = E_0^t e^{-\gamma_2 z} \mathbf{a}_y \quad \text{and} \quad \tilde{\mathbf{H}}_2 = H_0^t e^{-\gamma_2 z} \mathbf{a}_y \tag{30.59}$$

At the interface between medium 1 and medium 2, the tangent component of the electric and magnetic fields must be equal, and therefore we can relate the transmitted and reflected field at the interface ($z = 0$), as shown below:

$$\tilde{\mathbf{E}}_1 = \tilde{\mathbf{E}}_2; \quad E_0^i + E_0^r = 0 \tag{30.60}$$

$$\tilde{\mathbf{H}}_1 = \tilde{\mathbf{H}}_2; \quad H_0^i + H_0^r = 0 = \frac{E_0^i}{\eta} - \frac{E_0^r}{\eta} \tag{30.61}$$

This enables us to define the reflection (Γ) and transmission (τ) coefficients at the boundary in terms of the constitutive properties of each medium:

$$E_0^r = E_0^i \frac{\eta_1 - \eta_2}{\eta_1 + \eta_2} = \Gamma E_0^i \tag{30.62}$$

$$E_0^t = E_0^i \frac{2\eta_2}{\eta_1 + \eta_2} = \tau E_0^i \tag{30.63}$$

$$\tau = \frac{E_0^t}{E_0^i} = \frac{2\eta_2}{\eta_1 + \eta_2} \quad \text{and} \quad \Gamma = \frac{E^r}{E_0^i} = \frac{\eta_1 - \eta_2}{\eta_1 + \eta_2} \tag{30.64}$$

Therefore, Equation 30.59 becomes

$$\tilde{\mathbf{E}}_1 = E_0^i (e^{-\gamma_1 z} + \Gamma e^{\gamma_1 z}) \quad \text{and} \quad \tilde{\mathbf{H}}_1 = H_0^i (e^{-\gamma_1 z} - \Gamma e^{\gamma_1 z}) \tag{30.65}$$

$$\tilde{\mathbf{E}}_2 = \tau E_0^i e^{-\gamma_2 z} \quad \text{and} \quad \tilde{\mathbf{H}}_2 = \tau H_0^i e^{-\gamma_2 z} \tag{30.66}$$

30.2 Transmission Lines

30.2.1 Overview

The term "transmission line" usually refers to two or more parallel conducting lines that allow the transmission of electrical power or energy between systems. Common examples of transmission lines include coaxial cables, high-power lines, and telephone lines. The transmission lines discussed in this chapter only allow a *transverse electromagnetic* (TEM) *wave* to propagate. A TEM is unique in that the electric and magnetic fields are perpendicular to each other and the direction of propagation. This differs from the behavior that occurs in Waveguides, which will be discussed in the subsequent chapter. It also allows us to use a bulk parameter (R, L, C, G per unit length) or distributed model for the electromagnetic analysis.

30.2.2 Analysis

It is generally assumed that any multiconductor line must be analyzed as a transmission line if its length is greater than a tenth of the wavelength ($\lambda/10$). The analysis begins by considering a per unit meter bulk parameter equivalent of a two conductor line shown in Figure 30.2.

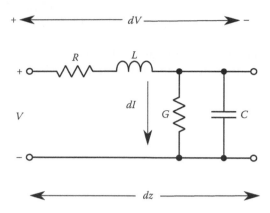

FIGURE 30.2   Bulk parameter representation of a transmission line.

The differential voltage (dV) and current drop that occurs over the length dz are given by

$$\frac{dV}{dz} = -IR - L\frac{dI}{dt} \tag{30.67}$$

$$\frac{dI}{dz} = -VG - C\frac{dV}{dt} \tag{30.68}$$

or in the phasor domain as

$$\frac{dV}{dz} = -I(R + j\omega L) \tag{30.69}$$

$$\frac{dI}{dz} = -V(G + j\omega C). \tag{30.70}$$

In the phasor domain, a single equation that depends only on V can be developed by taking the derivative of (30.69) with respect to z and substituting Equation 30.70 for dI/dz, as shown below:

$$\frac{d^2V}{dz^2} = -\frac{dI}{dz}(R + j\omega L) \tag{30.71}$$

$$\frac{d^2V}{dz^2} = (G + j\omega C)(R + j\omega L)V \tag{30.72}$$

Equation 30.72 is usually written as

$$\frac{d^2V}{dz^2} - \gamma^2 V = 0 \tag{30.73}$$

where $\gamma = \sqrt{(G + j\omega C)(R + j\omega L)} = \alpha + j\beta$, α being the attenuation constant and β the propagation constant. An analogous equation for the current waveform on a transmission line can be developed by taking the derivative of (30.70) with respect to z and substituting Equation 30.69. The solutions for the voltage and current waveforms on a transmission line are therefore given in the phasor domain as

$$V = (V_0^+ e^{-\gamma z} + V_0^- e^{\gamma z}) \tag{30.74}$$

$$I = (I_0^+ e^{-\gamma z} + I_0^- e^{\gamma z}) \tag{30.75}$$

and time domain as

$$V(z,t) = (V_0^+ e^{-\alpha z} \cos(\omega t - \beta z + \theta_{V^+}) + V_0^- e^{\alpha z} \cos(\omega t + \beta z + \theta_{V^-})$$

$$I(z,t) = (I^+ e^{-\alpha z} \cos(\omega t - \beta z + \theta_{I^+}) + I^- e^{\alpha z} \cos(\omega t + \beta z + \theta_{I^-})$$

The characteristic impedance of the region Z_0 can be found for a positively propagating wave by taking the derivative of the voltage ($V = V_0^+ e^{-\gamma z}$) with respect to z and solving for ($Z_0 = V/I$), as shown below:

$$\frac{dV}{dz} = -I(R + j\omega L) = -\gamma V_0^+ e^{-\gamma z} = -\gamma V \tag{30.76}$$

$$Z_0 = \frac{R + j\omega L}{\gamma} = \sqrt{\frac{R + j\omega L}{G + j\omega C}}. \tag{30.77}$$

For waves traveling in the $-z$-direction, the relationship between current and voltage is

$$\frac{V^-}{I^-} = -Z_0. \tag{30.78}$$

30.2.3 Lossless Transmission Lines

Several of the relationships for lossy transmission lines can be simplified if the line is assumed lossless. They include the phase velocity v_p, characteristic impedance Z_0, the propagation constant γ, and the phasor and temporal description of the voltage and current waveform (only voltage waveform shown) among others.

$$Z_0 = \sqrt{\frac{L}{C}} \tag{30.79}$$

$$\gamma = j\beta = j\omega \sqrt{LC} \tag{30.80}$$

$$v_p = \frac{\omega}{\beta} = \sqrt{\frac{1}{LC}} \tag{30.81}$$

$$V = V_0^+ e^{-j\beta z} + V_0^- e^{j\beta z} \tag{30.82}$$

$$V(z,t) = V_0^+ \cos(\omega t - \beta z) + V_0^- \cos(\omega t + \beta z) \tag{30.83}$$

There is also a relationship between the per unit length capacitance C and inductance L, the phase velocity v_p, and the constitutive parameters of the region:

$$LC = \varepsilon\mu \quad \text{and} \quad v_p = \frac{1}{\sqrt{\mu\varepsilon}}. \tag{30.84}$$

30.2.4 Terminated Transmission Lines

Transmission lines used to transmit electromagnetic waves are usually terminated in loads, which cause a reflected wave to occur. The loads are typically modeled as lumped elements at $z = 0$, as shown in Figure 30.3. The load impedance (Z_L) is the ratio of the voltage to current at the load and given as

$$Z_L = \frac{V(z=0)}{I(z=0)} = \frac{(V_0^+ e^{-\gamma 0} + V_0^- e^{\gamma 0})}{(I_0^+ e^{-\gamma 0} + I_0^- e^{\gamma 0})} = \frac{V_0^+ + V_0^-}{I_0^+ + I_0^-} \tag{30.85}$$

Therefore, by rearranging the terms in Equation 30.78, Z_L and V_0^- can be expressed in terms of Z_0, V_0^+, V_0^- as

$$Z_L = Z_0 \frac{V_0^+ + V_0^-}{V_0^+ - V_0^-} \tag{30.86}$$

$$V_0^- = \frac{Z_L - Z_0}{Z_L + Z_0} V_0^+ \tag{30.87}$$

This allows us to define the reflection coefficient at the load (Γ_L) as

$$\Gamma_L = \frac{V_0^-}{V_0^+} = \frac{Z_L - Z_0}{Z_L + Z_0} \tag{30.88}$$

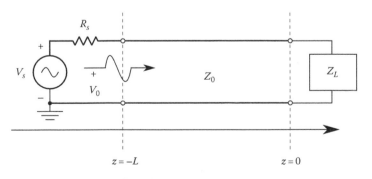

FIGURE 30.3   Schematic representation of a transmission line.

Γ is generally a complex number that describes both the magnitude and the phase of the reflected wave. Some of the simplest cases are when the transmission line is terminated into a short circuit, open circuit, or it's characteristic impedance Z_0, as shown below:

$\Gamma_L = -1$: maximum negative value of
reflection coefficient when the line is short-circuited

$\Gamma_L = 0$: no reflection on positive traveling
wave when the line is terminated into Z_0;
this is known as a matched condition

$\Gamma_L = +1$: maximum positive reflection when the line is open-circuited

or at any point on the line as

$$\Gamma = \Gamma_L e^{j2\gamma z} \tag{30.89}$$

Equations 30.9 and 30.10 can therefore be expressed as

$$V = V_0^+ (e^{-\gamma z} + \Gamma_L e^{\gamma z}) \tag{30.90}$$

$$I = I_0^+ (e^{-\gamma z} - \Gamma_L e^{\gamma z}). \tag{30.91}$$

Another common quantity that describes how effectively a transmission line is delivering power to a load is the voltage standing wave ratio (VSWR). The VSWR is defined as the ratio of the maximum voltage amplitude to the minimum voltage amplitude that occurs on a transmission line. Figure 30.4 shows how the peak value of the sinusoidally varying voltage waveform varies versus position z. The maximum and minimum values of the voltage waveform are $V_{max} = V_0^+(1+|\Gamma_L|); V_{min} = V_0^+(1-|\Gamma_L|)$. Therefore, the standing wave ratio is a strictly positive number that varies between 1 for a perfectly matched load to ∞ for either an open- or a short-circuit load. Equation 30.27 defines the VSWR of a transmission line terminated into a load Z_L:

$$VSWR = \frac{V_{max}}{V_{min}} = \frac{1+|\Gamma_L|}{1-|\Gamma_L|} \tag{30.92}$$

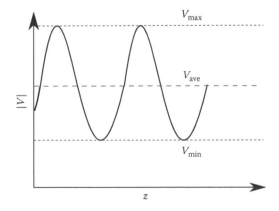

FIGURE 30.4 VSWR on a transmission line.

The input impedance (Z_{in}) at any distance from the load ($-l$) can be found using Equations 30.25 and 30.26, as shown below:

$$Z_{in}(-l) = \frac{(V_0^+ e^{+\gamma l} + V_0^- e^{-\gamma l})}{(I_0^+ e^{+\gamma l} + I_0^- e^{-\gamma l})} = Z_0 \frac{Z_L + Z_0 \tanh(\gamma l)}{Z_0 + Z_L \tanh(\gamma l)} \tag{30.93}$$

or for a lossless transmission line the equation becomes

$$Z_{in}(-l) = Z_0 \frac{Z_L + jZ_0 \tan(\beta l)}{Z_0 + jZ_L \tan(\beta l)}. \tag{30.94}$$

30.2.5 Smith Chart

Prior to the advent of high-speed digital computers, the most common technique employed to analyze transmission line behavior required a graphical solution using a Smith chart (shown in Figure 30.5). The Smith chart was created by Phillip H. Smith as a tool for electrical and electronic engineers to assist them in solving problems with transmission lines and matching networks. Use of the Smith chart is still very popular today for both problem solving and as a graphical aid demonstrating how the behavior of an electrical system changes as a function of frequency. Even though a Smith chart can be used for the analysis of a lossy transmission line, only the lossless case will be discussed here.

The Smith chart is a polar plot of the complex reflection coefficient (Γ) and the normalized line impedance (Z_{in}) at all points on the transmission line. A derivation of the Smith chart begins by defining the normalized load impedance z_L and calculating the complex reflection coefficient Γ_L, as shown below:

$$z_L = \frac{Z_L}{Z_0} \tag{30.95}$$

$$\Gamma_L = \frac{z_L - 1}{z_L + 1} \tag{30.96}$$

The reflection coefficient any distance from the load z is given in Equation 30.24, and for a lossless line it becomes

$$\Gamma_z = \frac{z_{in} - 1}{z_{in} + 1} = \Gamma_L e^{j2\beta z} \tag{30.97}$$

and may be expressed in terms of rectangular components as

$$\Gamma(z) = \Gamma_{Re} + j\Gamma_{Im} \tag{30.98}$$

The normalized input impedance (z_{in}) can be solved for in terms of $\Gamma(z)$, as shown below:

$$z_{in} = r + jx = \frac{1 + \Gamma_{Re} + j\Gamma_{Im}}{1 - \Gamma_{Re} - j\Gamma_{Im}}. \tag{30.99}$$

By multiplying the last term in (30.34) through by the complex conjugate ($1 - \Gamma_{Re} + j\Gamma_{Im}$) and rearranging the terms (30.34), r and jx are given as

$$r = \frac{1 - \Gamma_{Re}^2 - \Gamma_{Im}^2}{(1 - \Gamma_{Re})^2 + \Gamma_{Im}^2} \tag{30.100}$$

Smith chart

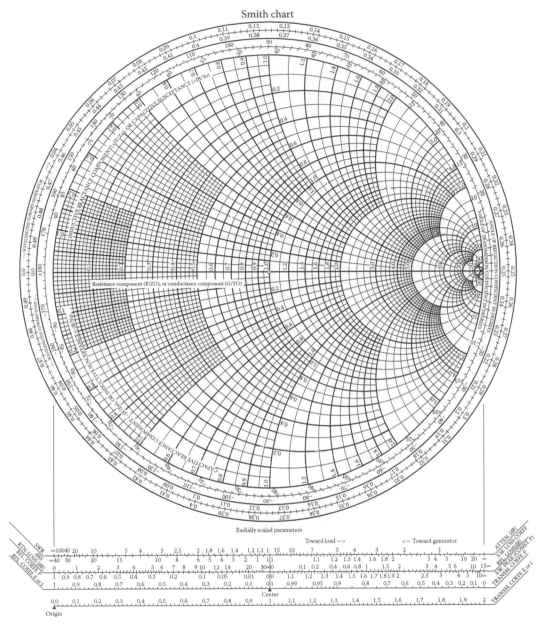

FIGURE 30.5 The Smith chart.

$$jx = \frac{j2\Gamma_{Im}}{(1 - \Gamma_{Re})^2 + \Gamma_{Im}^2} \tag{30.101}$$

Equations 30.35 and 30.36 can easily be formulated as circular functions, which, when plotted, constitute the Smith chart:

$$\left(\Gamma_{Re} - \frac{r}{r+1}\right)^2 + \Gamma_{Im}^2 = \left(\frac{1}{1+r}\right)^2 \tag{30.102}$$

$$\left(\Gamma_{Re} - 1\right)^2 + \left(\Gamma_{Im} - \frac{1}{x}\right)^2 = \left(\frac{1}{x}\right)^2 \tag{30.103}$$

The real part of the normalized load impedance r forms complete circles and the imaginary part jx circular arcs.

Figure 30.6 shows a simple example of the normalized load impedance of $z_L = 0.8 + j1.2$ plotted on the Smith chart with the reflection coefficient shown. Moving away from the load toward the

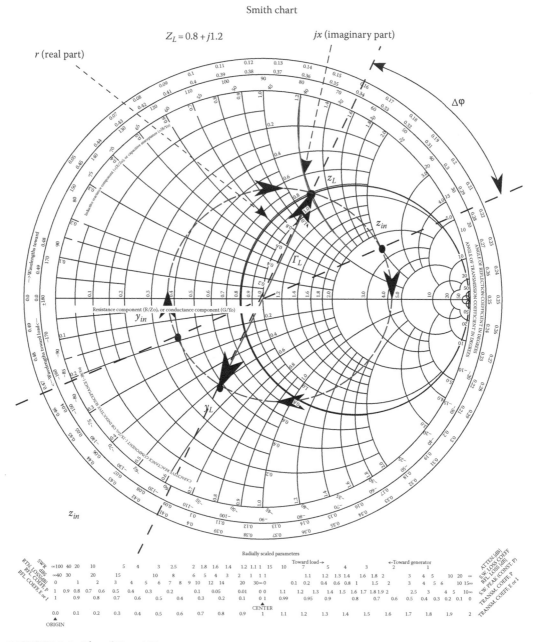

FIGURE 30.6   Plot of Z_L and Γ_L.

generator, a distance of 0.058 wavelengths (λ) is equivalent to moving in the clockwise direction around a circular arc centered at the origin, as indicated on Figure 30.6. Note the information on the rim of the Smith chart. The magnitude of the reflection coefficient stays constant but the phase changes $\Delta\phi = 2\beta z$. Since the original position corresponded to a reading on the outer rim of the Smith chart of 0.16λ, we need to add 0.058λ to get the final point located at $d = 0.218\lambda$. The input impedance at that point is indicated z_{in}. It is also important to note that the normalized admittance (y) that corresponds to each point in the normalized impedance domain is located at a conjugate point 180° away.

There are several features of a Smith chart that are important to note:

1. Moving in the clockwise direction around the Smith chart corresponds to moving away from the load or toward the generator.
2. Likewise, moving in the counterclockwise direction around the Smith chart corresponds to moving away from the generator or moving toward the load.
3. Moving one full circle around the Smith chart corresponds to moving a half-wavelength ($\lambda/2$), or 180°, in either direction.

30.2.6 Impedance Matching

The goal of impedance matching is to couple all of the transmitted power to the load. There exist a number different methods used to create impedance-matching networks. One of the most common is stub matching. Stub matching requires a short segment of the transmission line be cut to a predetermined length (ds) and placed in parallel with the transmission line a given distance from the load (dL). The stub may be terminated in either a short or open circuit, as shown in Figure 30.7.

The length of the stub and distance from the load are determined by first plotting the normalized load admittance on the Smith chart. Next, move around the arc of constant $|\Gamma|$ toward the generator until you reach the $r = 1 \pm s$ circle. Since there are two points where the circles intersect, you are free to choose which distance form the load you prefer. Once this point is determined, the value of the normalized stub admittance will be equal to $y_{stub} = -s$ and connected in parallel to the transmission line.

This is best understood by a simple example. Assume you have load 0. The normalized load impedance is $z_L = 0.4 + j1$ and normalized load admittance $y_L = 0.345 - j1.842$, as shown in Figure 30.8. By working in the admittance domain, we move away from the load until we reach the point $y_L = 1 + j1.867s$ and add a shunt stub having a value $y_{sub} = -j1.867$, making the total normalized admittance and normalized impedance equal to 1. This effectively sets the reflection coefficient ($\Gamma = 0$) to the left of the stub equal to 0.

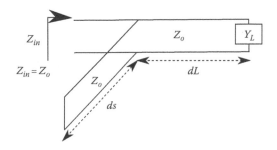

FIGURE 30.7 Impedance matching using a short-circuit stub.

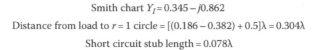

Smith chart $Y_l = 0.345 - j0.862$

Distance from load to $r = 1$ circle $= [(0.186 - 0.382) + 0.5]\lambda = 0.304\lambda$

Short circuit stub length $= 0.078\lambda$

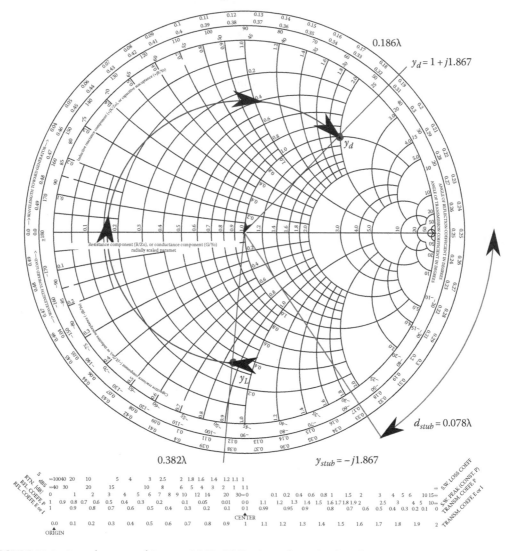

FIGURE 30.8 Impedance matching $y_L = 0.345 - j8862$ using a short-circuit stub.

30.3 Waveguides

A waveguide refers to any structure that directs electromagnetic energy down a specific path. In most cases, the cross-sectional area of a waveguide is invariant with the direction of propagation. However, there are exceptions to this generalization including the natural waveguide formed by the earth–ionosphere boundary. Transmission lines are a specific type of waveguide that allow a TEM wave to propagate and can therefore be treated in terms of bulk or distributed parameters (i.e., R, L, C, G). However, the waveguides discussed in this chapter operate in the transverse magnetic (TM) or transverse electric (TE) mode and act as high-pass filters.

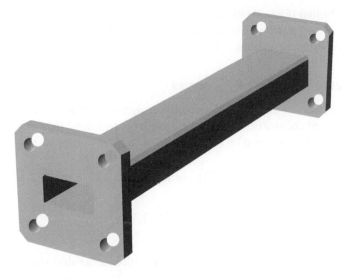

FIGURE 30.9 Rectangular waveguide section with flanges.

Some of the most common types of waveguides include the hollow conducting cylinders, dielectric waveguides, and optical fibers. The analysis of waveguides begins by formulating the problem in terms of Maxwell's equations and applying the appropriate boundary conditions.

30.3.1 Hollow Conducting Cylinders

The analysis described in this section may be applied to any hollow conducting cylinder; however, we will only consider waveguides having a rectangular cross section. Figure 30.9 shows a typical rectangular waveguide section. To begin the analysis, consider the cross-sectional diagram of a waveguide, as shown in Figure 30.10. The waveguide is filled with a dielectric material having constitutive parameters (ε, μ). The electric and magnetic fields may therefore be expressed in terms of $\tilde{\mathbf{E}} = \mathbf{E}_0(\mathbf{x}, \mathbf{y})e^{-\gamma z}$, and $\tilde{\mathbf{H}} = \mathbf{H}_0(\mathbf{x}, \mathbf{y})e^{-\gamma z}$ where γ is the propagation coefficient in the z direction. By applying Maxwell's equations to the expressions for $\tilde{\mathbf{E}}$ and $\tilde{\mathbf{H}}$, we obtain

$$E_x = -\frac{1}{\zeta^2}\left(\gamma\frac{\partial E_z}{\partial x} + j\omega\mu\frac{\partial H_z}{\partial y}\right) \tag{30.104}$$

$$E_y = -\frac{1}{\zeta^2}\left(\gamma\frac{\partial E_z}{\partial y} - j\omega\mu\frac{\partial H_z}{\partial x}\right) \tag{30.105}$$

$$H_x = -\frac{1}{\zeta^2}\left(-j\omega\varepsilon\frac{\partial E_z}{\partial y} + \gamma\frac{\partial H_z}{\partial x}\right) \tag{30.106}$$

$$H_y = -\frac{1}{\zeta^2}\left(j\omega\varepsilon\frac{\partial E_z}{\partial x} + \gamma\frac{\partial H_z}{\partial y}\right) \tag{30.107}$$

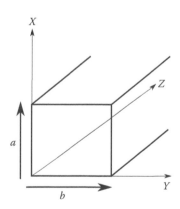

FIGURE 30.10 Diagram of rectangular waveguide.

where $\zeta^2 = \gamma^2 + \beta^2$ and $\beta^2 = \omega^2\mu\varepsilon$.

The propagation coefficient γ and six of the field components are not known. They may be determined by solutions to the differential equations given in (30.104) through (30.107) and applying the appropriate electromagnetic boundary conditions. In general, the propagation constant is given by $\gamma = \sqrt{j\omega\mu(\sigma + j\varepsilon\omega)}$, but for the remainder of this section, we will assume the waveguide is lossless and $\gamma = j\beta$.

30.3.2 Transverse Electric $\text{TE}_{m,n}$ Waves

$\text{TE}_{m,n}$ waves or modes have no electric field and only a magnetic field in the direction of propagation (H_z). To solve for the field components, we use the boundary condition $\partial H_{tan}/\partial n = 0$ on all conducting surfaces and construct solutions based on (30.25) for H_z as shown below:

$$\nabla^2 \tilde{H} + \omega^2 \mu\varepsilon \tilde{H} = 0 \tag{30.108}$$

$$\frac{\partial^2 H_z}{\partial x^2} + \frac{\partial^2 H_z}{\partial y^2} + \frac{\partial^2 H_z}{\partial z^2} + \omega^2 \mu\varepsilon H_z = 0 \tag{30.109}$$

The solution to Equation 30.109 is found using the separation of variables shown below:

$$H_z = X(x)Y(y)Z(z)$$

$$X''YZ + XY''Z + XYZ'' + \omega^2 \mu\varepsilon XYZ = 0$$

$$\frac{X''}{X} + \frac{Y''}{Y} + \frac{Z''}{Z} + \omega^2 \mu\varepsilon = 0$$

$$-k_x^2 - k_y^2 - k_z^2 + \omega^2 \mu\varepsilon = 0$$

where
$$X''/X = -k_x^2$$
$$Y''/Y = -k_y^2$$
$$Z''/Z = -k_z^2$$

Each of the differential equations has a solution of the same for form: $X(x) \sim \sin(k_x x); \cos(k_x x)e^{\pm jk_x x}$; $Y(y) \sim \sin(k_y y)\cos(k_y y); e^{\pm jk_y y}$; $Z(z) \sim \sin(k_z z), \cos(k_z z)e^{\pm jk_z z}$. The exact solution is dependant on the dimensions of the guide, as shown below:

$$X(x) = \cos\left(\frac{m\pi x}{a}\right); \quad Y(y) = \cos\left(\frac{n\pi y}{b}\right) \tag{30.110}$$

$$H_z = H_0 \cos\left(\frac{m\pi x}{a}\right)\cos\left(\frac{n\pi y}{b}\right)e^{\pm jk_z z} \tag{30.111}$$

The propagation constant k_z is given as

$$k_z^2 = \omega^2 \mu\varepsilon - \left(\frac{m\pi}{a}\right)^2 - \left(\frac{n\pi}{b}\right)^2 \tag{30.112}$$

$$k_z = \pm\sqrt{\omega^2 \mu\varepsilon - \left(\frac{m\pi}{a}\right)^2 - \left(\frac{n\pi}{b}\right)^2} \tag{30.113}$$

The radian frequency at which the argument $\sqrt{\omega^2\mu\varepsilon - (m\pi/a)^2 - (n\pi/b)^2}$ becomes negative is known as the cutoff frequency (ω_c or f_c). Waves having a frequency lower than cutoff frequency will not propagate. The cutoff frequency in radians per second and hertz are given below:

$$\omega_c = \frac{1}{\mu\varepsilon}\sqrt{\left(\frac{m\pi}{a}\right)^2 + \left(\frac{n\pi}{b}\right)^2} \tag{30.114}$$

$$f_c = \frac{1}{2\mu\varepsilon}\sqrt{\left(\frac{m}{a}\right)^2 + \left(\frac{n}{b}\right)^2} \tag{30.115}$$

and m, n are known as the mode numbers. It is important to note that for a $\text{TE}_{m,n}$ wave, the lowest order mode is the $\text{TE}_{1,0}$ mode corresponding to a cutoff frequency of $fc_{1,0} = 1/2a\sqrt{\mu\varepsilon}$.

This also allows us to reduce all relative equations to a more simplified form in terms of the cutoff frequency for the respective mode.

The remainder of the field components can be found by solving Maxwell's equations with the given value of H_z and are shown below:

TE case

$$E_x = \frac{j\omega u n\pi}{\zeta^2 b} H_0 \cos\left(\frac{m\pi x}{a}\right) \sin\left(\frac{n\pi y}{b}\right) e^{-jk_z z} \tag{30.116}$$

$$E_y = \frac{-j\omega u m\pi}{\zeta^2 a} H_0 \sin\left(\frac{m\pi x}{a}\right) \cos\left(\frac{n\pi y}{b}\right) e^{-jk_z z} \tag{30.117}$$

$$H_x = \frac{\gamma m\pi}{\zeta^2 a} H_0 \sin\left(\frac{m\pi x}{a}\right) \cos\left(\frac{n\pi y}{b}\right) e^{-jk_z z} \tag{30.118}$$

$$H_y = \frac{\gamma n\pi}{\zeta^2 b} H_0 \cos\left(\frac{m\pi x}{a}\right) \sin\left(\frac{n\pi y}{b}\right) e^{-jk_z z} \tag{30.119}$$

where $\zeta^2 = \omega^2\mu\varepsilon - k_z^2$.

The wave impedance for TE modes is given as

$$Z_{\text{TE}} = -\frac{E_y}{H_x} = \frac{E_x}{H_y} = \frac{\sqrt{\frac{\mu}{\varepsilon}}}{\sqrt{\left(1 - \left(fc/f\right)^2\right)}} \tag{30.120}$$

30.3.3 Transverse Electric $\text{TM}_{m,n}$ Waves

$\text{TM}_{m,n}$ waves or modes have no magnetic field and only an electric field in the direction of propagation (E_z). The solution for these modes follows in the same manner as the TE case. We begin by starting with the Helmholtz equation for the electric field and then apply the appropriate boundary conditions:

$$\nabla^2\tilde{\mathbf{E}} + \omega^2\mu\varepsilon\tilde{\mathbf{E}} = 0 \tag{30.121}$$

$$\frac{\partial^2 E_z}{\partial x^2} + \frac{\partial^2 E_z}{\partial y^2} + \frac{\partial^2 E_z}{\partial z^2} + \omega^2 \varepsilon \mu E_z = 0 \tag{30.122}$$

The solution to Equation 30.109 is found using the separation of variables shown below:

$$E_z = X(x)Y(y)Z(z)$$

$$X''YZ + XY''Z + XYZ'' + \omega^2 \mu \varepsilon XYZ = 0$$

$$\frac{X''}{X} + \frac{Y''}{Y} + \frac{Z''}{Z} + \omega^2 \mu \varepsilon = 0 \tag{30.123}$$

$$-k_x^2 - k_y^2 - k_z^2 + \omega^2 \mu \varepsilon = 0$$

where

$$X''/X = -k_x^2$$
$$Y''/Y = -k_y^2$$
$$Z''/Z = -k_z^2$$

Each of the differential equations has the same form of the solution as the TE mode: $X(x) \sim \sin(k_x x); \cos(k_x x)e^{\pm jk_x x}$; $Y(y) \sim \sin(k_y y)\cos(k_y y); e^{\pm jk_y y}$; $Z(z) \sim \sin(k_z z), \cos(k_z z)e^{\pm jk_z z}$. The exact solution is found by enforcing the boundary conditions $E_{tan} = 0$ at all conducting surfaces. The final form of the solution is shown below:

$$E_z = E_0 \sin\left(\frac{m\pi x}{a}\right)\sin\left(\frac{n\pi y}{b}\right)e^{\pm jk_z z} \tag{30.124}$$

The remainder of the equations are found by solving Maxwell's equations using the E_z and are given as TM case:

$$E_x = \frac{-jk_z}{h^2}\left(\frac{m\pi}{a}\right)E_0 \cos\left(\frac{m\pi x}{a}\right)\sin\left(\frac{n\pi y}{b}\right)e^{-jk_z z} \tag{30.125}$$

$$E_y = \frac{-jk_z}{h^2}\left(\frac{n\pi}{b}\right)E_0 \sin\left(\frac{m\pi x}{a}\right)\cos\left(\frac{n\pi y}{b}\right)e^{-jk_z z} \tag{30.126}$$

$$H_x = \frac{j\omega\varepsilon}{h^2}\left(\frac{n\pi}{b}\right)E_0 \sin\left(\frac{m\pi x}{a}\right)\cos\left(\frac{n\pi y}{b}\right)e^{-jk_z z} \tag{30.127}$$

$$H_y = \frac{-j\omega\varepsilon}{h^2}\left(\frac{m\pi}{b}\right)E_0 \cos\left(\frac{m\pi x}{a}\right)\sin\left(\frac{n\pi y}{b}\right)e^{-jk_z z} \tag{30.128}$$

where

$$h^2 = \left(\frac{m\pi}{a}\right)^2 + \left(\frac{n\pi}{b}\right)^2 \tag{30.129}$$

and

$$k_z = \omega\mu\varepsilon\sqrt{1 - \left(\frac{f_c}{f}\right)^2} \tag{30.130}$$

The wave impedance for TM modes is given as

$$Z_{TM} = -\frac{E_y}{H_x} = \frac{E_x}{H_y} = \frac{\sqrt{\frac{\mu}{\varepsilon}}}{\sqrt{\left(1 - \left(\frac{f_c}{f}\right)^2\right)}}$$

(30.131)

For both TE and TM modes, the phase velocity along the guide may approach infinity, while the velocity of energy propagation is always less than the speed of light in the medium. Both of these quantities along with the wavelength along the guide are given below:

$$v_p = \frac{1}{\sqrt{\mu\varepsilon}\sqrt{\left(1 - \left(\frac{f_c}{f}\right)^2\right)}}$$

(30.132)

$$v_e = \frac{1}{\sqrt{\mu\varepsilon}} \sqrt{\left(1 - \left(\frac{f_c}{f}\right)^2\right)}$$

(30.133)

$$\lambda_{guide} = \frac{\lambda_u}{\sqrt{\mu\varepsilon}\sqrt{\left(1 - \left(\frac{f_c}{f}\right)^2\right)}}, \quad \text{where } \lambda_u = \frac{1}{f\sqrt{\mu\varepsilon}}.$$

(30.134)

30.3.4 Dielectric Waveguides

Hollow metallic waveguides are useful as waveguides over a frequency range of ~1–40 GHz. Beyond these frequencies, the dimensions of the guide become prohibitively small and cannot be easily fabricated. Dielectric slab waveguides overcome this problem but do have evanescent fields existing outside the guide. Typical applications for these structures include planar light guides used in optical integrated circuits.

The investigation of wave propagation in dielectric materials is based on a solution of Maxwell's equations with the appropriate boundary conditions applied. A difficulty arises in the analysis due to the existence of evanescent or non-propagating fields outside the dielectric material. However, the behavior may be understood in an uncomplicated manner by first considering an obliquely incident wave striking a dielectric interface and applying Snell's law. In Figure 30.11, a wave is shown incident from medium 1 toward region 2. The incident wave results in a transmitted and reflected wave. The incident (θ_i), reflected (θ_r), and transmitted (θ_t) angles are related by Snell's law of refraction 30.137, as shown below:

$$\theta_i = \theta_r$$

(30.135)

$$\sqrt{\mu_1\varepsilon_1}\sin(\theta_i) = \sqrt{\mu_2\varepsilon_2}\sin(\theta_t)$$

(30.136)

Considering only nonmagnetic materials where the incident wave is in a dielectric having $\varepsilon_{r1} > \varepsilon_{r2}$, then transmitted wave is at a larger angle than the incident wave ($\theta_t > \theta_i$). Furthermore, there is an incident angle at which the entire wave is reflected, the critical angle θ_c. The critical angle can be expressed in terms of the permittivities of each region as

$$\theta_c = \arcsin\left(\frac{\sqrt{\varepsilon_{r2}}}{\sqrt{\varepsilon_{r2}}}\right)$$

(30.137)

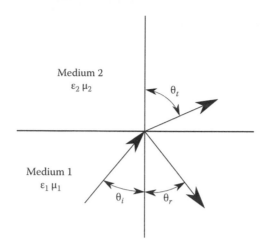

FIGURE 30.11 Wave obliquely incident on boundary of two different dielectric materials.

A more traditional method of expressing Snell's law is to do so in terms of the index of refraction of the medium *n*:

$$n = \sqrt{\varepsilon_r} \tag{30.138}$$

Snell's law then becomes

$$n_1 \sin(\theta_i) = n_2 \sin(\theta_t) \tag{30.139}$$

and

$$\theta_c = \arcsin\left(\frac{n_1}{n_2}\right) \tag{30.140}$$

For dielectric slab waveguide (Figure 30.12), only waves at angles greater than the critical angle will propagate if constructive interference is present. Additionally, you will have both TE and TM modes that include TE_0 and TM_0 modes. It can be shown that for only one mode to propagate, the following inequality must exist:

$$\frac{a}{\lambda_0} = \frac{1}{2}\left(\frac{1}{\sqrt{n_1^2 - n_2^2}}\right) \tag{30.141}$$

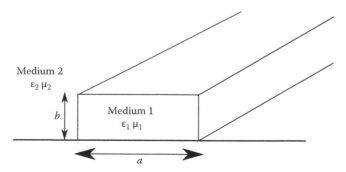

FIGURE 30.12 Dielectric waveguide.

30.3.5 Fiber Optics

Fiber optic technology is rapidly replacing copper as the backbone of the telecommunication industry. Optical fiber has enormous advantages over existing copper wire in long-distance applications due to its much lower attenuation and immunity to electromagnetic interference. There are three primary frequency-transmission windows used by industry centered at 850, 1300, and 1550 nm wavelengths.

There are two main types of optical fibers used in telecommunications: (1) multimode optical fiber and (2) single-mode optical fiber. Multimode optical fiber typically has a larger core (~50 μm) that results in less expensive transmitters, receivers, and connectors. The major problem associated with multimode fiber is the fact that it introduces intermodal dispersion, which limits the bandwidth and length of the link. Additionally, because of its higher dopant content, multimode fibers are more expensive and have higher attenuation than single-mode fibers. The core of a single-mode fiber is smaller (~10 μm) but they require more expensive components and interconnection methods, but do allow much longer, higher-performance links.

A typical step-index fiber consists of an inner core of radius *a* surrounded by an outer cladding and jacket, as shown in Figure 30.13. The inner core and surrounding cladding create a cylindrical dielectric waveguide that guides incident light down the fiber at all angles less than the acceptance angle. This is commonly referred to as the cone of acceptance and is shown as the gray region. A common term that quantifies how effectively any type of optical fiber accepts light is the "numerical aperture" (NA).

The NA of a given optical fiber determines this maximum acceptance angle of light entering the fiber (Figure 30.14). Assuming that angle θ_c is the critical angle for light in the core striking the interface between the inner core and cladding, it can be defined by

$$\sin(\theta_c) = \frac{n_c}{n_f} \tag{30.142}$$

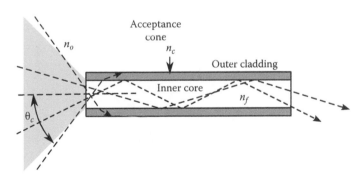

FIGURE 30.13   Cut-away view of a typical fiber optic cable.

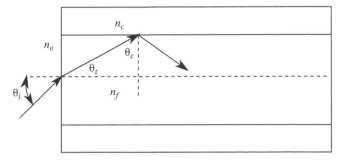

FIGURE 30.14   Cross-sectional view of fiber for determining acceptance angle.

and by inspection, it is obvious that $\sin(\theta_c) = \cos(\theta_t)$. Furthermore,

$$\sin^2(\theta_t) + \cos^2(\theta_t) = 1 \tag{30.143}$$

$$\sin(\theta_t) = \sqrt{1 - \cos^2(\theta_t)} \tag{30.144}$$

$$\sin(\theta_t) = \sqrt{1 - \sin^2(\theta_c)} \tag{30.145}$$

$$\sin(\theta_t) = \sqrt{1 - \left(\frac{n_c}{n_f}\right)^2}. \tag{30.146}$$

By applying Snell's law to the air–fiber interface we see that

$$n_o \sin(\theta_i) = n_f \sin(\theta_t) \tag{30.147}$$

$$n_o \sin(\theta_i) = n_f \sqrt{1 - \left(\frac{n_c}{n_f}\right)^2} \tag{30.148}$$

$$\sin(\theta_i) = \frac{\sqrt{n_f^2 - n_c^2}}{n_0} \tag{30.149}$$

$$NA = \frac{\sqrt{n_f^2 - n_c^2}}{n_0}. \tag{30.150}$$

As was true in the dielectric slab waveguide, the TE_0 and the TM_0 mode propagate in fiber optic cable due to the existence of an evanescent external electric field. Each of these modes has no associated cutoff frequency. For step-index fiber optic cables, the free space wavelength required for only one mode to propagate is given by

$$\lambda_0 > \frac{2\pi a \left(\sqrt{n_f^2 - n_c^2} \right)}{k_{01}} \tag{30.151}$$

where k_{01} is the first root of the zeroth order Bessel function ($k_{01} = 2.405$). It should be noted that single mode propagation requires keeping ($n_f - n_c$) small for realistic core diameters.

Additionally, for multimode step-index fiber, the number of modes is given as

$$N = 2 \left(\frac{\pi a}{\lambda_0} \right)^2 \left(n_f^2 - n_c^2 \right)^2 \tag{30.152}$$

30.4 Antennas

Antennas are a class of devices specifically developed to radiate or receive electromagnetic waves. They do so by converting alternating currents into electromagnetic waves when radiating and electromagnetic waves into currents when receiving. Some typical applications for antennas are wireless phones, airport radars, all AM and FM broadcast transmissions, and radio astronomy. Antennas are used in a variety of environments, but we will restrict our analysis to propagation in free space

Antennas usually consist of an arrangement of conductors. Several of the more common types are shown in Figure 30.15. They generate a radiating electromagnetic field in response to an applied alternating electric current or develop induced currents in response to an incident electromagnetic wave.

30.4.1 Differential or Hertzian Dipole

Analysis of antennas typically begins by considering the simple, electrically short dipole ($l \ll \lambda$), commonly referred to as a differential or Hertzian dipole (Figure 30.16), centered at the origin and oriented in the $\boldsymbol{a}_z$ direction. It consists of a wire of length l or dl electrically driven at the center. The radiated electric and magnetic field propagates in the r-direction and is given as the following:

$$\tilde{E}_\theta(r,\theta) = \frac{j\beta Il \sin(\theta)}{4\pi r} \sqrt{\frac{\mu}{\varepsilon}} e^{-j\beta r} \tag{30.153}$$

$$\tilde{H}_\phi(r,\theta) = \frac{j\beta Il \sin(\theta)}{4\pi r} e^{-j\beta r} \tag{30.154}$$

The waves propagate in phase, and the ratio of the electric to magnetic field is equal to the characteristic impedance of free space (η_0):

$$\eta = \frac{\tilde{E}_\theta}{\tilde{H}_\phi} = \sqrt{\frac{\mu}{\varepsilon}} \tag{30.155}$$

It is interesting to note that the field strengths from these finite structures fall off as ~$(1/r)$ rather than ~$(1/r^2)$ as static fields or quasi-static fields would. This behavior is true for all propagating waves from finite antennas.

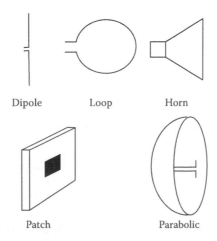

Dipole   Loop   Horn

Patch   Parabolic

FIGURE 30.15 Several typical antenna classes.

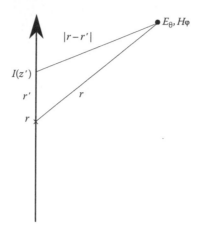

FIGURE 30.16 Dipole antenna of arbitrary length.

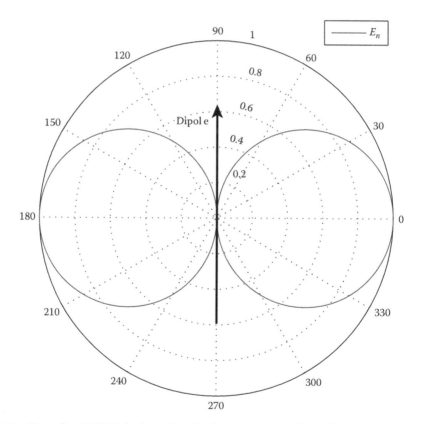

FIGURE 30.17 Normalized E-field plot from short dipole antenna oriented in z-direction.

An additional noteworthy feature of the Hertzian dipole is that no electric or magnetic fields radiate along the antenna's axis. The relative field strengths are a maximum at $\theta = 90°$ and are azimuthally symmetric. A normalized polar plot of the field strength magnitude is shown in Figure 30.17.

The Hertzian dipole is important to understand since all radiating antennas can be approximated as a large number of short (or differential) dipoles. This allows the total field to be calculated by summing or integrating the total field resulting from all differential dipoles.

30.4.2 Dipole of Length L

The general expression for the fields from a dipole of arbitrary length centered at the origin and oriented in the a_z direction can be found by integrating the value of the differential field over the length of the antenna. In order to do this, the current distribution must be known with a high degree of accuracy. For a center feed dipole having a given current distribution, $I(Z)$, the differential magnetic field is given as

$$I(z') = I_0 \sin\left(\beta\left(\frac{l}{2} - |z'|\right)\right) \tag{30.156}$$

$$dH_\phi = \frac{j\beta I(z')e^{-j\beta|r-r'|}}{4\pi|r-r'|}\sin\left(\beta\left(\frac{L}{2} - |z'|\right)\right) \tag{30.157}$$

where r is the distance from the origin to the point of observation of the field and r' is the distance from the origin to the differential current element (30.16). The total magnetic field is given by

$$H_\phi = \frac{j\beta I_0 a_\phi}{4\pi}\int_{-L/2}^{L/2}\frac{e^{-j\beta|r-r'|}}{|r-r'|}\sin\left(\beta\left(\frac{L}{2} - |z'|\right)dz'\right) \tag{30.158}$$

$$H_\phi = \frac{jI_0 e^{-j\beta r}a_\phi}{2\pi r}\frac{\left(\cos\left((BL/2)\cos(\theta)\right) - \cos(BL/2)\right)}{\sin(\theta)} \tag{30.159}$$

30.4.3 Half-Wave Dipole

Another very common antenna is the half-wave dipole antenna. Its total length in practice is usually slightly less than $\lambda/2$ in order to eliminate the imaginary part on the driving point impedance ($L = 0.485\lambda$). The impedance becomes $Z_{in} \sim 73\ \Omega$ and the current distribution is approximated as resulting in a magnetic field:

$$H_\phi = \frac{jI_0 e^{-j\beta r}a_\phi}{2\pi r}\frac{\left(\cos\left((\pi/2)\cos(\theta)\right)\right)}{\sin(\theta)} \tag{30.160}$$

30.4.4 Antenna Characteristics

30.4.4.1 Power Density and Total Radiated Power

Determining the total radiated power from any antenna system requires the integration of the time-average power density P over a surface enclosing the antenna. The time-average power density is defined as

$$P(r,\theta,\phi) = \frac{1}{2}\text{Re}\left(E \times H^*\right) \tag{30.161}$$

The radiated power density is commonly normalized and plotted versus (θ, ϕ):

$$P_n(r,\theta,\phi) = \frac{P(r,\theta,\phi)}{P_{max}} \tag{30.162}$$

and the total radiated power may therefore be found by integrating the power density:

$$P_{rad} = \oint_S P(r,\theta,\phi) \cdot dS \tag{30.163}$$

$$P_{rad} = \oint_S P(r,\theta,\phi) r^2 \sin(\theta) d\theta d\phi \tag{30.164}$$

$$P_{rad} = P_{max} \oint_S P_n(r,\theta,\phi) r^2 \sin(\theta) d\theta. \tag{30.165}$$

30.4.4.2 Beam Solid Angle and Directivity

A major goal of antenna design is to create antennas that transmit most of their power in a particular direction. Since antennas have identical transmitting and receiving patterns, they are preferentially sensitive to radiation incident from the same direction.

To quantify the degree of beam focusing that occurs for a given antenna system, we introduce the concept of a pattern solid angle (Ω_p). This is defined by

$$\Omega_p = \oiint_S P_n(r,\theta,\phi) \sin(\theta) d\theta d\phi \tag{30.166}$$

The directivity of an antenna is defined as the ratio of the maximum value of the power radiated in a given direction to the average power radiated by the antenna per steradian:

$$D = \frac{P_{max}(r,\theta,\phi)}{P_{ave}} \tag{30.167}$$

$$D = \frac{4\pi}{\Omega_p} \tag{30.168}$$

$$\text{where } P_{ave} = \frac{P_{rad}}{4\pi}$$

The radiation resistance of an antenna can be determined using the total power radiated according to Equation 30.18:

$$\frac{1}{2} I_0^2 R_{rad} = P_{rad} \tag{30.169}$$

$$R_{rad} = \frac{2P_{rad}}{I_0^2} \tag{30.170}$$

30.5 Summary

This chapter included the study of four basic areas:

1. The general solution of Maxwell's equations for plane waves for a variety of typically encountered scenarios was investigated. The behavior of waves normally incident on a surface was characterized including the effects of lossy materials.
2. A characterization of the behavior of electromagnetic wave propagation in transmission lines with terminating impedances was presented. It included a discussion of Smith charts and typical methods used for impedance matching including stub tuning.
3. An overview of the electromagnetic behavior in waveguides via a solution of the Helmholtz equation was discussed. A section was also included describing the behavior of light in optical fiber.
4. Lastly, a discussion of basic antenna characteristics was provided including a mathematical description of the currents and fields resulting from a dipole of arbitrary length. The radiated electromagnetic field, power density, total radiated power, directivity, and radiation resistance were found for several typical antennas.

31

Transmission Line Time-Domain Analysis and Signal Integrity

Edward Wheeler
Rose-Hulman Institute of Technology

Jianjian Song
Rose-Hulman Institute of Technology

David R. Voltmer
Rose-Hulman Institute of Technology

31.1 Introduction

Transmission lines are two or more uniform, parallel conductors able to transmit signals and power along their length. Two conductors connecting a source to a load is the basic picture taken in this chapter. The study of transmission lines is important in a variety of fields, from the performance of communication systems to signal integrity in printed circuit boards [EEM03,ME04,HSDD09].

Understanding the characteristics of transmission lines is necessary where wave behaviors such as propagation, reflection, and interference are important factors. In the time domain, this can be quantified by comparing propagation time to rise or fall times of the signals propagating. Whenever propagation times are not much shorter than rise times, wave behavior can no longer be ignored. Consider the system shown in Figure 31.1. If the rise time (t_r) associated with the source voltage is much longer than the propagation delay time (t_d) for the line connecting the source and load, then wave behavior is not important. In this case, all time variations are gradual compared to the propagation delay, and understanding transmission lines is not an important factor in understanding system behavior. These lines could simply be treated as nodes connecting the source and load. If the electrically short line's resistance, capacitance, or inductance were important, then the lines could be represented in a lumped element model. On the other hand, at higher speeds, where t_t is not much longer than t_d, reflections can play an important role. The source could see waves that have been reflected from the load and which could be re-reflected from the source, and so on. For this case, wave behavior is clearly important and a sound understanding of transmission lines becomes essential to understanding how the system works.

Speaking more generally, study of quasi-statics shows that transmission line models are not needed if the electrical length of the conductors is short. In the frequency-domain, electrical length is proportional

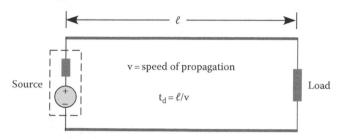

FIGURE 31.1 Transmission line connecting source to load.

to the ratio of conductor length (ℓ) to wavelength (λ). [EETR60] In the time domain, it is proportional to the ratio propagation delay-to-signal rise time. For lines that are electrically short—that is, when $\lambda \gg \ell$ or when $t_r \gg t_d$—wave effects are negligible and transmission line models are unnecessary.

31.2 EM Fundamentals

We begin our discussion with some electromagnetic fundamentals that can be omitted without loss of continuity.

Transverse electromagnetic (TEM) waves, in which the electrical and magnetic waves are perpendicular to the direction of propagation, are the principal type of electromagnetic wave that propagates along transmission lines. The TEM character of the fields is important as it allows a unique definition of voltage and current [PAEMF61].

TEM waves traveling in the $+\mathbf{a}_z$-direction can be written in terms of a transverse vector which varies in the x–y plane and a wave function g:

$$
\left.\begin{array}{c} \mathbf{E}(x,y,z) \\ \mathbf{H}(x,y,z) \end{array}\right\} = \left\{\begin{array}{c} \mathbf{E}_t(x,y)g\left(t-\dfrac{z}{v}\right) \\[2mm] \mathbf{H}_t(x,y)g\left(t-\dfrac{z}{v}\right) \end{array}\right.
$$

These waves are solutions of Maxwell's equations. Below, the del operator is split into two parts, one along z and the other acting in the x–y plane. Using this notation with Faraday's law,

$$
\nabla \times \mathbf{E} = \left(\nabla_t + \mathbf{a}_z \frac{\partial}{\partial z}\right) \times \mathbf{E} = -\frac{\partial \mathbf{B}}{\partial t}
$$

Substituting the wave functions assumed for **E** and **H**,

$$
\left(\nabla_t + \mathbf{a}_z \frac{\partial}{\partial z}\right) \times \mathbf{E}_t(x,y)g\left(t-\frac{z}{v}\right) = -\mu\,\frac{\partial\left[\mathbf{H}_t(x,y)g\left(t-\dfrac{z}{v}\right)\right]}{\partial t}
$$

$$
\nabla_t \times \mathbf{E}_t(x,y)g\left(t-\frac{z}{v}\right) - \frac{1}{v}\mathbf{a}_z \times \mathbf{E}_t(x,y)g\left(t-\frac{z}{v}\right) = -\mu\,\frac{\partial\left[\mathbf{H}_t(x,y)g\left(t-\dfrac{z}{v}\right)\right]}{\partial t}
$$

In this expression, since both ∇_t and $\mathbf{E}_t$ are directed in the tangential plane, which implies their cross product must be along the z-direction, $\nabla_t \times \mathbf{E}_t(x,y)g(t - (z/v))$ must too be in the z-direction. On the other hand, the right-hand side is purely in the transverse direction in the x–y plane. Therefore, equating tangential and longitudinal components, it is clear that $\mathbf{E}_t(x,y)$ satisfies the electrostatic equation:

$$\nabla_t \times \mathbf{E}_t(x,y)g(t-(z/v)) = 0$$

$$\nabla_t \times \mathbf{E}_t(x,y) = 0$$

Following a similar tack with Ampere's law, one readily finds that the transverse magnetic field satisfies the magnetostatic equation:

$$\nabla_t \times \mathbf{H}_t(x,y) = J_z$$

Therefore, if the waves are TEM in character, the tangential fields can be found from statics. This is of utmost importance since the resulting tangential fields—being those found from the electrostatic and magnetostatic equations—are *independent* of frequency. Their direction and relative amplitude are determined solely by the cross-sectional geometry and material properties. Because the transverse fields are independent of frequency, unique relationships exist between $\mathbf{E}_t$ and the line voltage (V), and between $\mathbf{H}_t$ and the line current (I).

This result provides the foundation for developing transmission line models in terms of V and I rather than in terms of field quantities. Due to a unique relation existing between the $\mathbf{E}_t$ and V and between $\mathbf{H}_t$ and I, for TEM waves traveling on the transmission line, voltage and current may just as well be used to represent the traveling waves as the electric and magnetic vectors. Moreover, because the relative amplitudes of the electric and magnetic fields are determined by the intrinsic impedance, and since the relationships between $\mathbf{E}_t$ and V and between $\mathbf{H}_t$ and I follow solely from the cross-sectional geometry and material properties, a unique and unambiguous relation can be defined between the voltage and current waves. This relation between the voltage and current waves is denoted as the transmission line's characteristic impedance (Z_c).

$$Z_c = \frac{V(t-(z/v))}{I(t-(z/v))} = \frac{V_m\left[g(t-(z/v))\right]}{I_m\left[g(t-(z/v))\right]}$$

In addition, TEM wave solutions exist for the lossless lines considered here and for lossy lines when the loss is confined to dielectric loss. TEM wave solutions are strictly no longer possible if the finite conductivity of the conductors is considered since a z-component of the electric field is required to sustain a current. For good conductors, however, the z-component is sufficiently small that the fields are still very nearly transverse and the wave is essentially TEM in character. It should also be noted that the waves are precisely TEM in character only for lines that are perfect electrical conductors (PEC) and that are surrounded by a homogeneous dielectric—a coaxial cable and a stripline for example. For a microstrip, the transmission line model is an excellent approximation [FME92]. Multiple transmission lines are treated by techniques similar to single transmission lines [MTL08].

31.3 Transmission-Line Modeling

Once it is established that transmission lines can be described in terms of traveling voltage and current waves, circuit-based models are a natural development. Starting by applying Faraday's law for the stationary dashed closed loop in Figure 31.2, a relation between current and the rate of change in voltage can be established, where φ is the magnetic flux penetrating the closed path of the E-field integral:

$$\oint E \cdot dl = -\frac{\partial\phi}{\partial t}$$

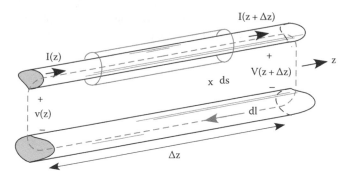

FIGURE 31.2 Transmission line.

$$V(z + \Delta z) - V(z) + RI = -\frac{\partial}{\partial t}\left[\left(\frac{\phi}{I}\right)I\right]$$

The ratio of magnetic flux to current is simply the self-inductance of the transmission line section and is not a function of time. Taking the limit as Δz grows small, a differential equation is obtained relating the spatial rate of change for voltage to current and where $\mathscr{R}$ is the resistance per unit length and $\mathscr{L}$ is the inductance per unit length:

$$V(z + \Delta z) - V(z) = -RI - L\frac{\partial I}{\partial t}$$

$$\lim_{\Delta z \to 0}\left[\frac{V(z + \Delta z) - V(z)}{\Delta z}\right] = \lim_{\Delta z \to 0}\left(-\frac{R}{\Delta z}I - \frac{L}{\Delta z}\frac{\partial I}{\partial t}\right)$$

$$\frac{\partial V}{\partial z} = -\mathscr{R}I - \mathscr{L}\frac{\partial I}{\partial t}$$

A similar relation can be found between the spatial rate of change in current to voltage by applying charge conservation to the cylindrical region in Figure 31.2 around the top conductor. The current out of a closed surface is equal to the rate that the charge stored within it decreases:

$$\oiint_{\text{surface}} J \cdot ds = -\frac{\partial q}{\partial t}$$

$$I(z + \Delta z) - I(z) + GV = -\frac{\partial}{\partial t}\left[\left(\frac{q}{V}\right)V\right]$$

The ratio of charge to voltage is simply the capacitance of the transmission line section and is not a function of time. Taking the limit as Δz grows small, a differential equation is obtained relating the spatial rate of change for current to voltage and where $\mathscr{G}$ is the conductance per unit length and $\mathscr{C}$ is the capacitance per unit length:

$$I(z + \Delta z) - I(z) = -GV - C\frac{\partial V}{\partial t}$$

$$\lim_{\Delta z \to 0}\left[\frac{I(z+\Delta z)-I(z)}{\Delta z}\right] = \lim_{\Delta z \to 0}\left(-\frac{G}{\Delta z}V - \frac{C}{\Delta z}\frac{\partial V}{\partial t}\right)$$

$$\frac{\partial I}{\partial z} = -\mathcal{G}V - \mathcal{C}\frac{\partial V}{\partial t}$$

Applying KVL and KCL to the circuit model shown in Figure 31.3 results in these two relations and therefore closely models TEM wave propagation on lossy transmission lines where $\mathcal{R}$ represents Joule losses in the conductors, which dominate at lower frequencies; and where $\mathcal{G}$ represents dielectric loss, which often dominates at higher frequencies.

For lossless lines considered in this chapter, $\mathcal{R}$ and $\mathcal{G}$ are taken to be zero as shown in Figure 31.4.

The equations describing the relation between the voltage and current waves on lossless transmission lines can be combined to obtain wave equations for V and I:

$$\frac{\partial V}{\partial z} = -\mathcal{L}\frac{\partial I}{\partial t} \quad \frac{\partial I}{\partial z} = -\mathcal{C}\frac{\partial V}{\partial t}$$

$$\frac{\partial^2 V}{\partial z^2} = -\mathcal{L}\frac{\partial}{\partial t}\frac{\partial I}{\partial z} = -\mathcal{L}\frac{\partial}{\partial t}\left(-\mathcal{C}\frac{\partial V}{\partial t}\right)$$

$$\frac{\partial^2 V}{\partial z^2} = \mathcal{L}\mathcal{C}\frac{\partial^2 V}{\partial t^2}$$

Solving for V, we obtain the solution that has been assumed, where the minus sign signifies travel in the $+\mathbf{a}_z$ direction and the plus sign indicates travel in the $-\mathbf{a}_z$ direction:

$$V\left(t\pm\frac{z}{v}\right) \quad \text{where the speed of propagation, } v = \frac{1}{\sqrt{\mathcal{L}\mathcal{C}}}$$

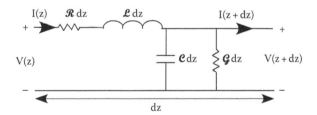

FIGURE 31.3 Lossy TL circuit model.

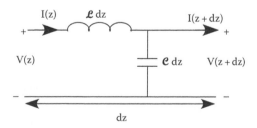

FIGURE 31.4 Lossless TL circuit model.

Given this voltage wave, the corresponding current wave can be found using one of the transmission line equations:

$$\frac{\partial I}{\partial z} = -\mathcal{C}\frac{\partial V}{\partial t}$$

$$\frac{\partial}{\partial z}\left\{I_m\left[g\left(t\pm\frac{z}{v}\right)\right]\right\} = -\mathcal{C}\frac{\partial}{\partial t}\left\{V_m\left[g\left(t\pm\frac{z}{v}\right)\right]\right\}$$

Letting α be the temporal-spatial argument of the wave functions, one obtains, using the chain rule,

$$\frac{\partial}{\partial z}\left\{I_m\left[g(\alpha)\right]\right\} = -\mathcal{C}\frac{\partial}{\partial t}\left\{V_m\left[g(\alpha)\right]\right\}$$

$$\pm\frac{I_m}{v}\frac{\partial g}{\partial\alpha} = -\mathcal{C}V_m\frac{\partial g}{\partial\alpha}$$

$$\frac{V_m}{I_m} = \mp\frac{1}{\mathcal{C}v} = \mp\sqrt{\frac{\mathcal{L}}{\mathcal{C}}} = \mp Z_c$$

The ratio between the voltage and current waves is the transmission line's characteristic impedance:

$$Z_c = \frac{V\left(t\pm(z/v)\right)}{I\left(t\pm(z/v)\right)} = \frac{V_m\left[g\left(t\pm(z/v)\right)\right]}{I_m\left[g\left(t\pm(z/v)\right)\right]} = \frac{V_m}{I_m} = \mp\sqrt{\frac{\mathcal{L}}{\mathcal{C}}}$$

The relation involves a minus sign, which simply reflects the fact that the direction of the current is reversed for waves traveling in the $-\mathbf{a}_z$ direction, as indicated in Figure 31.5.

The remainder of the chapter will discuss applications and examples illustrating the analysis of systems involving transmission lines. Aside from the common coaxial transmission line, two types of transmission lines should be mentioned and are shown in Figure 31.6.

Coaxial cables find widespread use in communication systems and a variety of test and diagnostic equipment. The microstrip and stripline configurations in Figure 31.6 are particularly amenable to planar deposition techniques, and find wide use in printed circuit boards and in integrated circuits.

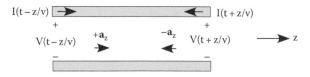

FIGURE 31.5   Waves traveling in $\pm\mathbf{a}_z$ directions.

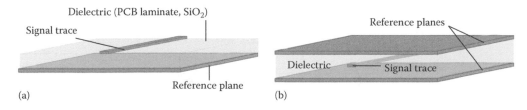

FIGURE 31.6   Transmission lines: (a) microstrip and (b) stripline.

31.4 Reflection and Transmission at Boundaries

Two types of waves are possible on a transmission line: waves moving away from the source toward the load and those moving in the opposite direction. When a single source excites a transmission line, it initiates a wave moving away from the source toward the load, here taken to be in the positive direction. Since the positive-going voltage wave is incident on the load, it is denoted as V_i. A corresponding incident current wave or $I_i = V_i/Z_c$ accompanies the incident voltage wave. The waves propagate along the transmission line unaltered until a change (or discontinuity) in impedance is encountered. Such a discontinuity would be created, for example, if a transmission line were joined to a second transmission line with a different characteristic impedance. At a discontinuity, part of the wave is reflected from the discontinuity and a part is transmitted across the discontinuity. An infinitely long transmission line would have no discontinuities and there would be no reflected wave.

To better understand the cause and nature of reflections consider Figure 31.7, which shows the boundary connecting two transmission lines. A voltage–current wave propagates to the right toward the junction; this is known as an *incident* wave and is characterized by $V_i/I_i = Z_{c1}$. The second transmission line carries the wave *transmitted* across the boundary and has a characteristic impedance value of $Z_{c2} = V_t/I_t$. There is also a reflected wave, in transmission line 1 traveling in the negative direction. This is the wave that has been *reflected* at the boundary. This wave is characterized by $V_r/I_r = -Z_{c1}$. At the junction, the total voltage across each transmission line must be equal at the boundary (KVL holds).

$$V_t = V_i + V_r$$

$$TV_i = V_i(1 + \Gamma)$$

$$T = 1 + \Gamma$$

In the relations above, the reflection coefficient, $\Gamma = V_r/V_i$, has been defined as the ratio of reflected voltage to incident voltage, and the transmission coefficient, $T = V_t/V_i$, has been defined as the ratio of transmitted voltage to incident voltage. Another relationship can be found using the fact that the current is continuous at the boundary (KCL holds):

$$I_i - I_r = I_t$$

$$\frac{V_i}{Z_{c1}} - \frac{V_r}{Z_{c1}} = \frac{V_t}{Z_{c2}}$$

$$\frac{1}{Z_{c1}} - \frac{\Gamma}{Z_{c1}} = \frac{T}{Z_{c2}}$$

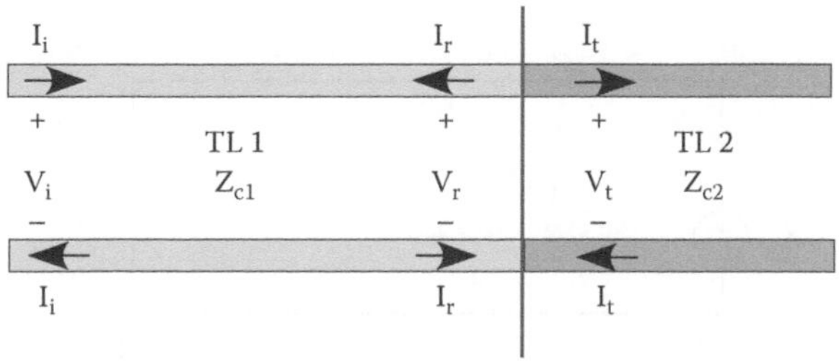

FIGURE 31.7 Reflection and transmission at the boundary between two transmission lines.

The result is two equations involving Γ and T, which can be solved for Γ and T:

$$\Gamma = \frac{Z_{c2} - Z_{c1}}{Z_{c2} + Z_{c1}}$$

$$T = \frac{2Z_{c2}}{Z_{c2} + Z_{c1}}$$

The reflection coefficient (and reflected voltage) vanishes when $Z_{c1} = Z_{c2}$. In this case, the lines are matched and no reflected wave exists. Lines that are not matched result in reflected waves that propagate back toward the source to the left.

What is done in other situations, perhaps in which a line is terminated by lumped elements or in which lumped elements are present at the boundary between two transmission lines? In such cases, we may add to the principles outlined above (continuity of voltage and current) the fact that the V–I relationships of the lumped elements must hold. These situations will be discussed below.

31.5 Transient Analysis

How do voltages and currents move from their initial to final values? That is, what is their transient signature? The general question regarding the time evolution of voltage and current waveforms is a central question in signal integrity. To begin, consider the system in Figure 31.8 with a source and load connected by a transmission line. We seek a technique by which v(t) can be found at any position along the line.

Considering the source to be off (zero) for t < 0, perhaps the first question to be answered is what happens just as the source is turned on? What happens at t = 0⁺? Since no time has elapsed, the wave cannot have traveled any finite distance. For example, the load resistance will experience no effects from the source turning on until the wave has had time to propagate to the load. Before $t = t_d$, conditions at the load will be just as they were when the source was off. At t = 0⁺, the effect of the source turning on is confined to just one location, and that is where the source connects to the line, at z = 0.

At z = 0, the ratio between the voltage and current waveforms being introduced to the line is determined by Z_c. Moreover, this ratio is independent of frequency. For all frequencies, the ratio of voltage to current waves is Z_c. Therefore, if one is only interested in what happens at z = 0, and there are no reflected waves to be considered, the transmission line can be replaced with a resistance R = Z_c, as shown in Figure 31.9. This will allow the determination of the positive-traveling waveform being injected into the transmission line at z = 0. This wave will subsequently travel down the transmission line as traveling step function. At $t = t_d$, the step's edge will encounter the resistive load. But for now, we are only interested in one instant in time (t = 0⁺) and one location in place (z = 0).

So, clearly, the initial voltage wave at t = 0⁺ and z = 0 can be determined by voltage division.

$$V^+\Big|_{z=0,\, t=0^+} = \frac{Z_c}{R_s + Z_c} V_s\Big|_{t=0^+}$$

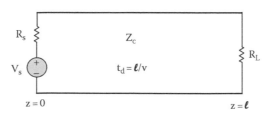

FIGURE 31.8 System of source, line, and load.

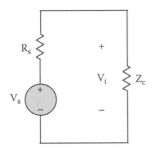

FIGURE 31.9 System at z = 0.

In general, the voltage injected into the transmission line at z = 0 is given by this equation. One must be careful when t > 2t$_d$; however, since the voltage being injected from the source may not be the total voltage, or even the total positive-going voltage. This is due to the fact that additional reflections can be present from the source. When t > 2t$_d$, it is possible that waves that have been reflected from the load are present. These waves can then be re-reflected from the source that can add to or subtract from the positive-going waves.

For waves traveling in the +**a**$_z$-direction on a transmission line, the ratio is Z$_c$. For waves traveling in the −**a**$_z$-direction on a transmission line, the ratio is −Z$_c$. The important concept to keep in mind is that the characteristic impedance is the ratio of traveling voltage and current waves; *it is not the ratio of total voltage to total current*. Total voltage and current are a combination of positively and negatively traveling waves. The resulting total voltage and total current are not related simply as Z$_c$ or −Z$_c$. When both incident and reflected waves are present, they can constructively or destructively interfere with each other, and their relationship is not simply Z$_c$. It is not true for t > 2t$_d$ that the ratio of total voltage to total current is Z$_c$, since there might be reflected waves present.

With that caution in mind, it is nevertheless true that, even when t > 2t$_d$, the voltage (the positive-going wave) being injected into the line at z = 0 in Figure 31.9 can be expressed as a voltage divider. This does not include any positive-traveling voltage wave resulting from being reflected from the load and then again from the source. It is just the voltage being injected into the line from the source:

$$V^+(t)\Big|_{z=0} = \frac{Z_c}{R_s + Z_c} V_s(t)$$

The incident voltage waveform from the source traveling on the transmission line is therefore

$$V_i(t,z)\Big|_{z=0} = V^+\left(t - \frac{z}{v}\right) = \frac{Z_c}{R_s + Z_c} V_s\left(t - \frac{z}{v}\right)$$

Again, it should be stressed that this is not the total voltage, but is rather only the voltage waveform that is injected from the source. To determine the total voltage on the line, the effects of reflection must be included as discussed below.

Upon being injected onto the transmission line, the voltage wave travels to the resistive load. What happens when it encounters the resistor can be found by invoking the resistance's v–i relation, Ohm's law, which gives the ratio that must exist between the total voltage and the total current across a resistor:

$$V = IR_L \rightarrow V_i + V_r = \left(\frac{V_i}{Z_c} - \frac{V_r}{Z_c}\right) R_L$$

$$1 + \Gamma_L = R_L\left(\frac{1}{Z_c} - \frac{\Gamma_L}{Z_c}\right) \rightarrow \Gamma_L = \frac{R_L - Z_c}{R_L + Z_c}$$

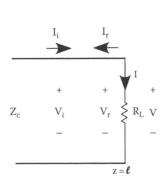

FIGURE 31.10 Reflection from load.

When the wave reflected from the load reaches the source, part of this reflected wave is then reflected from the source and subsequently travels to the load, where a part is reflected, and so on (Figure 31.10). The expression for the source reflection coefficient is similar to that of the load:

$$\Gamma_s = \frac{R_s - Z_c}{R_s + Z_c}$$

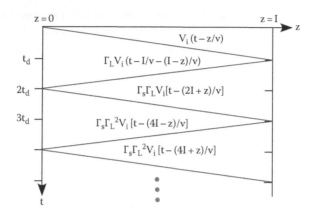

FIGURE 31.11   Generic bounce diagram.

At any particular time and position, the total voltage is the sum of the incident and all the reflected waves that are present at the particular time and location in question. This challenging bookkeeping task is eased by the use of bounce diagrams as illustrated in Figure 31.11.

Bounce diagrams are plots of time versus location in which the reflected waves are individually tracked. In Figure 31.11, V_i is the initial wave injected from the source and its propagation toward the load is indicated by the slanted line that arrives at the load in one delay time.

Example 31.1

Find $V_{in}(t)$ and $V_L(t)$.

For the system shown in Figure 31.12, V_i, t_d, Γ_s, and Γ_L can readily be found.

$$V_{i\,(z=0,\ t=0^+)} = \frac{50}{150+50}\,2.88\ \text{V} = 0.72\ \text{V}$$

$$t_d = \frac{0.1\,\text{m}}{2(10^8)\,\text{m/s}} = 0.5\ \text{ns}$$

$$\Gamma_s = \frac{150-50}{150-50} = \frac{1}{2} \quad \Gamma_L = \frac{25-50}{25-50} = -\frac{1}{3}$$

With these parameters, a bounce diagram for the system can be completed. Since the source is a step function and therefore constant for $t > 0$, the reflected waves are particularly simple, and all that is required is that their wave fronts be tracked, which the bounce diagram readily does (Figure 31.13).

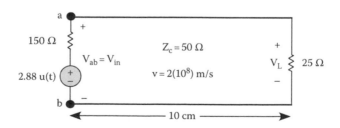

FIGURE 31.12   System with resistive load and 10 cm TL.

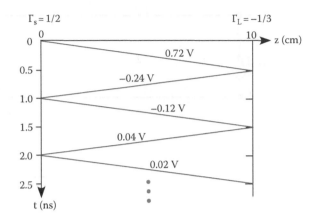

FIGURE 31.13 Bounce diagram.

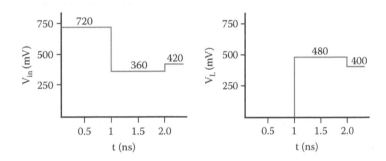

FIGURE 31.14 $V_{in}(t)$ and $V_L(t)$.

Having constructed the bounce diagram, voltage vs. time plots are readily developed for any location along the line as the sum of all voltage waves that have passed the location. This is a function of time as more reflections must be included with increasing time. See Figure 31.14 for $V_{in} = V_{(z=0)}$, and for $V_L = V_{(z=10cm)}$.

Example 31.2

Pulses are a signal type common in digital systems and much of our previous work can be utilized if we recognize that constant-valued pulses can be expressed as the difference of step functions. Given the system and the measured input voltage, $V_{ab}(t)$, as shown in Figure 31.15, determine the transmission line's characteristic impedance and propagation speed. Also find $\mathcal{L}$, $\mathcal{C}$, and R_L.

Z_c can be determined from the z = 0 equivalent circuit and the measured value of V_{ab} for 0 < t < 1 ns. The speed of propagation can be determined from the time at which the reflected pulse is seen at the source, having traveled to the load and then back to the source. Knowing Z_c and v fixes $\mathcal{L}$ and $\mathcal{C}$ since these four quantities are related through two equations. Knowing the reflected pulse determines R_L:

$$10\,V = \frac{Z_c}{50\,\Omega + Z_c}\,20\,V \rightarrow Z_c = 50\,\Omega \quad v = \frac{2(0.3\,m)}{3(10^{-9})\,s} = 2(10^8)\,m/s$$

$$\left.\begin{array}{l} Z_c = \sqrt{\mathcal{L}/\mathcal{C}} \\ v = \sqrt{1/\mathcal{L}\mathcal{C}} \end{array}\right\} \rightarrow \mathcal{L} = 0.25\,\mu H/m, \quad \mathcal{C} = 100\,pF/m$$

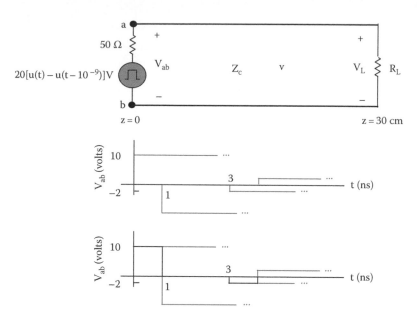

FIGURE 31.15 System with a pulse input.

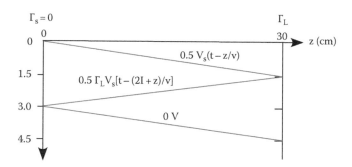

FIGURE 31.16 Bounce diagram for system with pulse input.

Since $Z_c = R_s$, and the source reflection coefficient $\Gamma_s = 0$, the source is said to be matched, which is meant to convey that the source impedance is equal, or matched, to the line's characteristic impedance. From the resulting bounce diagram in Figure 31.16, the reflected wave seen at the source is purely due to reflection from the load and no portion is "re-reflected" from the source (since $\Gamma_s = 0$):

$$\Gamma_L = \frac{R_L - 50\,\Omega}{R_L + 50\,\Omega} = \frac{-2}{10} \rightarrow R_L = 33\frac{1}{3}\,\Omega$$

31.6 Multiple Transmission Line Sections

Consider the system in Figure 31.17 with two transmission line segments of differing impedances but with identical propagation speeds. This is a common occurrence with PCBs where the dielectric is unchanged (resulting in the same speed of propagation) but where the microstrip trace geometry changes resulting in a discontinuity for Z_c. Since the trace-reference separation often remains unchanged

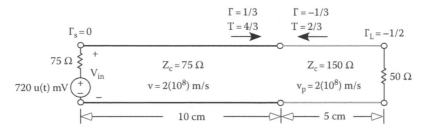

FIGURE 31.17 System with multiple TL sections.

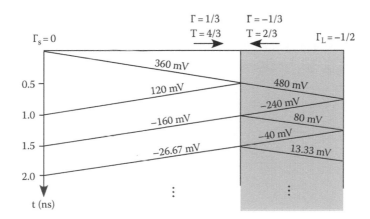

FIGURE 31.18 Bounce diagram for a multiple TL section system.

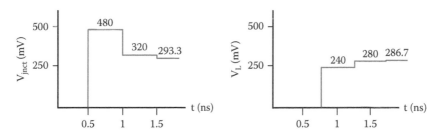

FIGURE 31.19 $V_{jnct}(t)$ and $V_L(t)$.

in PCBs, the change in characteristic impedance is most often due to a change in trace width, e.g., wider traces result in higher $\mathcal{C}$ and lower $\mathcal{L}$ giving a lower Z_c.

The resulting bounce diagram is shown in Figure 31.18.

Plots of the line voltage at the junction between the two lines ($z = 10\,$cm) and at the 50 Ω load are shown in Figure 31.19.

31.7 Transmission Line Junctions

When a junction between two transmission lines involves lumped elements, whether due to intentional loads or to unintended parasitics, the reflection and transmission coefficients at the junction can readily be determined by enforcing continuity of current and voltage and the element v–i relations.

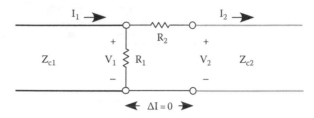

FIGURE 31.20   Transmission line junction with discontinuity.

Example 31.3

Consider the junction in Figure 31.20 and assume the wave is coming from TL 1 on the left and with the meaning of the effective reflection and transmission coefficients as given below. Assuming that lumped element analysis is valid at the junction, Γ and T can be found by enforcing KVL and KCL at the junction. The assumption underlying the analysis is that the junction's length can be neglected, $\Delta\ell = 0$:

$$
\text{Definitions} \begin{cases}
V_1 = (1+\Gamma)V_i \\
V_2 = TV_i \\
I_1 = \dfrac{V_i}{Z_{c1}} - \dfrac{\Gamma V_i}{Z_{c1}} \\
I_2 = \dfrac{TV_i}{Z_{c2}}
\end{cases}
$$

$$
\text{Continuity of V} \quad \{ V_1 = V_2 + R_2 I_2
$$

$$
\text{Continuity of I} \quad \left\{ I_1 - \dfrac{V_1}{R_1} = I_2 \right.
$$

From the six equations, V_1, V_2, I_1, and I_2 can be eliminated resulting in two equations giving Γ and T in terms of Z_{c1}, Z_{c2}, R_1, and R_2 (V_i is common to all terms and can be divided out):

$$
V_1 = V_2 + R_2 I_2 \rightarrow 1 + \Gamma = T + R_2 \frac{T}{Z_{c2}}
$$

$$
I_1 - \frac{V_1}{R_1} = I_2 \rightarrow \frac{1}{Z_{c1}} - \frac{\Gamma}{Z_{c1}} - \frac{1+\Gamma}{R_1} = \frac{T}{Z_{c2}}
$$

For $Z_{c1} = R_1 = R_2 = 50\ \Omega$ and $Z_{c2} = 100\ \Omega$, $\Gamma = -1/7$ and $T = 4/7$.

31.8 Reactive Loads

The preceding analysis involved transmission line junctions and resistances where the v–i relations were algebraic. When energy storage elements such as capacitances or inductances are present, v–i relations become differential equations. The electric and magnetic energies stored within these elements cannot be changed instantaneously but require a finite interval of time for any finite change in energy:

$$
I = C\frac{dV}{dt}
$$

$$
V = L\frac{dI}{dt}
$$

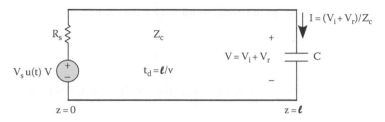

FIGURE 31.21 Capacitor-terminated transmission line.

Let's begin simply with a single capacitance terminating a transmission line with a characteristic imped-ance Z_c that is fed by a step-function source as shown in Figure 31.21.

At $z = 0$, the step-function V_s excites a voltage step at the input of the line, which can be calculated by using voltage division. This voltage wave is accompanied by a current wave:

$$V_i\big|_{z=0} = \frac{Z_c}{R_s + Z_c} V_s$$

$$I_i\big|_{z=0,\,t=0^+} = \frac{Z_c}{R_s + Z_c} \frac{V_s}{Z_c}\bigg|_{t=0^+}$$

The voltage and current waves propagate toward the capacitive load with velocity v_p and arrive after a propagation time delay:

$$t_d = \frac{\ell}{v_p}$$

The total voltage and current at the capacitance must be related via the capacitance's v–i relation. We assume here that the capacitance is uncharged prior to the arrival of the wave from the source:

$$I = C\frac{dV}{dt}$$

$$\frac{V_i - V_r}{Z_c} = C\frac{d(V_i + V_r)}{dt}$$

The total voltage is the sum of the incident (which, in this case, is a time-invariant step from the step source) and the reflected wave. The sum of the incident and reflected voltages must satisfy the capacitance's element relation:

$$V_i = \frac{Z_c}{R_s + Z_c} V_s$$

The resulting differential equation reads

$$\frac{V_i - V_r}{Z_c} = C\frac{dV_r}{dt}$$

$$Z_c C\frac{dV_r}{dt} + V_r = V_i$$

This differential equation is subject to the initial condition for the capacitance's voltage:

$$V\big|_{z=\ell} = (V_i + V_r)\big|_{z=\ell} = 0$$

The resulting solution of the first-order differential equation for the reflected voltage wave is

$$V_r = V_i(1 - 2e^{-(t-t_d/Z_cC)})u(t - t_d)$$

The preceding analysis uses the principles outlined above—that current and voltage continuity together with element v–i relationships can be used to determine reflections. Let us now approach this same problem with an alternative analysis. The result will be a time-dependent reflection coefficient for the capacitor, which must be the case since the ratio of the incident to reflected voltages is time-dependent. We will find nothing new but may discover a different way of looking at the problem (Figure 31.22).

So, beginning with the system shown in Figure 31.21, one can recognize that the voltage across the capacitance cannot change instantly. This requires that, at the instant of incidence, there must be a reflected voltage step equal to the incident voltage step; the total voltage is composed of the sum of the incident and reflected waves, and by KVL this must be equal to the capacitor terminal voltage according to

$$V = V_i + V_r$$

Simultaneously, the incident current step initiates a reflected current step; the incident is directed toward the load into the upper terminal of the capacitor, the reflected is directed away from the load out of the upper terminal:

$$I = I_i - I_r$$

$$I = \frac{V_i}{Z_C} - \frac{V_r}{Z_C}$$

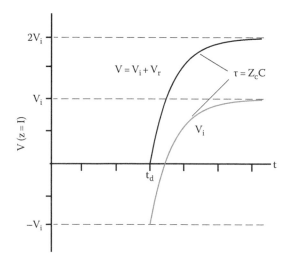

FIGURE 31.22   Total and reflected load voltage for a capacitive termination.

But rather than a linear constant relating the terminal voltage and current as in a resistor, the capacitor terminal voltage and current are related by a differential equation:

$$I = C\frac{dV}{dt}$$

$$\frac{V_i - V_r}{Z_C} = C\frac{d(V_i + V_r)}{dt}$$

Both sides of this equation can be divided by V_i, where the reflection coefficient is defined as usual as the ratio of the reflected and incident voltages at the capacitor terminals:

$$\frac{1 - (V_r/V_i)}{Z_c} = C\frac{d(1 + (V_r/V_i))}{dt} \quad \text{or} \quad \frac{1 - \Gamma}{Z_c} = C\frac{d(1 + \Gamma)}{dt}$$

This reflection coefficient must vary with time if it is to satisfy the differential equation. Even though the incident voltage is a step function, the reflected voltage is time varying as indicated by the time-varying reflection coefficient:

$$\frac{d\Gamma}{dt} + \frac{\Gamma}{Z_c C} = \frac{1}{Z_c C}$$

The initial condition requires that $V_c(t_d) = 0$. Solving the above differential equation subject to this initial condition is sufficient for us to solve for a time-varying reflection coefficient:

$$\Gamma = \left[1 - 2e^{-(t-t_d)/Z_c C}\right]u(t - t_d)$$

where the $(t - t_d)$ term accounts for the time required for the incident wave to travel from the source to the capacitance. The incident voltage is from the voltage divider at the source:

$$V_i\big|_{z=\mathcal{L},t=t_d} = \frac{V_s Z_c}{R_s + Z_c}$$

The reflected voltage at the capacitive load is the product of the incident voltage and the reflection coefficient. The total voltage at the capacitive load is the sum of the incident and reflected voltages. It can readily be seen that the voltages are the same as those found above:

$$V_r = \Gamma_c V_i = \frac{V_s Z_c}{R_s + Z_c}\left[1 - 2e^{-(t-t_d)/Z_c C}\right]u(t - t_d)$$

$$V = V_i + V_r = V_i + \Gamma_c V_i$$

$$V = \frac{V_s Z_c}{R_s + Z_c}\left[1 - 2e^{-(t-t_d)/Z_c C}\right]u(t - t_d)$$

The physical basis of the mathematical relations is summarized in the following points:

1. The reflection coefficient has an initial value of −1 since the uncharged capacitor initially, at the instant the incident wave arrives, acts as a short circuit.
2. The reflection coefficient has a final value of +1 since, for sufficiently long times, the capacitor is fully charged and acts as an open circuit.
3. The transition of the reflection coefficient from −1 to +1 proceeds with a time constant of $\tau_c = Z_cC$.

Based upon these physical principles, we can sidestep the mathematics associated with an inductive load with no initial current flow. Since the current in an inductance cannot change instantaneously, the inductance initially appears as an open circuit with a reflection coefficient of +1. For this step source, the steady-state current will be constant and the response of an inductance will be like a short circuit; so after a suitably long time, the reflection coefficient will be −1. The time constant of this transition is that of an inductance L and Z_c, $\tau = L/Z_c$. Plots of the terminal voltages of an inductive load on a transmission line are shown in Figure 31.23.

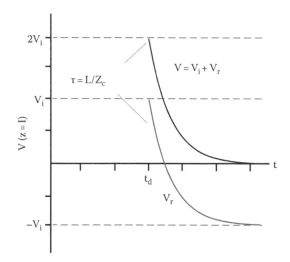

FIGURE 31.23   Total and reflected load voltage for an inductive termination.

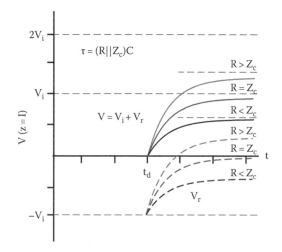

FIGURE 31.24   Total and reflected load voltage for a parallel RC termination.

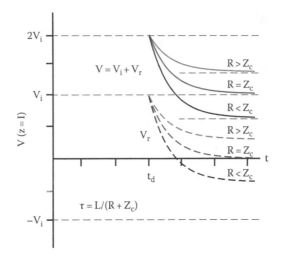

FIGURE 31.25 Total and reflected load voltage for a series RL termination.

With a step input, loads that contain a single capacitor or inductor and several resistors can be analyzed in a similar manner. Consider a capacitance in parallel with a resistance. This combination will initially act as a short circuit (the capacitor acting as a short will dominate) and will end acting as just the resistance (the capacitance acting as an open). The time constant will be $(R\|Z_c)C$. Typical responses appear in Figure 31.24.

Consider, again with a step input, an RL series load. In this case, the load will initially act as an open (the inductance acting as an open will dominate) and will end as just the resistance (the inductance acting as a short). The time constant will be $L/(R + Z_c)$. Typical responses appear in Figure 31.25.

31.9 Nonlinear Loads

Finally, what are the effects of a nonlinear element such as a diode or a logic circuit element as a load? The question cannot be determined with the techniques used in the previous sections since the techniques are valid for linear systems only. Linear elements and solution methods are based upon a constant relationship between the voltage and current. Nonlinear elements have varying proportionality between the terminal current and voltage. Consequently, we can understand their behavior more readily by using their voltage–current characteristics and using graphical methods. Additional details of this method, i.e., the method of Bergeron, are available online [TIBD96].

To explain the process, consider a linear source and load. The source V_S with a source resistance R_S excites a transmission line characterized by Z_c and time delay $t_d = \ell/v_p$ and terminated by load R_L as shown in Figure 31.26.

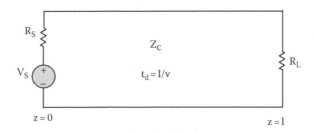

FIGURE 31.26 Step excitation of terminated transmission line.

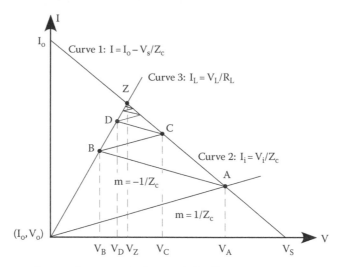

FIGURE 31.27 Bergeron diagram for step-excited transmission line.

We must determine the incident voltage from the source toward the load, V_i. The terminal characteristic of the source is a linear I–V relationship, curve 1 in Figure 31.27:

$$I = I_o - \frac{V}{R_S}$$

The input terminal characteristic of the input of the transmission line is given by its characteristic impedance and is shown as curve 2 in Figure 31.27. Note the slope of the V_i–I_i curve is $1/Z_c$:

$$V_i = I_i Z_c$$

When the source is connected to the transmission line input, the terminal voltages and currents must be equal; these conditions are satisfied at point A where the two curves intersect. Voltage and current at point A and time $t = 0$ are $V_A = V_i$ and $I_A = I_i$, respectively. In this example of a step function, the voltage and current were both zero for $t < 0$, so the initial point of curve 2 is the origin of the I–V plane. In general, the starting point for the curve 2 is (I_o, V_o) with curve 2 connected to curve 1 with slope $1/Z_c$ as for the step function.

The transient voltage V_i propagates to the load and arrives at time $t = t_d$. The incident voltage excites a reflected voltage when it is incident upon the load R_L. At the resistive load, the total voltage and current are related by $R_L = V_L/I_L$, which is represented by curve 3 in Figure 31.27. This load voltage, composed of both incident and reflected voltages, must lie on curve 3 at some point B where curves 1 and 3 intersect and must be determined. Therefore, the load voltage at $t = t_d$ is expressed as

$$V_L = V_B = V_i + V_r = V_A + V_r$$

and can be rewritten as

$$V_r = -(V_A - V_B)$$

Likewise, the load current is expressed as

$$I_L = I_B = I_i - I_r = I_A - I_r$$

and can be rewritten as

$$I_r = I_A - I_B.$$

Recall that the ratio of the reflected voltage and current waves is given by the characteristic impedance, the negative already accounted for in the equation above ($I_L = I_i - I_r$):

$$Z_c = \frac{V_r}{I_r} = -\frac{V_A - V_B}{I_A - I_B}$$

The "two-point" form of a straight line is expressed as

$$I = mV + I_0 = \frac{I_A - I_B}{V_A - V_B} V + I_0$$

where m is the slope of the line and I_0 is the I-axis intercept. Comparing the forms of the two previous equations, we see a line with slope $m = -1/Z_C$ that passes through points A and B. We can find point B as the intersection of the load line (curve 3) and a line of slope $-1/Z_C$ that passes through point A. The initial transient pulse V_A initiates this load voltage $V_L = V_B$ at time $t = t_d$.

This process continues ad infinitum as the wave undergoes repeated reflections. The voltages at the intersection points of the lines with slope $m = 1/Z_C$ and the source line are the input voltages of the transmission line at $t = 2nt_d$ intervals. The voltages at the intersection points of the lines with slope $m = -1/Z_c$ are the output voltages of the transmission line at $t = (2n + 1)t_d$ intervals. These voltages can be plotted with respect to time as shown in Figure 31.28. Note that after many reflections, the input and output voltages asymptotically approaches point Z. This is expected since each reflection is smaller than the previous. The transients grow smaller and, after many reflections, it is as if the source was connected directly to the load. Since the source resistance is less than the characteristic impedance in Figure 31.27 and has a greater slope, all reflections occur in the region below the source load line with monotonic voltage changes at both the input and output of the transmission line.

In the above example, the Bergeron method was demonstrated for a case that was simplified by the linear nature of the source and the load impedances. However, this technique brings the power to treat nonlinear elements with a positive sloped line ($m = 1/Z_c$) intersecting the source I–V curve at points that represent the line input voltages—points A, C, E, and so on. A negative sloped ($m = -1/Z_c$) line,

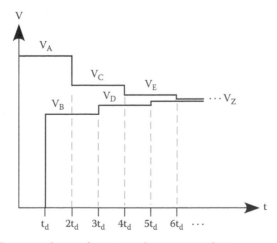

FIGURE 31.28 Input and output voltages of step-excited transmission line.

representing negatively traveling waves, intersects the load I–V curve at points that represent the line output voltages—points B, D, F, and so on.

The following example is for a transmission line with a matched source and a diode load. The source switches from a low state to a high state at t = 0 and so can be represented as a step function. $V_s = 5u(t)$ and the source resistance $R_s = 100\,\Omega$. The line has $Z_c = 50\,\Omega$ with a delay time of $t_d = 10\,\mu s$. The diode can be approximated by a 0.7 V source in series with a 5 Ω resistance. The diode-terminated transmission line and the resulting Bergeron diagram are shown below (Figures 31.29 and 31.30).

The time-domain plot of the resulting input and output voltages is given in Figure 31.31.

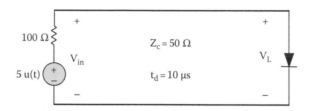

FIGURE 31.29   Diode-terminated transmission line.

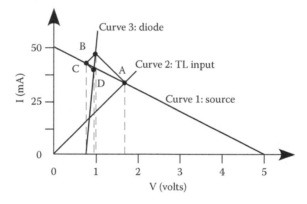

FIGURE 31.30   Bergeron diagram for diode-terminated transmission line.

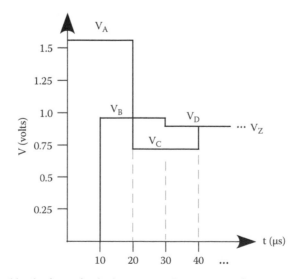

FIGURE 31.31   Input and load voltages for diode-terminated transmission line.

Note that the source impedance is greater than the characteristic impedance for this example and the reflections encircle the final point with the resulting voltage differences alternating in sign. When the source is matched to the line, there is only one reflection from the load, and point B and point Z are identical.

31.10 Conclusion

Whenever the propagation delay for a transmission is not negligible compared to the rise or fall time of the signal being propagated, wave effects such as reflection and interference can be of crucial importance in determining overall system performance. A sound understanding of transmission line behavior can be of crucial importance in a variety of contexts, from the performance of communication systems to signal integrity in printed circuit boards.

References

[EEM03] N. Ida, *Engineering Electromagnetics* (2nd edition), Springer-Verlag, New York, 2003.

[EETR60] R.B. Adler, L.J. Chu, and R.M. Fano, *Electromagnetic Energy Transmission and Radiation*, John Wiley & Sons, Inc., New York, 1960.

[FME92] R.E. Collins, *Foundations for Microwave Engineering* (2nd edition), McGraw-Hill Inc., New York, 1992.

[HSDD09] S.H. Hall and H.L. Heck, *Advanced Signal Integrity for High-Speed Digital Designs*, John Wiley & Sons, Inc., Hoboken, NJ, 2009.

[ME04] D.M. Pozar, *Microwave Engineering* (3rd edition), Wiley-Interscience, Hoboken, NJ, 2004.

[MTL08] C.R. Paul, *Analysis of Multiconductor Transmission Lines* (2nd edition), Wiley-Interscience, New York, 2008.

[PAEMF61] R. Plonsey and R.E. Collins, *Principles and Applications of Electromagnetic Fields*, McGraw-Hill Book Company, Inc., New York, 1961.

[TIBD96] Texas Instruments *Bergeron Method: A Graphic Method for Determining Line Reflections in Transient Phenomena*, Texas Instruments Inc., Dallas, TX, 1996 (available online).

Index

C

F

W

Z